20 JAN 2005

A16460

66.01(03)

HANDBOOK OF PETROCHEMICALS PRODUCTION PROCESSES

HANDBOOK OF PETROCHEMICALS PRODUCTION PROCESSES

Robert A. Meyers, Ph.D.   Editor-in-Chief

McGRAW-HILL

New York   Chicago   San Francisco   Lisbon   London   Madrid
Mexico City   Milan   New Delhi   San Juan   Seoul
Singapore   Sydney   Toronto

The McGraw·Hill Companies

Cataloging-in-Publication Data is on file with the Library of Congress

Copyright © 2005 by The McGraw-Hill Companies, Inc. All rights reserved. Printed in the United States of America. Except as permitted under the United States Copyright Act of 1976, no part of this publication may be reproduced or distributed in any form or by any means, or stored in a data base or retrieval system, without the prior written permission of the publisher.

1 2 3 4 5 6 7 8 9 0   DOC/DOC   0 1 0 9 8 7 6 5 4

ISBN 0-07-141042-2

The sponsoring editor for this book was Kenneth P. McCombs, the editing supervisor was Stephen M. Smith, and the production supervisor was Pamela A. Pelton. It was set in Times Roman by Paul Scozzari of McGraw-Hill Professional's Hightstown, N.J., composition unit. The art director for the cover was Handel Low.

Printed and bound by RR Donnelley.

McGraw-Hill books are available at special quantity discounts to use as premiums and sales promotions, or for use in corporate training programs. For more information, please write to the Director of Special Sales, McGraw-Hill Professional, Two Penn Plaza, New York, NY 10121-2298. Or contact your local bookstore.

This book is printed on acid-free paper.

Information contained in this work has been obtained by The McGraw-Hill Companies, Inc. ("McGraw-Hill") from sources believed to be reliable. However, neither McGraw-Hill nor its authors guarantee the accuracy or completeness of any information published herein, and neither McGraw-Hill nor its authors shall be responsible for any errors, omissions, or damages arising out of use of this information. This work is published with the understanding that McGraw-Hill and its authors are supplying information but are not attempting to render engineering or other professional services. If such services are required, the assistance of an appropriate professional should be sought.

CONTENTS

Contributors   xix
Preface   xxi
Acknowledgments   xxiii

Part 1   Acetic Acid

Chapter 1.1.   Chiyoda Acetic Acid Process ACETICA®   *Yasuo Hosono and Minoru Tasaki, P.E.*   1.3

Introduction / *1.3*
Chemistry / *1.4*
Process Features / *1.7*
Process Description / *1.7*
Product Specifications / *1.11*
Process Yield and Emissions / *1.11*
Economics of the Chiyoda ACETICA Technology / *1.11*
Scope of Chiyoda's Package of Services / *1.12*
Experience / *1.12*
References / *1.13*

Part 2   Aniline

Chapter 2.1.   DuPont/KBR Aniline Process   *Eric W. Wong and Ronald Birkhoff*   2.3

Introduction / *2.3*
Aniline Market Overview / *2.4*
Process Chemistry / *2.5*
Process Description / *2.5*
Technology Features / *2.6*
Operating Requirements / *2.7*
Product Quality / *2.7*
Wastes and Emissions / *2.7*
References / *2.8*

Part 3   1,3-Butadiene

Chapter 3.1.   BASF Butadiene Extraction Technology   *Robert Brummer*   3.3

Introduction / *3.3*
Process Perspective / *3.3*
Process Description / *3.4*
Economics / *3.7*
Environmental Considerations / *3.8*
Summary of Process Features / *3.8*

Chapter 3.2. UOP KLP 1,3-Butadiene from Acetylene Process
Steve Krupa, Tim Foley, and Stephen McColl   3.11

Introduction / *3.11*
Butadiene / *3.11*
The KLP Process / *3.12*
Process Chemistry / *3.12*
Commercial Experience / *3.13*
Economics and Operating Costs / *3.13*

Part 4   Cumene

Chapter 4.1. ABB Lummus Global Cumene Production via CD*Cumene*® Technology   *Stephen Pohl and Sanjeev Ram*   4.3

Introduction / *4.3*
Process Perspective / *4.4*
Process Chemistry / *4.4*
Process Description / *4.4*
Process Economics / *4.7*
Summary of Process Features / *4.9*

Chapter 4.2. UOP Q-Max™ Process   *Gary A. Peterson and Robert J. Schmidt*   4.11

Introduction / *4.11*
Process Chemistry / *4.12*
Description of the Process Flow / *4.14*
Feedstock Considerations / *4.15*
Process Perfomance / *4.18*
Case Study / *4.18*
Commercial Experience / *4.19*
Bibliography / *4.19*

Part 5   Ethylbenzene

Chapter 5.1. Lummus/UOP Liquid-Phase EB*One* Process and CDTECH *EB*® Process   *Stephen Pohl and Sanjeev Ram*   5.3

Introduction / *5.3*
Process Perspective / *5.4*
Process Chemistry / *5.4*
Process Description / *5.5*
Economics / *5.9*
Summary of Process Features / *5.12*

Chapter 5.2. Polimeri Europa Ethylbenzene Process   *Fabio Assandri and Elena Bencini*   5.13

Introduction / *5.13*
Description of the Process Flow / *5.16*
Process and Catalyst Advanced Features / *5.20*

Process Performance / 5.21
Commercial Experience / 5.21

Chapter 5.3. ExxonMobil/Badger Ethylbenzene Technology
Brian Maerz and C. Morris Smith   5.23

Introduction / 5.23
Ethylbenzene Manufacturing / 5.23
Properties of Ethylbenzene / 5.25
EBMax Process Catalysts / 5.26
Process Chemistry and EBMax Catalyst Performance / 5.28
Process Description / 5.31
Process Design Customization and Optimization / 5.33
EBMax Process Designs for Dilute Ethylene Feedstocks / 5.33
Technology Conversion and Capacity Expansion with EBMax / 5.34
Ethylbenzene Product Quality / 5.35
Raw Materials and Utilities Consumption / 5.37
Catalyst Requirements / 5.37
EBMax Plant Design / 5.38
Reference / 5.38

Part 6   Ethylene

Chapter 6.1. ABB Lummus Global SRT® Cracking Technology for the Production of Ethylene   *Sanjeev Kapur*   6.3

Introduction / 6.3
Development and Commercial History / 6.4
Process Chemistry / 6.5
Cracking Heater / 6.8
Ethylene Process Flow Schematic / 6.11
Refinery and Ethylene Plant Integration / 6.15
Recent Technology Advances / 6.16
Commercial Operations / 6.19
Economic Aspects / 6.19

Chapter 6.2. Stone & Webster Ethylene Technology   *Colin P. Bowen*   6.21

Introduction / 6.21
Economic Drivers / 6.21
Development History: Pyrolysis / 6.23
Development History: Recovery / 6.27
Process Description / 6.29
Megaplant Design Issues / 6.44
Project Execution Aspects / 6.47
References / 6.49

Chapter 6.3. KBR SCORE™ Ethylene Technology   *Steven Borsos and Stephen Ronczy*   6.51

Development and History / 6.51
Selective Cracking Furnace Technology / 6.52
Optimum Recovery-Section Design / 6.57
Future Developments / 6.63

Part 7 Methanol

Chapter 7.1. Lurgi MegaMethanol® Technology   *Alexander Frei*   7.3

History / 7.3
MegaMethanol Technology / 7.4
Process Description / 7.5
Latest Lurgi Methanol Project References / 7.17

Part 8 Oxo Alcohols

Chapter 8.1. Johnson Matthey Oxo Alcohols Process™   *Jane Butcher and Geoff Reynolds*   8.3

Introduction / 8.3
Process Description / 8.3
Process Flowsheet / 8.7
Benefits of the Johnson Matthey Technology / 8.9
Feed Specifications / 8.10
Process Economics / 8.12
Capital Costs / 8.12
Operational Experience / 8.13
Reference / 8.13

Part 9 Phenols and Acetone

Chapter 9.1. Polimeri Europa Cumene-Phenol Processes   *Maurizio Ghirardini and Maurizio Tampieri*   9.3

Introduction / 9.3
Cumene Technology / 9.3
Phenol Technology / 9.8

Chapter 9.2. Sunoco/UOP Phenol Process   *Robert J. Schmidt*   9.13

Introduction / 9.13
Cumene Production / 9.13
Phenol Production / 9.14
Sunoco/UOP Cumene Peroxidation Route to Phenol Production / 9.15
Overall Process Description/Chemistry / 9.15
Process Flow and Recent Technology Advances / 9.16
Conclusion / 9.28
References / 9.29

Chapter 9.3. KBR Phenol Process   *Alan Moore and Ronald Birkhoff*   9.31

Introduction / 9.31
History / 9.31
Markets / 9.32
Process Chemistry / 9.34
Process Description / 9.36
Feedstock and Product Properties / 9.41
Production Yields / 9.43

Utility Requirements / 9.43
Product Storage and Shipping / 9.43
Environmental Features / 9.44
Safety / 9.45
Operating Economics / 9.46
Investment/Economies of Scale / 9.46
Acetone Netback / 9.48
Technology Advantages / 9.49
Bibliography / 9.50

Chapter 9.4. QBIS™ Process for High-Purity Bisphenol A *Ed Fraini, Don West, and George Mignin* 9.51

Overview / 9.51
Commercial Experience / 9.56
Wastes and Emissions: Expected Performance / 9.57

Part 10 Propylene and Light Olefins

Chapter 10.1. Lurgi MTP® Technology *Waldemar Liebner* 10.3

Introduction / 10.3
Process Overview / 10.3
Detailed Process Description / 10.4
Products, By-Products, Wastes, and Emissions / 10.9
Technical and Commercial Status / 10.10
Process Economics / 10.11
Bibliography / 10.13

Chapter 10.2. UOP/Hydro MTO Process *Peter R. Pujadó and James M. Andersen* 10.15

Introduction / 10.15
MTO Technology / 10.18
Economic Basis / 10.20
Investment Estimates / 10.20
Economic Comparisons / 10.23
Economic Sensitivity / 10.25
Conclusions / 10.26
References / 10.26

Chapter 10.3. UOP Oleflex™ Process *Joseph Gregor and Daniel Wei* 10.27

Introduction / 10.27
Process Description / 10.27
Dehydrogenation Plants / 10.29
Propylene Production Economics / 10.31

Chapter 10.4. ABB Lummus Global Propylene Production via Olefins Conversion Technology *Catherine A. Berra and James T. C. Wu* 10.35

Introduction / 10.35
Development and Commercial History / 10.36

Process Chemistry / *10.37*
Process Description / *10.37*
Process Economics / *10.38*
Summary of Process Features / *10.40*
Conclusion / *10.41*

Chapter 10.5. Propylene via CATOFIN® Propane Dehydrogenation Technology *V. K. Arora* 10.43

Introduction / *10.43*
Process Chemistry / *10.44*
Process Description / *10.44*
Process Economics / *10.47*
Feedstock and Utility Consumption / *10.47*
Product Quality and By-Products / *10.47*
Catalyst and Chemical Consumption / *10.49*
Environmental Emissions / *10.49*
Summary of Technology Features / *10.49*

Part 11   Styrene

Chapter 11.1. Lummus/UOP "Classic" Styrene Technology and Lummus/UOP SMART[SM] Styrene Technology *Stephen Pohl and Sanjeev Ram* 11.3

Introduction / *11.3*
Process Perspective / *11.4*
Process Chemistry / *11.5*
Process Descriptions / *11.6*
Economics / *11.9*
Summary of Process Features / *11.11*

Chapter 11.2. Stone & Webster (Badger) Styrene Technology *Vincent Welch* 11.13

Introduction / *11.13*
Styrene Industry / *11.13*
Use of Styrene Monomer / *11.14*
Properties / *11.15*
Styrene Manufacturing / *11.15*
Process Chemistry / *11.16*
Process Description / *11.18*
Product Specification / *11.23*
Operating Economics / *11.23*

Chapter 11.3. Polimeri Europa Styrene Process Technology *Leonardo Trentini and Armando Galeotti* 11.25

Introduction / *11.25*
Process Chemistry / *11.26*

Description of the Process Flow / 11.27
Process and Mechanical Design Advanced Features / 11.32
Process Performance / 11.33
Commercial Experience / 11.34

Part 12 Terephthalic Acid

Chapter 12.1. E PTA: The Lurgi/Eastman/SK Process
Frank Castillo-Welter   12.3

Introduction / 12.3
Chemistry Overview and Product Specification / 12.4
Process Description / 12.4
Highlights and Benefits of E PTA Technology / 12.9
Economics of E PTA Technology / 12.10
Commercial Experience / 12.11

Part 13 Xylenes

Chapter 13.1. ExxonMobil PxMax^{SM} *p*-Xylene from Toluene
Terry W. Bradley   13.3

Introduction / 13.3
Process Chemistry / 13.4
Process Description / 13.5
Operating Performance / 13.5
PxMax Retrofit and Debottleneck Applications / 13.7
Aromatics Complex and PxMax Unit Description / 13.8
Case I: Grassroots PxMax Unit / 13.8
Case II: Retrofit of Selective TDP to PxMax / 13.11
Case III: Retrofit of Nonselective TDP to PxMax / 13.12
Conclusion / 13.13

Chapter 13.2. ExxonMobil XyMax^{SM} Xylene Isomerization
Terry W. Bradley   13.15

Introduction / 13.15
Process Chemistry / 13.16
Process Description / 13.17
Operating Performance / 13.18
XyMax Cycle Length / 13.20
Commercial Experience / 13.21

Chapter 13.3. UOP Parex™ Process for *p*-Xylene Production
Scott E. Commissaris   13.23

Introduction / 13.23
Parex versus Crystallization / 13.23
Process Performance / 13.26

Feedstock Considerations / 13.26
Description of the Process Flow / 13.26
Equipment Considerations / 13.28
Case Study / 13.29
Commercial Experience / 13.29
Bibliography / 13.30

Part 14 Polyethylene

Chapter 14.1. Basell *Spherilene* Technology for LLDPE and HDPE Production   *Maurizio Dorini and Gijs ten Berge*   14.3

General Process Description / 14.3
Process Chemistry and Thermodynamics / 14.3
Spherilene Process Perspective / 14.7
Process Description / 14.8
Products and Applications / 14.10
Process Economics / 14.12

Chapter 14.2. Borstar LLDPE and HDPE Technology
Tarja Korvenoja, Henrik Andtsjö, Klaus Nyfors, and Gunnar Berggren   14.15

Process Description / 14.15
Advanced Process Control / 14.21
Capacities and Locations of Borstar PE Plants / 14.23
Borstar PE Products / 14.23
Process Economics / 14.29

Chapter 14.3. Chevron Phillips Slurry-Loop-Reactor Process for Polymerizing Linear Polyethylene   *Mike Smith*   14.31

History / 14.31
Process Description / 14.32
Slurry-Loop Reactor / 14.35
Polymer Finishing and Packaging / 14.37
Utilities / 14.37
Technical Advantages of the Chevron Phillips Slurry-Loop Process for PE / 14.38
Summary / 14.44

Chapter 14.4. ExxonMobil High-Pressure Process Technology for LDPE   *Charles E. Schuster*   14.45

Introduction / 14.45
Reaction Mechanism / 14.46
Process Overview/Description / 14.48
LDPE versus LLDPE / 14.53
Product Capability/Grade Slate / 14.54
LDPE Markets / 14.54
Strengths of ExxonMobil Technology / 14.56
Summary / 14.57
Reference / 14.58
Disclaimer / 14.58

Chapter 14.5. Polimeri Europa Polyethylene High-Pressure Technologies *Mauro Mirra*   14.59

Introduction / *14.59*
Polimeri Europa Trademarks / *14.60*
Chemistry and Thermodynamics / *14.61*
High-Pressure Reactor Technologies / *14.63*
Detailed Process Description / *14.65*
Reactor Safety Discharge System / *14.67*
Process Performance / *14.69*
Plant Battery Limits / *14.69*

Chapter 14.6. Basell *Hostalen* Technology for Bimodal HDPE Production   *Dr. Reinhard Kuehl and Gijs ten Berge*   14.71

General Process Description / *14.71*
Process Chemistry / *14.71*
Hostalen Process Perspective / *14.74*
Process Description / *14.75*
Product Range and Applications / *14.82*
Process Economics / *14.85*

Chapter 14.7. Basell *Lupotech* G Technology for HDPE and MDPE Production   *Cyrus Ahmadzade and Gijs ten Berge*   14.87

General Process Description / *14.87*
Process Chemistry and Thermodynamics / *14.87*
Lupotech G Process Perspective / *14.88*
Process Description / *14.89*
Product Specifications / *14.92*
Process Economics / *14.93*

Chapter 14.8. Basell *Lupotech* T Technology for LDPE and EVA-Copolymer Production   *André-Armand Finette and Gijs ten Berge*   14.95

General Process Description / *14.95*
Process Chemistry and Thermodynamics / *14.95*
Lupotech T Process Perspective / *14.102*
Process Description / *14.104*
Product Specifications / *14.110*
Process Economics / *14.111*

Chapter 14.9. UNIPOL™ PE Gas-Phase Process: Delivering Value to the PE Industry   *Mardee McCown Kaus*   14.113

Introduction / *14.113*
History / *14.114*
General Process Description / *14.115*
Process Perspective / *14.119*
Product and By-Product Specifications / *14.120*
Wastes and Emissions / *14.122*
Process Economics / *14.124*

Chapter 14.10. NOVA Chemicals SCLAIRTECH™ LLDPE/HDPE Swing Technology   *Keith Wiseman*   14.131

Introduction / *14.131*
Chemistry and Catalysis / *14.132*
Process Overview / *14.134*
Advantages of the SCLAIRTECH Technology Platform / *14.138*
Economics / *14.140*
Product Capability / *14.140*
Commercial Installations / *14.143*
Summary / *14.143*
Acknowledgment / *14.144*
Disclaimer / *14.144*

Part 15   Polyethylene Terephthalate

Chapter 15.1. UOP Sinco Solid-State Polymerization Process for the Production of PET Resin and Technical Fibers
Stephen M. Metro and James F. McGehee   15.3

Introduction / *15.3*
Melt-Phase Polymerization / *15.5*
SSP Process Chemistry / *15.6*
Crystallization of PET / *15.8*
Sticking Tendency of PET / *15.10*
Detailed Process Description / *15.10*
Reactions of the Catalytic Nitrogen Purification System / *15.13*
Oxidation of PET / *15.14*
Process Variables / *15.14*
Feed Properties / *15.14*
Product Properties / *15.16*
Product Yield / *15.16*
Wastes and Emissions / *15.16*
Utilities / *15.16*
Equipment Considerations / *15.17*
Commercial Experience / *15.17*
References / *15.18*

Part 16   Polypropylene

Chapter 16.1. Basell *Spheripol* Technology for PP Production
Maurizio Dorini and Gijs ten Berge   16.3

General Process Description / *16.3*
Process Chemistry and Thermodynamics / *16.3*
Spheripol Process Perspective / *16.13*
Process Description / *16.13*
Process Economics / *16.17*
Products and Applications / *16.17*

Chapter 16.2. Basell *Spherizone* Technology for PP Production
Riccardo Rinaldi and Gijs ten Berge   16.21

General Process Description / *16.21*
Process Chemistry and Thermodynamics / *16.21*

Spherizone Process Perspective / *16.30*
Process Description / *16.30*
Economics / *16.37*
Products and Applications / *16.37*

Chapter 16.3. Borstar Polypropylene Technology *Jouni Kivelä, Helge Grande, and Tarja Korvenoja* 16.41

Introduction / *16.41*
Features of the Borstar PP Process Technology / *16.43*
Process Description / *16.44*
Production Cycle and Grade Transitions / *16.47*
Advanced Process Control / *16.48*
Catalyst / *16.49*
Environment / *16.49*
Operating Requirements / *16.50*
Products / *16.50*

Chapter 16.4. UNIPOL™ Polypropylene Process Technology *Barry R. Engle* 16.57

General UNIPOL PP Process Description / *16.57*
Process Chemistry / *16.60*
Process Perspective / *16.62*
Products and By-Products / *16.64*
UNIPOL PP Product Attributes Summary / *16.65*
Wastes and Emissions / *16.67*
Process Economics / *16.68*

Chapter 16.5. Chisso Gas-Phase Polypropylene Process *Takeshi Shiraishi* 16.71

Technology Background and History / *16.71*
Polymerization Mechanism and Polymer Type / *16.71*
Process Features / *16.74*
Process Description / *16.76*
Safety and Environmental Considerations / *16.78*
Product Capabilities / *16.78*
Economics / *16.79*
Reference Plants / *16.79*

Part 17 Polystyrene

Chapter 17.1. BP/Lummus Technology for the Production of Expandable Polystyrene *Robert Stepanian* 17.3

Introduction / *17.3*
Operating Plants / *17.3*
Process Chemistry / *17.3*
Process Description / *17.4*
Feedstock/Product Specifications / *17.4*
Waste and Emissions / *17.5*

Process Economics / *17.5*
Summary of Process Features / *17.7*
Reference / *17.7*

Chapter 17.2. BP/Lummus Technology for the Production of General-Purpose and High-Impact Polystyrenes   *Robert Stepanian*   17.9

Introduction / *17.9*
Operating Plants / *17.10*
Process Chemistry / *17.10*
Process Description / *17.10*
Feedstock and Product Specifications / *17.12*
Waste and Emissions / *17.15*
Process Economics / *17.15*
Summary of Process Features / *17.15*
References / *17.17*

Chapter 17.3. Polimeri Europa General-Purpose Polystyrene Process Technology   *Francesco Pasquali and Riccardo Inglese*   17.19

Introduction / *17.19*
Process Chemistry / *17.20*
Description of the Process Flow / *17.22*
Process Advanced Design Features / *17.23*
Process Performance / *17.24*
Plant Capacity / *17.24*
Commercial Experience / *17.25*
The Edistir GPPS Product Portfolio / *17.25*

Chapter 17.4. Polimeri Europa Expandable Polystyrene Process Technology   *Dario Ghidoni and Riccardo Inglese*   17.27

Introduction / *17.27*
Process Chemistry / *17.28*
Description of Process Flow / *17.30*
Process Advanced Design Features / *17.31*
Process Performance / *17.32*
Plant Capacity / *17.32*
Commercial Experience / *17.32*
The Extir EPS Product Portfolio / *17.32*

Chapter 17.5. Polimeri Europa High-Impact Polystyrene Process Technology   *Francesco Pasquali and Franco Balestri*   17.35

Introduction / *17.35*
Process Chemistry / *17.36*
Description of Process Flow / *17.37*
Process Advanced Design Features / *17.38*
Process Performance / *17.39*
Plant Capacity / *17.39*
Commercial Experience / *17.40*
The Edistir HIPS Product Portfolio / *17.40*

Part 18 Vinyl Chloride and Polyvinyl Chloride

Chapter 18.1. Vinnolit Vinyl Chloride and Suspension Polyvinyl Chloride Technologies   *Ulrich Woike and Peter Kammerhofer*   18.3

Company Introduction / *18.3*
Process Perspective / *18.4*
Vinnolit Vinyl Chloride Monomer (VCM) Process / *18.4*
Vinnolit Direct Chlorination Process / *18.7*
Vinnolit Oxychlorination Process / *18.12*
Vinnolit Thermal Cracking Process of 1,2-Dichloroethane to Vinyl Chloride / *18.18*
Vinnolit Suspension Polyvinyl Chloride (S-PVC) Process / *18.21*
Abbreviations and Acronyms / *18.33*
References / *18.35*

Chapter 18.2. Chisso Polyvinyl Chloride Suspension Process Technology and Vinyl Chloride Monomer Removal Technology   *Seiichi Uchida*   18.37

Chisso Polyvinyl Chloride Suspension Process Technology / *18.37*
Chisso Vinyl Chloride Monomer Removal Process Technology / *18.43*

Index follows Chapter 18.2

CONTRIBUTORS

Cyrus Ahmadzade *Basell Polyolefine GmbH, Frankfurt, Germany* (Chap. 14.7)
James M. Andersen *UOP LLC, Des Plaines, Illinois* (Chap. 10.2)
Henrik Andtsjö *Borealis Polymers O/Y, Porvoo, Finland* (Chap. 14.2)
V. K. Arora *ABB Lummus Global, Bloomfield, New Jersey* (Chap. 10.5)
Fabio Assandri *"C. Buonerba" Research Centre, Mantova, Italy* (Chap. 5.2)
Franco Balestri *"C. Buonerba" Research Centre, Mantova, Italy* (Chap. 17.5)
Elena Bencini *"C. Buonerba" Research Centre, Mantova, Italy* (Chap. 5.2)
Gunnar Berggren *Borealis Polymers O/Y, Porvoo, Finland* (Chap. 14.2)
Catherine A. Berra *ABB Lummus Global, Houston, Texas* (Chap. 10.4)
Ronald Birkhoff *Kellogg Brown & Root, Inc. (KBR), Houston, Texas* (Chaps. 2.1, 9.3)
Steven Borsos *Kellogg Brown & Root, Inc. (KBR), Houston, Texas* (Chap. 6.3)
Colin P. Bowen *Stone & Webster, Inc., Houston, Texas* (Chap. 6.2)
Terry W. Bradley *ExxonMobil Chemical Company, Baytown, Texas* (Chaps. 13.1, 13.2)
Robert Brummer *ABB Lummus Global, Bloomfield, New Jersey* (Chap. 3.1)
Jane Butcher *Johnson Matthey Catalysts, Billingham, England* (Chap. 8.1)
Frank Castillo-Welter *Lurgi Oel Gas Chemie GmbH, Frankfurt am Main, Germany* (Chap. 12.1)
Scott E. Commissaris *UOP LLC, Des Plaines, Illinois* (Chap. 13.3)
Maurizio Dorini *Basell Polyolefine GmbH, Frankfurt, Germany* (Chaps. 14.1, 16.1)
Barry R. Engle *Dow Chemical Company, Danbury, Connecticut* (Chap. 16.4)
André-Armand Finette *Basell Polyolefine GmbH, Frankfurt, Germany* (Chap. 14.8)
Tim Foley *UOP LLC, Des Plaines, Illinois* (Chap. 3.2)
Ed Fraini *Dow Chemical Company, Midland, Michigan* (Chap. 9.4)
Alexander Frei *Lurgi Oel Gas Chemie GmbH, Frankfurt am Main, Germany* (Chap. 7.1)
Armando Galeotti *"C. Buonerba" Research Centre, Mantova, Italy* (Chap. 11.3)
Dario Ghidoni *"C. Buonerba" Research Centre, Mantova, Italy* (Chap. 17.4)
Maurizio Ghirardini *Polimeri Europa, San Donato Milanese, Italy* (Chap. 9.1)
Helge Grande *Borealis Polymers O/Y, Porvoo, Finland* (Chap. 16.3)
Joseph Gregor *UOP LLC, Des Plaines, Illinois* (Chap. 10.3)
Yasuo Hosono *Chiyoda Corporation, Yokohama, Japan* (Chap. 1.1)
Riccardo Inglese *"C. Buonerba" Research Centre, Mantova, Italy* (Chaps. 17.3, 17.4)
Peter Kammerhofer *Vinnolit GmbH & Co. KG, Ismaning, Germany* (Chap. 18.1)
Sanjeev Kapur *ABB Lummus Global, Houston, Texas* (Chap. 6.1)
Mardee McCown Kaus *Univation Technologies LLC, Houston, Texas* (Chap. 14.9)

Jouni Kivelä *Borealis Polymers O/Y, Porvoo, Finland* (Chap. 16.3)
Tarja Korvenoja *Borealis Polymers O/Y, Porvoo, Finland* (Chaps. 14.2, 16.3)
Steve Krupa *UOP LLC, Des Plaines, Illinois* (Chap. 3.2)
Dr. Reinhard Kuehl *Basell Polyolefine GmbH, Frankfurt, Germany* (Chap. 14.6)
Waldemar Liebner *Lurgi AG, Frankfurt am Main, Germany* (Chap. 10.1)
Brian Maerz *Badger Licensing LLC, Cambridge, Massachusetts* (Chap. 5.3)
Stephen McColl *UOP LLC, Des Plaines, Illinois* (Chap. 3.2)
James F. McGehee *UOP LLC, Des Plaines, Illinois* (Chap. 15.1)
Stephen M. Metro *UOP LLC, Des Plaines, Illinois* (Chap. 15.1)
George Mignin *Dow Chemical Company, Midland, Michigan* (Chap. 9.4)
Mauro Mirra *Polimeri Europa, Milan, Italy* (Chap. 14.5)
Alan Moore *Kellogg Brown & Root, Inc. (KBR), Houston, Texas* (Chap. 9.3)
Klaus Nyfors *Borealis Polymers O/Y, Porvoo, Finland* (Chap. 14.2)
Francesco Pasquali *"C. Buonerba" Research Centre, Mantova, Italy* (Chaps. 17.3, 17.5)
Gary A. Peterson *UOP LLC, Des Plaines, Illinois* (Chap. 4.2)
Stephen Pohl *ABB Lummus Global, Bloomfield, New Jersey* (Chaps. 4.1, 5.1, 11.1)
Peter R. Pujadó *UOP LLC, Des Plaines, Illinois* (Chap. 10.2)
Sanjeev Ram *ABB Lummus Global, Bloomfield, New Jersey* (Chaps. 4.1, 5.1, 11.1)
Geoff Reynolds *Johnson Matthey Catalysts, Billingham, England* (Chap. 8.1)
Riccardo Rinaldi *Basell Polyolefine GmbH, Frankfurt, Germany* (Chap. 16.2)
Stephen Ronczy *Kellogg Brown & Root, Inc. (KBR), Houston, Texas* (Chap. 6.3)
Robert J. Schmidt *UOP LLC, Des Plaines, Illinois* (Chaps. 4.2, 9.2)
Charles E. Schuster *ExxonMobil Chemical Company, Baytown, Texas* (Chap. 14.4)
Takeshi Shiraishi *Chisso Corporation, Tokyo, Japan* (Chap. 16.5)
C. Morris Smith *Badger Licensing LLC, Cambridge, Massachusetts* (Chap. 5.3)
Mike Smith *Chevron Phillips Chemical Company LP, Kingwood, Texas* (Chap. 14.3)
Robert Stepanian *ABB Lummus Global, Bloomfield, New Jersey* (Chaps. 17.1, 17.2)
Maurizio Tampieri *Polimeri Europa, San Donato Milanese, Italy* (Chap. 9.1)
Minoru Tasaki, P.E. *Chiyoda Corporation, Yokohama, Japan* (Chap. 1.1)
Gijs ten Berge *Basell Polyolefine GmbH, Frankfurt, Germany* (Chaps. 14.1, 14.6, 14.7, 14.8, 16.1, 16.2)
Leonardo Trentini *"C. Buonerba" Research Centre, Mantova, Italy* (Chap. 11.3)
Seiichi Uchida *Chisso Corporation, Tokyo, Japan* (Chap. 18.2)
Daniel Wei *UOP LLC, Des Plaines, Illinois* (Chap. 10.3)
Vincent Welch *Badger Licensing LLC, Cambridge, Massachusetts* (Chap. 11.2)
Don West *Dow Chemical Company, Midland, Michigan* (Chap. 9.4)
Keith Wiseman *NOVA Chemicals Corporation, Calgary, Alberta, Canada* (Chap. 14.10)
Ulrich Woike *Vinnolit GmbH & Co. KG, Ismaning, Germany* (Chap. 18.1)
Eric W. Wong *Kellogg Brown & Root, Inc. (KBR), Houston, Texas* (Chap. 2.1)
James T. C. Wu *ABB Lummus Global, Houston, Texas* (Chap. 10.4)

PREFACE

This handbook is a reference and guide to present-day real-world petrochemicals (intermediates, monomers, and plastics) production methods, product properties, and economics. There is no other single source of such information. The focus is on the most economically important petrochemicals (which together represent a $200 billion market): acetic acid, aniline, 1,3-butadiene, cumene, ethylbenzene, ethylene, methanol, oxo alcohols, phenols and acetone, propylene and light olefins, styrene, terephthalic acid, xylenes, low-density/linear low-density and high-density polyethylenes and copolymers, polyethylene terephthalate, polypropylene, polystyrene, and vinyl chloride and polyvinyl chloride.

In order to present the global technology base adequately, it was decided to ask major licensers to contribute to this handbook the processes and economics of their most advanced and utilized licensable technologies (the reports were to be in a common format, to allow for side-by-side comparison). This resulted in a total of 53 technologies offered by 18 of the largest firms in the petrochemicals licensing business. These firms also represent a large fraction of global petrochemicals production. The information given in this handbook will let engineers make a first evaluation of licensable processes for new production and will allow engineering students to perform class exercises that compare the various characteristics of today's most-used technologies.

Each licenser was asked to follow the chapter format below as closely as possible:

General process description: including feed definition and product yield and a simplified flow diagram.

Process chemistry and thermodynamics: for each major processing unit as applicable.

Process perspective: developers, locations, and specifications of all test and commercial plants, and near-term and long-term plans.

Detailed process description: process flow diagram with mass and energy balances for major process variations, and feeds and details on unique or key equipment.

Product and by-product specifications: detailed analyses of all process products and by-products as a function of processing variations and feeds.

Wastes and emissions: process solid, liquid, and gas wastes and emissions as a function of processing variations and feeds.

Process economics: installed capital cost by major section, total capital investment, operating costs, annualized capital costs with the basis, and price range for each product if applicable.

This handbook is a companion to the *Handbook of Petroleum Refining Processes,* Third Edition (McGraw-Hill), published last year. That handbook, with the same format as this one, consists of 61 licenser technology chapters for the processing of petroleum to gasoline and other fuels as well as some petrochemical intermediates. The two handbooks together provide a comprehensive set of technologies for converting crude oil to fuels, intermediates, commodity chemicals, and the major plastics.

The reader of this handbook is specially directed to the section on gas-to-liquid technologies in the *Handbook of Petroleum Refining Processes,* which includes methanol production and also conversion to ethylene and propylene, as well as to the section on aromatics complexes for additional information on licensable petrochemical technologies.

Robert A. Meyers, Ph.D.

ACKNOWLEDGMENTS

A distinguished group of 72 engineers prepared the 53 chapters of this handbook, and I thank them for their good work. I also wish to thank the 18 petrochemicals process research and development and licensing firms that developed and market the technologies presented.

Many thanks to my wife Ilene for encouraging my efforts in bringing together these inventive concepts in handbook form and also for her direct participation in the preparation of the index.

PART 1

ACETIC ACID

CHAPTER 1.1
CHIYODA ACETIC ACID PROCESS ACETICA®

Yasuo Hosono and Minoru Tasaki, P.E.
Chiyoda Corporation,
Yokohama, Japan

INTRODUCTION

Acetic acid is an important raw material used for the production of vinyl acetate, acetic anhydride, polymer-grade terephthalic acid (PTA), etc. The growth rate in per capita consumption of acetic acid in Asia is typically 7 to 8 percent per annum, the requirement of acetic acid for the production of PTA being the most important and leading reason for the increase. Although there are several methods of producing acetic acid, such as oxidation of acetaldehyde, direct oxidation of ethylene, etc., synthesis from methanol and carbon monoxide, the methanol carbonylation process, is the world's leading process.

This process description refers to the Chiyoda ACETICA® process, developed and owned by Chiyoda, to produce acetic acid by the carbonylation of methanol and carbon monoxide. The ACETICA process for producing acetic acid is based on the presence of a proprietary heterogeneous rhodium catalyst in the active rhodium complex that is chemically immobilized on a polyvinylpyridine resin. To minimize attrition of the solid catalyst material, a bubble column reactor is employed.

This immobilized catalyst system has the following advantages over the conventional liquid rhodium catalyst:

- Handling of the catalyst is easy because it need not be recovered by separating the rhodium from the reaction liquid.
- A high level of reactor productivity can be obtained because the concentration of rhodium can be increased without the limitations imposed by solubility.
- The formation of by-products can be suppressed because operation is conducted under a low water concentration.
- The corrosive environment is moderated because the concentration of hydrogen iodide is decreased.

1.4 ACETIC ACID

The methanol carbonylation reaction between methanol and carbon monoxide is conducted at moderate temperatures (170 to 190°C) and pressures (3.0 to 4.5 MPa). The principal reactions are listed below. (Details are given in the following section, "Chemistry.")

Carbonylation: $CH_3OH + CO \rightarrow CH_3COOH$

Esterification: $CH_3OH + CH_3COOH \leftrightarrow CH_3COOCH_3 + H_2O$

Etherification: $2CH_3OH \leftrightarrow CH_3OCH_3 + H_2O$

The process consists of the following units:

- Feed/absorption/reaction unit
- Distillation unit
- Product-treating unit
- Methyl iodide (MI)–generation unit
- Iodide-removal unit
- Waste-treating unit

CHEMISTRY

Principal Reactions

The reaction unit of the Chiyoda ACETICA process operates at a moderate temperature range of 170 to 190°C and pressure range of 3.0 to 4.5 MPa in the presence of a heterogeneous rhodium (Rh) catalyst and methyl iodide as a promoter in a bubble-column reactor. The usual acetic acid yields based on methanol and CO consumption are greater than 99 and 92 percent, respectively.

The basic chemistry of carbonylation is similar to that of a homogeneous catalyst in the conventional processes. The net reaction in the carbonylation of methanol is

Carbonylation: $CH_3OH(l) + CO(g) \rightarrow CH_3COOH(l)$

$$\Delta H = -138 \text{ kJ/gmol methanol}$$

ACETICA requires no additional water to stabilize the active rhodium complex. Therefore, the reaction solution contains less than 8 percent water according to the following equilibrium reactions:

Esterification: $CH_3OH + CH_3COOH \leftrightarrow CH_3COOCH_3 + H_2O$

Etherification: $2CH_3OH \leftrightarrow CH_3OCH_3 + H_2O$

Hydrogen iodide is formed in the reaction solution by hydrolysis of the metal iodide:

$$CH_3I + H_2O \leftrightarrow CH_3OH + HI$$

Certain amounts of methanol and acetic acid may react with the methyl iodide as shown below:

$$CH_3I + CH_3COOH \leftrightarrow CH_3COOCH_3 + HI$$

$$CH_3I + CH_3OH \leftrightarrow CH_3OCH_3 + HI$$

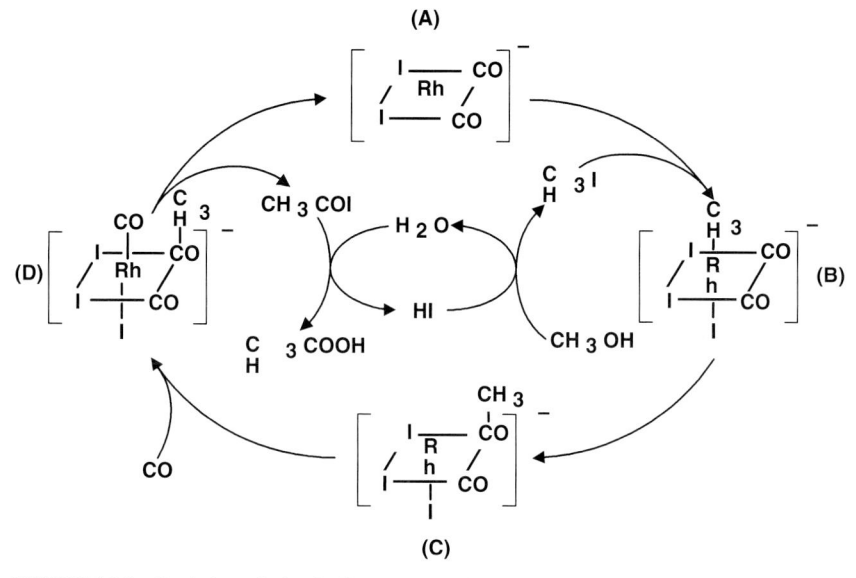

FIGURE 1.1.1 Catalytic cycle for rhodium carbonylation.

The detailed pathway of the Rh complex catalyst supported on solid resin can be depicted as two interacting cycles, as shown in Fig. 1.1.1.

Methyl iodide is added oxidatively to the rhodium-dicarbonyl-diiodide complex [RhI$_2$(CO)$_2$](A) to generate a rhodium-methyl complex (B). This rhodium-methyl complex rapidly undergoes a methyl migration to a neighboring carbonyl group in the acetyl form (CH$_3$CO) and reacts with CO to generate the rhodium-acetyl complex (D). Reductive elimination of the acetyl iodide (CH$_3$COI) then liberates the original rhodium complex (A). The hydration of acetyl iodide is very rapid in the presence of water and results in the formation of acetic acid and hydrogen iodide to complete the cycle.

The figure indicates that the concentrations of dissolved CO, CH$_3$I, catalyst, and H$_2$O affect the reaction rate of carbonylation and catalyst stability.

Catalyst Preparation Reactions

Promoter (CH$_3$I). The ACETICA process uses methyl iodide as the promoter for the carbonylation reaction, as described earlier. The method of producing CH$_3$I in a methyl iodide generator is

$$I_2 + CO + H_2O \rightarrow 2HI + CO_2$$

$$HI + CH_3OH \rightarrow CH_3I + H_2O$$

Heterogeneous Catalyst (Rhodium Immobilized on Resin). The ACETICA process is based on a heterogeneous rhodium catalyst. The nitrogen atoms of the resin pyridine groups become positively charged after quaternization with methyl iodide.

The active rhodium complex, [Rh(CO)$_2$I$_2$], is immobilized by the ion exchange on the quaternized polyvinylpyridine resin.

Because the ion-exchange equilibrium favors the solid phase, almost all Rh in the reaction mixture is immobilized on the resin support.

By-Product Formations

Gaseous By-Products (CO$_2$, H$_2$, CH$_4$). Gaseous by-product formation reactions, such as water-gas shift and methane formation, also occur in the carbonylation reactor, but losses of CO and methanol by these reactions amount to less than 1 percent of the total feed:

Water-gas shift reaction: $CO + H_2O \leftrightarrow CO_2 + H_2$

Methanation: $CH_3OH + H_2 \leftrightarrow CH_4 + H_2O$

$CO + 3H_2 \leftrightarrow CH_4 + H_2O$

Liquid By-Products. Very small amounts of liquid by-products are produced at the initial stage of plant operation, and their concentrations come to equilibrium, except for some heavy by-products such as propionic acid. These heavy by-products are eliminated with small purges from the purification unit.

Iodide-Removal Reaction

To minimize iodide loss, most of the iodide compounds in the crude product are recovered and returned to the synthesis section from the distillation section. Small amounts of iodide compounds remain in the distilled product, and these are removed by a proprietary adsorbent.

PROCESS FEATURES

In comparison with conventional systems, the key features of the Chiyoda ACETICA process are

- *Higher productivity.* Unlike homogeneous catalyst systems, the concentration of Rh is not restricted by solubility limitations of a liquid catalyst. With ACETICA, the Rh concentration can be increased, enabling higher reaction levels.
- *Few mechanical problems.* The ACETICA process employs a unique bubble-column reactor that has no mechanical agitator, meaning fewer problems with leakage and maintenance.
- *Moderate corrosive system.* The heterogeneous catalyst system enables the use of less water in the system, resulting in a lower concentration of hydrogen iodide, which is the leading contributor to corrosion. Zr is used in the reactor system. However, Ti or hastelloy is acceptable for use in the rest of the system, thereby reducing investment costs.
- *Higher product purity and lower by-product formation.* The high degree of catalyst activity under a low water concentration suppresses the formation of by-products, resulting in higher product purity.
- *Competitive economics.* A moderate corrosive system environment enables the use of lower-grade materials. In addition, the higher yield of acetic acid and the recovery of the reaction heat in the reactor system reduce operating costs.
- *More flexible for large-scale capacity.* Because there are no moving parts or mechanical equipment in the reactor, ACETICA offers more flexible application to larger plants having an annual capacity such as 200,000 to 500,000 tons.
- *Easy catalyst handling.* There is no need to recover the rhodium by separating it from the reaction liquid.
- *Open licensing policy.* Chiyoda offers ACETICA technology based on an open license policy.
- *Chiyoda's full service as EPC contractor.* Not only as a technology supplier but also as an EPC contractor, Chiyoda can supply all its accumulated expertise, including the design, procurement, and construction details of an actual acetic acid plant, as well as the license and the basic process package.

PROCESS DESCRIPTION

A simplified diagram of the process flow of the ACETICA process is shown in Fig. 1.1.2.

Feed/Absorption/Reaction Unit

The CO compressor compresses carbon monoxide received by pipeline at the battery limit to the reaction level. After the moisture is removed, the CO is sparged into the carbonylation reactor.

To enhance the absorbing efficiency, fresh methanol is split into two streams, each of which is fed to a separate countercurrent high-pressure (HP) absorber and low-pressure (LP) absorber. Part of the methanol feed contacts the reactor offgas in the HP absorber, which mostly contains unconverted CO, methyl iodide, methyl acetate, and so on. The other part of the methanol feed stream contacts the light gases in the LP absorber, which were

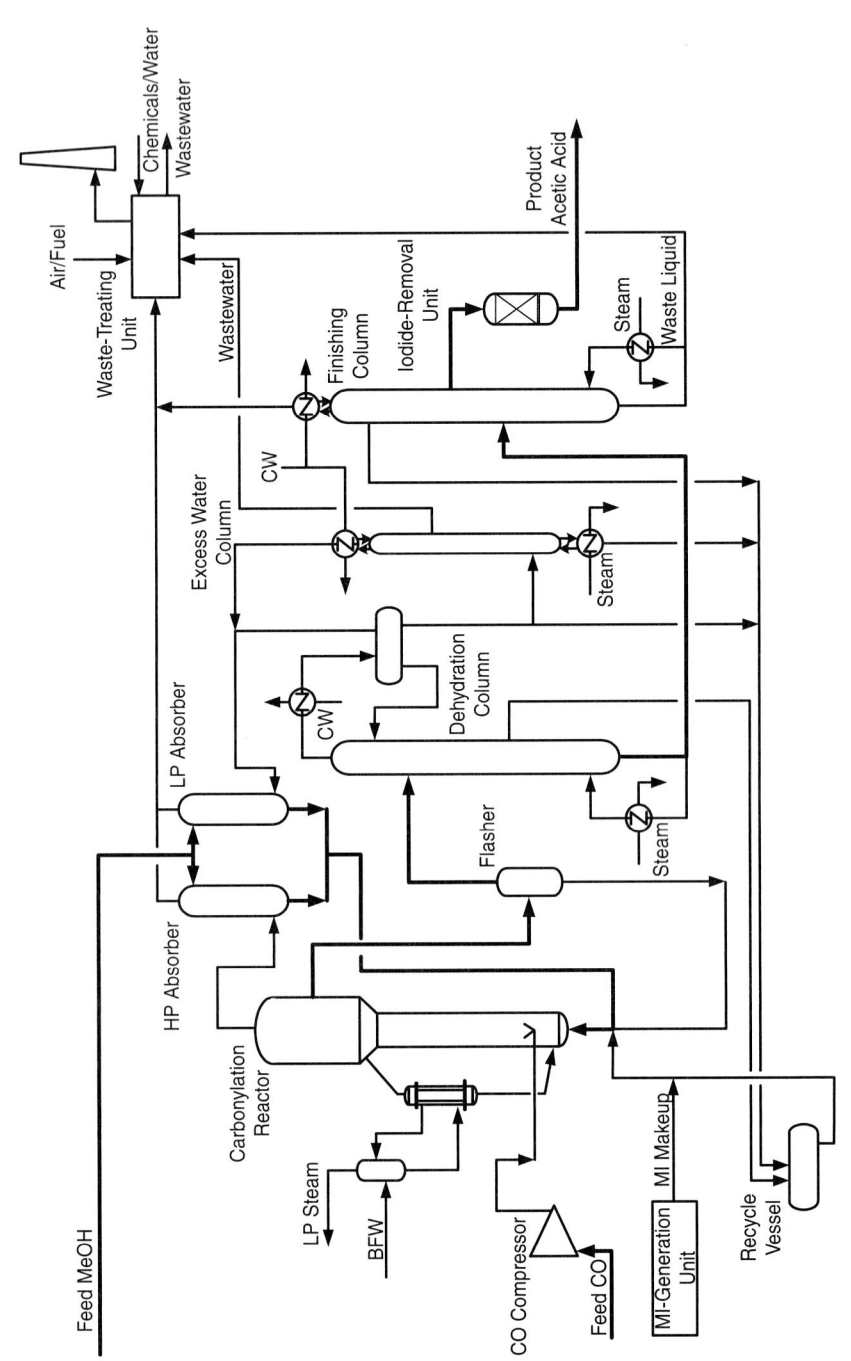

FIGURE 1.1.2 Simplified process flow diagram.

released at low pressure from the distillation unit. The main purpose of this absorption system is to maximize recovery of valuable methyl acetate and methyl iodide, which otherwise would exit the system with the vented gas, resulting in unnecessarily high chemical consumption and yield loss. The methanol feed streams exiting the absorbers are recombined and mixed with the recycled liquid from the recycle vessel and makeup methyl iodide from the MI generator unit. The combined and recycled stream is then charged to the bottom of the carbonylation reactor riser section.

The carbonylation reactor, using a three-phase bubble-column system, consists of a riser, separator, downcomer, and reactor cooler. The reaction conditions are as follows:

Temperature: 170–190°C

Pressure: 3.0–4.5 MPa

Methanol, CO, and recycled liquid from the distillation unit are introduced at the bottom of the carbonylation reactor riser section. The compressed CO is fed through a sparger for even distribution. These feeds and catalyst, together with the circulation, flow from the downcomer and up the riser, where almost all the CO and methanol are converted into acetic acid by carbonylation. The difference in density between the gaseous CO-rich riser and CO-depleted downcomer drives the circulation. The agitation provided by the high velocities of the liquid and gaseous reactants rapidly dissolves the CO into a liquid phase that reacts with methanol so that the system is not mass-transfer-controlled.

Unreacted CO and other gaseous by-products are vented from the top of the separator to the HP absorber. A portion of the reactor liquid effluent (crude acetic acid) is disengaged by gravity from the solids in the recycle slurry at the separator and sent to the downstream distillation unit. The remaining catalyst-liquid mixture is routed through the reactor downcomer, cooled by the reactor cooler to remove the large heat of reaction, and returned to the bottom of reactor riser to maintain circulation. The reactor cooler generates the low-pressure steam by recovering the reaction heat, which is used as a heat source of the distillation unit. Catalyst retention in the reactor is almost 100 percent.

A drawing of the bubble-column reactor is shown in Fig. 1.1.3.

Distillation Unit

The purposes of the distillation unit are

- Production of purified acetic acid from crude acetic acid by purification
- Maintenance of the stable condition in terms of water content and impurities
- Recovery of valuable hydrocarbons, including methyl iodide

Discharged liquid from the carbonylation reactor, containing crude acetic acid product, methyl acetate, water, carbon monoxide, etc., is flashed and vaporized in the flasher to separate vapor and liquid streams. A major portion of the acetic acid, unreacted methanol, methyl acetate, methyl iodide, water, and some heavy impurities such as propionic acid are flashed into the vapor phase and fed to the dehydration column. At the same time, some heavy impurities are recycled from the bottom liquid stream containing acetic acid to the carbonylation reactor.

The vapor stream from the flasher is sent to the dehydration column to remove dissolved gases, light organic components, and water. The overhead stream is condensed and passed to the dehydration column receiver. In the receiver, uncondensed volatile materials consisting mainly of CO with minor amounts of vaporized methyl iodide are sent to the LP absorber for recovery and to prevent yield loss. A part of the condensed liquid is sent to the excess water column. The remainder of this stream is returned to the carbonylation reactor.

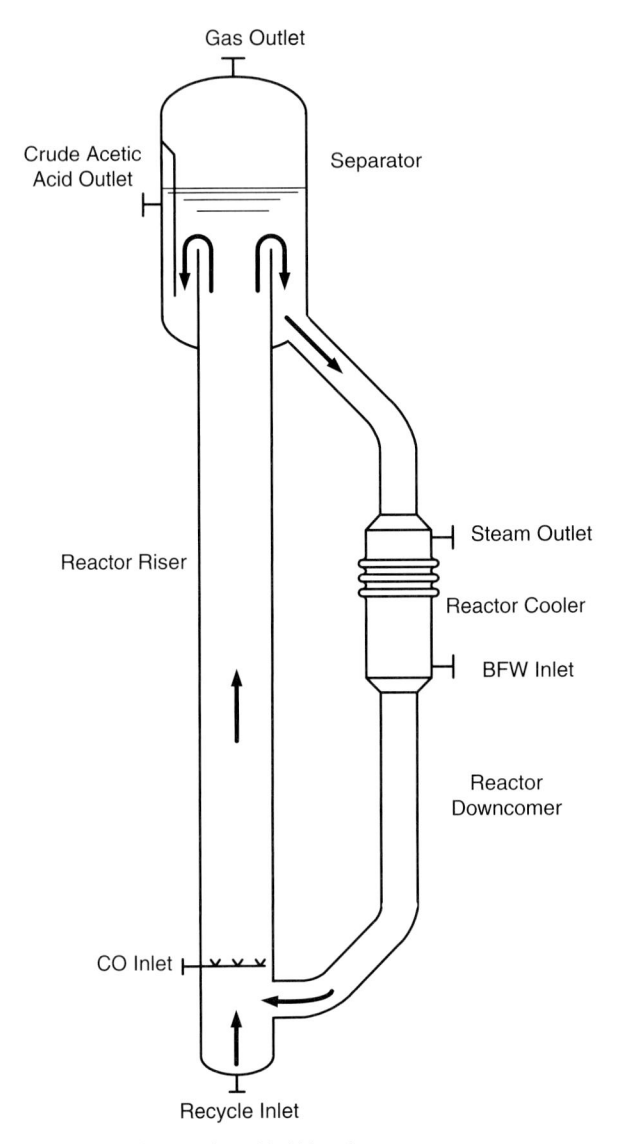

FIGURE 1.1.3 Drawing of bubble-column reactor.

The side stream containing hydrogen iodide, water, and acetic acid is withdrawn from the column. The dehydrated acetic acid from the bottom of the column is sent to the finishing column.

A part of the stream from the dehydration column receiver is fed to the excess water column to remove excess water and some of the accumulated impurities and to maintain a constant concentration of water and impurities in the reactor. The overhead vapor is cooled, and any uncondensed vapor is fed to the LP absorber for recovery. The water from the bottom of the column is also returned to the carbonylation reactor.

The dehydrated acetic acid is fed to the finishing column, where the heavy by-products (predominantly propionic acid) are removed with small amounts of acetic acid in the bottom

TABLE 1.1.1 Typical Product Quality

Property	Value	Unit
Purity	>99.9	Mass %
Water	<0.1	Mass %
Formic acid	<0.05	Mass %
Aldehydes	<0.005	Mass %
Propionic acid	<100	Mass ppm
Iodide	<3	Mass ppb
Specific gravity	1.049–1.056	(20°C/4°C)
Distillation	117.5–119.0	°C
Color	<10	APHA

draw-off and sent to the incinerator. The purified product acetic acid drawn from the middle of the upper distillation section is cooled and sent to the product-treating unit. The overhead acetic acid, containing small amounts of water and light organic compounds, is returned as a recycle stream. Small amounts of uncondensed light gaseous components containing traces of thermally cracked by-products are removed as vent gas to the waste-treating unit.

Iodide-Removal Unit

Product acetic acid still contains traces of iodide compounds, which might harm the catalyst in a vinyl acetate plant. In this section, any remaining traces of iodide compounds are adsorbed by a proprietary adsorbent to reduce the level of the iodide contents to less than 3 ppb.

PRODUCT SPECIFICATIONS

The typical expected product quality is listed in Table 1.1.1.

The Chiyoda ACETICA process is capable of processing a wide range of feedstock specifications, including optimization of a carbon monoxide production and/or purification system. This is done in conjunction with the optimization of a total methanol/CO/acetic acid complex including the operation parameters for the ACETICA process. Product specifications can be adjusted to meet downstream requirements.

PROCESS YIELD AND EMISSIONS

The expected raw materials and major utility consumption per ton of product acetic acid are listed in Table 1.1.2. All waste gases and liquids are processed by the incinerator, and pollution-causing materials are converted to nonpoisonous compounds.

ECONOMICS OF THE CHIYODA ACETICA TECHNOLOGY

The estimated annual production cost of the Chiyoda ACETICA process based on the yield of a 200,000-t/yr plant located in Asia is summarized in Table 1.1.3.

TABLE 1.1.2 Raw Materials and Major Utilities Consumption

Raw materials/utilities	Consumption
Methanol	0.538 (t/t)
Carbon monoxide	0.510 (t/t)
Electricity*	129 (kWh/t)
Cooling water*	137 (m^3/t)
Steam	1.6 (t/t)

*Including CO compressor.

TABLE 1.1.3 Estimated Annual Production Cost

Description	Cost (US$/t product)
Raw materials	237
Utilities	38.9
Capital-related operation cost	20.0
Labor-related operation cost	3.0
Capital-related cost (depreciation, etc.)	55.6
Total production cost	354.5

SCOPE OF CHIYODA'S PACKAGE OF SERVICES

Chiyoda is an engineering company able to supply total EPC package services as well as the ACETICA license. Chiyoda's scope of services ranges from licensing through to EPC services to assist clients in successfully constructing projects under a single point of responsibility, including the following:

- Open license of ACETICA technology
- Basic engineering package
- Detailed engineering
- Procurement of equipment and materials
- Construction
- Training and startup assistance
- Supply of proprietary resin
- Arrangement of rhodium supply/recycle service
- Financing arrangements for EPC project

EXPERIENCE

In 1999, Chiyoda built a pilot ACETICA plant in China with a capacity of approximately 10 t/yr and successfully demonstrated long-term continuous operation. In addition, Chiyoda conducted a cold-flow test of the bubble-column reactor system together with CFD modeling to confirm stable operation of the reactor system. As a result of these demonstration projects, the Chinese authority has approved and accepted Chiyoda's noncommercial process technology for a client in China. The Chiyoda ACETICA process has the commercial experience listed in Table 1.1.4.

TABLE 1.1.4 Chiyoda ACETICA Process Commercial Experience

Client	Country	Capacity	Contract year
CMC International Tendering Company/Guizhou Crystal Organic Chemical Group Co., Ltd.	China	36,000 t/yr	2002

REFERENCES

1. Yoneda, N., Minami, T., Hamato, K., Shiroto, Y., Hosono, Y., *J. Jpn. Petrol. Inst.* 46(4):229, 2003.
2. Yoneda, N., Minami, T., Shiroto, Y., Yasui, M., Matsumoto, T., Hosono, Y., *J. Jpn. Petrol. Inst.* 46(4):240, 2003.

PART · 2

ANILINE

CHAPTER 2.1
DUPONT/KBR ANILINE PROCESS

Eric W. Wong and Ronald Birkhoff
Kellogg Brown & Root, Inc. (KBR)
Houston, Texas

INTRODUCTION

The production of aniline by the catalytic hydrogenation of mononitrobenzene (MNB) is the dominant production route practiced today and accounts for over 90 percent of the aniline production in commercial-scale production units. Kellogg Brown & Root, Inc. (KBR), is the exclusive licensor of the DuPont aniline technology.

Aniline was first produced commercially for the dyestuff industry in the 1850s by the reduction of MNB with iron in the presence of aqueous HCl (the Béchamp process). In the 1930s, aniline was first produced by the catalytic hydrogenation of MNB. This route has proven to be the most economical for the large-scale production of aniline and has displaced other production routes. A small portion of aniline is produced via the ammonolysis of phenol. This production route is economical for specific integrated production sites with phenol feedstock availability and a need for the coproduct produced but has not received wide acceptance. Today, over 90 percent of the aniline production is from the catalytic hydrogenation of MNB.

Since over 95 percent of MNB is used for the production of aniline, almost all aniline production is integrated with MNB manufacture. From the 1930s, MNB has been produced commercially by the batch nitration of benzene with nitric acid in the presence of sulfuric acid. By the 1950s, most of the large-scale aniline plants were integrated with a continuous MNB plant. Isothermal nitration technology was the principal commercial process until the 1980s, when adiabatic nitration and dehydrating nitration were commercialized, resulting in reduced energy consumption associated with the reconcentration of the spent sulfuric acid. Today, all modern large-scale MNB plants have adopted either one of these two production technologies.

DuPont has over 60 years of experience in MNB and aniline production. In the 1940s, DuPont pioneered the concept of adiabatic nitration of benzene to MNB, and in the 1950s, DuPont developed a proprietary noble metal catalyst for the liquid-phase hydrogenation of MNB to aniline. DuPont has built and operated two continuous MNB/aniline plants. The first plant began operation in the early 1960s at Gibbstown, New Jersey, and the second plant was started up in Beaumont, Texas, in the early 1970s. This technology has been refined and optimized over the years to become the most efficient aniline production process in the industry. Aniline technology has been improved steadily in the areas of process yields, energy consumption, waste minimization, product quality, and higher operating efficiency. These process improvements have resulted in reduced capital cost and a lower cost of production.

ANILINE MARKET OVERVIEW

The worldwide consumption of aniline was approximately 3 million tons in 2003. An overall profile of the world's consumption of aniline by end use is shown in Fig. 2.1.1.

About 80 percent of the aniline produced is for the production of MDI (p,p-methylene diisocyanate), a key intermediate in the production of polyurethanes. MDI-based polyurethanes are used in rigid and semirigid foams, elastomers, coatings, and resins. The polyurethane products are used primarily in the construction, housing, appliance, and automotive industries. The second largest end use of aniline is as an intermediate for rubber-processing chemicals, such as vulcanization accelerators, antioxidants, antiozonants, stabilizers, and inhibitors. Although aniline has been an important intermediate for dyes and pigments for over 150 years, today this represents only a small fraction of the total market. Aniline is also used in agricultural chemical applications such as herbicides, fungicides, and defoliants. Other uses of aniline include specialty fibers, photographic chemicals, pharmaceuticals, amino resins, and explosives.

Western and eastern Europe account for over 45 percent of the worldwide production capacity, whereas the Americas, primarily the United States, account for over 33 percent of the production capacity. The rest of the production capacities are located in China, Japan, Korea, and India. Worldwide aniline consumption is projected to grow by an average of more than 4 percent per year until 2010, primarily due to the anticipated high growth in MDI demand.

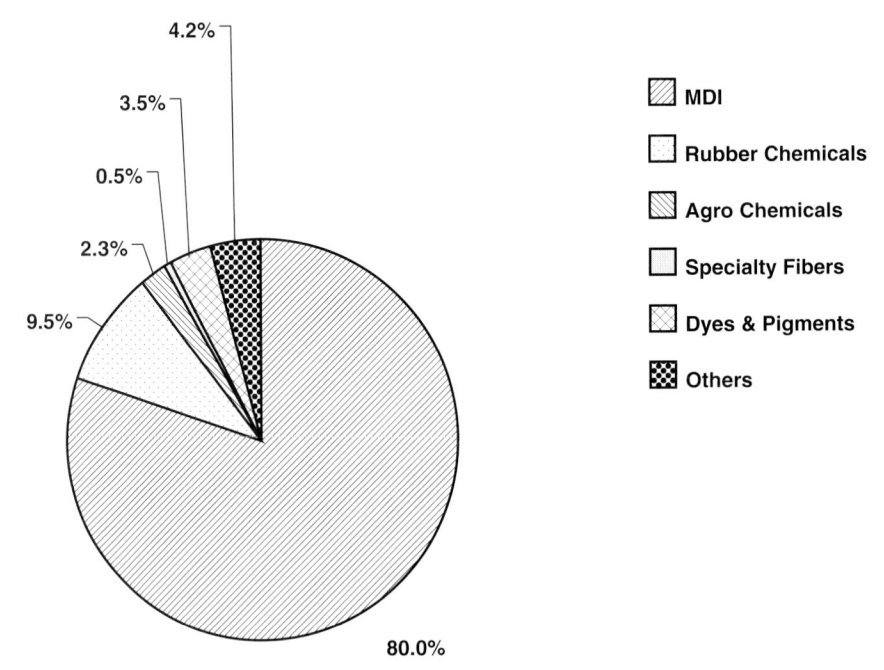

FIGURE 2.1.1 Worldwide end use of aniline.

PROCESS CHEMISTRY

The formation of MNB by the nitration of benzene with nitric acid in a nitric and sulfuric acid medium is a highly exothermic reaction.

Benzene + HNO₃ → (H₂SO₄) → Mononitrobenzene-NO₂ + H₂O

Benzene   **Nitric Acid**   **Mononitrobenzene**   **Water**

The reaction path produces an intermediate nitronium ion, NO_2^+, that is produced from nitric acid in the presence of sulfuric acid. The nitronium ions react with benzene to form MNB. The process yield is high: 98+ percent based on nitric acid and 99+ percent based on benzene. Minor impurities formed in the reaction include dinitrobenzene, dinitrophenols, and trinitrophenol (picric acid).

MNB is catalytically hydrogenated with a noble metal catalyst in the liquid phase to produce aniline. This reaction is also exothermic.

Mononitrobenzene-NO₂ + 3H₂ → (Catalyst) → Aniline-NH₂ + 2H₂O

Mononitrobenzene   **Hydrogen**   **Aniline**   **Water**

DuPont's proprietary catalyst and efficient reaction system achieve 99+ percent yield to aniline from MNB.

PROCESS DESCRIPTION

Mononitrobenzene (MNB) Production

Benzene is nitrated with mixed acid (nitric and sulfuric) at high selectivity and conversion to produce MNB. The nitration product is phase-separated, and the spent acid is reconcentrated by removing the water of reaction. Both the dehydrating nitration and the adiabatic nitration processes use the heat of nitration for reconcentration of the sulfuric acid.

The water produced by nitration of benzene, as well as the water that enters the nitration system with the nitric acid, leaves the nitration/acid-concentration system and is sent to waste treatment. The crude MNB product is washed in a series of contactors to remove residual acid and nitrophenolic impurities formed during the reaction. The product is then distilled, and the unreacted benzene is recovered and recycled. A block flow diagram of the MNB process is shown in Fig. 2.1.2.

Aniline Production

MNB is fed together with hydrogen into a liquid-phase plug-flow hydrogenation reactor that contains a proprietary noble-metal-on-carbon catalyst. The catalyst has a high selectivity, and

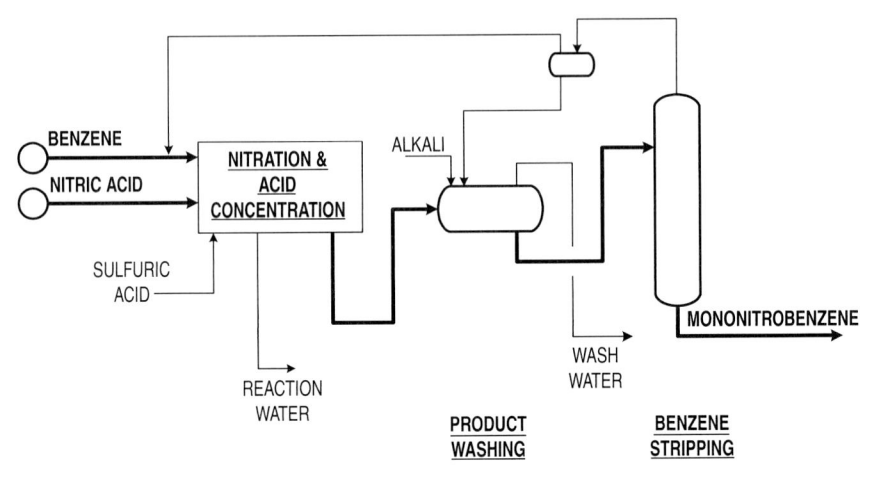

FIGURE 2.1.2 Schematic flow diagram of MNB production process.

the MNB conversion per pass is essentially 100 percent. The reaction conditions are optimized to achieve quantitative yields; i.e., essentially 100 percent of the MNB is converted to aniline in a single pass, and the reactor effluent is free of MNB. A small amount of excess hydrogen from the reactor effluent is vented, and the reactor product is sent to a dehydration column to remove the water of reaction, followed by a purification column to produce high-quality aniline product. A block flow diagram of the DuPont/KBR aniline process is shown in Fig. 2.1.3.

TECHNOLOGY FEATURES

The DuPont/KBR aniline process has been optimized to produce high-quality aniline product at a low manufacturing cost. Key features of this technology are

- *High yields.* The process consistently achieves 99.5 percent efficiency of mononitrobenzene utilization.
- *High-purity product.* Aniline produced from this technology maintains constant high product purity without the variations due to changes in catalyst activity that are typical of vapor-phase fixed-bed hydrogenation technologies. The aniline product from this process is the highest quality in the industry, with less than 0.1 ppmwt MNB.
- *Superior catalyst system.* The proprietary DuPont hydrogenation catalyst and the liquid-phase hydrogenation system provide excellent catalyst life. The DuPont system also avoids the complexity of a catalyst-regeneration system typical of vapor-phase technologies. The high-selectivity catalyst achieves essentially quantitative yields with minimal by-product formation, thus resulting in a very simple product purification system.
- *Low capital cost.* The DuPont/KBR hydrogenation system is mechanically simple and compact. Unlike vapor-phase fluidized-bed or fixed-bed technologies, parallel reactor trains or multiple reactor stages are not required. The DuPont/KBR hydrogenation reactor does not require mechanically complex reactor internals. A single-pass hydrogen system is sufficient for complete conversion of MNB; therefore, large hydrogenation gas recycle systems, typical of vapor-phase technologies, are not required.

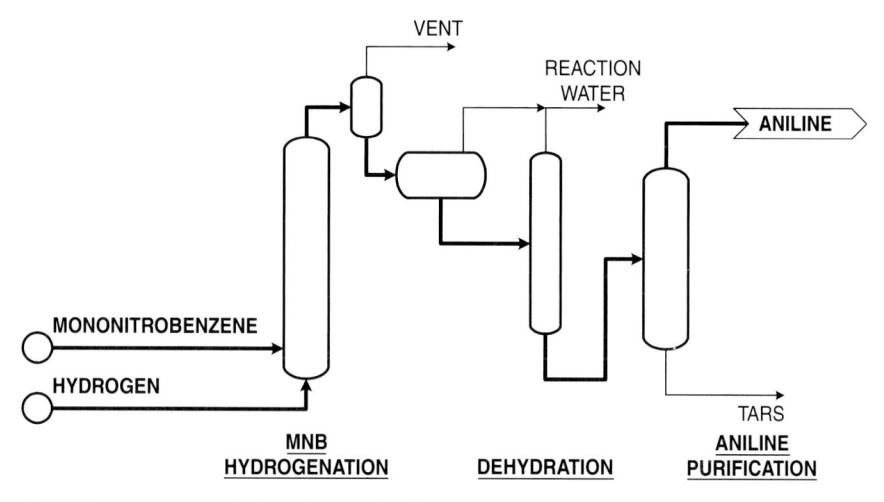

FIGURE 2.1.3 Schematic flow diagram of aniline production process.

- *Low environmental emissions.* The process includes waste-minimization features that reduce the total quantity of process wastewater. Furthermore, the quantity of biologically toxic aqueous discharge is reduced to a practical minimum. The waste management system is designed to meet and exceed the most stringent environmental regulations.
- *High on-stream factor.* Since shutdown for catalyst regeneration is not required, the on-stream time of the aniline plant is very high. DuPont maintains a high on-stream time by preventive maintenance of major equipment, typically achieving a 98+ percent service factor at the Beaumont aniline plant.

OPERATING REQUIREMENTS

The raw materials and utilities consumption for a typical aniline plant based on the DuPont/KBR process are shown in Table 2.1.1.

PRODUCT QUALITY

The DuPont/KBR aniline process consistently produces a very high quality aniline product suitable for all MDI production technologies and other chemical applications. Typical product specifications are as shown in Table 2.1.2.

WASTES AND EMISSIONS

The aqueous effluent from MNB plants contains nitrophenols that are toxic to biologic treatment systems. KBR has developed an integrated aqueous management system that reduces the nitrophenol-containing aqueous effluent. The treatment methods for this par-

TABLE 2.1.1 Raw Materials and Utility Requirements

Raw materials utilization	Per kg aniline product
Benzene (100%)	0.846 kg
Nitric acid (100%)	0.686 kg
Hydrogen	0.067 kg

Utilities	Per kg aniline product
Steam	(0.15 kg) export
Cooling water (9°C rise)	0.1 m^3
Electricity	0.04 kWh

Catalysts and chemicals	Per metric ton aniline product
Chemicals and catalyst	US$4

TABLE 2.1.2 Typical Product Specifications

Product	Specification
Aniline	99.95 wt %
Nitrobenzene	0.1 wt ppm
Water	300 wt ppm
Color, APHA	30
Freeze point (dry basis)	−6.0°C

ticular stream include extraction, incineration, or thermolysis. The resulting aqueous effluent from an MNB/aniline plant can be readily treated in a biologic treatment system.

Depending on local environmental regulations, the vents from an MNB/aniline plant may require incineration and NO$_x$ treatment. The organic heavies from the aniline plant can be either blended into a liquid fuel for the plant utilities or incinerated.

REFERENCES

1. F. R. Lawrence and W. J. Marshall, in W. Gerhartz (ed.), *Ullmann's Encyclopedia of Industrial Chemistry*, Vol. A2, VCH Verlagsgesellschaft, Weinheim, 1985, p. 303.
2. B. Amini, in M. Howe-Grant (ed.), *Kirk-Othmer Encyclopedia of Chemical Technology*, 4th ed., John Wiley & Sons, New York, 1991, p. 426.
3. E. W. Wong and J. E. Wallace, *Efficient Technology for the Production of Aniline*, Second Asian Petrochemicals Technology Conference, Seoul, Korea, May 7–8, 2002.

P · A · R · T · 3

1,3-BUTADIENE

CHAPTER 3.1
BASF BUTADIENE EXTRACTION TECHNOLOGY

Robert Brummer
ABB Lummus Global
Bloomfield, New Jersey

INTRODUCTION

Butadiene (BD) is a major petrochemical commodity used in the production of rubbers and plastics, such as styrene butadiene rubber (SBR), polybutadiene rubber (PR), and styrene butadiene latex (SBL). The main butadiene source is by-product from naphtha crackers, where the butadiene is contained within a mixed C_4 stream and must be separated before use in other processes. Since many of the C_4 compounds in the mixed C_4 stream have relative volatilities very similar to 1,3-butadiene, it would be nearly impossible to achieve the required separation using conventional distillation. However, in some solvents these same compounds have different solubilities compared with 1,3-butadiene, allowing the 1,3-butadiene to be separated by extractive distillation.

In general, the more unsaturated the hydrocarbon (the more double and triple bonds it contains), the more soluble it is in the solvent. Therefore, butanes and butenes are less soluble than 1,3-butadiene. Similarly, C_4 acetylenes are more soluble than butadiene. While methyl acetylene and 1,2-butadiene have solubilities in the solvent similar to 1,3-butadiene, there are sufficient relative volatility differences to use conventional distillation to meet the required specifications for methyl acetylene and 1,2-butadiene in the final 1,3-butadiene product.

PROCESS PERSPECTIVE

The BASF butadiene extraction process is unique because it uses *N*-methylpyrrolidinone (NMP) as solvent. Other technologies use different solvents such as dimethylfornamide

TABLE 3.1.1 Butadiene Extraction Projects

Company	Plant location	Startup or status	Capacity MTA
SECCO (BP/SPC)	Caojing, China	Under design	90,000
SABINA (Shell/BASF/Atofina)	Port Arthur, Texas	Under constr.	410,000
Haldia Petrochemicals, Ltd.	Haldia, India	2000	75,000
BSL Petrochemieverb.	Boelen, Germany	1999	120,000
Indian Petrochemical Co., Ltd.	Boroda, India	1996	37,000
Xinjang Dushanzi Petrochemical	Dushanzi, China	1995	34,000
Beijing Eastern Chemical Works	Beijing, China	1995	29,000
Arabian Petrochemical Co.	Al Jubail, Saudi Arabia	1993	130,000
BASF	Mannheim, Germany	1993	105,000
Huntsman	Port Neches, Texas	1991	430,000
Korea Kumho Petrochemical Co., Ltd.	Yeo Cheon, South Korea	1989	45,000
EniChem Eslatomeri	Ravenna, Italy	1989	120,000
OMV AG	Schwechat, Austria	1983	47,000
Korea Kumho Petrochemical Co., Ltd.	Yeo Cheon, South Korea	1980	75,000
Huntsman Chemical Corp., Ltd.	Wilton, U.K.	1979	100,000
Japan Synthetic Rubber Co., Ltd.	Yokkaichi, Japan	1976	148,000
BP Köln	Köln, Germany	1974	150,000
Basell	Wesseling, Germany	1973	170,000
Petroquimica Unaio S.A.	Capuava, Brazil	1972	75,000
Oxeno	Marl, Germany	1972	180,000
Atofina	Gonfreville, France	1972	68,000
Napththachimie SA	Lavera, France	1972	114,000
Japan Synthetic Rubber Co., Ltd.	Kashima, Japan	1971	120,000
Grupul Industrial de Petrochimie	Petesti, Rumania	1969	18,000
Imperial Chemical Industries, Ltd.	Wilton, U.K.	1968	80,000
BP Köln	Köln, Germany	1968	95,000

(DMF) and acetonitrile (ACN). Table 3.1.1 lists the plants using the BASF butadiene extraction process and the plants currently in design/construction.

Figure 3.1.1 is a simplified block flow diagram of the process. There are three main process steps: extractive distillation, degassing, and distillation. Two stages of extractive distillation result in a raffinate-1 stream containing butenes from the first stage overhead and a crude butadiene stream from the second stage overhead. The solvent (NMP) loaded with acetylenes is recovered in the solvent degassing system and recycled to the extractive distillation system. The crude butadiene is further purified using conventional distillation technology to yield a 1,3-butadiene product of greater than 99.7 percent purity.

PROCESS DESCRIPTION

Figure 3.1.2 is a simplified process flow diagram. The mixed C_4 feed stream is fed into the first extractive distillation column, which produces an overhead butene stream (raffinate-1) that is essentially free of butadiene and acetylenes.

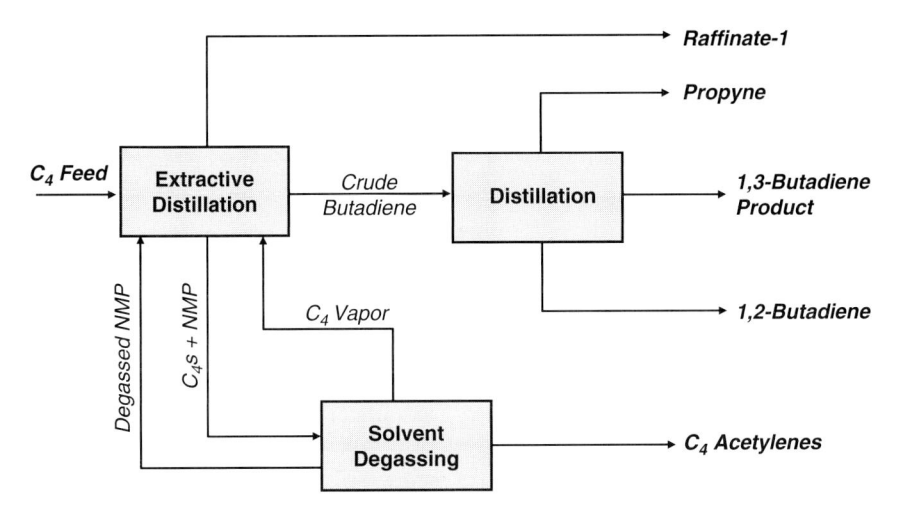

FIGURE 3.1.1 Simplified block flow diagram.

The bottoms stream from this column, containing butadiene, the acetylenic compounds, and some butenes dissolved in the solvent, is stripped free of butenes in the top half of the rectifier. A side stream containing butadiene and small amount of acetylenic compounds (vinyl and ethyl-acetylene) is withdrawn from the rectifier and fed into the second extractive distillation column. The C_4 acetylenes, which have higher solubilities in NMP than 1,3-butadiene, are removed by the solvent in the bottoms and returned to the top of the lower section of the rectifier. A crude butadiene stream from the overhead of the second extractive distillation column is fed into the propyne column, the first distillation column of the butadiene purification train.

Both extractive distillation columns have a number of trays above the solvent addition point to allow for removal of solvent traces from the overheads (C_4 hydrocarbons). The bottoms of the rectifier, containing butadiene, C_4 acetylenes, and C_5 hydrocarbons in NMP, are preheated and fed into the degasser (solvent stripping column). In this column, solvent vapors are used as the stripping medium to remove all hydrocarbons from NMP.

The hot-stripped solvent from the bottom of the degasser passes through the heat economizers (a train of heat exchangers) and is fed to the extractive distillation columns. A slipstream from the bottoms of the degasser is sent for solvent purification via batch vacuum distillation in an agitated kettle to prevent the buildup of high-boiling polymeric residues.

The hydrocarbons leaving the top of the degasser are cooled down in a column by direct contact with solvent (NMP and water) and fed to the bottom of the rectifier via a recycle gas compressor. Hydrocarbons having higher solubilities in the solvent than 1,3-butadiene accumulate in the middle zone of the degasser and are drawn off as a side stream. This side stream, after dilution with raffinate-1, is fed to a water scrubber to remove a small amount of NMP from the exiting gases. The scrubbed gases, containing the C_4 acetylenes, are purged to disposal, whereas the aqueous bottoms of the scrubber are returned to the middle zone of the degasser.

In the propyne column, the propyne (C_3 acetylene) is removed as overhead and sent to disposal. The bottoms are fed to the second distillation column (the 1,3-butadiene column), which produces pure butadiene as overhead and a small stream containing 1,2-butadiene and C_5 hydrocarbons as bottoms.

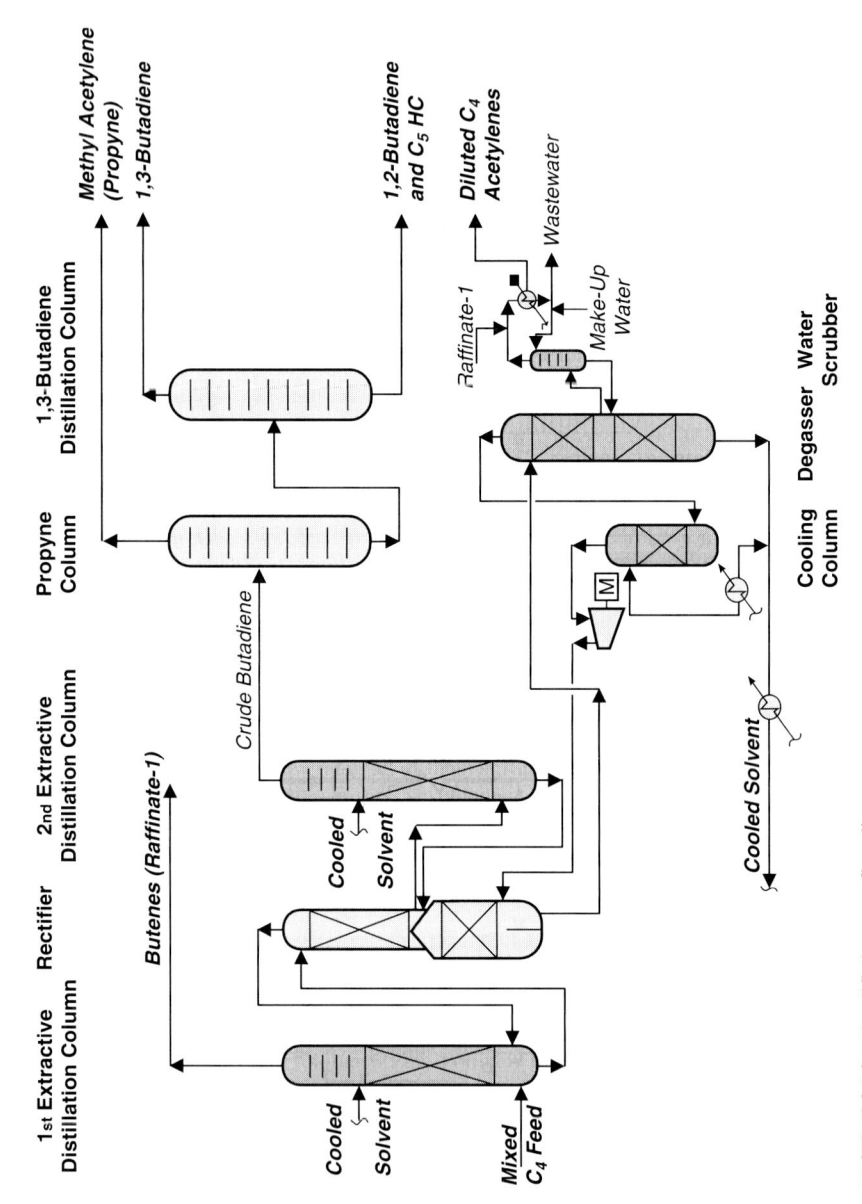

FIGURE 3.1.2 Simplified process flow diagram.

ECONOMICS

Feedstocks

Typical feed composition ranges for C_4 streams from naphtha crackers are shown in Table 3.1.2. By adjusting operating parameters such as solvent flow, reflux rates, and product draw-off rates, a well-designed butadiene unit can efficiently handle C_4 feeds across this range.

Product Quality and Yields

Typical product specifications for 1,3-butadiene produced in a BASF butadiene extraction unit are shown in Table 3.1.3. Usually more than 98 percent of the 1,3-butadiene contained in the feed is recovered as product, but this can vary slightly with the feed composition.

The butenes (raffinate-1) from the first extractive distillation column overhead generally will contain less than 0.2 wt % 1,3-butadiene and less than 10 wt ppm solvent.

Utility Consumption

The overall normal utility requirements will depend on equipment selection (e.g., compressor, pump efficiencies, etc.). Table 3.1.4 shows the range of utility requirements per metric ton of 1,3-butadiene product.

Chemical Consumption

The approximate costs of solvent makeup and chemicals consumed per metric ton of 1,3-butadiene product while processing a typical feedstock are summarized in Table 3.1.5.

TABLE 3.1.2 Typical C_4 Feed Specifications

1,3-Butadiene	35–55 wt %
Butanes	Up to 14 wt %
Butenes	40–51 wt %
Total acetylenes	Up to 2 wt %

TABLE 3.1.3 Typical Butadiene Produce Quality

1,3-Butadiene	99.7 wt %
Propadiene	<5 wt ppm
1,2-Butadiene	<20 wt ppm
Acetylenes	<20 wt ppm
NMP	<5 wt ppm

TABLE 3.1.4 Utilities per Metric Ton Product

Steam (medium pressure)	1.5–2.5 t
Cooling water (based on 25–30°C temperature rise)	100–200 m^3
Electrical energy	100–200 kWh

TABLE 3.1.5 Cost of Chemicals Consumed per Metric Ton Product

Solvent (NMP) makeup	$0.60
Other chemicals	$2.50

ENVIRONMENTAL CONSIDERATIONS

There are three categories of effluents from a butadiene extraction unit: hydrocarbons, wastewater, and solvent residue.

Intermittent hydrocarbon vents generally are sent to the flare or to an OSBL waste-gas incinerator, depending on local regulations. The two continuous hydrocarbon streams—one is a mixture of methylacetylene and 1,3-butadiene and the other is a mixture of C_4 acetylenes diluted with mixed C_4's—are always handled carefully because they contain acetylenes. They can be sent to a flare or an OSBL waste-gas incinerator depending on local regulations. In some plants, they are diluted with fuel gas and used as fuel. Alternatively, the acetylenes can be hydrogenated, and the stream can be mixed with the raffinate by-product.

A portion of the wastewater is recycled to the process, reducing the total amount of effluent. The balance is sent to a stream stripper, where the hydrocarbons are removed. The net wastewater, which contains traces of NMP solvent, is sent to conventional biologic treatment. An advantage of the NMP solvent is that it can be treated easily by biologic degradation.

The quantity of solvent residue is reduced by concentrating it in an ISBL semicontinuous still. Intermittently, the concentrated residue, which is made up of inorganic salts, polymers, and NMP, is diluted with water and discharged to containers that are transported to an incinerator.

There are some special considerations in any butadiene extraction plant regarding safety and polymer formation. The safety concern mostly deals with the issue of handling streams containing C_3 and C_4 acetylenes. In the BASF butadiene extraction process, these acetylene-containing streams are always kept diluted with other hydrocarbons so that the concentration of acetylenic compounds is in a safe range. This dilution is maintained by the operating parameters of the plant and by process controls, alarms, and shutdown systems.

Polymer formation is minimized by the use of inhibitors and by special design considerations. Pretreatment of the plant for rust and oxygen removal is important for polymer prevention. Oxygen must be kept out of the plant at all times during operation because even small amounts can increase the likelihood of polymer formation. Plants are designed to minimize stagnant areas in piping and equipment and to allow flushing of stagnant areas where they cannot be avoided.

SUMMARY OF PROCESS FEATURES

- NMP is a nontoxic, noncarcinogenic solvent that is treated easily by conventional biologic degradation. The maximum permissible concentration in air is higher than that for comparable solvents. Moreover, due to its low vapor pressure, only very low concentrations occur in air.
- Unsafe accumulations of acetylenic compounds do not occur in any part of the plant, even in the event of a disturbance.

- All industrial C_4 hydrocarbon mixtures can be processed regardless of their butadiene content, and a high-purity butadiene product is obtained with higher recovery.
- NMP is miscible with water in all proportions. The solvent used in butadiene extraction is a mixture of NMP and water. The degassing temperature therefore can be kept low.
- NMP and its mixtures with water are not corrosive, so no corrosion inhibitor is required. Carbon steel is used for all equipment and materials.
- Due to the good stability of NMP to hydrolysis and thermal decomposition, and because of NMP's high boiling point, solvent losses can be kept very low. NMP traces in the wastewater can be degraded easily in a biologic treatment plant.

CHAPTER 3.2
UOP KLP 1,3-BUTADIENE FROM ACETYLENE PROCESS

Steve Krupa, Tim Foley, and Stephen McColl
UOP LLC
Des Plaines, Illinois

INTRODUCTION

The presence of dienes and/or acetylenes in light olefinic streams is often undesirable, and these reactive contaminants must be removed without affecting the concentrations of desired components in the stream. The dienes and/or acetylenes can be removed by selective hydrogenation to produce only the desired product, typically a diene/acetylene-free olefinic product or an acetylene-free diene product.

Light olefin streams are produced by steam cracking, dehydrogenation of paraffins, or fluid catalytic cracking (FCC). Hydrogenation of dienes and/or acetylenic compounds in these feedstocks can be accomplished selectively under mild conditions. UOP offers the KLP™ process to selectively remove acetylenes from crude butadiene feedstocks. For the selective hydrogenation of dienes, and the Huels selective hydrogenation process licensed by UOP, see Chapter 8.2 of the *Handbook of Petroleum Refining Processes,* 3d ed., edited by Robert A. Meyers (McGraw-Hill, 2004).

BUTADIENE

Butadiene is a major petrochemical commodity. The compound of commercial interest is 1,3-butadiene, CH_2=CHCH=CH_2. It is a colorless gas normally stored and handled as a liquid under refrigeration and pressure. 1,2-Butadiene has little commercial importance and is a minor component in cracking unit C_4's. It is removed from 1,3-butadiene by distillation. Most of the world's butadiene is produced as a coproduct of ethylene that is obtained by the steam cracking of naphtha (C_5–C_8). Smaller amounts are produced in the United States by the oxidative dehydrogenation of *n*-butenes. Because of its two double bonds, the butadiene molecule is very reactive and readily forms polymers. A large fraction of butadiene production is consumed in the manufacture of styrene-butadiene rubber (SBR) and other elastomers, most of which are used in tires or other products associated with the automobile or appliance industry.

THE KLP PROCESS

The KLP process was originally developed and commercialized by the Dow Chemical Company in 1984 as an economic means to avoid safety issues associated with processing acetylene-rich streams. In the KLP process, acetylenes in the crude butadiene stream are selectively hydrogenated to their corresponding diene or olefin under mild conditions in the liquid phase. By coupling the KLP process with the first stage of extractive distillation, the second stage of extractive distillation (acetylene separation from butadiene) is eliminated. UOP acquired the KLP process and technology from Dow in 1990 and now offers it for licensing on a worldwide basis. These commercial operations include both new units and revamps.

The C_4 cut from a steam cracking unit contains up to 60 percent butadiene as well as small amounts (typically 0.5 to 2.0 wt %) of C_4 acetylenes that need to be separated from the main butadiene product. Acetylenic compounds, such as vinyl acetylene and ethyl acetylene, can be selectively removed using the KLP process prior to butadiene extraction. The KLP process is used to convert essentially 100 percent of the *alpha*-acetylenes to mono-olefins and butadiene. The KLP process is highly selective, and there is no yield loss of butadiene; in fact, there is typically a yield gain. With acetylenes removed, the extraction of butadiene can be accomplished in a single-stage unit. The KLP process can be readily integrated into existing extraction units to increase capacity or for debottlenecking of the existing extraction unit.

Acetylene removal from butadiene streams is required to meet strict polymer-grade 1,3-butadiene specifications. Alternative technologies remove acetylenes via extractive distillation.

Stand-alone two-stage extraction processes have a lower butadiene yield than those including the KLP process. In the stand-alone extraction unit, 1,3-butadiene is lost via the drag streams used to remove acetylenes (see Fig. 3.2.1). The KLP process eliminates this drag steam and converts undesired acetylenes into the desired 1,3-butadiene product. Additionally, incorporation of the KLP process reduces utility costs of 1,3-butadiene production and avoids the need to handle concentrated acetylene drag streams. The concentrations and temperatures of these acetylene-rich streams must be controlled carefully to prevent potential explosions.

The UOP KLP process is a simple fixed-bed catalytic process that selectively hydrogenates *alpha*-acetylenic compounds in crude butadiene streams. Extremely high selectivity yields essentially 100 percent conversion of acetylenes with no net loss of butadiene by hydrogenation.

PROCESS CHEMISTRY

Liquid feed, free of catalyst poisons, is charged to the reactor, where the acetylenes are removed over a fixed bed of catalyst. A two- or three-reactor system can be used. In the two-reactor system, one is online and one is in standby or regeneration. In the three-reactor system, two reactors are online in series, with the third being regenerated. The reactors are operated at moderate pressures and low temperatures to ensure H_2 solubility and to

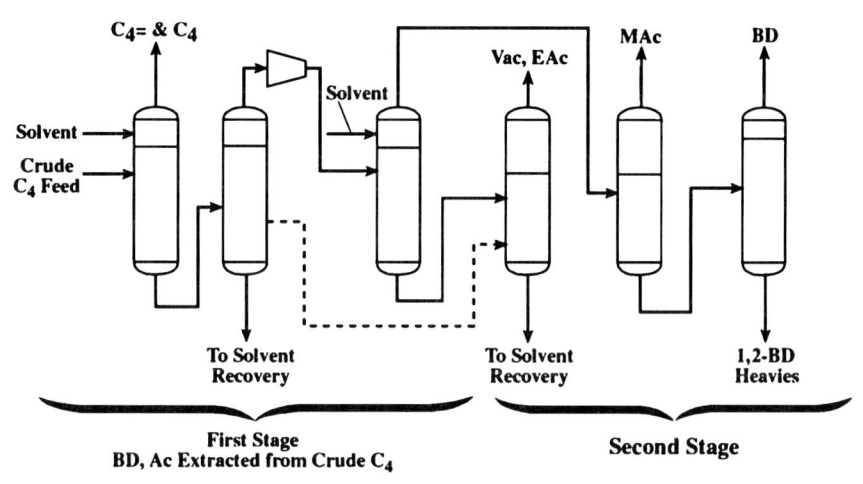

FIGURE 3.2.1 Generic two-stage butadiene extraction unit.

avoid thermal polymerization reactions. The reactor effluent is fractionated to remove any small amounts of green oil present, and the product typically is sent to a butadiene extraction unit. A flow diagram of the KLP process integrated with a butadiene extraction unit is shown in Fig 3.2.2 on p. 3.14.

The main KLP reactions are presented in Fig. 3.2.3 on p. 3.14. The reactions take place in the liquid phase over a fixed bed of catalyst. A modern KLP unit has two reactors online in series. Temperature, pressure, and the H_2 addition are all controlled to promote complete acetylene conversion while minimizing side reactions. The KLP-60™ catalyst is a copper-nickel formulation on a spherical alumina support.

COMMERCIAL EXPERIENCE

A total of nine KLP units have been licensed, and eight are in operation. The first unit came on-stream in 1984. The total licensed capacity for KLP is 850 kilometric tons per annum (kMTA).

A new catalyst rejuvenation procedure was commercialized in 2003 that has further reduced unit costs and simplified operation. The new rejuvenation procedure uses a solvent-wash design. This eliminates the need for a fired heater and associated equipment otherwise required with a carbon-burn catalyst regeneration design.

ECONOMICS AND OPERATING COSTS

The utility requirements for a KLP unit processing 200 kMTA are summarized in Table 3.2.1 on p. 3.14.

3.14   1,3-BUTADIENE

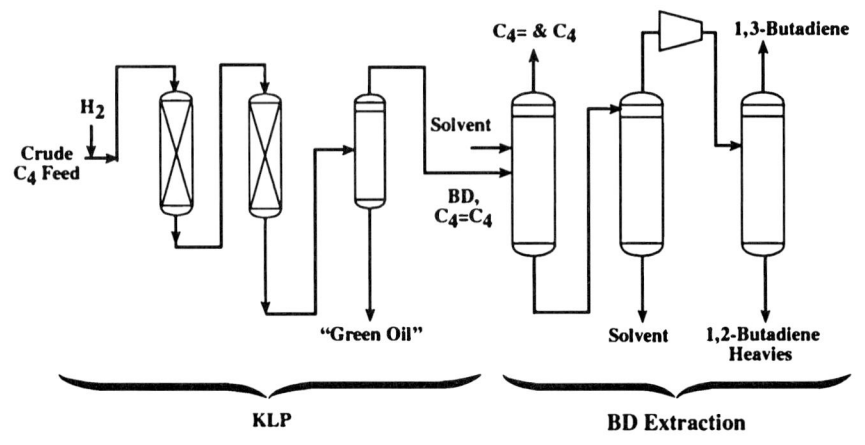

FIGURE 3.2.2 Integration of KLP unit with one-stage butadiene extraction unit.

$$H\text{-}\underset{\underset{H}{|}}{\overset{\overset{H}{|}}{C}}\text{-}C\equiv C\text{-}H + H_2 \longrightarrow H\text{-}\underset{\underset{H}{|}}{\overset{\overset{H}{|}}{C}}\text{-}\overset{\overset{H}{|}}{C}=\overset{\overset{H}{|}}{C}\text{-}H_2$$

Methyl Acetylene   **Propylene**

Ethyl Acetylene → 1-Butene

Vinyl Acetylene → 1,3-Butadiene

FIGURE 3.2.3 Acetylene hydrogenation reactions in the KLP process.

TABLE 3.2.1 Butadiene Extraction Economics

	KLP/one-stage	Two-stage extraction
Crude C_4 feed, t/yr	200,000	200,000
Acetylenes, t/yr	2,000	2,000
Butadiene, t/yr	92,500	92,500
Other C_4's, t/yr	105,500	105,500
Butadiene loss in drag streams/fractionation, t/yr	(1,000)	(2,500)
Net butadiene product, t/yr	93,000	90,000
Butadiene product value, $US MM/yr @ $420/t	39.1	37.8
Fixed investment, ISBL, $US MM	26	28
Total investment cost (TIC), $US MM	37	40
Utilities, $US MM/yr	2.4	2.7
Other operating costs, $US MM/yr	3.4	3.6

PART 4

CUMENE

CHAPTER 4.1
ABB LUMMUS GLOBAL CUMENE PRODUCTION VIA CD*CUMENE*® TECHNOLOGY

Stephen Pohl and Sanjeev Ram
ABB Lummus Global
Bloomfield, New Jersey

INTRODUCTION

Cumene processes were developed originally between 1939 and 1945 to meet the demand for high-octane aviation gasoline during World War II. Today, almost all the world's supply of cumene is used to produce phenol and its coproduct acetone. Worldwide cumene production in 2002 was about 8.5 million metric tons per annum (MTA) and is expected to reach 10 million MTA by 2006 (according to CMAI). Many consumer and industrial products, such as plywood and composition board bonded with phenolic resins, nylon-6, epoxy, and polycarbonate resins and solvents, have origins that can be traced back to cumene.

High-purity cumene is produced by the catalytic alkylation of benzene with propylene in either the liquid or vapor phase. Zeolite catalyst technology is now used in almost all new cumene plants, replacing both vapor-phase solid phosphoric acid (SPA)– and liquid-phase aluminum chloride–catalyzed processes. The SPA technology has two major limitations: Cumene yield is limited to about 95 percent, and the catalyst cannot be regenerated and so must be discarded. Aluminum chloride–based plants generate waste streams, suffer from corrosion, and make a less pure cumene product. Zeolite catalyst technology is also being used to revamp/replace these technologies in existing plants because zeolites are noncorrosive, regenerable, environmentally friendly, and give both a high yield and a high-purity product.

The CD*Cumene*® process from CDTECH, a partnership between ABB Lummus Global, Inc., and Chemical Research & Licensing, a CRI International company, uses proprietary zeolite catalyst packaged in a patented catalytic distillation (CD) structure (catalyst bales). The CD*Cumene* process creates no hazardous or acidic effluents and consumes no chemicals other than feedstocks.

PROCESS PERSPECTIVE

The CD*Cumene* process is based on the unique catalytic distillation system developed by CDTECH that combines reaction and fractionation in a single-unit operation. The application of catalytic distillation technology to cumene production allows the alkylation reaction to take place isothermally and at low temperature. Reaction products are removed continuously from the reaction zones by distillation, resulting in high selectivity to cumene. These factors limit the formation of by-product impurities, enhance product purity and yields, and result in expected reactor run lengths of up to 6 years. The exceptional quality of the cumene product—attained without clay treatment—easily surpasses current requirements of phenol producers.

In general, the catalytic distillation system has wide applications to various processes and has 20 years of proven commercial application for the production of MTBE. The same concept has been applied and tailored specifically for the production of cumene and ethylbenzene. One world-scale cumene plant using the CD*Cumene* process and three world-scale ethylbenzene plants using the CDTECH *EB*® process are currently operating. The first commercial CD*Cumene* plant started up in April 2000, as shown in Table 4.1.1.

PROCESS CHEMISTRY

The manufacture of cumene by the CD*Cumene* process starts with the alkylation of benzene with propylene to yield a mixture of alkylated and polyalkylated benzenes. The mixture is then processed to recover cumene (isopropylbenzene) in a distillation train. The polyalkylated benzenes are recovered and transalkylated with benzene for maximum cumene yield.

Alkylation reactions:

$$\underset{\text{Benzene}}{C_6H_6} + \underset{\text{Propylene}}{C_3H_6} = \underset{\text{Cumene}}{C_6H_5CH(CH_3)_2}$$

$$\underset{\text{Cumene}}{C_6H_5CH(CH_3)_2} + \underset{\text{Propylene}}{C_3H_6} = \underset{\text{Diisopropylbenzene}}{C_6H_4[CH(CH_3)_2]_2}$$

Transalkylation reaction:

$$\underset{\text{Diisopropylbenzene}}{C_6H_4[CH(CH_3)_2]_2} + \underset{\text{Benzene}}{C_6H_6} = \underset{\text{Cumene}}{2C_6H_5CH(CH_3)_2}$$

Based on the reaction stoichiometry, 1 t of cumene is produced from 0.65 t of benzene and 0.35 t of propylene.

PROCESS DESCRIPTION

Figure 4.1.1 is a simple flow schematic of the CD*Cumene* process.

Alkylation

The CD alkylation reaction system consists of two reaction stages: the alkylator and the finishing reactor. Propylene is introduced as a vapor at the bottom of the reaction zone in the

TABLE 4.1.1 CDTECH Cumene Awards

Client/location	Design capacity, MTA	Startup date
Formosa Chemicals and Fibre Corp., Taiwan (revamp)	540,000	2004
Formosa Chemicals and Fibre Corp., Taiwan	270,000	2000
AO Orgsteklo, Russia	170,000	On hold

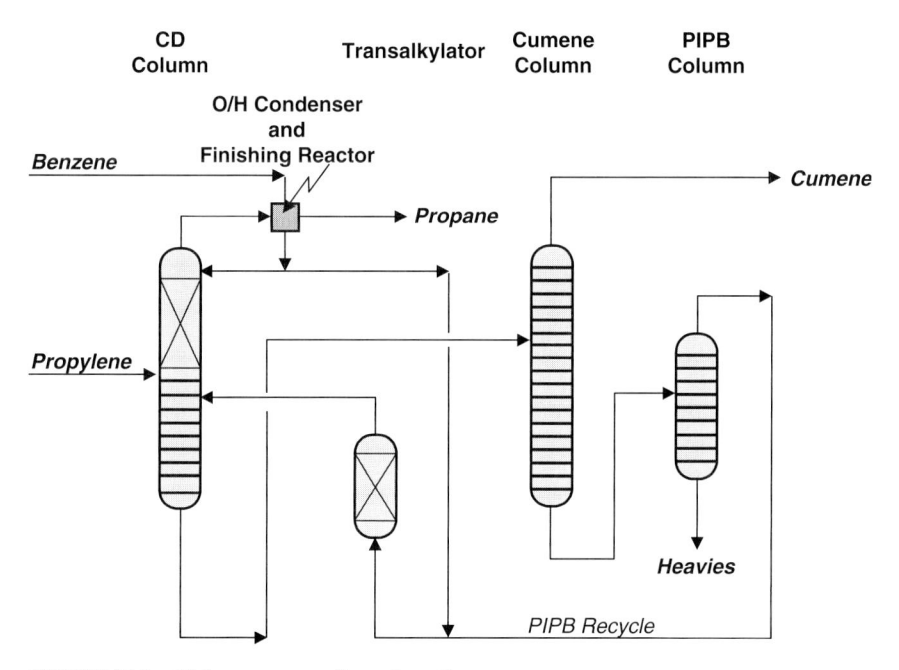

FIGURE 4.1.1 CD*Cumene* process flow schematic.

upper section of the CD column; benzene is fed in as a liquid above the reaction zone. Typical feedstock compositions are shown in Table 4.1.2. The benzene is alkylated with propylene to form cumene over a zeolite catalyst packaged in a special manner to allow the reactor to operate in two phases, vapor and liquid. Successive alkylations occur to a small extent, producing some diisopropylbenzene (DIPB) and very small amounts of higher isopropylated benzenes. Other coupling reactions occur to a minor extent, yielding high-boiling materials. The reactions occur in the catalytic zone of the column, whereas distillation occurs throughout the column, resulting in a countercurrent flow of vapor and liquid throughout the column.

Reaction products are removed continuously from the catalytic zone by distillation, whereas any unconverted reactants and other lights are taken overhead. The heat of reaction is immediately removed by vaporizing benzene, which is significant in that it provides a near-isothermal operation at the optimal reaction temperature. As propylene vapor is injected into the reactor, it comes in contact with the liquid benzene coming from above and is absorbed in the liquid phase. At equilibrium, most of the propylene is present in the vapor phase. As the small amount of propylene in the liquid comes in contact with the catalyst, it immediately reacts to form cumene. This moves the propylene concentration in the liquid phase away from equilibrium. The vapor-liquid equilibrium then causes propylene to be "injected" from the vapor phase into the liquid phase, thus restoring the equilibrium.

TABLE 4.1.2 Typical Feedstock Specifications

Benzene	
Sp. gr. 15.5/15.5°C	0.882–0.886
Solidification pt.	5.45°C min. (anhydrous basis)
Distillation range (including 80.1°C)	1.0°C max.
Benzene purity	99.9 wt % min.
Nonaromatics	1500 wt ppm max.
Toluene	500 wt ppm max.
Chloride	1 wt ppm max.
Total sulfur	1 wt ppm max.
Water	Saturated; no free water
Propylene	
Grade	Chemical/polymer
Propylene	95–99.9 mol % typical
Ethylene	0–10 mol ppm typical
Butylenes	0–50 mol ppm typical
Acetylenes and dienes	10 mol ppm max.
Propane	Balance
Total sulfur	1 wt ppm max.
Arsine	Not detectable
Other basic compounds	Not detectable
Water	No free water

Note: These specifications may be relaxed, if required. Refinery-grade propylene or dilute propylene streams with purity as low as 10 mol % can be used provided the content of other olefins and related impurities is within specifications.

All the alkylation reactions are highly exothermic. The heat of reaction is efficiently removed by the vaporization of benzene so that the alkylation can proceed at isothermal condition while maintaining high catalyst productivity. As a result of the isothermal operating condition, the selectivity to cumene is high. The internal recycle of benzene required for a high overall yield therefore is minimized.

The proprietary CD*Cumene* catalyst bales containing the alkylation catalyst are stacked on top of each other inside the alkylator like structured packing, enabling the alkylation reaction and the distillation of reactants and products to take place simultaneously. Several beds of stacked bales are used. The high-activity zeolite catalyst is contained in glass fiber, which is then rolled in a wire mesh into a spiral configuration. The unique structure of the bales gives them the required void fraction for the vapor to flow upward through the reactor. The glass-fiber packing acts as a barrier and prevents the vapors from coming in direct contact with the catalyst. The bales are about 1 ft in diameter and height and are easily handled during loading operations.

Light alkanes and noncondensables, introduced with the propylene feed, are distilled with benzene in the CD column overhead. Benzene is condensed and returned to the CD column as reflux. Light gases are recovered either as propane product or as fuel. The CD column bottom section strips benzene from cumene and heavier, and the benzene is returned to the reaction zone. The bottoms stream from the stripping section of the CD column, consisting mainly of cumene and polyisopropylbenzene (PIPB), is sent to distillation.

Most of the propylene is reacted in the alkylator, and any unconverted propylene contained in the alkylator overhead effluent is then fully converted in the finishing reactor. The finishing reactor contains a fixed bed of loose zeolite catalyst and operates adiabatically in the liquid phase. Since the propylene flow to the reactor is small, the benzene-to-propylene ratio is very high, resulting in excellent catalyst selectivity and stability.

TABLE 4.1.3 Typical Cumene Product Specifications

Purity	99.95 wt % typical
Sp. gr., 20/20°C	0.864 min.
Distillation range (including 152.5°C)	1°C max.
Corrosion, copper	No iridescence, black or gray
Bromine index	2 typical
Appearance	Clear water white liquid
Color, Pt-Co scale	15 max.
Color, APHA	10 max.
Acid wash color	2 max.
Normal propylbenzene	250 ppm max.
Light aliphatics	Not detectable
Phenolics	Not detectable
Cumene hydroperoxide	Not detectable
Peroxides	Not detectable
Sulfur content	Not detectable (dependent on feed sulfur content)

The zeolite catalysts employed in the process are highly active, selective, and stable, resulting in low catalyst volume and long run lengths of 2 to 6 years (proven in commercial plants).

Transalkylation

A mixture of the di- and triisopropylbenzenes recovered from the PIPB column is mixed with benzene and sent to the transalkylator, where a portion of the polyalkylated benzenes is transalkylated into cumene. Higher-alkylated benzenes also participate in the transalkylation reactions. The reaction is thermally neutral and is limited by approach to chemical equilibrium. The transalkylation product mixture is sent to the bottom section of the CD column, where cumene and heavier are recovered at the bottom for direct feed to distillation.

Distillation

Cumene and other alkylation products in the bottom section of the CD column are sent to the cumene column, where the cumene product is fractionated overhead. Typical product specifications are shown in Table 4.1.3. Components heavier than cumene—di- and trialkylated benzenes, as well as other heavier components—are distilled to the bottom of the column. Both the di- and triisopropylbenzenes are recovered in the PIPB column overhead for conversion in the transalkylator. The residue of the process is withdrawn from the PIPB column bottom and may be used as a fuel blending component.

PROCESS ECONOMICS

The investment required for a cumene plant based on the CD*Cumene* process is estimated to be less than half that required for a comparable acid-based unit and significantly less than that required by other fixed-bed zeolite processes. The factors contributing to the low investment cost are process simplicity, low equipment piece count, mild operating conditions, and all-carbon-steel construction.

Due to the noncorrosive nature of the process, all equipment is constructed of mild carbon steel. No stainless steel, higher alloys, or brick and glass linings are required.

All operations are conducted at low temperatures and low pressures, and no special fabrication is needed for any equipment. The alkylation reactor is similar to a simple packed column, and the product distillation train is similar to other commercial cumene processes except that equipment sizes are smaller. The CD*Cumene* process is not sensitive to moisture, so there is no need for benzene drying. Reactor effluent is fed directly into distillation, and treatment systems such as acid neutralization or catalyst filtration are not required.

In contrast, aluminum chloride–based processes are more complicated, need extensive alloy and other exotic materials, and must have facilities for aluminum chloride handling and disposal of spent catalyst. The SPA-based process has a much lower yield and must have handling and disposal facilities for phosphoric acid "syrup" and spent SPA catalyst.

Raw Materials and Utility Consumption

The benefits of reduced feedstock requirements (high yield), high energy utilization efficiency, low capital cost, and low maintenance cost due to the absence of severe corrosion problems result in extremely favorable overall economics relative to alternative technologies. Typical feedstock and utility consumptions per metric ton of cumene are shown in Tables 4.1.4 and 4.1.5. Yield on feed is 99.7 wt %.

Catalysts

The proprietary, patented zeolite catalysts used in the alkylator, transalkylator, and finishing reactor have been tested extensively. The selectivity of the catalysts essentially eliminates *n*-propylbenzene production, which in turn eliminates the costly handling of *n*-propylbenzene in the downstream phenol unit and ensures the highest-quality phenol. The alkylation catalyst is very monoselective (i.e., makes little dipropylbenzene), which allows operation at a low benzene recycle rate, thereby reducing energy consumption. The transalkylation catalyst has very high transalkylation activity and converts all the higher-polypropylated benzenes to cumene.

Both the alkylation and transalkylation catalysts are extremely stable. In addition to extended run lengths, the commercial plants also have demonstrated that operating conditions remain constant throughout the run length with no changes in selectivity patterns. The benefits of long run length are infrequent plant shutdowns for catalyst changeover and low catalyst cost per ton of cumene produced.

TABLE 4.1.4 Raw Materials per Metric Ton Product

Propylene, kg	353
Benzene, kg	650

TABLE 4.1.5 Utilities per Metric Ton Product

Electricity, kWh	8
Heat (import), 10^6 kcal	0.5
Steam (export), t	1.0
Cooling water, m^3	12

The catalysts age through deposition of carbon or poisons and are regenerated by a controlled burn to remove this deposit. The extended time between regenerations (2 to 6 years) eliminates the need for spare reactors or mandatory regeneration equipment; the generally preferred procedure is to remove the catalysts and regenerate them in an outside facility. The regenerated catalysts are as active and selective as fresh catalysts.

The catalysts are indifferent to most poisons associated with other alkylation processes, notably water and sulfur. Since the catalysts can operate in the presence of water, there is no need to dry the feedstocks, and water-stripping equipment is not required. Any water present in the benzene or propylene feeds will either pass out with the vent gas or be condensed and removed by the vent condenser of the alkylator overhead system.

Process Effluents

The inert components of the propylene feed will appear as noncondensables in the alkylator overhead system vent. This alkylation offgas is directed to a flare or a fired heater. The column overhead systems are designed to minimize losses of hydrocarbons using dual-range pressure controllers with nitrogen. The vents normally are directed to a flare.

The process creates a small liquid wastewater stream, mainly from the dissolved water present in feed benzene, although much of this water leaves with the vent gas. Any condensed water is removed as a separate phase in the alkylator overhead system. This condensate is handled easily within the plant facility.

The zeolite catalyst is environmentally inert. At the end of its life it can be disposed of in an ordinary landfill.

SUMMARY OF PROCESS FEATURES

- The process has a net make of heavy ends of less than 0.3 kg/100 kg of cumene. This corresponds to a product yield of 99.7 percent.
- Exceptional-quality cumene (typical purity of 99.95 percent or higher and ultralow bromine index) is produced without drag streams or clay treatment. This is the direct result of the highly selective catalytic distillation operation.
- The zeolite catalyst can be disposed of safely as landfill. This eliminates the problems of corrosion and acid waste disposal with earlier catalysts. Low-pressure operation minimizes fugitive emission and lowers the potential severity of emergency discharges.
- Almost all the heat input into the process, including the exothermic heat of reaction, is recovered as useful steam for export.
- All equipment is constructed of mild carbon steel, and no special fabrication is needed, resulting in low capital cost.
- Dilute propylene feedstock (containing up to 90 mol % of propane and other inerts) can be accommodated with only minor modifications.
- Reaction temperature is controlled easily by the setting of the alkylation column overhead pressure. The unit is inherently stable and forgiving of upsets. Reaction heat is removed by the vaporization of benzene, so the risk of a runaway reaction is virtually nonexistent.
- The cumene process can be integrated easily with a phenol plant for optimization of energy and products.

CHAPTER 4.2
UOP Q-MAX™ PROCESS

Gary A. Peterson and Robert J. Schmidt
UOP LLC
Des Plaines, Illinois

INTRODUCTION

The Q-Max™ process converts benzene and propylene to high-quality cumene by using a regenerable zeolitic catalyst. The Q-Max process represents a substantial improvement over older cumene technologies and is characterized by its exceptionally high yield, superior product quality, low investment and operating costs, reduction in solid waste, and corrosion-free environment.

Cumene is produced commercially through the alkylation of benzene with propylene over an acid catalyst. Over the years, many different catalysts have been proposed for this alkylation reaction, including boron trifluoride, hydrogen fluoride, aluminum chloride, and phosphoric acid. In the 1930s, UOP introduced the UOP catalytic condensation process, which used a solid phosphoric acid (SPA) catalyst to oligomerize light olefin by-products from petroleum thermal cracking into heavier paraffins that could be blended into gasoline. During World War II, this process was adapted to produce cumene from benzene and propylene to make a high-octane blending component for military aviation gasoline. Today, cumene is no longer used as a fuel, but it has grown in importance as a feedstock for the production of phenol.

Although SPA is a highly efficient and economical catalyst for cumene synthesis, it has two important limitations:

1. Cumene yield is limited to about 95 percent, because of the oligomerization of propylene and the formation of heavy alkylate by-products.
2. The catalyst is not regenerable and must be disposed of at the end of each catalyst cycle.

In recent years, producers have been under increasing pressure to improve cumene product quality so that the quality of the phenol produced downstream (as well as acetone and *alpha*-methylstyrene, which are coproduced with phenol) could be improved. Twenty-five years ago, most phenol was used to produce phenolic resins, and acetone was used primarily as a solvent. Today, both phenol and acetone are used increasingly in the production of polymers such as polycarbonates and nylon. Over the years, improvements to the SPA process managed to keep pace with the demand for higher cumene product quality, but

4.12 CUMENE

producers still sought an improved cumene process that would produce a better-quality product at higher yield.

Because zeolites are known to selectively perform many acid-catalyzed reactions, UOP began searching for a new cumene catalyst that would overcome the limitations of SPA. UOP's objective was to develop a regenerable catalyst that would increase the yield of cumene and lower the cost of production. More than 100 different catalyst materials were screened, including mordenites, MFIs, Y-zeolites, amorphous silica-aluminas, and *beta*-zeolite. The most promising materials were modified to improve their selectivity and then subjected to more rigorous testing. By 1992, UOP had selected the most promising catalyst based on *beta*-zeolite for cumene production and then began to optimize the process design around this new catalyst. The result of this work is the Q-Max process and the QZ-2000 catalyst system. More recently, UOP has introduced QZ-2001 as a new alkylation catalyst, also based on *beta*-zeolite, that offers improved stability and operation at very low B/P feed ratio.

PROCESS CHEMISTRY

The synthesis of cumene from benzene and propylene is a modified Friedel-Crafts alkylation, which can be accomplished by many different acid catalysts. The basic alkylation chemistry and reaction mechanism are shown in Fig. 4.2.1. The olefin forms a carbonium ion intermediate, which attacks the benzene ring in an electrophilic substitution. The addition to the olefin double bond is at the middle carbon of propylene, in accordance with Markovnikov's rule. The addition of the isopropyl group to the benzene ring weakly activates the ring toward further alkylation, producing diisopropylbenzene (DIPB) and heavier alkylate by-products.

The QZ-2001 catalyst functions as strong acid. In the QZ-2001 catalyst, the active surface sites of the silica-alumina structure act to donate the proton to the adsorbed olefin. Because the QZ-2001 catalyst is a strong acid, it can be used at a very low temperature. Low reaction temperature reduces the rate of competing olefin oligomeriza-

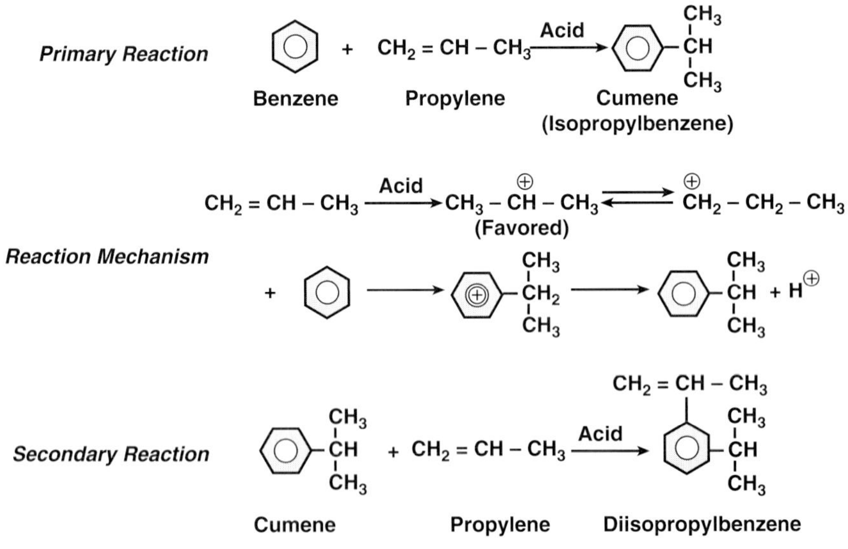

FIGURE 4.2.1 Alkylation chemistry.

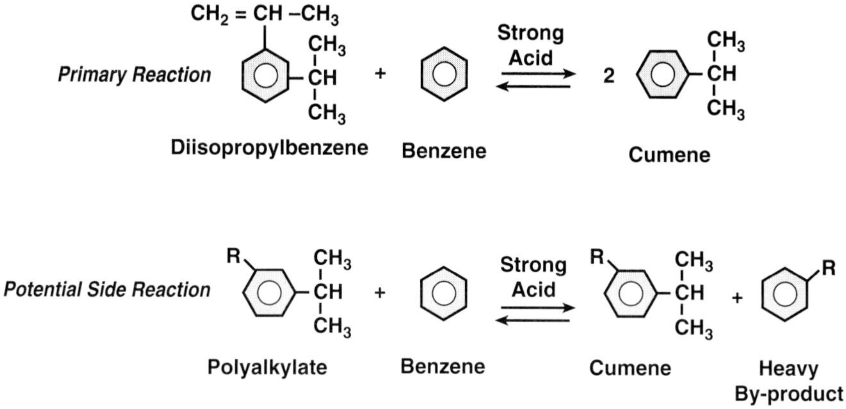

FIGURE 4.2.2 Transalkylation chemistry.

tion reactions, resulting in higher selectivity to cumene and lower production of heavy by-products.

Transalkylation of DIPB

Transalkylation is the acid-catalyzed transfer of one isopropyl group from DIPB to a benzene molecule to form two molecules of cumene (Fig. 4.2.2). The QZ-2000 catalyst is recommended for transalkylation due to its very high transalkylation activity. The Q-Max process is designed with an alkylation reactor section, which produces about 85 to 95 wt % cumene and 5 to 15 wt % DIPB. After recovery of the cumene product by fractionation, the DIPB is reacted with recycle benzene at optimal conditions for transalkylation to produce additional cumene. With the alkylation and transalkylation reactors working together to take full advantage of the QZ-2001/QZ-2000 catalyst system, the overall yield of cumene is increased to 99.7 wt %.

Side Reactions

In addition to the principal alkylation reaction of benzene with propylene, all acid catalysts promote the following undesirable side reactions to some degree (Fig. 4.2.3):

- *Oligomerization of olefins.* The model for acid-catalyzed alkylation is diffusion of the olefin to an active site saturated with benzene followed by adsorption and reaction. One possible side reaction is the combination of the propyl carbonium ion with propylene to form a C_6 olefin or even further reaction to form C_9, C_{12}, or heavier olefins.
- *Alkylation of benzene with heavy olefins.* Once heavy olefins have been formed through oligomerization, they may react with benzene to form hexylbenzene and heavier alkylated benzene by-products.
- *Polyalkylation.* The addition of an isopropyl group to the benzene ring to produce cumene weakly activates the ring toward further substitution, primarily at the *meta* and *para* positions, to make DIPB and heavier alkylates.
- *Hydride-transfer reactions.* Transfer of a hydrogen to an olefin by the tertiary carbon on cumene can form a cumyl carbonium ion that may react with a second benzene molecule to form diphenyl propane.

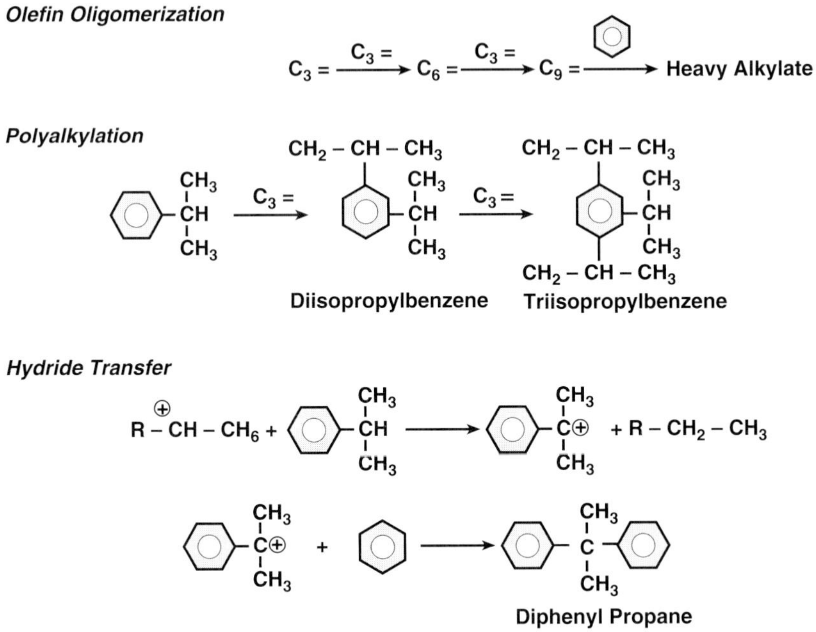

Olefin Oligomerization

$$C_3= \xrightarrow{C_3=} C_6= \xrightarrow{C_3=} C_9= \longrightarrow \text{Heavy Alkylate}$$

Polyalkylation

Diisopropylbenzene   Triisopropylbenzene

Hydride Transfer

Diphenyl Propane

FIGURE 4.2.3 Possible alkylation side reactions.

In the Q-Max process, the reaction mechanism of the QZ-2000 catalyst and the operating conditions of the unit work together to minimize the impact of these side reactions. The result is an exceptionally high yield of cumene product.

DESCRIPTION OF THE PROCESS FLOW

A representative Q-Max flow diagram is shown in Fig. 4.2.4. The alkylation reactor is typically divided into four catalyst beds contained in a single reactor shell. The fresh benzene is routed through the upper midsection of the depropanizer column to remove excess water and then sent to the alkylation reactor via a sidedraw. The recycle benzene to both the alkylation and transalkylation reactors comes from the overhead of the benzene column. A mixture of fresh and recycle benzene is charged downflow through the alkylation reactor. The fresh propylene feed is split between the four catalyst beds. An excess of benzene is used to avoid polyalkylation and to help minimize olefin oligomerization. Because the reaction is exothermic, the temperature rise in the reactor is controlled by recycling a portion of the reactor effluent to the reactor inlet, which acts as a heat sink. In addition, the inlet temperature of each downstream bed is reduced to the same temperature as that of the first bed inlet by injecting a portion of cooled reactor effluent between the beds.

Effluent from the alkylation reactor is sent to the depropanizer column, which removes any propane and water that may have entered with the propylene feed. The bottoms from the depropanizer column are sent to the benzene column, where excess benzene is collected overhead and recycled. Benzene column bottoms are sent to the cumene column,

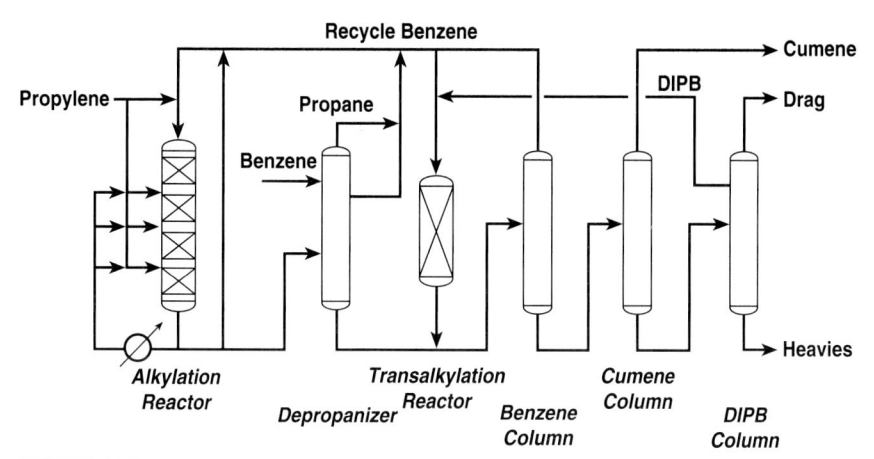

FIGURE 4.2.4 Process flow diagram.

where the cumene product is recovered overhead. The cumene column bottoms, which contain mostly diisopropylbenzene, are sent to the DIPB column. The DIPB stream leaves the column by way of a sidecut and is recycled to the transalkylation reactor. The DIPB column bottoms consist of heavy aromatic by-products, which are normally blended into fuel oil. Steam or hot oil provides the heat for the product fractionation section.

A portion of the recycle benzene from the top of the benzene column is combined with the recycle DIPB from the sidecut of the DIPB column and sent to the transalkylation reactor. In the transalkylation reactor, DIPB and benzene are converted to additional cumene. The effluent from the transalkylation reactor is then sent to the benzene column.

The QZ-2001/QZ-2000 catalyst system utilized in the alkylation and transalkylation reactors is regenerable. At the end of each cycle, the catalyst is typically regenerated *ex situ* via a simple carbon burn by a certified regeneration contractor. However, the unit can also be designed for *in situ* catalyst regeneration. Mild operating conditions and a corrosion-free process environment permit the use of carbon-steel construction and conventional process equipment.

FEEDSTOCK CONSIDERATIONS

Impact of Feedstock Contaminants on Cumene Purity

In the Q-Max process, the impact of undesirable side reactions is minimal, and impurities in the cumene product are governed primarily by trace contaminants in the feeds. Because of the high activity of the QZ-2001 catalyst for alkylation, it can be operated at very low temperature, which dramatically reduces the rate of competing olefin oligomerization reactions and decreases the formation of heavy by-products. Thus, with the Q-Max process, cumene product impurities are primarily a result of impurities in the feedstocks. Table 4.2.1 lists the common cumene impurities of concern to phenol producers, and Fig. 4.2.5 graphically shows the reactions of some common feedstock contaminants that produce these impurities.

- *Cymene and ethylbenzene.* Cymene is formed by the alkylation of toluene with propy-

4.16 CUMENE

TABLE 4.2.1 Common Cumene Impurities

Trace contaminant	Concern in downstream phenol unit
Nonaromatics	Form acids and other by-products in phenol unit, yield loss
Ethylbenzene	Forms acetaldehyde, an acetone contaminant
n-Propylbenzene	Forms propionaldehyde, an acetone contaminant
Butylbenzenes	Resist oxidation, an *alpha*-methylstyrene contaminant
Cymenes	Form cresols, phenol contaminants
Polyalkylates	Form alkylphenols, yield loss

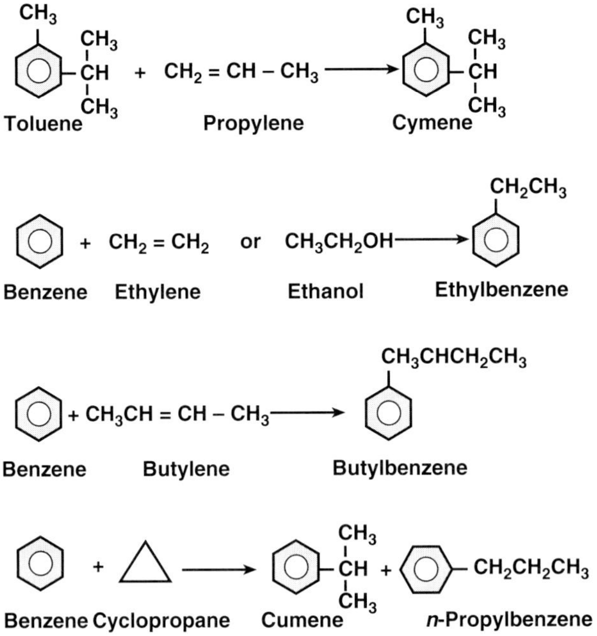

FIGURE 4.2.5 Reactions of feed impurities.

lene. The toluene may already be present as an impurity in the benzene feed, or it may be formed in the alkylation reactor from methanol and benzene. Ethylbenzene is primarily formed from ethylene impurities in the propylene feed. However, as with cymene, ethylbenzene can also be formed from ethanol. Small quantities of methanol and ethanol are sometimes added to the C_3's in a pipeline to protect against hydrate freezing. Although the Q-Max catalyst is tolerant of these alcohols, removing them from the feed by a water wash may be desirable to achieve the lowest possible levels of ethylbenzene or cymene in the cumene product.

- *Butylbenzene.* Although butylbenzene is produced primarily from traces of butylene in the propylene feed, it may also be created through the oligomerization of olefins. However, the very low reaction temperature of the Q-Max process reduces oligomerization, resulting in minimal overall butylbenzene formation.

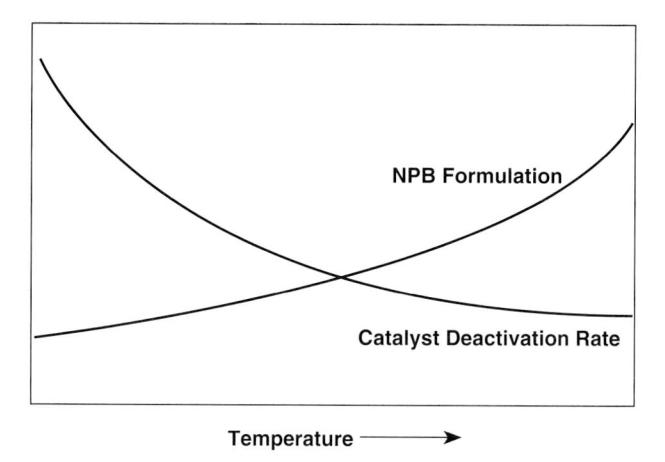

FIGURE 4.2.6 Effect of reactor temperature.

- *n-Propylbenzene.* The *n*-propylbenzene (NPB) is produced from trace levels of cyclopropane in the propylene feed. The chemical behavior of cyclopropane is similar to that of an olefin: It reacts with benzene to form either cumene or NPB. The tendency to form NPB rather than cumene decreases as the reaction temperature is lowered. Unfortunately, the catalyst deactivation rate increases with lower reaction temperature (Fig. 4.2.6). Because of the exceptional stability of the QZ-2001/QZ-2000 catalyst system, a Q-Max unit can be operated for extended cycle lengths and still maintain an acceptable level of NPB in the cumene product. For example, with a typical FCC-grade propylene feed containing normal amounts of cyclopropane, the Q-Max process can produce a cumene product containing less than 250 wt ppm NPB and maintaining an acceptable catalyst cycle length.

Impact of Catalyst Poisons on Catalyst Performance

A list of potential Q-Max catalyst poisons is found in Table 4.2.2. All the listed compounds are known to neutralize the acid sites of zeolites. Good feedstock treating practice or proven guard-bed technology easily handles these potential poisons.

Water in an alkylation environment can act as a Brønsted base to neutralize some of the stronger zeolite acid sites first. However, as a result of the inherently high activity of the Q-Max catalyst, water does not have a detrimental effect at the typical feedstock moisture levels and normal alkylation and transalkylation conditions. The Q-Max catalyst can process feedstocks up to the normal water saturation conditions, typically 500 to 1000 ppm, without any loss of catalyst stability or activity.

TABLE 4.2.2 Handling Potential Catalyst Poisons

Poison	Source	Removal
Basic nitrogen	Trace levels in feedstocks	Guard bed
Ammonia	Common impurity in FCC propylene	Water wash or guard bed
Arsine (AsH_3)	Common impurity in FCC propylene	Guard bed

Sulfur does not affect Q-Max catalyst stability or activity at the levels normally present in the propylene and benzene feeds processed for cumene production. However, trace sulfur in the cumene product, for example, might be a concern in the downstream production of certain monomers (e.g., phenol hydrogenation for caprolactam). Within the Q-Max unit, the majority of sulfur compounds associated with propylene (mercaptans) and those associated with benzene (thiophenes) are converted to products outside the boiling range of cumene. However, the sulfur content of the cumene product does depend on the sulfur content of the propylene and especially benzene feeds. Sulfur at the levels normally present in propylene and benzene feeds considered for cumene production will normally result in cumene product sulfur content that is within specifications (for example, <1 wt ppm).

Successful operation with a wide variety of propylene feedstocks from different sources has demonstrated the flexibility of the Q-Max process. Chemical-grade, FCC-grade, and polymer-grade propylene feedstocks can all be used to make high-quality cumene product.

PROCESS PERFORMANCE

The Q-Max unit has high raw material utilization and an overall cumene yield of at least 99.7 wt % based on using typical propylene and benzene feedstock. The remaining 0.3 wt % or less of the overall yield is in the form of a heavy aromatic by-product.

The cumene product quality summarized in Table 4.2.3 is representative of a Q-Max unit processing commercially available, high-quality feedstocks. The quality of the cumene product from any specific Q-Max unit is strongly influenced by the specific contaminants present in the feedstocks.

Propane entering the unit with the propylene feedstock is unreactive in the process and is separated in the fractionation section as a propane product.

CASE STUDY

A summary of the investment cost and utility consumption for a new Q-Max unit producing 200,000 MTA of cumene from extracted benzene and chemical-grade propylene is shown in Table 4.2.4. The estimated erected cost for the Q-Max unit assumes construction on a U.S. Gulf Coast site in 2002. The scope of the estimate includes basic engineering, procurement, erection of equipment on the site, and the initial load of QZ-2000 catalyst.

TABLE 4.2.3 Representative Cumene Product Quality

Cumene purity, wt %	≥ 99.97
Bromine index	≤ 10
Sulfur, wt ppm	≤ 0.1
Specific impurities, wt ppm:	
Ethylbenzene	≤ 30
n-Propylbenzene	≤ 250
Butylbenzene	≤ 20
Cymene	≤ 5
Diisopropylbenzene	≤ 10
Total nonaromatics	≤ 20

TABLE 4.2.4 Investment and Operating Cost for 200,000-MTA Q-Max Unit

Feedstock requirements:	
Extracted benzene (99.8 wt %)	132,300 MTA
Chemical-grade propylene (95 wt %)	74,240 MTA
Utility consumption per MT cumene produced:	
Electric power	12.3 kWh
High-pressure steam	0.81 MT
Medium-pressure steam	0.20 MT
Low-pressure steam credit	−0.31 MT
Cooling water	3.1 m^3
Erected cost estimate	$14.2 million

The utility requirements for a Q-Max unit depend on the project environment (i.e., feed, product specifications, and utility availability). Q-Max units are often integrated with phenol plants where energy use can be optimized by generating low-pressure steam in the Q-Max unit for utilization in the phenol plant.

COMMERCIAL EXPERIENCE

The first Q-Max unit went on-stream in 1996. Since that time, UOP has licensed a total of 10 Q-Max units throughout the world having a total plant capacity of 2.6 million MTA of cumene. Seven Q-Max units have been commissioned and three more are in various stages of design or construction. Capacities range from 37,000 to 700,000 MTA of cumene produced. Several of these units have been on-stream for more than 7 years without performing a single catalyst regeneration.

BIBLIOGRAPHY

Jeanneret, J. J., D. Greer, P. Ho, J. McGeehee, and H. Shakir: "The Q-Max Process: Setting the Pace for Cumene Production," *DeWitt Petrochemical Review* Conference, Houston, March 1997.

Schmidt, R. J., A. S. Zarchy, and G. A. Peterson: "New Developments in Cumene and Ethylbenzene Alkylation," AIChE Spring Meeting, New Orleans, March 2002.

PART 5

ETHYLBENZENE

CHAPTER 5.1
LUMMUS/UOP LIQUID-PHASE EB*ONE* PROCESS AND CDTECH *EB*® PROCESS

Stephen Pohl and Sanjeev Ram
ABB Lummus Global
Bloomfield, New Jersey

INTRODUCTION

Ethylbenzene (EB) is used almost exclusively as an intermediate in the production of styrene monomer (SM). It is produced by liquid-phase or vapor-phase alkylation of benzene with ethylene. Commercial production started in the 1930s and has grown to over 23 million metric tons per annum (MTA).

Until 1980, almost all EB was manufactured with an aluminum chloride catalyst using a Friedel-Crafts reaction mechanism. A few EB production units employed a different Friedel-Crafts catalyst, boron trifluoride. Small amounts of EB also were recovered as a by-product from mixed xylenes streams using a very energy-intensive distillation process. In 1980, the first commercial facility using a zeolite catalyst started production. The ease of operation of zeolite-based processes and the absence of the maintenance and environmental problems associated with the Friedel-Crafts catalysts have allowed zeolite catalysts to completely displace the older catalysts in all modern production facilities.

The first zeolitic process was based on a vapor-phase reactor at a temperature of over 400°C. In this temperature range, reactions such as isomerization/cracking, isomerization, and hydrogen transfer produce a number of by-products that contaminate the EB product. Efforts were made to reduce by-product formation by changing reaction conditions, but it was not until the advent of liquid-phase and mixed-phase processes operating at temperatures lower than 270°C that zeolite-catalyzed processes were truly capable of producing high-purity ethylbenzene. The first high-purity zeolite-based EB plant, based on technology developed by UOP and ABB Lummus Global, started up in 1990.

Two zeolitic EB technologies are discussed in this chapter: EB*One* and CDTECH

EB®. The EB*One* technology is a joint development of ABB Lummus Global and UOP, whereas the CDTECH *EB* process was developed by Lummus and Chemical Research & Licensing (CR&L) based on catalytic distillation technology, which combines catalytic reaction and distillation in a single operation.

The liquid-phase Lummus/UOP EB*One* process employs a highly selective and robust catalyst that has proven run lengths of over 6 years. This process has significant capital and operating cost advantages compared with older vapor-phase zeolite- and $AlCl_3$-based processes, as well as competing liquid-phase EB processes.

The CDTECH *EB* process uses a zeolite catalyst specially packaged in bales. The catalyst bales are installed in the alkylator in a manner similar to structured packing, enabling the alkylation reaction and the distillation of reactants and products to take place simultaneously. The alkylation reactor operates in a mixed liquid-vapor phase, making it a natural match for dilute ethylene feedstock. The dilute ethylene source can be refinery fluid catalytic cracking (FCC) offgas or steam crackers with ethylene concentrations in the range of 10 to 85 mol %.

Both processes are characterized by high yields, superior product purity with very low xylene content, low energy consumption, and long catalyst run lengths that eliminate the need for spare reactors and regeneration equipment that are required with a vapor-phase zeolite process. In addition to being well suited for new plants, the EB*One* and CDTECH *EB* processes are adaptable for revamp of both $AlCl_3$ and vapor-phase fixed-bed zeolite process-based plants with minimal new investment and a potential increase in capacity.

PROCESS PERSPECTIVE

The Lummus/UOP EB*One* technology was first commercialized in 1990 at Nippon Styrene Monomer Company (NSMC), Japan. Commercial operation has demonstrated the sturdiness of the zeolite catalysts both chemically and mechanically. The catalysts have gone through regeneration without loss of mechanical strength or process performance.

Two other EB plants, at Mitsui Chemicals and Chiba Styrene Monomer, also in Japan, were subsequently started up successfully in 1994. As shown in Table 5.1.1, this process has been selected for 26 projects with design capacities ranging from 100,000 to 750,000 MTA EB. Currently, 16 operating plants use this technology.

Since 1990, the CDTECH *EB* technology has been selected for six projects worldwide. As shown in Table 5.1.2, three plants are in operation and one is under construction. The largest CDTECH *EB* plant designed to date uses a very dilute ethylene feedstock containing less than 40 mol % ethylene (the balance consisting mainly of hydrogen, methane, and ethane).

PROCESS CHEMISTRY

As shown in Fig. 5.1.1, EB is made by the alkylation of benzene with ethylene in the presence of zeolite catalyst. Successive alkylations also occur to a minor extent, producing diethylbenzene and higher-ethylated benzenes, collectively termed *polyethylated benzene* (PEB). Very small levels of other coupling reactions also occur, yielding materials such as diphenylethane (DPE) and high-boiling compounds. All alkylation reactions are exothermic.

The PEB produced by successive alkylations is transalkylated (transfer of ethyl groups) with benzene to produce additional EB. These reactions are slower than alkylation and are limited in extent by equilibrium. Higher-ethylated benzenes also participate. The heat of reaction is essentially zero, and the reaction conditions are effectively isothermal.

TABLE 5.1.1 Zeolite Fixed-Bed Plant Awards

Client	Location	MTA	Status
Asahi Kasei (revamp)	Japan	360,000	Engineering
BP-SPC (SECCO)	China	530,000	Engineering
Asahi Kasei	Japan	360,000	Construction
Kaucuk	Czech Republic	300,000	Operating
Jilin (Revamp)	China	160,000	Operating
Shell/CNOOC	China	640,000	Engineering
Dow	Germany	Confidential	Operating
Repsol	Spain	380,000	Operating
P.T. SMI	Indonesia	212,000	Operating
SADAF	Saudi Arabia	530,000	Operating
PT TPPI	Indonesia	530,000	On hold
Jilin	China	106,000	Operating
Lyondell (ARCO)	The Netherlands	726,000	Operating
Nova (Huntsman)	United States	749,000	On hold
TPI	Thailand	212,000	Operating
BP Amoco (Hüls)	Germany	400,000	Operating
Yangzi/BASF	China	127,000	Operating
Angarsk	Russia	129,000	On hold
Maoming	China	106,000	Operating
Seraya Chemicals	Singapore	360,000	Operating
Supreme	India	106,000	On hold
Supreme (Polychem)	India	64,000	On hold
Chiba Styrene Monomer	Japan	265,000	Operating
Tabriz Petrochemicals	Iran	100,000	Operating
Mitsui Chemicals	Japan	254,000	Operating
NSMC	Japan	210,000	Operating

TABLE 5.1.2 CDTECH *EB* Plant Awards

Client	Location	MTA	Awarded	Status
Chevron Phillips Chemical (CPC)	Saudi Arabia	760,000	2002	Engineering
Dow	United States	Confidential	2000	Construction
Nova	Canada	477,000	1996	Operating
PASA	Argentina	140,000	1995	Operating
Pemex	Mexico	106,000	1992	On hold
Mitsubishi	Japan	260,000	1992	Operating

PROCESS DESCRIPTION

Benzene is alkylated with ethylene to yield a mixture of alkylated benzenes. This mixture is distilled to recover product EB, excess benzene for recycle, and higher-ethylated benzenes (polyethylbenzenes, or PEBs). The PEBs are transalkylated with benzene to form additional EB.

Lummus and UOP were the first to introduce the concept of separate alkylation and transalkylation reactors for a zeolite-based process. This permits operation at lower benzene recycle than alkylation-only processes. Thus the process can operate at an optimal benzene recycle rate with consequent savings in energy input and equipment cost. Figure 5.1.2 is a simplified flow diagram of the EB*One* process.

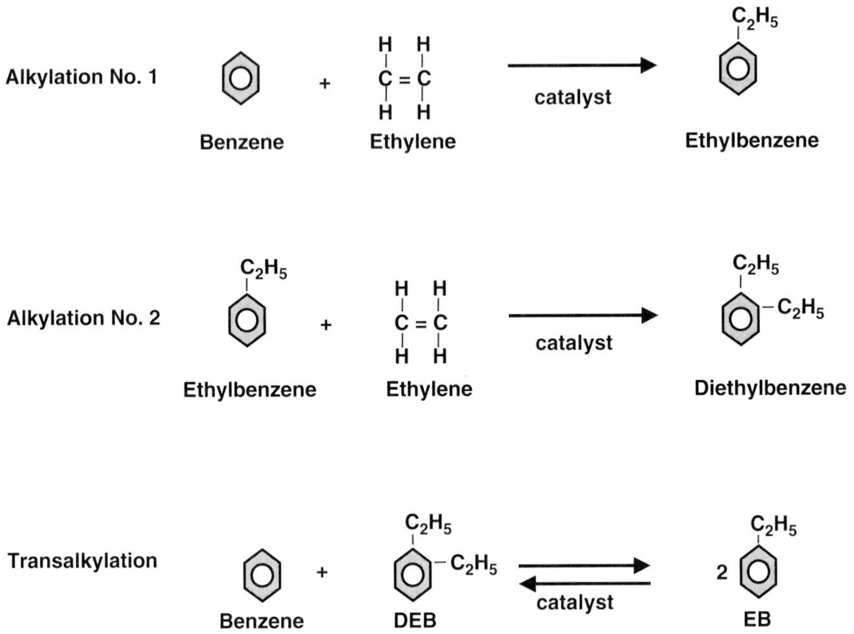

Alkylation No. 1: Benzene + Ethylene (H₂C=CH₂) →[catalyst] Ethylbenzene (C₆H₅–C₂H₅)

Alkylation No. 2: Ethylbenzene + Ethylene →[catalyst] Diethylbenzene

Transalkylation: Benzene + DEB ⇌[catalyst] 2 EB

FIGURE 5.1.1 Process chemistry.

The liquid-phase alkylation reactor consists of multiple beds of zeolite catalyst operating adiabatically. Process conditions are selected to keep the aromatic reaction mixture in the liquid phase. Excess benzene is used, and ethylene is injected before each bed. Multiple ethylene injection points improve selectivity and enhance catalyst stability. In the alkylation reactor, ethylene reacts completely, leaving only the inert constituents of the feed, such as ethane. These inerts pass through the reactor and are removed from the plant at a convenient point.

The transalkylation reactor also consists of multiple beds of zeolite catalyst. Conditions are selected to keep the reactor contents in liquid phase.

The alkylation and transalkylation effluents are fed to the benzene column, where benzene is taken as the overhead product for recycle to the reactors. The benzene column bottoms feed the EB column. Here EB is taken as the overhead product. The EB column bottoms feed a small PEB column where the diethylbenzene and triethylbenzene are distilled overhead for recycle to the transalkylator. The small bottoms flow of heavy compounds (flux oil) is used as fuel.

The reboilers of the distillation columns may use hot oil, high-pressure steam, or direct firing. Overhead vapors are condensed in waste heat boilers, generating valuable steam useful in a downstream SM or propylene oxide/styrene monomer (PO/SM) plant. The EB unit has considerable flexibility to meet a variety of local site conditions in an efficient manner. If no steam export is required, the net heat import can be reduced considerably.

CDTECH *EB* Process Flow Scheme

As shown in Fig. 5.1.3, the CDTECH *EB* process flow scheme differs only in the alkylation reactor system.

Reaction Section

Alkylation Reactor | **Transalkylation Reactor**

Distillation Section

Benzene Column | **EB Column** | **PEB Column**

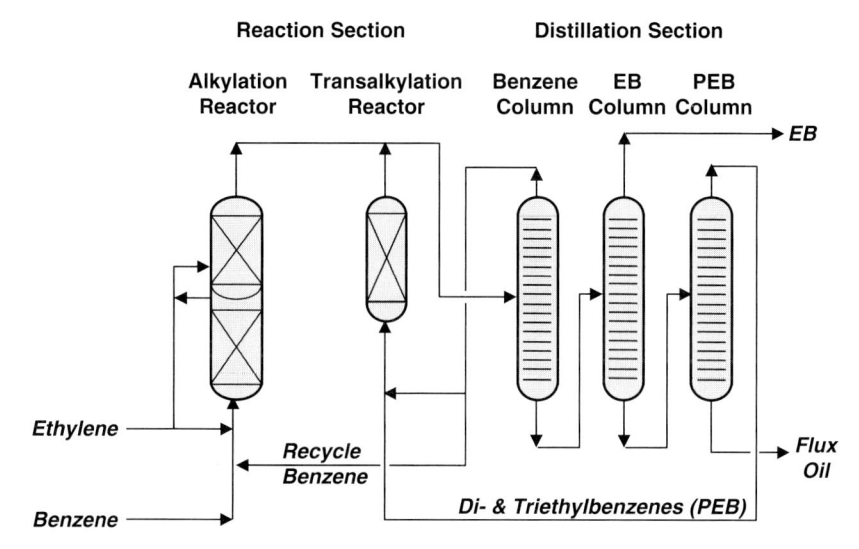

- Ethylene
- Benzene
- Recycle Benzene
- Di- & Triethylbenzenes (PEB)
- EB
- Flux Oil

FIGURE 5.1.2 EB*One* process flow scheme.

At the heart of the CDTECH *EB* process is the patented catalytic distillation concept that combines catalytic reaction and distillation in a single operation. Ethylene is introduced as a vapor at the bottom of the reaction zone. Reaction occurs in the catalytic zone of the column, and distillation occurs throughout the column, resulting in a countercurrent flow of vapor and liquid throughout the column.

Reaction products are removed continuously from the catalytic zone as a result of distillation, whereas any unconverted reactants and other lights are taken overhead. The heat of reaction is removed immediately by vaporizing benzene, which is significant in that it provides a near-isothermal operation at the optimal reaction temperature. As ethylene vapor is injected at multiple points in the reactor, it comes in contact with the liquid benzene coming from above and is absorbed in the liquid phase. At equilibrium, most of the ethylene is present in the vapor phase. As the small amount of ethylene in the liquid comes in contact with the catalyst, it immediately reacts to form ethylbenzene. This moves the ethylene concentration in the liquid phase away from equilibrium. The vapor-liquid equilibrium then causes ethylene to be "injected" from the vapor phase to the liquid phase, thus restoring the equilibrium.

The high-activity zeolite catalyst is contained in glass fiber, which is rolled in a wire mesh into a spiral configuration. The unique structure of the bales gives them the required void fraction for the vapor to flow upward through the reactor. The glass-fiber packing acts as a barrier and prevents the vapors from coming in direct contact with the catalyst. The bales are about 1 ft in diameter and height and are handled easily during loading operations (Fig. 5.1.4).

These proprietary CDTECH catalyst bales containing the alkylation catalyst are stacked on top of each other inside the alkylator like structured packing, enabling the alkylation reaction and the distillation of reactants and products to take place simultaneously. Several beds of stacked bales are used.

The typical range of operating parameters for the two processes is shown in Table 5.1.3. Process turndown to about 70 percent of capacity can be achieved with no economic penalty. Beyond 70 percent turndown, the distillation section will encounter some losses in energy efficiency; however, the reaction system can be turned down to capacities below 50 percent without any adverse processing or economic impact.

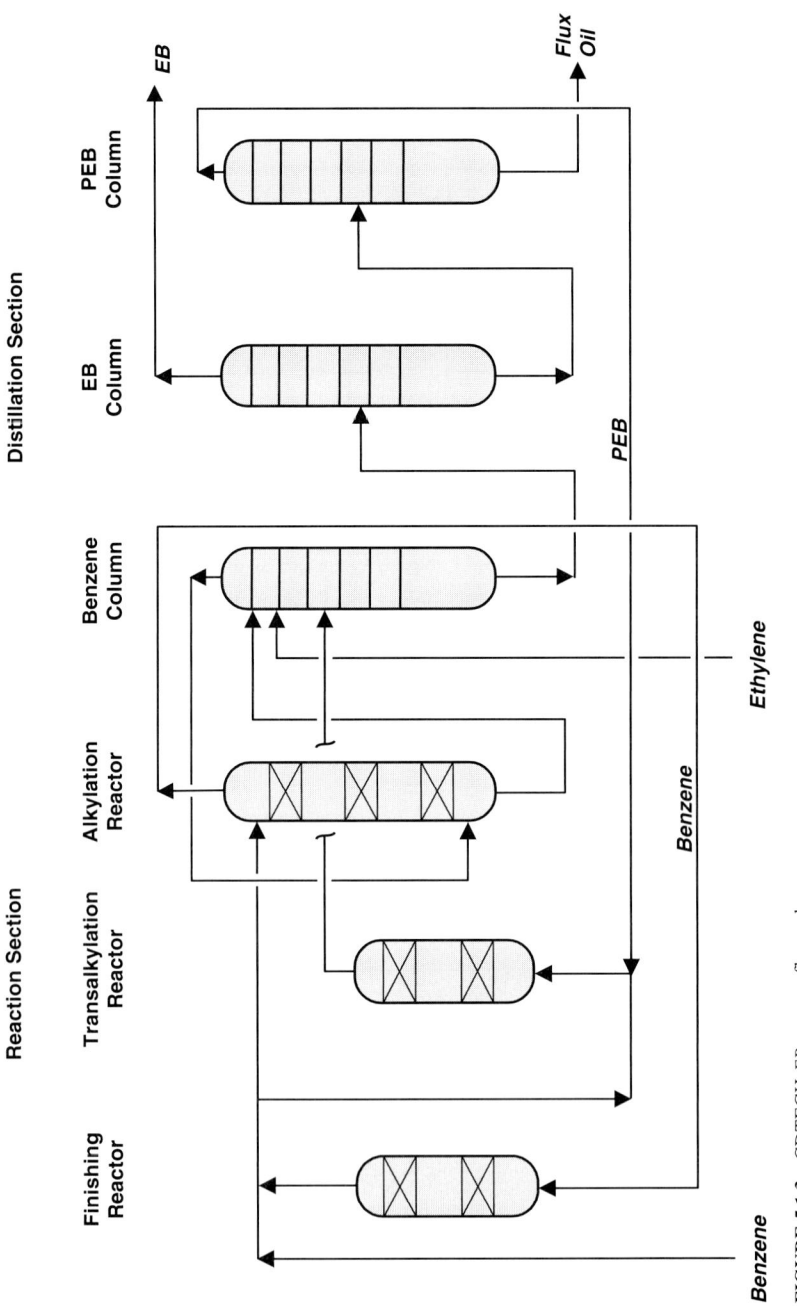

FIGURE 5.1.3 CDTECH *EB* process flow scheme.

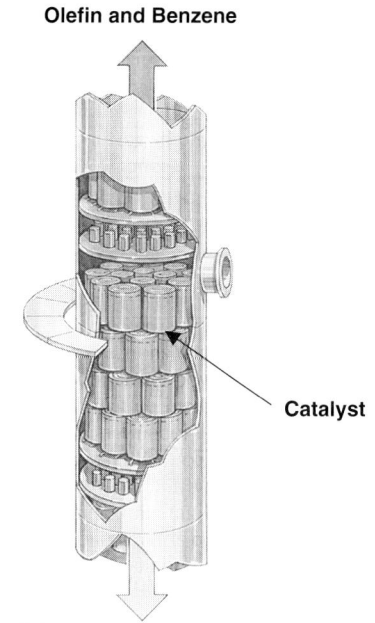

FIGURE 5.1.4 CDTECH *EB* reactor.

TABLE 5.1.3 Typical Range of Operating Parameters

	EB*One*	CDTECH *EB*
Alkylation B/E, mol	2.0–4.0	2.0–4.0
TA P/E, mol	2.0–4.0	2.0–4.0
Alkylation temp. window, °C	185–270°C	200°C top, 240°C btm
Alkylator pressure, bar(g)	30–40	20–25
Ethylene consumption, kg/kg EB	0.264	0.264
Benzene consumption, kg/kg EB	0.738	0.738

ECONOMICS

Investment Costs

The Lummus/UOP EB*One* and CDTECH *EB* processes are simple and use all-carbon-steel equipment. Investment is considerably lower than for other commercially available processes, typically in the range of US$30–45/MTA EB product on an ISBL Gulf Coast basis.

Feedstocks

The two technology options can process the complete range of ethylene feedstocks. Both technologies are equally capable of processing polymer-grade ethylene, which has a typical specification of 99.9 mol % ethylene, 50 ppm maximum of other unsaturates, and 10 ppm maximum of propylene.

TABLE 5.1.4 Typical Dilute Ethylene Feed Specification (after Pretreatment)

Ethylene	As received
Hydrogen	As received
Nitrogen	As received
Methane	As received
Ethane	As received
Propylene and heavier	30 vppm max
Water	No free water
Acetylene	5 vppm max
Carbon monoxide	As received
Carbon dioxide	0.5 vol % max
Oxygen	10 vppm max*
Total sulfur	10 vppm max
Basic nitrogen	Not detectable
Arsenic	5 wppb max
Heavy metals	Not detectable
Alkaline	Not detectable

*Occasional excursions to higher concentrations are acceptable.

The CDTECH *EB* process is also able to handle dilute ethylene feedstock with ethylene feed compositions ranging from 10 percent to polymer grade (100 percent). The dilute ethylene source can be refinery FCC offgas or dilute ethylene streams from steam crackers with ethylene concentrations in the range of 30 to 85 mol %. Table 5.1.4 shows a typical dilute ethylene feed specification for the CDTECH *EB* process. In order to meet this specification, FCC offgas typically requires some form of pretreatment, which may involve water scrubbing, distillation, absorption, and acetylene hydrogenation. Feed benzene has the typical specifications shown in Table 5.1.5.

Yields and Product Quality

The EB*One* and CDTECH *EB* alkylation and transalkylation reactions are highly selective. Minimum residue formation and 100 percent ethylene conversion result in high yields that approach theoretical. The processes have an overall yield of 99.7 percent. Since reactor operating conditions are mild, the high product yields and complete ethylene conversion are maintained throughout the catalyst run length.

The ethylbenzene product is an ideal feed for downstream SM and PO/SM plants. Because of the low-temperature alkylation/transalkylation reactors and the selective zeolite catalysts, insignificant amounts of xylenes are produced. This feature eliminates the requirement for expensive separation equipment to prevent buildup of xylenes in the styrene unit and permits the highest-purity SM (99.9 percent plus) to be produced. Although the absolute ethylbenzene purity depends on the impurities in the benzene feedstock, the net impurities make in both ethylbenzene processes is summarized in Table 5.1.6. Ethylbenzene product purities of up to 99.95 wt % have been attained commercially.

Catalysts

The proprietary zeolite catalysts used in both the alkylation and transalkylation reactors are extremely stable. In addition to extended run lengths, the commercial plants also have demonstrated that operating conditions remain constant throughout the run with no changes in selectivity patterns. The benefits of long run length are infrequent plant shutdowns for catalyst changeover and low catalyst cost per ton of ethylbenzene produced. The

TABLE 5.1.5 Typical Benzene Feed Specifications

Distillation range	1.0°C max, including 80.1°C
Freezing point, dry	5.4°C min
Toluene	500 wt ppm max
Nonaromatics	1500 wt ppm max

TABLE 5.1.6 Typical Ethylbenzene Product Specifications

Ethylbenzene	99.90 wt % min
Color, APHA max	15
Sp. gr. (15.5°C/15.5°C)	0.869–0.872
Acidity	Neutral
Benzene	Balance
Toluene	450 wt ppm max
Cumene	100 wt ppm max
Nonaromatics	125 wt ppm max
Diethylbenzenes	5 wt ppm max
Xylenes	10 wt ppm max

design run length typically is chosen to match the turnaround schedule of the downstream SM or PO/SM unit, within the range of 2 to 6 years.

The catalysts age through deposition of carbon or poisons and are regenerated by a controlled burn to remove this deposit. The extended time between regenerations eliminates the need for spare reactors or regeneration equipment. The generally preferred regeneration procedure is to remove the catalysts and regenerate them in an outside facility. The regenerated catalysts are as active and selective as fresh catalysts.

The catalysts are indifferent to most poisons associated with other alkylation processes, notably water and sulfur. Since the catalysts can operate in the presence of water, there is no need to dry the feedstocks, and water-stripping equipment is not required. Any water present in the benzene or ethylene feeds will either pass out with the vent gas or be condensed and removed by the vent condenser of the benzene column.

Process Effluents

The inert components of the ethylene feed will dissolve in the alkylator effluent and appear as noncondensables in the benzene column vent, which is directed to a flare or a fired heater. The column overhead systems are designed to minimize losses of hydrocarbons using dual-range pressure controllers with nitrogen. The vents normally are directed to a flare.

A small liquid wastewater stream is derived from the process, mainly from the dissolved water present in feed benzene. Much of this water leaves with the vent gas. Any condensed water is removed as a separate phase in the benzene column overhead system and is handled easily within the plant facility.

The zeolite catalyst is environmentally inert. At the end of its life, it can be disposed of in an ordinary landfill.

Raw Materials and Utility Consumption

The EB*One* and CDTECH *EB* processes feature consistently high product yields over the entire catalyst life cycle. Table 5.1.7 summarizes raw materials and utility consumptions for both processes. Additional utility savings can be realized via heat integration with a downstream styrene unit.

TABLE 5.1.7 Raw Material/Utilities per Metric Ton Product

Ethylene, kg	264
Benzene, kg	738
Utilities, $US	1

All the exothermic heat of reaction produced by alkylation is recovered. The alkylation and transalkylation reactions occur in the liquid phase, and thus the combined benzene feed does not have to be vaporized. The benzene-to-ethylene ratio is moderate. These features, in combination with a fractionation system designed to recover heat efficiently through steam generation, result in a low net utility consumption. Many energy options are available to optimize utility consumption within the plant complex and to respond to any specific utility pricing economics.

SUMMARY OF PROCESS FEATURES

EBOne Technology

- 99.7 wt % yield and 100 percent ethylene conversion throughout the run length
- Low investment
- Uninterrupted catalyst run lengths of 2 to 6 years
- Extended catalyst life (up to four regeneration cycles with no performance deterioration)
- Up to 99.95 wt % ethylbenzene product purity with essentially no xylene formation

The Lummus/UOP EBOne technology is a true liquid-phase process because both the alkylation and transalkylation reactors operate in the liquid phase. This results in xylene make of less than 10 wt ppm in the ethylbenzene product. The alkylation reactor's mild operating conditions permit the use of carbon-steel construction throughout the process and avoid potential maintenance problems. Furthermore, reactor feed vaporization and additional heat recovery equipment are not required in an all-liquid-phase alkylation process. All the ethylene feedstock is reacted completely in the liquid benzene, thus eliminating offgas recovery equipment.

CDTECH EB Technology

The CDTECH EB process is able to process a wide range of ethylene feedstocks from polymer grade down to a highly dilute ethylene mixture. The long run lengths (up to 6 years) demonstrate the robust nature of the catalyst system. High yield (99.7 wt %) and excellent product quality (99.95 wt % purity) are attained throughout the run.

The CDTECH EB process has the lowest alkylator operating pressure and benzene-to-ethylene (B/E) ratio of all commercially proven technologies. The low B/E ratio makes the CDTECH EB process especially attractive for revamping an existing unit to a much higher capacity.

Since the CDTECH EB alkylation reactor operates as a two-phase system, it is uniquely able to process very dilute feedstocks from FCC units or ethylene crackers. If the dilute ethylene feedstock contains significant amounts of light gases such as hydrogen and methane, the reactor configuration is somewhat modified in order to achieve high ethylene conversion and reject the lights from the process.

The combination of process simplicity, low-temperature operation, and all-carbon-steel equipment results in a significantly lower investment compared with competing technologies.

CHAPTER 5.2
POLIMERI EUROPA ETHYLBENZENE PROCESS

Fabio Assandri and Elena Bencini
"C. Buonerba" Research Centre
Mantova, Italy

INTRODUCTION

Ethylbenzene (EB) is one of the largest-volume petrochemicals intermediate in use today. Nearly all of EB production is used for manufacturing styrene monomer. For this reason, EB and styrene plants generally are located in the same production site, where they also can take advantage of an integration of steam production (on EB plant) and consumption (on styrene plant).

EB production technology has evolved considerably in the last 25 years. Prior to 1980, EB was produced almost exclusively via Friedel-Crafts liquid-phase alkylation with aluminium chloride catalyst. Despite the lower quality of EB produced and the need to dispose of the spent catalyst as a waste stream, many of the aluminium chloride–based EB plants built in the 1970s are still operating today in a satisfactory way.

Starting in the 1980s and until middle of the 1990s, a new gas-phase alkylation process based on zeolite catalyst has been applied for most of the new installations, overcoming the main drawbacks of aluminium chloride technology. Finally, starting in the beginning of 1990s, the development of zeolite catalysts has led to the new liquid-phase alkylation processes, which are the only ones today applied for new installations due to best product quality and lower catalyst consumption.

Polimeri Europa (formerly Montedison and then EniChem) has experienced in part this process technology development. It started the production of EB in Mantova, Italy, in the 1960s with aluminium chloride–based plants and developed such technology internally up to the construction of a new EB production unit at the end of 1980s.

In a parallel way, starting in the 1990s, Polimeri Europa also has developed, with the support of EniTecnologie (ENI Corporate Research), proprietary zeolite-based catalysts for liquid-phase ethylbenzene and cumene production that show outstanding performance in comparison with similar types of catalysts used for the same alkylation processes.

Since then, continuous research and development of zeolite and process technology have led Polimeri Europa, as final result, to

- Retrofit to zeolite its 400-kMTA cumene plant in Porto Torres, Italy
- Successfully run its zeolite on a commercial-scale EB unit

Today, Polimeri Europa is one of the few companies that, with its proprietary zeolite-based catalysts (PBE family) and its proprietary process technology, is in a position to offer an up-to-date and flexible EB production technology. Its points of strength are the high selectivity and excellent stability of the PBE catalysts, together with a design experience, based also on its industrial experience as an EB and cumene producer, always oriented to reliability, safety, and great attention to details.

Process Chemistry

Alkylation of benzene with ethylene is an exothermic reaction that occurs in liquid phase and in the presence of a zeolitic catalyst. The zeolite works as a solid acid, which includes both Brønsted and Lewis acid sites. The reaction is commonly considered as an electronic substitution on the aromatic ring, proceeding via a carbonium ion type of mechanism. Ethylene is protonated by the acid site to form the active intermediate.

The main alkylation reaction leads to production of EB according to the following scheme:

$$C_6H_6(l) + C_2H_4(g) \rightarrow C_6H_5CH_2CH_3(l)$$

$$\Delta H = -27.19 \text{ kcal/mol}$$

Moreover, due to the presence of catalyst acid sites, ethylene can further react with the EB already produced, leading to production of diethylbenzene, which is the main by-product of the process:

$$C_6H_5CH_2CH_3(l) + C_2H_4(g) \rightarrow C_6H_4(CH_2CH_3)_2(l)$$

$$\Delta H = -25.82 \text{ kcal/mol}$$

Together with diethylbenzene, other polyethylbenzenes (mainly with three or four ethyl groups) are also produced, but in much smaller amounts. Other side reactions lead to the production of butylbenzene and polycyclic compounds, mainly diphenylethanes, and to negligible quantities of light paraffins and other alkylated aromatic compounds.

The alkylation reaction is carried out in presence of an excess of benzene, to be recycled, while ethylene always reaches 100 percent of conversion. The main process parameter that affects selectivity is the benzene-ethylene molar ratio, which typically ranges between 6 and 2. The lower the benzene-ethylene molar ratio, the higher is the quantity of the diethylbenzene and butylbenzene produced, whereas, on the other hand, the higher the benzene-ethylene molar ratio, the higher is the quantity of the benzene that will need to be recycled.

Reaction temperature needs to be balanced at the optimal value that allows complete conversion of ethylene with the lowest formation of high-boiling alkylated and ethylene oligomers.

All polyethylbenzenes produced (mainly diethylbenzene) are not a loss of material for the whole process because they can be recovered by a transalkylation reaction carried out in a dedicated plant section according to the following scheme:

$$C_6H_4(CH_2CH_3)_2(l) + C_6H_6(l) \leftrightarrow 2\ C_6H_5CH_2CH_3(l)$$

$$\Delta H = -1.3 \text{ kcal/mol}$$

Also, this reaction is catalyzed by the acid sites of a zeolite. The relevant mechanism seems to be different depending on the zeolite structure. For medium-pore zeolites, the operating mechanism involves the dealkylation of diethylbenzenes followed by realkylation of benzene to give ethylbenzene. On the contrary, for large-pore zeolites, the reaction involves a hydride-transfer mechanism and a bulky diarylethane intermediate.

Main reaction parameters are the same as for alkylation: temperature, residence time, and ethyl-phenyl group molar ratio, which typically ranges between 6 and 3. Diethylbenzene conversion increases if temperature or residence time increases until it reaches a value close to chemical equilibrium. In the same conditions, ethylbenzene selectivity generally decreases, even if it strongly depends on the whole set of operating conditions.

The Catalyst

Zeolites are crystal aluminosilicates with a regular microporous structure that can be of different types depending on chemical composition and preparation procedure. Zeolite structure is responsible for the very high catalyst selectivity. In fact, we speak about reagent selectivity or transition-state selectivity, respectively, when a reagent or an intermediate has a kinetic diameter too big if compared with micropore dimensions. In both cases, reaction does not occur. On the other hand, product selectivity is related to incompatible kinetic diameter of the product that has to isomerize before going out the pore.

Other important parameters for zeolite catalytic activity are the silica-to-alumina ratio (SAR) and the particle size. Catalytic activity generally decreases with an increase in the silica-to-alumina ratio—in other words, decreasing the number of acid sites. The same effect on catalytic activity is obtained when the mean particle diameter is increased.

The acid strength required for transalkylation is higher than that required for alkylation; thus, for a given catalyst, the best conditions for an acceptable rate for the transalkylation reaction should be more severe than for alkylation. More conveniently, a different and specifically designed catalyst should be adopted for transalkylation.

Catalyst Deactivation

Zeolitic materials can be deactivated either by poisoning of the active sites or by blockage of the pores. The first deactivation mode is caused by poisons that, even in small quantities, can contaminate the raw materials. The second is a consequence of coke formation, which is an undesiderable side phenomenon commonly associated with the reaction of organic compounds over heterogeneous catalysts.

Known poisons for acid zeolites are, of course, basic compounds, such as amines or other nitrogen compounds, that can be present in the benzene and/or ethylene feed to the reactors. In fact, they neutralize the acid sites of the zeolite and in such a way deactivate the catalyst.

Other poisons for zeolites are oxygen and oxidated compounds, particularly in the alkylation section, where they can favor the oligomerization of ethylene and therefore the formation of coke precursors. For similar reasons, it is also important to minimize the amount of olefins and especially dienes in the benzene feed.

The generic term *coke* signifies a complex mixture of carbon compounds, polyaromatic or not, both external (on the outer surface) and internal (inside the pores). The formation of carbonaceous deposits and their action on the zeolites depend on their pore structure, the nature of the reactants involved, and the experimental conditions, such as reaction temperature, pressure, and feed composition. In any case, the presence of a coke molecule in a channel makes all the active sites in the channel inaccessible to the reactants, whereas a deposit of coke on the external surface can reduce the open area of the channels and so the catalytic activity. Deactivation by coke formation is reversible; removal of coke by an optimized regeneration process generally allows recovery of catalyst activity. Burning of the coke is one of the most widely used methods for this purpose, but it is important to pay attention to avoid high-temperature spots (caused by the high exothermicity of the oxidation), which can structurally damage the catalyst.

DESCRIPTION OF THE PROCESS FLOW

An EB production plant consists of four main sections: pretreatment, alkylation, transalkylation, and distillation. Figures 5.2.1 to 5.2.3 show the general flowsheet for a typical unit.

In the pretreatment section (see Fig. 5.2.1), fresh benzene coming from OSBL is fed to a drying column to remove water and molecular oxygen. Noncondensable impurities from the ethylene feed and light nonaromatic compounds produced in the reaction are vented/purged in this column, too. Dried benzene is then sent to a guard bed for removal of possible poisons for the catalyst.

In the alkylation section (see Fig. 5.2.2), the reaction of ethylene with benzene takes place in one or more fixed-bed reactors with a multibed arrangement. Due to reaction exothermicity, some external intrabed refrigerators are required, which allows complete heat recovery (preheating the feed and/or producing steam). The operating pressure of the reactors is high enough to maintain liquid-phase conditions. The reactor operating conditions are selected so as to achieve optimal performance (minimum by-product formation and long catalyst life). The excess benzene, recycled from the distillation section together with fresh treated benzene, is first preheated and then fed at the beginning of the alkylation train. The ethylene, coming from OSBL, is compressed (where necessary) to the reactor pressure and then fed to each alkylation bed.

The alkylation zone effluent, which consists mainly of unreacted benzene, ethylbenzene, and by-products such as diethylbenzene, polyethylbenzene, butylbenzene, diphenylethanes, and higher-boiling components, is fed to the distillation system.

In the transalkylation section (see Fig. 5.2.2), the diethylbenzene produced in the alkylation zone and separated in the distillation section reacts with excess benzene to produce additional EB. The benzene-diethylbenzene mixture is first preheated and then sent to a dedicated fixed-bed reactor, where the isothermal reaction takes place in liquid phase. Thus the pressure is high enough to maintain the liquid-phase conditions. The transalkylator effluent, which consists mainly of ethylbenzene, unreacted benzene and diethylbenzenes, and by-products such as diphenylethanes and higher-boiling components, is fed to the distillation section.

When coke deposition on the catalyst, for both the alkylation and the transalkylation sections, leads to an irreversible decrease in activity, thermal regeneration of the catalyst is required. With the Polimeri Europa process, the catalyst lifetime, before thermal regeneration is required, is very high; *ex situ* thermal regeneration is then preferred from an economic standpoint, and no additional equipment is required for *in situ* regeneration.

The distillation section (see Fig. 5.2.3) consists of a three-column train. From the first column, fed with alkylation and transalkylation effluents, the benzene is removed from the top to be recycled back to the alkylation and transalkylation sections. From the top of the second column is separated pure EB to be sent to product storage. In the third column, operated under vacuum, is carried out the separation of the heaviest by-product. Transalkylable polyethylbenzenes (mainly diethylbenzene) are removed from the top of the column and recycled to the transalkylation section, whereas high-boiling by-products (flux oil) are extracted from the bottom.

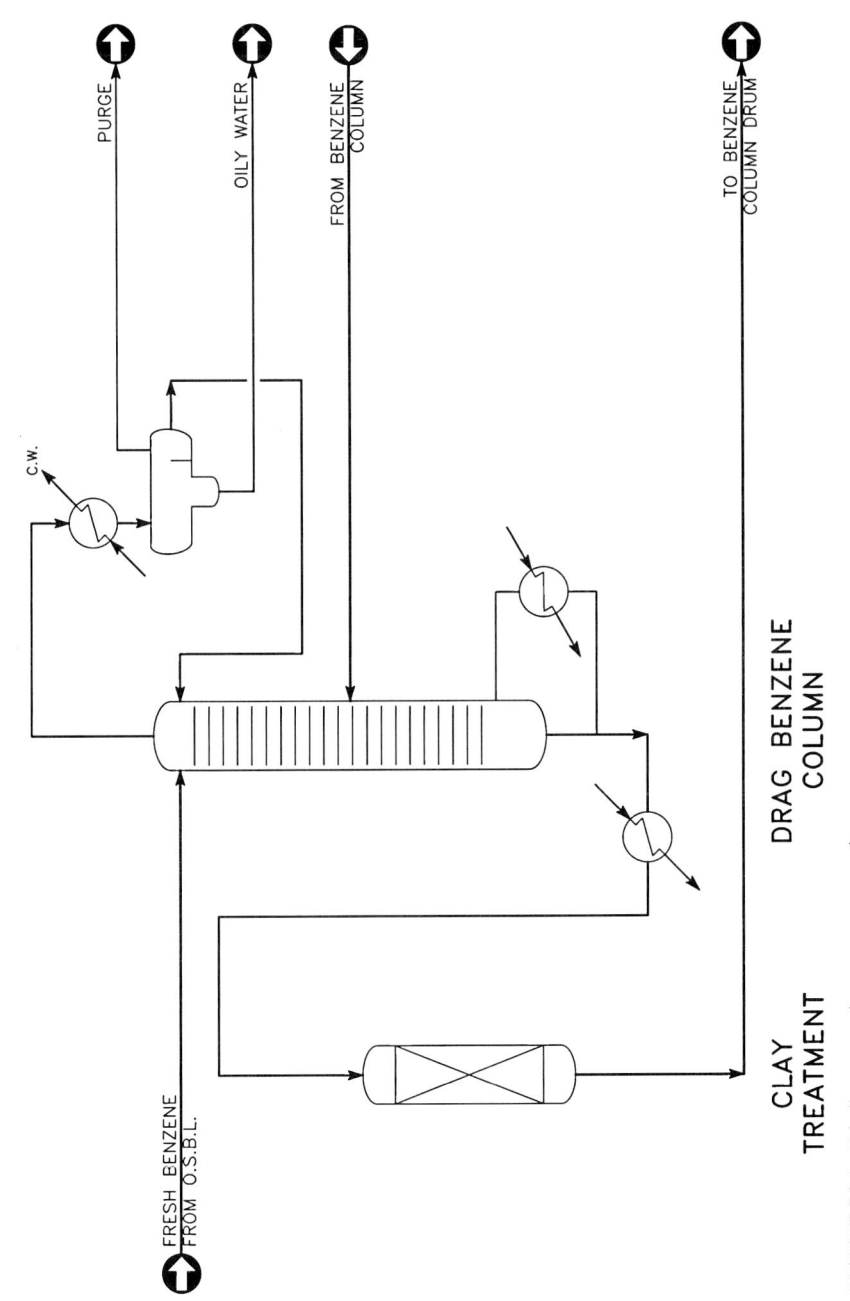

FIGURE 5.2.1 Ethylbenzene unit, pretreatment section.

5.17

FIGURE 5.2.2 Ethylbenzene unit, alkylation and transalkylation sections.

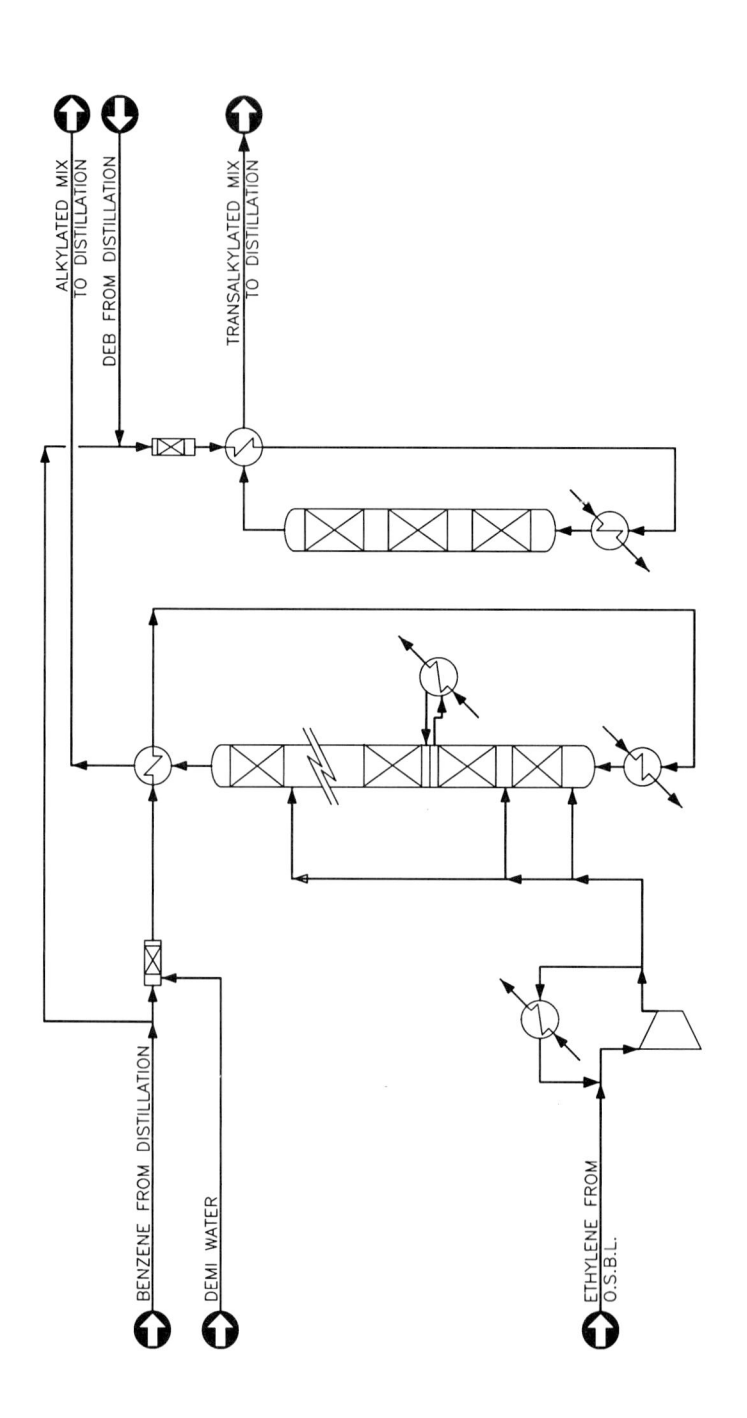

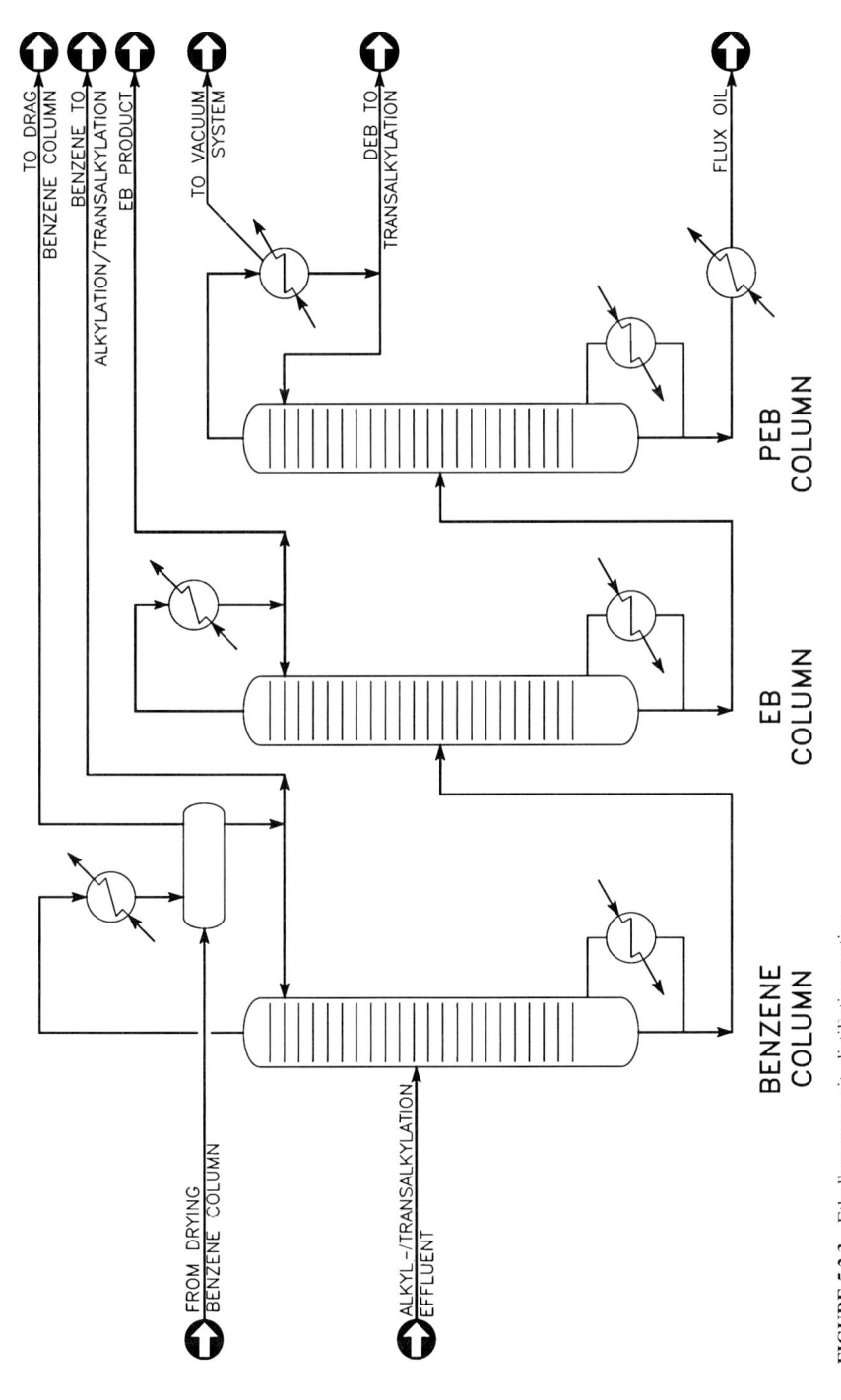

FIGURE 5.2.3 Ethylbenzene unit, distillation section.

PROCESS AND CATALYST ADVANCED FEATURES

As with every EB production technology based on liquid-phase zeolite, the following well-known advantages, over aluminium chloride technology, are available:

- Low investment and maintenance costs because all equipment in the reaction section is of carbon steel; no need for special alloy or surface lining.
- Low operating cost because of heterogeneous catalysis instead of a homogeneous one; no need for washing and neutralizing the reactor effluent.
- Low environmental impact because no wastewater is generated by the process.
- High selectivity due to much higher performance of the catalyst, resulting in high ethylbenzene purity.

From this common base, every improvement in process performance can only be obtained with either a good and updated process technology and a solid know-how in zeolite catalyst development and production. Polimeri Europa is strongly involved in both these fields of activity.

Process Technology

Process technology has been set up by Polimeri Europa in its Styrenics Research Centre "C. Buonerba," located in Mantova, Italy. Starting in the beginning of the 1990s, a huge number of different zeolite catalysts have been tested, covering a wide range of process operating conditions, for both alkylation and transalkylation reactions. Experimental activities have led to the choice of zeolite catalysts tailored for each section and to the definition of relevant best-process scheme and operating conditions. A great effort also has been made in investigating catalyst deactivation due to the different possible poisons and to set up suitable pretreatment sections to avoid sudden losses of activity.

For all these purposes, several lab-scale pilot plants have been used, together with a large pilot plant (500-kg/h flow rate), all of them still in operation. A second pilot plant is under construction. In addition to process development, Polimeri Europa expertise in FEM analysis has been applied in the design of key equipment.

One of the most critical aspects, from the process point of view, is the mixing of ethylene with benzene before the catalyst bed. With the help of CFD techniques, an original design has been settled on for this position that is capable of maximizing process performances.

Catalysts: PBE Family

With reference to catalyst development, the Polimeri Europa corporate Research Centre ("G. Donegani"), at the forefront in this area, has developed and patented PBE1 zeolite for cumene and ethylbenzene synthesis, with the support of EniTecnologie (ENI Corporate Research). More recently, PBE2 catalysts have been set up for DEB transalkylation.

Generally, catalysts of the PBE family show very high selectivity for alkylation and transalkylation and, above all, a very high stability, at the highest level among common zeolites and zeolite-based catalysts. More particularly, the preparation procedures allow optimal values for extrazeolite porosity and degree of interconnectivity, which results in very high catalyst stability and very low ethylene oligomerization and deactivation due to coke deposition. This behavior allows lower alkylation temperatures and thus lower operating and design pressures. Of course, Polimeri Europa also has developed and experi-

TABLE 5.2.1 Ethylbenzene Unit: Typical Process Performances

EB purity	>99.95%
Ethylene (kg/kg EB)*	0.265
Benzene (kg/kg EB)*	0.741
Hot oil or HP steam (kcal/kg EB)	540
Electricity (kWh/kg EB)	0.02
LP steam generated (kg/kg EB)	1.5
Catalyst life cycle (between regenerations)	2–4 years

*100 percent basis.

enced, for its PBE catalysts, industrial-scale production and regeneration. At present, hundreds of tons of PBE catalyst have been produced and loaded in industrial reactors since 1996.

In addition, the proprietary catalyst regeneration procedure, which is able to prevent any catalyst modifications (such as zeolite dealumination and/or catalyst sintering), has been proven by Polimeri Europa at industrial scale.

PROCESS PERFORMANCE

For an ethylbenzene unit, a typical set of process parameters is summarized in Table 5.2.1.

COMMERCIAL EXPERIENCE

A first ethylbenzene unit (650 kMTA) licensed by Polimeri Europa will be started up in 2005.

CHAPTER 5.3
EXXONMOBIL/BADGER ETHYLBENZENE TECHNOLOGY

Brian Maerz and C. Morris Smith
Badger Licensing LLC
Cambridge, Massachusetts

INTRODUCTION

Badger Licensing LLC, an affiliate of Stone & Webster and ExxonMobil Chemical Company, offers the EBMax^{SM} process for the production of ethylbenzene by the alkylation of benzene with ethylene. In the EBMax process, ethylbenzene is produced under full or partial liquid-phase conditions over proprietary ExxonMobil zeolite catalysts contained in fixed-bed reactors. EBMax units can be designed to accept either a polymer-grade or dilute ethylene feedstock. The purity of the ethylbenzene produced by the EBMax process is very high, usually in excess of 99.9 percent, with less than 100 ppm of impurities boiling in the range of ethylbenzene. The main by-product of the process, a heavy residue stream, is produced at a rate of only 3 to 4 kg per metric ton of ethylbenzene produced.

An ethylbenzene unit typically is located adjacent to a styrene monomer (SM) unit and is tightly integrated with the SM unit through the ethylbenzene intermediate product, utilities systems, and minor gas and liquid by-product streams. Stone & Webster (Badger®) offers the ATOFINA/Badger styrene process through Badger Licensing for the production of SM by the dehydrogenation of ethylbenzene. Most of the SM manufactured worldwide is produced by the dehydrogenation of ethylbenzene. Styrene also is produced as a by-product of propylene oxide manufacture by the ethylbenzene hydroperoxide route. Three large EBMax units currently in service feed propylene oxide units.

ETHYLBENZENE MANUFACTURING

Approximately 26 million metric tons of ethylbenzene is produced annually, nearly all of which is consumed in the production of SM. Prior to 1980, ethylbenzene was produced

TABLE 5.3.1 Typical Specification for Polymer-Grade Ethylene

Ethylene	99.9 vol % min
Methane + ethane + N_2	1000 ppm vol max
Propylene and higher olefins	10 ppm vol max
Acetylene	5 ppm vol max
Carbon monoxide	5 ppm vol max
Carbon dioxide	5 ppm vol max
Hydrogen	10 ppm vol max
Total sulfur	2 ppm wt max
Oxygen	2 ppm vol max
Water	5 ppm vol max
Total nitrogen compounds	0.1 ppm wt max

almost exclusively from aluminum chloride–catalyzed liquid-phase alkylation processes. In that year, the first commercial-scale ethylbenzene unit based on alkylation over a zeolite catalyst was commissioned, a 400,000–metric ton per annum (MTA) plant located in Bayport, Texas. The new zeolite-based vapor-phase alkylation process, developed by The Badger Company and Mobil Research and Development Corporation, quickly became the leading licensed technology for ethylbenzene production. Only a few aluminum chloride technology ethylbenzene units have been constructed since 1980. ExxonMobil and Badger commercialized the EBMax process for the production of ethylbenzene under liquid-phase conditions in 1995. Today, ExxonMobil/Badger zeolite-catalyzed processes account for approximately 55 percent of the total ethylbenzene produced worldwide, with over half of this percentage being produced by the EBMax process.

Polymer-grade ethylene is used as the sole feedstock source in all but a few ethylbenzene units. A typical specification for polymer-grade ethylene is given in Table 5.3.1.

Polymer-grade ethylene is available by pipeline in many locations. Extensive ethylene pipeline networks are located in the Gulf Coast region of the United States, in western Europe, and elsewhere. Ethylene also may be fed directly from an ethylene unit. A few plants are equipped to receive ethylene shipped as a refrigerated liquid.

Ethylbenzene also can be produced from a dilute ethylene stream obtained from the distillation section of an ethylene plant or from refinery offgas in which ethylene is a minority constituent. Examples of the former include deethanizer overhead streams, in which ethane would be the main impurity, or streams withdrawn from a demethanizer, which would contain significant quantities of methane and hydrogen. A dilute ethylene stream from the distillation section of an ethylene plant otherwise may be considered equivalent to polymer-grade ethylene from a trace impurities standpoint, provided that acetylene has been removed upstream. Two ExxonMobil/Badger units produce ethylbenzene from ethylene plant dilute streams.

Refinery offgas normally contains significant quantities of higher olefins along with heteroatom-bearing (nitrogen, sulfur, oxygen, etc.) compounds and metals that may adsorb on zeolite alkylation catalysts, reducing their activity. Some of these impurities cannot be removed by regeneration. As a consequence, refinery offgas must be pretreated prior to being fed to a zeolite-catalyzed ethylbenzene unit. This is particularly true for liquid-phase technology units because impurities are adsorbed more strongly at the low temperatures that are characteristic of all liquid-phase alkylation processes. Two ExxonMobil/Badger units produce ethylbenzene using refinery offgas as a source.

So-called nitration-grade, or refined, benzene is used in the production of ethylbenzene. Approximately half the benzene produced worldwide is consumed in ethylbenzene manufacture. The purity specification for benzene is typically 99.8 percent by weight; however, the actual purity generally exceeds 99.9 percent on a water-free basis, with nonaromatic

TABLE 5.3.2 Typical Specification for Benzene

Benzene	99.8 wt % min
Nonaromatic hydrocarbons	1500 ppm wt max
Toluene	300 ppm wt max
Distillation range	1.0°C incl. 80.1°C
Solidification point	5.4°C
Specific gravity, 15.56/15.56°C	0.882–0.886
Appearance	Clear, free of sediment
Thiophene	1 ppm wt max
Carbon disulfide	1 mg/L max
H_2S/SO_2	None
Acid wash color	1 max
Color	20 Pt/Co max
Acidity	None detected
Bromine index	10 mg/100 g max
Copper corrosion	Pass (1a or 1b)
Total chlorides (as Cl_2)	3 ppm wt max
Total nitrogen compounds	1 ppm wt max

compounds and toluene being the primary hydrocarbon impurities. A typical benzene specification is given in Table 5.3.2.

Benzene usually is obtained directly from a nearby aromatics unit or shipped by barge or other means of bulk conveyance. In the former case, at least a small portion of the total benzene supply generally is obtained from one or more external sources. In the latter case, it is not unusual for benzene to be obtained from many different sources. Nitrogen-containing and other basic compounds are of particular interest from an impurities perspective because these are strongly adsorbed by the process catalysts at liquid-phase operating temperatures. Such compounds are not adsorbed as strongly under the conditions that are characteristic of vapor-phase alkylation. Three aromatics extraction processes currently in widespread use employ nitrogen-containing solvents, and extraction processes using other types of solvents frequently add amines at certain locations within the process to control corrosion.

Ethylbenzene plants that are integrated with dehydro technology SM plants sometimes include a small distillation column to produce a salable toluene by-product from the benzene/toluene fraction that is produced in the SM unit. The benzene that is recovered as a distillate from this column usually is recycled to the ethylbenzene unit. This recycle benzene stream can contain ppm quantities of nitrogen compounds that originate with the corrosion-control chemicals introduced upstream of the main condenser in the SM dehydro section. Feedstock purification techniques are described later.

PROPERTIES OF ETHYLBENZENE

Ethylbenzene is a colorless liquid with a gasoline-like odor. It is soluble in alcohol, benzene, and other aromatic liquids but is only slightly soluble in water. Ethylbenzene vapor forms explosive mixtures with air. It is a respiratory, eye, and skin irritant. The threshold limit value for ethylene is 100 ppm in air. Physical properties for ethylbenzene are given in Table 5.3.3. The boiling point of pure ethylbenzene at various pressures is given in Table 5.3.4.

TABLE 5.3.3 Physical Properties of Ethylbenzene

Formula	C_8H_{10}
Molecular weight	106.167
Critical temperature	344°C
Critical pressure	35.8 kg/cm^2(g)
Critical volume	0.374 m^3/kmol
Accentric factor	0.3035
Normal boiling point	136.2°C
Melting point	−95°C
Specific gravity at 15.56/15.56°C	0.873
Autoignition temperature	430°C
Lower limit of flammability	1.0%
Upper limit of flammability	6.7%
Flash point	15°C

TABLE 5.3.4 Boiling Points of Pure Ethylbenzene at Various Pressures

Pressure, mmHg	Boiling point, °C
5	14.1
50	57.6
100	74.1
200	92.6
760	136.2

EBMAX PROCESS CATALYSTS

The EBMax process is based on a unique family of ExxonMobil zeolite catalysts first offered for commercialization in liquid-phase benzene alkylation applications by Badger in the early 1990s. The unique structural features of the EBMax alkylation catalyst and the resulting performance characteristics make EBMax the clear choice for the commercial production of ethylbenzene.

- Adsorption of benzene is strongly favored over ethylene adsorption. This effectively eliminates the formation of ethylene oligomers, which in turn reduces heavies formation and virtually eliminates catalyst aging caused by fouling of the catalyst by heavy compounds.
- The absence of oligomerization permits a significant reduction in (and, in principle, the elimination of) the quantity of excess benzene that must be fed to the alkylation reactor to maintain catalyst stability. This reduces the size and cost of reaction and distillation equipment used for the recovery of unreacted benzene from the alkylation effluent and also reduces energy consumption in the reaction and distillation sections of the plant.
- The absence of oligomerization reduces the formation of ethylbenzene impurities and heavy by-products. The alkylator produces insignificant quantities of cumene and *n*-propylbenzene and very small quantities of diphenyl compounds.
- The EBMax alkylation catalyst produces very small quantities of polyethylbenzenes, reducing the size and cost of equipment required for polyethylbenzene recovery and the quantity of catalyst required for the transalkylation of polyethylbenzenes.

The EBMax alkylation catalyst has unique structural features not seen before in a molecular sieve.[1] ExxonMobil scientists resolved the structure of the catalyst by using high-resolution microscopy and x-ray diffraction analysis, as depicted in Fig. 5.3.1. The catalyst crystallizes in thin sheets or plates. The crystalline structure consists of two independent, noninterconnecting channel systems, each accessible through 10-member-ring apertures. One system consists of 10-member-ring channels. The other possesses large supercages, the diameters of which are defined by 12-member rings. The alkylation reaction is believed to take place primarily within surface pockets defined by the top and bottom halves of these supercages (see Fig. 5.3.1). It is thought that the 10-member-ring system does not contribute significantly to ethylbenzene production.

The catalyst supercage structure (7.1 Å × 7.1 Å × 18.2 Å) is depicted in Fig. 5.3.2, which shows the two interconnected 12-ring pockets; in the catalyst, these supercages are found in a hexagonal array pattern. The pocket structure is uniquely selective for aromatics adsorption and, as a result, strongly promotes aromatics alkylation reactions over olefin oligomerization reactions.

A key feature of the EBMax alkylation catalyst is its ability to operate with stability at low benzene concentrations with minimal production of polyethylbenzenes and ethylene oligomers. The unique crystalline structure is believed to be responsible for its high monoalkylation selectivity (i.e., low polyethylbenzene production relative to ethylbenzene).

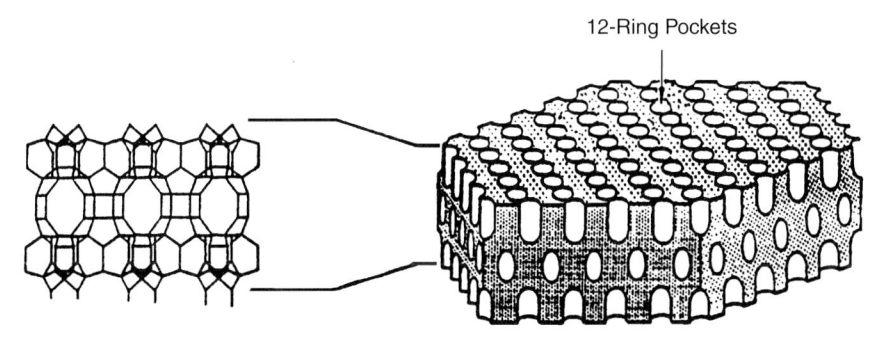

FIGURE 5.3.1 Catalyst chrystalline structure.

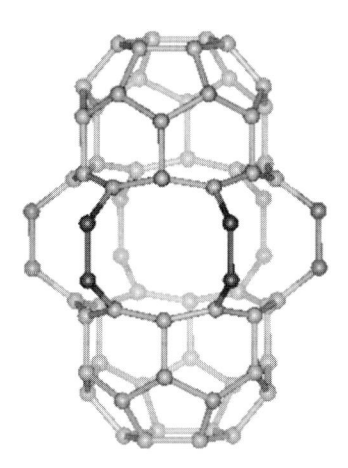

FIGURE 5.3.2 Catalyst supercage structure.

Badger has operated an EBMax pilot plant at its Weymouth, Massachusetts, research and development (R&D) facility for over 8 years. The pilot plant has logged over 35,000 hours of operation in EBMax service during that period, operating over a wide range of operating conditions. Without exception, ExxonMobil's EBMax alkylation catalyst has operated with complete stability at all conditions tested. The catalyst has demonstrated its ability to operate efficiently and with complete stability at benzene-to-ethylene feed ratios that are very close to the stoichiometric requirement. As noted earlier, the ability to design for operation with a minimal excess of benzene fed to the alkylator is desirable from both capital investment and energy-consumption standpoints.

The EBMax process uses a proprietary ExxonMobil catalyst for the transalkylation of polyethylbenzenes with benzene. Like the alkylation catalyst, the EBMax transalkylation catalyst does not require that a large excess of benzene be fed to the reactor to ensure stable operation and high reaction selectivity. Pilot plant tests and commercial plant operation both indicate that the presence of a large excess of benzene in the reactor feed does not measurably benefit per-pass conversion, selectivity, or catalyst stability. The EBMax transalkylation catalyst produces only small quantities of heavy compounds and ppm quantities of impurities boiling in the range of ethylbenzene. This makes it possible to operate the alkylation system at very low benzene-to-ethylene feed ratios because the production of greater quantities of polyethylbenzenes in the alkylation reactor does not result in a significant increase in the yield loss to heavies for the process as a whole.

PROCESS CHEMISTRY AND EBMAX CATALYST PERFORMANCE

In the EBMax process, benzene is reacted with ethylene in an alkylation reactor containing multiple fixed catalyst beds. The alkylation reaction is exothermic, releasing 960 kcal of energy per kilogram of ethylene reacted. Polyethylbenzenes formed in the alkylation reactor are recovered in the distillation section of the plant and fed to a transalkylation reactor, where they react with benzene, forming ethylbenzene. There is no net release of energy in the transalkylation reaction. The principal reactions are described below.

In the alkylation reactor, ethylene reacts with benzene, the latter of which is fed in excess of the stoichiometric requirement, forming ethylbenzene:

$$\underset{\text{Benzene}}{C_6H_6} + \underset{\text{Ethylene}}{C_2H_4} \rightarrow \underset{\text{Ethylbenzene}}{C_6H_5\text{-}C_2H_5}$$

In the operating temperature range that is characteristic of liquid-phase alkylation, ethylene consumption may be considered to be complete (in excess of 99.99 percent).

Benzene is strongly adsorbed by the ExxonMobil EBMax alkylation catalyst. As a consequence of the catalyst's strong affinity for benzene, the formation of by-products that originate with ethylene oligomerization reactions (cycloparaffins, cumene, higher alkylbenzenes, and heavies) is practically eliminated. Ethylene oligomers are also believed to be precursors in the formation of heavy compounds that can foul (coke) the catalyst, thereby reducing its activity. EBMax commercial plant operating experience shows that there is no measurable loss of catalyst activity due to ethylene oligomerization or catalyst fouling over many years of continuous operation.

The ethylbenzene produced is subject to further ethylation:

$$\underset{\text{Ethylbenzene}}{C_6H_5\text{-}C_2H_5} + \underset{\text{Ethylene}}{C_2H_4} \rightarrow \underset{\text{Diethylbenzene}}{C_6H_4\text{-}(C_2H_5)_2}$$

$$\underset{\text{Diethylbenzene}}{C_6H_4\text{-}(C_2H_5)_2} + \underset{\text{Ethylene}}{C_2H_4} \rightarrow \underset{\text{Triethylbenzene}}{C_6H_3\text{-}(C_2H_5)_3}, \text{ etc.}$$

These compounds are collectively referred to as *polyethylbenzenes*. Polyethylbenzene production relative to ethylbenzene is a function of catalyst selectivity and the quantity of excess benzene that is fed to the alkylation reactor. At a given ethylene feed rate, increasing the benzene rate to the alkylation reactor will reduce the quantity of polyethylbenzenes formed. The impact of alkylator polyethylbenzene production on the ethylbenzene yield of the process is described below. The intrinsically low rate of polyethylbenzene formation over the EBMax catalyst makes it practical to operate the alkylation reactor with only a modest excess of benzene.

Ethylene will react with other alkylbenzenes present in the feed to the alkylation reactor, forming the corresponding ethyl-alkylbenzene. For example:

$$\underset{\text{Toluene}}{C_6H_5\text{-}CH_3} + \underset{\text{Ethylene}}{C_2H_4} \rightarrow \underset{\text{Ethyltoluene}}{C_6H_4\text{-}(CH_3)(C_2H_5)}$$

All the toluene contained in the benzene feed to the unit will be present in the ethylbenzene product in the form of either toluene or ethyltoluene.

A small yield loss is incurred in the alkylation reactor to heavy by-products that cannot be converted easily to ethylbenzene in the transalkylation reactor. These include tetraethylbenzenes and heavier polyethylbenzenes and diarylalkanes, the latter of which are formed by complex hydride transfer reactions. For example:

$$\underset{\text{Ethylbenzene}}{C_6H_5\text{-}C_2H_5} + \underset{\text{Benzene}}{C_6H_6} \rightarrow \underset{\text{Diphenylethane}}{C_6H_5\text{-}C_2H_4\text{-}C_6H_5} + \underset{\text{Hydrogen}}{H_2}$$

Extremely small quantities of light paraffins are also formed by these reactions.

In the transalkylation reactor, polyethylbenzenes are converted to ethylbenzene by reaction with benzene over a single bed of proprietary ExxonMobil transalkylation catalyst. For example:

$$\underset{\text{Diethylbenzene}}{C_6H_4\text{-}(C_2H_5)_2} + \underset{\text{Benzene}}{C_6H_6} \rightarrow \underset{\text{Ethylbenzene}}{2(C_6H_5\text{-}C_2H_5)}$$

Both diethylbenzenes and triethylbenzenes are converted easily to ethylbenzene. Per-pass polyethylbenzene conversion is in the range of 60 to 70 percent. Benzene in excess of that required for polyethylbenzene conversion is fed to the transalkylator to maintain per-pass conversion at the desired level and to limit the formation of heavy compounds.

Other polyalkylated species present in the reactor feed are also subject to transalkylation. For example, ethyltoluene formed by the reaction of toluene with ethylene in the alkylation reactor will be partially converted to toluene and ethylbenzene:

$$\underset{\text{Ethyltoluene}}{C_6H_4\text{-}(CH_3)(C_2H_5)} + \underset{\text{Benzene}}{C_6H_6} \rightarrow \underset{\text{Toluene}}{C_6H_5\text{-}CH_3} + \underset{\text{Ethylbenzene}}{C_6H_5\text{-}C_2H_5}$$

Compounds boiling between ethylbenzene and triethylbenzene that cannot be converted to ethylbenzene will be contained in the polyethylbenzene recycle stream and converted to heavy compounds. For example, the small quantities of butylbenzenes formed in the EBMax alkylator will be converted to heavy compounds by complex hydride transfer reactions:

$$\underset{\text{Butylbenzene}}{C_6H_5\text{-}C_4H_9} + \underset{\text{Ethylbenzene}}{C_6H_5\text{-}C_2H_5} \rightarrow \underset{\text{Diphenylethane}}{C_6H_5\text{-}C_2H_4\text{-}C_6H_5} + \underset{\text{Butane}}{C_4H_{10}}$$

Some of the nonaromatic compounds that enter the EBMax unit with the benzene feedstock, those boiling closest to benzene in particular, will be concentrated in the benzene recycle and will be present in the feed to both the alkylation and transalkylation reactors. These compounds may form other nonaromatic compounds and heavy compounds. While some of these nonaromatic compounds may escape the EBMax unit with the ethylbenzene product or the lights column vent, at least a portion must be removed by reaction because the nonaromatics boiling closest to benzene are not removed by fractionation to any significant extent.

Nearly all the polyethylbenzenes produced in the alkylator are converted to ethylbenzene in the transalkylator; however, it is desirable to select an alkylation catalyst that minimizes the production of these compounds for the following reasons:

- A small fraction of the polyethylbenzenes is converted to heavies in the transalkylation reactor; therefore, minimization of their production in the alkylator will minimize the yield loss to heavies in the transalkylation reactor.

- Minimization of polyethylbenzene production will reduce the size of the transalkylation reactor and the quantity of transalkylation catalyst required.

- An alkylation catalyst with inherently low polyethylbenzene production will permit operation with a reduced excess of benzene in the alkylation reactor feed. As a consequence, the quantity of unreacted benzene that must be recovered from the reaction system effluent will be reduced. This reduces the size of the benzene recovery column, reboiler, condenser, overhead drum, reflux pumps, and reactor feed pumps. The energy consumption associated with benzene recovery also will be reduced. Most of the energy consumed in the distillation section of an ethylbenzene unit is consumed in benzene recovery.

- The diameter of the alkylation reactor can be reduced if the benzene feed rate to the alkylator is reduced.

The EBMax alkylation catalyst possesses an intrinsically high monoalkylation selectivity; therefore, its superior performance in this regard reduces capital investment and energy consumption in both the reaction and distillation sections of the ethylbenzene unit.

The EBMax alkylation catalyst does not require the presence of an excess of benzene in the reactor feed to promote stable operation; however, the presence of an excess of benzene does serve to reduce the quantity of polyethylbenzenes produced in the alkylation reactor. Polyethylbenzenes are converted to ethylbenzene in the transalkylation reactor, but since small quantities of heavies are produced in the transalkylator, it is desirable to minimize the production of polyethylbenzenes produced in the alkylator. The benzene feed rate to the alkylator relative to the ethylene feed rate—the benzene-to-ethylene (B/E) feed ratio—is one of the principal design variables in the economic optimization of an EBMax design. A reduction in the B/E ratio reduces capital investment and energy consumption, whereas an increase in the B/E ratio reduces raw materials consumption.

All zeolite catalysts are susceptible to poisoning by basic compounds. As noted previously, nitrogen-containing compounds, including certain solvents used in benzene extraction, and amines used for corrosion control or their decomposition products may be present in trace quantities in the benzene feedstock to the ethylbenzene unit. If not removed, these compounds will be adsorbed by the catalyst's active sites, reducing catalyst activity. Activity loss from trace benzene impurities typically is seen as a gradual reduction in catalyst bed temperatures in the first alkylation catalyst bed in series, proceeding in a bandwise fashion through the bed. Catalyst activity can be restored by regeneration (burning). Given the infrequency with which regeneration is required, most EBMax units are not designed for *in situ* regeneration of the process catalysts.

Most liquid-phase ethylbenzene units incorporate benzene pretreatment systems for removal of basic compounds. Solid adsorbents such as mole sieve, acid-activated clay, acid resins, and alumina typically are employed in benzene pretreatment. The treatment may be applied directly to the benzene feed from battery limits or to an internal benzene-rich stream. Adsorbent performance often can be enhanced if the treated stream is substantially water-free.

PROCESS DESCRIPTION

All EBMax plants include a reaction section that consists of an alkylation and transalkylation reaction system and a distillation system consisting of four columns for the recovery of unreacted benzene, recovery of ethylbenzene product, rejection of heavy compounds, and rejection of lights. A simplified flow diagram is presented in Fig. 5.3.3. A high-temperature heat source, such as 600-psig [42-kg/cm^2(g)] steam or hot oil is used to preheat the feeds to the reaction systems and reboil the distillation columns. For an EBMax unit that is integrated with an ATOFINA/Badger SM unit, steam generated in the alkylation system and in the overhead condensers of the distillation columns is used as dilution and vaporizing steam in the dehydro unit and as steam to the reboilers of the styrene distillation columns.

Plants that use a dilute ethylene feedstock require additional equipment for recovery of benzene and other aromatics from the inert diluents that constitute the offgas from the process. For EBMax plants that will receive an ethylene feedstock containing large quantities of diluents, the alkylation system is designed for operation under partial liquid-phase conditions, in which case an ethylene-rich vapor phase and an aromatics-rich liquid phase flow concurrently downward through the catalyst beds of the alkylation reactor. Unless otherwise indicated, the process descriptions that follow refer to an EBMax plant fed with polymer-grade ethylene.

In the alkylation system, ethylene is reacted with benzene in the liquid phase over ExxonMobil EBMax alkylation catalyst to form ethylbenzene and small quantities of polyethylbenzenes. An EBMax alkylation reactor typically consists of multiple liquid-filled chambers. Benzene is preheated to reaction temperature and introduced at the bottom of the reactor, flowing through the chambers in series. A portion of the total ethylene feed is

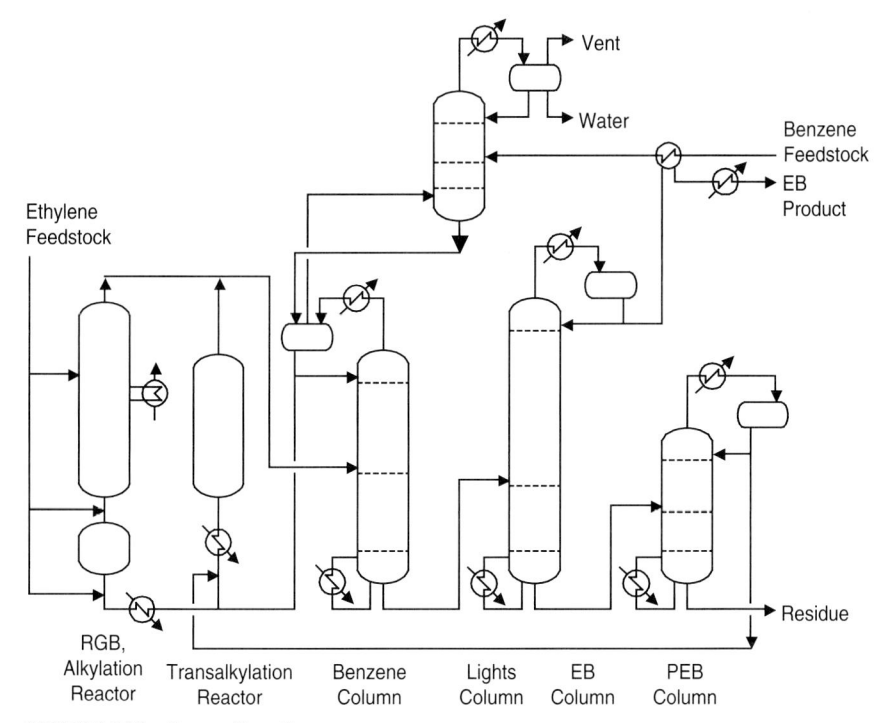

FIGURE 5.3.3 Process flow diagram.

introduced upstream of each catalyst bed. The heat of reaction can be recovered stage-wise in steam-generating exchangers.

Commercial experience indicates that the first bed of alkylation catalyst can be subject to a gradual loss of activity due to nitrogen poisons that are present in the benzene feed to the reactor. Catalyst contained in the alkylation beds that follow does not age under normal circumstances. Badger can place the first alkylation catalyst bed in a separate vessel (a reactive guard bed, or RGB) that can be bypassed when removal of its contained catalyst for regeneration is required. In this configuration, ethylbenzene production can be maintained continuously at design rates for extended periods.

Polyethylbenzenes are recovered from the crude ethylbenzene in the distillation section of the EBMax plant. In the transalkylation system, the polyethylbenzenes are converted to ethylbenzene by reaction with benzene in a liquid-filled vessel. Polyethylbenzenes and benzene are externally mixed and preheated prior to feeding to the reactor. The reactor contains a single fixed bed of ExxonMobil transalkylation catalyst.

In the distillation section, unreacted benzene, polyethylbenzenes, and heavier compounds are separated from the alkylation and transalkylation reactor effluents to produce the ethylbenzene product. The distillation section consists of four columns, three of which generate useful steam in their overhead condensers.

The benzene column recovers unreacted benzene from both the alkylation and transalkylation reactor effluents. The column operating pressure is set based on the utilities scheme adopted in the design of the EBMax unit. In a typical plant, the benzene column is reboiled with high-pressure steam, with low-pressure steam generated in the overhead condenser. A portion of the column overhead vapor is fed to a small lights column for rejection of the inerts present in the ethylene feedstock. The net liquid overhead is mixed with the dry, pretreated benzene feedstock and pumped to the reaction section of the plant. The benzene is split immediately downstream of the feed pumps, with the largest portion flowing to the alkylation system. The benzene feed to the reaction system typically contains only a small fraction of ethylbenzene. The bottoms from the benzene column consists of ethylbenzene and heavier compounds along with trace amounts of benzene.

As mentioned in the preceding paragraph, a portion of the benzene column overhead vapor is fed to a lights column for rejection of C_4 and lighter compounds. This column also can be used to dry the benzene feedstock from battery limits after it has been preheated. The wet benzene feed contacts the vapor feed from the benzene column in the bottom section of the lights column. The dry bottoms stream is pumped to the benzene column reflux drum. The vapor overhead from the column is condensed against cooling water, and the water that enters the lights column system with the benzene feedstock is separated from the condensed hydrocarbons in the reflux drum. In an integrated EB/SM unit, aromatics recovery is practiced, and the small water stream from the reflux drum can be handled in existing equipment. It should be noted that the yield loss to lights is extremely small and for practical purposes can be ignored.

The benzene column bottoms is fed to the ethylbenzene column for recovery of an ethylbenzene product of high purity. The column operating pressure typically is set to permit reboiling with high-pressure steam and generation of low- or intermediate-pressure (dehydro dilution) steam in the overhead condenser. The column is designed to produce an ethylbenzene product that is essentially free of diethylbenzenes. Diethylbenzenes undergo dehydrogenation to divinylbenzenes in the SM reaction system and will copolymerize with styrene monomer in the distillation section of the SM plant, forming a cross-linked polymer that can foul or plug equipment. The ethylbenzene column distillate normally is cooled against the benzene feed to the ethylbenzene unit and then fed directly to the SM unit. The column bottoms, still containing percent levels of ethylbenzene, is fed to the polyethylbenzene column.

The ethylbenzene column bottoms is fed to the polyethylbenzene column for recovery of polyethylbenzenes and rejection of heavy compounds. In most plants, the column is designed for operation under vacuum to permit reboiling with high-pressure steam. In plants that are integrated with an ATOFINA/Badger SM unit, steam generated in the overhead condenser is used to strip hydrocarbons from the SM reaction section process condensate. The column is operated to produce a polyethylbenzene stream substantially free of heavies while minimizing the triethylbenzene content of the bottoms stream.

PROCESS DESIGN CUSTOMIZATION AND OPTIMIZATION

All EBMax plants incorporate the basic unit operations in the processing arrangement described earlier; however, the operating conditions in the reaction and distillation sections will differ from one plant to the next depending on how the design for each plant has been optimized for integration with the utilities systems and with the SM unit that it will feed. For example, in instances in which an EBMax unit feeds a propylene oxide/styrene monomer unit, there frequently is only limited use for the steam that normally would be generated in the EBMax unit. Such designs would incorporate heat integration within the reaction section and between the reaction and distillation sections of the EBMax unit to minimize steam production and also may incorporate heat integration between columns in the distillation section to further reduce steam production. Minimization of low-pressure steam production also may be required when EBMax units are integrated with units producing SM by the dehydro route if the SM plant includes heat-recovery schemes that minimize low-pressure steam consumption. In such cases, operating conditions in the ethylbenzene unit's distillation columns can be set to produce steam at a higher pressure that can be used as dilution steam in the dehydrogenation reactors.

EBMax unit designs are optimized to balance capital and operating costs according to the economic criteria that are specific to each project, also taking into account the feedstock supply conditions and any constraints imposed by the utilities systems. As noted previously, the alkylator B/E feed ratio, which also effectively sets the relative PEB/EB ratio, is one of the principal design variables in the economic optimization of an EBMax unit. In most cases, economic criteria for grassroots plants favor designing at a low B/E ratio. The impact of the B/E ratio on equipment cost is significant, whereas its impact on raw materials consumption is small by comparison. Other factors also may determine selection of the B/E ratio, e.g., the pressure at which the ethylene feedstock is available at the plant's battery limits. If the ethylene supply pressure is low, it may be optimal to design for operation at a higher B/E ratio in order to maintain liquid-phase conditions in the alkylation system (ethylene partial pressure is reduced and, with it, the bubble-point pressure of the reaction mix). While designing for operation at a higher B/E ratio will increase the cost of the alkylation and benzene recovery systems, the elimination of an ethylene compressor from the design will more that offset these cost increases. Of course, operational considerations also favor a design without such a compressor.

EBMAX PROCESS DESIGNS FOR DILUTE ETHYLENE FEEDSTOCKS

Under unusual circumstances, it may be possible to obtain dilute ethylene, e.g., containing 20 to 30 percent ethane by volume, as a feedstock to the EBMax unit. Such a source may satisfy the entire ethylene requirement or may constitute only a fraction of it. The EBMax unit may be designed for operation with such a feedstock with the following departures from a typical design:

- A small absorber is added to recover aromatics from the process offgas, which consists primarily of ethane, methane, and other nonreactive gasses contained in the ethylene

feedstock. The polyethylbenzene stream (polyethylbenzene column distillate) normally is used as the absorber oil. The bottoms from the absorber are sent directly to the transalkylation reactor. Absorbed inerts do not affect transalkylation performance.
- The operating pressure of the alkylation system is increased to maintain liquid-phase conditions.
- The B/E feed ratio of the alkylation system may be increased for this purpose as well.
- The operating temperature of the alkylation system may be reduced for the same purpose.
- The operating pressure of the benzene column is increased, and the overhead system is designed to minimize the carryover of aromatics to the absorber.

Design and operation of equipment downstream of the benzene column are not affected by the use of a dilute ethylene feedstock. Reaction chemistry is not affected by the presence of inert diluents, and ethylene and benzene consumption are not increased to any appreciable extent.

For feedstocks containing higher concentrations of inerts, particularly those containing significant quantities of methane and/or hydrogen, it is not practical to maintain liquid-phase conditions in the alkylation system; therefore, the alkylation reactor must be designed for concurrent downward flow of an aromatics-rich liquid phase and an ethylene-rich vapor phase through the catalyst beds. The alkylation reactor size and alkylation catalyst quantity are increased in this case; however, the simple fixed-bed configuration shared by all EBMax alkylation system designs is retained. There is no need to contain the alkylation catalyst in a complex baling system such as that employed by so-called catalytic distillation processes because the presence of an ethylene-rich vapor phase in the reactor will not age the EBMax alkylation catalyst, nor will it lead to excessive polyethylbenzene or ethylbenzene impurities formation.

For dilute ethylene designs, the increased capital cost and utility consumption in the EBMax unit typically must be justified by an increase in production in the ethylene unit (debottlenecking) or a reduction in operating costs in that unit, in other words, by a reduction in the transfer price of the contained ethylene. In a grassroots complex that includes both ethylene and ethylbenzene/styrene plants, integration of the two units through the ethylene feedstock and utilities systems (steam, refrigeration, etc.) can be actively considered.

TECHNOLOGY CONVERSION AND CAPACITY EXPANSION WITH EBMAX

Aside from grassroots ethylbenzene plants, the EBMax technology can be used to increase the production capacity of existing vapor-phase zeolite technology ethylbenzene plants or to replace the aging reaction sections of aluminum chloride technology units. In both cases, major modifications to existing distillation systems generally are not required, making it possible to effect the technology conversion during a scheduled styrene unit turnaround. New EBMax reaction systems typically are constructed in available adjacent plot space while the existing unit is operating and are tied into the existing distillation equipment during the turnaround.

In vapor-phase technology upgrades, the existing alkylation reactors generally are reused in alkylation and transalkylation services, reducing capital investment. A capacity increase in the range of 50 to 100 percent can be achieved for such revamps with no major modifications to the distillation section of the plant. The following factors make such expansions possible:

- The quantity of excess benzene fed to the alkylation system relative to the ethylene fed is reduced dramatically. Vapor-phase technology units typically operate at B/E feed

ratios of 18:1 or more by weight. In an EBMax revamp, this ratio is reduced to less than 7:1. Major modifications to the benzene column thus are not required.
- Separation requirements in the ethylbenzene column are reduced. In vapor-phase technology plants, the ethylbenzene column is designed to separate ethylbenzene from *ortho*-xylene or cumene. In the EBMax technology, the requirement is reduced to the separation of ethylbenzene from diethylbenzenes. No modifications to the ethylbenzene column are required.
- The monoalkylation selectivity of the EBMax alkylation catalyst is very high—only small quantities of polyethylbenzenes are produced. Furthermore, polyethylbenzene production does not increase with time on-stream. The existing polyethylbenzene column, originally designed to satisfy the requirements of vapor-phase technology operation, can be reused without modification in an EBMax plant operating at twice the ethylbenzene production rate.

In addition to the increased production capacity, the purity of the ethylbenzene product is increased. Xylenes are effectively eliminated as an ethylbenzene impurity after conversion, easing separation requirements in the distillation section of dehydro technology SM units. The toluene content of the ethylbenzene product is reduced because none is made in the reaction systems. The yield loss to the ethylbenzene unit residue is also reduced substantially after conversion to EBMax technology.

Most of the aluminum chloride technology ethylbenzene units currently operating are more than 20 years old. It is not unusual for maintenance costs to be high in the reaction section of these aging plants due to the corrosive nature of the aluminum chloride catalyst. Disposal of the aluminum chloride waste also can be a concern. These factors, along with a desire to improve yield, provide significant incentives for reaction section technology replacement.

Most aluminum chloride technology units operate with only a small excess of benzene in the feed to the reaction system—benzene-to-ethylene molar feed ratios typically are in the range of 2.0 to 2.5:1. As noted previously, the EBMax alkylation catalyst can operate with complete stability at very low B/E ratios. Therefore, it is possible to replace the existing aluminum chloride reaction section equipment with EBMax reaction equipment without the addition of a second benzene column (sometimes referred to as a *prefractionator*). In most plants, no modifications to the existing benzene column are required. The existing ethylbenzene and polyethylbenzene columns also can be reused without modification.

The purity of the ethylbenzene product will be marginally improved after conversion from aluminum chloride to EBMax reaction technology, with reduced nonaromatics and cumene content. Raw material consumption should be reduced by 2 to 4 kg/t of ethylbenzene product.

ETHYLBENZENE PRODUCT QUALITY

The purity of the EBMax ethylbenzene product is very high. Distillation column operation and benzene feedstock purity are the principal factors that affect ethylbenzene product quality. EBMax units that have been designed to produce a sharp separation in the benzene column have reported product purities in excess of 99.98 percent by weight when operating with a benzene feedstock of high purity. A typical EBMax ethylbenzene product composition is given in Table 5.3.5.

The design and operation of the benzene column affect the benzene content of the ethylbenzene product. Benzene and toluene are produced as a by-product of ethylbenzene dehydrogenation in quantities that are large relative to the residual benzene content of the ethylbenzene product. Therefore, a benzene content of approximately 1000 ppm in the ethylbenzene product is generally acceptable. Dehydro units include a column that

TABLE 5.3.5 Typical EBMax Ethylbenzene Product Composition

Ethylbenzene	99.9 wt %
Nonaromatics	50 ppm wt
Benzene	500 ppm wt
Toluene	100 ppm wt
Xylenes	10 ppm wt
Cumene	30 ppm wt
n-Propylbenzene	15 ppm wt
Ethyltoluenes	50 ppm wt
Diethylbenzene	1 ppm wt

separates the benzene/toluene fraction from the crude SM, returning this fraction to a nearby aromatics unit. Alternatively, an integrated EB/SM plant also may include a small column that separates the benzene from the toluene, the former being recycled to the ethylbenzene unit. Benzene and toluene content in some cases must be reduced to very low levels when the product ethylbenzene feeds a propylene oxide/styrene monomer unit. In such plants, the ethylbenzene purity often exceeds 99.98 percent by weight.

The toluene content of the ethylbenzene product is determined mainly by the toluene content of the benzene feedstock. The net production of toluene in the EBMax reaction systems is very small (on the order of 20 ppm relative to ethylbenzene) and is not measurable in a commercial unit. As noted previously, a portion of the toluene contained in the feed to the alkylator will alkylate, forming ethyltoluenes, and a portion of the ethyltoluenes contained in the transalkylator feed will be converted to toluene. The reflux ratio in the ethylbenzene column directly affects the concentration of toluene relative to ethyltoluene in the ethylbenzene product. Ethyltoluenes will be partially converted to vinyltoluenes in SM dehydrogenation units. Generally, there is no specification on the maximum quantity of vinyltoluenes that may be present in the styrene monomer product.

Xylenes are produced in insignificant quantities in the transalkylation reactor and as such do not affect the purity of the ethylbenzene product or the product from the downstream SM unit.

In addition to the propylbenzenes that are produced from any propylene contained in the ethylene feedstock, very small quantities of propylbenzenes—less than 50 ppm total of cumene and n-propylbenzene—are produced in the EBMax reactors. At these concentrations, propylbenzenes do not affect the design of the SM unit. In dehydro technology plants, cumene is partially converted to *alpha*-methylstyrene in the reactors. The cumene, along with a portion of the *alpha*-methylstyrene, will be present in the SM product. The design and operation of the styrene finishing column affect the *alpha*-methylstyrene content of the SM product.

In a dehydro technology SM unit, most of the nonaromatics contained in the ethylbenzene feed will be contained in the benzene/toluene by-product from the unit. If a salable toluene by-product will be produced by fractionation, then the concentration of nonaromatic compounds in the benzene feed to the EBMax unit must be limited, or their concentration in the benzene column bottoms must be reduced by improving the fractionation (i.e., reducing the benzene content of the bottoms). In most cases, nonaromatics are a potential problem in this regard only if their total concentration in the benzene feedstock exceeds 1000 ppm by weight.

Diethylbenzenes are an undesirable impurity in the feed to dehydro technology SM units because they form divinylbenzenes in the reactors. As noted previously, divinylbenzenes can cause problems in the distillation section of an SM unit if present in excessive

TABLE 5.3.6 Raw Materials and Utilities Consumption

Item, units	Unit/kg EB
Benzene, kg	0.265
Ethylene, kg	0.739
High-pressure steam, kg	1.00
Dilution steam for SM dehydro unit, kg	0.74
Steam for SM distillation, kg	0.61
Power, kWh	0.0075
Cooling water, m^3	0.0059

concentrations. Divinylbenzenes are also produced from ethylbenzene in the SM reactors in quantities such that the relative significance of 10 to 20 ppm of diethylbenzenes in the ethylbenzene feed is not that great. However, the ethylbenzene column generally is designed based on a more stringent specification to minimize the concentration of diethylbenzenes during upsets.

RAW MATERIALS AND UTILITIES CONSUMPTION

The consumption of feedstocks and utilities for a typical EBMax unit is given in Table 5.3.6. Feedstock purities of 100 percent are assumed in the consumption calculations. The EBMax unit is assumed to be integrated with an ATOFINA/Badger or other unit that produces SM by the dehydrogenation route.

In most designs, ethylene compression inside the battery limits of the EBMax unit is not required if the supply pressure is in excess of 26 kg/cm^2(g). High-pressure steam typically is available at a pressure of 41 kg/cm^2(g). Steam for the dehydro and distillation sections of the SM unit normally is supplied at pressures of 4 and 2 kg/cm^2(g), respectively. As discussed in previous sections, EBMax units are designed based on requirements for integration with utilities systems, the SM unit, and the source of ethylene supply. In most cases, variances from the typical EBMax design described earlier will have little or no impact on raw materials consumption. However, designs frequently are adjusted to match specific steam consumption and production requirements. For example, steam production can be minimized by preheating EBMax reactor feeds and reboiling columns in the ethylbenzene distillation section with reactor effluent streams, and high-pressure steam consumption can be eliminated by incorporating a hot-oil system in the EBMax unit design.

CATALYST REQUIREMENTS

The total quantity of process catalyst is computed based on plant capacity, reactor operating conditions, and the target for continuous operation between maintenance shutdowns. The ExxonMobil EBMax catalysts used in the alkylation reactor and transalkylation reactor will operate for many years before replacement is required. Under normal circumstances, the catalysts require regeneration only infrequently—10 years is typical for catalyst in the alkylation reactor, and 5 years is typical for the transalkylation catalyst. Regeneration is accomplished *ex situ* by a controlled burn. Catalyst regeneration services are available in a number of locations worldwide through third-party providers.

EBMAX PLANT DESIGN

A typical EBMax unit consists of approximately 40 pieces of process equipment. All the major equipment is fabricated from carbon steel. In-line ethylene mixing devices, catalyst support grids, and the tubes in one small cooling water exchanger are fabricated from stainless steel. Distillation towers incorporate conventional valve trays in most plant designs.

Single-train EBMax units with capacities in excess of 1 million MTA are easily constructed. One EBMax unit with a capacity of 900,000 MTA of ethylbenzene entered service late in 2001. Another EBMax unit with a capacity of 1.3 million MTA of ethylbenzene entered service early in 2004. In more than one location, a single EBMax unit provides ethylbenzene to two or more SM units. Once again, the ability of the EBMax catalysts to operate with only a small excess of benzene in the reactor feeds is an important attribute because the alkylation reactor and benzene columns are two of the largest vessels in an ethylbenzene plant. Even for a plant that is designed to produce 1 million MTA of ethylbenzene, the size of both the alkylation reactor and the benzene column are not exceptionally large, with diameters of less than 10 and 15 ft, respectively.

REFERENCE

1. M. E. Leonowicz, J. A. Lawton, S. L. Lawton, M. K. Rubin, "MCM-22: A Molecular Sieve with Two Independent Multidimensional Channel Systems," *Science* 264:1910, 1994.

PART 6

ETHYLENE

CHAPTER 6.1
ABB LUMMUS GLOBAL SRT® CRACKING TECHNOLOGY FOR THE PRODUCTION OF ETHYLENE

Sanjeev Kapur
ABB Lummus Global
Houston, Texas

INTRODUCTION

Ethylene ($H_2C=CH_2$) is the lightest olefinic hydrocarbon, and it does not occur freely in nature. It represents the largest building block for a variety of petrochemicals such as plastics, resins, fibers, solvents, etc. Ethylene is produced primarily from the thermal cracking of hydrocarbon feedstocks derived from natural gas and crude oil. Alternate methods for ethylene production are constantly being explored; however, the economics of large-scale investment, along with technology risks, have prevented these technologies from making any significant impact.

Ethylene production and consumption are expected to grow at a rate faster than the world economy despite the economic uncertainties surrounding the petrochemical industry. This will be a result mainly of increased demand growth from developing economies. Total world ethylene capacity in 2002 was close to 110 million metric tons per annum (MTA). Approximately 60 percent of the ethylene is consumed in the production of polyethylene (low density, linear low density, and high density), nearly 13 percent for each of polyvinyl chloride (PVC) and ethylene oxide, and about 7 percent in the production of ethylbenzene.

The hydrocarbon feedstocks used for the production of ethylene are shown in Fig. 6.1.1. Naphtha cracking represents about 45 percent of production capacity, whereas nearly 35 percent of capacity is produced from ethane cracking. Most of the capacity addition in the Middle East will be based on ethane cracking, and therefore, the feedstock mix will change in coming years. This change also will have an impact on production of coproducts such as propylene, butadiene, and benzene. Figure 6.1.2 shows the product distribution for ethylene capacity in the United States from various feeds.

6.4 ETHYLENE

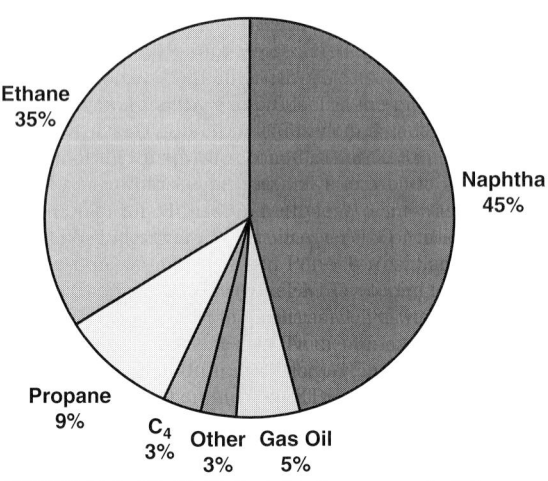

FIGURE 6.1.1 Worldwide feed slate (percentage of ethylene production capacity).

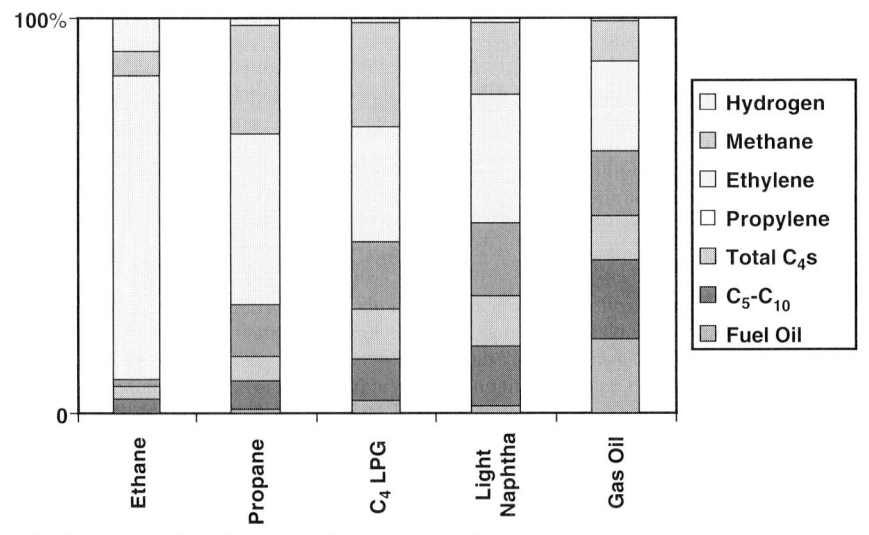

FIGURE 6.1.2 U.S. product distribution from various feeds.

DEVELOPMENT AND COMMERCIAL HISTORY

ABB Lummus Global (Lummus) designed and built the first ethylene plant in 1942. By the mid-1950s, total world capacity (excluding the Soviet bloc) was around 1.6 million MTA, and the average world-scale plant produced less than 70,000 thousand metric tons per annum (MTA). The majority of this production was in the United States. In comparison, world capacity at the end of 2002 was close to 110 million MTA, with an average world-scale plant size of 700,000 MTA. The large-capacity plants being built today range in size from 1 million to 1.5 million MTA, depending on feedstock. While more than 25 percent of world ethylene production capacity is in the United States, current growth areas are in the Middle East and the Far East, and growth is driven by feed availability and demand patterns.

Ethylene plants are capital- and energy-intensive operations, and historically, developments in this technology were driven by attempts at reduction in capital and operating costs. The literature from the early to middle 1960s indicates that the focus of ethylene plant designers was on issues related to reduction in plant cost and improving operating cost through selectivity improvements. In the 1960s, Lummus introduced the first generation of vertical coil cracking heaters and thereby reduced the residence time to below 1 s. Plant sizes increased to 200,000 MTA. In the 1970s, the ethylene industry—like many others during the "oil crisis"—went through an extensive redesign of its flowsheet and lowered the specific energy requirements by 40 to 50 percent. The thermal efficiency of cracking heaters improved by about 10 percent, from approximately 82 to around 92 percent. Cracking heater radiant coil designs improved to shorten the typical residence time closer to 350 to 400 ms. The trend continued in the 1980s, with selectivity improvements and cracking coil designs with a residence time between 150 to 250 ms. The late 1980s saw a shift in emphasis toward reduction in capital cost, and this trend continues today. Large-capacity plant sizes in the 1990s reached ethylene production levels of over 1 million MTA, and single cracking heater capacities increased to above 200,000 MTA ethylene.

Developments in machinery and equipment have contributed significantly to the improvement in performance efficiency and size of plants and therefore the economics of ethylene production. With the introduction of centrifugal compressors in the 1960s, single-train plant sizes started increasing rapidly. Thermodynamic efficiencies improved substantially over the next two decades, and the efficiency and capacity of separation devices improved continually.

The environmental performance of ethylene plants has improved significantly over the years as well. Waste effluents are minimized by improvements in energy efficiency, recycling and recovery of vent and purge streams, closed vent and drain systems, and reduced stack gas emissions. Plants are designed and operated to minimize flaring during startup, normal operations, and plant-wide upsets. The designs limit noise emissions inside the plant operations area and at the fence limit of the plant.

PROCESS CHEMISTRY

Feedstock selection is a key consideration for the economics of ethylene production. The yield of coproducts and the coproducts' values influence the rate of return and therefore drive the economics. The quantities of coproducts and the plant design vary with the properties of the selected feedstock. Conventional feedstocks include ethane, propane, butane, pentanes, and naphthas. Other possible feeds include refinery offgas, natural gasoline liquids, wide-boiling condensate fractions, atmospheric gas oils, vacuum gas oils, and hydrotreated or hydrocracked vacuum gas oils. Table 6.1.1 provides typical yields for the most commonly used feedstocks.

The main route to ethylene production is by the thermal cracking of hydrocarbons. This reaction is carried out in tubular coils located in the radiant zone of fired heaters. Steam is added to reduce the partial pressure of the hydrocarbons in the radiant coils. The reactions that result in the transformation of mostly saturated hydrocarbons to olefins are highly endothermic and require reactor temperature in the range of 750 to 900°C depending on the feedstock and design of the reactor coils.

TABLE 6.1.1 Typical Yields

	Ethane	Propane, low conversion	Propane, high conversion	80/20 ethane/propane	C₄ LPG, low conversion	C₄ LPG, high conversion	Light naphtha	Full-range naphtha	Gas oil
Hydrogen	4.20	1.20	1.43	3.68	1.21	1.25	0.96	0.91	0.49
Methane	5.50	21.77	25.82	9.46	24.15	25.42	16.46	14.48	9.17
Acetylene	0.59	0.28	0.51	0.58	0.79	0.85	0.47	0.46	0.17
Ethylene	53.52	32.05	39.98	51.02	27.65	29.65	31.11	28.10	21.87
Ethane	29.74	4.99	3.99	24.73	2.44	2.41	4.19	3.35	3.95
MA + Pd	0.07	0.33	0.46	0.14	1.91	1.96	0.89	0.91	0.49
Propylene	1.27	17.32	11.06	3.05	15.25	13.31	17.13	15.46	15.27
Propane	0.21	14.85	4.95	1.01	1.02	0.52	0.53	0.36	0.53
Butadiene	1.86	1.86	4.32	2.35	3.75	3.99	5.64	7.45	4.86
Butylene	0.17	1.16	0.74	0.27	6.68	5.81	5.28	4.83	5.95
Butane	0.22	0.10	0.07	0.18	1.96	0.33	0.98	0.75	0.12
C₅+ or C₅s	2.65	4.09	6.69	3.53	1.75	1.70	2.84	3.61	3.65
C₆–C₈ NA					1.08	1.19	3.62	4.60	2.62
Benezene					3.92	4.39	4.03	4.57	4.59
Toluene					1.08	1.18	2.49	3.69	2.71
Xy + EB					0.37	1.42	0.73	1.71	1.48
Styrene					0.31	0.34	0.44	0.73	0.71
C₉–200 oC					1.48	1.62	0.39	1.65	3.87
Fuel oil					3.20	3.66	1.82	2.38	17.50

Thermal cracking reactions also produce valuable by-products, including propylene, butadiene, benzene, gasoline, and hydrogen. Some of the less valuable coproducts include methane and fuel oil. An important parameter in the design of commercial thermal cracking coils is the optimal selectivity to produce the desired product slate for maximizing economic returns. The product distribution is affected by feed specifications, reactor coil residence time, severity of operation, and hydrocarbon partial pressure.

The thermal cracking reaction chemistry is quite complex, and the degree of complexity increases when cracking heavier hydrocarbons at higher operating severity. The enormous multiplicity of chemical reactions coupled with nonuniform heat transfer makes the development of a comprehensive pyrolysis theory a major challenge for chemists and mathematicians. Most of the concepts applied to thermal cracking reactors and the prediction of yields are mathematical models with a certain degree of empiricism in kinetic modeling. The Pyrolysis Yield Prediction System (PYPS^{SM}) model developed by Lummus is one well-proven industrial model.

Rice and Herzfeld proposed the concept of a free-radical mechanism as the basis for the decomposition of hydrocarbons by thermal cracking. This concept is accepted to represent a sequence of reactions correctly and therefore rationalizes the distribution of the product slate. The concept postulates three basic types of reactions:

1. *Chain initiations,* or the initial formation of radicals:

$$C_nH_{2n+2} \rightarrow C_mH_{2m+1}\cdot + C_{(n-m)}H_{2(n-m)+1}\cdot$$

2. *Chain propagations* through reaction of radicals with molecules:

$$C_nH_{2n+2} + C_mH_{2m+1}\cdot \rightarrow C_nH_{2n+1}\cdot + C_mH_{2m+2}$$

$$C_nH_{2n+1}\cdot \rightarrow C_mH_{2m} + C_{(n-m)}H_{2(n-m)+1}\cdot$$

3. *Chain terminations* causing the disappearance of radicals:

$$C_nH_{2n+1}\cdot + C_mH_{2m+1}\cdot \rightarrow C_nH_{2n} + C_mH_{2m+2}$$

$$C_nH_{2n+1}\cdot + C_mH_{2m+1}\cdot \rightarrow C_nH_{2n+2} + C_mH_{2m}$$

$$C_nH_{2n+1}\cdot + C_mH_{2m+1}\cdot \rightarrow C_{n+m}H_{2(n+m)+2}$$

During the chain-initiation step, two radicals are produced by the cleavage of the C-C bonds of paraffin molecules. Chain propagation involves many different reactions, including hydrogen abstraction, addition, radical decomposition, and radical isomerization. Chain termination is the reverse of chain initiation, causing the disappearance of radicals. The free-radical mechanism describes the primary decomposition of saturated hydrocarbons to olefins, methane, and hydrogen. This helps predict the product distribution of single hydrocarbons at lower feed conversions with reasonable accuracy; however, at higher conversions, secondary reactions involving decomposition of primary products gain significance. These reactions can be described by the free-radical mechanism to some extent, but it does not adequately explain the formation of mono- and polycyclic aromatics.

Heavier liquid feeds are composed principally of paraffins, cycloparaffins, aromatic ring structures, and occasionally, olefins. The paraffinic hydrocarbons have a high hydrogen-to-carbon ratio and decompose to produce high yields of gaseous products and low yields of heavy aromatic oils. The cycloparaffins decompose by ring rupture accompanied by the formation of alkyl radicals. These compounds require higher activation energy levels compared with paraffins. The hydrogen-to-carbon ratio is lower than paraffins but higher than aromatics, and therefore, the yield of gaseous products and heavy liquids is intermediate (i.e., between paraffins and aromatics). Aromatic ring structures are highly heat-stable and are refractory compounds. The hydrogen-to-carbon ratios are low, and olefin yields from aromatic feeds are negligible. These compounds are tar precursors and cause deposition of coke in the cracking coils through condensation reactions.

The terms *severity* and *conversion* are used to measure the extent of cracking. Conversion of a single component is measured by the rate of its disappearance relative to its concentration in the feed. However, when a mixed feed is cracked, the single compound is also formed as a product of cracking from other compounds. The measured conversion of a single compound in a mixed feed provides only an approximation of the true conversion. For liquid feeds, such as naphtha, it is impractical to calculate true conversion. Instead, several other indicators are measured/calculated to define severity of operation. These include propylene-to-ethylene ratio, propylene-to-methane ratio, cracking severity index, kinetic severity factor, molecular collision parameter, average molecular weight of the complete product slate, hydrogen content in the C_5+ product, etc.

Coke Formation and Deposition

The on-stream availability of the thermal cracking reactor is determined by fouling of either the cracking coils or the cracked effluent transferline exchangers (TLEs). Coke is produced as a side product of thermal cracking and deposits on the radiant coil walls and inside the tubes of TLEs. This limits the heat transfer and increases the pressure drop, thus reducing the olefin selectivity. The run length is normally determined by the tube metal temperature increase of the radiant coil, the outlet temperature of the TLE, or the increased pressure drop. Therefore, kinetic models that predict the rate of coke laydown are of particular interest to designers and operators.

Coke is believed to be formed by two mechanisms—catalytic and condensation. The metal surface of the cracker coil catalyzes the growth of a filamentary type of coke and contains metal granules (Fig. 6.1.3). The second type of coke is formed by condensation, polymerization, and/or agglomeration of heavies in the gaseous phase. The coking behavior of various feedstocks differs in cracking coils and TLEs, and these can be influenced by contaminants in the feed, dilution steam, and coil surface. The precise modeling of coking in commercial cracking coils is a highly complex process, and the mechanism is not fully understood. Simplified models using empirical constants have been employed to describe the coking phenomenon in cracking coils and TLEs. Radiant coils see the quick formation of a coke layer in the first few hours of operation due to a highly active tube surface, resulting in a quicker rise in tube metal temperature. After that, the temperature continues to rise at a relatively gradual increment for the duration of the run. TLEs see a similar phenomenon. Figure 6.1.4 shows TLE coking for various feeds. Plants cracking ethane and propane experience coking at the inlet (on the tubesheet) of the TLEs due to gas-phase reactions at higher inlet temperatures and to the discontinuities in flow distribution. This inlet fouling normally results in high-pressure drop, limiting the run length. Liquid-feed cracking results in fouling on the TLE tubes near the outlet. This is caused by the condensation of tarry materials, which form a thin, oily layer that gradually polymerizes. The rate of fouling reduces significantly when the coke/polymer layer reaches the hydrocarbon dew point.

CRACKING HEATER

The cracking heaters are a very important part of an ethylene plant. The cracking heaters represent 20 to 30 percent of the equipment cost. The majority of energy in an ethylene plant is consumed in the cracking heaters to provide the heat of reaction and the sensible/latent heat to bring the reactants to the desired reaction temperature. The design of the cracking heaters determines the product slate, which sets the profitability of an ethylene plant.

ABB LUMMUS GLOBAL SRT CRACKING TECHNOLOGY 6.9

Catalytic Coke Layer

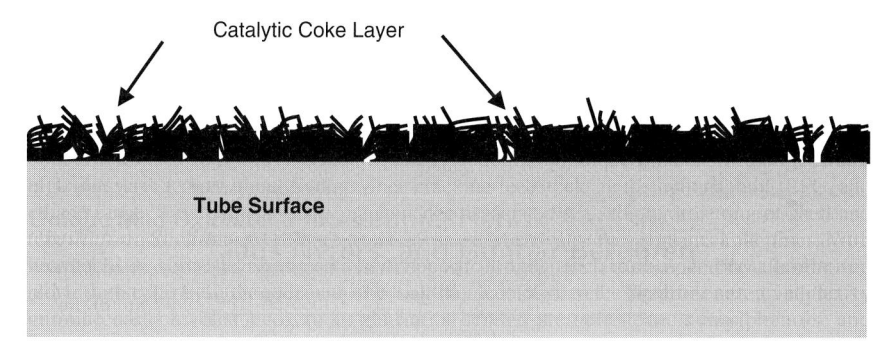

Tube Surface

Catalyzed by Ti, Cr, Fe and Ni

Ti < Cr < Fe < Ni

Fe appears to play the predominant role

Coking tendency: Acetylenes > Olefins > Paraffins

FIGURE 6.1.3 Catalytic coke.

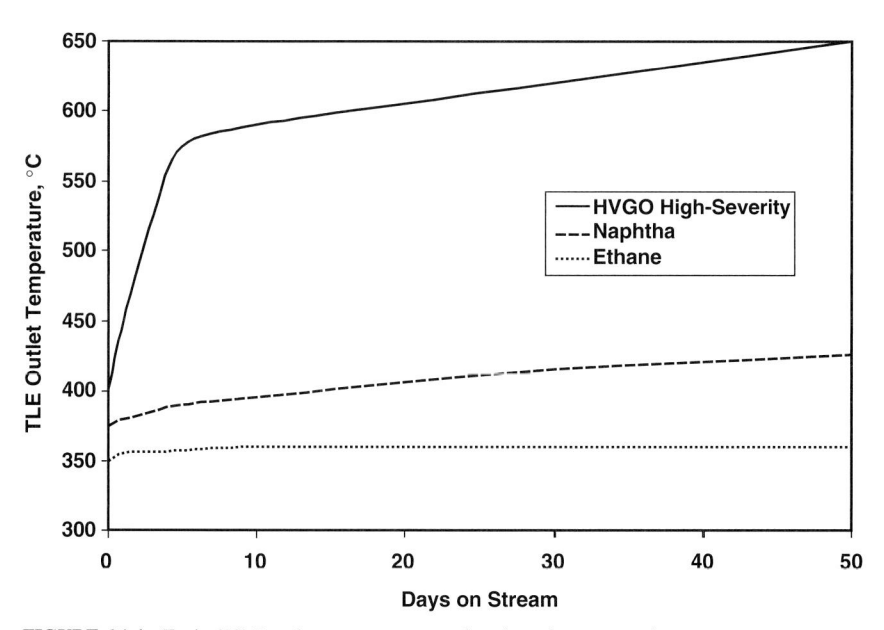

FIGURE 6.1.4 Typical TLE outlet temperature as a function of on-stream time.

Thermal cracking of hydrocarbons takes place in tubular coils placed in the center of a fired radiant box. The cracked effluents leave the radiant coils at a temperature of 750 to 900°C, depending on feedstock, cracking severity, and selectivity. In order to maintain the overall process efficiency, it is required to efficiently recover the heat in the cracked effluents. The cracked effluents also need to be quenched quickly to stop secondary reactions that result in yield degradation. This is achieved by the TLEs, which cool the furnace effluents to nearly 350 to 450°C at clean conditions, and this heat is used to generate very high pressure steam ($\sim$125 bar). The higher steam pressure results in higher tube metal temperature and therefore minimizes condensation of tarry materials. For ethane/propane cracking, the cracked effluents are further cooled in secondary TLEs or, for heavier hydrocarbon feeds, directly quenched with recycled heavy oil from the gasoline fractionator.

Figure 6.1.5 shows the configuration of a typical cracking heater. It consists of a convection section in the upper offset arrangement and a radiant section at the lower end. The vertical radiant coils are located close to the center plane of the radiant box and are suspended on a hanging system from the top of the radiant box. The hanging system allows the radiant coils to expand without causing any additional stresses on the radiant coils, and current designs are mostly based on a counterweight philosophy for higher reliability. The radiant coils are centrifugally cast from 25Cr/35Ni or 35Cr/45Ni materials for their carburization and creep resistance. These materials have a maximum service temperature of up to 1150°C. The typical composition of these materials is shown in Table 6.1.2. The optimal radiant coil design for a particular plant is determined by its feedstock, product/by-product slate, and economic criteria. The SRT heater design is characterized by a configuration that maximizes the coil surface-to-volume ratio at the coil inlet, where coking tendencies are low. This is done by using small-diameter parallel tubes in the inlet pass or passes. At the outlet, where coking tendencies are high, large-diameter tubes are used. These minimize the rate of coke laydown and the impact coke laydown has on the coil pressure drop and therefore on the selectivity over the heater cycle. The use of this design concept results in essentially constant yields throughout the heater cycle.

The convection section recovers the flue-gas heat by preheating the hydrocarbons and dilution steam. In addition, heat is recovered through boiler feed water preheating and

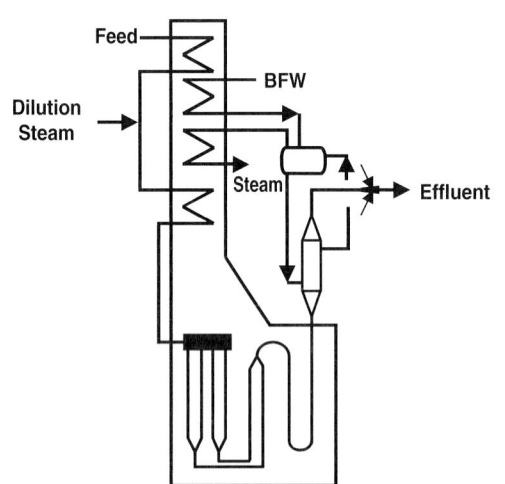

FIGURE 6.1.5 Cracking heater.

TABLE 6.1.2 Typical Composition of Radiant Coil Material

Element	HP, 40 Mod	Pompey HP, 40 W	Pompey Manaurite XM	Kubota Manaurite XTM	KHR 45ACR
Cr	23.5/26.5	24/27	23/28	34/37	30/35
Ni	34/37	33/37	33/38	43/48	40/46
Si	1.5/2.0	1.5/2.0	1.0/2.0	1.0/2.0	2.0 max
Mo	1.25 max	1.5 max	1.0/1.5	1.0/2.0	2.0 max
C	0.37/0.45	0.37/0.50	0.37/0.50	0.40/0.45	0.40/0.60
W	—	3.8/5.0	—	—	—
Additions	W, Nb	—	Nb, Ti, Zr	Nb,Ti + others	Ti + others + others

superheating of very high pressure steam. The high-temperature coil or part thereof is bare due to excessive tube metal temperatures, and normally all the other convection coils have fins to improve the heat-transfer coefficient.

The SRT design uses the economical single-stage TLE concept with multiple tubes based on the conventional approach and also a new approach based on bathtub or "quick quench" designs. In the inlet cone of conventional TLEs, the residence time is about 45 ms. While this is not significant in less selective coils having residence times on the order of 500 ms, this would result in a measurable yield degradation for the new high-selectivity coils. The bathtub and "quick quench" approaches reduce the inlet cone residence time by up to 35 ms, to as low as 10 ms.

The radiant box is equipped with mostly multiple floor-fired burners and also, sometimes, wall burners to provide the required heat for cracking reactions. The burners are designed to produce a heat-flux profile that maximizes utilization of tube heat transfer surface and keeps the tube metal temperatures low. The flame pattern is stable and stiff to avoid flame impingement on the radiant coils, which could result in hot spots and therefore shorter run lengths due to excessive localized coking. It also could result in excessive carburization and creep of radiant coil material resulting in tube deformation. The burner designs need to meet the required environmental emissions for NO_x, carbon monoxide, particulates, etc.

Present-day single cracking heater capacities vary between 100,000 and 200,000 MTA ethylene production, with designs available to achieve capacities as high as 300,000 MTA.

ETHYLENE PROCESS FLOW SCHEMATIC

The ethylene plant forms the core facility of a world-scale, fully integrated petrochemical complex producing a variety of products. Ethylene plants are very complex, with some process sections operating at very high temperatures (nearly 900°C) and others at very low temperatures (lower than −140°C), process-side pressures ranging from almost atmospheric up to 40 bar, and the steam system at pressures close to 125 bar. The capacity of most plants designed since 2000 ranges between 800,000 and 1,300,000 MTA. These plants are highly capital-intensive and require a developed infrastructure around the facility to handle the logistics of feeds, products, utilities, etc. An ethylene plant can be based on cracking gas feeds only, either as a single component or as a combination of ethane, propane, and butane. A liquid cracking facility normally can crack more than one feed from a combination of feeds ranging from LPGs to vacuum gas oils. Some facilities are designed with the flexibility to handle a combination of feeds ranging from gas to liquid hydrocarbons.

Process Description

Oil Quench and Water Quench. Figure 6.1.6 is a simplified flow scheme of a liquid-feed cracker. The major difference in a liquid cracker flow scheme versus a gas cracker flow scheme is the fuel oil fractionation. Liquid-feed crackers require a gasoline (primary) fractionation step to remove heavier fuel components, whereas the cracked effluents from a gas cracker enter a water quench tower directly. The cracked gas leaves the primary TLEs at between 350 to 600°C, depending on the type of feedstock and the runtime conditions. TLEs are clean at the beginning of a cycle and cool the cracked gas to lower temperatures; temperatures rise as TLEs foul. For gas crackers, cracked gas is further cooled to nearly 200°C in secondary TLEs. For liquid-feed crackers, this is achieved by direct quench with quench oil.

The mixed-phase flow of cracked gas and quench oil is separated in the bottom of the gasoline (primary) fractionator. The gasoline fractionator operates at a bottoms temperature of 190 to 230°C and an overhead temperature of 95 to 120°C. The fuel oil components are separated and stripped and routed to the fuel-handling facilities. This tower is refluxed with the heavy gasoline cut from the water quench tower. The separated quench oil at the bottom of gasoline fractionator provides heat to generate dilution steam. Some of the heat is also recovered by preheating various process streams. A part of the cooled quench oil is used for direct quench of cracked gas in the quench fittings, and the rest is used as pump-around liquid for the lower portion of the gasoline fractionator. The quality of the circulating quench oil and its bottoms temperature are carefully controlled to keep its viscosity within a certain range. The cracked components of quench oil are unstable and tend to polymerize and agglomerate, resulting in higher viscosity. The fuel oil yield, and therefore the residence time of quench oil in the system, has an impact on the accumulation of polymerization products. Naphthas and lighter feeds yield lower fuel oil components, whereas gas oils and heavier feeds yield much higher quantities of fuel oil components.

The cracked gas is cooled to nearly 40°C by direct quenching with circulating water in the water quench tower. This tower condenses the heavy gasoline components and provides reflux for the gasoline fractionator. The dilution steam is also condensed in this tower. The condensed gasoline and dilution steam are separated in the base of the tower (or in an oil/water separator connected to the bottom). The separation is controlled by the proper maintenance of pH values through injection of neutralizing agents. The bottom temperature is maintained at 80 to 85°C to recover heat in various low-level heating services. The condensed process water is stripped of acid gases and some of the dissolved hydrocarbons before routing to the dilution steam generator or the feed saturator (for gas feeds). This helps in conserving and recycling water, thus keeping net discharge to a minimum to control the quality of dilution steam water.

Cracked Gas Compression, Acid Gas Removal and Drying. The cracked gas leaving the water quench tower is compressed to 32 to 37 bar in a four- or five-stage centrifugal compressor. The permissible discharge temperatures that will limit machine fouling determine the number of stages. (The diolefins in cracked gas tend to polymerize at high temperatures, resulting in deterioration of compressor performance.) The condensed water and hydrocarbons are separated from the cracked gas between the stages. The water is returned to the quench water system. The hydrocarbons from the first three stages are sent to the pyrolysis gasoline stripper, whereas the condensate from the last stage is sent to the condensate stripper (deethanizer). The compressor normally is driven by a single extraction/condensing turbine.

Acid gases (carbon dioxide and hydrogen disulfide) are removed after the third or fourth stage of compression. Scrubbing with a dilute caustic soda solution is economical

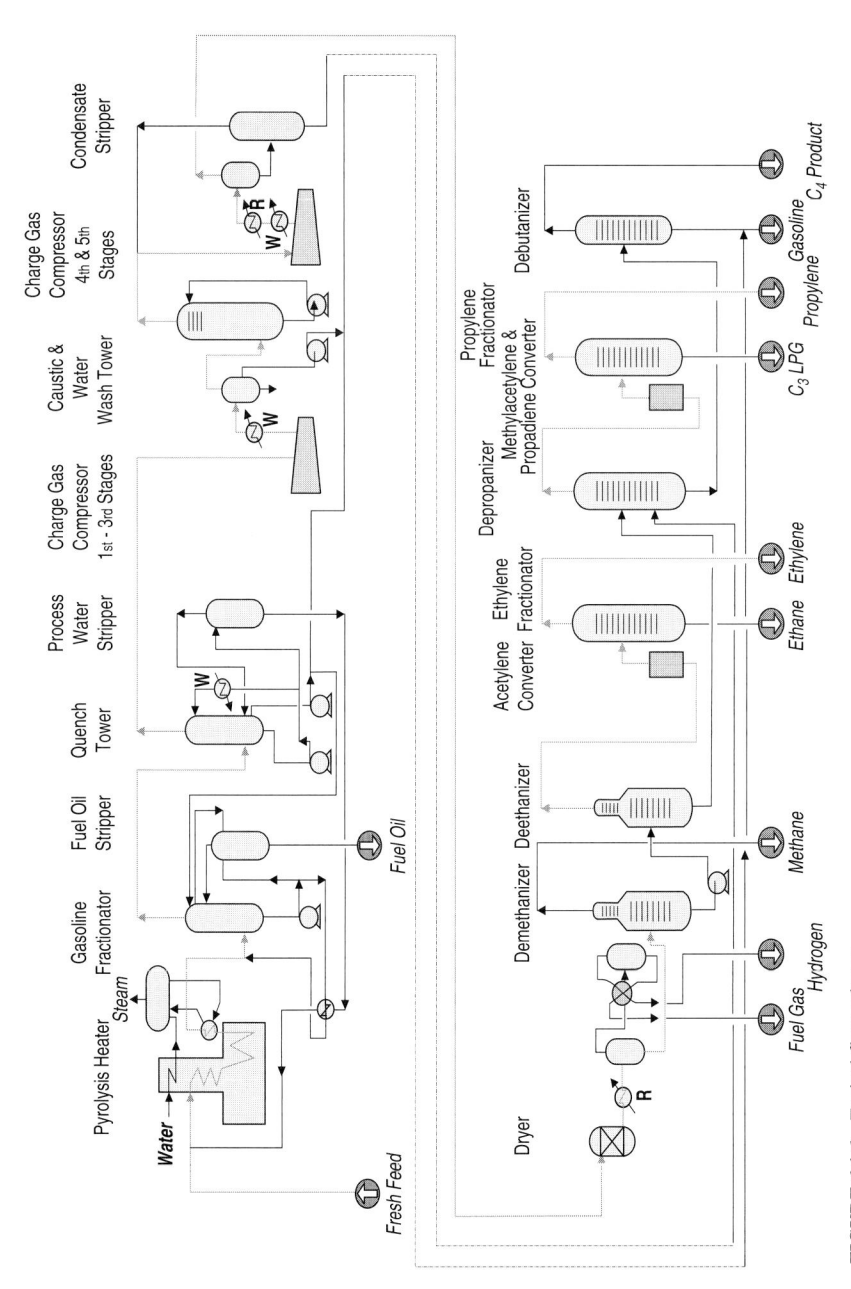

FIGURE 6.1.6 Typical flow scheme.

6.13

because normally feedstock sulfur is low. A two- or three-stage caustic wash tower is provided for maximum use of caustic. The solution circulating over the top contains 8 to 12 percent sodium hydroxide, and overall caustic use is controlled at between 80 and 90 percent. A water wash section is provided on top of the wash tower to remove any entrained caustic from the charge gas to avoid any damage to the compressor system. The cracked gas leaving the wash tower contains less than 1 ppm acid gases to meet final product specifications. The spent caustic from the caustic wash tower needs subsequent treatment before it can be discharged safely.

Trouble-free operation of the caustic wash system depends on maintaining the proper hydrocarbon dew point, the concentration of circulating caustic solution, and the overall caustic use. In order to avoid an aldol condensation phenomenon, chemical inhibitors can be injected in the cracked gas.

The complete removal of water is essential in preparation for cryogenic separation and is achieved with adsorption over molecular sieves. Typically, the system consists of two dryers, with one in normal operation while the other is being regenerated. The dryers are designed for an on-stream time of 24 to 36 hours between successive regenerations. High-pressure methane, heated with steam to 225°C, normally is used as the regeneration medium because this stream is dry and easily available within the ethylene plant. The cracked gas normally is chilled with propylene refrigerant to limit the size of the dryer. The chilling temperature is kept above the hydrate formation temperature, which depends on the system pressure.

Chilling Train and Demethanizer. The chilling train involves the successive chilling of the charge gas at essentially constant pressure in heat exchange with various stages of boiling refrigerant and with cold streams that need to be vaporized and/or reheated. The stage condensates are separated and fed to the appropriate stage of the demethanizer tower. Hydrogen (typically 95 mol % purity) is withdrawn from the lowest-temperature stage separator and can be further purified, after reheat, in a pressure-swing adsorption system or a wash type of system. Since 1980, Lummus's standard conventional flow scheme has used a low-pressure demethanizer. This approach lowers the overall energy requirements due to a significant reduction in the thermodynamic irreversibilities (due to higher temperature differential) of the fractionation system and lowers the investment at the same time. The overhead from the demethanizer tower essentially consists of methane ($\sim$ 95 mol %) with hydrogen, carbon monoxide, and some ethylene.

Deethanizer, Acetylene Hydrogenation, and Ethylene Fractionator. The C_2's and heavies from the demethanizer bottoms are fractionated in the deethanizer column into a C_2 stream in the overhead and a bottom product of C_3's and heavier hydrocarbons. The C_2 specification in the bottom product is maintained to meet the propylene product specification. This tower typically operates at 20 to 25 bar.

The C_2 stream is treated to either remove or convert acetylene to meet the desired ethylene product specification. Due to the limited demand for acetylene, most modern-day plants convert acetylene selectively into ethylene. This reaction is carried out in the vapor phase over palladium catalyst. The reaction is exothermic, and in order to control the hydrogenation selectivity, multiple beds with intermediate cooling normally are provided. The reactor effluents typically contain less than about 1 ppm acetylene. The efficient control of acetylene hydrogenation is critical to continuously meeting product specification. The reaction chemistry also results in side reactions that form olefins with carbon numbers ranging from 4 to 20 or more, which slowly deactivate the catalyst and require increased reactor temperatures to maintain catalyst activity. After several months of operation, catalyst is regenerated to restore catalyst activity.

The ethylene-ethane mixture is fractionated to recover polymer-grade ethylene product. The ethane, drawn from the bottom of tower, normally is recycle-cracked to extinction. This tower normally is provided with a "pasteurization" section for removal of light ends.

Purification of Propylene and Heavier Products. The deethanizer bottoms mixed with condensate stripper bottoms are sent to the depropanizer for separation of C_3's from C_4 and heavier hydrocarbons. The bottom product of the depropanizer tends to polymerize due to the presence of butadienes and heavier diolefins. This requires that the tower be operated at a lower bottoms temperature, which in turn requires a lower-temperature condenser that uses refrigerant instead of cooling water, thus increasing the energy requirements for separation. Energy-efficient units typically consist of a two-tower system with different operating pressures in each. The higher-pressure depropanizer is condensed with cooling water to minimize the refrigeration requirement in the lower-pressure depropanizer.

The depropanizer bottoms are further processed in the debutanizer to separate C_4 product from pyrolysis gasoline. The debutanizer typically operates at 4 to 6 bar and is condensed against cooling water.

The C_3 stream from the depropanizer overhead is treated to convert methylacetylene and propadiene to achieve the desired product purity and also to avoid excessive buildup of these compounds in the separation tower. This reaction is carried out in either the vapor or liquid phase over a palladium-based catalyst. The vapor-phase system is similar to the acetylene hydrogenation system described earlier. The liquid-phase system is normally a single-bed system and is simpler to operate. The exothermic heat of reaction is removed by maintaining liquid recycle. For polymer-grade propylene, the MAPD content in the reactor effluent is not critical because it leaves with the propane. Therefore, reactor operation is optimized to maintain higher selectivity.

Separation of the propylene-propane mixture is more difficult than fractionation of the ethylene-ethane mixture due to their low relative volatility. Since low-level heat is available in the quench water, this separation can be accomplished at higher pressure without any mechanical energy required other than reflux or liquid transfer pumps (in the case of a two-tower system). A low-pressure propylene fractionator can be considered where energy costs justify this approach. Similar to the ethylene fractionator, a "pasteurizing" section can be incorporated for the rejection of lights to meet product specification. The propane stream from the bottom of the tower normally is recycle-cracked to extinction.

Refrigeration System. The refrigeration needs of a conventional ethylene plant are provided by ethylene, propylene, and methane refrigeration (or methane/hydrogen turboexpanders or methane recycle over charge gas compressor) systems. The separation of cracked gas through condensation and fractionation requires refrigeration over the entire temperature range, from ambient to $-140°C$. In some cases, an even lower temperature is achieved using the Joule-Thomson expansion of process streams. The ethylene and propylene compressors normally are driven by steam turbines.

REFINERY AND ETHYLENE PLANT INTEGRATION

An integrated refinery and ethylene plant complex offers significant opportunities to upgrade various hydrocarbons into higher-value products. These integration opportunities improve the gross margin of the combined operation, therefore improving overall return on capital. The potential hydrocarbon feeds from a refinery to an ethylene plant include gaseous streams containing ethane, propane, and butanes; liquid streams, from C_4 LPGs to naphthas; and raffinates, gas oils, vacuum gas oils, hydrocracker distillate, etc. In addition, olefinic streams such as FCC offgases, coker offgases, and propylene streams can be integrated with the ethylene plant recovery section. Various streams from the ethylene plant, like pyrolysis gasoline and fuel oil, can be blended with the refinery product slate. Excess hydrogen from the ethylene plant can supply the refinery's needs.

In addition to process stream integration, there are significant opportunities for utility integration that can result in higher energy efficiencies and lower investments for the combined operation. One such example is the steam system. An ethylene plant typically has excess low-pressure steam, whereas refinery operations are large sinks for low-pressure steam. Therefore, integrating the two steam systems will result in a combined higher energy efficiency.

RECENT TECHNOLOGY ADVANCES

The production of ethylene and propylene by steam cracking is a fairly mature technology. While improvements in the configuration of the cracking heater and product-recovery sections continue to enhance process efficiency, the process chemistry and fundamental flowsheet have remained relatively unchanged. Olefin producers have faced a number of challenges in recent years. The first is to reduce greenhouse gas emissions by reducing the fuel fired in the thermal cracking step and by lowering the energy consumption in the product recovery section. The second is to lower the investment level associated with new facilities. The third challenge is to improve the rate of return on these investments by improving the operating margins through an enhanced product slate.

To help meet these challenges, Lummus has an ongoing research and development program that is focused on changing the process chemistry downstream of cracking heaters and on a fundamental reconfiguration of the flowsheet. Some improvements from this program include

- The application of metathesis chemistry to the ethylene plant flowsheet
- The application of CD*Hydro*® catalytic distillation hydrogenation technology
- The integration of gas turbines with the cracking heaters
- The application of binary and tertiary refrigeration technologies
- Large-capacity cracking heaters
- Low NO_x burner concepts and heater designs developed using computational fluid dynamics (CFD)

Application of Metathesis Chemistry

Applying metathesis chemistry to an ethylene plant can improve the operating margin by increasing the value of the by-product slate. In this application, the metathesis reaction is used to react butylenes with ethylene to produce propylene. Metathesis also reduces the fuel fired in the cracking heaters because this reaction is essentially energy-neutral, lowering the energy consumption by up to 13 percent and the production of greenhouse gases by an equivalent amount. This technology is discussed in more detail in Chap. 10.4.

Application of CD*Hydro* Catalytic Distillation Hydrogenation Technology

In a conventional steam cracker flowsheet, the selective hydrogenation of C_2 and heavier acetylenes and dienes and the saturation of C_4 and heavier olefins are accomplished in a

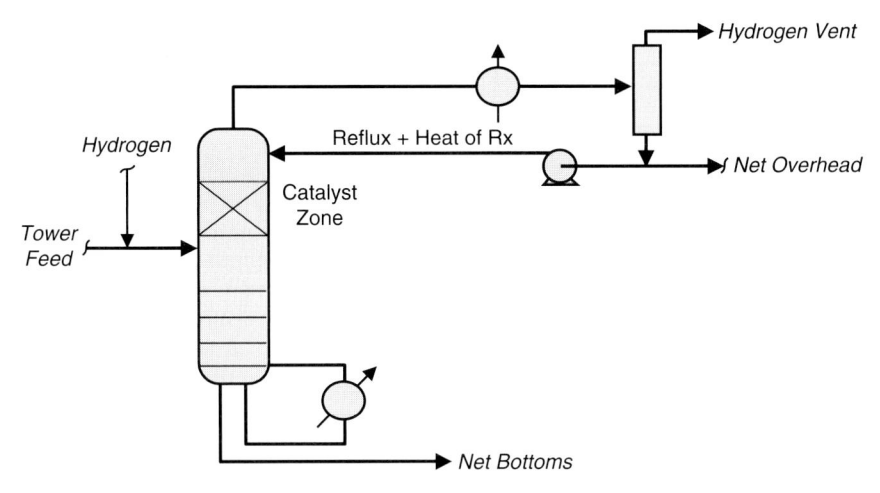

FIGURE 6.1.7 CD*Hydro* tower.

series of fixed-bed reactor systems. As shown in Fig. 6.1.7, the CD*Hydro* process combines the hydrogenation step with the distillation step, thereby eliminating separate fixed-bed reactor systems. The CD*Hydro* process offers a number of advantages when compared with conventional fixed-bed reactor systems. Dimers and oligomers are fractionated from the catalyst zone by the wash effect of liquid flowing down. This prevents catalyst fouling and the loss of catalyst activity that is experienced in conventional fixed-bed systems, eliminating the need for a spare reactor. Green oil is fractionated out with the bottom product, eliminating the need for a separate removal facility. Higher selectivities are achieved because the catalyst fouling rate is minimized. The reaction and distillation steps are combined, which eliminates the equipment associated with the fixed-bed reactor system, including the reactors, feed heaters, and effluent coolers. This results in reduced investment cost and improved operating efficiency.

Integration of Gas Turbines with Cracking Heaters

The integration of gas turbines with cracking heaters—using the oxygen-rich hot exhaust gas as combustion air and to produce power—significantly reduces the specific energy for ethylene production. This option is very attractive for plant locations with high energy costs and when steam and power export sinks are available. The selection of a gas turbine for a given application needs to account for various process, mechanical, operational, and cost considerations. Gas turbine integration saves between 10 and 20 percent of overall energy requirements for an ethylene plant. Figure 6.1.8 is a typical flow scheme for gas turbine integration. Eleven plants designed by Lummus based on the integration concept are operating successfully.

Application of Binary and Tertiary Refrigeration Technologies. Lummus has developed a patented scheme of mixed refrigerants that provides the complete range of refrigeration from ambient to $-140°C$, eliminating the need for three separate refrigeration systems required in the conventional scheme. This reduces the number of rotating equipment pieces in the ethylene plant, resulting in investment savings without any energy penalties and in increased reliability with a reduction in maintenance costs. The binary system combines the methane and ethylene

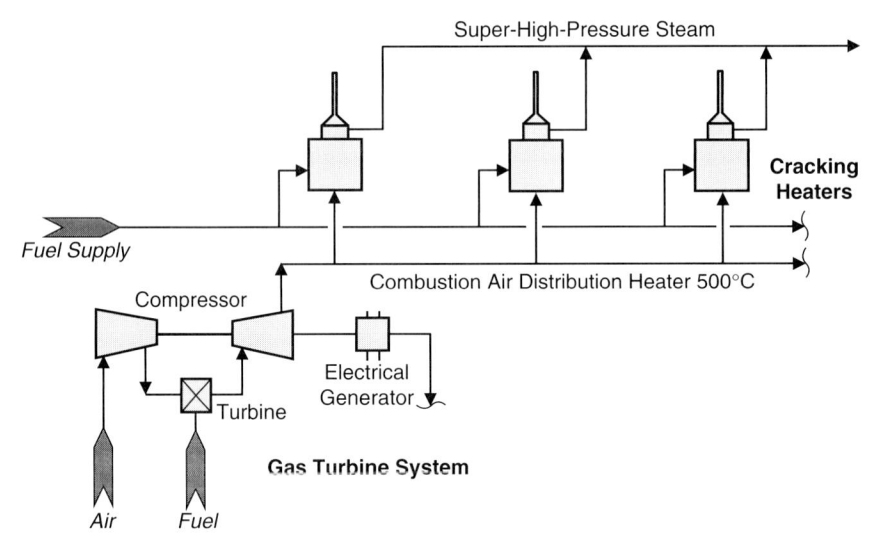

FIGURE 6.1.8 Cracking heater/gas turbine integration.

refrigeration systems. The tertiary system combines the methane, ethylene, and propylene refrigeration systems. Both these schemes meet the needs of changing loads in the ethylene plant due to frequent changes in feed slate and cracking severity. These systems have been applied in four plants, with two in successful operation and two currently under design/construction.

Large-Capacity Cracking Heaters

A single SRT cracking heater can be built up to a capacity of 300,000 MTA of ethylene. The current average size of a new cracking heater ranges from 100,000 to 200,000 MTA. The high-capacity heaters are designed to use plot space more efficiently and save on investment for piping and pipe racks. Modern-day large-capacity plants can be built with five to six operating heaters to meet capacities of up to 1.5 million MTA and therefore saving significantly on overall investment.

Low NO$_x$ Burner Concepts and Heater Designs Using Computational Fluid Dynamics

Historically, heaters have been designed using detailed process-yield models with uniform-temperature firebox models or simplified zonal firebox models. It was assumed that the burners themselves did not have a major impact on the performance calculations and simply provided the heat.

Recent trends in the design and operation of ethylene plants have led to larger and more intensely fired cracking heaters. In the last decade, the capacity of a typical cracking heater has almost doubled, and it is anticipated that this may increase by another 40 percent in the next 5 years. A more intensely fired heater operates closer to its metallurgical limits, which increases the importance of the heat-flux profile that is generated by the burners and its influence on metal temperatures and process performance. In order to design these more intensely fired cracking heaters, it is necessary to consider the actual individual burner performance. Further complicating the design is the requirement for low NO_x. Low-NO_x wall and hearth burners are designed to entrain flue gas from the firebox in order to dilute the flame and retard combustion. This lowers the peak flame temperature and the subsequent formation of thermal NO_x. It also affects the heat-release patterns within the firebox and the overall process and heater performance.

Computational fluid dynamics (CFD) is an advanced tool that can be used to predict the influence of burner design and operation on cracking coil performance. Using CFD to model cracking heaters is not new; several examples can be found in the literature, but for those cases, only the firebox aerodynamics were simulated with CFD. Several years ago Lummus began an intensive effort to develop a CFD model of a cracking heater, and the model that has been developed is unique in several respects. First, it combines the firebox model with a pyrolysis model for reactions taking place within the cracking coil. Second, by taking advantage of modern computing power, the individual burners can be modeled to include the very small fuel jets that define burner performance and hence heat-release profiles within the heater. Finally, the model is large enough to incorporate fully 50 percent of the heater as opposed to a small "slice." In addition to predicting the overall firebox aerodynamics, it shows the influence of burner design, operation, and location on the coil temperature profile and the reactions taking place. The integrated model is built up from individually validated components. It is currently used in all new and retrofit furnace optimizations, as well as in the troubleshooting of existing heaters.

COMMERCIAL OPERATIONS

More than one-third of the world's ethylene capacity is based on Lummus's SRT heater technology. These plants process a variety of hydrocarbon feedstock, ranging from ethane to vacuum gas oils with end points above 550°C. Table 6.1.3 shows the typical range of operating parameters. Several of these plants are designed to handle multiple feeds to improve economics of operation based on seasonal and opportunity feed availabilities. They also include designs that take advantage of the synergistic integration of the ethylene plant with refinery, downstream petrochemicals, and natural gas separation units.

ECONOMIC ASPECTS

The feedstocks for manufacturing ethylene are either natural gas components or various petroleum fractions. The economics therefore are strongly influenced by forces affecting the natural gas and petroleum markets. Ethane feed produces mostly ethylene, with very small quantities of coproducts; heavier feeds produce a significant amount of coproducts. The combined by-products volume of plants processing naphtha or heavier feeds exceeds the production rate of ethylene by a factor of 2. Therefore, coproduct prices and demand affect the economics of an ethylene plant significantly. High-value coproducts improve gross margin and smooth the peaks and troughs of product/coproduct demand and price fluctuations.

The cost of raw materials and energy/utilities and the market value of the products are site-specific and can vary significantly, affecting economic decisions. Total capital investment is lowest for ethane feed plants and increases successively for heavier feeds. However, the gross margin also will increase due to extra revenue from the greater volumes of coproducts. This increment may influence positively the rate of return on employed capital and therefore justify higher capital investment. Another significant fac-

TABLE 6.1.3 Typical Range of Operating Parameters

Cracking heater outlet temperature	750–900°C (1380–1650°F)
Cracking heater outlet pressure	1.5–2.8 bar (22–40 psia)
Ethane conversion	65–75%
Propane conversion	80–95%
Propylene/ethylene ratio	0.4–0.7 (for liquid feeds)
Dilution steam/hydrocarbon ratio	0.25–1 (by weight depending on feed)
Charge gas compressor discharge	30–38 bar (425–550 psia)
Demethanizer	7–32 bar (100–465 psia)
Deethanizer	20–27 bar (300–400 psia)
Depropanizer	10–18 bar (150–270 psia)
Debutanizer	4–6 bar (60–90 psia)
Ethylene fractionator	8–20 bar (110–300 psia)
Propylene fractionator	8–20 bar (110–300 psia)

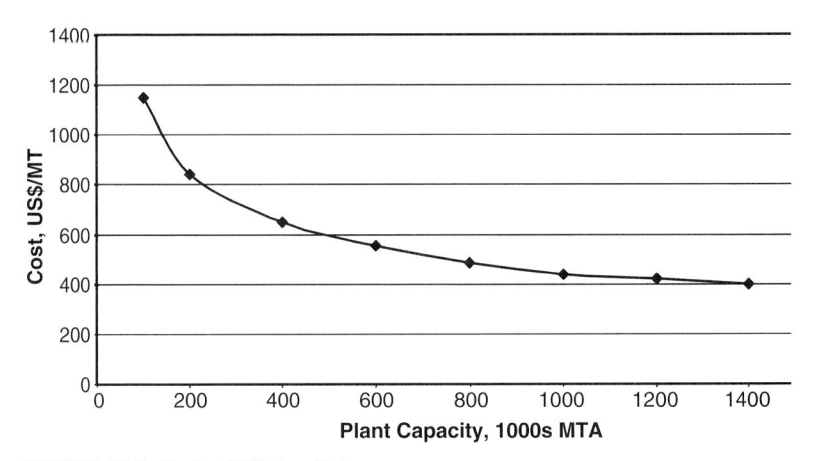

FIGURE 6.1.9 Typical ISBL installed cost.

tor affecting capital investment is plant capacity. Economies of scale favor larger-capacity plants, and the trend toward large-capacity plants continues. The maximum capacity of a single train is typically limited by compressor frame size.

Typical inside battery limit (ISBL) installed cost is presented in Fig. 6.1.9. The cost of outside battery limit (OSBL) support facilities typically varies between 30 and 50 percent of ISBL investment, depending on infrastructure and storage facilities.

CHAPTER 6.2
STONE & WEBSTER ETHYLENE TECHNOLOGY

Colin P. Bowen
Stone & Webster, Inc.
Houston, Texas

INTRODUCTION

Stone & Webster (S&W) ethylene technology enjoys a leading worldwide reputation. At the time of this writing, S&W technology has been selected for over 120 grassroots units; since 1990, S&W technology has contributed approximately one-third of total world additional capacity, including the world's largest unit to date [Nova/Dow E-3, Alberta, Canada; it currently operates at almost 1.5 million metric tons per annum (MMTA) of ethylene]. The 16 grassroots projects awarded to the company since 1990 represent a total ethylene capacity of almost 10 MMTA. S&W is also the world leader in major ethylene revamps, many of which have been applied to units designed by others. Revamp projects of approximately 5 MMTA of additional capacity have been implemented since 1990. S&W-designed ethylene plants have established a worldwide reputation for exceptionally high operational reliability, rapid startup, and superior performance.

Ethylene production technology covers a wide range of unit operations, thermal and catalytic reactions, heat- and mass-transfer systems, and very high and very low operating temperatures and pressures. This chapter will cover a range of topics related to the many aspects of this key technology.

ECONOMIC DRIVERS

Ethylene production, which is now approaching 100 MMTA, continues to grow at rates well above those of gross domestic product (GDP). This is due to further improvements in ethylene derivative performance and continuing ethylene-based polymer substitution of natural products. Ethylene production technology also has made significant advances. S&W has pioneered many developments in the field of ethylene technology and is a recognized leading licensor. The ethylene and derivatives market today is truly global; accordingly, all producers must seek to maintain economic competitiveness. In the field of ethylene production this translates to lower life-cycle cost and comprises several elements, most of which are linked to the technology employed. These are summarized below and will serve to provide an economic introduction for the various areas of technology that will be described in subsequent sections of this chapter.

1. *Feedstock cost.* This is the single largest cost item in ethylene production. *Net feedstock cost* (cost of feed − value of produced coproducts) is frequently regarded as a more valid economic term. To reduce feedstock-related costs, the process must be capable of cracking the selected feedstock range at optimal yields. Seasonal and opportunistic feedstock pricing requires that most plants be capable of cracking a relatively wide range of feedstocks to maintain profitability. The S&W technology covers the entire range of feedstocks.

2. *Selectivity.* This is a primary economic variable because it directly influences net feedstock cost. Cracking selectivity is principally determined by residence time, temperature profile, and hydrocarbon partial pressure, which, in turn, are functions of cracking furnace radiant coil geometry and imposed operating conditions. S&W recognized the importance of these variables in the 1950s, patented the low-residence-time USC (Ultra Selective Conversion) radiant coil dimensions in the late 1960s, and has continued to develop and demonstrate further selectivity advances.

3. *Efficiency.* Energy can represent up to 25 percent of today's production cost. Hence specific energy consumption (kcal/kg ethylene) is a key evaluation factor in comparing ethylene technologies. Energy consumption is divided approximately 50-50 between the cracking and recovery sections of the ethylene plant. In response to continually increasing energy costs, all technology licensors have employed standard energy-reduction measures, and specific energy consumption values have been reduced significantly within the last 25 years. Such measures involved reduced furnace heat losses, increased waste heat recovery, and other combustion-related enhancements. The largest energy component is the heat of cracking, which is essentially fixed. Hence the area for greatest energy savings is the recovery section. S&W has demonstrated a clear lead in this area of technology with its patented Advanced Recovery System (ARS).

4. *Operability/availability/reliability.* Although these performance aspects are less definable than the preceding factors, they are no less important in determining life-cycle cost. Costs of downtime, emergency flaring, maintenance, and equipment replacement can represent a significant proportion of total production cost. In particular, the ability to operate efficiently at full capacity for a predictable extended period (typically 5+ years) between scheduled shutdowns is most important. The S&W design pays particular attention to these issues by avoiding fouling conditions in critical services, by substituting or eliminating mechanically susceptible components (e.g., shell and tube quench exchangers, cryogenic pumps), and by designing equipment and control systems to minimize flaring during startup and shutdown.

5. *Investment costs.* Although this may not be the most important factor in determining life-cycle costs, it is certainly the most readily identifiable, particularly where projects are contracted on a lump-sum turnkey (LSTK) basis. Many factors contribute to investment cost, several of which are controlled by the owner (e.g., design codes and standards, operational and layout preferences, sparing philosophy, overcapacity margins, etc.). But the technology definitely contributes to total installed cost (TIC) both inside battery limit (ISBL) and outside battery limit (OSBL). A major factor that both the owner and the designer can use effectively is that of scale economies. Today's world-scale mega-units (1+ MMTA) achieve significant unit cost reduction over earlier plants (scale cost exponent approximately 0.65). Technology simplification, process design intensification, and design optimization can contribute to both investment and operating cost reductions. In these respects, S&W has taken a leading role, having built the world's

TABLE 6.2.1 Ethylene Production Cost Components[a, b]

Location	N.E. Asia / W. Europe	Middle East	United States
Feedstock	Naphtha	Ethane	Ethane
Feedstock cost ($/t feed)	320[c]	62[d]	317[e]
Net feedstock cost[f] ($/t C_2H_4)	55	68	266
Energy cost ($/t C_2H_4)	194	16	140
Fixed costs[g] ($/t C_2H_4)	66	56	51
Total production cost ($/t C_2H_4)	315	140	457
Contract sales price ($/t C_2H_4)		650–700	

[a] Amortization costs for capital investment are excluded.
[b] Cost basis: first quarter 2004.
[c] N.E. Asia/W. Europe naphtha cost = approx. $37.5/BBL (130% crude price).
[d] Middle East ethane cost = $1.25/MMBtu.
[e] United States ethane cost = $5.45/MMBtu natural gas + $1.0/MMBtu extraction cost.
[f] Net feedstock cost = feedstock cost − price of total nonethylene co-products.
[g] Fixed costs include labor, supervision, maintenance, insurance, overhead.

largest unit (Nova/Dow E-3, Alberta, Canada; its current capacity is approaching 1.5 MMTA) and the world's largest cracking furnaces (USC-32M, 0.23 MMTA) and having installed some 11 MMTA of capacity based on the proprietary ARS recovery process, thereby minimizing cold fractionation/refrigeration system duties and related ISBL/OSBL system costs.

The relative ethylene production cost components in the principal production regions are listed in Table 6.2.1. These data, based on 2004 first-quarter costs, emphasize the advantage of low feedstock and energy costs, the primary incentive for current Middle East projects.

Not shown in this table is the impact of integration of upstream feedstock supply and downstream derivatives marketing on overall economics. Such integration clearly can benefit the owner (or joint-venture partners) of the combined business scheme.

DEVELOPMENT HISTORY: PYROLYSIS

To date, most ethylene synthesis has been based on thermal pyrolysis. Many of the key pyrolysis parameters and their interrelationships were described in the late 1940s and early 1950s by two of S&W's leading technologists, Schutt and Zdonik.[1] This allowed greater understanding of cracking furnace design and yield prediction. Various radiant coil designs were then developed to improve capacity and selectivity. Most were based on horizontal serpentine designs and were limited by available metallurgies.

Then, in the mid-1960s, S&W introduced the revolutionary Ultra Selective Conversion (USC) coil design, which was based on vertical short residence-time coil and linear quench exchanger design principles. Ethylene yields were improved considerably, individual furnace capacities were increased, and a range of supporting design features were commercialized.

In conjunction with this technology breakthrough, S&W published a series of articles entitled, "Manufacturing Ethylene," by Zdonik, Green, and Hallee[2] that provided a definitive text on the theory and practice of cracking furnace design. The key principles identified and employed in the USC pyrolysis system are described below.

1. *Low residence time.* This is the fundamental key to greater cracking selectivity. In simple terms, this can be linked to the fact that ethylene is produced primarily by first-order dissociation of larger molecules. Second- and third-order reactions serve to further crack or recombine this ethylene to lighter or heavier, less valuable species. Residence time is a function of cracking coil volume. A related parameter, heat flux, is a function of cracking coil surface area. Hence, to minimize residence time without exceeding heat flux limits, the optimal cracking coil should have a small diameter with an adequate length. Since residence time is also determined by reactant vapor density, which is in turn a function of system pressure and hence pressure drop through the coil, the coil diameter-to-length ratio is an important design parameter. Early calculations typically were based on a length-to-inside-diameter (ID) ratio of approximately 500, representing the optimal balance between key design variables.

When USC coil designs were first patented and introduced, all other non-S&W cracking furnace designs were still based on much larger diameter serpentine-type coils. Both the kinetic and mechanical performance of the USC design was clearly superior. Hence, now that most of the USC patents have expired, most other designers today have copied many of the USC principles, including small coil diameters, flow control inlet venturis, close-coupled linear primary quench exchangers, and other secondary features.

2. *Low hydrocarbon partial pressure (HCPP).* Since the cracking reaction results in molecular expansion (e.g., one molecule of naphtha typically yields four molecules of different reaction products), in order to promote this effect, system pressure should be at minimum (Le Châtelier principle). The USC process employs a low coil outlet pressure, dictated by downstream pressure profile, and a low system pressure drop, together with the standard use of a dilution steam to further reduce HCPP. Early USC coils employed uniform diameters in order to promote low residence time. Recognizing the pressure-drop effects of molecular and thermal expansion and coil outlet zone coking effects, the coil design was changed to the "swaged" type, where the process flow ID increased stagewise through the coil and the downstream linear exchanger. This serves to improve average yield from start of run (SOR) to end of run (EOR), allow higher capacity flow, avoid loss of venturi criticality, and thus increase operating run length. A secondary benefit of the swaged concept is its low coke plugging susceptibility.

3. *High cracking conversion.* Maximum ethylene yield corresponds to high conversion (or severity) operation. Ethylene production pyrolysis is a high heat intensity process. Heat of cracking depends on feedstock and conversion. Ethane cracking can require up to 800 kcal/kg feed (approximately 75 percent conversion). Naphtha heat of cracking levels at maximum practical severity are in the 450-kcal/kg feed range. These energy levels correspond to approximately 40 percent of the total energy consumption within the ethane cracker and 30 percent within the naphtha cracker. Conversion is a function of both residence time and temperature (refer to definition of kinetic severity function below).

Definition of Kinetic Severity Function (KSF)

The severity of the pyrolysis reaction for naphtha-type feedstocks may be defined conveniently by measuring the content of n-pentane in the feedstock and in the furnace effluent. n-Pentane is selected as the reference component because it is commonly present in most light liquid feedstocks and because its degree of formation from either lighter or heavier components present in the reaction is negligible. Applying the first-order dissociation of n-pentane, the definition of severity is thus

$$KSF = \ln \frac{1}{1-a}$$

where $\ln$ = the natural logarithm
a = conversion of n-pentane

Applying this definition, the following severity table may be developed. Typical product maxima are listed for reference.

% conversion	KSF	
82	1.7	Lowest commercial severity; max. propylene
90	2.3	Max. butadiene
93	2.7	Max. combined olefins
98	3.9	Highest commercial severity; max. ethylene

Hence, in order to achieve minimum residence time, the cracking temperature must be high. Further kinetic optimization indicates that the process temperature profile should increase through the entire reaction sequence. Because the USC coil has a relatively high surface-to-volume ratio, this permits operation at high temperatures without exceeding the limiting tube metal temperature. Consequently, USC coil designs can achieve both high-selectivity (defined as combined yields of ethylene, propylene, and butadiene from a given feedstock at a given conversion) and high-severity operation.

Linear Quench Exchanger. Ethylene yield is determined not only by residence time in the cracking coil but actually by residence time in the entire pyrolysis system. Hence, to retain the high yield achieved within the coil without allowing it to deteriorate as a result of secondary/tertiary reactions, it is essential to "freeze" the reaction equilibrium as quickly as possible. Pre-USC designs employed conventional shell and tube type exchangers installed on manifolded coil outlets. Such designs created a high inlet adiabatic volume (thereby extending residence time and reducing selectivity), suffered from high coking effects on the inlet tube sheet, and resulted in increased coil outlet pressure, particularly at EOR conditions (thereby further affecting yield). Other operating and maintenance problems also detracted from the suitability of such transfer-line exchangers (TLEs) for this service. S&W recognized the need for a completely different type of quench exchanger to complement the multicoil USC designs while avoiding the problems just stated. The patented USX (Ultra Selective Exchanger) linear exchanger was developed to achieve these objectives. The USX was close-coupled to the cracking coil outlets and avoided all the TLE-type problems; inlet adiabatic volume was minimized; reaction temperature was quickly quenched, thereby "freezing" the yield; coking tendencies were greatly reduced; and SOR/EOR pressure drop was low. Depending on the feedstock type, the USX primary exchanger was supplemented by a conventional shell and tube type secondary exchanger (TLX) to maximize waste heat recovery. Later designs employed the next generation Selective Linear Exchanger (SLE) designed by S&W's partner Borsig. The SLE combines the heat-transfer duty of both the primary USX and the secondary TLX into a single element but retains all the USX performance benefits.

Feedstock Flexibility. The key principles stated earlier apply to both gas and liquid feedstocks. However, due to the different reaction conditions and reactant properties, certain design differences also apply. When cracking ethane or propane, the molecular expansion effect is less pronounced, but the heat of cracking is considerably higher than that for liquid feedstocks. These two kinetic variables therefore correspond to a somewhat different optimal cracking coil design. The USC coil for ethane/propane cracking is thus a larger version of the same coil employed for liquid cracking; its somewhat higher residence time has only a minimal yield debit but allows for high-conversion cracking at reasonable run lengths. The large-diameter USX double-pipe quench exchanger again matches the larger gas cracking coil outlet diameter.

6.26 ETHYLENE

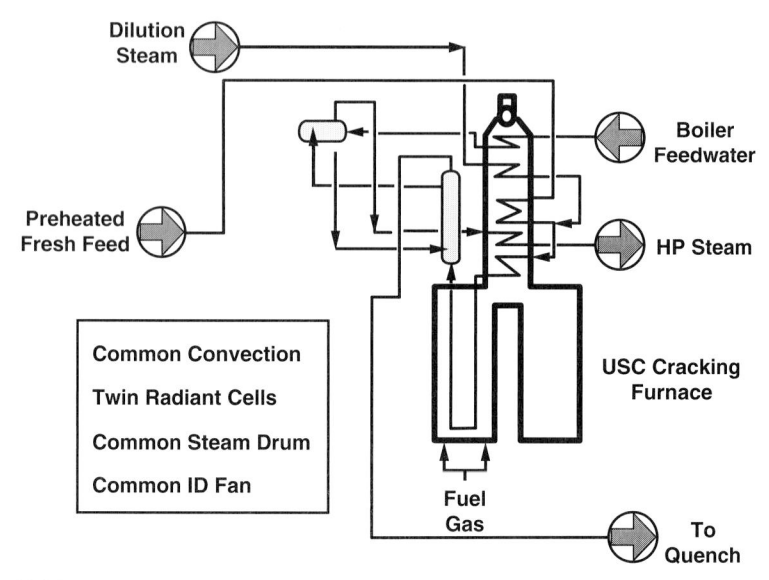

FIGURE 6.2.1 Twin-cell configuration.

Capacity. Today's ethylene plants are up to 100 times larger than the first commercial units. Furnace capacities have increased by a factor of 10 since introduction of USC designs in the late 1960s. Current megafurnace ethylene capacities are up to 250 kilotons per annum (KTA) (ethane feed) and 175 KTA (naphtha feed). Even larger furnaces are in the design stage; furnace size is usually determined by plant capacity and optimal furnace count (four to eight furnaces).

In order to benefit from the scale economics of larger furnaces while retaining the operational flexibility of multiple furnaces, large USC furnaces are designed with both features. This is achieved with the dual radiant cell zoned feed cracking design, as shown in Fig. 6.2.1. The dual radiant cell feature results in a more compact, cost-effective layout. It can be arranged to allow independent cell decoking and feedstock cracking. The cracking coils also may be arranged to allow different feeds to be cracked within a single radiant cell.

Independent cell decoking is possible by virtue of the ability to completely decoke the coils and SLE quench exchangers by online decoking (i.e., maintaining normal operating temperature during steam-air decoking). Such a system would be impossible if the decoking operations could not be conducted at normal operating temperatures. Since shell and tube type TLE exchangers require frequent offline (i.e., cold) mechanical/hydroblast decoking, independent cell decoking of such furnaces would be operationally incompatible.

Efficiency. All modern furnaces are designed for high (i.e., 92+ percent) thermal efficiency. Several USC furnaces have achieved even higher efficiencies by incorporating combustion air preheat and by designing for optimal preheating of ambient-temperature feedstocks in order to achieve very low stack gas temperatures. Such furnaces can achieve thermal efficiencies of up to 96 percent.

A further overall system efficiency credit is achieved by linking the furnace with a gas turbine and thereby recovering the gas turbine exhaust waste heat within the furnace. The gas turbine shaft power typically is directed to power generation. Such gas turbine integration (GTI) schemes require careful selection of all involved components and controls to ensure safety and operating compatibility under all operating conditions. GTI designs can achieve overall energy savings within the ethylene unit of as high as 15 percent.

Environmental. Increasing concerns over environmental emissions have had a major impact on both new and existing furnaces. The major design challenge is to meet the very strict NO_x emission levels set by most regional authorities. In order to comply, S&W new designs employ total floor firing with fewer relatively large low-NO_x burners. Such burners are designed with staged firing, flue gas recirculation, and optional steam injection to achieve NO_x levels as low as 70 mg/Nm³. Newer burner designs are under development to further reduce NO_x emissions. Most new furnaces are also designed with convection zone space for an integral selective catalytic reduction (SCR) system in order to further reduce net NO_x output if required.

Other emission standards that are imposed today include particulates, carbon monoxide, and volatile organic compounds (VOCs). Current USC furnace designs incorporate controlled recycle decoking to the firebox to minimize such emissions and standard VOC control components.

Safety and Control. The ethylene industry is continuously aware of furnace-related safety incidents; several major accidents have occurred as a result of a variety of equipment failures, design problems, and operator errors. S&W furnaces have a much better safety record than others due principally to their mechanical reliability and basic safety/control schemes. Current USC furnace designs employ a three-stage shutdown scheme: partial, total, and pilot. Such a system, coupled with the use of both local and remote instrumentation, will allow either manual or automatic shutdown corresponding to the relevant emergency condition. A further key safety system is the high-quality burner control/startup instrumentation scheme to avoid startup ignition risks.

DEVELOPMENT HISTORY: RECOVERY

The pyrolysis furnace effluent consists of an extremely wide range of components. The technical challenge of separating the primary products from the secondary coproducts in the most efficient manner continues to exercise the several ethylene licensors. S&W's approach to this separation technology employs a combination of fractionation, catalytic, and cryogenic techniques. The basic recovery scheme employed on early designs consisted of the following sequence:

- *Quench oil/water towers.* Fuel oil rejection, preliminary gasoline separation.
- *Cracked gas compression.* Achieves necessary system pressure for downstream low-temperature fractionation.
- *Acid gas removal, dehydrators.* Eliminates H_2S, CO_2, and H_2O.
- *Front-end demethanization.* Principal separation between C_2+ and lighter residue gas fractions with maximum recovery of ethylene and hydrogen.
- *Deethanization/C_2 hydrogenation/C_2 splitter.* Produces high-purity ethylene, recycles ethane.
- *Depropanizer/C_3 hydrogenation/C_3 splitter.* Produces high-purity propylene, recycles propane.
- *C_4+ processing.* Separates C_4, C_5, and gasoline fractions.

To support this basic scheme, a number of refrigeration and other ISBL (inside battery limit) utilities systems are required; external offsite systems also represent considerable investment (up to 40 percent of ISBL investment).

In order to achieve the necessary separations, the early designs consumed a relatively high energy level. Pre–energy crisis specific energy consumption values for a typical naphtha cracker were on the order of 9000 kcal/kg C_2H_4. In response to higher energy costs, considerable effort was directed in the 1970s and 1980s toward efficiency improvements; energy consumption levels, accordingly, were reduced to the 7500-kcal/kg C_2H_4 range. For ethane crackers, the corresponding energy consumption values were approximately 70 percent of the preceding naphtha cracker levels.

During this same time frame, various other recovery-area improvements were introduced. Advances in catalysis allowed more selective conversion of acetylenes and dienes to their olefinic analogues. Improved desiccant materials and contaminant-removal systems were introduced; tighter product specifications thus were achieved. Higher-efficiency compression systems and antifouling design measures improved energy consumption and extended operating cycles. High-efficiency, mechanically reliable turboexpanders became an integral part of most designs and provided the bulk of the cryogenic level refrigeration duty (below $-101°C$).

In conjunction with the preceding equipment and materials advances, a similar level of improvements in the area of instrumentation and control occurred in the 1980s. These technologies have progressed to the extent that online computer control/monitoring/optimization today is a standard component of most ethylene units.

A significant further reduction in energy consumption came about in the early 1990s when S&W introduced the Advanced Recovery System (ARS). This scheme, co-invented with Mobil Chemical, was based on several high-efficiency concepts, including integrated simultaneous heat and mass transfer, distributed distillation, higher-selectivity acetylene conversion, and open-loop heat-pump refrigeration schemes. The combined effects of the preceding combination of ARS elements reduced specific energy consumption of the naphtha cracker to the 5000-kcal/kg C_2H_4 range.

As ethylene plants became larger, it became economically viable to recover several of the higher-value secondary coproducts rather than merely converting them to ethylene/propylene or to combine them with fuel streams. S&W developed several extraction/recovery processes to separate such coproducts, e.g., acetylene, methyl acetylene/propadiene, cyclopentadiene, and isoprene. In conjunction with other licensors, S&W offers associated recovery techniques for butadiene, butenes, amylenes, aromatics, and so on.

Another area of proprietary S&W development was the ripple tray. First developed and patented in the 1950s, this simple but highly effective tray design (corrugated, nondowncomer, sieve tray) solved several of the heat/mass-transfer/fouling problems that other designs encountered. Today, the ripple tray is the preferred internal for the quench towers (high hydraulic capacity and heat transfer, low fouling; see Fig. 6.2.2), the caustic scrubber (high fouling tendency with associated CO_2 breakthrough), and other fouling services. To increase capacity and overcome fouling problems, S&W has retrofitted the ripple tray

- **Enhanced Heat Transfer**
- **Proven Fouling Resistance**
- **Maximum Hydraulic Capacity**

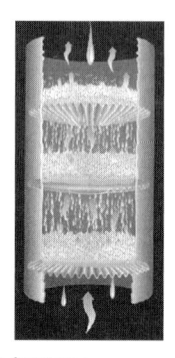

FIGURE 6.2.2 QO tower ripple tray advantages.

in revamp applications on many ethylene units, including non-S&W designs and other difficult fractionation services.

A key area in ethylene recovery technology is that of hydrogenation catalysis. Early catalysts were based on alumina/nickel combinations and were relatively insensitive to impurities and temporary poisons, e.g., mercury, arsenic, sulfur, and oxides. However, their selectivity for acetylene-olefin conversion and/or conjugated diene saturation was relatively low. In the late 1960s, the first generation of noble metal (Pd, Pt) catalysts was introduced. Conversion selectivity was improved, but sensitivity to poisons also increased. S&W developed an isothermal C_2 hydrogenation reactor system to allow for higher inlet acetylene levels. Various catalyst manufacturers subsequently enhanced performance by employing bimetallic noble metal systems, e.g., Pd/Ag and Pt/W, with associated guard bed materials.

Today's S&W design employs front-end vapor phase hydrogenation catalysis (higher selectivity, eliminates regeneration) with optional back-end liquid phase hydrogenation of residual methyl acetylene/propadiene. If appropriate, partial or complete saturation of C_4/C_5 fractions is also included. Most ethylene units are linked to an associated pyrolysis gasoline hydrogenation unit. Such mixed-phase systems reduce gum precursors by conjugated diene saturation (mixed phase, first stage) to meet gasoline blendstock specifications and optionally produce a sulfur-free, olefin-free aromatic concentrate (vapor phase, second stage) for BTX extraction.

Another emerging aspect of ethylene plant recovery design is the integration of the ethylene unit with its adjacent refinery. As the degree of refinery conversion processing increases, recovery of light olefins from FCC-type units becomes economically viable. Accordingly, it is frequently attractive to separate these refinery offgas olefins within the adjacent ethylene unit. This requires a series of contaminant removal systems to avoid safety risks and product specification deviations but provides a low-cost supplementary feedstock source. Other areas of refinery-petrochemical integration also have been devised to further improve combined operating economics, e.g., hydrocracker residue cracking, pyrolysis gasoline interchange, hydrogen management, recycle cracking of various raffinate streams, metathesis-type conversions, etc.

An emerging area of refinery-petrochemical technology is that of catalytic olefin manufacture. New developments in the areas of higher-activity FCC-type catalysts combined with many features of S&W residue FCC technology today achieve high product levels of ethylene, propylene, and butylenes from vacuum gas-oil (VGO)–type feedstocks. Linked recovery schemes between such catalytic olefin units and the standard ethylene unit allow for minimum overall investment and energy consumption.

The first-generation catalytic olefin process, Deep Catalytic Cracking (DCC), is described in Chap. 3.2 of the Third Edition of the *Handbook of Petroleum Refining Processes,* edited by Robert A. Meyers (McGraw-Hill, 2004).

PROCESS DESCRIPTION

This nonconfidential description of today's S&W ethylene process recognizes all the key design issues indicated in the proceeding discussion and incorporates the latest technologies to achieve the economic targets identified.

Ethylene unit design depends on the type of cracking feedstock. Ethylene plants generally fall into two categories, gas feeds and liquid feeds. Several features are similar, but there are sufficient differences between the two plant types that it is more convenient to provide separate descriptions. In general terms, the ethane cracker is a somewhat simpler design, contains approximately two-thirds of the items of equipment, and consumes approximately two-thirds of the energy of its naphtha cracking counterpart.

Gas-Feed Cracker Process Design

The basic block flow diagram for the ethane/propane cracking process is shown in Fig. 6.2.3.

Pyrolysis Furnace/Radiant Coil Selection. The principles of radiant coil design are outlined in the introductory section of this chapter. The S&W patented coil geometry represents the optimum from the standpoint of yield selectivity, operability, and capital cost. The special features of USC technology that produce this performance are

- Swaged coil design to better accommodate process-fluid thermal/molecular expansion and coke laydown effects.
- A single-inlet, single-outlet flow path through the entire coil and primary quench exchanger; this corresponds to optimal selectivity and mechanical reliability.
- Positive and uniform flow control to each radiant coil during all operating and decoke modes.
- The optimal combination of residence time and coil capacity (and hence optimal surface-area-to-volume ratio) results in conservative thermal requirements, relatively low metal temperatures, and long coil life.

The simple mechanical configuration of the USC radiant coil provides extended, reliable operation and eliminates the high maintenance associated with more complex coil designs. S&W licensees report an average radiant coil life of 7 years (for all feeds; ethane feed operation typically 5 years), indicative of the robust mechanical coil design.

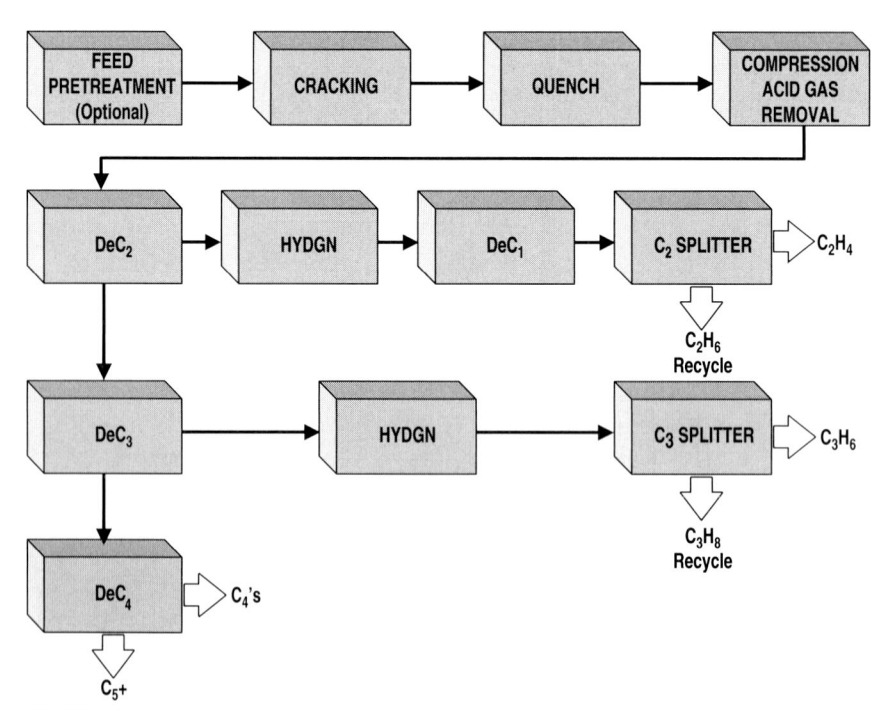

FIGURE 6.2.3 Ethane/propane cracking block flow diagram.

S&W offers a family of radiant coils employing a standard design philosophy; length and diameter are optimized to the client's individual needs. The longest coil, the M-coil, has the lowest selectivity, the highest feed capacity, and the longest run length and results in the lowest capital cost furnace per unit quantity of ethylene produced. The shortest coil, the U-coil, achieves the best selectivity but has the lowest feed capacity, shorter run lengths, and a relatively higher unit cost. The W-coil is intermediate in all respects.

The most appropriate coil for any given application will depend on a variety of economic considerations, e.g., capital cost, feed cost, and relative product values. These will, in turn, relate to the design feedstock range, the design severity range, and specific requirements for run length.

Certain basic relationships govern technology selection under these circumstances. Selectivity is linked to residence time, but for lighter feedstocks, the effect of residence time on selectivity is diminished. In addition, since severity of cracking is reduced, the yield advantage associated with selectivity also reduces. The USC range of coil designs therefore has been developed with residence times of 0.15 to 0.55 s to best meet the current technical and economic requirements; the lower end of the range is most applicable to liquids cracking, and the higher end is most applicable to gas cracking. Designs outside this range have inherent difficulties—coils with residence times shorter than this target range suffer rapid selectivity loss and poor run lengths due to accelerated coking; coils with longer residence times unduly sacrifice selectivity.

The USC M-coil is usually the most appropriate selection for ethane/propane cracking for the following reasons:

- High capacity/low investment
- Long run lengths between decokes
- Wide conversion range
- Mechanical simplicity, reliability, and survivability

A separate consideration is the relative cracking conversions to be employed for the ethane and propane feedstocks. These feeds may be either cocracked or cracked independently. For operational and investment reasons, it is preferable to standardize on a single furnace type. However, it is probably advantageous to crack the two feedstock types at individual optimal conversions, particularly if propylene coproduct production is to be maximized (corresponding to lower-conversion propane cracking).

Some of the important mechanical features of the USC furnace radiant coil design are

- A single-row radiant coil layout with conservative coil spacing minimizes circumferential temperature gradients.
- The simple mechanical configuration of the USC radiant coil with a single coil inlet for each coil outlet is a rugged, reliable design. There are no multileg fittings within the firebox because such fittings are prone to failure due to thermal stresses and increase creep deformation due to their heavier weight. This translates into long USC radiant coil life.
- USC radiant coils have been fabricated from a range of high-temperature chromium-nickel alloys. Early designs employed combinations of HK and HP alloys. The latest designs now use higher-performance HP materials containing various proportions of anticarburizing and welding additives (e.g., tungsten, niobium). Tube metal maximum design temperatures range from 1100 to 1150°C depending on alloy type.
- The latest developments in extended-surface coil geometries and catalytic suppression coatings also have been employed on recent USC M-coils.

Burner Design. Furnaces are equipped with partial, total, and pilot shutdown. This enables rapid restart of the furnace in the event of upset or emergency situations. Total floor firing is employed for the USC-M furnace. Advantages of this design include simpler control, easier burner maintenance, and tighter NO_x emission control.

Piloted floor-mounted low-NO_x staged fuel recirculation type gas burners are selected to minimize NO_x emissions over the full range of available fuels. S&W, in conjunction with John Zink and REI (Reaction Engineering, Inc.), recently has developed advanced CFD techniques for more accurately predicting firebox conditions and heat-transfer effects. This modeling technique results in more uniform radiant section performance and lower overall emissions.

The latest generation of S&W furnaces incorporates a system that routes decoke effluent directly back to the firebox of the decoking furnace. This saves capital cost and contributes to effluent minimization.

Quench Exchangers. The basic quench exchanger configuration used in the USC-M process is the Ultra Selective Exchanger (USX), followed by the Transfer Line Exchanger (TLX). The USX primary exchanger is a double-pipe design close-coupled to the outlet of the radiant coil. The USX quenches the cracked gas effluent to below the temperature where the cracking reaction ceases. Additional heat recovery then takes place in the TLX exchanger. The recovered heat from both the USX and TLX exchangers generates high-pressure steam that is subsequently superheated in the integrated superheater zone of the furnace.

The USX/TLX quench exchanger system is an integral component of the S&W pyrolysis system. This patented design is particularly valuable in ethane/propane cracking service, where quench exchanger inlet fouling has been a major operating problem for non-USC designs. Fouling in the exchanger inlet is reduced significantly by the unique USX design. Other advantages that the USX/TLX quench system achieves include

- Low transfer-line volume, hence rapid quenching. This reduces secondary reactions that degrade olefin products and produce accelerated coking.
- Anticoking design features due to streamlined flow and large internal diameter. The stagewise diameter increase (swaging) throughout the radiant coil/USX quench exchanger system avoids the risk of coke plugging.
- Multiple operating cycles between offline decokes; based on feedback from clients, depending on site operating and maintenance practices at the design ethane conversion level, this furnace should not require offline decoking other than during planned furnace inspection/plant turnaround, e.g., every 2 to 3 years.
- Ease of decoking because of its simple yet rugged construction. Online decoking by steam-air typically requires no more than 24 hours of process interruption.
- The oversized coke-collection header at the outlet of the USX eliminates TLX tubesheet fouling. This header, upstream of the shell and tube TLX exchanger, has been proven to be effective in preventing spalled coke particles from fouling the hot-end tubesheet at the TLX.

Furnace Control Systems. The furnace typically incorporates four major control systems to maintain stable operation:

- Flow distribution
- Conversion
- Furnace draft
- Steam drum level

Uniform flow of hydrocarbon and dilution steam to each furnace zone will be controlled by a separate flow-control valve. Uniform flow to each radiant coil is ensured by the use of critical flow venturis.

Conversion is controlled by adjusting the firing rate of the burners in each furnace using the furnace average coil outlet temperature (COT) to regulate fuel gas flow. The COT is reset to control the conversion based on a correlation with the measured composition of the cracked gas.

Furnace draft is controlled by a damper at the induced draft (ID) fan inlet; the ID fan typically is driven by a variable-speed electric motor.

Accurate control of the steam drum level is critical to maintain thermosyphon stability to the quench exchangers. A three-element control typically is used, where the BFW flow rate will be reset by the level controller and the flow of the SHP steam generated from the steam drum. The TLX outlet temperature is controlled by operator-adjusted BFW flow to the TLX.

Particular attention is paid to the protection of the cracking furnace from unnecessary thermal stresses caused by plant trips. In order to provide the maximum freedom from tube thermal cycling and to eliminate potential problems of coke spalling and possible deterioration of insulation and refractory caused by repeated furnace shutdowns, a pilot and staged shutdown arrangement is employed.

Quench Water Tower System. The quench water tower receives the cracking furnace effluent, condenses the dilution steam from the cracked gas, rejects the heavy contaminants, and recovers the process waste heat against other suitable process streams. Gas cracking traditionally has resulted in highly contaminated quench water. This contamination by tars, polymer oils, and coke fines has resulted in costly dilution steam generation feed water cleanup and high effluent treating costs. Cracking of gas feeds produces heavy hydrocarbons and pyrolysis tar, which have a density similar to that of water. The continuous injection of sulfur into furnace feeds as a coking and carbon monoxide control method complicates the control of pH in the bottoms of the quench tower and the resulting oil-water separation. Once the water and hydrocarbon have emulsified, it becomes very difficult to separate them. S&W has minimized this problem by designing the quench water tower system with two discrete quench water circulation loops corresponding to a saturation section and a condensation section.

In the bottoms saturation section, the cracked gas is cooled and saturated with quench water to a few degrees above its adiabatic dew point. All the dilution steam remains in the vapor phase. The temperature is reduced by this quenching and results in condensation of the tar and the heavy hydrocarbons. The saturation circulating water is withdrawn to the first oil-water separator (OWS-1).

The purpose of OWS-1 is to effect a three-phase separation of the process water stream. The heavy phase, which contains tar and coke particles, is separated from the bulk stream and fed into the tar disposal separator, from which the tar is withdrawn periodically. After separation of the pyrolysis tar within OWS-1, the residual water and light hydrocarbon phases are separated by conventional interface weir compartments. The aqueous phase is recirculated to the saturator section; the gasoline phase is pumped forward for further processing.

The saturated furnace effluent passes from the saturator section into the upper tower (condensation) section via chimney trays. In this section the dilution steam is condensed from the cracked gas by the primary quench water. Since the heavy hydrocarbons have been removed in the saturator section, quench water quality in the upper section is relatively clean and causes minimal fouling problems in the cooling circuit. Cooling of the quench water is via process-process heat recovery followed by air and/or cooling-water exchange. Quench water is returned to the top of the quench tower at the closest possible approach to the heat-sink temperature available. The cracked gas is cooled thereby as much as possible so that the subsequent compression power is minimized. A quench-water slipstream is used as feedwater to the dilution steam system.

This arrangement of the quench-water circuits minimizes fouling of the heat-recovery exchangers and results in a reliable and consistent quench-water and dilution steam system operation. It eliminates one of the most frequent problems with high-severity gas cracking—the handling and cleaning up of oily quench water.

Dilution Steam Generation. The feedwater for dilution steam generation is withdrawn from the clean primary quench-water circuit. This feed water is saturated with acid gas and light hydrocarbons. The water is filtered initially and treated to remove emulsified and dissolved oils and then is routed to the pretreat stage. The pretreat process recently developed by S&W employs solvent-based liquid-liquid extraction to remove any residual tar/oil (free, dissolved, emulsified) from the dilution steam feedwater. The solvent is recovered and recycled with a minimum purge and makeup.

The purified dilution steam feedwater is then pumped forward to a dual-function stripper tower reboiled against MP steam. Dissolved solvent is stripped and recycled to the upstream quench tower. Dilution steam is withdrawn as a vapor sidedraw. Net bottoms blowdown is routed to offsite recovery or treatment.

This system is preferred to the feed-gas saturation system employed by others because it provides a more reliable method of producing dilution steam and decoke steam without fouling risk.

Compression. Most ethane/propane crackers employ a four-stage cracked gas compressor, interstage coolers, knockout drums, acid gas removal system, dehydrators, and front-end hydrogenation systems. Selection of the number of stages for the compression system depends on the characteristics of each specific process application, and evaluation of the system depends principally on the following variables:

- Overall suction-discharge compression ratio, i.e., selected operating pressures for the demethanizer system and the coil outlet of the cracking furnaces
- Available cooling water temperature
- Interstage discharge temperatures with or without water injection
- Interstage pressure drop allowance
- Location of the acid gas removal system

The caustic tower is usually located on the penultimate stage discharge because lower-pressure operation tends to reduce fouling, overall power requirements, and capital cost.

A multisection caustic tower is included to minimize caustic use and to produce a spent caustic stream that contains less than 1 percent free caustic. The cracked gas is contacted by circulating strong and weak caustic solutions to remove the acid gases. The wash-water section at the top of the tower minimizes caustic carryover to the compressor. The cooler reduces the overhead temperature, thereby reducing the water content in the overheads. Ripple trays are used in the caustic tower to prevent plugging. The spent caustic is fed to the degassing drum for final hydrocarbon de-entrainment, prior to downstream disposal treatment.

Deethanization. The compressed cracked gas is dried before being routed to the recovery section for purification and fractionation of the ethylene product. The initial processing step is removal of the acetylene. Prior to introduction of the cracked gas to the acetylene hydrogenation reactor, it is prefractionated to remove any C_3+ heavy components. Such heavy ends would interfere with the hydrogenation reaction and prematurely require catalyst regeneration/replacement.

S&W employs a dual-pressure front-end deethanizer to make this separation. The overhead vapor from the high-pressure (HP) deethanizer is routed to the hydrogenation reactor.

Bottoms temperature of the HP deethanizer is controlled to avoid fouling by leaking C_2's to the low-pressure (LP) deethanizer. The LP tower is operated as a simple stripper; overhead vapor is recycled to the cracked gas compressor, and bottoms C_3+ product is delivered to downstream processing.

The deethanizer system thus protects the downstream purification units and avoids the fouling problems associated with processing of the highly unsaturated C_3+ fraction.

Acetylene Hydrogenation. The modern front-end hydrogenation system is the process of choice for a number of well-proven reasons. The rate of reaction is partly moderated by the CO in the cracked gas from pyrolysis but mainly depends on acetylene concentration and temperature. The process is highly controllable by catalyst bed inlet temperature only.

There is little overreaction, which would result in the production of dimers or homologues (green oil). The catalyst deactivates slowly and hence does not require regeneration facilities. Selectivity is high and is predictable for several years. Several palladium-based front-end catalysts are available from experienced manufacturers.

The reactor trip system is simple and inherently safe and is based on the principle of blocked-in feed flow and depressurization. In addition, the catalyst support pores, which control the activity, will not admit hydrocarbon species at elevated temperatures. Thus any temperature excursion due to ethylene-ethylene reactions is automatically quenched, and further reaction is prevented, before dangerous temperatures are experienced.

Front-end reactor systems produce a consistent and much higher final ethylene purity than older back-end hydrogenation systems. Because it is unnecessary to provide a separate purity hydrogen reactant stream from the demethanizer overheads, the front-end acetylene reactor scheme can be readily started up and thereby reduces overall startup time.

The front-end hydrogenation scheme consists of two reactors in series with interstage cooling followed by a single secondary dehydrator. Should it be necessary to remove one of the reactors from service, the operating condition of the remaining unit can be adjusted (at lower net selectivity) to maintain effluent acetylene level.

Demethanization. The acetylene-free cracked gas contains ethylene, ethane, methane, and hydrogen. The methane and hydrogen (together with any residual carbon monoxide) are stripped from the C_2 fraction in the demethanizer system. The C_2 product stream from the bottom of the demethanizer is routed to the downstream ethylene fractionator. The demethanizer is an energy-intensive section of the plant; considerable optimization of the refrigeration systems is employed to minimize net energy consumption. Initial chilling of the feed gas is against cascade propylene-ethylene refrigerants. Lowest-level refrigeration, necessary for the final ethylene-hydrogen separation, is derived from the isentropic expansion of the hydrogen-rich residue gas through a two-stage expander-recompressor set. In the unlikely event of expander outage, the process gas is expanded isenthalpically; ethylene production and specification thus are maintained, although at a higher net leakage of ethylene into the residue gas.

Ethylene Fractionation. Ethylene-ethane separation is achieved in an open-loop heat-pumped fractionator. The operating pressure for this ethylene-ethane fractionation system is set to achieve an optimal balance of investment cost against operating cost. The system operates at relatively low pressure, thus increasing the relative volatility and decreasing the required reflux ratio and the number of trays required.

The C_2 splitter system achieves significant energy savings by using an open-loop heat-pump system. The heat pump provides both the reboiling heat input and the reflux refrigeration. The tower overheads are reheated against the vapor streams to the reboiler and the tower reflux. Low-level propylene refrigeration is used in parallel with the bottom reboiler to provide any necessary trim condensation. Most of the heat of compression is removed against cooling water in the heat-pump compressor discharge desuperheaters rather than being rejected to refrigeration.

This critical fractionation system achieves high thermodynamic efficiency, simple and reliable operation, and an optimized configuration. This system reduces energy consumption significantly compared with the previous high-pressure, indirectly refrigerated system.

The combination of front-end acetylene hydrogenation and open-loop ethylene/ethane fractionation will allow the production of the highest-purity ethylene and the lowest energy consumption. The hydrogen and carbon monoxide concentrations in the product ethylene are almost unmeasurable (<0.1 ppm).

Refrigeration Systems. The heat-pump compressor circuit is optimally integrated with the C_2 refrigerant circuit. Plant refrigeration duties are provided by cascaded C_2/C_3 refrigeration. Various levels of process-process refrigeration recovery are also employed. Respective suction/discharge pressures are optimized against available condensing media and equipment characteristics. All refrigeration compressors are equipped with dry-gas seals to prevent lube/seal oil contamination.

C_3/C_4 Processing. The LP deethanizer bottoms stream consists of C_3+ components with a relatively high concentration of unsaturates. Depending on the specification requirements for the C_3 and C_4+ coproducts, this stream will be depropanized/hydrogenated/fractionated accordingly. Assuming that a desired coproduct is polymer-grade propylene, the C_3 fraction will be hydrogenated to convert the bulk of the methyl acetylene/propadiene to propylene and then superfractionated to meet coproduct specifications (99.6 percent C_3H_6). Propane bottom product then will be recycled and recombined with fresh propane feed.

The depropanizer bottoms probably will be exported as an untreated C_4 coproduct stream. Depending on specific needs, it may be further fractionated into a C_4 fraction (with high butadiene content) and a C_5+ fraction.

Conclusion. The S&W ethane propane cracking/recovery scheme offers many technical, operational, and investment advantages. It embodies several generations of process design evolution and the latest equipment and catalyst developments, integrated to achieve an optimal overall result. Many of the features described are demonstrated in the Nova/Dow E-3 cracker, the world's largest and most efficient ethane cracker. This design results in a safe, reliable, and economic plant, achieving the lowest life-cycle costs.

Liquid-Feed Cracker Process Design

The basic block flow diagram for the liquids cracking process is shown in Fig. 6.2.4.

Pyrolysis Furnace. As explained in the preceding pyrolysis development section, when cracking liquids feedstocks, the optimal furnace design typically will combine the features of high selectivity, high severity, high efficiency, and high capacity. Many of the design principles for liquids cracking furnaces are similar to those outlined earlier for gas feed crackers.

The precise configuration of the liquids cracker furnace section will depend on the feedstock range [light naphtha (LN)/atmospheric gas-oil (AGO)/hydrocracker residue (HCR)/vacuum gas-oil (VGO)] and the selected recycle cracking arrangement (ethane/propane/C_4's).

Most large crackers will consist of a liquids cracking furnace section and associated ethane/propane recycle furnace(s). Recycle furnace design features are described in the

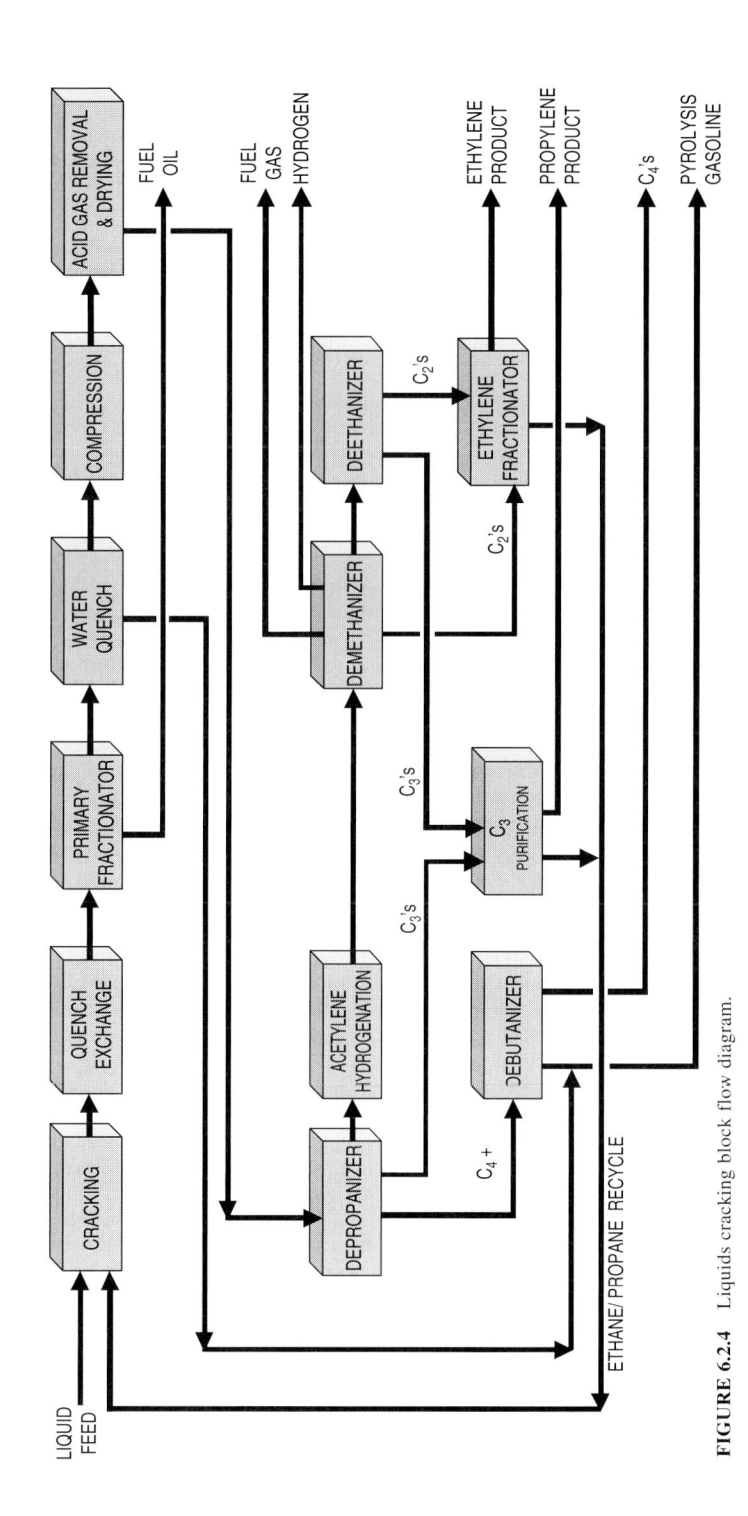

FIGURE 6.2.4 Liquids cracking block flow diagram.

6.37

preceding section. Certain furnace designs can combine both fresh feed and recycled feed cracking within a single furnace. Today's maximum-capacity liquids cracking furnaces can produce up to 175 KTA C_2H_4 in a single unit. Such designs employ a dual-cell zone cracking/zone decoking feature and can even be further subdivided to allow for four separate feedstocks to be cracked within a single furnace (*quadracracking*). Hence the number of furnaces installed in modern world-scale plants typically ranges from four to eight.

The gravity and boiling range of the liquid feedstock will dictate the steam-to-hydrocarbon ratio employed in the furnace (0.4 for LN, up to 1.0 for HCR and VGO). The complexity of the convection-section vaporization system similarly will be dictated by feedstock characteristics. Heavy feedstocks have a greater fouling tendency; hence operating temperatures (convection-radiant crossover temperature, radiant coil outlet temperature) are reduced correspondingly. Heavy-feedstock cracked effluent is also more susceptible to fouling; hence quench exchanger outlet design temperature is elevated, and the prequenched effluent usually is further cooled by direct quench with recycle quench oil. This minimizes transfer-line temperature stress and quench oil tower heat contact duty.

Radiant Coil Selection. To achieve high-selectivity, high-severity operation, the radiant coil residence time and coil outlet pressure are minimized. Typical radiant coil dimensions for a high-selectivity furnace are in the range of 40 to 55 mm ID/18 to 27 m length. Accordingly, modern world-scale furnaces may contain up to 250 radiant coil passes.

The venturi nozzle located at the inlet of each radiant coil ensures uniform distribution of feed to all radiant coils. This flow element creates critical flow, where flow is a function of the upstream pressure only, and is completely independent of the downstream pressure. Thus flow to each coil remains equal at all times, even if the pressure drop through the individual radiant coils varies due to uneven coking rates. The venturi provides uniform flow distribution as long as the ratio of absolute outlet pressure to absolute inlet pressure is 0.90 or less. Each radiant pass then feeds an individual SLE quench exchanger.

The internal diameter of the radiant coil is swaged (i.e., increased stagewise) approximately midway through the coil to enhance the yield/capacity and reduce the risk of spalled coke blocking the coil. An optimal residence-time coil gives the advantage of being able to crack liquid feedstocks at high conversions and high selectivity while still maintaining exceptionally long run lengths.

The radiant coils typically are manufactured from cast HK-40 and HP modified (Cr-Ni-Nb) alloys. Average coil life reported by S&W licensees exceeds 7 years. The end-of-run peak metal temperature (typically 1100–1150°C) or the inlet venturi pressure ratio will limit the furnace run length. Coil manufacturers and designers are continually developing improved metallurgies and geometric configurations to extend the operating envelope of the radiant coil.

Burner Design. Each radiant cell is 100 percent floor fired using state-of-the-art low-NO_x burners. The burners are designed to fire the plant residue fuel gas and/or natural gas import at 10 percent excess air. Each burner is equipped with an individual pilot burner. Draft required for combustion will be supplied by an induced draft (ID) fan located downstream of the convection-section breeching. Bridge-wall (radiant-to-convection transition) arch pressure is controlled by fan speed/damper operation to be slightly subatmospheric.

Quench Exchangers. The furnace effluent quench system generally consists of two stages. Primary quenching occurs in close-coupled, roof-mounted SLE exchangers, one per radiant pass. The SLE is a double-pipe linear exchanger that generates SHP steam [up to 130 bar(g)]. Coke laydown in the SLE is minimal, online decoking is very effective, and mechanical cleaning is essentially eliminated. For heavy liquids feed cracking, the cracked gas from the SLEs is further cooled in a quench fitting by direct injection of quench oil.

Boiler feed water (BFW) is supplied to the steam-generating quench exchangers from a high-pressure steam drum under thermosyphon. BFW is preheated in the upper zones of the convection section. Balance of convection heat recovery preheats/vaporizes feed and superheats both dilution steam and SHP steam. The steam drum is equipped with internal separators to produce high-quality steam (>99.9 percent vapor fraction), minimizing turbine and superheater fouling. Furnace export SHP steam is supplied to recovery-area compressor turbines.

The pyrolysis of hydrocarbons forms a large quantity of coke that deposits in the radiant coil, thus increasing the average metal temperature and pressure drop, particularly in the outlet tube. The furnaces must be decoked periodically using steam and air. This procedure oxidizes and spalls the coke from the coil and quench exchanger surfaces. The furnaces are designed to allow for the injection of the decoke effluent into the radiant firebox for incineration. Waste gases from various recovery section services are also incinerated in the firebox.

Quench Oil Tower System. Most S&W liquids crackers employ a high-temperature (220°C+) quench oil tower design (Fig. 6.2.5). This system recovers heat from the cracked gas at optimal levels by using circulating quench oil, pan oil, and quench water at successive temperature levels. Hot quench oil is pumped from the bottoms of the quench oil tower and used to supply heat to the dilution steam reboilers. A slipstream from the tower bottom to the heavy fuel oil stripper is fed to the quench fitting on the recycle furnace effluent. Viscosity of the quench oil is controlled by recycling the middle boiling fuel oil components, via the stripper overheads recycle, back into the quench oil. This recycle allows for the operation of the quench oil tower at a hotter bottoms temperature, thus maximizing use of the furnace effluent heat. The fuel oil product from the stripper bottoms is pumped through a cooler to tankage.

In the middle section of the quench oil tower, a pan oil loop removes heat at a somewhat lower temperature level; this heat is used in a number of services. A slipstream from the pan oil loop and a small drawoff from the rectifying section are stripped with dilution steam in the light fuel oil stripper to remove C_8 and lighter material. The stripper bottoms stream typically is mixed with the fuel oil product from the

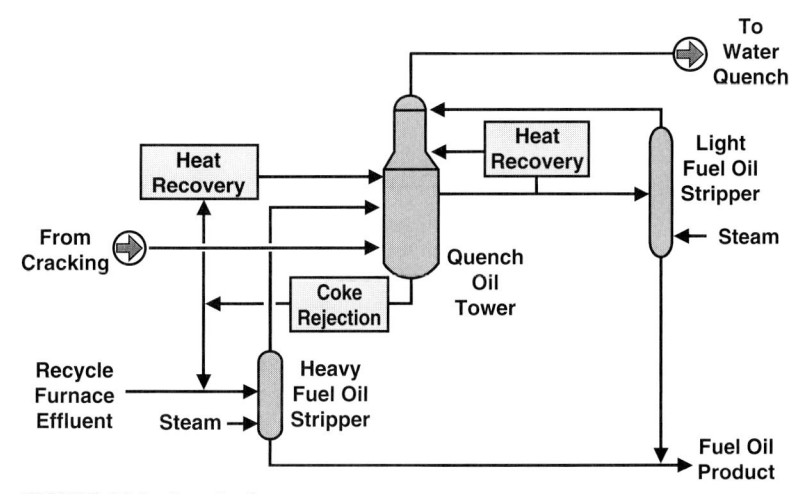

FIGURE 6.2.5 Quench oil system.

bottom of the heavy fuel oil stripper. The overhead of the stripper is recycled back to the quench oil tower.

The rectifying section of the quench oil tower is designed to ensure the split between fuel oil and gasoline fractions. Reflux is provided by hydrocarbon condensate from the oil-water separator located in the base of the quench water tower. The temperature of the overheads from the quench oil tower is controlled to avoid steam condensation in the upper section of the tower and to maximize heat recovery.

S&W proprietary ripple trays are the optimal quench oil tower internals by virtue of their high-capacity, high-heat-transfer, and nonfouling characteristics. They are also used in the fuel oil stripper to minimize the effects of fouling.

The overheads of the quench oil tower proceed to the quench water tower. A primary water loop recovers heat at a temperature level that can be used in a number of reboilers and heaters in the recovery section of the plant. Excess heat is removed by air and/or cooling water in trim coolers. A secondary water loop removes an additional amount of heat from the quench water against cooling water. The quench water tower overheads are then fed to the cracked gas compression system.

The quench water tower also uses ripple trays. The oil water separator in the bottom of the tower achieves the necessary gasoline water separation to prevent fouling in the downstream water circuits and dilution steam generation system. The hydrocarbon condensate is used as reflux to the quench oil tower. Net hydrocarbon condensate is fed to the distillate stripper.

Dilution Steam Generation. Feed water for the dilution steam system is withdrawn from the circulating quench water loop. Filters and a coalescer remove most of the entrained hydrocarbons and coke fines. The dilution steam feedwater is then heated against pan oil before entering the low-pressure water stripper. The volatile hydrocarbons in the feedwater to the quench water tower are stripped using dilution steam. The resulting bottoms stream is pumped and heated against pan oil and condensate and then fed to the dilution steam generator (DSG). To mitigate the effects of fouling, the DSG is designed as a flash drum without any tray internals. Quench oil and medium-pressure steam are used to raise dilution steam in the DSG reboilers. Blowdown from the bottom of the drum is cooled against cooling water prior to disposal. Dilution steam from the DSG is superheated against superheated MP steam.

Compression. The cracked gas compression system in most liquids crackers uses a five-stage compressor. The overhead from the quench water tower is fed to the first-stage compressor suction. The first-stage compressor discharge is cooled by cooling water and fed to the second-stage suction drum. This drum separates the liquid into a water stream and a hydrocarbon stream. The water stream is returned to first-stage suction drum and pumped to the quench water tower. The hydrocarbon stream is pumped directly to the distillate stripper. The vapor proceeds to the second stage of the cracked gas compressor.

The debutanizing distillate stripper processes the excess hydrocarbon condensate from the oil-water separator in the quench water tower bottoms section and the condensate from the first four stages of compression. The debutanized bottoms constitute a portion of the raw pyrolysis gasoline stream. The overhead vapor recycles back to the quench water tower. Pan oil is used to reboil the distillate stripper.

Successive compressor stages are cooled against cooling water; condensed liquid is returned to the preceding-stage suction drum. The cracked gas leaving the fourth-stage discharge drum is heated against quench water and is then fed to the bottom of a three-section caustic tower. In the bottom two sections, the gas is contacted countercurrently with caustic solution, the concentration of which decreases from top to bottom. The third (upper) section of the caustic tower is a cooled-water wash section to protect downstream equipment. Ripple trays are used throughout this tower to provide good gas-liquid contact while resisting fouling. A heater against quench water in the caustic tower feed superheats the cracked gas to avoid condensation of hydrocarbons and provide optimal reaction temperature. Makeup caustic from offsite is diluted with cooled, deaerated water before being injected into the tower. Boiler feed water is also added to the wash-water loop to account for vaporization within the tower.

The overhead from the caustic tower is cooled against propylene refrigeration. Condensed water from the downstream knockout drum is returned to the fourth-stage suction drum. Condensed hydrocarbon liquid is pumped, coalesced, and dried prior to feeding the high-pressure depropanizer. Vapor from the knockout drum is dried and fed to the depropanizer feed/overhead exchanger before entering the high-pressure depropanizer.

The cracked gas drier system in a world-scale plant typically consists of three vessels containing 3A molecular sieves, two in operation and one on regeneration. Start-of-run cycle time exceeds 48 hours. Each dehydrator consists of a main section and a guard section. A moisture analyzer is located between the primary and guard beds to monitor water breakthrough.

Depropanization. The S&W dual-pressure depropanizer system is designed to bypass C_4 and heavier components around the front-end hydrogenation and cold fractionation section of the plant. In this distributed distillation system, the high-pressure depropanizer makes a sharp separation between C_2 and lighter components and C_4 and heavier components. Reflux is adjusted to control the tower overhead butadiene content. Bottoms temperature is controlled below fouling level by splitting the C_3 components between the bottoms and overhead streams. The bottoms of the high-pressure depropanizer are cooled against cooling water and fed to the low-pressure depropanizer.

The fifth stage of the cracked gas compressor is a heat pump for the high-pressure depropanizer system. The overhead vapor is reheated in the feed/overhead exchanger and compressed by the fifth stage of the cracked gas compressor to a pressure consistent with the separation of methane and ethylene against $-101°C$ ethylene refrigeration in the demethanizer tower condenser. The compressor discharge flows through the C_2 hydrogenation reactors before being cooled against cooling water. The gas is dried in the secondary drier, partially condensed against demethanizer prefractionator bottoms and propylene refrigeration, and fed to the HP depropanizer reflux drum. Reflux is flashed to the tower; any excess liquid feeds the demethanizer prefractionator. Vapor from the reflux drum feeds the demethanizer chilling train. The low-pressure depropanizer splits C_3 material from the C_4+ components. The tower overhead stream is condensed against propylene refrigeration. The net overheads are pumped to the downstream C_3 processing system. The tower bottoms are fed to the debutanizer. Both depropanizers are reboiled against waste heat from the pan oil system.

Acetylene Hydrogenation. The front-end acetylene hydrogenation system is located in the discharge of the fifth stage of the cracked gas compressor. The discharge feeds the hydrogenation system via a reactor feed cooler/heater. This allows the inlet temperature to the system to be adjusted depending on the age and activity of the catalyst and the inlet CO level. Three catalyst beds are used. Cooling between beds and at the outlet of the third reactor is provided via cooling water. A spare reactor is not necessary because *in situ* regeneration is not used. Catalyst life of 5 to 7 years is typical. Excess hydrogen in the feed gas, combined with the high hydrogen partial pressure, requires lower operating temperatures and allows for higher space velocities than with a typical back-end hydrogenation system. Green oil formation is extremely low, is typically a low-molecular-weight material, and is returned to the high-pressure depropanizer via the reflux drum.

Demethanization. The cold fractionation area of the ethylene recovery unit is the most energy-intensive due to the required refrigerated fractionation necessary to separate polymer-grade ethylene from methane and ethane at very high recovery levels. The Advanced Recovery System (ARS) solves this efficiency challenge by using a combination of distributed distillation services and simultaneous chilling/prefractionation devices. Distributed distillation fractionates to produce only a single overhead or bottoms stream to tight specification, thereby reducing required reflux/reboil energy. Simultaneous chilling/prefractionation of demethanizer feed minimizes refrigeration duty, prestrips methane (thereby reducing demethanizer duty), and simultaneously prefractionates ethylene and ethane. The principle of this scheme is shown diagrammatically in Fig. 6.2.6 and is described below.

The vapor from the depropanizer reflux drum is chilled sequentially against propylene refrigeration and process refrigeration credit streams. The liquid from the final flash drum feeds the demethanizer prefractionator. The net liquid from the depropanizer reflux drum is fed directly to the demethanizer prefractionator.

The vapor from the final chilling train flash drum is chilled successively against ethylene refrigerant levels to knock out C_3+ material. The flashed liquid feeds the demethanizer prefractionator, and the flashed vapor feeds a dephlegmator or similar simultaneous heat- and mass-transfer device. The dephlegmator is a vertical-plate fin exchanger that provides adequate fractionation to separate C_2 and lighter components from the C_3-rich bottoms liquid. The overhead vapor from this dephlegmator device is further chilled and rectified in a second similar device operating at lower temperature. Overhead vapor from the second stage consists of hydrogen and methane only; bottoms liquid consists of C_2's with a minimum amount of methane.

The demethanizer prefractionator is reboiled using warm propylene refrigerant vapor. The prefractionator uses the overheads stream from the deethanizer as a condensing medium. The demethanizer prefractionator overheads consist of C_2 and lighter components; the prefractionator bottoms feed the deethanizer.

The vapor from the prefractionator and the liquid from the second-stage dephlegmator device are fed to the low-temperature demethanizer. These feeds are essentially C_2's and lighter material. Therefore, the bottoms from this column do not need deethanization and feed the C_2 splitter directly. The low-temperature demethanizer is reboiled with propylene

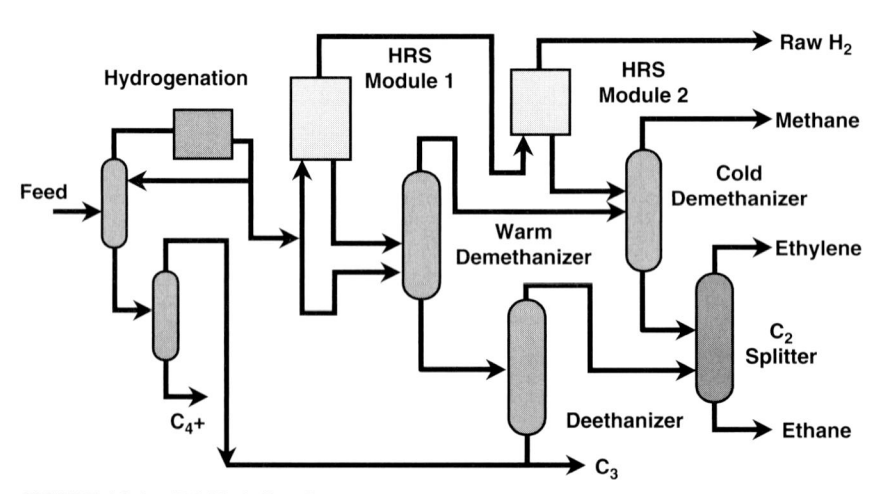

FIGURE 6.2.6 ARS block flow diagram.

vapor, and the reflux is condensed against $-101°C$ temperature ethylene refrigerant. The demethanizer bottoms product is subcooled against propylene refrigerant and proceeds directly to the C_2 splitter.

Residue gas vapor from the final dephlegmator device feeds the two-stage autorefrigerated hydrogen recovery system. Methane-rich liquid from the first hydrogen drum is flashed to a pressure consistent with the reactivation system pressure. Vapor from the second drum is hydrogen product that ultimately feeds the methanator or pressure swing adsorption (PSA) unit. The residue gas and hydrogen streams are reheated successively in the dephlegmator device and the refrigerant subcoolers. The hydrogen then passes to the methanator or PSA unit, where high-purity hydrogen is produced.

The demethanizer net overhead vapor is preheated against ethylene refrigeration, combined with the excess residue gas, and expanded in the first stage of the residue gas expander. The stream is reheated in the dephlegmator device before being expanded in the second stage. After the second stage of expansion, the stream is reheated, recompressed, mixed with the reactivation gas from the hydrogen-recovery system, and delivered to the fuel-gas system.

Deethanization. The bottoms from the demethanizer prefractionator are separated into an overhead C_2 stream and a bottoms C_3 stream in the nonfouling deethanizer. The tower overheads are condensed totally with propylene refrigerant, and the bottoms are reboiled against quench water. The bottom product is cooled before feeding the optional C_3 hydrogenation system. The total liquid overhead product is subcooled against propylene refrigeration. A portion of this stream is used as refrigerant in the upstream demethanizer train before entering the C_2 splitter.

Ethylene Fractionation. The C_2 splitter is a low-pressure tower integrated with the C_2 refrigerant/heat pump compressor and works as an open-loop heat pump system. The C_2 splitter fractionates the feeds from the deethanizer reflux drum and from the demethanizer bottoms. The fractionation in the C_2 splitter separates the ethylene/ethane feed into a high-purity ethylene overhead vapor and an ethane bottoms stream for recycle cracking. The purer top feed sets the reflux requirements, thereby reducing energy consumption.

The overhead of the C_2 splitter is reheated and fed to the fourth stage of the C_2 refrigerant/heat-pump compressor. The compressor fourth-stage discharge vapor is desuperheated against cooling water and split into three streams. One stream is removed from the system as ethylene vapor product. The other streams are both condensed, one in the C_2 splitter reboiler and the other in the C_2 heat pump trim condenser. The condensed ethylene flows to the C_2 splitter reflux drum. Part of the liquid from the C_2 splitter reflux drum is subcooled in the C_2/C_3 refrigerant subcooler and the ethylene product chiller and then is routed to the ethylene liquid product atmospheric storage tanks as the final ethylene liquid product. The remaining liquid is subcooled and returned to the top tray of the C_2 splitter as reflux. The tower is reboiled by compressed ethylene vapor in the C_2 splitter reboiler. Ethane is withdrawn from tower bottoms and is recycled via the demethanizer refrigeration recovery system before being routed to the recycle furnace for cracking.

Propylene Processing. The C_3 hydrogenation system is a liquid-phase single-stage reactor system. The feed to the system is the low-pressure depropanizer overhead liquid and the deethanizer bottoms. The system is designed to selectively hydrogenate the methyl acetylene (MA) and propadiene (PD) in the feed to the propylene recovery section. Hydrogen from the methanator or PSA unit feeds the reactor on flow control. The effluent from the reactor is cooled against cooling water and fed to the C_3 splitter. A spare reactor is provided. The standby reactor is put in service during catalyst reactivation of the primary reactor. Depending on the specification of the propylene product and the recycle propane, the C_3 hydrogenation system may be omitted.

The propylene recovery system typically consists of two towers in series, a stripper and a rectifier. C_3 feed is introduced to the stripper, the overheads of which feed the rectifier. The rectifier overheads are condensed against cooling water. The light ends introduced with hydrogen in the hydrogenation system are stripped out in the upper pasteurization section of the tower to make the required propylene specification. A vent condenser minimizes the recycle to the cracked gas system. The polymer-grade propylene product is taken off as a side draw from the rectifier column. Both towers are reboiled with quench water. Reflux is delivered to the rectifier pasteurization section and subsequently transferred from rectifier bottoms to the stripper top tray.

C_4+ Processing. The low-pressure depropanizer bottoms stream is fed to the debutanizer. The debutanizer separates C_4's from higher-boiling material. The C_4 components exit overhead as a raw C_4 stream. C_4's product may be fractionated/extracted to provide a range of C_4 products (butadiene, isobutylene, butene-1) or partially hydrogenated to provide a mixed-butane stream for catalytic or thermal recycle cracking. The debutanizer C_5+ bottoms product is combined with gasoline from the distillate stripper bottoms, cooled against cooling water, and delivered to the offsite hydrogenation unit. The debutanizer is reboiled with low-pressure steam and condensed against cooling water.

Refrigeration Systems. The refrigeration demands of the process are supplied by various levels of cascaded propylene and ethylene refrigerant. These are further supplemented by refrigeration from matched internal process streams. Lowest-level refrigerant is supplied from the turboexpander. Propylene is condensed against cooling water. Refrigerant is supplied to the demethanizer chilling train by a three-stage ethylene machine. The ethylene is condensed using propylene refrigerant. Process-process heat integration is restricted to streams within a given system, thereby avoiding potential operating instability within the plant.

MEGAPLANT DESIGN ISSUES

The capacity of the next generation of ethylene plants will be in the range of 1 to 1.8 MMTA. The primary incentive to construct such large units is investment-scale economies. Such plants will still be single-train units but will be designed with different temperature/pressure profiles in order to minimize both equipment sizes and piping account cost. Key areas of design philosophy are outlined below.

The optimal furnace count in today's ethylene units is in the range of four to eight. Twin radiant cell designs offer the scale economies of large units with the optional ability to operate as 4 × 25 percent zones and decoke as 2 × 50 percent zones. Latest design intensification schemes (e.g., the patented symmetric USC-SU design), coupled with proven large-scale furnace designs, translate into individual furnace capacities of up to 300 KTA ethylene for gas crackers and 230 KTA ethylene for liquids crackers. Figure 6.2.7 shows a recently constructed furnace employing the preceding features. Hence a 1.5-MMTA cracker might consist of no more than five dual-cell/zone-operation furnaces with an optional recycle cracking furnace to process light recycles. The mechanical and operating challenges of such units already have been addressed successfully.

The quench tower area of the plant involves some of the largest equipment. Inevitably, the quench towers will be site fabricated unless exceptional project conditions so permit. It will be important to minimize the cost of heat removal systems by judicious selection of exchanger type, cleaning arrangements, and compact piping circuits. Plate and frame exchangers and air coolers will be considered. Parallel dilution steam generators may be

FIGURE 6.2.7 Two large-capacity twin-cell quadracracking USC-SU furnaces recently installed at the Formosa Plastics, Taiwan, facility as part of a major expansion revamp. Feedstock combinations include naphtha, LPG, and recycled ethane/propane.

required for liquid crackers because the number of exchangers servicing one dilution steam generator unit may be limited by hydraulic considerations. The dimensions and weights of the quench oil and quench water towers for a 1.8-MMTA cracker will be less than those of many site-fabricated towers in today's refineries and petrochemical units.

A key limit in determining plant size is the availability of proven rotating equipment. The preference is to accommodate the high ethylene capacity rates in single-train available size casings without having to resort to parallel compressor lines or increased suction pressure. Discussion with vendors has indicated that rates can be accommodated in a single-compressor train without having to increase the compressor suction pressure. (Increased cracked gas compressor suction pressure adversely affects furnace yields and requires more fresh feed.)

The S&W ethylene process scheme employs refrigeration and heat-pump compressors with casing sizes that do not exceed those in proven service, even up to 1.8 MMTA ethylene capacity. The critical service is thus the cracked gas compressor/driver. Key design considerations include

- Aerodynamic capacity of the first-stage inlet
- Ability to incorporate the necessary number of impellers per stage in the given rotor shaft span with large-diameter nozzle spacing allocation
- Optimal stage revolutions per minute (RPM)
- Ability to provide proven variable-speed driver of adequate power
- Compressor and turbine mechanical design issues (e.g., tip speeds, bearing types, impeller dimensions, etc.)

Based on the preceding considerations, the cracked gas compressor design for a representative liquids cracking unit probably would employ a single-train, dual-driver arrangement. Total shaft power would be approximately 85 MW. The critical first-stage design would employ a dual-stage casing to accommodate the high-suction flow. Such an arrangement solves the challenges of high inlet flow at adequately low suction pressure while allowing driver flexibility and optimal revolutions per minute. The first-stage suction flow for a 1.5-MMTA naphtha cracker with a 4-psig [0.275-bar(g)] suction pressure would be about 300,000 ACFM (140 m^3/s), and the total train power requirement would be about 115,000 BHP (85 MW). The required power would be split between two turbines. From a steam balance perspective, the use of multiple drivers, as implemented in several previous designs, allows increased flexibility.

In conjunction with compressor/turbine manufacturers, S&W has confirmed that a combination of available compressor casing sizes and variable-speed steam turbines can be assembled to achieve the necessary flow/head requirements. The arrangement would power stage 1 (dual stage, single casing) plus stages 2 and 3 (single casing) with a single turbine. Stages 4 and 5 (single casing) would be powered by a separate, higher-RPM driver.

A related consideration for megaplant compression equipment is the size and routing of interstage piping. S&W has optimized equipment designs, piping pressure profiles, and layout arrangements to minimize equipment piece count and associated piping system cost.

Depending on the specific location, site access, and availability of local fabrication facilities and heavy-lift equipment, the choice of shop- versus site-built vessels will play an important role and hence will be identified very early on in any such project. If the site is close to a port or a canal where large process equipment vessels can be brought into the plant via barges, then shop fabrication of large vessels becomes feasible and cost-effective. In certain cases, process design conditions may be adjusted to allow shop rather than site fabrication. For example, purities of internal recycles may be relaxed to minimize sizes of critical superfractionators, vane separators could be installed in drums, and high-capacity trays can be used in towers to minimize vessel dimensions.

The recovery area of the plant also presents new challenges. Chilling train plate-fin exchangers will require multiple shells or cores with inherent higher manifolding costs and layout problems. Core-in-kettle or core-in-tower units could be maximized. Driers or reactors may have to be operated in parallel because of regeneration gas limitations. Isolation valves may be required to minimize downtime to facilitate major equipment servicing requirements. The enormous size of relieving volumes with multiple releases could require staggered relief valve setpoints to minimize instantaneous flare loads.

Generally, the accepted 0.65 scale exponent applies for the sizing of individual equipment. However, for piping systems, the cost of fittings and valves becomes disproportionately high, availability of vendors decreases, and layout and control issues become critical. Pressure drops through the piping and associated equipment will be increased to minimize overall cost impact. Equipment location and orientation will have to be reevaluated. Several novel concepts have been employed to minimize piping complications, e.g., integrated intercooler separators, core-in-kettle exchangers, and integrated towers/condensers/reboilers.

As plants become larger, the importance of associated project factors also increases. Among such issues are the logistics of the coproduct disposition from the liquid crackers, the size of supporting offsite systems, the safe containment of large hydrocarbon inventories, and the special commissioning and maintenance requirements for large containment systems.

Many of the concepts just described have been demonstrated in the world's largest single-train plant, the Nova/Dow E-3 unit, in Alberta, Canada, to which previous references have also been made. This S&W design has been in service since 2000; original design capacity was 1.275 MMTA based on ethane feed; current operating rates approach 1.5 MMTA (Fig. 6.2.8).

FIGURE 6.2.8 Nova/Dow E-3 ethane cracker, Alberta, Canada.

PROJECT EXECUTION ASPECTS

The preceding sections described the key areas of S&W ethylene technology and the related economics. Ethylene units are probably the largest single petrochemical units constructed today; the various stages in implementation of such projects require extensive planning, organization, and control. Key issues involved are summarized below. Projects of this type usually are awarded on either a lump-sum turnkey (LSTK) basis or proceed under the client's closer supervision following successive development phases (e.g., approve/select/define/execute) with corresponding approval milestones. In all cases, the final plant performance and project cost depend primarily on the decisions taken during the initial phase of the project. Accordingly, the quality of the technology, its supporting data base, and design team expertise are vital to overall project success.

The project schedule involves the following key phases that proceed sequentially with some parallel activities:

1. *Process design.* This is the initial phase, which sets the basic definitions of plant configuration, layout, equipment sizing, operational performance, and key mechanical design conditions. Much of this information is developed using "intelligent" software systems that permit further development and definition in subsequent design phases. The process flow diagram (PFD) and associated heat and material balances are essentially fixed during this phase.

2. *Basic engineering design.* The process design is then expanded in further detail using linked design software. Specific codes, standards, regulations, etc., are embodied in the design. The designer usually assists the client at this stage to obtain the necessary environmental permits. A major portion of the basic design effort is completion of the piping and instrumentation diagram (P&ID), which defines equipment details, piping configurations/specifications/dimensions, and instrumentation and safety systems. Mechanical design details of all equipment items are finalized The plot plan with associated piping layout arrangement is essentially fixed. All the preceding activities are linked via hazard, reliability, and operating (HAZROP) studies that focus on the key issues of safety, constructability, operations, maintenance, accessibility, etc. Such studies involve a range of relevant design and operating/maintenance/construction staff. To achieve a minimum project schedule during this phase, long-lead equipment items may be specified and placed on order. At the conclusion of the basic engineering phase, it should be possible to define plant cost to ±25 percent. Assuming that this cost estimate meets target objectives, the project will receive full approval to proceed.

3. *Detailed engineering design.* This is the most extensive design phase, where all mechanical design definitions are finalized. A major portion of this effort is associated with the design, layout, and stress analysis of all piping systems. Precise sizing and orientation of all equipment, nozzles, manways, valving, instrumentation, etc., are defined. Key to much of this activity is the use of three-dimensional (3D) computer-aided design (CAD) and modeling, which is in turn linked to the intelligent design data base developed in the preceding process and basic engineering phases. Again, during this project phase, further HAZROP studies take place prior to releasing design documentation for fabrication/construction. In today's competitive environment, it is quite common for much of this detailed engineering effort to be allocated to lower-cost satellite design offices. The linked design data base system including the 3D layout model makes such work scope allocation both effective and efficient and allows the headquarters design office to maintain design oversight. During the final stage of detailed engineering design, a further cost estimate (target ±10 percent) frequently is developed for final confirmation/approval.

4. *Procurement.* Having established the process and mechanical design documentation, this is then combined with the associated equipment specifications developed by S&W over many projects, together with any client-related instructions. This design statement forms the basis for the purchase inquiry issued to prequalified vendors. The selected supplier then proceeds with the fabrication of that equipment under routine inspection by S&W and client. As appropriate, various levels of oversight, expediting, and testing are also applied. Depending on the type and location of the project, much of the equipment and material may be procured locally, often with controlling input from the client. In such cases, S&W will provide varying levels of procurement assistance.

5. *Project controls.* Throughout the engineering and procurement phases, it is essential that close monitoring of costs, schedule, man-hours, document records, etc., is maintained. A related area of oversight is quality assessment and control (QA/QC), which is conducted by an independent project team member group. Key to effective project control is the concept of work breakdown structure. This allows each activity to be precisely defined, assigned, monitored, and costed.

6. *Construction.* Although the construction phase follows the preceding engineering/procurement phases in terms of overall schedule, successful projects involve construction management personnel from the outset. In this manner, various construction-related issues can be incorporated without the need to recycle engineering activities. The field effort usually represents the longest phase in the overall project schedule (±20 months), the highest man-hour-related costs (up to 25 percent of total cost), and the area where all upstream engineering and downstream operations requirements interact. Site preparation usually begins at the outset of the project. Piling and underground piping typically are completed within 12 months, and then the major erection effort starts. Erection sequence generally is major towers, furnaces, piperack structures, piping installation, rotating equipment, other equipment, and instrumentation. Key to successful completion is the planned testing, handover, and commissioning of integral plant systems in order that a logical commissioning sequence can be carried out.

7. *Commissioning/startup.* Following the phased mechanical completion of the ethylene unit and its supporting utilities systems, a carefully planned and executed commissioning phase is the key to final project success. Various precommissioning activities take place in advance of hydrocarbon introduction. An initial activity involving both the oper-

TABLE 6.2.2 Typical Project Statistics for World-Scale Unit (1000-KTA Ethylene plus Propylene)

Engineering man-hours	520,000
Construction man-hours	7,000,000
Piping installation length	173 km
Concrete volume poured	15,000 m^3
Steel weight installed	6450 tons
Construction taskforce	3,000 persons
Equipment installed	383 items
Manufacturers involved	125 companies

ations and construction staff is the "punching" of the entire unit to verify complete and correct construction per design documents. The offsite steam supply should be available as early as possible. This allows pressure testing, system flushing, and other precommissioning activities. The steam system within the furnace area can then be cleaned chemically. The plant is dried by air circulation from the cracked gas compressor (seals removed). A vital precommissioning activity is the cryogenic system drying. Catalysts and desiccants are loaded carefully and, if appropriate, tested for uniform pressure drop. All rotating equipment is run-in. Instrumentation/control systems are loop-tested. Furnace refractory is dried under controlled firing conditions; furnace piping expansion tolerances are checked. Cooling water circuits are checked for debris; chemical treatment is applied and monitored frequently during startup.

Staged precommissioning and commissioning will start at least 6 months before actual startup. Prior to startup, the unit is flushed with inert gas, pressure-tested, and finally inventoried with hydrocarbons. All main-train machinery is tested under recycle or vacuum conditions.

The precise startup sequence is developed during the engineering phase, coordinated with the client's operating staff, and implemented under S&W supervision. Operations and maintenance staff are trained extensively in advance (schoolroom, simulators, host plant training).

A correctly designed, built, and commissioned unit will start up smoothly. The shortest time for a S&W grassroots unit initial startup is only 15 h. Within a few days, the unit achieved full design capacity.

8. *Project statistics.* An ethylene unit probably represents one of the most complex plant projects in today's petrochemical industry. An indication of the extent of the engineering, procurement and construction scope is shown in Table 6.2.2.

REFERENCES

1. Schutt and Zdonik, *Oil and Gas Journal*, February 13, April 2, May 14, June 25, July 30, September 10, 1956.
2. Zdonik, Green, and Hallee, *Manufacturing Ethylene* (articles reprinted from *Oil and Gas Journal*, 1966–1970). Tulsa, Okla.: Petroleum Publishing Co., 1970.

CHAPTER 6.3
KBR SCORE™ ETHYLENE TECHNOLOGY

Steven Borsos and Stephen Ronczy
Kellogg Brown & Root, Inc. (KBR)
Houston, Texas

DEVELOPMENT AND HISTORY

KBR was formed by the merger of the engineering and construction firms M. W. Kellogg Company (Kellogg) and Brown & Root in 1998. In the same year, KBR signed a license agreement with ExxonMobil Chemical Company that resulted in the combining of experiences in technology, design, and operations in steam cracking plants. The combination of and improvements to these three formerly separate technologies are the basis of KBR's SCORE™ (Selective Cracking, Optimum REcovery) technology.

KBR's olefins experience dates back to the early days of the olefins industry. Kellogg built the world's first naphtha cracker in 1951 for ICI in Wilton, England, and C. F. Braun (Braun), which was later incorporated into Brown & Root, built the world's first gas oil cracker in 1952 for Esso in Baton Rouge, Louisiana.

Development of the respective technologies continued into the 1960s as Kellogg became the first designer to incorporate dephlegmators and plate-fin heat exchangers into the ethylene plant flowsheet in 1965. In 1967, Braun pioneered the development of the front-end acetylene converter and introduced the front-end deethanizer flowsheet for ethane cracking.

Kellogg continued flowsheet development with introduction of the front-end demethanizer flowsheet. Further refinements during the 1970s improved operational reliability, reduced plant costs, improved energy efficiency, and increased plant size. These developments culminated in construction of the world's two largest single-train ethylene plants at the time, one at Shell's facility in Deer Park, Texas, and the other at Exxon's complex in Baytown, Texas, both of which are still in operation today.

In the 1980s, Braun introduced the Low Cost Ethylene Technology (LCET) design, which made extensive use of low-pressure-unit operations and process integration such as heat-pumped distillation. Further refinements of Braun's LCET design and of Kellogg's front-end demethanizer technology lead to major improvements in recovery-section energy efficiency through the 1970s and 1980s.

As the energy price shocks subsided in the mid-1980s, focus again returned to reducing capital investment and overall project costs. Kellogg used its project-execution capabilities to trim schedules. Green-field ethylene projects were completed in as little as 29 months, from contract signing to ethylene product in the pipeline. Kellogg was a pioneer in short-contact-time pyrolysis research in its research and development (R&D) facilities, leading to construction of the first Millisecond test furnace in Tokuyama, Japan, in 1973. The first commercial-scale Millisecond furnaces were built in Channelview, Texas, in 1982, and in 1985, the first all-Millisecond ethylene plant was commissioned in Chiba, Japan. Ultimately, Kellogg designed and built over 100 Millisecond cracking furnaces worldwide.

In the 1970s Exxon Chemical Company began developing short-contact-time cracking furnaces. Exxon used its in-house development, as well as its extensive commercial operating experience, to refine its furnace design. This culminated in Exxon's commitment in 1987 to its short-contact-time LRT-II furnace design. Finally, in 1998, the technology from these three industry pioneers was combined in KBR's SCORE ethylene technology.

SELECTIVE CRACKING FURNACE TECHNOLOGY

KBR's SCORE selective cracking (SC) pyrolysis technology offers the olefins industry a broad selection of proven furnace coil designs. These designs range from serpentine coil designs (SC-4) to two-pass coils (SC-2) and short-residence-time, high-selectivity straight-tube configurations (SC-1). The SC-1 furnace design represents an advance over prior technologies by providing the highest selectivity to ethylene available to the industry from the shortest-contact-time cracking furnaces. The SC-1 design combines a process and mechanical configuration that allow the cracking reactions to take place in approximately 0.1 s. The mechanical configuration is simple, making the furnace very robust, reliable, and easy to operate, and control features provide enhanced operating flexibility.

Impact of Short-Contact-Time Cracking

The primary benefit of KBR's SC-1 short-residence-time pyrolysis furnace is its superior selectivity, or yield, to valuable light olefins. Decreasing residence increases ethylene yields across the entire range of feeds—from ethane to vacuum-gas oils.

Short contact time favors the primary reactions that produce high-value light olefins, which produce more ethylene per unit of feedstock. Short-contact-time cracking restricts the secondary cracking and polymerization reactions that destroy light olefins and form fuel components. The result is a higher-value product slate. Because fewer low-value by-products are processed in the recovery section, the plant requires less capital investment and less energy input per unit of ethylene produced.

The relative impact of pyrolysis contact time on ethylene yield is shown in Fig. 6.3.1 for a variety of feeds. For ethane, short-contact-time cracking will produce up to 4 percent more ethylene per unit of feed than will the typical four-pass serpentine coil often used in this service. When cracking naphtha and gas-oil-range liquid feeds, this advantage increases to up to 12 percent more ethylene relative to typical designs used.

Table 6.3.1 shows typical single-pass SC-1 yields for various feeds.

Economics of Short-Contact-Time Cracking

Benefit of Short-Residence-Time Cracking: Ethane Feed. The change in ethylene yield as residence time changes for ethane feed is shown in Fig. 6.3.2. In this plot,

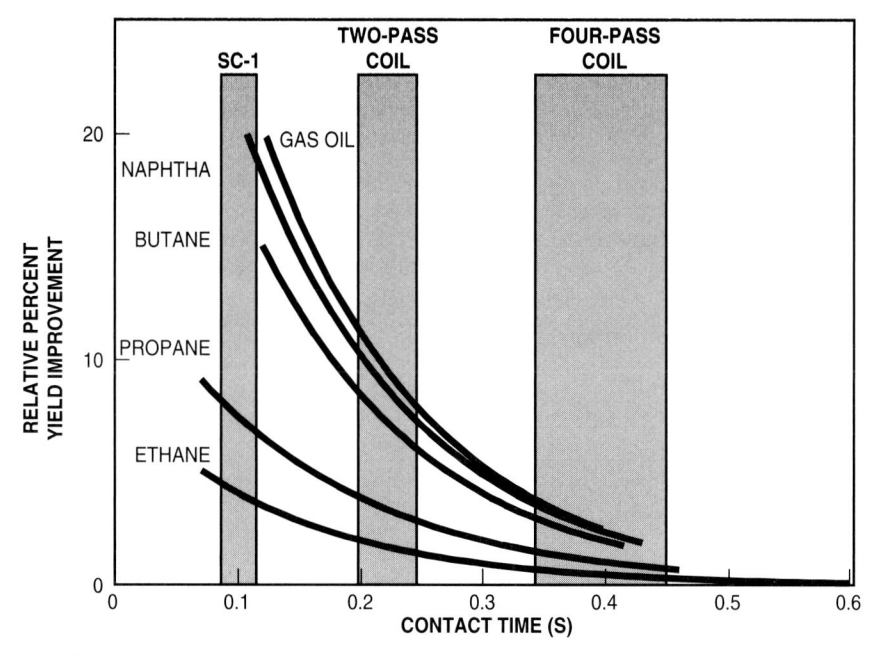

FIGURE 6.3.1 Impact of contact time on ethylene yield.

TABLE 6.3.1 Typical SC-1 Yield Slates (High Severity; wt %)

	Ethane	Propane	Light naphtha	Atmospheric gas oil
H_2	4.7	1.9	1.2	0.7
CH_4	5.3	22.6	17.1	10.2
C_2H_2	0.9	0.9	1.0	0.6
C_2H_4	57.3	38.9	33.6	27.2
C_2H_6	25.0	2.6	3.3	2.6
C_3H_4	0.1	1.3	1.2	0.8
C_3H_6	1.1	13.2	15.1	13.9
C_3H_8	0.2	8.0	0.3	0.5
C_4H_6	2.5	2.8	5.4	5.8
C_4H_8	0.2	1.1	3.9	4.1
C_4H_{10}	0.1	-0-	0.3	0.1
RPG	2.44	6.34	15.4	16.6
FO	0.1	0.3	2.1	16.8
$CO + CO_2 + H_2S$	0.06	0.06	0.1	0.1
Total	100.0	100.0	100.0	100.0

ultimate ethylene yield (the ultimate production of ethylene from ethane feed, including recycle to extinction of the uncracked ethane in the furnace effluent) is plotted at three different contact times across a broad range of single-pass ethane conversions. The ultimate yield of ethylene increases as the residence time in the furnace coil decreases.

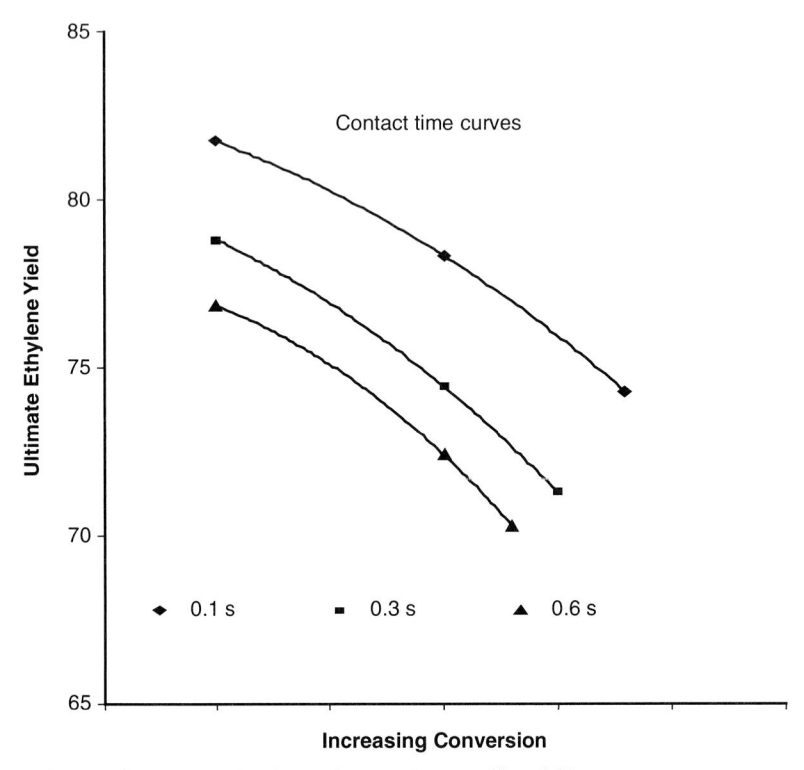

FIGURE 6.3.2 Impact of residence time on ethane cracking yield.

Translating this difference in selectivity into a comparison of overall plant material balances based on a constant ethane feed rate at 75 percent ethane conversion gives the results shown in Table 6.3.2. The gross margin (value of products minus feeds) shown in Table 6.3.2 is derived from historic values for ethane feed, ethylene product, and fuel. For a plant with an ethylene capacity of 1 million t/yr based on ethane cracking, the increase in gross margin from the plant for an SC-1 coil relative to a four-pass coil is approximately $15 million to $20 million per year.

Benefit of Short-Residence-Time Cracking: Naphtha Feed. As shown in Fig. 6.3.1, the yield advantage for short contact time increases as feeds become progressively heavier. A comparison of ethylene and propylene yields from an SC-1 and a two-pass coil furnace for typical full range of naphthas is given in Fig. 6.3.3 as a function of severity (methane yield). Cracking residence time has little impact on propylene yield, and all furnace designs are able to produce similar amounts of propylene from a unit of feed. However, the short-contact-time coil is more selective to ethylene, and the total amount of valuable light olefins (ethylene plus propylene) is higher for the SC-1 furnace. Since the ethylene yield is higher, the propylene-to-ethylene ratio (P:E ratio) tends to be lower for the SC-1 coil. Some incorrectly interpret this as an inability of short-contact-time furnaces to produce propylene when, in fact, these furnaces actually produce more ethylene at the same propylene yield, resulting in a lower P:E ratio.

A comparison of the overall material balances for short-contact-time and two-pass coil furnace plants, both low-severity (P:E = 0.6) and high-severity (P:E = 0.45) cracking of

TABLE 6.3.2 Overall Material Balance: Ethane Cracking at 75% Conversion

	SC-1	Four-pass coil
Feed (KTA)	1355	1355
Products (KTA)		
Ethylene	1050	1000
C_3 and heavier	127	130
Fuel gas/fuel oil	178	225
Gross margin, $MM/yr	284	268
Difference, $MM/yr	+16	

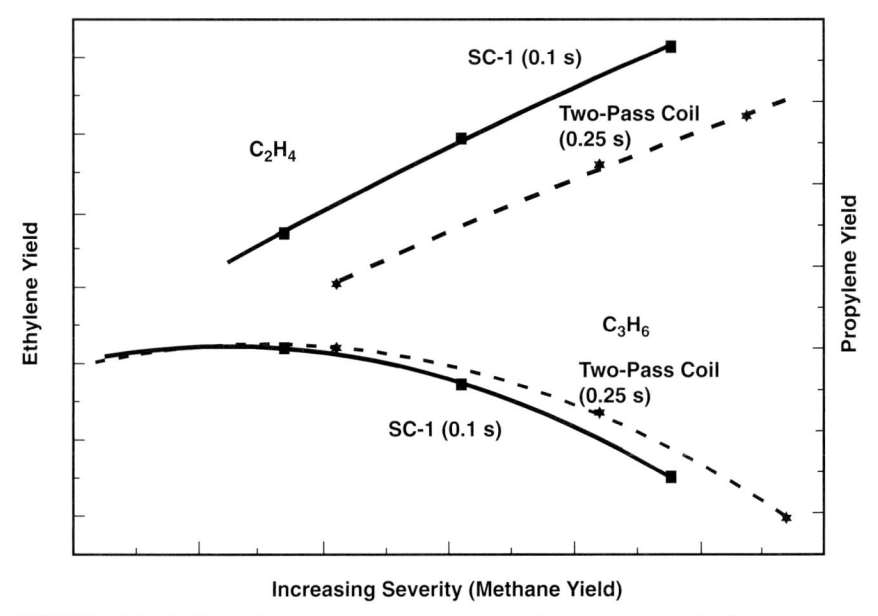

FIGURE 6.3.3 Residence time impact on ethylene and propylene yield (naphtha feed).

light naphtha, is given in Table 6.3.3. The SC-1 coil is more selective to olefins than the two-pass coil across this entire range of operating severity. Using historic values for the naphtha feed and the various by-products, the gross margin for the plants can be estimated. As shown, a 1-million-t/yr plant with short-contact-time furnaces operates with a gross margin of approximately $15 million to $25 million per year more than a plant with two-pass coils.

Features of the SC-1 Furnace

A simplified elevation drawing of an SC-1 furnace is provided in Figure 6.3.4. A typical layout of the SC-1 furnace includes two parallel rows of radiant tubes contained within the firebox. This design is more compact and uses less plot space than a twin-cell furnace with the same capacity. In fact, it is possible to place seven SC-1 furnaces into the same plot space as six twin-cell furnaces of the same per-furnace capacity.

TABLE 6.3.3 Overall Material Balances, Naphtha Cracking

	Low severity (P:E = 0.60)		High severity (P:E = 0.45)	
	SC-1	Two-pass coil	SC-1	Two-pass coil
Naphtha (KTA)	3012	3012	2806	2806
Products (KTA)				
Ethylene	1035	1000	1045	1000
Propylene	575	549	453	426
Mixed C_4's	410	382	277	264
(Butadiene)	(180)	(158)	(158)	(134)
Gasoline	503	569	431	492
(BTX)	(211)	(234)	(243)	(294)
Fuel gas/oil	580	603	600	624
Gross margin, $MM/yr	499	480	455	435
Difference, $MM/yr	+19		+20	

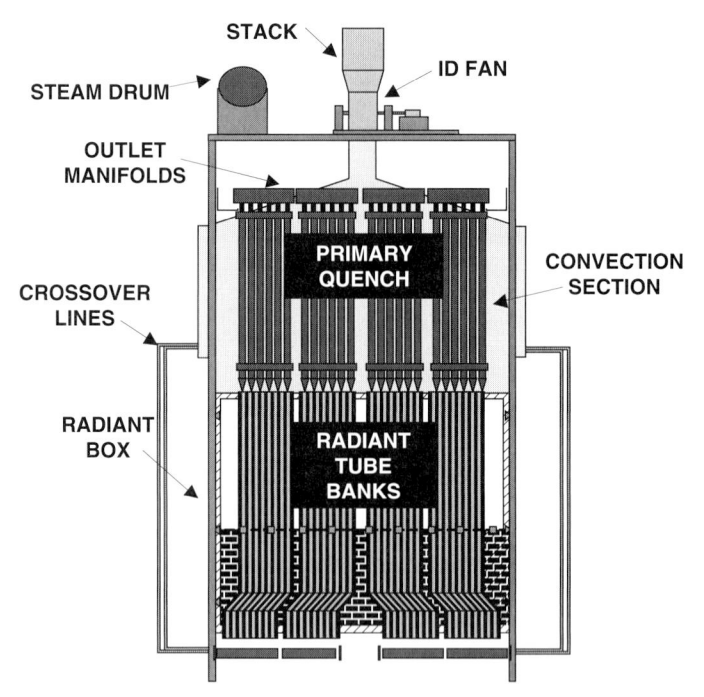

FIGURE 6.3.4 SC-1 furnace elevation.

The radiant tubes make one upflow pass through the radiant box. The radiant tubes have no return bends, eliminating a major point of mechanical stress. This extremely simple and robust mechanical design allows each tube to operate at the same temperature yet flex independently. The single-pass, upflow configuration allows easy access to the inlet manifolds for maintenance. With this design, no welds are located within the fired zone of the furnace. The primary quench exchangers are close-coupled to the radiant tube outlets, resulting in a short unfired, adiabatic residence time. This contributes to the very high selectivity to light olefins.

The burners are 100 percent floor fired, resulting in a simple fuel piping arrangement and easy access for burner maintenance. Additionally, a proprietary low-NO_x burner design can be used when NO_x emissions from the furnaces must be controlled.

SC-1 furnace technology uses a proprietary double-pipe quench system that generates high-pressure steam without fouling when cracking feedstocks in the light naphtha, LPG, and ethane range. This eliminates the maintenance associated with the hydrojetting of shell and tube TLXs and eliminates the need to dispose of wastewater that contains coke fines. With heavier feeds such as gas oils, coke will deposit in the quench exchanger, but it is removed along with coke from the radiant coil during the decoking cycle. Therefore, even with heavy gas oil feedstocks, heat recovery for generation of high-pressure steam is possible.

The SC-1 furnace uses a single-pass, small-diameter tube to obtain a reaction time in the range of 0.1 s. Because of the small tube diameter, SC-1 furnaces generally are designed to operate with run lengths in the 30- to 35-day range depending on feedstock and the severity of cracking. However, the decoking operation is proportionately short also, in the 18- to 24-h range. This provides a high on-stream efficiency of 96 to 98 percent, which is similar to furnaces with larger-diameter radiant tubes. The primary quench exchangers are also cleaned when the radiant tubes are decoked owing to system geometry, and they continue to generate high-pressure steam during the decoking operation.

By the use of proprietary control technology, SCORE furnaces have the ability to simultaneously crack different feeds in different areas of the same furnace, each at its optimal severity. This allows a very large capacity furnace (with the economic advantage of economy of scale) to also have exceptional flexibility. For example, a single SCORE furnace can crack an ethane/propane recycle stream at high conversions (to limit recycle flows), whereas the balance of the furnace may be cracking naphtha at low severity to maximize propylene yield. Such flexibility, which has been demonstrated routinely in operating plants, is the result of both mechanical features and the unique control system that is used.

The SCORE control system also can be used for a special online decoking technique that reduces furnace offline time typically associated with conventional steam/air decoking. This technique allows one or more flow passes of a furnace to be decoked, whereas the remaining flow passes remain in production cracking hydrocarbon feed. This reliable technique, which has been demonstrated in many plants, significantly extends furnace run cycles between offline steam-air decoking operations. Therefore, a full spare furnace is not required to maintain full plant production during online decoking.

The combined knowledge of KBR and ExxonMobil brought together as SCORE technology creates the ability to build very large capacity furnaces in the range of 200,000 to 250,000 t/yr of ethylene in a single firebox. This is accomplished by combining commercially proven features of both technologies. The single-pass radiant coil and primary quench system is modular in concept, and it is relatively straightforward process to add more capacity. The largest short-contact-time furnace currently in operation is at ExxonMobil's BOP-X Baytown, Texas, plant and has a capacity of over 200,000 t/yr of ethylene in a single firebox. A number of furnaces with individual capacities of over 200,000 t/yr of ethylene are in various stages of design and construction.

OPTIMUM RECOVERY-SECTION DESIGN

The use of high-selectivity furnaces inherently reduces capital cost and energy consumption owing to reduced plant traffic in the plant recovery section. SCORE technology also includes *o*ptimum *re*covery-section (ORE) technology for the specified plant feed slate. KBR offers three recovery-section flowsheets, all of which are proven in commercial operation. The front-end deethanizer flowsheet is KBR's primary offering for ethane and ethane-propane (E/P) feeds. The front-end depropanizer is KBR's primary offering for propane and heavier feeds and is also flexible to ethane feeds. The front-end demethanizer design can be used for propane, naphtha, and gas oil feeds and for ethane feeds where flexibility to heavier feeds is a design requirement.

Each of these recovery-section designs has been optimized through the use of PINCH and lost-work analysis and the implementation of preseparation and distributed distillation concepts to reduce energy consumption and capital costs. Throughout the optimization process, KBR has paid special attention to maintaining flexibility and operability of the recovery system.

The front-end deethanizer and front-end depropanizer flowsheets each provide excellent economies of scale and low cost for large-capacity projects, and both designs have high efficiency and feedstock flexibility. The low-pressure tower designs in these flow schemes provide significantly lower tower weights and do not impose the fabrication, delivery, and erection problems that are associated with the higher-pressure towers found in other designs. Compressor loads are optimized by using low-pressure distillation techniques. Loads are shifted from the large propylene refrigeration compressor to the smaller ethylene refrigeration compressor and to the high-pressure section of the process gas compressor. Megacrackers therefore can use machines that are similar in size. In addition, these flow schemes make extensive use of heat-pump systems, eliminating pumps, drums, and structures normally associated with conventional distillation systems. A consequence of the reduction of recovery-section operating pressure and hydrocarbon inventory is, directionally, a safer plant.

Features of the Front-End Deethanizer

A block flow diagram for the front-end deethanizer is shown in Fig. 6.3.5. This design is used when the primary feed is ethane or ethane-propane (E/P). The front-end deethanizer is flexible and can process a feed slate that contains a large percentage of propane.

Quench Section. For light feeds, fuel oil and gasoline production is low, and a fractionator is not needed. The furnace effluent can be routed directly to the water quench tower. Some of the heat rejected to the circulating water is used as process heat, and the remainder is rejected to cooling water or to the atmosphere through air coolers.

Process Gas Compressor and Deethanizer. The process gas compressor raises the pressure of the cracked gas in preparation for separation into various products. Only four stages of compression typically are required because of the overall low pressures used in the SCORE recovery-section design. The low-pressure front-end deethanizer provides efficient separation and is inherently low fouling. The cracked gas compressor is used as a heat pump for the separation process and to provide refrigeration. As a result, low-temperature ethylene refrigeration and methane refrigeration are not required, saving significant cost.

Front-End Acetylene Hydrogenation. The hydrogen already present in the process gas stream is used for the hydrogenation of the acetylene to ethylene. The acetylene converter is located within the process gas compressor system to stabilize flow through the reactor under all operating conditions. The front-end converter system simplifies the plant and reduces cost because no hydrogen purification system is required.

The presence of carbon monoxide in the reactor feed moderates the reaction kinetics and enhances the selectivity of the hydrogenation reaction to ethylene rather than ethane.

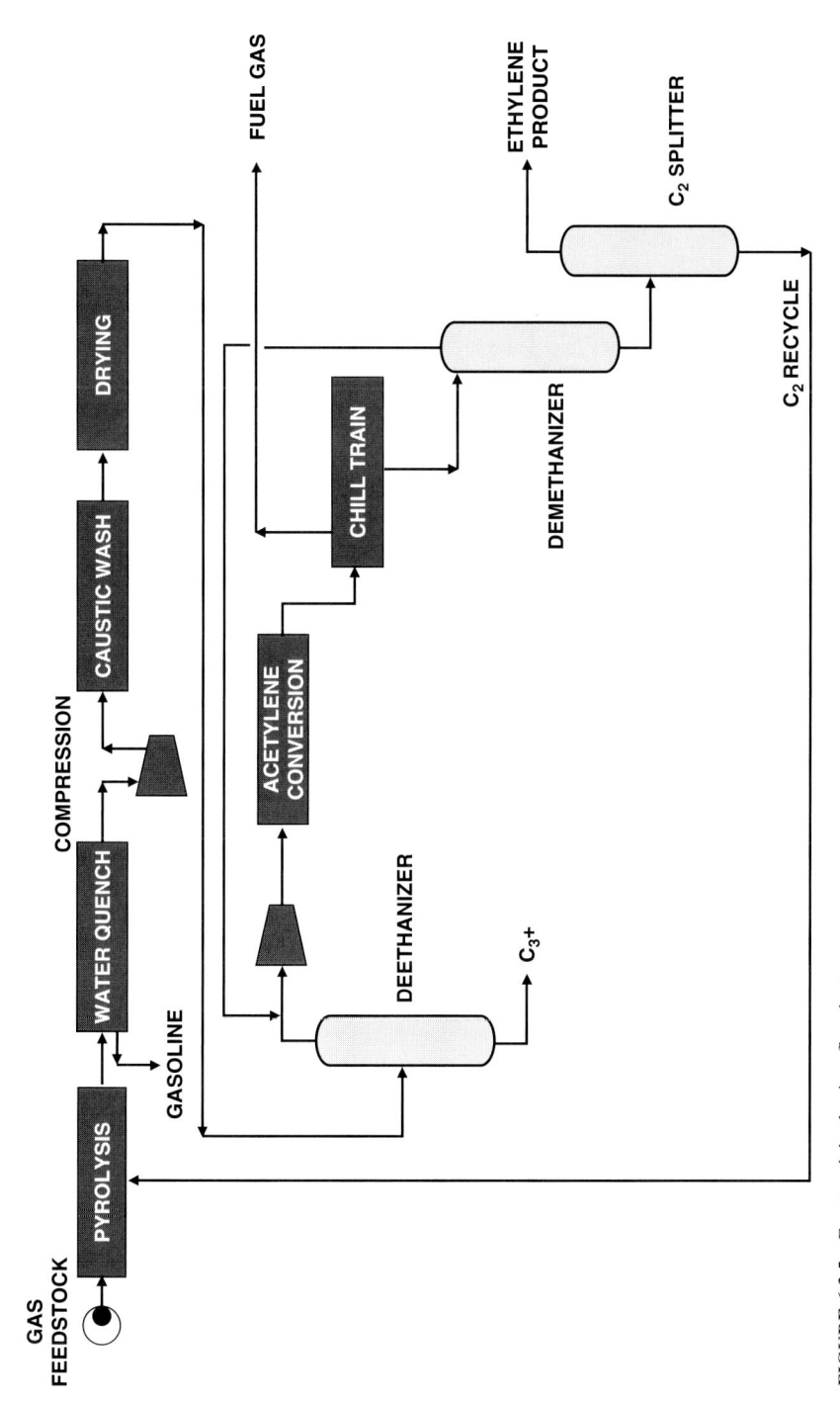

FIGURE 6.3.5 Front-end deethanizer flowsheet.

6.59

The moderated reaction kinetics and the natural presence of chain terminators act to minimize green oil formation, and catalyst fouling is very low. The operating cycle for the front-end converters is typically 5 years or longer. As a result, a spare reactor and *in situ* regeneration systems are not required, providing significant capital cost savings.

Demethanizer System. The demethanizer feed system is an efficient, heat-integrated system using state-of-the-art heat-transfer equipment to minimize the requirement for ultracold refrigeration. There is no need for low-temperature ethylene or methane refrigeration or for a cryogenic reflux pump. Tail gas, consisting of the hydrogen and methane, is recovered for use as fuel. Modifications can be incorporated into the design to provide a hydrogen product.

Ethylene Purification. The C_2 splitter is a heat-pump unit integrated with the ethylene refrigeration system. This eliminates the reflux drum and reflux pumps and significantly reduces the propylene refrigeration compressor loads. The C_2 splitter operates at a much lower pressure than traditional C_2 splitters and therefore requires fewer trays and a much thinner tower shell. This is a significant benefit for large-capacity plants because it facilitates fabrication, transportation, and erection.

Features of the Front-End Depropanizer

A block flow diagram for the front-end depropanizer is shown in Fig. 6.3.6. This design is used when propane or heavier material is the primary cracker feed. Flexibility for multiple feeds, including a large percentage of ethane, also can be provided with the front-end depropanizer design.

Fractionation and Quench. Gasoline and fuel-oil production become significant when the olefins plant feed slate contains naphthas or gas oils. In these situations, a pyrolysis fractionator is used to recover the fuel oil and heavy gasoline products.

The combined effluent from the furnaces is routed to the bottom of the fractionator. Circulating quench oil is used to cool the furnace effluent and condense the fuel oil. Maintenance of the viscosity of the circulating quench oil is critical to the performance of the fractionator system. KBR's proprietary fractionator design provides a means to remove all the pyrolysis tar from circulating quench oil. Since the circulating quench oil contains no tar, low viscosities can be maintained at high operating temperatures, thus improving the heat-recovery efficiency. Waste heat recovered by the circulating oil is used as a low-level heat source within the olefins plant to generate dilution steam and preheat furnace feeds. The fractionator is refluxed with heavy gasoline condensed in the water quench tower to maintain the gasoline product specifications. Fractionator overhead vapors are routed to the water quench tower.

Process Gas Compressor and Front-End Depropanizer. As with the front-end deethanizer design, only four stages of compression normally are required. The discharge temperature from each of the stages is low, and compressor fouling is minimized.

The depropanizer system provides efficient separation and low-fouling operation. This system also uses the process gas compressor as a heat pump for the separation process.

Front-End Acetylene Hydrogenation. As with the front-end deethanizer design, this configuration maintains flow through the reactor under wide variations in operating conditions because it is incorporated into the process gas compression system. The acetylene converter also hydrogenates a significant portion of the methyl acetylene and propadiene (MAPD) that is contained in the cracked gas to propylene. Depending

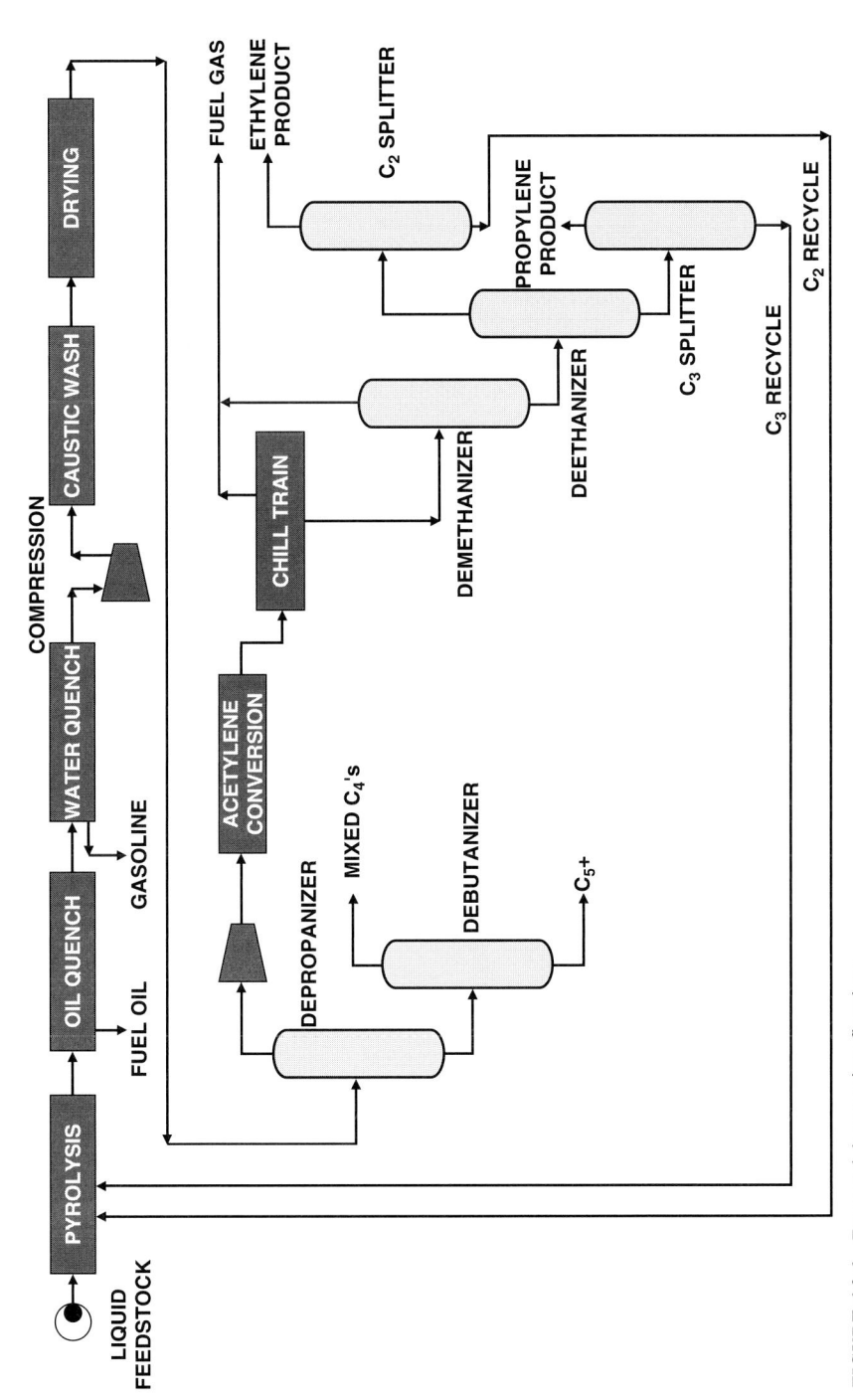

FIGURE 6.3.6 Front-end depropanizer flowsheet.

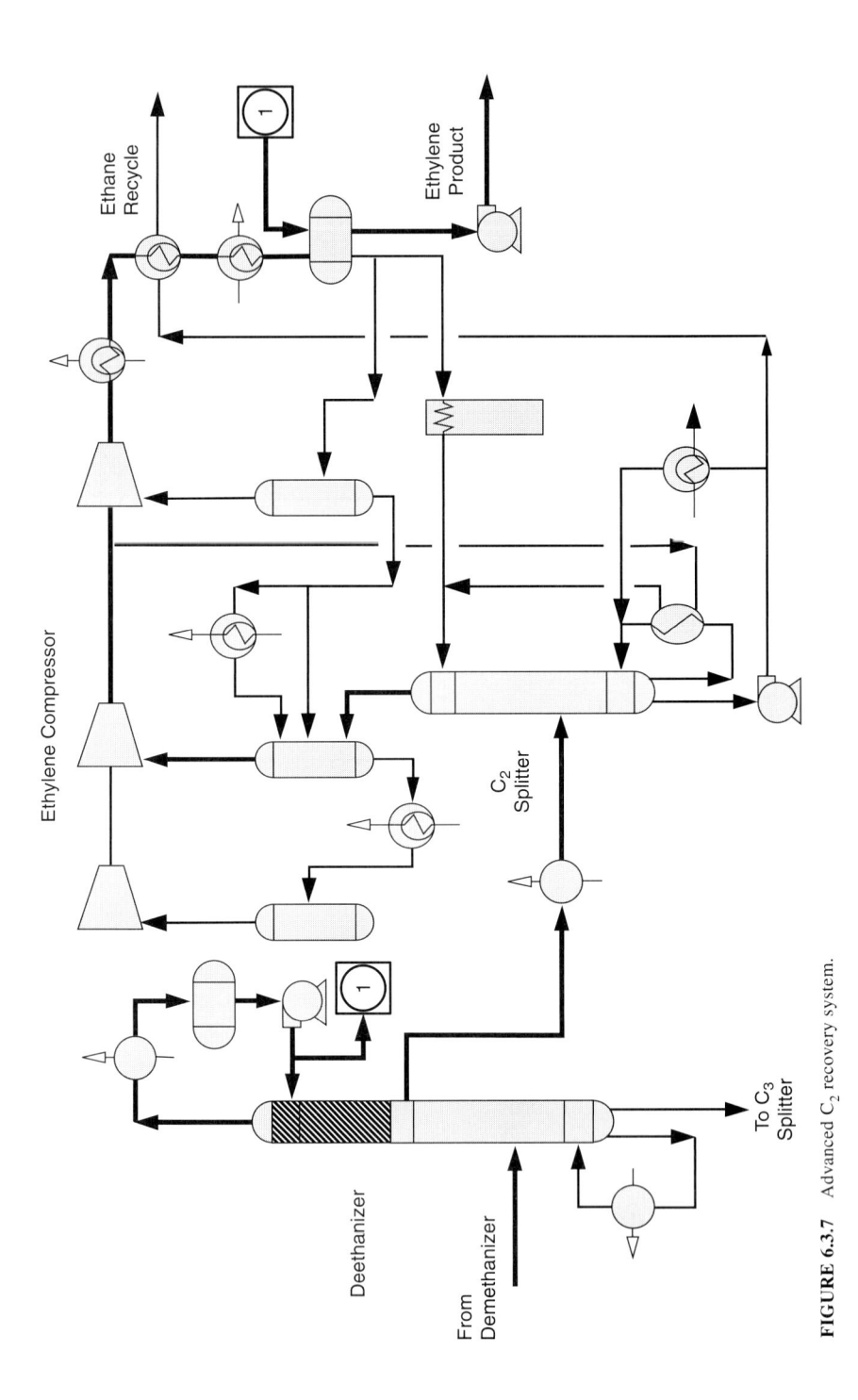

FIGURE 6.3.7 Advanced C_2 recovery system.

on product requirements and the feed slate, the partial hydrogenation of the MAPD may eliminate the need for a downstream MAPD converter.

Demethanizer System. The demethanizer feed system chills and condenses the cracked gas in a stagewise manner. A raw hydrogen stream, generally of 85 to 96 percent purity, normally is taken off the feed system as a product.

Ethylene Purification. An innovation of the SCORE front-end depropanizer recovery system is KBR's patented Advanced C_2 Recovery System, shown in Fig. 6.3.7. In this design, the deethanizer can produce up to 40 percent of the polymer-grade ethylene product. The flow to the downstream C_2 splitter is reduced by a corresponding amount, resulting in capital and energy savings. The Advanced C_2 Recovery System facilitates the design of very large capacity liquid cracking plants because the C_2 splitter is smaller and lighter than more conventional designs, which facilitates fabrication, transportation, and erection.

FUTURE DEVELOPMENTS

KBR continuously investigates using alternative recovery-system designs and new separation technologies to ensure the competitiveness of our technology, especially as it relates to the design, construction, and operation of very large plants. Throughout these investigations, KBR's goal is to find ways to reduce capital cost and energy consumption while still paying the utmost attention to safety, reliability, and environmental requirements.

PART 7

METHANOL

CHAPTER 7.1
LURGI MEGAMETHANOL® TECHNOLOGY

Alexander Frei
Lurgi Oel Gas Chemie GmbH
Frankfurt am Main, Germany

HISTORY

One of Lurgi's core process technologies, synthesis of methanol—an important chemical raw material—has become established all over the world, known as the Lurgi Low-Pressure Methanol Process. Basically, the methanol process consists of synthesis gas generation, methanol synthesis, and methanol distillation. Gas may be generated starting from various different feedstocks (such as natural gas, naphtha, heavy residues, and coal).

Methanol production on an industrial scale was introduced in 1923 when BASF Ludwigshafen commissioned a methanol synthesis plant on the basis of a chromium/zinc catalyst. Since this type of catalyst was not highly active, it was necessary to apply operating pressures of between 300 and 400 bar and operating temperatures of between 350 and 400°C. Based on this type of high-pressure methanol catalyst, Lurgi has built six plants in various countries that use synthesis gas produced from natural gas or steam reforming of light naphtha or by partial oxidation of heavy residual oil, the total capacity amounting to 181,000 metric tons per annum (MTA).

In order to improve the process economics, in the 1960s, Lurgi initiated the development of a process that allows processing of methanol at low pressures and temperatures. Development work at Lurgi's research and development (R&D) center focused on a suitable catalyst. The first LP methanol tests were run in 1969. Since Lurgi itself is not a catalyst manufacturer, it started a cooperation with Süd-Chemie AG in 1970 for fabrication of the catalyst. In the same year, Lurgi built a semicommercial demonstration unit with a 100-tube methanol reactor. The design and technology of the methanol reactor were based on a water-cooled reactor, already service-proven in Lurgi's Fisher-Tropsch synthesis. In the demonstration plant, extensive research and experimental work was conducted for virtually all types of synthesis gas, irrespective of origin, i.e., produced by steam re-forming of natural gas or naphtha or by partial oxidation of heavy residue oils. The influence of harmful components on the methanol catalyst also was investigated in-depth in the demonstration unit. The operating results of the demonstration unit convinced Lurgi's clients so much that the first three contracts for Lurgi LP Methanol Process plants were signed in that same year, representing a total capacity of more than 300,000 MTA. Given that each of these plants was based on a different feedstock—heavy residue of a refinery, natural gas, feed range from natural gas to naphtha—Lurgi was able to prove that the new LP Methanol Process technology can cope with all those feedstocks. Thanks to the know-how and experience gained with the demonstration unit, all three plants went onstream without problems in early 1973.

Meanwhile, the methanol catalyst has been improved significantly in terms of selectivity, conversion rates, and durability through intense joint research efforts of Lurgi and Süd-Chemie AG. Today, Lurgi has highly active and stable catalysts available for varying process conditions so as to largely suppress the formation of by-products.

The economics of the process also have been improved continuously with respect to equipment, feedstock requirements, and capacity. Lurgi has so far built (or is building) 41 LP Methanol Process plants with a total capacity of 20.1 million MTA. This represents about a third of the world installed capacity.

Vast natural gas and oil-associated gas reserves are available in remote areas at low and stable cost. Combining low-cost feedstock with large single-train synthesis technology will be the strategy of the next decades in order to achieve a remarkable production cost reduction.

Lurgi has developed its MegaMethanol® technology on the basis of the syngas technologies available in the 1990s, i.e., conventional steam re-forming and combined re-forming together with a new synthesis concept. Lurgi's MegaMethanol process is an advanced technology for converting natural gas to methanol at low cost in large quantities. It permits the construction of highly efficient single-train plants of at least double the capacity of those built to date. This paves the way for new downstream industries that can use methanol as a competitive feedstock.

MEGAMETHANOL TECHNOLOGY

Lurgi's MegaMethanol technology has been developed for world-scale methanol plants with capacities larger than 1 MMTA. To achieve such a capacity in a single train, a special process design is needed, incorporating advanced but proven and reliable technology, cost-optimized energy efficiency, low environmental impact, and low investment cost.

The main process features to achieve these targets are

- Oxygen-blown natural gas re-forming, either in combination with steam methane re-forming or as pure autothermal re-forming
- Two-step methanol synthesis in water- and gas-cooled reactors operating along the optimal reaction route
- Adjustment of syngas composition by hydrogen recycle

The configuration of the re-forming process—autothermal or combined re-forming—mainly depends on the feedstock composition, which may vary from light natural gas (nearly 100 percent methane content) to oil-associated gases. The aim is to generate an optimal synthesis gas, characterized by the stoichiometric number given below:

$$SN = \frac{H_2 - CO_2}{CO + CO_2} = 2.0\text{--}2.1$$

The Lurgi MegaMethanol process, based on re-forming of gaseous hydrocarbons, especially natural gas, consists of the following essential process steps:

- Desulfurization
- Pre-re-forming (optional)
- Saturation
- Pure autothermal or combined re-forming
- Methanol synthesis
- Methanol distillation

The syngas generation section of a MegaMethanol plant using combined re-forming is shown in Fig. 7.1.1.

PROCESS DESCRIPTION

The synthesis gas production section of a conventional methanol plant accounts for more than 50 percent of the capital cost of the entire plant. Thus, optimizing this section yields a significant cost benefit. Conventional steam methane re-forming is economically applied in small and medium-sized methanol plants only, with the maximum single-train capacity being limited to about 3000 metric tons per day (MTD). Oxygen-blown natural gas re-forming, either in combination with steam re-forming or as pure autothermal re-forming, is today considered to be the best-suited technology for large syngas plants. The reason for this appraisal is that the syngas generated through oxygen-blown technology becomes available in stoichiometric composition and under very high pressure. Hence very high quantities can be produced in a single train using reasonably small equipment.

Desulfurization

Catalyst activity is seriously affected even by traces of catalyst poisons in the gas feedstock. Among others, sulfur compounds in particular lower the catalyst activity considerably.

In order to protect the re-former and synthesis catalysts from sulfur poisoning, the feedstock must be desulfurized. Desulfurization operates at approximately 350 to 380°C in the desulfurization reactor. The feedstock is routed through zinc oxide beds, where hydrogen sulfide is adsorbed according to the following equation:

$$H_2S + ZnO \leftrightarrow ZnS + H_2O$$

If the feedstock contains organic sulfur compounds such as mercaptans or thiophenes, hydrogenation is required prior to desulfurization. This is often accomplished in a separate reactor, where the feedstock, after adding a small amount of hydrogen-rich methanol synthesis purge gas, is hydrogenated over cobalt-molybdenum catalysts. A residual sulfur content of less than 0.1 ppm is obtainable and can be tolerated for the downstream processes.

Saturation

After desulfurization, the natural gas feedgas is cooled and then enters the top of a saturator for saturation, with water supplying a major part of the H_2O required for the re-forming reactions. After makeup with process condensate and process water from distillation, hot circulation water is fed to the top section of the saturator. Circulation water is withdrawn from the bottom of the saturator by a recirculation pump and then is heated by a circulation water heater in the re-formed gas cooling train before it is refed to the saturator.

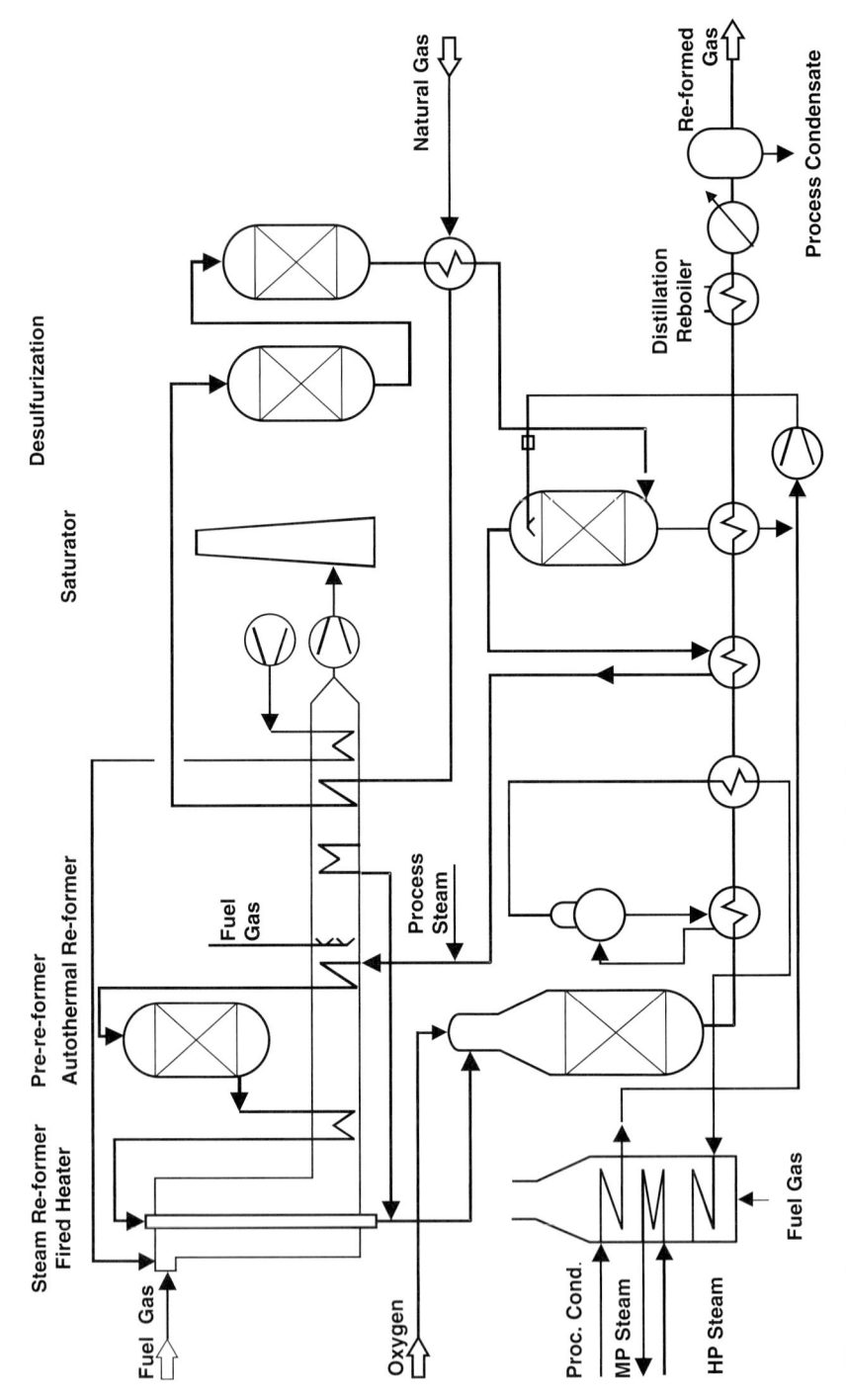

FIGURE 7.1.1 Syngas generation using combined re-forming for the MegaMethanol process.

Pre-re-forming (Optional)

If the feedstock contains fractions of higher hydrocarbons, the steam re-former catalyst can be affected by carbon deposits owing to cracking reactions when operated at low steam-carbon ratios. This should be avoided by pre-re-forming the feedstock in a pre-re-former. The conversion of higher hydrocarbons in an adiabatic reactor produces a gas rich in methane and hydrogen that is perfectly suitable for further steam re-forming. The fixed-bed-type pre-re-former is arranged between the process feed superheater and the steam re-former. The desulfurized feedstock, with process steam added, is routed through the catalyst bed, where almost all higher hydrocarbons and a small percentage of the methane are re-formed with steam according to the following equations:

Steam re-forming of higher hydrocarbons:

$$C_n H_m + nH_2O \leftrightarrow nCO + (m/2 + n)H_2$$

Methanation:

$$CO + 3H_2 \leftrightarrow CH_4 + H_2O$$

Water-gas shift reaction:

$$CO + H_2O \leftrightarrow CO_2 + H_2$$

Pre-re-formed gas is produced at about 380 to 480°C, the overall heat balance being slightly endothermic or exothermic depending on the content of higher HCs in the feedstock. The pre-re-formed gas only contains a few parts per million of hydrocarbons higher than methane and allows reduction of the steam-to-carbon ratio for the steam re-former to 1.8. A low steam-to-carbon ratio and superheating of the pre-re-formed gas upstream of the steam re-former inlet reduces the size of the steam re-former significantly. Additionally, the amount of waste heat is reduced by saving steam re-former underfiring when combined re-forming is used.

Autothermal Re-forming

Pure autothermal re-forming can be applied for syngas production in MegaMethanol plants whenever light natural gas is available as feedstock to the process. The desulfurized and optionally pre-re-formed feedstock is re-formed with steam to synthesis gas at about 40 bar using oxygen as the re-forming agent. The process offers great operating flexibility over a wide range to meet specific requirements. Re-former outlet temperatures are typically in the range of 950 to 1050°C. The synthesis gas is compressed to the pressure required for methanol synthesis in a single-casing synthesis gas compressor with integrated recycle stage.

With the help of a proprietary and proven three-dimensional computational fluid dynamics (CFD) model (Fig. 7.1.2), gas flows and temperature profiles are simulated with the objective of designing burner and reactor as an integrated unit. Autothermal processes produce the heat required for gasification through partial combustion of the feedgas to be converted in the reactor. Oxygen is usually added for this purpose. Suitable feedstocks for autothermal catalytic re-forming are light natural gases or steam re-formed gases with a high residual methane content.

The principal chemical reactions involved in the process are those of complete combustion of methane

$$CH_4 + 2O_2 \leftrightarrow CO_2 + 2H_2O$$

partial oxidation of methane

$$CH_4 + O_2 \leftrightarrow CO + H_2 + H_2O$$

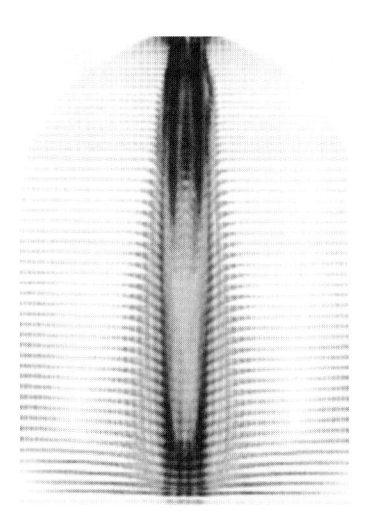

FIGURE 7.1.2 Autothermal reactor simulation, performance by 3D CFD model.

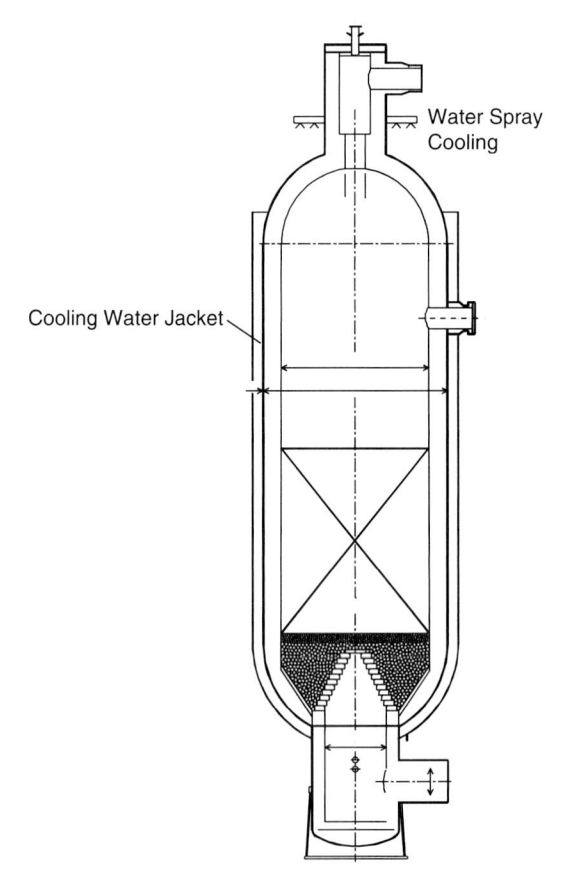

FIGURE 7.1.3 Autothermal re-former.

as well as methane and higher hydrocarbon re-forming and a CO shift reaction, as described under "Steam Re-forming."

The combustion (partial oxidation) is highly exothermic, the re-forming reaction being endothermic. In order to attain the desired product gas quality, the outlet temperature of the reactor is selected and controlled by metering in the required amount of oxygen to maintain the heat balance between exothermic and endothermic reactions. The autothermal re-former (ATR) is a refractory-lined pressure vessel. A cross section of an ATR is illustrated in Fig. 7.1.3. The lower, cylindrical part of the ATR contains re-forming catalyst. The entire ATR shell is protected by high-temperature-resistant brick lining. A mixer in the ATR top section provides fast and uniform mixing of the gas-steam mix with preheated oxygen. The combustion and prereaction zone is located above the catalyst bed. The catalyst employed is a steam re-forming catalyst. Because of the operating conditions, establishing a stable flame contour is essential to protecting the reactor walls and catalyst from excessive temperatures. Even when using pure methane as feedstock for autothermal re-forming, it is necessary to condition the synthesis gas because its stoichiometric number is below 2.0. The most economical way to achieve the required gas composition is to add hydrogen, withdrawn from the methanol synthesis purge stream by a membrane unit or a pressure swing adsorption (PSA) unit.

Compared with its competitors, Lurgi has the most references and experience for this re-forming technology. This process has been implemented in Lurgi plants since the 1950s.

Combined Re-forming

The combination of oxygen-blown autothermal re-forming and conventional steam methane re-forming, the so-called combined re-forming process, has an advantage in that it yields synthesis gas of optimal composition and at a high pressure. A cross section of Lurgi steam re-former is shown in Fig. 7.1.4. In a steam re-forming process, hydrocarbons and steam are catalytically converted into hydrogen and carbon oxides. The composition of the product gas is reached according to the following equilibrium reactions:

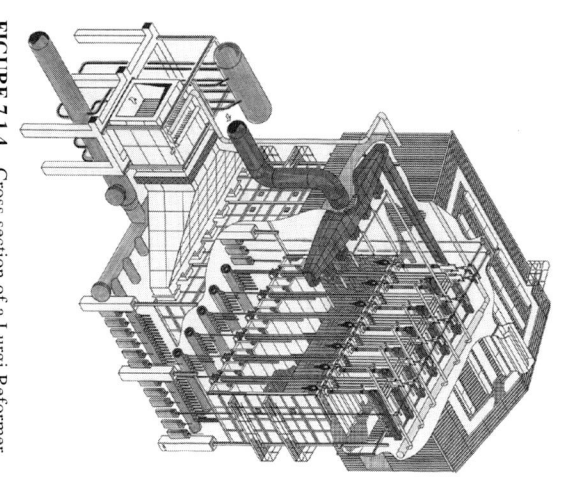

FIGURE 7.1.4 Cross section of a Lurgi Reformer.

METHANOL

$$C_nH_m + nH_2O \leftrightarrow nCO + (m/2 + n)H_2$$

$$CH_4 + H_2O \leftrightarrow CO + 3H_2$$

$$CO + H_2O \leftrightarrow CO_2 + H_2$$

The overall reaction is highly endothermic, so reaction heat has to be provided externally. The synthesis gas is characterized by a relatively low pressure and a surplus of hydrogen. By adding carbon dioxide, the composition of the synthesis gas can be adjusted to be more favorable for methanol production.

The tubular steam re-former is the most expensive single item of a methanol plant when combined re-forming is used.

Against this background, Lurgi has developed the Lurgi Reformer™ that allows for the construction of very large re-formers of up to 1000 catalyst tubes in a single cell. The largest Lurgi Reformer implemented so far contains 720 tubes.

The Lurgi Reformer is a top-fired re-former, and as such, it exhibits all the advantages of this typical design:

- Multiple-tube rows, resulting in a lower number of burners and lower heat loss
- Almost uniform wall temperature over the entire heated tube length
- Easier burner adjustment and reduced burner maintenance because of the reduced number of burners
- Less NO_x formation by more accurate fuel and combustion air equipartition of the burners
- Easier noise abatement

The features of this re-former design are

- Advanced inlet pigtail design
- Internal insulation at the top of the re-former catalyst tube
- Counterweight tube support system
- Flexitube outlet system

The main advantage of the combined re-forming process over similar process alternatives is the patented feedgas bypass of the steam re-former. The optimal stoichiometric synthesis gas composition can be achieved by choosing the appropriate bypass ratio. Since less than half the feed gas is routed through the steam re-former, the overall process steam requirements are also roughly halved compared with other processes that use an autothermal re-former downstream of the steam re-former without such a bypass. The lower process steam consumption translates into reduced energy requirements and lower capital costs.

Methanol synthesis gas generation by means of combined re-forming is a well-proven technology. For natural gases or oil-associated gases with a methane content above 80 percent and for methanol syntheses with capacities above 1500 MTD of methanol, this process route offers potential capital investment savings compared with the conventional steam re-forming process.

The composition of the generated synthesis gas is characterized by the stoichiometric number

$$SN = \frac{H_2 - CO_2}{CO + CO_2}$$

In the steam methane re-forming process, the given C:H ratio of the natural gas and the hydrogen added by steam decomposition leads to a stoichiometric number that is higher than optimal for methanol production. The surplus of hydrogen would have to be compressed and behaves as an inert gas in the synthesis loop. It increases the size of the equipment and has to be discharged with the purge gas. Thus it can only be used as fuel gas for firing. In autothermal re-forming, the heat required for re-forming is generated by partially combusting feedgas. Since the ATR is a pressure vessel with an inner lining, its limitations with regard to pressure and temperature are much less stringent than those of a steam re-former tube. The temperature is only limited by the thermal stability of the catalyst and of the interior lining. On the other hand, if autothermal re-forming were to be applied alone, the heat required for re-forming largely would be generated by free oxygen, and the resulting synthesis gas would exhibit a substoichiometric ratio.

Therefore, Lurgi combined the two processes in such a way that only the amount of natural gas is routed through the steam re-former that is required to generate a final synthesis gas with the desired stoichiometric number of about 2.05. Thus the synthesis gas demand per ton of methanol is reduced by approximately 25 percent compared with steam methane re-forming. Depending on the composition of the natural gas, only about 30 percent of the hydrocarbons are converted in the steam methane re-former, and hence the steam methane re-former in the combined re-forming is only about one-quarter of the size of a re-former in the conventional steam methane re-forming process. This means considerable savings in cost and energy. Owing to the higher pressure in the re-forming section, the compression energy is reduced, and compression to synthesis pressure is possible in a single-stage compressor.

The cost saved in comparison with conventional steam methane re-forming is higher than the investment for the air separation unit. Further energy savings of about 8 percent are achieved by combined re-forming.

An important secondary effect is that instrument air, plant air, and nitrogen required in the plant are obtained as by-products from the air separation unit so that no separate utility units are to be considered.

Waste Heat Recovery

Flue-Gas Cooling Section. When combined re-forming is used for the syngas production, the flue gases leaving the steam re-former tubes are routed through a duct to the flue gas waste heat recovery section.

In order to use the sensible heat of the flue gas, several heat exchanger tube banks are arranged in series. The feed preheater, which appropriately serves to heat the natural gas/process steam mixture, is arranged at the outlet of the re-former. Subsequently, superheating of high- and medium-pressure steam is achieved. Before the flue gases are exhausted, further cooling is obtained by simultaneously preheating combustion air. The draft necessary for re-former firing and transport of the flue gases through the waste-heat section is provided by a flue-gas fan that delivers the flue gas to the stack.

Re-formed Gas Cooling Section. The re-formed gas leaving the base of the ATR at approximately 950 to 1050°C represents a considerable source of heat with a potential for energy recovery. Directly at the outlet of the ATR the gas enters the waste-heat boiler to generate saturated high-pressure steam. The waste heat boiler is a horizontally arranged cooler with fixed tube sheet. It is connected to the high-pressure steam drum by risers and downcomers using a natural circulation system. The re-formed gas leaving the waste-heat boiler is cooled in the natural gas preheater while simultaneously preheating the natural gas. Re-formed gas leaving the natural gas preheater is then routed to the circulation water heater, where the circulating water for the saturator is preheated. Further cooling is performed in the distillation section where the re-formed gas heat is utilized for reboiler duties. Final cooling to syngas compressor suction temperature is achieved in the final cooler. During the cooling process, the gas temperature drops below the water dew point. Separation of the process condensates is achieved subsequently and the saturated re-formed gas is routed to the methanol synthesis unit.

Methanol Synthesis

In the Lurgi MegaMethanol process methanol is synthesized from hydrogen, carbon monoxide, and carbon dioxide in the presence of a highly selective copper-based catalyst. The principal synthesis reactions are as follows:

$$CO + 2H_2 \leftrightarrow CH_3OH$$

$$CO_2 + 3H_2 \leftrightarrow CH_3OH + H_2O$$

These reactions are highly exothermic, and the heat of reaction must be removed promptly from its source. This is accomplished most effectively in the Lurgi methanol reactors described below.

Efficient conversion in the methanol synthesis unit is essential to low-cost methanol production. In addition, optimal use of the reaction heat offers cost advantages and energy savings for the overall plant.

Nowadays, two types of catalytic fixed-bed reactors are used in industry: steam-raising reactors and gas-cooled reactors. From the very beginning of the low-pressure technology era, Lurgi has been equipping its methanol plants with a tubular reactor in which the heat of reaction is transferred to boiling water.

The Lurgi Water-Cooled Methanol Reactor (WCR) is basically a vertical shell-and-tube heat exchanger with fixed-tube sheets. The catalyst is accommodated in tubes and rests on a bed of inert material. The water-steam mixture generated by the heat of reaction is drawn off below the upper tube sheet. Steam pressure control permits an exact control of the reaction temperature. The quasi-isothermal reactor achieves very high yields at low recycle ratios and minimizes the production of by-products.

A significant improvement in synthesis technology has been achieved by combining the WCR with a downstream Gas-Cooled Reactor (GCR). The so-called combined converter synthesis is shown in Fig. 7.1.5. The excellent heat transfer in the WCR allows this reactor to operate with a high concentration of reaction components in the inlet gas. This highly concentrated gas results from a drastically reduced recycle rate. Under these conditions, a very high methanol yield is achieved in the WCR. The methanol-containing exit gas of the WCR is fed to the downstream GCR. In the GCR, the reaction is accomplished at continuously reduced temperatures along the optimal reaction route. The optimal temperature profile is achieved by countercurrent preheating in the inlet gas to the WCR.

The latest generation of methanol catalysts makes it possible to select an outlet temperature of the GCR of about 220°C.

Methanol Synthesis Loop

Since economical conversion of the synthesis gas to methanol cannot be achieved in a single reactor pass, unreacted gases are circulated in a loop, thus increasing the conversion rate. Figure 7.1.6 shows a typical diagram of the synthesis loop with distillation section. Recycle gas and synthesis gas are mixed and preheated in the trim heater by cooling the reactor outlet gas. Preheated recycle gas and synthesis gas are routed to the GCR. On the tube side of the GCR, the reactor inlet gas is further heated to the inlet temperature of the WCR (approximately 240°C).

Gas-Cooled Reactor

Water-Cooled Reactor

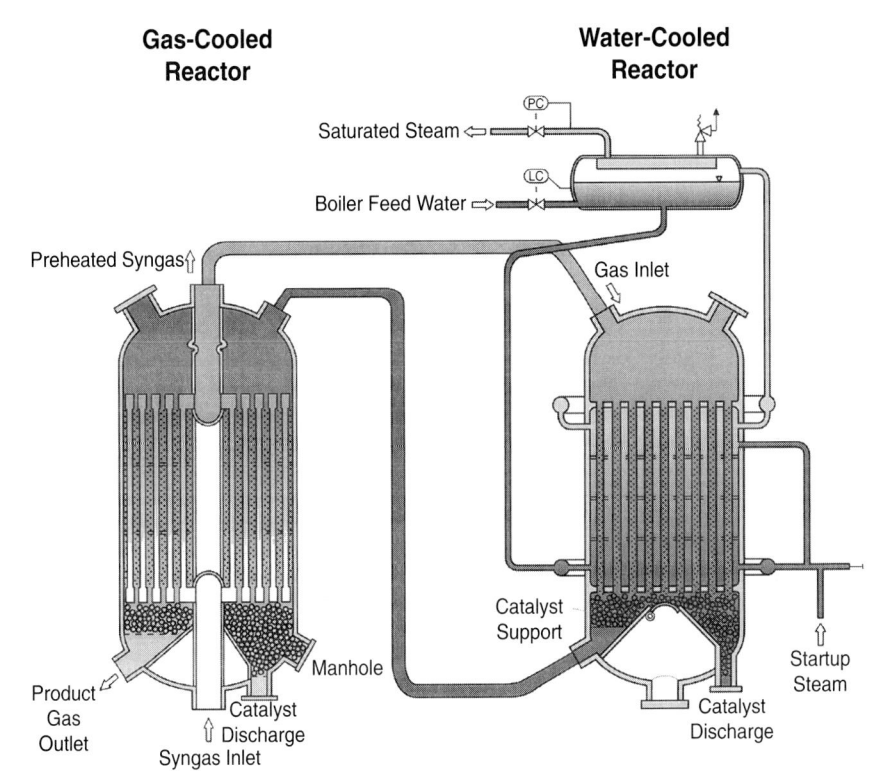

FIGURE 7.1.5 Combined converter synthesis.

Boiler water from the steam drum enters the reactor shell side at the bottom through a distributor and rises up to the outlet at the top due to a thermosyphoning effect. The steam-water mixture coming out of the reactor shell side is separated in the steam drum. Saturated middle-pressure (MP) steam is discharged from the steam drum via a pressure-control valve, and water circulates back to the reactor. The pressure control at the steam drum outlet controls the pressure in the shell side of the reactor and thus the boiling point of water, which in turn controls the reaction temperature.

The "preconverted" gas is routed to the shell side of the GCR, which is filled with catalyst, and the final conversion to methanol is achieved at continuously reduced temperatures along the optimal reaction route. The heat of reaction is used to preheat the reactor inlet gas inside the tubes.

Aside from methanol and water vapor, the reactor outlet gas contains nonreacted H_2, CO and CO_2, inerts such as CH_4 and N_2, and some parts per million of reaction by-products. This gas needs to be cooled from the reactor outlet temperature to about 40°C in order to condense and separate CH_3OH and H_2O from the gases. The hot gas is routed to the MP-BFW preheater, where the heat released is used to preheat MP boiler feed water for the steam drum. The gas stream is further used for preheating the recycle gas and the synthesis gas to the reactor in the trim heater.

At last the gas is cooled in the synthesis air cooler and the final cooler by cooling water. Condensed raw methanol contains, apart from methanol and water, dissolved gases and reaction by-products.

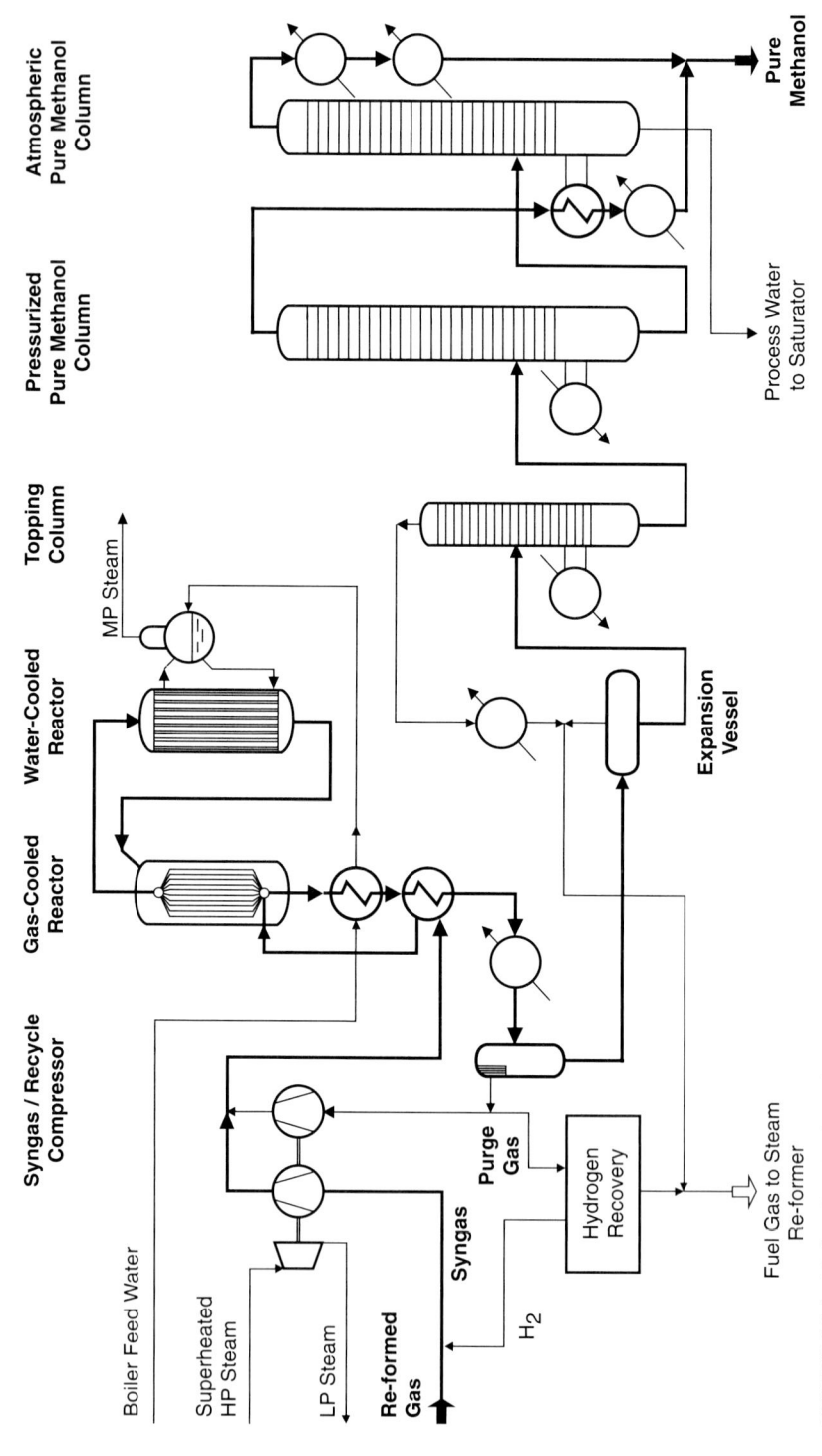

FIGURE 7.1.6 Methanol synthesis section with distillation.

Separation of raw methanol from nonreacted gases takes place in the methanol separator. Raw methanol leaves the vessel on a level control for distillation, where pure methanol is distilled from water and other impurities. The major portion of gas is recycled back to the synthesis reactors via a recycle gas compressor in order to achieve a high overall conversion. A small portion is withdrawn on pressure control as purge gas to avoid excessive accumulation of inerts in the loop.

The major portion of the purge gas is routed to the hydrogen recovery unit, where hydrogen is recovered. The hydrogen product is mixed with the synthesis gas, whereas the remaining gas is recycled as fuel gas. During startup and shutdown, the purge gas is routed to the flare. A small amount of purge gas is used for hydrogenation of the natural gas for desulfurization.

The most important advantages of combined-synthesis converters are

- *High syngas conversion efficiency.* At the same conversion efficiency, the recycle ratio is about half the ratio in a single-stage, water-cooled reactor.
- *High energy efficiency.* About 0.8 t of 50- to 60-bar steam per ton of methanol can be generated in the reactor. In addition, a substantial part of the sensible heat can be recovered at the gas-cooled reactor outlet.
- *Low investment cost.* The reduction in catalyst volume for the water-cooled reactor, the omission of a large feedgas preheater, and savings resulting from other equipment due to the lower recycle ratio translate into specific cost savings of about 40 percent for the synthesis loop.
- *High single-train capacity.* The design of two plants has confirmed that single-train plants with capacities of 5000 MTD and above can be built.

Methanol Distillation

The raw methanol produced in the methanol synthesis unit contains water, dissolved gases, and a quantity of undesired but unavoidable by-products that have either lower or higher boiling points than methanol. The purpose of the distillation unit is to remove those impurities in order to achieve the desired methanol purity specification. A three-column methanol distillation is shown in Fig. 7.1.7.

This is accomplished in the following process steps:

- Degassing
- Removal of low-boiling by-products
- Removal of high-boiling by-products

Dissolved gases are driven out of the raw methanol simply by flashing it at a low pressure into the expansion-gas vessel. Removal of light ends and remaining dissolved gases is carried out in a light ends column. Finally, the methanol is separated from the heavy ends in a pure methanol distillation section consisting of one or two columns. The split of the refining column into two columns allows for very high single-train capacities. Methanol purity remains unaffected, whereas the single-train capacity, consumption of steam, and investment cost depend on the distillation concept.

The first pure methanol column operates at elevated pressure and the second column at atmospheric pressure. The overhead vapors of the pressurized column heat the sump of the atmospheric column. Thus about 40 percent of the heating steam and, in turn, about 40 percent of the cooling capacity are saved. The split of the refining column into two allows for very high single-train capacities.

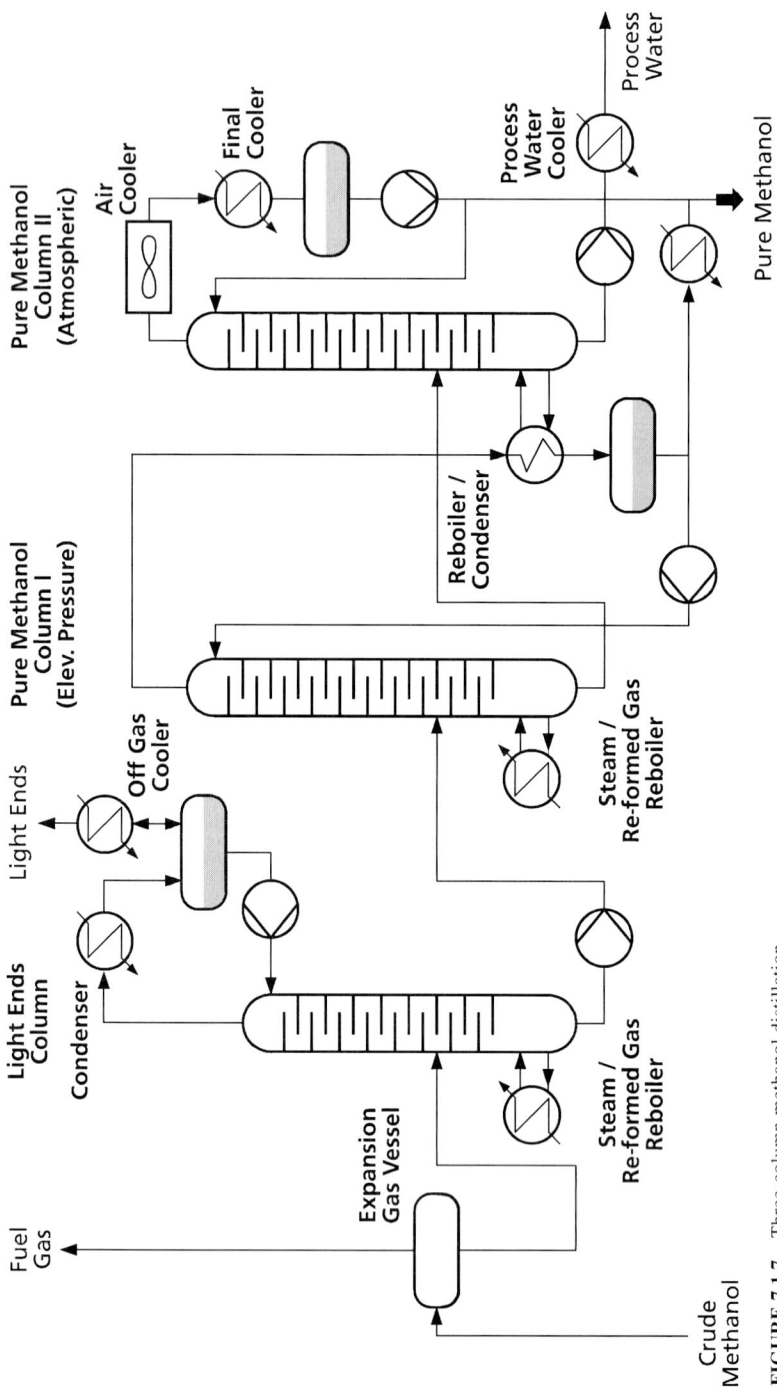

FIGURE 7.1.7 Three-column methanol distillation.

LATEST LURGI METHANOL PROJECT REFERENCES

- Methanex, United States, 1700 MTD, 1992
- Statoil, Norway, 2400 MTD, 1992
- CINOPEC, China, 340 MTD, 1993
- KMI, Indonesia, 2000 MTD, 1994
- NPC, Iran, 2000 MTD, 1995
- Sastech, South Africa, 400 MTD, 1996
- Titan, Trinidad, 2500 MTD, 1997
- PIC, Kuwait, 2000 MTD, 1998
- YPF, Argentina, 1200 MTD, 1999
- Atlas, Trinidad, 5000 MTD, 2000
- ZAGROS, Iran, 5000 MTD, 2000
- Methanex, Chile, 2400 MTD, 2002
- ZAGROS II, Iran, 5000 MTD, 2004
- Hainan Methanol, China, 2000 MTD, 2004
- QAFAC, Qatar, 6750 MTD, 2004

PART 8

OXO ALCOHOLS

CHAPTER 8.1
JOHNSON MATTHEY OXO ALCOHOLS PROCESS™

Jane Butcher and Geoff Reynolds
Johnson Matthey Catalysts
Billingham, England

INTRODUCTION

The Johnson Matthey Oxo Alcohols Process™ can be used to convert a wide range of olefins to the corresponding alcohols. An olefin is converted to the aldehyde by reaction with carbon monoxide and hydrogen. This reaction is called *hydroformylation*, or the *oxo reaction*. The aldehyde is then hydrogenated to the alcohol, and the alcohol is purified by distillation and optionally by hydrorefining. See Fig. 8.1.1.

The alcohols can be converted to esters, which are used to plasticize polyvinyl chloride (PVC). In some cases the esters are used in synthetic lubricant formulations. The alcohols produced from linear olefins can be converted to detergents. The intermediate aldehydes from linear olefins can be converted to long-chain amines with a range of applications in speciality chemicals manufacture.

The distinctive feature of the Johnson Matthey Oxo Alcohols Process compared with processes offered by other companies is that it uses an unliganded rhodium catalyst for the hydroformylation reaction. This was made possible by the development of a highly efficient rhodium harvesting process.[1] Use of unliganded rhodium means that the process can convert a wide range of linear and branched olefins (typically C_6 to C_{14}) to give aldehydes and alcohols with minimum by-product formation and therefore a low feedstock usage.

This allows refinery operators to upgrade refinery olefinic gas streams from gasoline or fuel value to highly valued chemical intermediates.

PROCESS DESCRIPTION

The Johnson Matthey Oxo Alcohols Process is best described by splitting the process into sections.

8.4 OXO ALCOHOLS

FIGURE 8.1.1 100-KTA INA plant designed and commissioned for an Asian customer showing (from left to right) the hydrogenation and hydroformylation sections and in the background the machinery hall, control building, distillation section, and storage.

Hydroformylation

A feed of olefin (fresh feed or recycle light ends) is reacted with a 50-50 mixture of carbon monoxide and hydrogen at a pressure of 245 bar(g) and a temperature in the range 95 to 180°C.

This produces an aldehyde containing one carbon atom more than the feed olefin. The specific temperature used depends on the particular olefin and the properties that are required in the product alcohol.

The essential chemistry of the hydroformylation step is

$$-CH=CH- + CO + H_2 \rightarrow -CH_2-\underset{\underset{CHO}{|}}{CH}-$$

The reaction is exothermic and is carried out in a train of equal-sized water-cooled reactors. The number of reactors is determined by the required plant capacity. (See Fig. 8.1.2.)

The reaction is catalyzed by a low concentration of a homogeneous unliganded rhodium catalyst. This is added as a rhodium salt dissolved in the feed olefin. In the presence of the carbon monoxide and hydrogen, the rhodium salt is converted into the actual catalytic species. Under the conditions used for the hydroformylation reaction, rhodium catalyzes the hydroformylation reaction but not the hydrogenation of the aldehyde. This reduces the formation of by-products compared with alternative hydroformylation technologies and gives a lower feedstock usage.

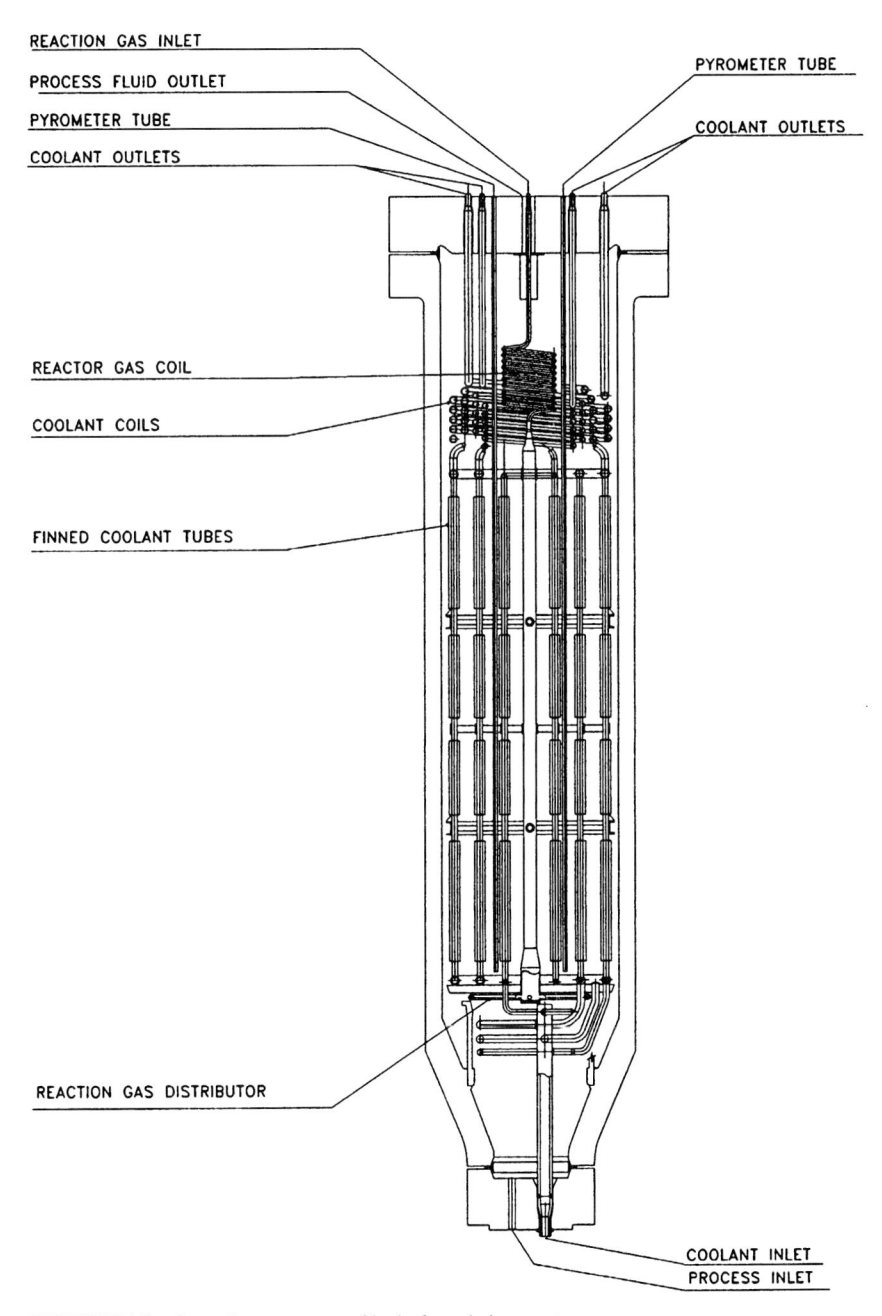

FIGURE 8.1.2 General arrangement of hydroformylation reactor.

Olefins with more than four carbon atoms have many isomeric forms. For a number of reasons (e.g., steric hindrance), some of the olefin isomers are difficult to hydroformylate. In addition to catalyzing the hydroformylation reaction, the rhodium is able to move the double bond to different positions in the molecule. By moving the double bond, it can be made accessible. Conversely, in the case of terminal olefins (alpha olefins), it is usually important to ensure that the double bond does not move. The conditions used in the hydroformylation reaction stage move the double bond or keep it fixed depending on the required properties of the alcohol.

Rhodium Catalyst Harvesting

Pressure letdown following the hydroformylation stage releases dissolved carbon monoxide and hydrogen, some of which is recycled to the hydroformylation section; the rest is sent to the fuel-gas system.

After letdown, the aldehyde-olefin mixture is passed through a series of ion-exchange resin beds in which the rhodium is extracted ("harvested"). Every 3 weeks the resin "loaded" with rhodium is removed from the plant and incinerated. At the same time, a fresh bed of resin is put online.

The incineration reduces dramatically the bulk volume of the rhodium residue to simplify shipping. Incineration also converts the rhodium into rhodium oxide, which is stable and is readily recovered and converted back into the rhodium salt by a specialist precious metal processor.

Hydrogenation

The lean aldehyde, so-called because the catalyst has been removed, is converted into the alcohol by hydrogenation in the vapor phase over a copper/zinc catalyst (Pricat CZ 29/2):

$$-CH_2-\underset{\underset{CHO}{|}}{CH}- + H_2 \rightarrow -CH_2-\underset{\underset{CH_2OH}{|}}{CH}-$$

In the standard design, this stage of the process is operated at 235 bar(g) pressure.

Lower-pressure and/or liquid-phase hydrogenation can be used depending on feedstock reactivity and the particular technology application.

The copper/zinc hydrogenation catalyst Pricat CZ 29/2 was developed to achieve efficient hydrogenation of aldehyde to alcohol with minimum conversion of the unconverted olefin to paraffin in two-pass operation (less than 10 percent of the feed olefin). The catalyst is highly selective, and by-product formation is minimized. The formulation of the catalyst ensures that it is resistant to sulfur, which is present in some olefin feedstocks. The life of the catalyst depends on the impurities present in the feed olefin and their concentration. Using olefins from a variety of suppliers, principally derived from polygas plants, can be expected to give catalyst lives of 12 to 18 months. Use of "Dimersol" octenes with low sulfur and chlorine can give lives up to 2 years.

The high-boiling by-products (heavy ends) produced in the overall process are mainly ethers together with esters and dimer alcohol.

In the higher-carbon-number olefins, some of the olefin isomers are totally unreactive. This unreactive olefin, together with any paraffin present in the feed olefin and paraffin produced in the process, is separated in a stream referred to as *secondary light ends*.

Distillation and Hydrorefining

The crude alcohol is refined, using three distillation columns, to produce four fractions:

- *A light ends stream containing paraffin and unconverted olefin.* This can be recycled in campaigns to the hydroformylation section.
- *A stream containing a small amount of unconverted aldehyde in alcohol.* This is recycled continuously to the hydrogenation stage of the plant.
- *The refined alcohol.* This can be sent directly to storage or hydrorefined.
- *A heavy ends stream that contains the by-product ethers and esters.* This stream can be used as a fuel or sold as a solvent.

In a final stage of the process, the alcohol can be hydrorefined by passing it over HTC 400 nickel catalyst in the presence of hydrogen. The process operates liquid phase at about 100°C and 10 to 30 bar(g). This removes trace levels of unsaturation and any remaining carbonyl.

In the case of some alcohols, the spread of the boiling points of the isomers can be so great that some aldehyde isomers cannot be separated easily from the alcohol by distillation. These "trapped" aldehyde species can be removed easily by hydrorefining.

Single- and Two-Pass Operation

The olefin isomers in any feedstock vary in their hydroformylation reactivity. Their reactivity can be divided into three bands: reactive, moderately reactive, and unreactive. Depending on the particular olefin, there can be different proportions of the three bands. Because the olefin feedstock is usually the most important cost component of the final alcohol, efficient conversion of olefin to alcohol is the most important variable cost parameter.

Feedstocks containing reactive olefins can be processed single-pass by converting about 95 percent of the feed to aldehyde. (The octenes product derived from the IFP "Dimersol" process is an example of such a feedstock.) The light ends from single-pass operation are sent to petrol blending or fuel.

Most olefins contain all three bands. The Johnson Matthey Oxo Alcohols Process has been designed to operate with two passes when using these types of feedstocks to achieve a high olefin efficiency.

In the first pass, the reactive olefin and part of the moderately reactive olefin are converted into aldehyde. This is typically 80 percent conversion of the olefin. (Limiting the conversion to 80 percent minimizes by-product formation.) The unconverted olefin (primary light ends) is recovered and stored as feed for a second pass.

The second-pass hydroformylation is operated under slightly more forcing conditions of temperature and catalyst concentration. Typically, 75 percent of the remaining olefin is converted to aldehyde in the second pass. The light ends collected from the second-pass operation (secondary light ends) are totally unreactive and can be used for petrol blending or fuel.

Because the alcohols from first- and second-pass operations are of different isomeric form, it is necessary to blend the two products to obtain a standard alcohol product.

PROCESS FLOWSHEET

Figure 8.1.3 shows a simplified flow diagram of the basic process.

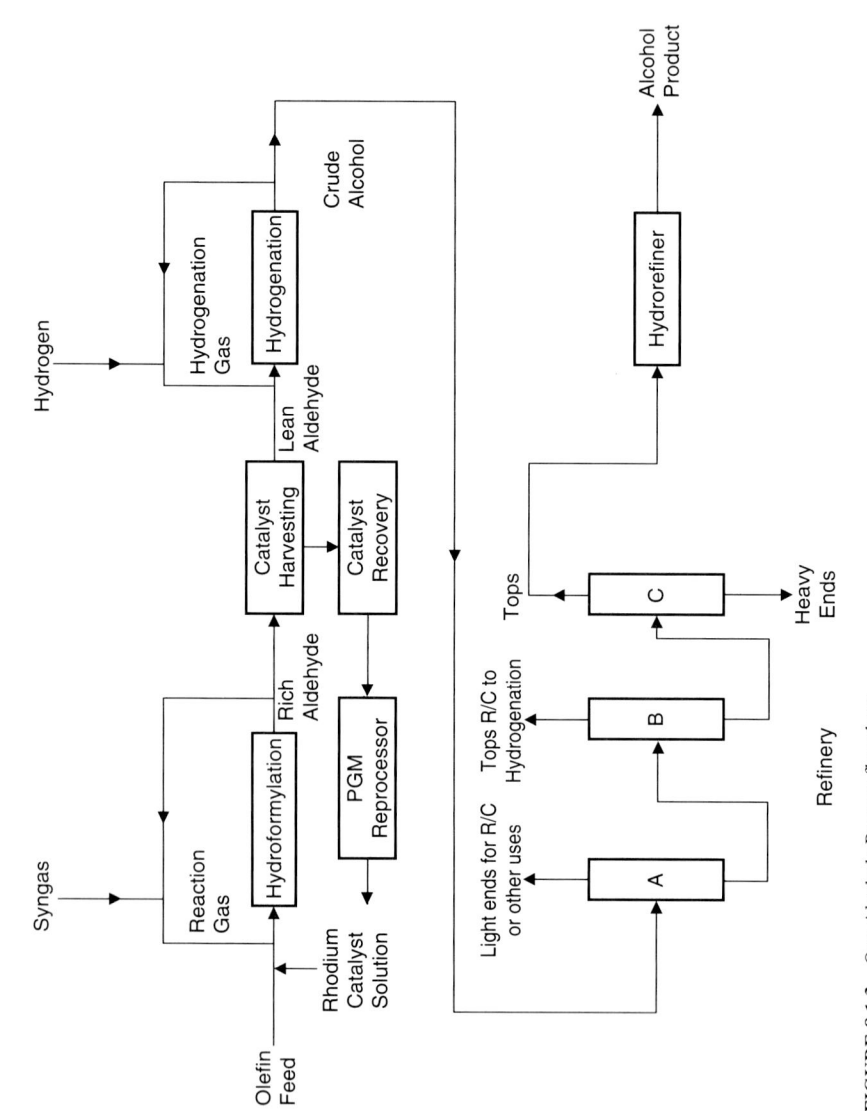

FIGURE 8.1.3 Oxo Alcohols Process flowsheet.

BENEFITS OF THE JOHNSON MATTHEY TECHNOLOGY

Feedstock Efficiency

The olefin feedstock is usually the most important variable cost component in any Johnson Matthey Oxo Alcohols Process. The technology has been developed to give a high conversion of olefin to alcohol with minimum by-product formation.

Feedstock usages for the most commonly produced alcohols (C_8–C_{10}) are in the range 0.87 to 0.93 t/t alcohol. Actual usages depend on the specific alcohol and type of olefin feedstock.

The benefit of this efficient high conversion rate is to minimize the feedstock usage and to eliminate the operating and capital costs associated with workup of by-products, as occurs in the cobalt-catalyzed processes.

Process Flexibility

A further benefit of the technology is the range of alcohols that can be produced from a large number of olefin feedstocks. The technology can be used to manufacture C_7 to C_{15} alcohols from their corresponding olefins (bearing in mind that one carbon atom is added to the carbon chain in the manufacturing process).

The olefin feedstocks can be derived from refinery gases (C_2", C_3" and C_4") using the UOP Polygas process, the IFP "Dimersol" process, or an ethylene growth-type process.

The following conversions can be made:

- Hexene → isoheptanol
- Heptene → iso-octanol
- Diisobutene → nonanol
- Octene → isononanol
- Nonenes → Isodecanol
- Propylene tetramer → tridecanol
- C_8–C_{10} alpha-olefins → C_9–C_{11} alcohols
- C_{12}–C_{14} alpha-olefins → C_{13}–C_{15} alcohols

From olefins, which can be processed single-pass, it is possible to manufacture rhodium-free aldehyde products that are sufficiently pure to be used as a feedstock in downstream processes without intermediate purification. Because of the very efficient and "clean" catalyst removal stage in the process, it is possible to produce aldehyde products that are phosphorus-free, in contrast to processes using a triphenyl phospine (TPP) ligand for catalyst removal. The benefit of this process flexibility is that olefin feedstocks that would be unsuitable for use in other process routes can be used in the Johnson Matthey Oxo Alcohols Process.

Seamless Product Changes

Changing from one product to another usually can be carried out, using a technique known as a *running switch*. The production change is made from one alcohol to another as a continuous operation without a break. The benefit of seamless product changes is to improve plant "occupacity" or, conversely, to reduce the capital cost of the plant for a given capacity.

Low Environmental Impact

The Johnson Matthey Oxo Alcohols Process has a very low environmental impact. There are several reasons for this:

- There is a high conversion of feed olefin to product alcohol.
- There is a high level of recovery of the rhodium catalyst used in the process.
- The spent copper/zinc (17 t/2 yr) and nickel catalysts (10 t/5 yr) are readily accepted for reprocessing by specialist metal processors.
- The ultimate waste streams from the process, secondary light ends and heavy ends, are readily acceptable as fuels or, in the case of the light ends, as re-former feedstock or for solvent use. Both streams are sulfur-free.
- The process has a very low water formation/usage (300 kg/h), and hence wastewater treatment costs are small.
- There are no gaseous emissions in normal operation other than combustion products from fired heaters, and these are sulfur-free.

The benefit of the low environmental impact of the technology is that it is likely to match or exceed local statutory environmental regulations now and well into the future.

FEED SPECIFICATIONS

General

A wide range of olefins—linear or branched, terminal or internal—can be converted to the corresponding aldehyde. In general, feedstocks contain a mixture of isomers and carbon numbers. A low degree of branching facilitates reactivity in the hydroformylation stage, and the distillation range (spread of carbon numbers) must be limited to avoid overlap with the distillation range of the derived alcohol.

Some of the more important impurities are referred to below. These must be kept below specified concentrations either to give a satisfactory level of reactivity in hydroformylation or to avoid premature failure of the hydrogenation catalyst.

Dienes and Peroxides

Dienes attach themselves around the rhodium catalyst ("chelation") and effectively lock it up, hence slowing down the conversion of mono-olefins to aldehydes. The diene is slowly hydrogenated to mono-olefin, and the reaction rate increases. Diene inhibition thus shows itself as an induction period to the reaction.

Diene, which is normally measured as a *maleic anhydride value,* is not normally present in the olefins produced by the "Polygas," "Dimersol," or Ethylene Growth processes. However, it can arise indirectly if the olefin is stored in contact with air so that peroxides are formed. Peroxides reduce reactivity, apparently by breaking down to form dienes. It is normal to inhibit peroxide formation in olefin storage by adding a suitable phenolic antioxidant.

Sulfur

Sulfur compounds can be divided into two groups:

1. Mercaptans, which act as ligands and behave as temporary poisons for the rhodium catalyst, resulting in an induction period. The mercaptan sulfur content of a feedstock should be less than 1 ppm. Mercaptans are not normally present in the more commonly used types of feedstock.
2. Other sulfur compounds are grouped together as *innocuous sulfur*, for which the specification is 50 ppm. While this concentration of sulfur will not affect the hydroformylation reaction, it will reduce the hydrogenation catalyst life.

Sulfur compounds are removed by the hydrogenation catalyst, which is self-protecting against sulfur.

Chlorine

Organic chlorine does not affect the hydroformylation process, but it permanently poisons the hydrogenation catalyst. It also can lead to a reduction of the copper/zinc catalyst pellet strength. The chlorine content must be less than 1 ppm. The octenes specification provided in Table 8.1.1 is given as an example, as is the specification for the derived isononanol in Table 8.1.2.

TABLE 8.1.1 Typical Specification for Octenes Feedstock

Appearance	Clear, free of suspended matter
Specific gravity (15.5/15.5°C)	0.728–0.734
Color (Saybolt)	+20 max
Sulfur, ppm wt/wt	5 max
Chlorine, ppm wt/wt	1 max
Peroxide content, ppm wt/wt	5 max
Olefin content, % vol/vol	98 min
Distillation range (ASTM D1078):	
Initial boiling pt., °C	115 min
Final boiling pt., °C	135 max
5–97%, °C	3 max
C_8 content, % wt/wt	98 min
Total acid number (mg KOH/g)	0.05 max
Residue on evaporation, mg/100 mL	5 max

TABLE 8.1.2 Typical Specification for Isononanol

Appearance	Clear, free of suspended matter
Specific gravity (20°C/20°C)	0.836–0.840
Color (platinum-cobalt)	10 max
Distillation range (ASTM D-1078):	
Initial boiling pt., °C	199 min
Dry pt., °C	210 max
Water, % wt/wt	0.05 max
Carbonyl (as C=O), % wt/wt	0.01 max
Hydroxyl number, mg KOH/g	380–396
Sulfuric acid, Pt-Co color	30 max
Acid value, mg KOH/g	0.01 max

TABLE 8.1.3 Consumptions of Raw Materials and Catalysts

	Ton/ton* INA	Ton/ton* IDA
Octenes	0.88	—
Nonenes	—	0.92
Synthesis gas (50-50)	0.238	0.22
Hydrogen	0.027	0.026
By-product light ends	−0.06	−0.08
By-product heavy ends	−0.06	−0.08
By-product fuel gas	−0.025	−0.026

Note: The consumption of rhodium catalyst, as well as the copper/zinc and nickel catalysts, depends on the nature of the feedstock olefins (e.g., sulfur content and branching).
*Equivalent to pound per pound (lb/lb).

TABLE 8.1.4 Utilities Consumption

Cooling water	2500 m^3/h
Demineralized water or condensate	400 m^3/yr
Intermediate-pressure steam [21 bar(g)]	10000 kg/h
Low-pressure steam [7 bar(g)]	150 kg/h
Plant air	250 Nm^3/h
Instrument air	300 Nm^3/h
Low-pressure nitrogen	750 Nm^3/h
Electricity	3130 kW

PROCESS ECONOMICS

The plant is easy to operate with a small workforce because operational control is relatively simple for a plant of this size. Usages are given in Table 8.1.3 for two of the most widely used alcohols, isononanol (INA) and isodecanol (IDA).

The utilities normally required and their expected maximum consumptions for a 100,000-t/yr INA plant are shown in Table 8.1.4.

The fuel requirements of the process are supplied partly by using the waste fuel gas and the heavy ends streams that are generated by the process. The remaining heat duty requires the following:

Vaporized LPG:
    Startup   450 kg/h
    Normal running   50 kg/h
Fuel oil   300 kg/h

In addition to the preceding, there is an additional minor usage of methanol.

CAPITAL COSTS

The approximate ISBL capital cost for a 100,000 t/yr isononanol plant using the Johnson Matthey Oxo Alcohols Process is US$75 million (year 2000). This cost is inclusive of all engineering and license fees. It should be emphasized that this capital cost estimate has been based on norms applicable to the United Kingdom and will be different depending on location and specific customer requirements.

OPERATIONAL EXPERIENCE

The Johnson Matthey Oxo Alcohols Process (formerly the ICI High Pressure Oxo Process) is based on over 30 years of operational experience and continuous development of the unliganded rhodium-catalyzed process at ICI's Billingham site. The strengths of the process include the flexibility of using a wide range of feedstocks, low operating cost, and very low environmental impact.

From 1961 to 1994, the flexibility of the process was exploited to manufacture a range of alcohols from C_7 to C_{15}, principally for plasticizer but also for lubricant and surfactant applications, using alpha-olefins and olefins derived from polygas processes. The ability to change quickly between different alcohol products was developed to maximize plant output and to allow a quick response to the changing needs of ICI's internal and external customers.

The strengths of the Johnson Matthey Oxo Alcohols Process have been recognized by many international operators. A license has been granted to a major Asian chemical producer. The resulting world-scale INA/IDA plant using the Johnson Matthey Oxo Alcohols Process came on stream in December 2001 and is operating at flowsheet rates (115,000 t/yr).

The catalyst harvesting technology has been licensed to a major European producer for retrofitting into an existing High Pressure Oxo Alcohols Process. This plant came on stream at the end of 2003 and is operating at flowsheet rate.

REFERENCE

1. Patent WO02/20541, "Recovery of metals by incineration of metal-containing basic ion exchange resin," 14.3.2002.

PART 9

PHENOLS AND ACETONE

CHAPTER 9.1
POLIMERI EUROPA CUMENE-PHENOL PROCESSES

Maurizio Ghirardini and Maurizio Tampieri
Polimeri Europa
San Donato Milanese, Italy

INTRODUCTION

The cumene-phenol process was developed in the 1940s, when UOP first introduced the cumene technology to increase the octane number of gasoline components and later BP and Hercules developed a commercial route to phenol and acetone via cumene peroxidation. Since then, cumene has become the major feedstock for the production of phenol, so the cumene hydroperoxidation process currently accounts for about 95 percent of the phenol produced worldwide. Therefore, it is reasonable to consider cumene and phenol processes together for new phenol capacities, as well as to evaluate each technology standing alone to revamp existing plants.

Polimeri Europa (former EniChem) has been running its cumene and phenol units since the early 1960s and has updated them during the years by developing significant improvements in single sections of the plants and also by introducing innovative technologies, as in the case of the cumene process.

CUMENE TECHNOLOGY

UOP developed the cumene process using SPA catalyst in the 1940s. The SPA catalyst is a composite of phosphoric acid and a solid binder material, such as kieselguhr or diatomaceous earth. Until the mid-1990s, the SPA process was the dominant technology for cumene production, with more than 40 plants licensed worldwide.

However, the SPA technology has relevant limitations, mainly related to product purity and yield, due to propylene oligomer and diisopropylbenzene (DIPB) formation, as well as to operational and environmental issues, such as plant corrosion problems and frequent catalyst disposal (usually not less than once a year).

An alternative technology, based on liquid-phase alkylation with AlCl$_3$, originally developed for ethylbenzene production, found some applications for cumene in the 1980s, but the environmental problems that arose with the SPA technology, namely, plant corrosion and catalyst disposal, were even worse.

It was in the late 1980s that, exploiting zeolite-based catalysts, new cumene processes were developed. They have been applied since the mid-1990s and have allowed producers to overcome the above-mentioned environmental issues and to increase substantially the process efficiency.

Polimeri Europa started its cumene plant, based on a proprietary technology and on a proprietary SPA catalyst, in the mid-1960s in Porto Torres, Sardinia. During the 1980s, a cooperation among the research centers of the ENI Group led to the development of a proprietary beta-zeolite-based catalyst (PBE-1) that in 1996 was charged in a reactor of the plant and then in a second reactor 1 year later. At the same time, kinetic studies were completed to develop an alkylation and transalkylation reactor model that was able to fit the industrial results. On the basis of this technical background and the positive industrial results, in 2000 the whole Porto Torres plant was debottlenecked with PBE-1 catalyst from 290 to 400 kilotons per annum (KTA) of cumene just introducing minor modifications, apart from the transalkylation section.

In these last few years, a continuous improvement has been carried out both in catalyst performance and in process optimization so that currently Polimeri Europa can offer one of the best available technologies on the market for cumene production.

Process Chemistry

The alkylation of aromatic hydrocarbons is known as a *Friedel-Crafts reaction*. It is an electrophilic reaction; typical catalysts are Lewis acids (AlCl$_3$, BF$_3$), Brønsted acids (H$_2$SO$_4$, H$_3$PO$_4$, HF), and acidic zeolites.

In the case of cumene, the alkylating agent is propylene, and the substrate is benzene. The reaction is exothermic, rapid, and irreversible. Although all the above-mentioned catalysts are active, AlCl$_3$, H$_3$PO$_4$, and zeolites are preferred commercially.

The reaction is carried out in the presence of excess benzene in order to enhance propylene selectivity by reducing propylene side reactions, such as further alkylation to diisopropylbenzenes (DIPBs) and oligomerization to hexenes, nonenes, or even heavier olefins. While DIPBs may be recovered by transalkylation, the oligomers represent a net loss of propylene. They affect the bromine index, which measures the olefin content in the final product. They also may react with aromatics to form heavier alkylbenzenes such as hexyl- and nonylbenzenes, which, through cracking reactions, lead to the formation of ethylbenzene, butylbenzenes, and biphenyl compounds.

In the presence of a strong acid and excess benzene, DIPBs can be converted back to cumene. The transalkylation reactions are equilibrium-limited, essentially neutral in terms of heat released, and slower than the alkylation reactions. Processes based on zeolite catalysts use transalkylation to improve overall cumene yield by separating the DIPB from the effluent of the bottom of the cumene column and recycling it into a transalkylation reactor filled with a zeolite catalyst, often the same as that used in alkylation. AlCl$_3$ also is active in transalkylation, whereas solid phosphoric acid is not effective as a transalkylation catalyst.

Another relevant by-product is *n*-propylbenzene (NPB), which forms both for nonselective alkylation of benzene with propylene and for cumene isomerization. Since the rate of NPB generation is strongly temperature-dependent, its formation can be controlled kinetically by a proper selection of reaction conditions.

Impurities in the benzene and propylene feedstocks also can lead to the formation of a number of by-products. Propane is the main impurity in the propylene feed, with concentrations typically in the range of 4 to 8 percent for chemical-grade and 30 to 50 percent for refinery-grade propylene. However, propane is inert during alkylation and usually is recovered in the cumene process as fuel or liquefied petroleum gas. Minor propylene impurities include ethylene, butylene, and cyclopropane, which by benzene alkylation may form ethylbenzene, butylbenzenes, and n-propylbenzene, respectively.

Among the impurities in the benzene feed, toluene can lead to the formation of cymenes (isopropyltoluenes), whereas thiophene can be readily alkylated to isopropylthiophene. Some of these by-products are distilled with cumene. Although their concentration is globally less than 500 ppm, they lead to the formation of typical contaminants in the phenol plant. Thus ethylbenzene and n-propylbenzene form aldehydes (acetone contaminants), cymenes lead to cresol formation (phenol impurities), and butylbenzenes are present in purified α-methylstyrene.

Process Features

The Polimeri Europa cumene process is based on PBE-1, a new zeolite-based catalyst specifically designed for the industrial production of cumene and ethylbenzene. The selectivity of PBE-1 catalyst to cumene is higher than that of common zeolites and zeolite-based catalysts. PBE-1 catalyst is equally effective for alkylation and transalkylation of polisoproplbenzenes to cumene. PBE-1 catalyst allows minimal by-product formation, giving high yield and high product purity.

The main properties of the PBE-1 catalyst include

- High activity for alkylation of benzene with propylene
- Low activity for propylene oligomerization, cumene isomerization, and cracking reactions giving product impurities
- High stability to allow for long cycle lengths (over 15 t of cumene per kilogram of catalyst)
- Easy regeneration and long catalyst life (estimated to be over 50 t of cumene per kilogram of catalyst)
- Increase in product purity
- No waste disposal problems
- No corrosion problems
- Ability to transalkylate polyalkylated products to cumene

The cumene process advantages of the PBE-1 catalyst include

- It uses a low-inventory, regenerable catalyst.
- It eliminates problems related to disposal and maintenance due to the presence of free phosphoric acid catalyst.
- It allows for the construction of plants completely of carbon steel with low investment costs.
- It avoids installation of equipment (pumps and drums) for the injection of anticorrosion additives.
- It avoids installation of a bromine index reduction unit.
- It allows low production costs.
- It provides a substantial increase in capacity when an existing cumene unit based on former SPA technology is revamped.

Process Description

Polimeri Europa cumene technology is based on alkylation and transalkylation on fixed catalyst beds of PBE-1 (Polimeri Europa proprietary catalyst). The fresh benzene feeds the pretreatment section (necessary to remove dissolved oxygen and solubility water, possibly containing chlorine) and then enters the distillation section, where the recycle benzene and propylene are pumped into the alkylation section.

The mixture produced in the alkylation section feeds the distillation section, which separates, in three columns, the propane (offgas) to be sent to the fuel gas network, the recycle benzene, the cumene product (purity higher than 99.9 wt %), the heavies, and the exhaust benzene. The last stream is recycled back to the aromatic unit.

The stream of heavies, composed of DIPBs and bottoms, is the feed of the transalkylation section, where a column separates the DIPBs from the bottoms; the DIPBs are fed to the reactor mixed with a stream of benzene coming from the distillation section, and the bottoms can be sent either to an aromatic production plant or to burn. The product of the transalkylation reactor is recycled to the distillation section to separate the cumene produced from the benzene and the not-converted DIPBs.

A simplified block scheme of the process is shown in Fig. 9.1.1. Cumene is produced from the distillation section at very high purity (see Table 9.1.1). The Polimeri Europa cumene process can be fed with a very wide range of propylene purity, ranging from deydro- and refinery- to chemical- and polymer-grade propylene, with a propane content of from 50 to 1 wt %.

Yields and Materials Balance

Table 9.1.2 shows the cumene plant materials balance.

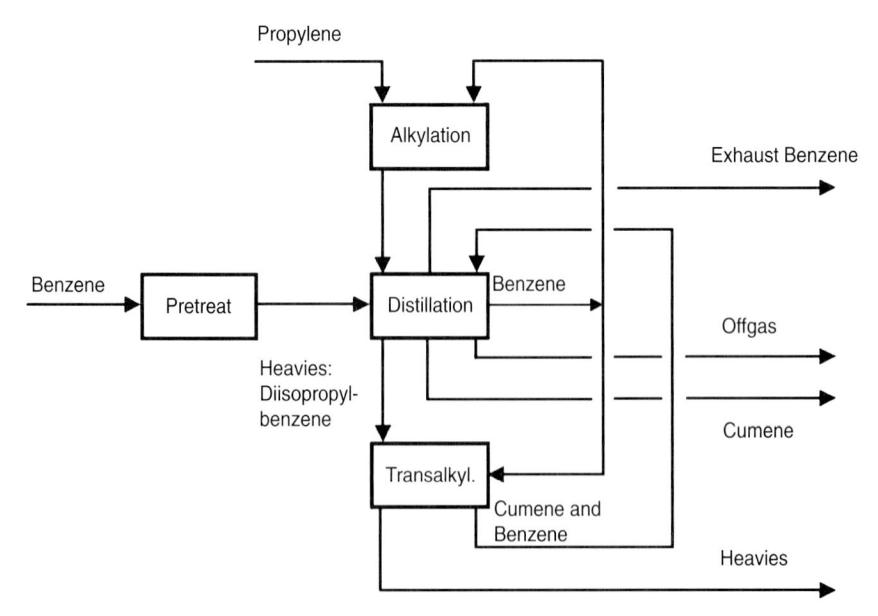

FIGURE 9.1.1 Polimeri Europa cumene process—simplified block scheme.

TABLE 9.1.1 Main Product Specification for Cumene

Cumene	99.94 wt % min
Bromine index	5 max
Ethylbenzene	100 ppmw max
n-Propylbenzene	300 ppmw max
Butylbenzene	100 ppmw max

TABLE 9.1.2 Cumene Plant Materials Balance

Material	Metric ton per metric ton cumene
Feed	
Benzene (as 100%)	0.652
Propylene (as 100%)	0.352
Product	
Cumene	1.000
By-products	
Heavies	0.005
Fuel gas (propane)	*
Hexaust benzene	†

*Function of propylene purity.
†Function of nonaromatics in benzene feed.

TABLE 9.1.3 Cumene Plant Utilities Consumption

Utilities	Per metric ton of cumene
Steam import [32 bar(g)]	1100 kg
Steam export [2.5 bar(g)]	266 kg
Cooling water	52 m^3*
Electricity	15 kWh

*No air coolers are assumed.

Process Economics

Table 9.1.3 reports the cumene plant utilities specific consumption. A 270-KTA ISBL cumene unit is estimated to cost approximately 20 million euros (NWE basis).

Wastes and Emissions

The process produces no liquid or vapor emissions, with the exception of the vacuum pump vents. Spent catalyst can be regenerated several times and at the end of its lifetime can be disposed of in a normal landfill.

PHENOL TECHNOLOGY

Prior to World War II, phenol was produced by recovery from coal tar and from benzene sulfonation and chlorination processes, followed by hydrolysis to phenol with salts as a by-product. In the late 1940s, BP and Hercules developed a route to phenol and acetone via cumene hydroperoxidation that quickly became the most important synthesis process for phenol. In the 1970s, a new process from toluene was industrialized by DSM consisting of toluene catalytic oxidation to benzoic acid and its subsequent oxydecarboxylation to phenol. However, its lower overall selectivity with respect to the cumene process makes it less attractive from a commercial point of view.

Polimeri Europa started a phenol plant in Mantua, Italy, in the 1960s based on cumene hydroperoxidation. In 1970, the plant reached a capacity of 100 KTA, which was doubled in 1979. During the 1980s, several improvements were introduced, such as extractive distillation to improve phenol quality, tar catalytic cracking to enhance the overall process efficiency, and the use of reactor models of the oxidation and cleavage reactions to minimize by-product formation. In 1990, plant capacity was increased to 300 KTA, and several minor improvements were introduced in the last few years to increase plant reliability.

Another phenol plant has been run by Polimeri Europa in Porto Torres, Sardinia, since the late 1960s, basically with the same cumene hydroperoxidation technology. However, in a few sections, technological applications different from those of the Mantua plant were adopted so that it was possible to verify in practice the best attainable efficiency. In the late 1980s, the plant reached a capacity of 100 KTA, and in 2000 it was revamped to 180 KTA, exploiting the oxidation and cleavage reactor models and introducing in each section the more efficient technology taken from the Porto Torres and Mantua experience.

The knowledge so acquired allows Polimeri Europa to make flexible and advanced technological proposals to build new phenol plants and to revamp existing ones. As an example, in 1995, a phenol plant in East Germany was revamped from 60 to 160 KTA under license of Polimeri Europa (EniChem at that time).

Process Chemistry

The phenol process is extremely complex, with several recycles and by-product upgradings to increase the overall efficiency. Three main sections can be defined.

Cumene Oxidation to Cumene Hydroperoxide. The cumene oxidation proceeds in liquid phase with air under pressure at temperatures ranging from 90 to 120°C. It is a free-radical chain reaction initiated by the presence of small amounts of cumene hydroperoxide (CHP). The radicals so generated produce CHP but also several by-products, such as dimethylphenylcarbinol, acetophenone, dicumylperoxide, methanol, and formic and acetic acids. Their formation is enhanced as CHP concentration and temperature increase; therefore, these two parameters must be controlled carefully, and optimization criteria usually are employed to maximize CHP selectivity. Also, the acidity of the mixture must be neutralized, usually with sodium carbonate or sodium hydroxide solutions, to avoid CHP cleavage and phenol formation even in limited extent (tenths of ppm) that would trap the radicals and reduce the oxidation efficiency.

Cumene Hydroperoxide Cleavage to Phenol and Acetone. After the oxidized product has been concentrated to recycle back the unreacted cumene, the resulting CHP solution is cleft in the presence of small amounts of acid catalyst, usually concentrated sulfuric acid, to give phenol and acetone at temperature ranging from

70 to 90°C. Besides the main reaction, other reactions take place, such as cleavage of dicumylperoxide, dehydration of dimethylphenylcarbinol to α-methylstyrene, addition of dimethylphenylcarbinol to phenol to form cumylphenol, and coupling and dehydration of acetone to mesityloxide. The reaction parameters, such as temperature, acid and water concentrations, and residence time, have to be controlled carefully, better with the aid of reactor models, to maximize the reaction efficiency.

Distillation Section. After neutralization of the solution, acetone and phenol are recovered in separate sections, where specific impurities are removed by distillation. α-Methylstyrene is recovered to be sold in part and to be hydrogenated in part to cumene and recycled back to oxidation, whereas acetophenone, cumylphenols, and other heavies may be burned or recovered in part to be sold or even cracked to obtain phenol and α-methylstyrene again.

Process Features

- Overall cumene consumption is only 1.330 kg/kg phenol.
- Proprietary oxidators designed to enhance selectivity to CHP and to minimize reactor volume ensure safer conditions.
- The purity of the phenol is 99.99 percent.
- Less than 150 ppm total gas chromatographic impurities are seen.
- Both phenol and acetone are suitable for bisphenol A–polycarbonate–grade applications.
- The proprietary cleavage reactor design ensures low selectivity to impurities and safer conditions.
- There is a low CHP inventory in the concentration section.
- No external effluent treatment is required apart from water.
- Operating experience of a more than 40 years is made available.

Process Description

Polimeri Europa phenol technology is based on cumene oxidation. The fresh cumene, along with recycled cumene coming from the cumene hydroperoxide (CHP) concentration and α-methylstyrene hydrogenation sections, is fed, together with air, into the cumene oxidation section, where the CHP is produced; the CHP reaches a concentration of about 30 wt % in the exit in a cumene mixture. Small amounts of a basic compound in aqueous solution are fed to the oxidizers to neutralize acidic by-products.

The offgases from the oxidation reactors pass through an offgas treatment section, where the hydrocarbons are recovered, before this stream is sent into the atmosphere. The CHP mixture produced in the oxidation section is sent to the CHP washing section to eliminate traces of basic compounds, and then it is concentrated to about 80 wt % in the CHP concentration section while the overhead product (mainly cumene) is recycled back to the cumene oxidation section.

The CHP concentrated mixture is cleft in the CHP cleavage section, with mineral acid as catalyst, to acetone, phenol, and by-products. The acid cleft product is neutralized in the cleft neutralization section to eliminate the salt content and then is stocked as feed to the distillation section.

The distillation section is composed, after a first raw phenol-acetone separation, of two subsections: the acetone purification section and the phenol purification section. In the acetone section, pure acetone (purity higher than 99.8 wt %) and an aqueous phase are produced. In the phenol section, pure phenol (purity 99.96 wt % on dry product) via extractive distillation and a hydrocarbon-rich cut are produced. The hydrocarbon-rich cut from the main distillation feeds the hydrocarbon recovery section, where a cumene and α-methylstyrene stream and a purge of heavier compounds are produced. Heavies can be sent to burn.

The cumene and α-methylstyrene stream is the feed of the α-methylstyrene hydrogenation section, where the α-methylstyrene is converted into cumene and recycled as raw material to the oxidation section. If required, some α-methylstyrene can be recovered to be sold. Also if required, phenol production can be fed, totally or partially, to a hyperpurification section, which is able to yield a purity higher than 99.99 wt % (polycarbonate grade).

High-capacity phenol plants (higher than 200 KTA) can be implemented with a catalytic cracking section (not included in this description) before the hydrocarbon recovery section in order to minimize the amount of heavies produced.

A simplified block scheme of the process is shown in Fig. 9.1.2, whereas Tables 9.1.4 and 9.1.5 report typical specifications both for acetone and phenol purity.

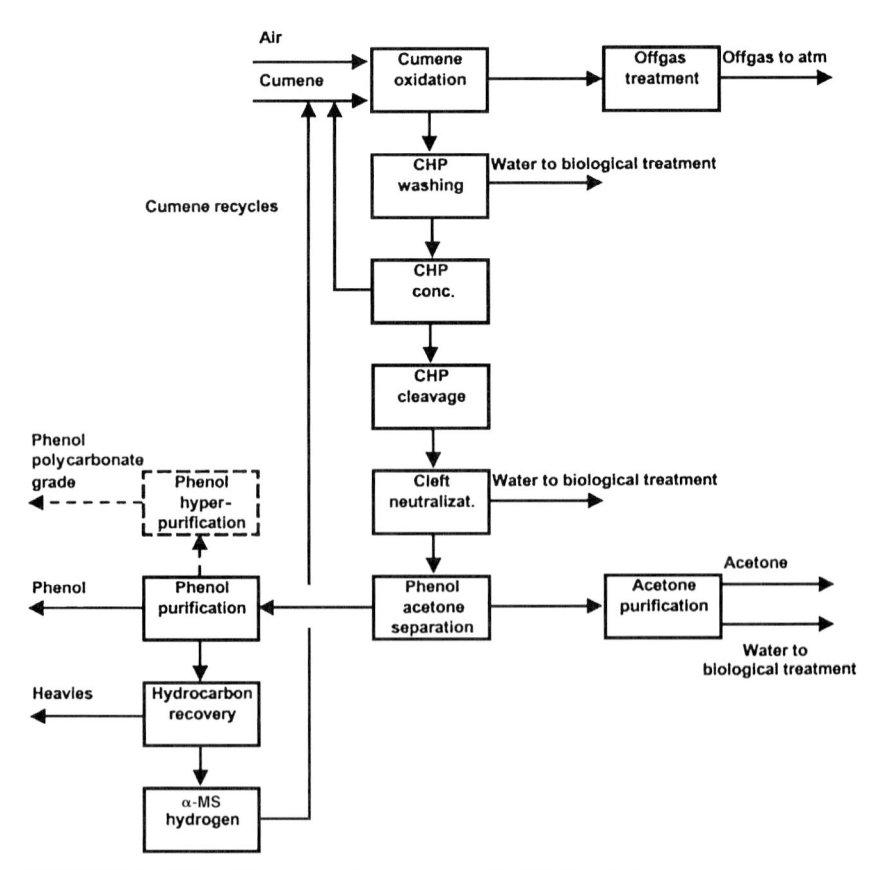

FIGURE 9.1.2 Polimeri Europa phenol process—simplified block scheme.

TABLE 9.1.4 Main Product Specification for Acetone

Acetone	99.8 wt % min (dry)
Water	0.2 wt % max
Benzene	10 ppmw max
Methanol	200 ppmw max
Permanganate test	180 min min

TABLE 9.1.5 Main Product Specification for Phenol

	Pure phenol	Hyperpure phenol
Phenol	99.96 wt % min (dry)	99.99 wt % min (dry)
α-Methylstyrene	100 ppmw max	10 ppmw max
Total organic impurities (cresols not included)	200 ppmw max	60 ppmw max
Cresols	120 ppmw max	60 ppmw max

TABLE 9.1.6 Phenol Plant Materials Balance

Material	Metric ton per metric ton of phenol
Feed	
Cumene (as 100%)	1.330
Minor chemicals	0.020
Product	
Phenol	1.000
By-products	
Acetone	0.626
Heavies	0.053
Process water	0.493

TABLE 9.1.7 Cumene Plant Utilities Consumption

Utilities	Per metric ton of cumene
Steam import [18 bar(g)]	1900 kg*
Steam import [5 bar(g)]	880 kg†
Cooling water	220 m^3
Electricity	180 kWh

*Hyperpure phenol requires 2660 kg.
†Hyperpure phenol requires 570 kg.

Yields and Materials Balance

Table 9.1.6 reports the phenol plant materials balance.

Process Economics

Table 9.1.7 reports the phenol plant utilities specific consumption. A 200-KTA ISBL phenol unit is estimated to cost approximately 95 million euros (NWE basis).

Wastes and Emissions

Wastes:

- Process water with organics and salts (normal biotreatment is suitable)
- Heavy organics that can be sent to a boiler or to an incinerator

Emissions:

- Exhaust air from cumene oxidation section (a recovery treatment of organics contained allows discharge to atmosphere)
- Vacuum pump emissions

CHAPTER 9.2
SUNOCO/UOP PHENOL PROCESS

Robert J. Schmidt
UOP LLC
Des Plaines, Illinois

INTRODUCTION

Using the cumene hydroperoxide route to produce phenol, the Sunoco/UOP process remains the process of choice for the production of phenol. Recent design innovations have improved process yields, economics/costs, and process safety. A number of these innovations are highlighted in this chapter. The Sunoco/UOP process is expected to continue to play a significant role in the future because phenol is a key intermediate feedstock for the growing bisphenol A and phenolic resins markets.

The history of the phenol market has been examined in great detail in the literature with much discussion regarding product usage, emerging markets, and process economics over the past two decades.[1] More than 90 percent of the world's phenol production is currently based on the cumene hydroperoxide route. The Sunoco/UOP (formerly AlliedSignal/UOP) phenol process represents state-of the-art technology for phenol production. It is based on autocatalytic cumene oxidation and dilute acid cleavage (cumene hydroperoxide decomposition) processing routes. Much of the recent advancement in this technology falls along the lines of improved yield, safety, and economy as a result of specific process and design improvements. The focus of this chapter is therefore to review the latest improvements to the Sunoco/UOP phenol technology made over the past decade.

CUMENE PRODUCTION

Cumene is the principal feedstock used in the production of phenol and is produced commercially through the alkylation of benzene with propylene over an acid catalyst. Over the years, many different catalysts have been proposed for this alkylation reaction, including boron trifluoride, hydrogen fluoride, aluminum chloride, and phosphoric acid. Cumene processes were

TABLE 9.2.1 Improved Cumene Product Quality

	Q-Max	SPA
Cumene purity, wt %	99.97	99.92
Butylbenzene, wt ppm	10	100–200
Bromine index	<1	50–100
Total nonaromatics, wt ppm	10	300
Ethylbenzene, wt ppm	20	5–50
n-Propylbenzene, wt ppm	250–300	200–300

TABLE 9.2.2 Common Cumene Product Impurities

Contaminant	Source	Concern in downstream phenol unit
Nonaromatics	Nonaromatics	Forms acids and other by-products, yield loss
Ethylbenzene	Ethylene, ethanol	Forms acetaldehyde, acetone contaminant
n-Propylbenzene	Cyclopropane	Forms propionaldehyde, acetone contaminant
Butylbenzenes	Butylenes, butanols	Resist oxidation, α-methylstyrene contaminant
Cymenes	Toluene, methanol	Form cresols, phenol contaminant
Polyalkylates	Side reaction	Form alkylphenols, yield loss

developed between 1939 and 1945 to meet the demand for high-octane aviation gasoline during World War II.[2,3] In 2000, about 95 percent of cumene demand was used as an intermediate for the production of phenol and acetone. A small percentage is used for the production of α-methylstyrene (AMS). The demand for cumene has risen at an average rate of 2 to 4 percent per year from 1970[4,5] to 2003. This trend is expected to continue through at least 2010.

Currently, almost all cumene is produced commercially by using zeolite-based processes. A limited number of remaining units consist of the fixed-bed, kieselguhr-supported phosphoric acid (SPA) catalyst process developed by UOP and the homogeneous AlCl$_3$ and hydrogen chloride catalyst system developed by Monsanto. Two new processes using zeolite-based catalyst systems were developed in the late 1980s. These included technology based on both a conventional fixed-bed system and catalytic distillation system.[6–9] The current state-of-the-art process for cumene production consists of the liquid-phase Q-Max™ process offered by UOP based on a ß-zeolite catalyst system.[10]

Over the past decade, great progress has been made in improving and optimizing catalyst formulations for improved yields and product quality for cumene alkylation. Table 9.2.1 shows the typical cumene product quality for the latest Q-Max ß-zeolite technology versus the older SPA technology. Note that a product purity of 99.97 wt % can be achieved. This is highly desirable for the downstream phenol process due to the reduced levels of impurity precursors leading to undesirable contaminants in the finished phenol and acetone products.

Table 9.2.2 shows typical cumene feed impurities that lead to undesirable contaminants in the phenol, AMS, and acetone products. The Q-Max process greatly reduces cumene impurities such as butylbenzene and nonaromatics that lead to AMS and phenol impurities. Thus the combination of the Q-Max process and the Sunoco/UOP phenol process results in the highest product quality and highest overall production yields available in the industry.

PHENOL PRODUCTION

The phenol market now spans over three centuries, with many production technologies and product uses having been developed over time. Nineteenth-century phenol was derived mainly from coal tar and used in the production of products such as disinfectants and salicylic acid. In the early twentieth century, phenol was produced primarily using benzene chlorination and sulfonation technologies to support the growing phenolic resin industry after World War I. From the mid-1940s through the mid-1960s, development of the cumene peroxidation process began taking hold such that by about 1990, more than 90 percent of the world's phenol demand was being satisfied by the cumene route. Refinement of the process over the past decade has been driven by several market forces, namely

1. The growing bisphenol A market, driven by the electronics and automotive industries' demand for epoxy and polycarbonate resins
2. The growing phenolic resin market, driven by housing and furniture industries for composite wood resins

Also, because competition in these areas for product differentiation and price has become so keen, phenol technology suppliers have been driven to improve production yield, process safety, and phenol-acetone product quality to remain competitive. The Sunoco/UOP phenol process represents the most advanced technology for phenol production via the cumene peroxidation route. Recent advances in this technology are highlighted below, along with more process details in the following section-by-section analysis.

SUNOCO/UOP CUMENE PEROXIDATION ROUTE TO PHENOL PRODUCTION

The Sunoco/UOP phenol process produces high-purity phenol and acetone by the cumene peroxidation route using oxygen from air. This process features low-pressure oxidation for improved yield and safety, advanced CHP cleavage for high product selectivity, an innovative direct product neutralization process that minimizes product waste, and an improved, low-cost product-recovery scheme. The result is a very low cumene feed consumption ratio of 1.31 wt cumene/wt phenol that is achieved without acetone recycle and without tar cracking. The process also produces an ultrahigh product quality at relatively low capital and operating costs. Extensive commercial experience validates these claims. These improvements are discussed for each of the key major sections of the process.

OVERALL PROCESS DESCRIPTION/CHEMISTRY

The main reactions for phenol and acetone production via cumene peroxidation are shown in Figure 9.2.1. Both reactions are highly exothermic. Oxidation of cumene to cumene hydroperoxide (CHP) proceeds via a free-radical mechanism that is essentially autocatalyzed by CHP. The decomposition reaction is catalyzed by strong mineral acid and is highly selective to phenol and acetone. In practice, the many side reactions that take place simultaneously with the preceding reactions are minimized by optimization of process conditions. Dimethylphenylcarbinol (DMPC) is the main oxidation by-product, and the DMPC/AMS reactions play a significant role in the plant, as shown in Fig. 9.2.2.

Figure 9.2.3 is a block flow diagram showing the major processing steps required to manufacture phenol. The major processing steps include (1) liquid-phase oxidation of cumene to cumene hydroperoxide (CHP), (2) concentration of CHP, (3) acid-catalyzed decomposition of concentrated CHP to phenol and acetone, (4) neutralization of acidic decomposition product, (5) fractionation of the neutralized decomposition product for recovery of acetone, phenol, AMS, and residue, (6) recovery of phenol and the effluent

FIGURE 9.2.1 Main reactions.

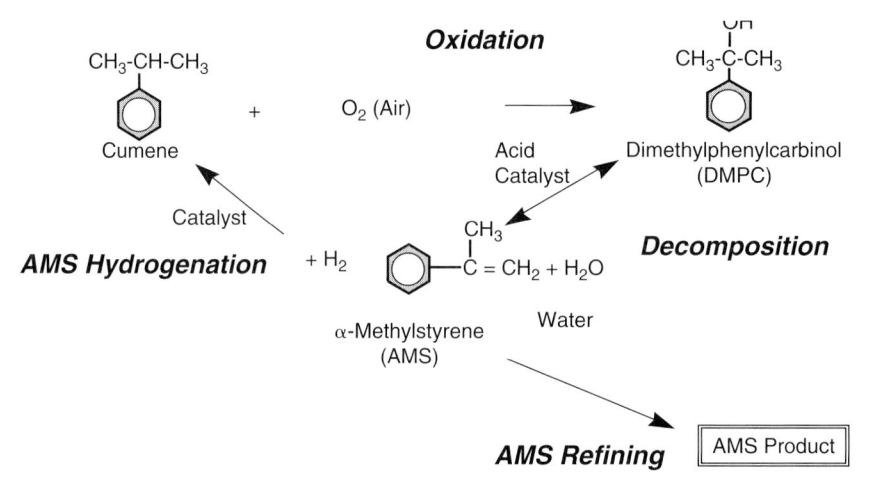

FIGURE 9.2.2 DMPC/AMS reactions.

wastewater via an extraction process to prepare it for further downstream treatment required to meet effluent quality specifications, and (7) hydrogenation of AMS back to cumene for recycling to synthesis or, optionally, refining of AMS for sale as a product.

The details of each of these main processing steps are discussed as follows, highlighting recent technological advances.

PROCESS FLOW AND RECENT TECHNOLOGY ADVANCES

Oxidation Section Process Flow

Figure 9.2.4 shows a typical series flow for a two-oxidation-reactor configuration for the low-pressure technology. Five or more reactors can be used in multiple reaction trains depending on the capacity of the unit, location, and processing objectives. This flexibility allows incremental investment staging to meet growing capacity demands.

The fresh cumene feed is pumped from the oxidation day tank to the combined feed surge drum. Once recycle cumene streams from other sections of the plant are combined, they flow through the feed prewash column, where organic acids are removed by scrubbing

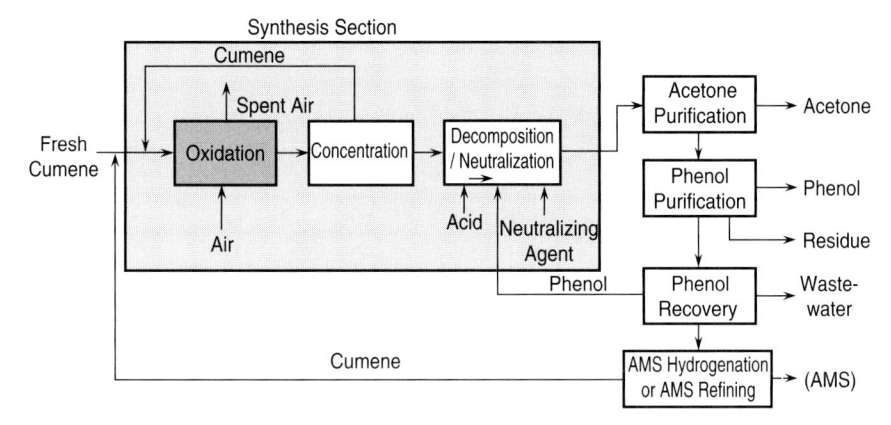

FIGURE 9.2.3 Sunoco/UOP phenol process.

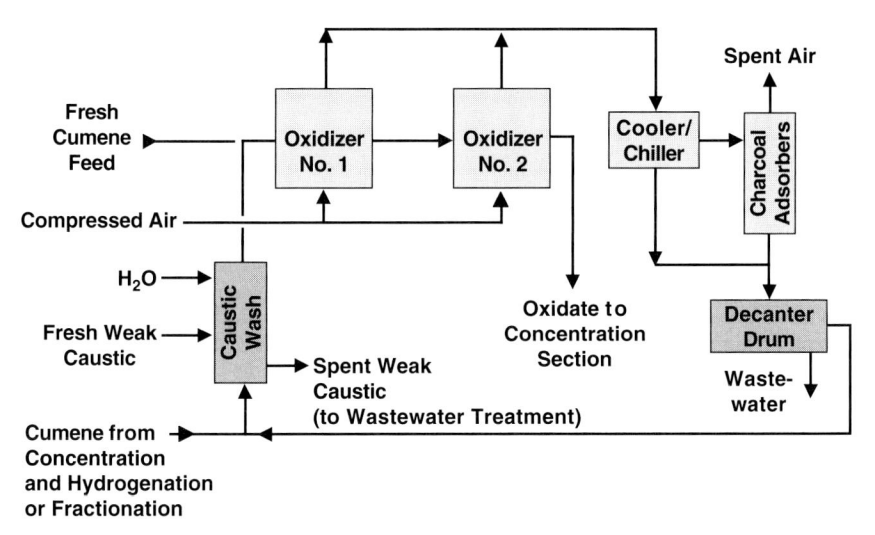

FIGURE 9.2.4 Oxidation section.

with weak caustic and water. The recycle cumene then joins with the fresh cumene feed in the combined feed surge drum. The combined feed is then pumped to oxidizer no. 1.

Cumene is also used for various utility-like purposes throughout the plant. Cumene is sent from the day tank to the phenol recovery section on a batch basis as makeup solvent. It is also used as pump-seal flush in various sections of the plant.

The two oxidizers are located in series with respect to liquid flow but in parallel with respect to air flow. The oxygen requirement for the oxidizers is supplied by atmospheric air. The air is first filtered and then compressed before going into the oxidizers through a sparger. The heat of reaction in the first oxidizer is balanced by adjusting the temperature of the cool cumene feed, so no other cooling is required.

For large phenol units, it is economical to recover the heat of reaction from the second oxidizer by heat integration with the concentration section. The hot oxidizer circulating

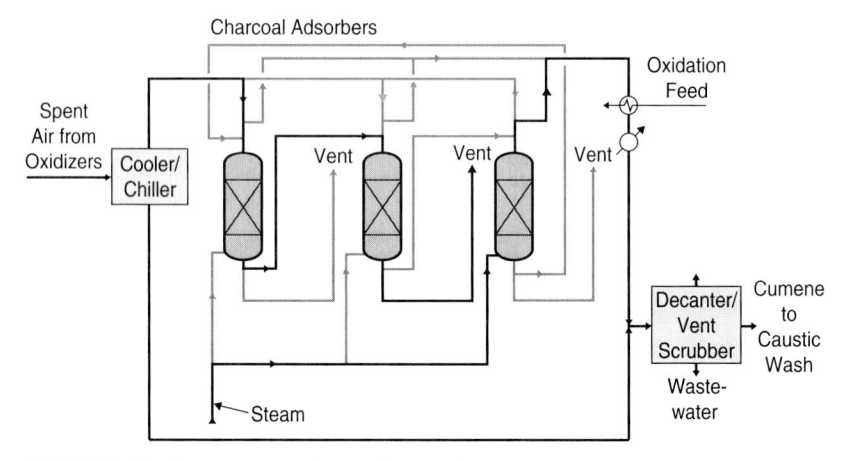

FIGURE 9.2.5 Cumene recovery from oxidation of spent air.

liquid stream is used to supply heating to the preflash column upper vaporizer. The net oxidate from oxidizer no. 2 (effluent from oxidation section) flows directly to the concentration section.

As can be seen in the flow diagram shown in Fig. 9.2.5, the spent air streams from both oxidizers are combined and routed through a water-cooled condenser, a chilled condenser, and an entrainment separator for maximum removal of hydrocarbon and cumene. From the entrainment separator, the air flows to the charcoal absorbers. Two of the adsorbers are always online in series flow while the third one is being regenerated with steam. The cleaned air from the charcoal absorbers is vented to safe atmospheric disposal. A catalytic incinerator usually is not needed to meet emission limits, but one can be provided if regulations stipulate incineration as the emission control method. The cumene collected by the charcoal adsorbers is recovered by desorption with low-pressure steam followed by condensing the steam and decanting the cumene and water phases. The cumene is then recycled to the feed prewash column.

Recent Advances in Oxidation Section Technology

Recent improvements to the oxidation section include (1) the use of high-efficiency charcoal adsorption to recover trace products from spent air, (2) use of an emergency water spray installation and elimination of oxidizer rupture disks, (3) a reduction in oxygen content of vent gas thus reducing air compressor capacity, (4) elimination of the requirement for caustic scrubbing of fresh feed from the zeolitic cumene unit, (5) use of a dilute caustic wash tower that replaces the feed wash mixer/settler system, (6) integration of the decanter with the concentration section vacuum system and elimination of the vent gas scrubber, (7) integration of the feed coalescer into the combined feed surge drum, and (8) use of common spares for oxidizer pumps and emergency coolers. All these improvements serve to reduce capital and operating costs of the process, making it one of the most effective phenol processes available.

With Sunoco/UOP low-pressure oxidation, the air stripping combined with partitioning of the acids to the condensate in the spent air cumene recovery system and the weak caustic scrubbing of the recycle cumene are so effective that no other method of acid removal is required.

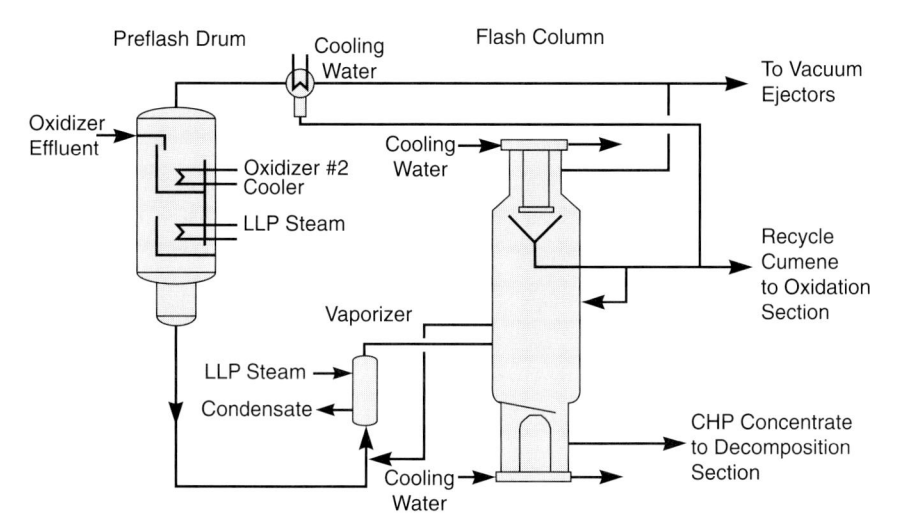

FIGURE 9.2.6 Concentration section.

Concentration Section Process Flow

An example of the typical flow for the concentration section of the process where CHP in the oxidizer reactor effluent is concentrated to a level of 75 to 85 wt % prior to decomposition to phenol and acetone is shown in Fig. 9.2.6. The oxidate from the last oxidizer flows to the concentration section to recover unreacted cumene. For large phenol units, it is economical to use a two-column concentration system, in which the heat of reaction from oxidizer no. 2 and very-low-pressure steam are used to vaporize cumene in the first (preflash) column, reducing the size of the main flash column. The preflash drum and flash column operate under vacuum to minimize the temperature necessary to concentrate the CHP. The vacuum typically is generated by an ejector system. Under vacuum in the preflash drum, cumene vaporizes in the upper vaporizer using heat from the second oxidizer cooler. Additional cumene vaporizes in the lower vaporizer with heat supplied by very-low-pressure steam. Final CHP concentration is achieved in the flash column vaporizer and flash column, both of which operate under deeper vacuum than the preflash drum. The preflash drum bottoms stream flows through the flash-column vaporizer, where additional cumene vaporizes using heat from low-pressure steam.

The CHP content of the flash-column overheads is minimized by rectification in the flash column using either screen trays or packing, whichever is more economical. The flash-column overheads, consisting primarily of cumene, are recycled to the oxidation section via the feed prewash column. The concentrated CHP collects in the integral receiver/cooler at the bottom of the flash column, where it is cooled to a safe temperature. The CHP concentrate from the flash-column bottoms reservoir is then pumped to the decomposition section.

A cumene quench tank is also provided in this section for automatic emergency quenching of various strategic sections of the concentration section, if necessary, to maintain safe operating temperatures in the event of an incipient CHP decomposition excursion.

Recent Advances in Concentration Section Technology

Recent improvements include (1) heat integration with the oxidation section, (2) a two-stage concentration section consisting of preflash and flash column, (3) elimination of overhead receivers, (4) use of a Packinox welded-plate exchanger in the flash-column condenser, and (5) use of power traps instead of level-controlled pumped condensate pots. All these improvements reduce capital costs for the process.

Decomposition Section Process Flow

The decomposition or cleavage section of the process involves the catalytic decomposition of concentrated CHP in the presence of parts per million levels of acid to crude phenol and acetone. The most effective technology for this section is a unique two-step process described in U.S. Patent (USP) no. 4,358,618[11] by Sifniades/Allied Corporation, patented in 1982. The process involves the use of a backmixed reactor section at low temperature/higher contact time for the main CHP decomposition step, followed by a plug-flow dehydration section at higher temperature/short contact time for conversion of dicumylperoxide (DCP) to AMS. The process represents a breakthrough in AMS yield improvement and is currently being used by all licensors as the process of choice for modern high-yield phenol technology.

An example of the most advanced decomposer technology is the process offered by Sunoco/UOP shown in Fig. 9.2.7. It consists of a very simple but elegant drum and loop reactor design where concentrated CHP from the concentration section flows into the decomposer drum, along with a metered amount of water, to maintain optimal reaction conditions in the decomposer recycle loop. Sulfuric acid is injected via injection pumps into the loop to provide the catalyst required for the decomposition of CHP to phenol and acetone. A circulation pump is provided to circulate the content of the decomposer. Sulfuric acid is injected into the circulating stream to such an extent that the decomposition of CHP and dehydration of dimethylphenylcarbinol (DMPC), a key by-product of the oxidation reaction, are precisely controlled. The level of unreacted CHP is monitored via calorimeters, to which part of the acid catalyst flow is routed.

The effluent from the decomposer is pumped to the dehydrator, in which the effluent is heated to a temperature where remaining DMPC is dehydrated and DCP converted to AMS at very high yield. This is a unique advantage of the Sunoco/UOP decomposition technology. The Sunoco/UOP process produces approximately 90 percent AMS yield from DMPC. This also results in higher phenol yield and thus lower cumene consumption and less residue (e.g., tar) formation.

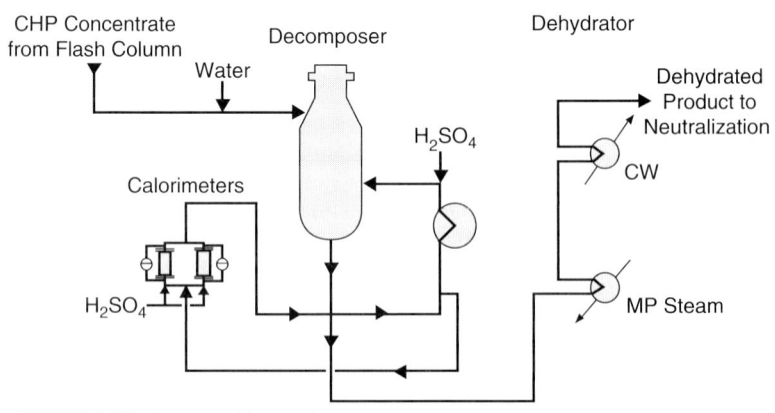

FIGURE 9.2.7 Decomposition section.

Recent Advances in Decomposition Section Technology

In addition to very high yields across the decomposition section, the Sunoco/UOP technology offers the following recent improvements: (1) implementation of advanced process control (APC), (2) a reduction in required recycle rate from 100:1 to 25:1, (3) elimination of water injection tank and pumps, (4) use of acid totes to eliminate the acid tank depending on unit size and client preference, and (5) design of the unit for safe containment in most probable relief situations and elimination of the catch tank. The major advantage of these improvements is reduced capital costs and improved process yields and economics. AMS yields as high as 85 to 90 percent across the decomposition section have been demonstrated, making the Sunoco/UOP technology one of the most selective offerings in the industry today.

Neutralization Section Process Flow and Recent Technology Advances

The acid catalyst that is added in the decomposition section must be neutralized to prevent yield loss due to side reactions and protect against corrosion in the fractionation section. The Sunoco/UOP phenol process uses a novel approach for neutralization: The acid catalyst is neutralized by injecting a stoichiometric amount of a diamine that does not need to be removed from the process, as shown in Fig. 9.2.8.

The diamine salt is soluble through the distillation train and leaves the unit with the residue stream. The diamine salt does not contribute significantly (if at all) to SO_x or NO_x when the residue is used as boiler fuel. The main advantages of direct diamine neutralization over conventional systems are (1) a new/simplified design that is easy to operate and reduces capital cost, (2) use of soluble salts that reduce reboiler fouling and lower maintenance costs, and (3) no need for water addition for neutralization, which in turn lowers wastewater production and reduces distillation utilities. By replacing older ion-exchange-resin technology with the new direct neutralization process, the phenolic wastewater production can be reduced by 45 percent or more at lower capital cost.

Acetone Refining Section Process Flow and Recent Technology Advances

Acetone is a major by-product of the CHP oxidation process for the production of phenol. The overall economics of the process are highly dependent on production of high-quality

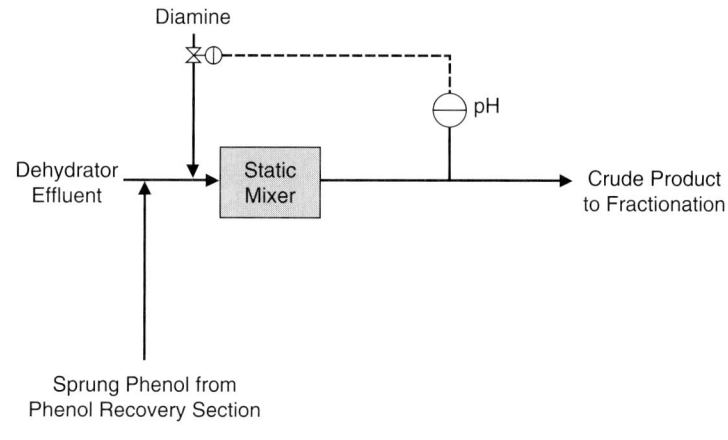

FIGURE 9.2.8 Direct diamine neutralization.

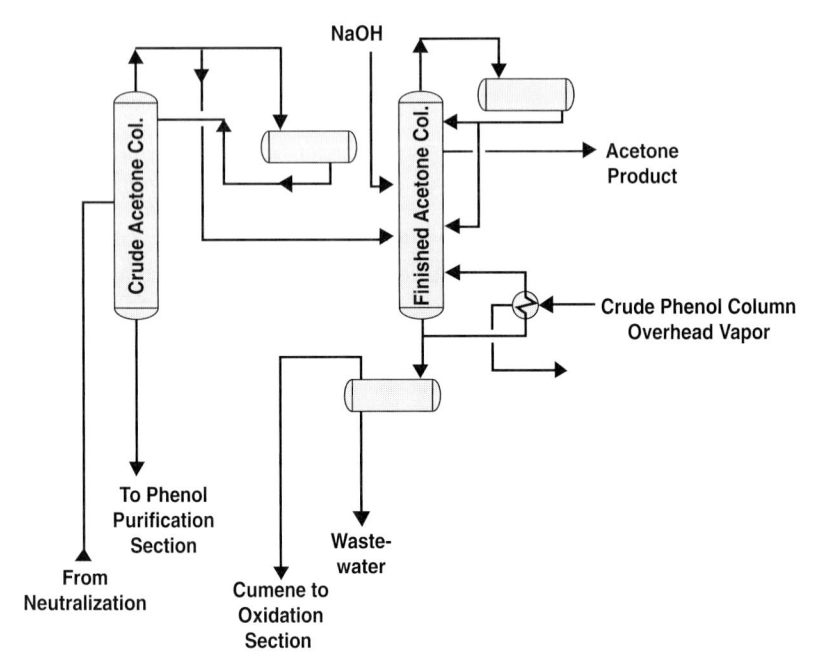

FIGURE 9.2.9 Acetone refining section.

acetone (e.g., 99.7 to 99.9 percent purity) for sales in the solvents and bisphenol A markets. Shown in Fig. 9.2.9 is a typical flowscheme for the Sunoco/UOP process.

Fractionation feed goes from the fractionation feed tank to the crude acetone column. Water is injected as necessary to the bottom of the crude acetone column to increase the volatility of the acetone and maintain the bottom temperature. The overhead of the column, consisting of acetone, water, and some cumene, flows to the finished acetone column (FAC).

The key impurities removed in the FAC are aldehydes, which historically have been analyzed with the permanganate fading test, and water. More recently, as product quality demands for acetone have increased, most phenol producers use gas chromatography (GC) as the definitive method for determining aldehyde content. The permanganate fading test is simply not effective for aldehydes unless the level is in the range of several hundred parts per million or more. Caustic is injected into the FAC column to catalyze the condensation of trace aldehydes. The heavier condensation products are less volatile and leave with the FAC bottoms. High-purity acetone flows by gravity from the FAC sidecut near the top of the column to the acetone product day tank. Table 9.2.3 shows the typical finished high-quality acetone that is achieved in the process. Note that this specification is sufficient to meet or exceed that typically required for bisphenol A manufacture.

Finally, the net bottom stream of the FAC flows to the FAC bottoms drum, where cumene and water are separated. The water goes to the sewer, whereas the cumene is recycled to the oxidation section. Recent improvements for the Sunoco/UOP process include the use of stabbed-in condensers and elimination of overhead receivers, where appropriate, to save capital cost.

Phenol Fractionation and Purification Section Process Flow

Once the crude phenol has been produced, it must be further fractionated to prepare a finished product that is of sufficient purity to meet downstream user specs. An example of the

TABLE 9.2.3 Typical Acetone Product Quality

Color, pt-co	5
Specific gravity, 20/20°C	0.790–0.792
Total GC impurities, wt ppm	250
Water content, wt %	0.3 (can design for 0.2)
Alkalinity, wt ppm as ammonia	< 10
Acidity, as acetic acid, wt %	0.001
Nonvolatile matter, g/100 mL	0.001
Water solubility	Passes
Reducing substances (permanganate test)	Passes (>4 h)
Distillation range, °C	0.5°C from start to dry, including 56.1°C

phenol purification is shown in Fig. 9.2.10 using AMS hydrogenation as a means of recycling the by-product AMS to maximize phenol production.

The bottoms material from the crude acetone column flows to the cumene/AMS column, where cumene and AMS are recovered overhead and sent to the cumene caustic wash column in the phenol recovery section. A chemical agent is injected into the bottom half of the column for the removal of carbonyl impurities such as acetol (α-hydroxyacetone) and mesityl oxide from the phenol. The bottoms from the column are routed through a chemical treatment reactor that provides residence time for the chemical treatment reactions.

The effluent from the chemical treatment reactor flows to the crude phenol column, where the heavy components distill to the bottoms and then flow into the residue stripper column for removal as the net residue by-product. This separate residue stripper column section allows the final stripping of phenol from the residue to be conducted at higher vacuum, which allows both the crude phenol column and residue stripper column to be reboiled with medium-pressure steam. *Thus no high-pressure steam is required for the phenol plant!* The residue product has flow and combustion properties similar to no. 6 fuel oil and typically is charged to a dedicated burner in a boiler furnace.

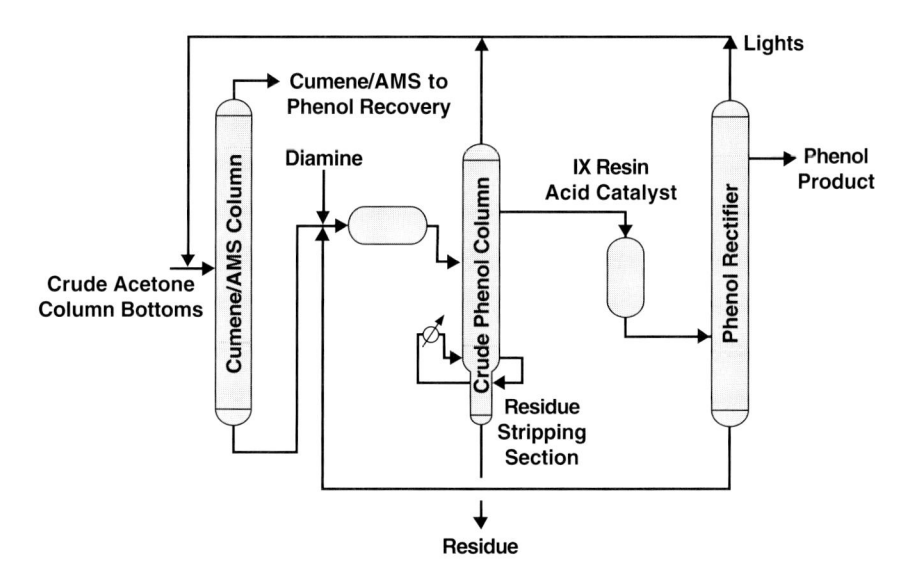

FIGURE 9.2.10 Phenol fractionation and purification scheme for AMS hydrogenation.

The crude phenol column has a top pasteurizing section to remove the small amount of light by-products generated during distillation. The main product from the column is taken off as a sidecut and flows to ion-exchange (IX) resin treaters, in which the IX resin catalyzes conversion of methylbenzofurans (MBF) and residual AMS to high-boiling components. MBF and AMS are otherwise difficult to remove by distillation. The effluent from the IX resin treaters goes to the phenol rectifier, where the heavy components and some phenol are distilled to the bottoms and recycled back to the crude phenol column. The phenol rectifier also has a top pasteurizing section for removal of small amounts of light by-products generated during distillation. Phenol product flows by gravity from the rectifier side cut to storage.

Recent Advances in Phenol Fractionation and Purification Section Technology

Improvements in the phenol fractionation and purification section include (1) replacement of the azeotropic phenol stripper with chemical/resin treating, (2) elimination of the acid neutralizer system, and (3) use of stabbed-in condensers and elimination of overhead receivers where appropriate. The main benefit of these changes has been reduced capital/utilities cost while maintaining the already very high overall phenol product quality. Further, equipment sizes and wastewater have been reduced significantly by implementation of the advanced Sunoco/UOP phenol recovery technology shown in Fig. 9.2.11.

In this section there are two process tanks: a phenolic wastewater tank containing various phenolic wastewaters from various sections of the plant and a sodium phenate tank containing sodium phenate (from phenol and caustic reaction) from other sections of the plant. This section performs several functions: removal of phenol from the cumene-AMS mixture prior to AMS hydrogenation or AMS purification, recovery of phenol from sodium phenate by "springing" with sulfuric acid, and recovery of phenol from the phenolic wastewater.

Phenol would inhibit the oxidation reaction if it were not removed from the cumene and AMS that ultimately will be recycled to the oxidation section. Due to the formation of azeotropes, removal of phenol from cumene-AMS mixture from the column of the same name to the low levels required is not possible via distillation. However, the phenol is readily removed by washing with sodium hydroxide (caustic), forming sodium phenate, which is soluble in water. This operation is conducted in the cumene caustic wash column, which is located in the phenol recovery section. The resulting sodium phenate flows to the sodium phenate tank. The organic phase from the cumene caustic wash column (a cumene-AMS mixture) flows through a sand filter to remove any entrained caustic and then to the AMS hydrogenation or AMS purification section. If there is a market for the AMS that is produced as a side product in the phenol process, it can to be recovered via distillation as a very high purity product for use in the ABS resin or adhesives markets. Figure 9.2.12 shows the typical fractionation scheme needed to recover high-quality AMS. To achieve a high yield of AMS (typically $>$ 93 percent recovery), the crude AMS first must be caustic washed to convert AMS impurities that coboil with the AMS into heavy by-products that can be removed by fractionation. The washed product is then fractionated to remove cumene and phenol, followed by fractionation in the AMS finishing column to remove the heavy by-products, resulting in a high-quality finished AMS product, as shown in Table 9.2.4.

The phenol removed from the cumene/AMS column overheads stream is recovered by mixing the sodium phenate stream from the sodium phenate tank with sulfuric acid, which converts the sodium phenate back to phenol. The phenol then forms a separate organic phase. This operation is commonly referred to as *springing* the phenol. The mixture flows to the phenol recovery unit separator, where the "sprung" phenol is recovered by decanting

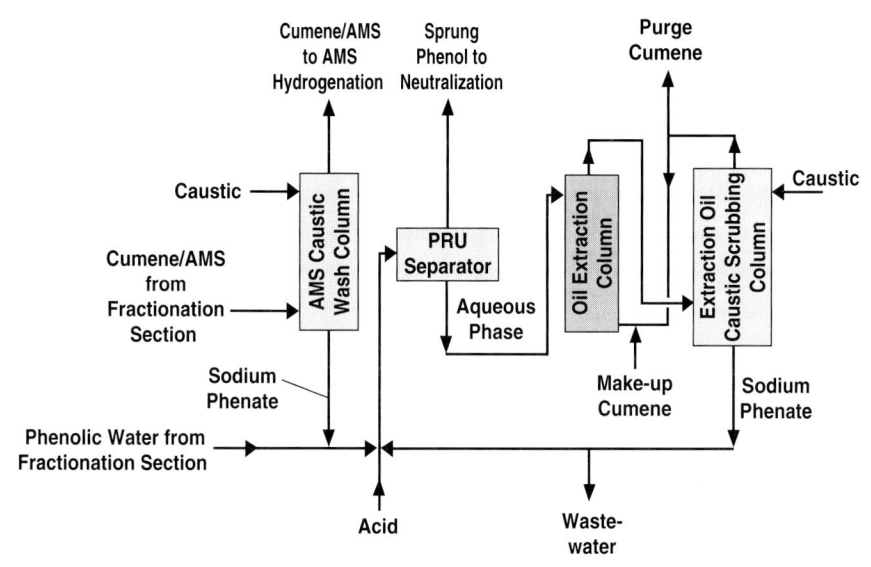

FIGURE 9.2.11 Phenol recovery section.

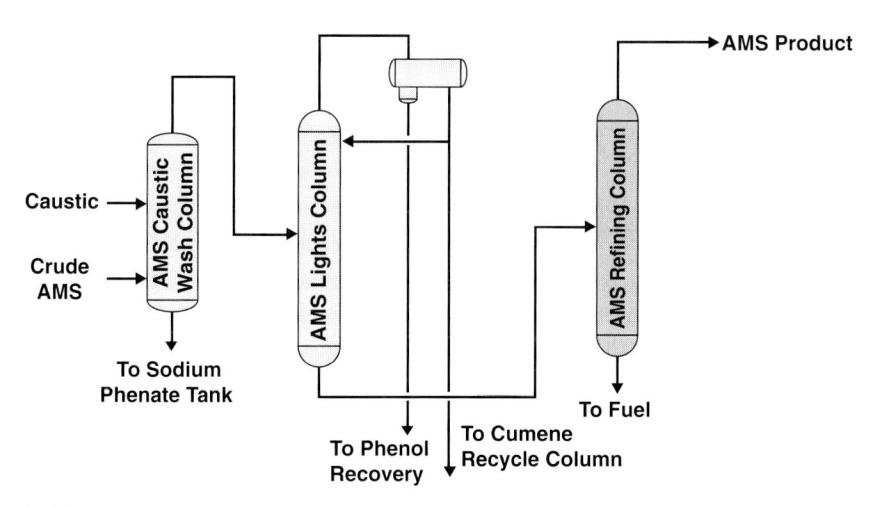

FIGURE 9.2.12 α-Methylstyrene (AMS) refining.

the phenol phase from the top of the vessel and recycling it to the neutralization section, where it joins the decomposition section effluent.

The aqueous phase from the phenol recovery unit combines with phenol-containing water from the phenolic wastewater tank and then flows to the oil extraction column for removal of residual phenol. Cumene is used as the extraction solvent to extract the phenol out of the water in a countercurrent extraction column. The extraction column bottoms are one of the net wastewater streams from the phenol unit. The cumene solvent from the top of the extraction column is then scrubbed of phenol in a countercurrent caustic scrubbing column, similar to the cumene/AMS caustic wash column. The resulting sodium phenate

TABLE 9.2.4 Typical AMS Product Quality

	Typical specifications	Typical quality
Appearance	Clear, water white liquid	Pass
Odor	AMS, characteristic	Pass
Color, pt-co	20, max	10
Specific gravity @ 15.5/15°C	0.912–0.915	0.914
AMS content, wt %	99.0 min	99.2–99.6 (up to 99.8 w/Q-Max)
Sulfur compounds	Free from $H_2S + SO_2$	Pass
Copper corrosion test	No iridescence, gray or black	Pass
Phenol content, ppm	20	5
Inhibitor level, PTBC, ppm	10–20	15
Distillation range, 5 to 95%, °C	0.5, including 165.4°C	0.4, including 165.4

TABLE 9.2.5 Typical Phenol Product Quality

	High-purity phenol	Ultrahigh-purity phenol
Color, pt-co	5	5
Water, wt ppm	<300	<300
Solidification point, °C	40.8	40.8
Total carbonyls, wt ppm	<30	<15
Total GC impurities, wt ppm	<100	<30
Cresols, wt ppm	<50	<25
2-MBF, wt ppm	≤5	≤1

solution then flows to the sodium phenate tank. The lean cumene is recycled to the extraction column.

Use of an IX resin catalyst that is better at controlling total carbonyls and 2-MBF in the phenol results in a very-high-purity phenol product in the range of less than 15 ppm total carbonyls and less than 1 ppm 2-MBF. This is compared with claims of less than 20 ppm total carbonyls and less than 5 ppm 2-MBF for some competing cumene hydroperoxide processes.

Table 9.2.5 shows the typical phenol product quality expected from the process for two different grades of phenol: (1) high purity suitable for most merchant market applications and many bisphenol A markets and (2) ultrahigh purity required by the most demanding bisphenol A and polycarbonate markets. Note that both products can be accommodated in the Sunoco/UOP phenol process by making appropriate adjustments to process operating and fractionation conditions.

AMS Hydrogenation Section Process Flow

The Sunoco/UOP phenol process uses AMS hydrogenation technology developed by Huels. The Huels MSHP™ process is a mild hydrogenation process based on a Pd-containing catalyst system that operates at moderate pressure. The process achieves nearly complete conversion of AMS with very high selectivity to cumene, resulting in a very low overall process cumene-phenol consumption ratio of 1.31 wt/wt. The simple process shown in Fig. 9.2.13 has been demonstrated to operate without significant catalyst deactivation over multiple years.

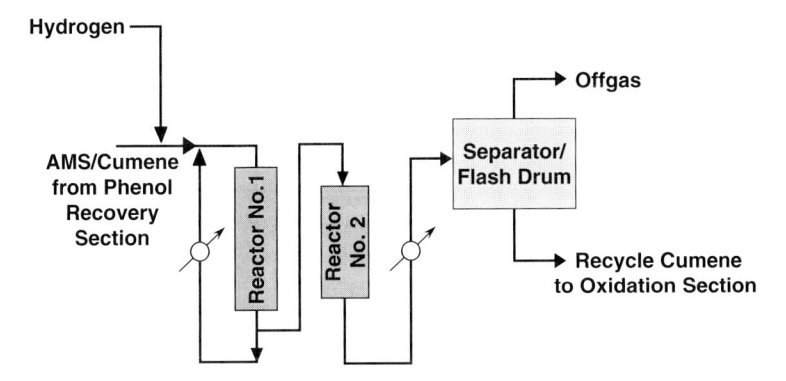

FIGURE 9.2.13 AMS hydrogenation section (Hüels MSHP™ process).

In this section the AMS in the cumene/AMS stream from the phenol recovery section is selectively hydrogenated to cumene by the Huels MSHP process. The fresh cumene/AMS feed is mixed reactor effluent recycle and hydrogen. The combined feed passes through hydrogenation reactor no. 1, where the bulk of the AMS is hydrogenated to cumene. The reaction is highly exothermic. The reactor effluent recycle is cooled before joining with the fresh feed. The net flow goes to hydrogenation reactor no. 2 as a finishing reactor to complete conversion of AMS to cumene. Product from the second reactor goes through the product cooler and then to the product separator. Hydrogen flow is once through, with only a slight excess over stoichiometric. The flow rate of feed hydrogen is regulated based on the flow of excess hydrogen and light gases from the product separator. Dissolved gases that come out of solution when the liquid flashes to low pressure are disengaged in the flash drum. The cumene liquid product is then recycled to oxidation. Recent advances in the Sunoco/UOP/Huels AMS hydrogenation technology include (1) elimination of the recycle hydrogen compressor and (2) elimination of the hydrogen flash drum and pumps. Both of these improvements save capital cost and utilities.

Tar Cracking

Tar cracking of heavy ends produced in the Sunoco/UOP phenol process is no longer required due to the improvement in process yield achieved over the last 10 years. By eliminating tar cracking, along with additional refinements in cumene and phenol fractionation technology, phenol product purity has improved such that total organic impurities have been reduced to a level of less than 30 ppm.

Phenol Process Safety

Safety considerations in the production of phenol and acetone from cumene include design and operating criteria for processing the intermediate CHP. CHP decomposes rapidly to phenol and acetone when exposed to strong acids, even at low temperatures. This reaction is highly exothermic and is the second reaction step in the process. At high temperatures, the rate of CHP decomposition catalyzed by weak acids also would become significant. In addition, CHP reacts with cumene to form dimethylphenyl-carbinol. This reaction occurs to some extent under normal conditions in the oxidation, concentration, and decomposition sections, but the rate becomes significant at higher temperatures. At still higher temperatures, CHP also decomposes thermally to form acetophenone and methane. CHP decomposition catalyzed by weak acids and the thermal CHP reactions would only become significant from a safety standpoint in the event that heat could not be removed. In such a case, the increasing temperature from the heat of reaction would result in a higher reaction rate, creating the potential for an uncontrolled reaction.

Thus the availability of heat exchangers, cooling medium, and pumps for cooling CHP mixtures is critical from a safety standpoint. A significant advantage of the Sunoco/UOP low-pressure oxidation technology in addition to high yields is the very mild operating temperature (e.g., typically 82 to 90°C) required for the process. The lower oxidizer operating temperature translates into a much longer allowable operator response time should intervention be required due to an upset to prevent oxidizer temperatures that are high enough to promote CHP thermal decomposition. The intervention response time may be as long as 24 hours or more to avoid elevated temperatures and high rates of CHP decomposition in the oxidizers. For high pressure processes offered by other licensors, the response time is much shorter, on the order of only a few hours, to prevent accelerated CHP thermal decomposition due to the higher initial process temperature (e.g., typically 95 to 100°C or more).

The Sunoco/UOP phenol process design and operating criteria are based on an industry-accepted 10,000-year probability guideline. Safety provisions include emergency coolers and pumps, reliable power supplies, and reliable cooling water supply. Further backup provisions include the ability to use firewater as once-though cooling water and the ability to use cool cumene to reduce (quench) temperature. For all cooling services designated as critical, if cooling becomes unavailable, it must be possible to reestablish cooling within 20 hours with 99.99 percent certainty. Analysis has shown that meeting this availability criterion typically requires a cooling water supply system with a minimum of three pumps, with two operating normally and the third in standby, and multiple independent power sources for the pumps and cooling tower fan. For all pumping services designated as critical, if pumping becomes unavailable, it must be possible to reestablish pumping within 20 hours with 99.99 percent certainty. Meeting this criterion typically requires multiple sources of power.

Options for multiple independent power sources can include, but are not necessarily limited to, an emergency electric power generator, a steam-driven turbine, a direct-drive engine, or multiple independent external electric power supplies. While the first three options are less reliable than normal electric power, the combination of multiple power sources provides a more robust system than a typical single external electric power supply. For example, if a diesel-powered emergency electric power generator is used, a probability of 97 percent is typical. Emergency generators are less reliable than normal electric power because of the probabilities associated with failure to start, failure to run, and unavailability due to testing and maintenance. Quantitative risk assessment typically is performed to verify the reliability of such systems.

CONCLUSION

A summary of various process and yield improvements is shown in Table 9.2.6 for the low-pressure Sunoco/UOP process versus the competitive high-pressure technology. Results show that higher yields can be achieved at lower cost with improved process safety as a result of mild oxidation reaction conditions and the advanced decomposition and product purification technology employed in the Sunoco/UOP process. With a cumene feed/phenol product consumption ratio (wt/wt) of 1.31, the process represents the highest yield and lowest-cost technology available in the industry today.

TABLE 9.2.6 Low- versus High-Pressure Oxidation

	Sunoco/UOP	High-pressure oxidation
Oxidation outlet pressure, kg/cm^2	0.2	5
Compressor stages	Single	Multiple
Compression utilities	Low	High (or turboexpander)
Number of oxidizers	2	3–6
CHP selectivity	95.7	~93
Tar cracking required	No	Yes
Capital cost including tar cracking	Base	Higher
Cumene/phenol consumption ratio, wt/wt	1.31	>1.35 w/o tar cracking

REFERENCES

1. M. Bentham et al. "Process Improvements for a Changing Phenol Market," *DeWitt Petrochemical Review* Conference, Houston, Texas, March 19–21, 1991.
2. G. Stefanidakis and J. E. Gwyn. In J. J. McKetta and W. A. Cunningham (eds.), *Encyclopedia of Chemical Processing and Design*, Vol. 2. New York: Marcel Dekker, 1977, p. 357.
3. W. Keim and M. Roper. In W. Gerhartz (ed.), *Ullmann's Encyclopedia of Industrial Chemistry*, Vol. A1. Weinheim: VCH Verlagsgesellschaft, 1985, p. 185.
4. Z. Sedaglat-Pour. "Cumene," *CEH Data Summary*, SRI International, Menlo Park, Calif., Mar. 1989, p. 638.5000A.
5. *Chem Mark Rep* 232(10):54, 1987.
6. J. D. Shoemaker and E. M. Jones, Jr. "Cumene by Catalytic Distillation," 1987 NPRA Annual Meeting, March 29–31, 1987.
7. E. M. Jones, Jr., and J. Mawer. "Cumene by Catalytic Distillation," AICHE Meeting, New Orleans, April 6–10, 1986.
8. W. P. Stadig. *Chem. Process.* 50(2):27, 1987.
9. J. D. Shoemaker and E. M. Jones, Jr. *Hydrocarbon Process* 66(6):57, 1987.
10. R. Schmidt, A. Zarchy, and G. Petersen. "New Developments in Cumene and Ethylbenzene. Alkylation," Paper 124b, 1st Annual Aromatic Producers Conference, AIChE Spring Meeting, New Orleans, March 10–14, 2002.
11. U.S. Patent no. 4,358,618 (November 9, 1982), S. Sifniades, A. A. Turnick, F. W. Koff (to Allied Corporation).

CHAPTER 9.3
KBR PHENOL PROCESS

Alan Moore and Ronald Birkhoff
Kellogg Brown & Root, Inc. (KBR)
Houston, Texas

INTRODUCTION

Phenol is a chemical that can be produced by several different processes. Today, the cumene hydroperoxide route is by far the dominant process, but various other routes have been used in the past, and new, novel routes are being investigated.

The cumene (isopropylbenzene)–based phenol process consists of a classic two-for-one overall reaction in which cumene is converted to phenol plus acetone in a fixed ratio. Marketplace demand, unfortunately, seldom requires both phenol and acetone in the same ratio as they are produced. This imbalance has resulted in the development of on-purpose acetone (e.g., acetone from isopropanol) or on-purpose phenol processes without a coproduct (e.g., benzene direct to phenol).

Phenol and acetone are key components in the cumene value chain, as illustrated in Fig. 9.3.1.

The focus of this chapter is on KBR's current phenol technology, which brings together more than 50 years of ongoing innovative process improvements. The KBR phenol technology is the world's leading phenol process, having been licensed to produce more than half the world's phenol capacity. The advantages of the process—high yields, high-purity products, low capital investment, low energy requirements, efficient environmental control, and safe operation—are described in later sections of this chapter.

HISTORY

Commercial synthesis of phenol began in the early part of the twentieth century to produce salicylic acid and phenolic resin/molding compounds. Early processes employed benzene chlorination (both direct and oxychlorination) and sulphonation.

Hercules, Inc., and BP Chemicals, Ltd. (Distillers Co.), began development of the process for producing phenol via the oxidation of cumene in late 1940s. KBR was engaged to scale up the pilot-plant data of Hercules and Distillers for a commercial plant design. The result of this effort was the world's first cumene-to-phenol plant for B. A. Shawinigan in Montreal, Canada, in 1952.

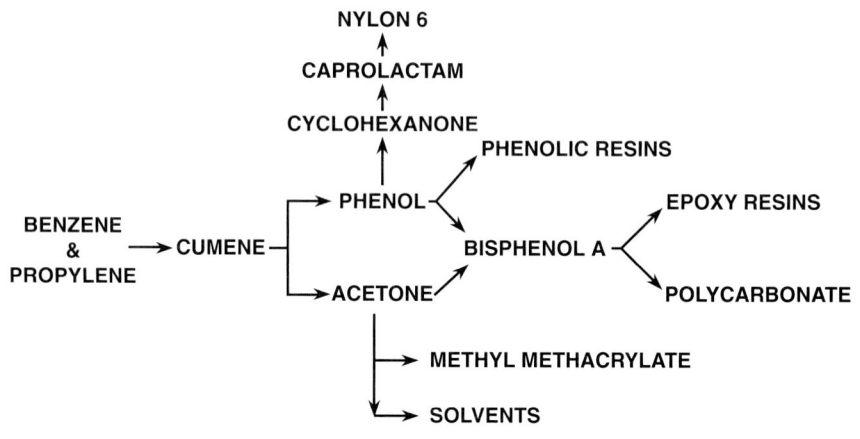

FIGURE 9.3.1 The cumene value chain.

Since its first commercialization in 1952, the Hercules/BP process has become the dominant technology in the world for the production of phenol and acetone. The technology was improved continuously by Hercules, BP/Distillers, and KBR. It has been licensed for more than half the phenol capacity throughout the world, including the largest and most efficient single-train plants. In 1991, KBR became the sole licensor of this technology and now owns and exclusively offers the KBR Phenol Process for worldwide licensing along with a complete range of engineering and construction services. Throughout this long, continuous participation in phenol technology, KBR has played a key role in the development of many phenol plant improvements relating to reduced capital and operating costs, improved plant safety and operability, improved yields, and reduced emissions.

Today, approximately 95 percent of synthetic phenol is produced via the cumene route and about 5 percent via the toluene oxidation process. A small amount of phenol is also recovered from coal gasification and other coal and petroleum processes.

MARKETS

Phenol Demand by Region

Because of complications involved in shipping, phenol production historically has been located near the site of consumption, i.e., in the most industrialized nations. With the industrial manufacturing base shifting to Asia, much of the recent growth in phenol demand and capacity has taken place in this region. Worldwide annual demand is predicted to grow from 6.93 million metric tons in 2001 to 8.2 million metric tons by 2006, with most of this growth (7 percent per year) occurring in Asia (Fig. 9.3.2).

Phenol Demand by Application

The dominant growth market for phenol, in Asia as well as the rest of the world, is bisphenol A (BPA). This growth is fueled in turn by the high growth rate for polycarbonate and to a lesser extent by epoxy resins. BPA's share of the total phenol consumption is predicted to increase from 35 percent in 2001 to 44 percent in 2006 (Fig. 9.3.3). The average annual growth rates, by region and by application, expected for phenol demand for 2001 to 2006 are summarized in Fig. 9.3.4.

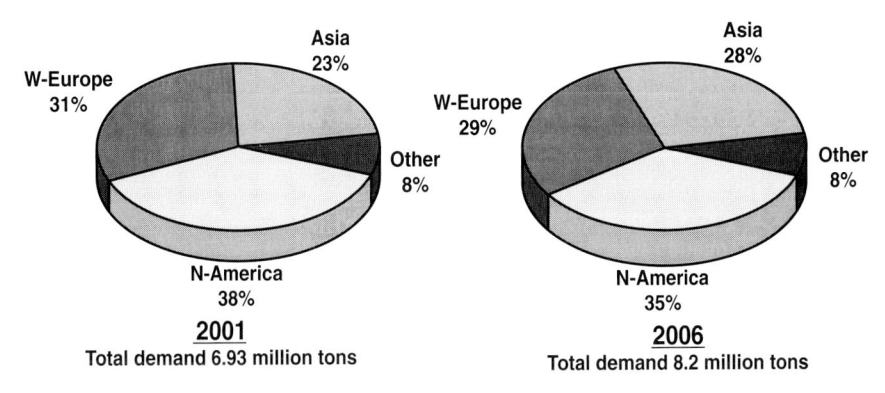

FIGURE 9.3.2 World phenol demand by region, 2001–2006.

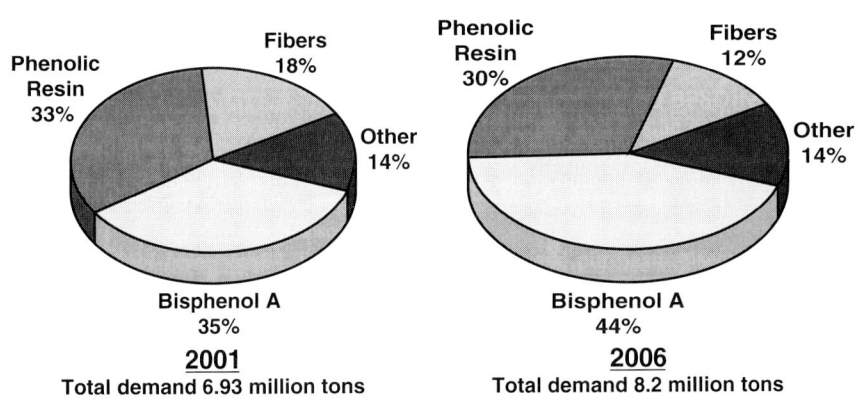

FIGURE 9.3.3 World phenol demand by end-use application, 2001–2006.

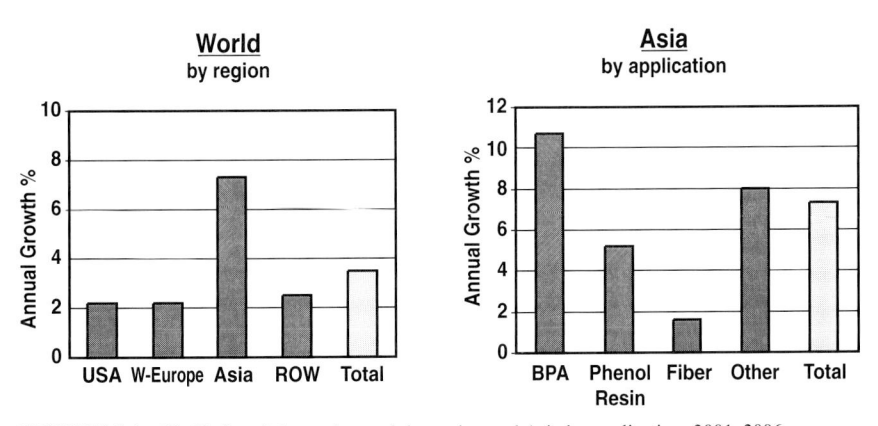

FIGURE 9.3.4 World phenol demand growth by region and Asia by application, 2001–2006.

9.34 PHENOLS AND ACETONE

PROCESS CHEMISTRY

The net overall chemistry of the cumene-based phenol process is the addition of one atom of oxygen to benzene to form phenol and the addition of one atom of oxygen to propylene to form acetone. These reactions are accomplished in two process units. First, cumene (isopropylbenzene) is produced by alkylation of benzene with propylene (in the cumene plant).

Benzene + $CH_3 - CH = CH_2$ → Cumene (with $CH_3 - CH - CH_3$ substituent)
Propylene

In the second step (i.e., in the phenol plant), cumene is converted to phenol and acetone by oxidation with air. The overall chemistry for the phenol unit is

Cumene + O_2 → Phenol + $CH_3 - CO - CH_3$ (Acetone)

In the phenol process, cumene and air are mixed in the oxidation reactor. The principal reaction in the oxidation reactor is the formation of cumene hydroperoxide (CHP) by the reaction of cumene and oxygen.

Cumene + O_2 → Cumene Hydroperoxide

The oxidation of cumene to cumene hydroperoxide is exothermic. The heat of reaction is 36 kcal/g·mol. Additional competing reactions occur, forming dimethyl benzyl alcohol (DMBA) and a small amount of acetophenone (AP).

Cumene + $1/2\ O_2$ → Dimethyl Benzyl Alcohol

Oxidation conditions are selected to minimize the formation of AP. The oxidizer yield of CHP from cumene is greater than 95 percent of theoretical. If α-methylstyrene (AMS) is not recovered as a by-product, the ultimate yield of cumene to phenol will be even higher due to the conversion of DMBA back to cumene via α-methylstyrene hydrogenation. In this case, the ultimate plant yield can exceed 98 percent of theoretical.

$$\text{Cumene} + O_2 \longrightarrow \text{Acetophenone} + \text{Methanol}$$

The function of the cleavage reaction is to decompose cumene hydroperoxide (CHP) into the phenol plant products—phenol and acetone. This is an extremely rapid exothermic reaction with a heat of reaction of 53 kcal/g·mol.

$$\text{Cumene Hydroperoxide} \xrightarrow{\text{acid}} \text{Phenol} + \text{Acetone}$$

The KBR advanced cleavage system is operated at optimal residence time and sulfuric acid catalyst conditions to minimize the formation of undesirable by-products. Under carefully controlled conditions of acid concentration, water and acetone content, temperature, and residence time, the conversion of CHP to phenol and acetone is greater than 99 percent of theoretical.

Additionally, within cleavage, DMBA is dehydrated to AMS.

$$\text{Dimethyl Benzyl Alcohol} \longrightarrow \alpha\text{-Methylstyrene} + \text{Water}$$

The AMS can be recovered and hydrogenated to cumene to improve the overall conversion of cumene to phenol.

$$\alpha\text{-Methylstyrene} + H_2 \longrightarrow \text{Cumene}$$

In this reaction, AMS combines directly with the stoichiometric amount of molecular hydrogen to form cumene, which is recycled to oxidation.

PROCESS DESCRIPTION

A typical phenol plant based on KBR technology consists of two principal process sections. In the front end or reaction area, cumene is oxidized to form cumene hydroperoxide (CHP), which is subsequently cleaved to produce phenol and acetone. The back end or product-recovery section employs fractionation to recover and purify the phenol and acetone products. The net effluent from the front-end reaction section consists of acetone, phenol, water, hydrocarbons such as cumene and AMS, and oxygenated organic compounds that must be separated. These components are highly nonideal and, in many cases, close boiling. KBR has developed a highly integrated system that separates the various materials at a minimum expenditure of energy and yet produces very-high-purity phenol (99.99 percent) and acetone (99.75 percent). The back end of the plant also contains the α-methylstyrene (AMS) recovery, purification, and hydrogenation sections, which permit the by-product AMS to be hydrogenated back to cumene and recycled to the front end for oxidation or, as an alternative, be recovered as a saleable by-product.

A more detailed breakdown of the processing steps employed in the preceding two areas is as follows:

Front end/reaction   *Back end/product recovery*
Oxidation   Acetone fractionation
Cumene concentration   Phenol fractionation and heavy ends removal
Cleavage reaction   AMS hydrogenation and purification
Neutralization

Additional environmental-related process steps, such as dephenolation and a process vent management system, normally are required to meet local regulations. Small quantities of waste streams containing light and heavy organic compounds are recovered and can be burned in an OSBL boiler for the generation of steam. Thus no waste organic streams require disposal attention. These issues are discussed in more detail in the environmental subsection of this chapter.

A block flow diagram for a KBR phenol plant illustrating the process connectivity of the steps just listed is shown in Fig. 9.3.5. A more detailed description of the functions of the various process sections follows.

Oxidation

The first step in the KBR process for producing phenol and acetone from cumene is the formation of the intermediate compound cumene hydroperoxide (CHP). This is accomplished by oxidizing cumene with air in KBR's medium-pressure dry oxidation process, which operates at moderate temperature and pressure conditions. Decomposition reactions proceed simultaneously with the oxidation reaction, with the net effect being a reduction in the yield of phenol and acetone and an increase in the quantity of undesired by-products. Since the rates of by-product formation and CHP decomposition both increase with CHP concentration, there is an optimal CHP concentration that sets the conversion level for the process.

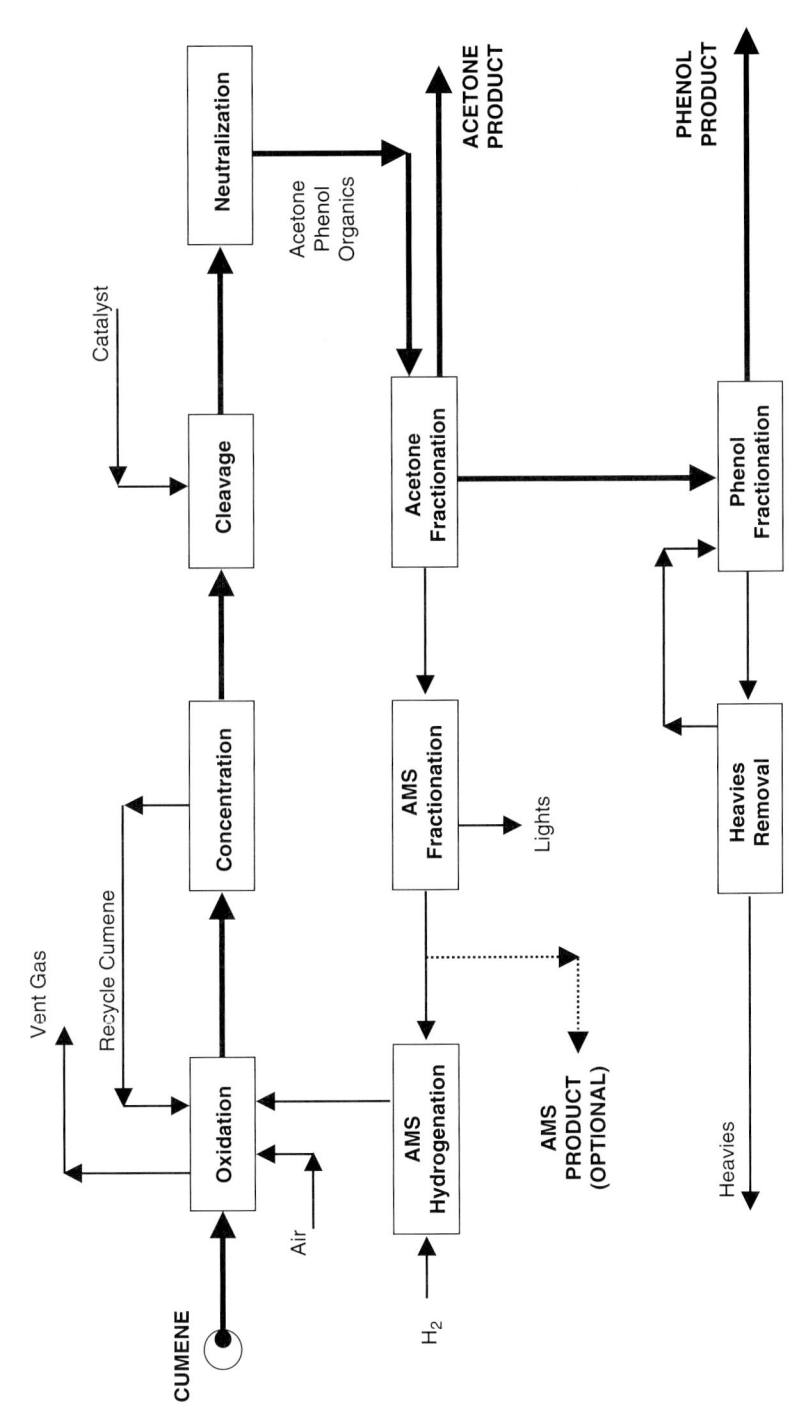

FIGURE 9.3.5 Block flow diagram of the KBR phenol process.

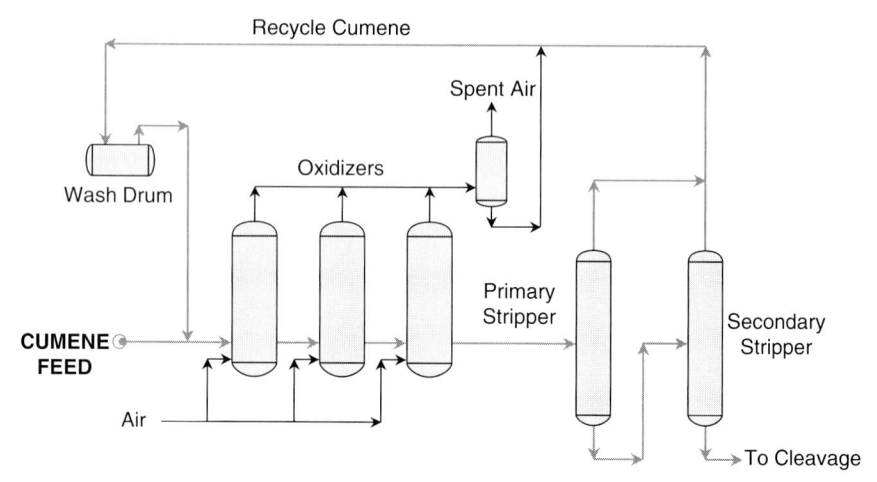

FIGURE 9.3.6 Oxidation and cumene concentration.

Proper operation of the oxidation section requires feed preparation and cumene recovery from the spent air (Fig. 9.3.6). The unit operations involved in these two sections are an integral part of the oxidation area. The recycled cumene may contain trace amounts of acids and phenol produced as a result of thermal decomposition of CHP in the stripping area. Caustic washing is required to remove the trace levels of phenol and/or organic acids, described earlier, which can act as retardants to the oxidation reaction.

Heat is evolved in the oxidizer reactors due to the exothermic nature of the oxidation reaction. A portion of this heat is removed by evaporation of water and organic liquid (primarily cumene) during saturation of the air as it rises through the liquid. The balance of the heat is removed by circulation of the reactor contents through external coolers.

The spent air stream from the oxidizers contains cumene that must be removed for both economic and environmental reasons. In the spent-air treatment section, the cumene is recovered by condensation, and the remaining volatile organic compounds (VOCs) in the spent air are incinerated in order to meet stringent environmental standards.

The yield in oxidation is primarily a function of temperature, CHP concentration, and the number of stages of oxidation. In the KBR process, the oxidation temperatures are between 90 and 98°C, the maximum CHP concentration is normally held to less than 25 percent, and the number of oxidation stages is three or four depending on the plant capacity. This provides an economically optimized design, with oxidation yields of between 95 and 96 percent of theoretical. Higher yields can be achieved but are not considered to be economically optimal.

Cumene Concentration

The oxidizer effluent typically contains about 22 to 28 wt % CHP, with the remainder being unreacted cumene and a small portion of oxidation by-products. Cumene is stripped from the oxidizer effluent by vacuum distillation because of the instability of CHP at high temperatures. Recovered cumene is recycled to the oxidation area. The concentrated oxidate, containing 80 to 85 percent CHP, is fed to cleavage. Figure 9.3.6 is a process schematic illustrating both the cumene oxidation and concentration systems.

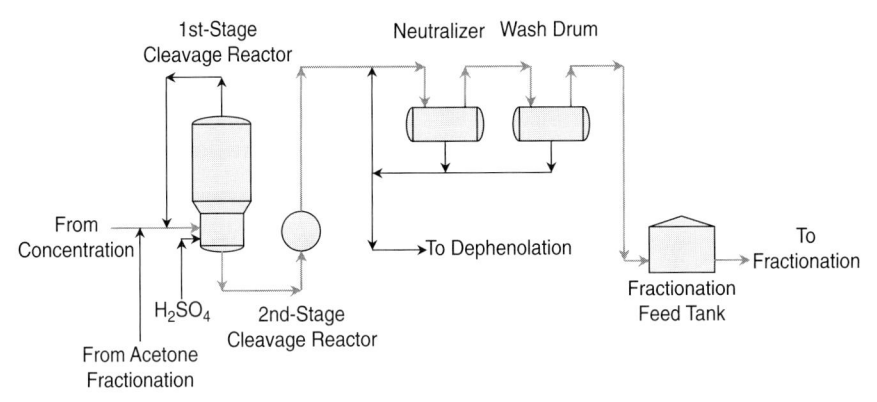

FIGURE 9.3.7 Cleavage reaction and neutralization.

Cleavage Reaction

Phenol and acetone are formed by the acid-catalyzed decomposition of CHP. This step is carried out in a two-stage cleavage system (Fig. 9.3.7) where the operating conditions are set to maximize yields of phenol, acetone, and AMS and minimize formation of by-products. A selectivity of phenol from CHP greater than 99 percent is achieved in the KBR advanced cleavage system.

The concentrated CHP solution from the cumene stripping section is fed to the first cleavage reactor. Recycle acetone is used to control the reaction temperature, as well as minimize the formation of undesirable by-products. In addition, dimethylbenzyl alcohol (DMBA), a by-product of oxidation, is dehydrated to AMS.

Net reactor product is pumped to the second-stage reactor to complete the reaction of CHP and dicumyl peroxide (DCP). The cleavage product is cooled before entering the neutralization section.

Neutralization

The cleavage effluent contains the sulfuric acid used as catalyst for the cleavage reaction plus formic and acetic acids formed by secondary cleavage reactions. To avoid corrosion problems in the downstream equipment, the acids must be extracted and neutralized. This is done in two vessels. The acids are extracted from the organic phase into an aqueous phase and then neutralized with sodium phenate. The cleavage and neutralization sections are shown schematically in Fig. 9.3.7.

Acetone Fractionation

After cleavage and neutralization, the mixed organics are fractionated and purified. The acetone fractionation system serves the purpose of (1) crude separation of acetone and hydrocarbons from phenol and heavies in the neutralizer product and (2) purification of acetone product.

The acetone fractionation train consists of two columns—the crude acetone column and the acetone product column—as illustrated in Fig. 9.3.8. In the crude acetone column, the neutralization product is fractionated to an overhead stream consisting of acetone, water, cumene, AMS, and other light materials and a bottoms stream consisting of phenol and heavier components. The vapor distillate is sent to the acetone product column for ace-

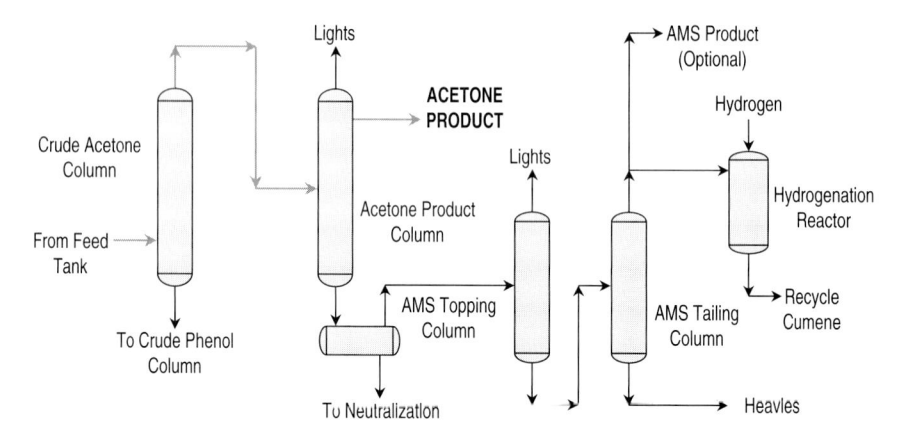

FIGURE 9.3.8 Acetone fractionation and AMS fractionation/hydrogenation.

tone purification. Azeotropes formed in this column are used to improve the efficiency of the separation.

The acetone product column operates under vacuum and produces specification acetone by removing light ends and separating water and hydrocarbons.

Phenol Fractionation and Heavies Removal

The phenol fractionation section is fed with the bottoms of the crude acetone column. This stream consists primarily of phenol, acetophenone, and heavy organics such as cumyl phenol, AMS dimers, and tars. The purpose of the phenol fractionation section is to isolate and purify the phenol product and to recover useful organics for recycle. This is achieved in a three-column fractionation train that includes the crude phenol column, the hydrocarbon removal column, and the phenol finishing column, as shown schematically in Fig. 9.3.9.

In the crude phenol column, the bulk of the phenol is recovered overhead along with virtually all the lighter organics. The bottoms stream of the crude phenol column is fed to the heavies removal column, where phenol is recovered in the overhead stream and recycled. The heavy ends stream is suitable for burning in the steam boiler. The liquid distillate of the crude phenol column is fed to the hydrocarbon removal column (HRC). This column separates hydrocarbons from phenol using water as an azeotropic agent in the top section. Phenol is dehydrated in the bottom section of the column.

The phenol purification resin bed improves phenol purity by converting the carbonyl impurities contained in the crude phenol to heavies that can be removed easily by distillation in the phenol finishing column. Final purification of the phenol product takes place in the phenol finishing column. The distillate purge (containing any residual water) is returned to the HRC. The bottoms stream is recycled to the crude phenol column. The final phenol product contains less than 30 ppm of organic impurities.

AMS Fractionation, Hydrogenation, and Purification

One of the major by-products of the phenol/acetone process is α-methylstyrene (AMS), which is formed by the dehydration of dimethylbenzyl alcohol (DMBA), an oxidation by-product in cleavage. In this section, trace amounts of phenol are removed from the crude

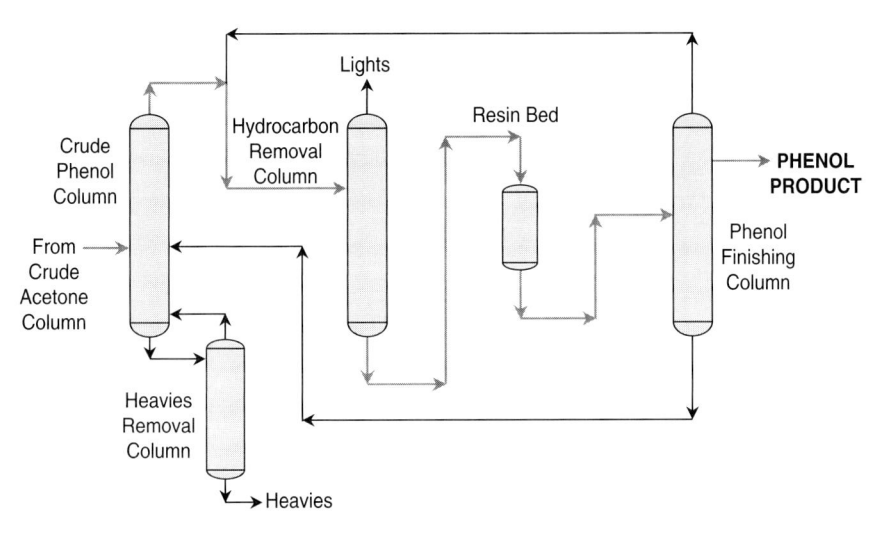

FIGURE 9.3.9 Phenol fractionation.

AMS, which is then fractionated and hydrogenated to cumene for recycle to oxidation, as illustrated in Fig. 9.3.8. Hydrogenation may be carried out over noble metal, nickel, or copper catalysts, generally in a fixed-bed reactor system.

Alternatively, AMS can be recovered as a by-product from the phenol plant, in which case the distillation is designed to produce high-purity AMS, typically 99.5 percent.

Dephenolation

The dephenolation section prepares effluent water for biologic treatment and recovers phenol from water streams for process economic reasons. Phenol is removed from the wastewater using a multistage solvent extraction column. Phenol is first extracted from the aqueous phase into the organic solvent. The aqueous phase is sent offsite to biotreatment. The phenol-rich solvent is regenerated with caustic.

Vent System and Emergency Relief Scrubber

The vent system is designed to collect vapor streams for recovery of phenol, acetone, and hydrocarbons and to ensure that the plant meets environmental regulations by first condensing these materials from the vapor. Any remaining noncondensables in the vent system vapors are then directed to the spent-air incinerator. The vent scrubber system has the capability of quenching and scrubbing relief vapors from the fractionation areas.

FEEDSTOCK AND PRODUCT PROPERTIES

Cumene

Table 9.3.1 shows a typical purchase specification for cumene. In actual practice, the modern zeolite cumene processes produce cumene of 99.95+ percent purity with a Bromine Index of less than 10. KBR's phenol process can handle a wider variety and levels of impurities in the cumene feed, if and when required, while still producing high-purity phenol and acetone.

TABLE 9.3.1 Typical Purchase Specification for Cumene

Property	Specification
Cumene content	99.9 wt % min
Appearance	Water white
Distillation range (including 152.5°C)	1.0°C max
Bromine index	50 max
ortho-Cymene	<15 wt ppm
Total cymenes	<100 wt ppm
Ethyl benzene	<300 wt ppm
n-Propylbenzene	<300 wt ppm
Butylbenzene	<200 wt ppm
Sulfur	<1 wt ppm

TABLE 9.3.2 Typical Phenol Specification

Property	Specification
Appearance	Clear
Purity (dry basis)	99.99 wt % min
Total organic impurities	<30 wt ppm
Water	<200 wt ppm
Color (APHA)	<5
Solidification point (dry material)	40.85°C min
Total carbonyls as mesityloxide	<10 wt ppm
2-Methylbenzofuran (2-MBF)	<2 wt ppm
Sulfuric acid discoloration (SAD) test	95% min
Iron	<0.2 wt ppm

Phenol Product

Table 9.3.2 shows a typical specification for phenol produced by a plant employing KBR technology. This phenol product, with its low carbonyl content, is especially suitable for the production of high-quality BPA for use in the manufacture of top-grade polycarbonates.

Acetone Product

Acetone product purity requirements are also very high due to the demands of downstream applications. The most important specifications are water, permanganate time, and acidity. If these specifications are met, other specifications such as total organic impurities are also met. The acetone finishing column (AFC) is designed to meet a water specification of less than 0.2 wt %. Permanganate time, a measure of aldehyde contained in the acetone product, and the methanol content, an important factor in BPA catalyst life, can be controlled by the operation of the AFC.

Acetone of very high quality is produced via the KBR process and, like phenol noted earlier, is especially suitable for the production of high-quality BPA for use in the manufacture of top-grade polycarbonates. Water content of less than 0.2 percent (wt) and a very high permanganate test time can be attained. Typical acetone quality is shown in Table 9.3.3.

TABLE 9.3.3 Typical Acetone Specification

Property	Specification
Purity	99.75 wt % min
Specific gravity, 20/20°C	0.7910–0.7930
Color, ASTM pt-co	5 max
Permanganate test	>4 h
Water	<0.25 wt %
Acidity, as acetic acid	<10 ppm wt

TABLE 9.3.4 Overall Production Yields

Material	Kilogram per kilogram of phenol
Raw materials	
Cumene	1.308
Hydrogen	0.001
Products	
Phenol	1.000
Acetone	0.617
α-Methylstyrene	*

*α-Methylstyrene is totally hydrogenated and internally recycled.

α-Methylstyrene Product (Optional)

α-Methylstyrene (AMS) can be recovered as a by-product with a typical purity of 99.5 wt percent. AMS is used as a modifier in the manufacture of heat-resistant ABS resins with applications in the automotive and electrical industries.

PRODUCTION YIELDS

The KBR phenol process is highly efficient—1.308 kg cumene consumed per kilogram of phenol produced, with total hydrogenation of the by-product AMS. Table 9.3.4 summarizes the unit consumptions of feedstock and the corresponding products from the phenol plant.

UTILITY REQUIREMENTS

The typical utilities required for a KBR phenol plant are summarized in Table 9.3.5.

PRODUCT STORAGE AND SHIPPING

Proper attention must be paid to the design of product storage tanks and loadout facilities for phenol and acetone to ensure safe operations and prevent contamination of the shipped product. Phenol product is stored in phenolic-lined carbon steel or 304 stainless steel tanks. Due to tighter iron specifications being imposed by downstream users of phenol,

TABLE 9.3.5 Typical Utility Consumptions

Utility	Quantity per ton of phenol
Steam	2.32 ton
Cooling water (10°C rise)	290 m^3
Electricity	170 kWh
Fuel gas	0.06 × 10^6 kcal
Nitrogen	0.51 N·m^3

304 SS is being used more extensively in both storage and shipping vessels. Bisphenol A catalyst and reaction systems are very sensitive to iron content (<0.2 ppm is specified), and the storage tanks and shipping vessels are the greatest potential source of contamination. Also, a concern for phenol product storage tanks and associated product lines is to keep the contents above the phenol freezing temperature of 40.9°C. Product lines typically are traced electrically and insulated. Storage tanks are insulated, traced, and fitted with external plate coils or an internal heating coil.

Acetone product is stored in inorganic zinc-coated carbon steel tanks fitted with an internal floating roof. The product storage tanks are nitrogen blanketed and provided with a vapor recovery system. Separate dedicated pumping systems normally are provided for ship and/or truck loading of the phenol and acetone products. Loading arms with vapor return lines are installed at the respective loading areas.

ENVIRONMENTAL FEATURES

KBR has developed state-of-the-art control technologies to minimize or eliminate releases and emissions to the environment. The following is a brief description of the waste sources and the control systems used.

Gaseous Emissions

KBR employs a centralized vent management system that includes collection and segregation headers, chillers, a scrubber, and other equipment. This system achieves very high recovery efficiencies while minimizing capital investment. The system collects and treats the spent air stream from the cumene oxidizers, as well as process vent streams in the oxidation and cumene concentration areas. The process vents are cooled and refrigerated to effectively remove and recover valuable hydrocarbons before directing to the spent air incinerator.

To meet current U.S. environmental regulations for new plants, a spent-air incinerator is employed to destroy the trace quantities of volatile organic compounds (VOCs) remaining in the spent air and other process vents after overhead cooling and separation of condensates.

Organic Liquid Waste

A small amount of organic waste is produced in the AMS fractionation and phenol fractionation areas of the process. These streams are composed of light and heavy oils separated from the process and typically are blended and used as fuel for steam generation.

Wastewater Effluent

The aqueous effluent streams from the phenol plant are segregated into two categories: process effluents and padded area rainwater runoff. These effluents can be discharged to offsite biotreatment facilities.

Process wastewater streams containing phenol are collected and directed to the dephenolation facilities, where pH is adjusted and the phenol content is reduced substantially before discharge to the biotreatment facilities. The system for removal of phenol from aqueous process waste streams uses a solvent extraction method referred to as *dephenolation*. Dephenolation reduces the phenol content of these wastewaters substantially to allow efficient biologic waste treatment. Caustic used to extract phenol from the solvent is used for cleavage effluent neutralization, thus avoiding the use of additional fresh caustic. KBR offers a very efficient dephenolation system that reduces the phenol concentration in the process wastewater upstream of biotreatment to below 50 ppm.

The phenol plant design includes segregated drainage systems. Two are closed systems handling and recycling process drainage, another handles phenolic effluents from equipment washing, while a fourth collects padded area runoff, such as rain and other surface water. This segregated sewer system minimizes loading on the waste treatment area and improves the overall waste management system efficiency.

SAFETY

KBR's integrated safety system for the phenol process has made an important contribution to the unmatched safety record of KBR-designed plants. In the area of oxidation, concentration, and cleavage, the design takes into account the highly reactive nature of CHP. A key KBR contribution in the area of plant safety has been the development of a comprehensive system for handling CHP, including CHP production, concentration, transfer, and cleavage.

Oxidation Area Safety

KBR's oxidation system offers numerous safety advantages, namely

- The oxidizer relief valve system is designed to handle thermal decomposition should all other precautions fail. The discharges from the individual oxidizer reliefs are directed to a blowdown drum for safe emergency discharge.
- Oxidation employs small oxidizer volumes and low CHP concentrations. Thus there is less physical volume of CHP.
- Specialized controls are provided to maintain operating temperature.
- Temperatures are monitored by a redundant system across each oxidizer. External cooling with variable cooling water flow rate ensures accurate control of each oxidizer's temperature.
- Special interlock control systems are incorporated that automatically initiate safe shutdown procedures in the case of upsets.
- Critical temperatures, pressures, levels, and flows are monitored continuously across the oxidizers and the spent-air system. Alarm conditions automatically initiate the proper safe shutdown procedure, which may include shutdown of cumene feed and oxidizer air and depressuring of the oxidizers.

Concentration Area Safety

As CHP is concentrated, the energy that could be released by the decomposition of a given volume of fluid increases. For increased safety in the concentration area, KBR employs these features in the design:

- Gravity-flow design is employed in the concentrators to minimize the number of concentrated CHP pumps.
- Pumps handling CHP are provided with minimum flow circulation lines to prevent deadheading of the pumps.
- All lines containing concentrated CHP are designed without pockets or deadlegs. Startup and recycle lines containing concentrated CHP are drained or continuously flushed to prevent stagnant holdup of CHP.

Cleavage Safety

The advanced cleavage reactor design employed by KBR allows the highly exothermic decomposition of CHP to phenol and acetone to be carried out safely and under control. Safety considerations of the reactor design include

- Heat of reaction is removed by boiling and refluxing acetone.
- Low liquid inventory and low residual CHP levels are maintained.
- Automatic proprietary shutdown devices safely stop flow into the reactor before serious concentrations of CHP can accumulate.

OPERATING ECONOMICS

Advancements in KBR's phenol technology have improved the overall operating economics for phenol/acetone production significantly. A historical perspective on these technology improvements is given in Fig. 9.3.10, which shows the change in selected parameters from the first cumene-based phenol plant engineered by KBR in the early 1950s to the modern day. The overall process yield and product purity have been improved while reducing energy requirements and plant emissions. While it is difficult to directly compare 1952 costs with 2004 costs, the reduction in the number of equipment items by 40 percent implies a reduction in capital cost as well.

The challenges for the phenol plant designer have been to reduce the total cost of production of phenol (both variable costs and fixed costs) while improving safety, operability, and product (phenol and acetone) purity and reducing emissions to the environment. KBR's ongoing program for improving the technology has, for example, reduced the phenol production cost by nearly 15 percent (equivalent to $85/metric ton) over the last 10 years, as demonstrated in Fig. 9.3.11.

INVESTMENT/ECONOMIES OF SCALE

Economies of scale also can reduce overall investment per unit ton of production significantly and hence the production cost. The first cumene-based phenol plant for B.A. Shawinigan in Montreal, Canada, had a capacity of 6000 metric tons per year. Current

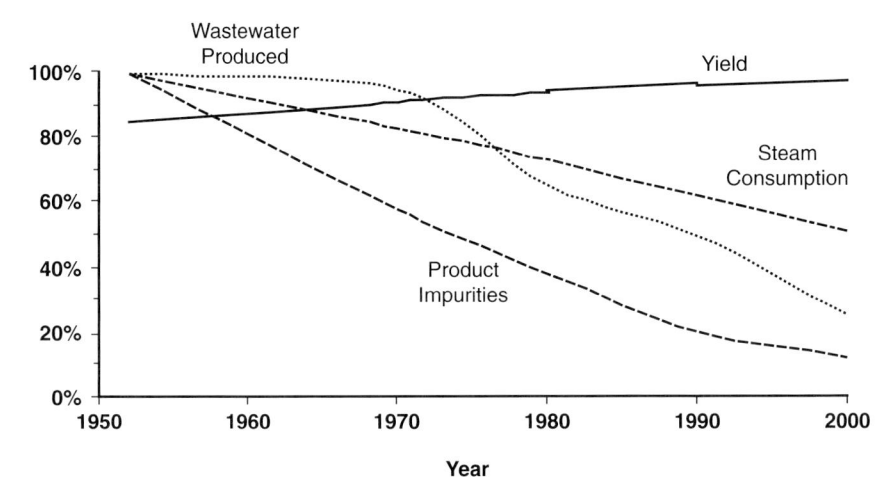

FIGURE 9.3.10 KBR technology improvements over 50 years.

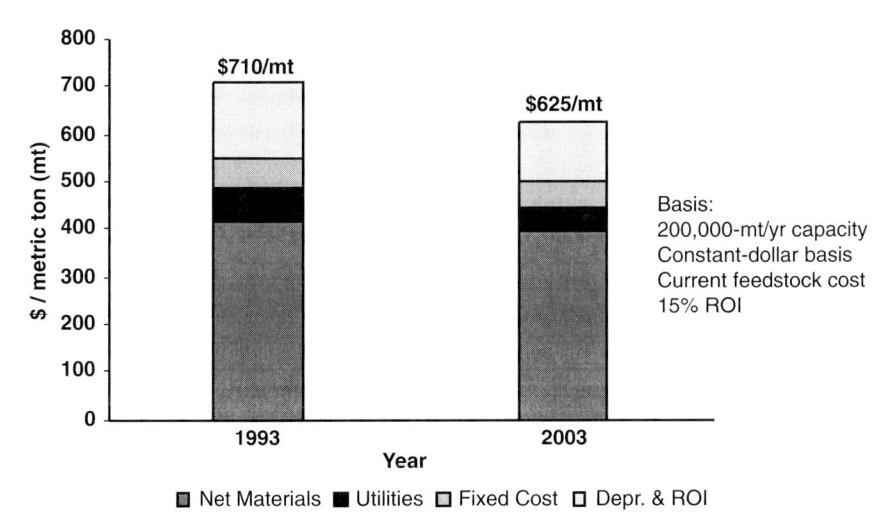

FIGURE 9.3.11 Phenol cost of production—effect of 10 years of technology improvements.

world-scale phenol plants have production capacities in excess of 400,000 metric tons per year—a near 70-fold increase in capacity in 50 years. The growth in typical world-scale plant sizes is shown in Fig. 9.3.12.

The impact of economies of scale on the production cost of phenol can be seen clearly in Fig. 9.3.13, which shows the production cost elements for three identical plant scenarios except for varying plant capacities of 100, 200, and 400 KTA. The "Net Materials" element includes the net impact of feedstock (cumene) consumption and buying price and acetone production and selling price. Variation in either of these prices can have a significant impact on the net phenol production cost, as illustrated in Fig. 9.3.14.

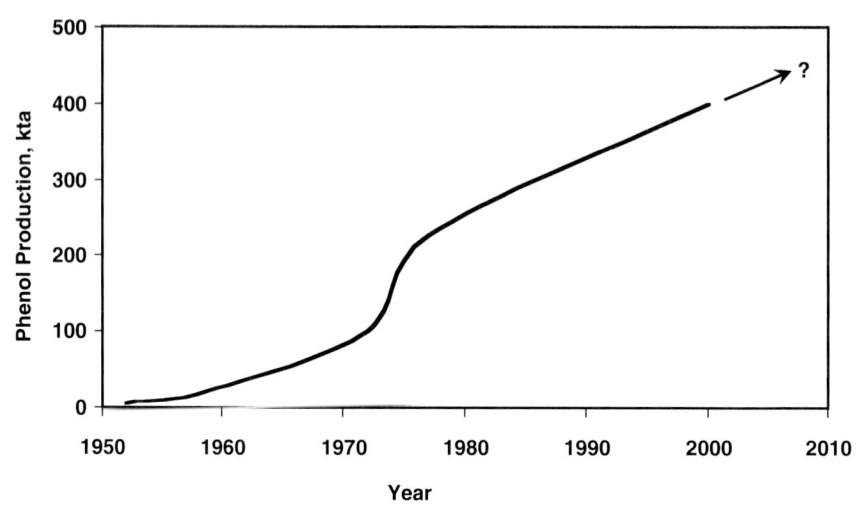

FIGURE 9.3.12 Typical world-scale phenol plant size.

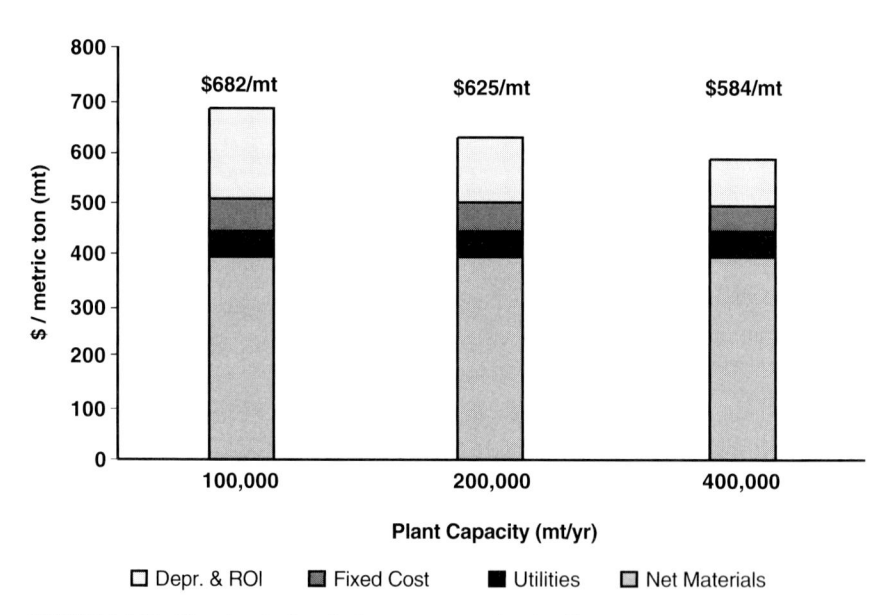

FIGURE 9.3.13 Phenol cost of production—economy of scale effect.

ACETONE NETBACK

The calculated phenol cost of production or the profitability of the phenol plant is strongly affected by the price received for the coproduct acetone. This has become an especially critical factor in the phenol production business in recent years. The amount of acetone

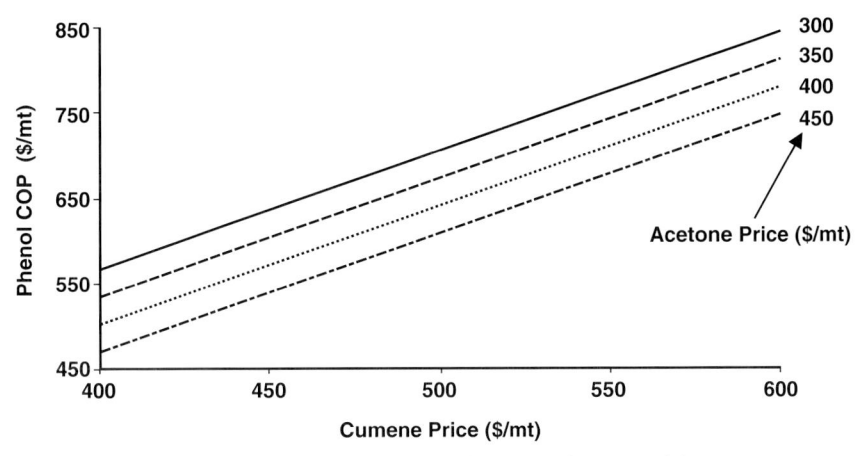

FIGURE 9.3.14 Phenol cost of production—impact of cumene and acetone pricing.

coproduced with phenol in the cumene oxidation process is fixed at 1 mole of acetone to 1 mole of (a weight relationship of 620 tons of acetone per 1000 tons of phenol. The growth rate of the phenol market is approximately 60 percent higher than that of acetone. Since plants are operated to meet the phenol market demand, the corresponding growth in acetone supply has resulted in excess acetone in the marketplace and a sharp reduction in the selling price. Today, in 2004, acetone is selling well below the price of propylene, from which it is made. This reduction in coproduct revenue results in an increase in the calculated cost of production for phenol, as illustrated in Fig. 9.3.14.

TECHNOLOGY ADVANTAGES

The key advantages of the KBR phenol process include

- High phenol and acetone yields, resulting from high-efficiency oxidation (95+ percent), advanced cleavage, and a highly integrated recovery and dephenolation system. Cumene consumption is reduced to less than 1.31 kg cumene per kilogram of phenol produced.
- High-quality phenol and acetone, meeting the stringent requirements of the most demanding bisphenol A manufacturers, are efficiently produced via this process.
- Low-energy-consumption phenol plants were pioneered by KBR in the early 1970s. Since then, KBR has continued to improve and refine the heat integration system to further reduce energy consumption.
- KBR-designed phenol plants have an unmatched record of safety, and KBR continues to develop new and improved safety systems, taking advantage of the continuing advances in electronics and computers.
- KBR-designed phenol plants remain at the forefront of pollution prevention, targeting the lowest emissions in the industry. Environmental controls and pollution prevention systems are a critical part of phenol plant design.
- Low annual maintenance costs.

BIBLIOGRAPHY

J. Wallace and R. Birkhoff, Phenol technology for the 21st century, 1st Asian Petrochemicals Conference, Taiwan, May 2001; also 3rd European Petrochemicals Technology Conference, Vienna, June 2001.

J. Wallace, Phenol: Past, present and future, 12th KBR Phenol Conference, Shanghai, 2002.

Nexant (Chem Systems), PERP Report, Phenol/Acetone/Cumene, 01/02-2, September 2002.

J. Wallace, Phenol, entry in *Kirk-Othmer Encyclopedia of Chemical Technology*, 4th ed. New York: Wiley, 1998.

J. Wallace, Integrated cumene and phenol facilities, WGI Technology Conference, Boston, Mass., October 2002.

M. Weber, O. Gerlich, et al., *U.S. Patent 6,307,112*, October 13, 2001.

A. P. Moore and J. Wallace, Integration of cumene, phenol and BPA plants, Indonesian Conference of Aromatics, November 1995

H. E. Gimpel, A. P. Moore, M. J. Van Sickels, and J. Wallace, *Phenol and Bisphenol A Plants*. Houston, Tex.: DeWitt Petrochemical Review, March 1995.

CHAPTER 9.4
QBIS™ PROCESS FOR HIGH-PURITY BISPHENOL A

Ed Fraini, Don West, and George Mignin
Dow Chemical Company
Midland, Michigan

OVERVIEW

Bisphenol A (BPA) is the basic monomer for the production of two major polymers: epoxy resins and polycarbonates. As such, it is the fastest growing end user of phenol and acetone, with a projected worldwide growth rate of 7 percent per year.

Dow's High-Purity Bisphenol A QBIS™ technology is an advanced process producing BPA suitable for polycarbonate and epoxy production at a low manufacturing cost. Key advantages of the Dow High-Purity Bisphenol A QBIS™ technology are as follows.

High Selectivity to BPA

This process employs a proprietary QCAT™ resin catalyst system that has demonstrated capability to consistently achieve *greater than 98 percent* efficiency of raw materials utilization. The QCAT catalyst used in the main reactors is a unique catalyst providing high asset utilization and long life. The proprietary QCAT catalyst used in the rearrangement reactor has been designed for improved selectivity for isomerization and color absorption.

Low Capital Cost

This process has fewer processing steps, resulting in a lower equipment count and a *lower capital cost* than other commercial resin-catalyzed processes. Because there are fewer pieces of equipment, this process requires a small plot plan and is easier to operate than competitive processes. A fixed promotion reactor system is used, so no promoter removal hardware is required.

High-Purity/High-Performance Product

Dow's BPA process is based on water crystallization that provides unique advantages with respect to product quality. The Dow process yields a very low level of color-forming impurities. It has demonstrated capabilities to produce high-quality epoxy resin and polycarbonate with one of the lowest yellowness index (YI) values in the industry. BPA product from this process is also well suited as a raw material for a melt-phase polycarbonate process, producing very low color prepolymer when compared with competitors' BPA products.

Product Form

The BPA can be produced in crystalline, flake, or prill form for easy loading and packaging.

Low Energy Consumption

The recycled phenol does not have to be distilled. It is of sufficient quality to use for the separation washes, as well as for reactor feed. Only one stage of phenol crystallization is used, thereby reducing the wash phenol and simplifying management of the plant phenol balance. Adduct crystallization is performed without an evaporative organic cosolvent, resulting in low overall steam consumption.

Low Maintenance

Maintenance costs are projected to be low at only 2 percent of ISBL (inside battery limit) capital costs per year due to the equipment selected and its robust operating design.

Low Environmental Emissions

The process is designed to meet or exceed the most stringent environmental regulations. The process uses no free mercaptan, simplifying vent and waste treatment.

Superior Technology Support

Dow continually enhances its BPA process and its catalyst offerings via a world-class research and development (R&D) organization to ensure that its BPA meets the needs of its own polycarbonate process, as well as those of its BPA licensees for both quality and cost. This ensures licensees a quality, low-cost process from engineering through commissioning and startup, as well as the opportunity for continuing technical improvements and support through years of operation.

Low Operating Manpower Requirements

The Dow BPA process is stable and simple to operate and has a high on-stream factor. A BPA plant employing the Dow technology therefore requires only a small operating staff. Plant staffing or manpower representative of a typical U.S. Gulf Coast (USGC) location is as follows: 11 operators (total covering all shifts), 2 foremen, 1 superintendent, and 1 laboratory supervisor.

World-Scale Capacity Design

The process scale is easily adjustable. Facilities using the Dow BPA process are designed for 90,000, 100,000, and 120,000 MTA. Figure 9.4.1 shows a process block flow diagram.

PROCESS DESCRIPTION

Bisphenol Reaction

BPA is produced by reacting 1 mol of acetone with 2 mol of phenol in an acid-catalyzed condensation reaction. The reaction is exothermic, and water is the main reaction by-product.

$$2 \text{ PHENOL} + \text{ACETONE} \longrightarrow \text{P,P - BISPHENOL A} + \text{WATER}$$

A QCAT cation-exchange resin catalyzes the BPA reaction. The catalyst is batch-promoted via a proprietary system to achieve the proper selectivity to the para, para isomer. The reaction is carried out in a significant excess of phenol, and the reactors run adiabatically.

The process minimizes the formation of rearrangeable and nonrearrangeable by-products.

By using a process recycle stream, the rearrangeable impurities reach equilibrium such that the reaction of phenol and acetone primarily produces high-purity *p,p*-bisphenol.

The plant has multiple fixed-bed reactors. When a reactor's catalyst activity becomes too low and the reactor is taken offline, the spent catalyst is removed and replaced with the QCAT catalyst before the reactor is brought back online.

Rearrangement Reaction

A portion of the reactor feed is passed through a rearrangement reactor and returned to the reactor feed tank. This reactor uses another unique QCAT cation resin catalyst to convert by-products into desired product while having a low reactivity to undesirable by-products. This enhances the overall raw material efficiency.

Adduct Crystallization

The outlet from the reactors is fed to the adduct or phenol crystallizers, where a crystal adduct that is a 1:1 molar ratio of phenol to bisphenol is produced. Crystallization is achieved by cooling the solution. The crystal slurry from the crystallizers is fed to a system using a proprietary separation system. The crystals are washed with phenol to remove contamination on the crystal surface. The adduct crystals recovered are very pure.

The crystals (phenol and BPA adduct) are then fed to the phenol separation section. The main filtrate and wash filtrate are recycled back to the reactor section via the drying/acetone recovery section, and the wash filtrate is recycled to the adduct crystallizers.

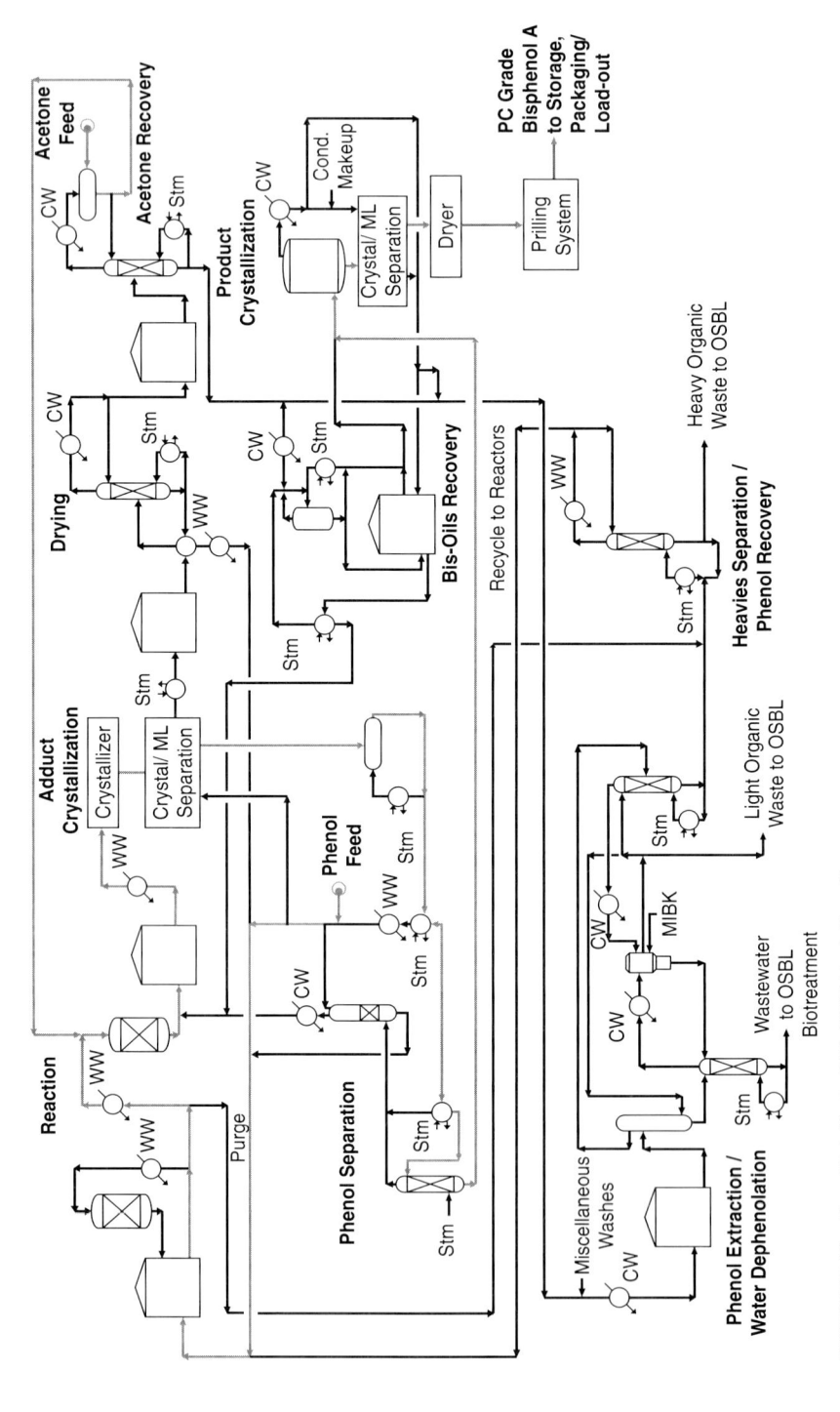

FIGURE 9.4.1 Dow High-Purity BPA QBIS block flow diagram.

Acetone Recovery/Water Separation

The filtrates from the proprietary separation system are fed to the drying column, where unreacted acetone and the by-product water are removed. The tower operates under vacuum and at a moderately high temperature.

The overheads from the drying column are fed to an acetone recovery column that separates acetone from water. A proprietary control scheme is used to ensure that the highest-purity acetone is recovered for recycle. The fresh acetone is combined with makeup acetone and fed to the main reactors. The water is sent to the phenol extraction system.

The bottom stream from the drying column is heat interchanged with the feed and then further cooled before being sent to the reaction section for further processing.

Phenol Separation

The phenol/BPA adduct crystals are melted, and the residual phenol is stripped off and recovered. The recovered phenol is of sufficient quality to be used with fresh phenol makeup to be partially recycled back as the crystal wash. The balance is recycled to the reactor section. The finished molten BPA is sent to the product crystallizers for further purification. The use of one phenol crystallization step makes the plant phenol balance easy to manage.

Product Water Crystallization

Water crystallization produces a product with less polycarbonate color-forming impurities than could be achieved otherwise with other types of crystallization processes. This is especially advantageous for polycarbonate plants using melt-phase technology.

In this unit operation, the molten BPA is mixed with water and fed to the water crystallizers. The slurry of crystals, water, and residual BPA is fed to a centrifuge for separation. The crystals are then fed to the drying train, and the filtrate and wash water are sent to the bis-oils recovery section.

Product Drying

The finished crystals are dried in a series of screw conveyors.

Prilling

The BPA crystals from the drying train are melted and pumped to the top of the prill tower. Solid prills are formed in the tower as the BPA flows countercurrent to a recirculating nitrogen stream. The nitrogen leaving the top of the prill tower passes first through a bag filter to remove any fines and is then cooled and recirculated back to the prill tower. The prills leaving the bottom of the tower pass through a vibrating screener. Oversize product is removed and recycled back to the process. The prilled product then can be conveyed to storage silos prior to packaging.

Bis-Oils Recovery

The purpose of this section is to separate and recover the BPA-related compounds in the wash from the product centrifuge. The water and BPA from the centrifuge are decanted

TABLE 9.4.1 Dow High-Purity Bisphenol A QBIS Technical Summary

Reaction	
Reactor type	Fixed bed
Catalyst type	QCAT resin
Promoter	Yes, fixed promoter
Selectivity, phenol %	98.7
Rearrangement catalyst	QCAT resin
Recovery	
Acetone distillation	Yes
Predistillation	No
Crystallization stages	2
Crystallization type	Adduct/water
S/L separation	Proprietary separation/centrifuge
M/L recycle to reactor	Yes
Rearrangement reactor	Yes
Rearrangement catalyst type	QCAT resin
Plot plan, $m \times m$	36×106

and separated in a separator tank. The aqueous phase is sent to the phenol extraction section. The organic phase is recycled back and combined with the feed to the drying/acetone recovery section.

Phenol Extraction/Water Dephenolation

This section recovers phenol from water in order to increase the overall phenol utilization and ensure low phenol concentration in the aqueous stream to the licensor's battery limit for the BPA plant. The phenol recovery section separates and recycles the phenol that is contained in the water streams from acetone recovery, bis-oils recovery, and intermittent reactor/vessel washes. The organic phase from the extractor is fed to another column where the stripping solvent is recovered and then combined with fresh solvent makeup and recycled back to the extractor. A light organic purge is taken from this stream.

Heavies Separation and Phenol Recovery

This section recovers phenol from the heavies (higher-molecular-weight phenolic compounds) streams. The phenol stream from the extractor system and a purge stream from the phenol recycle/mother liquor stream are fed to the heavies column. The purge stream reduces the buildup of nonrearrangeable heavies in the phenol recycle. In the heavies column, phenol is stripped, and the heavies are concentrated in the bottoms. The phenol is recycled back to the reactor section, and the heavies are stored for transport to OSBL incineration. Table 9.4.1 provides a technical summary of the process. Tables 9.4.2 and 9.4.3 show the raw materials requirements.

COMMERCIAL EXPERIENCE

Dow's extensive experience in the development of BPA technology and the engineering, design, and operation of BPA plants is unique in the industry. Dow has been producing BPA since the 1960s. At present, Dow is one of the world's leading producers of BPA, with

TABLE 9.4.2 Phenol Raw Materials Requirements

Property	Value	Units
Purity, minimum	99.9	wt %
Color, molten, maximum	10	
Solidification point, minimum	40.8	°C
Water, maximum	0.1	wt %
Iron, maximum	0.5	ppm
Mesityl oxide, maximum	20	ppm
Hydroxyacetone, maximum	20	ppm
Carbonyls, maximum	30	ppm

TABLE 9.4.3 Acetone Raw Materials Requirements

Property	Value	Units
Purity, minimum	99.7	wt %
Color hazen, maximum	5	
Water, maximum	0.3	wt %
Acidity (as acetic acid), maximum	0.002	wt %
Alkalinity (as NH_3), maximum	0.001	wt %
Iron, maximum	0.5	ppm
Permanganate test, minimum	2	h

a total production capacity of 300,000 MTA in four plants located in the United States and Germany. The location and year of startup for the four plants currently in operation are shown in Table 9.4.4.

Both the Freeport and Stade sites have undergone numerous expansions with technology advancements since their startup. The Stade site successfully demonstrated the High-Purity BPA QBIS technology in 2001.

Nan Ya Plastics Corp. was the first licensee of the Dow Chemical Company BPA Generation I technology for its 90,700-MTA grassroots BPA plant at Mailiao, Taiwan. The plant came on-stream in first quarter 1999 and has met all performance guarantees successfully.

Since the Nan Ya Plastics licensing activity and its startup, Dow has continued to improve its BPA process and now *directly* offers its High-Purity BPA QBIS technology for licensing.

PARABIS™ polycarbonate grade BPA is made and used by Dow to produce CALIBRE 1080 DVD™ optical media-grade polycarbonate resin. CALIBRE 1080 DVD resin is used commercially by a wide range of companies to produce optical media format products such as DVDs, CD-videos, CD-Rs, CD-RWs, CD-ROMs, and audio CDs. Dow is presently selling optical media-grade polycarbonate based on its PARABIS bisphenol A in North America, Europe, South America, and Asian Pacific regions (Table 9.4.5).

WASTES AND EMISSIONS: EXPECTED PERFORMANCE

The Dow BPA process produces low emissions and is designed to meet the most stringent environmental regulations.

TABLE 9.4.4 Location and Year of Startup of Four BPA Plants

Producer	Location	Startup year
Dow USA	Freeport, Texas	1972
Dow Europe	Stade, Germany	1977
Dow Europe	Stade, Germany	1987
Dow USA	Freeport, Texas	1991

TABLE 9.4.5 Dow High-Purity Bisphenol A QBIS Typical Product Quality

Property	Value	Units
Purity	99.92	wt %
o,p-Isomer	700	ppm wt
Other organic impurities	100	ppm wt
Phenol	<10	ppm wt
Iron, maximum	<0.1	ppm wt
Caustic color	20	APHA
Alcohol color	10	APHA

Gaseous Emissions

All the vents containing phenol and acetone are collected in a common header. Depending on local regulations, the vent stream can be sent to a scrubber, a small self-standing flare, an incinerator, or an existing flare. The vent collection and disposal system is for environmental reasons and not for emergency flaring situations.

Aqueous Effluent

A Dow-designed BPA plant achieves a low phenol concentration in the wastewater leaving the process unit. This goal is accomplished by extracting phenol from wastewater in the phenol extraction or water dephenolation section of the plant. By doing so, the overall phenol utilization factor is improved, and the loading on the downstream (OSBL) wastewater treatment unit is minimized. Project wastewaters normally are categorized into two general source types, process wastewaters and storm waters. Process wastewaters are those which come into contact with process streams or commercial products and thus are contaminated. In a Dow-designed BPA plant, there are three sources of process wastewaters:

- Continuously generated within the process unit
- Intermittently generated within the process unit (e.g., wash water during a catalyst changeout)
- Drain water (but excluding storm water runoff)

The continuous and intermittent streams generated by the process (i.e., excluding drain water) are collected in a tank and processed through the phenol extraction/water dephenolation section of the BPA plant before discharge to the OSBL wastewater treatment facility.

TABLE 9.4.6 Dow High-Purity Bisphenol A QBIS Process Economics

100K mtpa capital, MM$	
ISBL	$55.3 (Gulf Coast, factored, no license fees)
120K mtpa capital, MM$	
ISBL	$64.2 (Gulf Coast, factored, no license fees)
Consumption/lb BPA	
Phenol, lb	0.836
Acetone, lb	0.261
Utilities	
Steam (16 kg/cm^2), t/t BPA	2.3
Electric power, kWh/t BPA	180
Cooling water, m^3/t BPA	230 ($\Delta T = 6°C$)
Maintenance, % ISB	2.0
Manpower requirements	
Labor	11
Foremen	2
Superintendent	1
Lab supervisor	1
Expected on-stream time, h/yr	8250±

Organic Effluent

The Dow-designed BPA plant generates two small organic effluent streams, both of which are collected and pumped to the battery limit for disposal. Disposal may be dictated by local regulations, but incineration is recommended.

Solid Waste

The only solid waste from the Dow-designed BPA plant is the spent resin catalyst, an intermittent waste product. If properly cleaned (washed) in the reactor prior to removal, the spent catalyst can be used as landfill, depending on local regulations. However, Dow normally recommends incineration of the spent catalyst to minimize possible environmental contamination.

Process Economics

Table 9.4.6 illustrates the process economics.

PART · 10

PROPYLENE AND LIGHT OLEFINS

CHAPTER 10.1
LURGI MTP® TECHNOLOGY

Waldemar Liebner
Lurgi AG
Frankfurt am Main, Germany

INTRODUCTION

Conventional steam cracking technology accounts for most of the worldwide production of propylene. Since in this process propylene is a by-product of the main product ethylene, the expected annual growth rate for propylene, which is higher than for ethylene, cannot be satisfied by existing technologies only.

Producing propylene from methanol has been considered an interesting alternative to the traditional production of propylene from petroleum. Methanol is deemed to be a readily stored and managed intermediate product for use of hitherto unused natural gas. Thus the increasing demand for propylene on the world market also could be met by using inexpensive methanol.

Lurgi's new Methanol To Propylene (MTP®) process represents a simple, cost-effective, and highly selective technology yielding an excellent value-added product for the use of natural gas reserves via syngas and Lurgi MegaMethanol®.

PROCESS OVERVIEW

Lurgi's new MTP process is based on an efficient combination of the most suitable reactor system and a highly selective and stable zeolite-based catalyst. Süd-Chemie AG manufactures this catalyst commercially; it provides maximum propylene selectivity and has a low coking tendency, a very low propane yield, and also limited by-product formation. This, in turn, leads to a simplified purification scheme that only requires a reduced cold box system as compared with on-spec ethylene-propylene separation.

Based on Fig. 10.1.1 a brief process description reads: The methanol (MeOH) feed from the MegaMethanol plant is sent to an adiabatic DME prereactor, where methanol is converted to dimethylether (DME) and water. The high-activity, high-selectivity catalyst used nearly achieves thermodynamic equilibrium. The methanol-water-DME stream is

10.4 PROPYLENE AND LIGHT OLEFINS

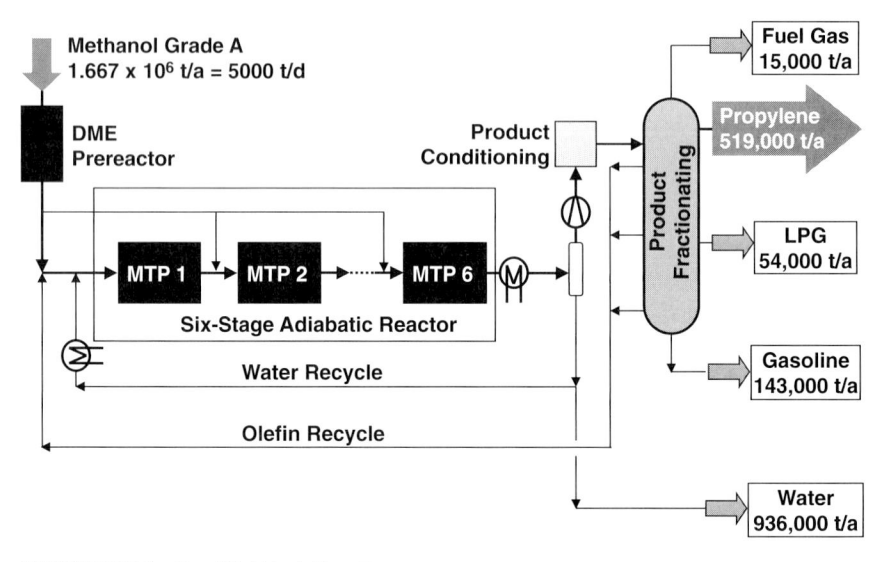

FIGURE 10.1.1 Simplified block flow diagram.

routed to the first MTP reactor stage, where steam is added. Methanol and DME are converted by more than 99 wt %, with propylene as the predominant hydrocarbon product. Additional reaction proceeds in the subsequent five MTP stages. Process conditions in the six MTP reactor stages are chosen to guarantee similar reaction conditions and maximum overall propylene yield. The product mixture is then cooled, and the product gas, organic liquid, and water are separated.

The product gas is compressed, and traces of water, CO_2, and DME are removed by standard techniques. The cleaned gas is then further processed, yielding chemical-grade propylene with a typical purity of more than 97 wt % or, if specified as polymer-grade, 99.6 wt %. Several olefin-containing streams are sent back to the main synthesis loop as an additional propylene source. To avoid accumulation of inert materials in the loop, small purges are required for light ends and the C_4/C_5 cut. Gasoline is obtained as an important by-product.

Water is recycled to steam generation for the process; the excess water resulting from the methanol conversion is purged. This process water can be used as raw-water supplement or for irrigation after appropriate and inexpensive biologic treatment. It can even be processed to potable water where needed.

An overall mass balance is depicted in Fig. 10.1.1 based on a combined MegaMethanol/MTP plant. At a feed rate of 5000 t of methanol per day (1.667 million tons annually), approximately 519,000 t of propylene is produced per year. By-products include fuel gas and LPG, as well as liquid gasoline and process water.

DETAILED PROCESS DESCRIPTION

The detailed process description is related to the block flow diagram of the MTP plant (Fig. 10.1.2). It describes the process flow of reaction, compression, and product separation, thus giving an overview of the MTP process.

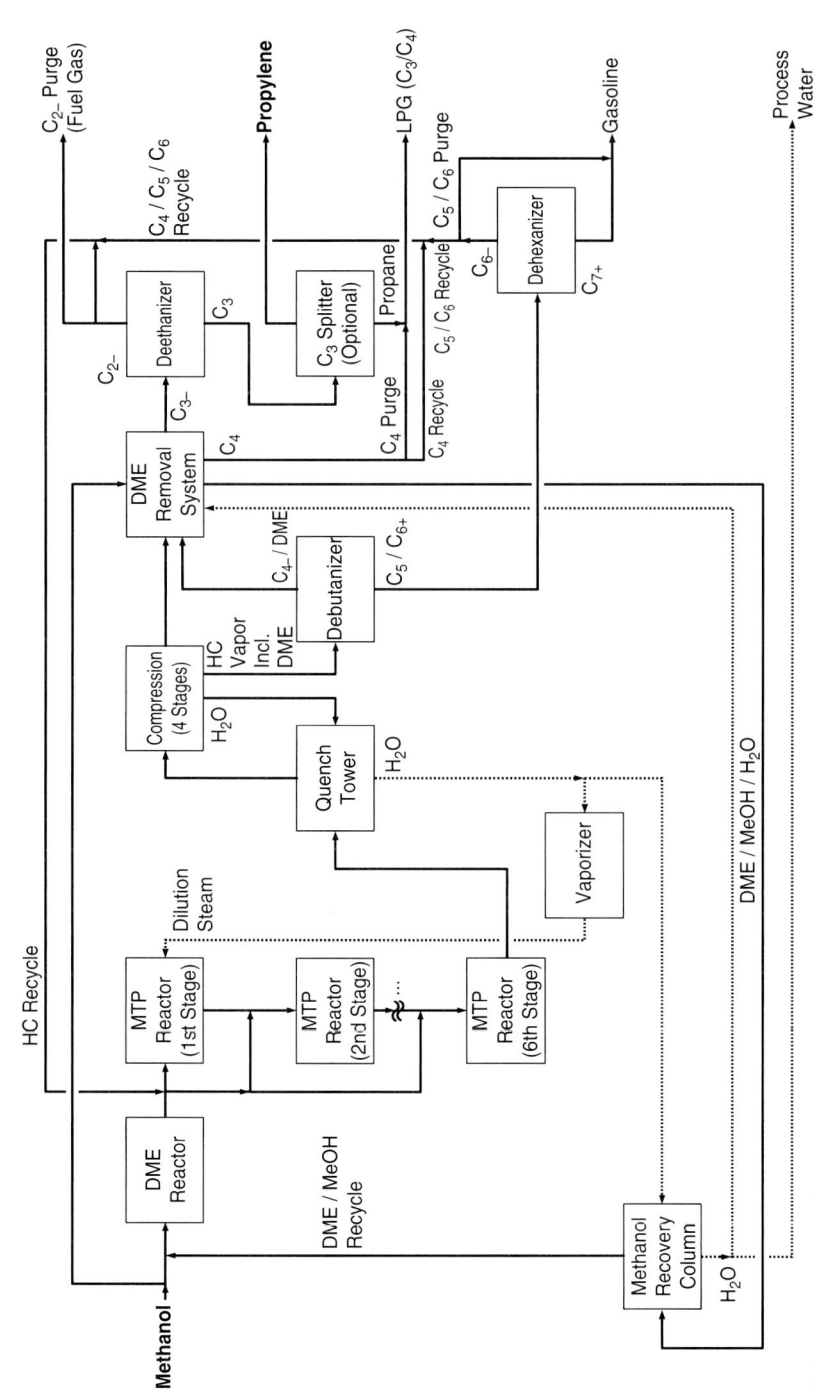

FIGURE 10.1.2 Detailed block flow diagram of the MTP process.

10.5

DME Reactor

The major part of the methanol feed is vaporized, superheated, and fed into the DME reactor. A smaller part of the feed methanol is used as solvent for purification within the DME removal section.

The DME reactor is a single-stage adiabatic reactor where most of the methanol vapor is dehydrated to dimethylether (DME) on an aluminum oxide catalyst (γ-Al_2O_3) according to the following equation:

$$2CH_3OH \rightarrow CH_3OCH_3 + H_2O$$

The catalyst features high activity and high selectivity, achieving almost thermodynamic equilibrium. The reaction is exothermic, and the reaction equilibrium is nearly independent of the operating pressure. The process is designed for a high conversion rate at moderate operating conditions.

Besides fresh methanol vapor, a recycle stream also consisting of methanol, DME, and some water is fed to the DME reactor. This stream is the overhead product of the methanol recovery column, which recovers methanol and DME from aqueous phases.

MTP Reactor

For a better approach to isothermal conditions, the MTP reactor was designed with six stages including six catalyst beds. The product of the DME reactor is divided into six streams, each feeding one of six stages of the MTP reactor. The feed to the first MTP reaction stage is mixed with a hydrocarbon recycle stream and some recycle steam. The mixture is further heated and then introduced into the MTP reactor.

In the MTP reactor, the DME-MeOH mixture is converted to olefins on a zeolite-based catalyst according to the following summary reaction:

$$nCH_3OCH_3 \rightarrow 2C_nH_{2n} + nH_2O \qquad n = 2, \ldots, 8$$

The MTP catalyst converts nearly all the DME-MeOH mixture with a high selectivity toward low-molecular-weight olefins. About 85 wt % of the carbon from the fresh feed (DME or MeOH) reacts to olefins in the range of C_2 to C_8 with the peak for propylene.

The quantity of fresh DME-MeOH fed to each catalyst bed is adjusted in such a way that the adiabatic temperature rise caused by the heat of the exothermic reaction is the same for each bed. This guarantees similar reaction conditions, resulting in maximum overall yield of low-molecular-weight olefins.

The above-mentioned high selectivity toward olefins requires relatively high operating temperatures over the catalyst beds and low operating pressures.

The intermediate reaction products from stages 1 to 5 are cooled and mixed with additional fresh DME-MeOH feed before entering the next reaction stage. During operation, small amounts of heavy hydrocarbons are formed that partly block the active sites of the catalyst. In order to minimize the carbonization process, steam is added to the feed of the first MTP reaction stage. The steam also serves as a heat sink for the exothermic reaction and thus supports control of the temperature rise over the catalyst.

The hydrocarbon recycle to the first MTP reaction stage increases the propylene yield by conversion of olefins with a carbon number lower or higher than 3 (propylene). In addition, the hydrocarbons serve as a heat sink for the exothermic reaction, again supporting temperature control over the catalyst.

The MTP reactor catalyst has to be regenerated when the overall conversion of the DME-MeOH feed falls below the economical limit. The regeneration is done *in situ* by controlled combustion of coke with an air-nitrogen mixture. In order to ensure continuous

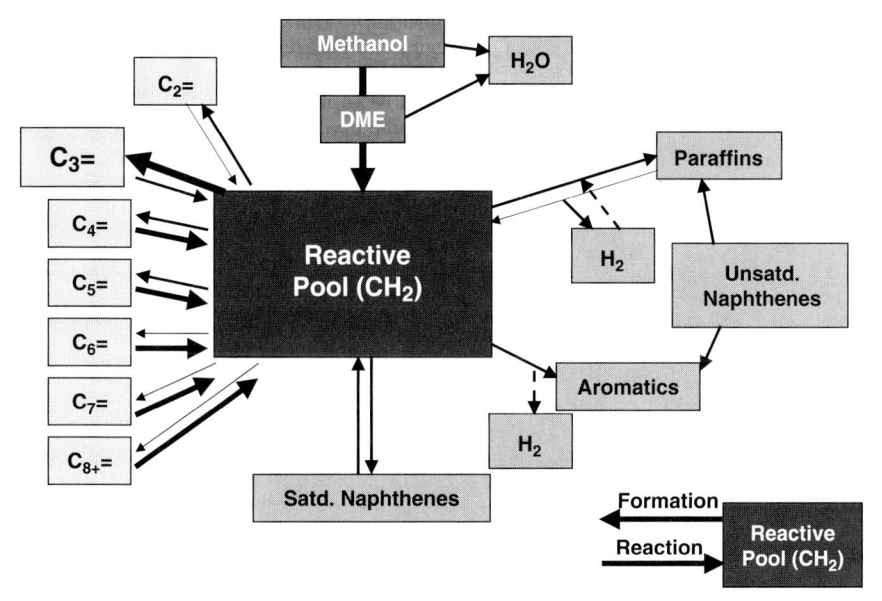

FIGURE 10.1.3 Simplified reaction model.

operation of the plant, the design consists of three MTP reactor trains. Two trains are in operation while one train is in regeneration or in standby mode.

A simplified reaction model for the methanol conversion is shown in Fig. 10.1.3. It is assumed that a so-called reactive pool exists on the zeolite surface that contains CH_2 fragments. It is very unlikely that isolated CH_2 species are actually formed, but for illustrative purposes, it can be assumed that this pseudospecies is formed by dehydration of methanol (or DME) and serves as a building block for all olefinic products in the MTP reaction. In addition, almost all reaction products can be converted on the catalyst, so this reactive pool is filled from several sources, with methanol-DME being the most prominent one.

The actual selectivities of the MTP reaction and the overall product distribution, i.e., the relative reaction rates from the reactive pool to each single product species, are an inherent feature of the specific catalyst used and depend on its pore and atomic structure. The catalyst used in the MTP process was optimized for maximum propylene yield and maximum total olefin selectivities.

Similar reactive pools are widely described in the literature for hydrocarbon reactions on zeolites, and their existence has been proven by isotope-exchange experiments.

Quench Tower and Water Stripper

The product leaving the reactor contains naphthenes, paraffins, aromatic components, and light ends, as well as olefins and process and reaction water. The hydrocarbon yield based on converted fresh DME-MeOH feed is 85 wt % olefins, about 9 wt % paraffins, less than 3 wt % aromatic compounds, and less than 2 wt % naphthenes, as well as light components (H_2, CO, CO_2). The reaction product is cooled in heat exchangers and finally in the quench tower, where the hydrocarbon product is separated from most of the water. The hydrocarbons leave the quench tower as overhead vapor, whereas the water is condensed and sent to the methanol recovery column.

The reactor product leaves the MTP reactor at a low pressure and has to be sent to the first compressor stage with a very small pressure drop. The use of an additional cooler and separator instead of the quench tower, with its virtually negligible pressure loss, therefore would cause problems with the maximum permissible pressure drop.

In addition to the components mentioned earlier, the reaction over the zeolite-based catalyst forms small amounts of organic acids such as acetic acid and propionic acid. Prior to withdrawal, process water, having a pH in the range of 3 to 4, is neutralized by adding caustic solution to the quench tower sump. The smaller portion of the water condensate is vaporized and recycled as dilution steam to the first MTP reaction stage, whereas the major portion of process water is routed to the methanol recovery column for recovery of methanol and DME. The stripped water containing about 3500 wt ppm of methanol is routed to battery limits as process water. It can be biotreated easily to supplement raw water or for use as irrigation water. A smaller portion of the process water from the methanol recovery column is recycled into the process for use as a solvent in the DME removal system.

Compression

The hydrocarbon vapor product from the quench tower is compressed by a multiple-stage turbocompressor. Between the compression stages, the product is cooled, and residual water, as well as hydrocarbon liquid, is separated from the vapor phase, which is compressed further. Residual water is recycled to the quench tower, whereas hydrocarbon liquid and hydrocarbon vapor are sent to a dryer each, which is not shown in the block flow diagram of the MTP plant. The dryer removes the remaining water by adsorption on mole sieves.

Separation

The dried hydrocarbon liquid is fed to the debutanizer column, whereas the dried hydrocarbon vapor is further processed in the DME removal system. The debutanizer distillation column separates light-boiling components C_{4-} and DME from C_{5+} hydrocarbons. The C_{5+} bottoms product is fed to a dehexanizer distillation column, where light naphtha components in the range of C_5 and C_6 are separated from a heavier gasoline fraction that contains C_{7+} hydrocarbons (paraffins, olefins, naphthenes) and the alcylated benzene derivatives toluene, xylene, and trimethylbenzene.

The C_5/C_6 overhead product (paraffins, olefins, naphthenes) is mainly recycled to the first MTP reaction stage for further conversion of higher olefins to propylene; a smaller portion is purged out the reaction loop. The C_5/C_6 purge stream is usually added to the C_{7+} fraction, forming the gasoline by-product.

The compressed and dried hydrocarbon vapor—including light olefins and DME—and the C_{4-}/DME overhead product from the debutanizer are both feed of the DME removal system. There, C_{3-} hydrocarbons are separated from C_4 hydrocarbons. In addition, DME is removed from the hydrocarbons. The overhead product C_{3-} is free of DME (<1 wt ppm) or any other oxygenate component. It is fed to the deethanizer distillation column.

The aqueous phase leaving the unit contains methanol and DME and is recycled to the methanol recovery column. The C_4 hydrocarbon fraction, now purified from most oxygenate components, is recycled to the MTP reaction system for further conversion of butenes to propylene. A smaller portion is purged out the reaction loop, forming a C_4 LPG by-product.

In the deethanizer distillation column, the C_{3-} hydrocarbon fraction is split into C_{2-} (containing ethylene, ethane and some light-ends methane, hydrogen, CO, and CO_2) and the C_3 product containing propylene (about 97 wt %) and propane (about 3 wt %) but no unsaturated components such as methylacetylene or propadiene (the analyzed concentration of these is less than 1 wt ppm). Alternatively, depending on the propylene specification requirement, the MTP process produces chemical-grade propylene as deethanizer bottoms product or polymer-grade propylene by use of an additional C_3 splitter distillation column that separates pure propylene (99.6 wt %) from the by-product propane. Usually the by-products propane and C_4 LPG are mixed to form a C_3- and C_4-containing LPG product.

The C_{2-} product from the deethanizer overhead consists of two streams of different composition. At first, the overhead vapors of the deethanizer are compressed in a single-stage compressor from the column operating pressure to an elevated pressure. Then the overhead product is partially condensed by means of propylene refrigerant. In the reflux drum the noncondensable vapor stream is separated from the liquid. The noncondensable vapor stream is withdrawn as first overhead product of the deethanizer. The condensed liquid is used in part as reflux to the column, whereas the remaining portion is taken off the column, forming the second overhead product of the deethanizer. The noncondensable vapor product is richer in light-end components (methane, hydrogen, CO, and CO_2). Therefore, it is purged off the reaction loop and used internally as fuel gas. The liquid overhead product is richer in ethylene. It is vaporized, routed to an adsorber filled with mole sieves for CO_2 removal (not shown in the block flow diagram of the MTP plant), and then recycled to the MTP reaction system for further conversion of ethylene to propylene. CO_2 is separated from the C_{2-} recycle stream to reduce the buildup of this component.

PRODUCTS, BY-PRODUCTS, WASTES, AND EMISSIONS

The products, by-products, wastes, and emissions listed below refer to a feed rate of 5000 t/d of methanol (Fig. 10.1.1).

Product

Propylene: 64,875 kg/h

Polymer grade: 99.6 wt %

By-Products

Gasoline: 17,875 kg/h

Composition:
- Olefins: about 20 wt %
- Paraffins: about 50 wt %
- Aromatics (methylated benzene derivatives): about 30 wt %
- No benzene
- No sulfur

An analysis of gasoline was made by a Statoil refinery laboratory in Norway. The analyzed gasoline was received from the demo unit at Tjeldbergodden, Norway. The value was stated as approximately "unleaded premium," which defines a typical gasoline with a research octane number (RON) of 95 and a motor research number (MON) of 85.

LPG: 6750 kg/h

Process water: 117,000 kg/h

for use as raw-water supplement or as irrigation water after biotreating; the fuel gas (Fig. 10.1.1) is used internally.

Wastes

The catalyst of the DME reactor is an aluminum oxide catalyst (γ-Al$_2$O$_3$) with an expected lifetime of 10 years, whereas the catalyst of the MTP reactor is a zeolite-based catalyst with a life of more than 1 year. Both catalysts are easily disposable as landfill after use.

Emissions

The only emissions of note are the "standard" flue gases from gas-fired heaters and boilers and the catalyst regeneration gas, which basically consists of nitrogen-diluted air with a somewhat elevated CO$_2$ content.

TECHNICAL AND COMMERCIAL STATUS

The technological status of MTP in the areas of process and catalyst can be summarized as follows: The basic process design data were derived from more than 9000 operating hours of a pilot plant at Lurgi's Research and Development Center. Besides optimization of reaction conditions, several simulated recycles also have been analyzed.

Parallel to this, Lurgi has opted to build a larger-scale demonstration unit to test the new process in the framework of a world-scale methanol plant with continuous 24/7 operation using methanol feedstock from an existing plant. The main purpose of the test was to demonstrate that the catalyst lifetime meets or exceeds the commercial target of 8000 hours on stream. After a cooperative agreement with Statoil ASA was signed in January 2001, the demo unit was assembled in Germany and then transported to the Statoil methanol plant at Tjeldbergodden, Norway, in November 2001 (Fig. 10.1.4). Later in 2002, Borealis joined the cooperation.

The demo unit was started up in January 2002, and the plant has been operated almost continuously since then. By April 2004 the on-stream time of the first catalyst batch had reached 8000 hours, and 3000 hours more had been used for tests with a new batch.

Cycle lengths between regenerations have been longer than expected. Deactivation rates of the methanol conversion reaction decreased with operation time. Propylene selectivity and yields were in the expected range for this unit, with only a partial recycle. To verify the full recycle of all light and heavier olefins, a bench-scale unit with a complete purification section will be installed at Lurgi's R&D center. This also will prove again the polymer-grade quality of the product, i.e., the absence of polymerization poisons. This was first demonstrated by producing polypropylene cups from a batch of MTP propylene that was distilled offline at the R&D center and then polymerized in Borealis' labs.

The gasoline product of the demo unit was analyzed in a Statoil refinery lab that reported the sample to have "premium gasoline quality." The catalyst development is completed, and the supplier commercially manufactures the catalyst, which is already used in a similar application.

Integrating all these favorable results and conditions allowed Lurgi to prepare commercial designs for middle- and large-scale MTP plants. These designs were the basis for thorough in-house costing and benchmarking. Client feasibility studies for large polypropylene complexes have been concluded. With that, Lurgi today offers MTP on commercial terms. By July 2004 negotiations for the first commercial-size MTP plant were in the final stage.

FIGURE 10.1.4 Demo unit for Tjeldbergodden, Norway.

PROCESS ECONOMICS

Since propylene itself is more an intermediate than an end product, an economics estimate was performed for a methanol-propylene-polypropylene complex. This just needs the addition of a block polypropylene synthesis to Fig. 10.1.1.

Thus the economic assessment included the MTP route with a polypropylene unit for the production of a more salable, higher-value end product. The case presented here is based on a feasibility study for a complex in the Middle East/Arabian Gulf region. It takes into account contingencies for the newly developed route. With that, the investment cost estimate is fairly high, and still an attractive return can be expected, as shown in Tables 10.1.1 and 10.1.2. The price basis for the investment cost estimate is the third quarter of 2002.

The remarkable facts here are the low production costs for the "intermediates" methanol and propylene and for the end product polypropylene. These leave room for healthy profit margins and for new applications such as MtPower® (the direct use of methanol or DME in power generation) and MtSynfuels®, the MTP-based route to diesel fuel and gasoline.

Given the fact that MegaMethanol plants are built for around US$300 to US$320 million in investment cost and that contingencies may be outweighed by additional integration savings or may not be needed in full, this route is seen as the most promising and most economical natural gas utilization.

TABLE 10.1.1 Production Cost: Integrated MegaMethanol-MTP-PP Complex

	Methanol	MTP	PP
Capacity, TPY	1,700,000	520,000	520,000
Investment cost EPC, million US$	350	215	165
Owner's cost, incl. capitalized interest, million US$	70	43	33
Feed cost, US$	Natural gas $0.5/MMBtu	Methanol $43/t	Propylene $174/t
Production cost, US$/t	42.9	210.1	261
Raw materials, US$/t	14.4	150.3	212.8
Utilities, US$/t	1.6	5.6	6.8
Operation and maintenance, US$/t	5.6	11.3	8.6
Plant OVHD and insurance, US$/t	6.0	12.1	9.2
Depreciation, US$/t	15.3	30.8	23.6
Credit for by-product naphtha, US$/t	—	−35.7	—
Cost of product at ROI = 0, US$/t	43	174	261

TABLE 10.1.2 ROI: Integrated MegaMethanol-MTP-PP Complex

	Methanol-MTP-PP
Investment cost EPC, million US$	730
Owner's cost incl. capitalized interest, million US$	146
Feed stock cost, US$	Natural gas $0.5/MMBtu
Production cost, million US$	154.3
Raw materials, million US$	49.7
Utilities, million US$	9.1
Operation and maintenance, million US$	19.9
Plant OVHD and insurance, million US$	21.2
Depreciation, million US$	54.4
Revenues, million US$	356.6
Naphtha (US$130/t), million US$	18.6
Polypropylene (US$650/t), million US$	338
Return on investment (ROI, %)	23.1

BIBLIOGRAPHY

1. MTP as part of the GTC technologies group (including MtSynfuels®): Gas to chemicals: Technologically advanced natural gas monetization, by Harald Koempel and Waldemar Liebner, Lurgi Oel · Gas · Chemie GmbH, Gas to Liquids Europe, Milan, Italy, July 2002.
2. MTP as R&D project: Demonstrating the methanol-to-propylene (MTP) process, by Martin Rothaemel and Jens Wagner, Lurgi Oel · Gas · Chemie GmbH, Staale Jensen, Ola Førre Olsvik, Statoil ASA, presentation at AIChE Spring National Annual Meeting, New Orleans, March 30–April 3, 2003.
3. MTP as economic undertaking: An economical route from natural gas to propylene, by Waldemar Liebner, Lurgi Oel · Gas · Chemie GmbH, PETCHEM 2003, Monte Carlo, Monaco, 25/26 September 2003.

CHAPTER 10.2
UOP/HYDRO MTO PROCESS

Peter R. Pujadó and James M. Andersen
UOP LLC
Des Plaines, Illinois

INTRODUCTION

The conversion of natural gas into syngas is the first step in utilizing natural gas for methanol and for the conversion of gas to liquids hydrocarbon products. Methanol as a base chemical offers limited opportunities for natural gas utilization unless linked to other derivative markets. Gas-to-liquids (GTL) technology using Fischer-Tropsch type of catalysts offers large market opportunities for natural gas utilization but is challenged by the economics of high capital costs with relatively low transportation fuel product values.

Syngas and methanol production technologies are achieving greater economies of scale. World-scale methanol production facilities have doubled in size compared to just a few years ago, and when combined with remote natural gas prices, these facilities offer substantially lower costs of production than those in existing plants. However, although methanol can be economically shipped from remote gas areas, the expected growth in demand for methanol for conventional uses does not support the addition of many new plants.

The conversion of methanol to fuel components was accomplished commercially in the Mobil MTG (methanol-to-gasoline) process[1-4] at a plant located near New Plymouth in New Zealand, but that plant has since been shut down on account of the relatively poor economics of gasoline production. However, Mobil did demonstrate that over a ZSM-5 (MFI) type zeolitic catalyst, methanol could be converted to a largely aromatic product, up to durene, but also with a significant proportion of olefins, principally propylene. Lurgi has recently developed a modified version of this process that minimizes the production of the gasoline fraction and maximizes production of propylene at about 70 percent; this process is known as MTP for methanol-to-propylene.[5]

Methanol can also be converted to ethylene and propylene via the UOP/Hydro MTO process, thus opening new opportunities for methanol utilization. Ethylene and propylene can then be used to satisfy the growing market demand for polyolefins or can be used in the production of other olefin derivatives. Remote gas strategies for MTO generally consider either the shipping of methanol from remote locations to countries with strong olefin

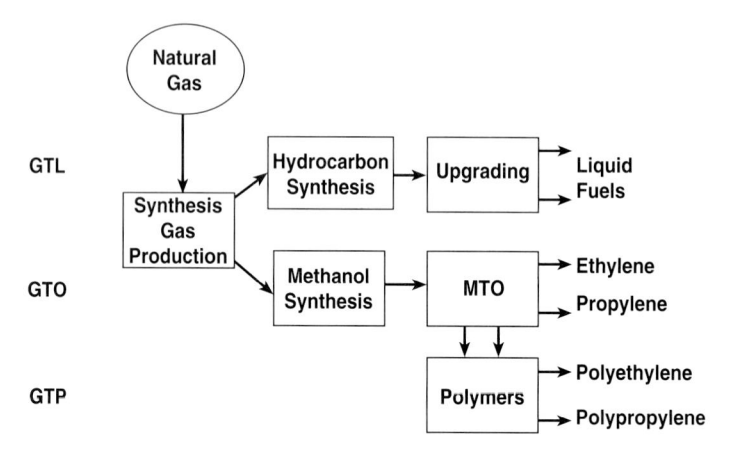

FIGURE 10.2.1 Conversion steps for natural gas to liquids and polyolefins.

demand or shipping of polymer pellets from fully integrated gas-to-polyolefin (GTP) facilities in remote locations.

Figure 10.2.1 illustrates various alternatives for the utilization of natural gas in the production of either liquid fuels or petrochemical derivatives. There is at present renewed interest and considerable activity in the planning and construction of large-scale GTL facilities. GTL technology is attractive because it offers great potential for the valorization of stranded gas by taking advantage of the large markets for fuel products. The implementation of GTL and other large-scale gas conversion projects is challenging because investments are high, technologies are often not well proven at the actual scale, there is competition with crude-oil-based products, and plant location is often in remote areas. One way to enhance the economics of GTL projects is to produce products with higher added value. This can include the recovery of normal paraffins for linear alkyl benzene (LAB) production, specialty lube oils, methanol, olefins, and polyolefins. The markets for some of these products can limit the opportunities for production in GTL facilities. The olefin and polyolefin markets, however, are exceptionally large, and these products offer very high added value.

Both GTL and GTP facilities incorporate sizable front-end syngas units for the processing of natural gas, as illustrated in Fig. 10.2.1. These units are the major contributors to the relatively high investments required for these complexes. It follows that the integration of these facilities could offer substantial synergistic savings. Potential savings for integrated GTL/GTP complexes may derive from

- Shared syngas plant
- Shared utility systems with by-products utilization
- Shared wastewater treatment facilities
- Shared administration, laboratory, and maintenance facilities
- Minimal intermediate-product storage facilities needed

There are other advantages for integrated facilities, such as advantages of back integration for polyolefin production:

- Lower cash costs of production
- Elimination of costs for intermediate-products shipping and handling
- Consistency and better control over feedstock quality

When one is considering an integrated facility to convert natural gas to polyolefins (GTP), there are three main process technologies involved. These technologies must fit together at world-scale capacities for an ideal integration. Each of these technologies is discussed further in the next paragraphs.

Syngas/methanol process technology is available from several well-known licensors. Until recently, world-scale capacity for methanol production was considered 2500 to 3000 metric tons per day (MT/D). Now there are a number of projects underway with capacities of 5000 MT/D, and licensors of syngas/methanol technologies are discussing capacities as high as 10,000 MT/D. Most of the new capacity that has recently come, or will soon be coming, on-stream reflects a growing trend in which methanol production has been shifting from heavily industrialized countries to locations with access to lower-priced natural gas.

The combination of large-scale production facilities with low-priced natural gas feedstock substantially reduces the fundamental costs of methanol production, as shown in Fig. 10.2.2.

Many smaller plants exist today in industrialized locations. The cash cost of production for these plants is typically more than $100/MT of methanol, due primarily to the cost of natural gas. If capital charges are added to provide, say, a 20 percent return on capital (ROC), then the delivered price of methanol is almost $180/MT. Large-scale plants in remote locations experience a considerable advantage due to low cash costs and economies of scale, even after accounting for the costs of shipping to distant markets. For these remote units the cash costs of production can be less than $50/MT. These units can deliver methanol at about $110/MT, even after adding capital charges and costs for shipping. New mega-scale projects enjoy an even greater advantage and can achieve attractive project economics with methanol delivered at prices less than $90/MT. This enables new applications for methanol such as fuel cells or conversion to olefins and offers large market growth potential.

Polyolefins are widely produced using technologies available from several licensors and may include flexibility to produce several grades of homopolymer and copolymer products. World-scale capacity for polyethylene processes is generally considered in the range of 300 to 350 kilometric tons per annum (kMTA). World-scale capacity for polypropylene processes is generally considered in the range of 250 to 300 kMTA.

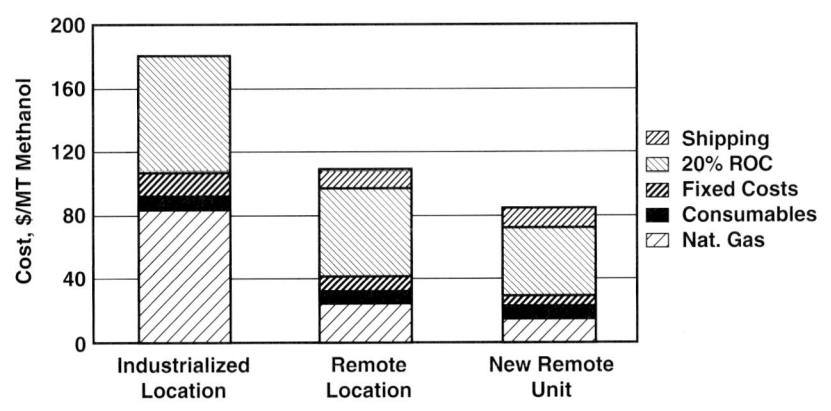

- "Industrialized Location" is based on 1500 MT/D capacity with $2.50/million Btu gas
- "Remote Location" is based on 3000 MT/D capacity with $0.75/million Btu gas
- "New Remote Unit" is based on 6000 MT/D capacity with $0.50/million Btu gas

FIGURE 10.2.2 Examples of methanol production costs.

MTO TECHNOLOGY

The remaining technology piece of the integrated GTP plant is an MTO unit capable of converting methanol to light olefins: ethylene and propylene. The UOP/Hydro MTO process provides the key link between natural gas and polyolefin production. It provides more profitable means to valorize remote gas and offers new opportunities for natural gas utilization. The MTO process is an innovative route for the production of olefins from natural gas. It offers yield flexibility that can deliver propylene as well as ethylene and satisfy the propylene demand that cannot be met by conventional ethylene plants alone.[2]

The conversion of methanol to olefins requires a selective catalyst that operates at moderate to high temperatures. The reaction is exothermic so heat can be recovered from the reaction. Methanol first goes through a dimethylether (DME) intermediate, and the reaction proceeds with further dehydration to yield ethylene and propylene. A limited amount of butenes and higher olefins is produced as well. Depending on the design and operation of the MTO unit, the overall yields of ethylene plus propylene can be almost 80 or 90 percent, based on the carbon content of the methanol feed.

Carbon or coke accumulates on the catalyst and requires removal to maintain catalyst activity. The coke is removed by combustion with air in a catalyst regenerator system. A fluidized-bed reactor and regenerator system is ideally suited for the MTO process. The reactor operates in the vapor phase at temperatures between 350 and 550°C and pressures between 1 and 3 bar gage. A slipstream of catalyst is circulated to the fluidized-bed regenerator to maintain high activity. The UOP/Hydro MTO process can be operated on "crude" or undistilled methanol as well as on pure methanol. The choice of feedstock quality generally depends on project-specific situations because there can be advantages in either case. Figure 10.2.3 illustrates a simple flow diagram for the UOP/Hydro MTO process. After the oxygenate recovery section, the effluent is further processed in the fractionation and purification section to separate the key products from the by-product components. Ethylene and propylene are produced as polymer-grade products and sent to storage.

The highly selective MTO-100 catalyst is based on SAPO-34, a template-based, silicoaluminophosphate molecular sieve with a chabazite structure and a unique pore size of about 3.8 Å (Fig. 10.2.4). The pore size controls the size of the olefins that emerge from the catalyst pores. Larger olefins diffuse out at a slower rate. Smaller olefins predominate in the reactor product. If, on the contrary, the reaction were conducted over an

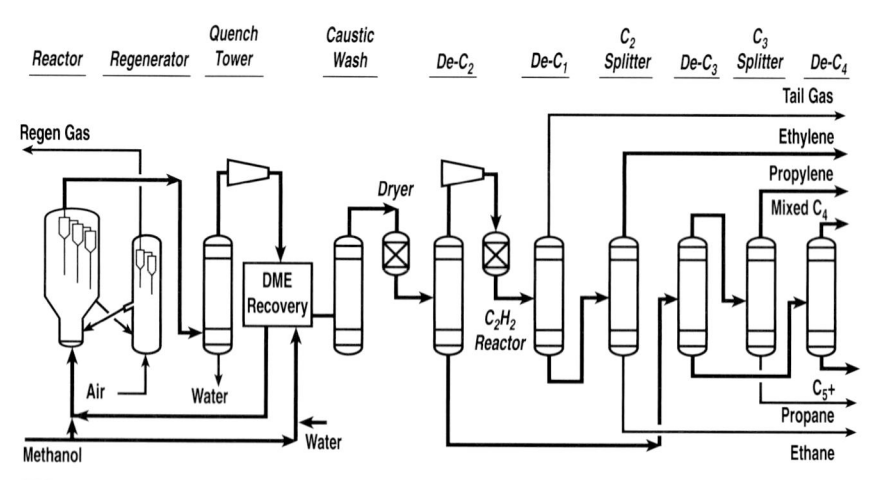

FIGURE 10.2.3 MTO process flow scheme.

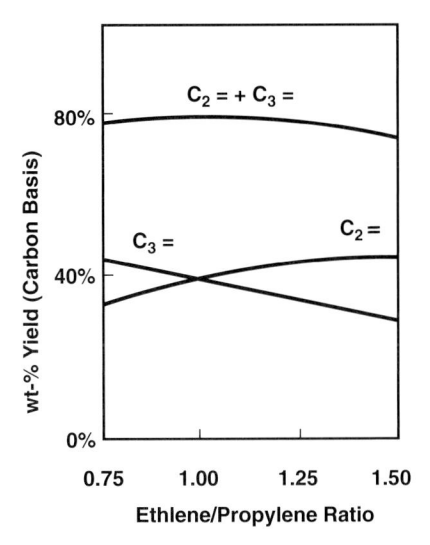

FIGURE 10.2.4 SAPO-34 structure and MTO light olefin yields.

MFI zeolite with a pore size of about 5.1 to 5.6 Å, the product would comprise much larger molecules, all the way to aromatics.

In a typical operation, up to 80 percent of the methanol feed (on a percent carbon basis) is converted to ethylene and propylene, with approximately 10 percent going to butenes. The overall carbon yield of light olefins can be increased to almost 90 percent by converting the C_4+ coproducts, mostly to propylene.

The UOP/Hydro MTO process offers a wide range of flexibility for altering the relative amounts of ethylene and propylene products by adjusting the operating severity in the reactor. The MTO process can be designed for an ethylene-to-propylene product ratio between 0.75 and 1.5. The overall yield of light olefins (ethylene plus propylene) changes

TABLE 10.2.1 MTO Mass Balance

600,000 MTA light olefins (ethylene + propylene)

	Feedstocks, MT/D	Products, MT/D
Methanol*	5204	
Ethylene		882
Propylene		882
Mixed butanes		272
C_5+ hydrocarbons		100
Fuel gas		88
Other (water, CO_x, coke, etc.)		2980
Total	5204	5204

*5204 MT/D of methanol requires about 155 million SCF/day (4.2 million N·m³/day) of natural gas, assuming MTO by-products are used as fuel.

slightly over this range with the highest yields achieved with about equal amounts of ethylene and propylene, roughly in the 0.8 to 1.25 range. This envelope provides the lowest methanol requirements, but the ratio can be adjusted to reflect the relative market demand and pricing for ethylene and propylene.

An example material balance is shown in Table 10.2.1 for the production of 600,000 MTA of light olefins with equal amounts of ethylene and propylene. Approximately 3 tons of methanol is required per ton of light olefins. This represents a carbon-based yield of almost 80 percent.

Because of the high olefin yields and low light-ends make, the MTO process does not require an ethylene refrigeration system. Although it is not yet commercialized, additional projected cost savings have been achieved by optimizing the reactor design and performing value engineering and pinch analysis. Several design packages have been prepared to determine the design requirements and costs for MTO projects. These studies have included design and cost requirements for off-sites and utility systems associated with stand alone MTO as well as integrated GTP projects.

ECONOMIC BASIS

To arrive at a meaningful economic comparison, we have made a number of pricing assumptions. Investment costs were adjusted to reflect a remote location. Allowances for the costs for off-sites and utilities were assumed to be equivalent to 35 percent of the inside-battery-limits (ISBL) estimated erected costs. This is expected to be a reasonable approximation for integrated facilities.

Product prices were estimated to roughly correspond to a crude oil price of $18/bbl (1 bbl ~ 0.159 m^3). The natural gas price was assumed at $0.50/million Btu (1 million Btu ~ 1.055 GJ), reflecting a price for remote gas utilization. GTL liquid products were assumed to have an aggregate value equivalent to $5/bbl above the crude oil price, and gas products were valued at the equivalent to the local fuel value.

Polyolefin yields were assumed at 98 wt % of the monoolefin feed rate. Polyolefin product prices were based on averages of historical spot and contract prices for western Europe roughly corresponding to crude oil at $18/bbl.

Shipping costs were assumed to estimate net-back revenues after marine transportation from a remote location to industrialized markets such as western Europe or the United States. These costs can vary significantly depending on project and market locations as well as fuel costs. Handling fees and import duties can also impact the net-back revenues.

Fixed costs of production were based on an allowance of 5 percent of the inside-battery-limits estimated erected cost (ISBL EEC) to cover the costs of labor and supervision, overhead, maintenance, taxes and insurance, and interest on working capital. Please refer to Table 10.2.2 for details.

INVESTMENT ESTIMATES

Investment costs were estimated based on scaling the estimated erected costs for the process units. These costs were determined by comparing cost information shown in various papers and publications as well as UOP in-house information. The basis and assumptions used in developing these costs are further explained in the paragraphs that follow.

The estimated costs for the options considered in this chapter are compared in Fig. 10.2.5. The investment costs for GTL and GTL integrated with methanol production are similar at about $1.2 billion. GTP also requires a similar investment. The addition/integration of olefin and polyolefin production facilities increases the investment cost to about $2.0 billion.

TABLE 10.2.2 GTL/GTP Economic Basis

Item	Cost
ISBL erected cost	Remote location basis
Off-sites and utilities	35% of ISBL assumed
Other costs*	Included
On-stream factor	340 days/yr
Project life	20 yr (17 operating)
Crude oil (corresponding)	$18/bbl
Natural gas feed	$0.50/million Btu
GTL products	$23/bbl
Polyethylene	$800/MT
Polypropylene	$705/MT
Methanol	$120 and $85/MT
Shipping	$12/MT liquids
	$40/MT polyolefins
Fixed operating costs	5% of ISBL (erected)

*Other costs include catalysts, license fees, and allowances for other miscellaneous owner's costs.

GTL Investment. A GTL complex for the production of 50,000 BPSD requires the conversion of about 450 million SCF/day of natural gas and has an estimated capital cost of about $1.25 billion. This cost assumes an all-inclusive plant cost in a remote location of $25,000/BPSD. The syngas facilities are assumed to account for about 60 percent of the ISBL costs for a GTL complex.[3]

GTL/Methanol Investment. The methanol synthesis and purification sections account for about 28% of the ISBL cost of a conventional methanol plant.[4] A world-scale methanol plant has a capacity of about 5000 MT/D (~1.7 million MTA). By scaling up the costs of the methanol synthesis and purification facilities to the world-scale capacity, it is estimated that the capital costs of these sections would be approximately $80 million. Integrating these facilities with the same syngas facilities used in the 50,000-BPSD GTL case described above would require about 38 percent of the syngas for methanol

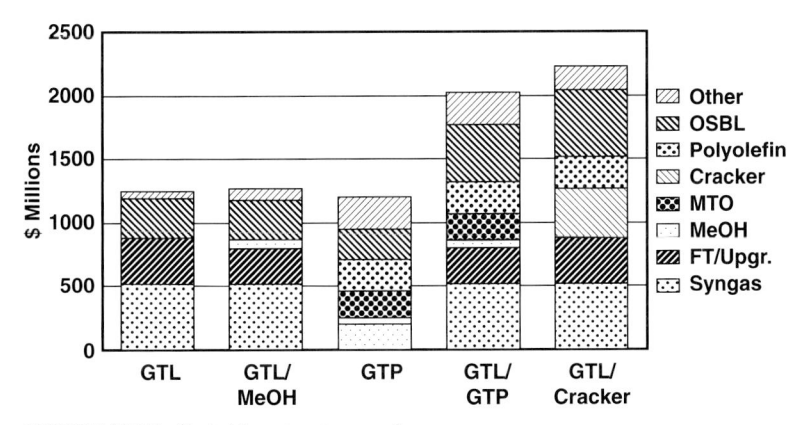

FIGURE 10.2.5 Capital investment comparison.

production. Such a complex would produce 5300 MT/D of high-purity methanol plus 31,000 BPSD of GTL liquid products. The size and cost of the Fischer-Tropsch (FT) synthesis and product upgrading facilities would be reduced for the lower GTL capacity. After adding the costs for outside-battery-limits (OSBL) allowance, catalysts, license fees, and other costs, the overall plant costs for the GTL/methanol facility would be only slightly higher than those for the GTL facility.

GTP Investment. The size of a world-scale GTP complex is mainly set by the capacity of the methanol and polyolefin units. World-scale polyolefin units have capacities of about 300,000 MTA. If equal amounts of polyethylene and polypropylene were desired, the MTO unit would require about 1.8 million MTA of methanol to support 600,000 MTA of polyolefin production. The methanol purification section can be greatly simplified for an integrated GTP complex because crude methanol can be used for feedstock to the MTO unit. This results in significant savings in the methanol plant, and the amount of intermediate-product storage is minimized for an integrated facility. The estimated cost for such a GTP complex is $1.21 billion. This includes the costs for OSBL allowance, catalysts, license fees, and other costs and is based on a remote location with an assumed location factor of 15 percent above the cost for a U.S. Gulf Coast location.

GTL/GTP Investment. The addition of an MTO unit and polyolefin units allows the methanol to be converted to olefins and then polyolefins. The conversion to polyolefins is necessary because it would be very costly to ship olefins from remote locations. Polyolefins are economically shipped over long distances. The estimated cost for the integrated GTL/GTP facility is about $2 billion. This includes the costs for OSBL allowance, catalysts, license fees, and other costs and is based on a remote location with an assumed location factor of 15 percent above the cost for a U.S. Gulf Coast location.

GTL/Cracker/Polyolefin Investment. The most common route to polyethylene and polypropylene production today is through steam cracking of naphtha. The GTL liquid products include naphtha boiling-range product. This naphtha is attractive for steam cracking applications because of its high concentration of normal paraffin components. This offers high ethylene yields for naphtha cracking. The naphtha can be shipped from remote sites to industrialized locations with naphtha crackers, and this is the most commonly envisioned outlet for the GTL naphtha product. Since this chapter discusses the potential advantages of integrating polyolefin production with GTL, it is appropriate to include the integration through conventional cracking. The naphtha portion of GTL products can vary considerably depending on the catalyst and operating conditions in the Fischer-Tropsch unit. For the purposes of this chapter it was assumed the naphtha cut accounts for 28 vol % of the total FT liquids. This would provide 14,000 BPSD of naphtha, but this amount by itself is too small to support a world-scale naphtha cracker. The resulting economics would be poor for such a project, so a larger cut of the FT liquids would be necessary.

We assumed that 56 percent of the FT liquids would be used as cracker feedstock. We estimate that this 28,000 BPD of feedstock, rich in normal paraffins, could support the production of 442,000 MTA of polyethylene and 166,600 MTA of polypropylene. This ethylene yield is significantly higher than cracker yields based on conventional feedstocks. The estimated erected cost of the ISBL facilities for such a cracker is $380 million, and the corresponding cost for the polyolefin ISBL facilities is $360 million. The total investment for this complex is estimated at slightly above $2.2 billion. This includes the costs for OSBL allowance, catalysts, license fees, and other costs and is based on a remote location with an assumed location factor of 15 percent above the cost for a U.S. Gulf Coast location.

ECONOMIC COMPARISONS

The integration of methanol production with GTL offers enhanced economics provided the methanol sales price is at around $120/MT or more. However, methanol consumption through conventional applications (i.e., formaldehyde, MTBE, chloromethanes, acetic acid, etc.) offers limited opportunities for remote gas utilization. Current demand for methanol is around 30 million MTA, and it is forecasted to grow to 37 million MTA in 10 years. This additional methanol demand would support the installation of only two or three world-scale methanol plants and consume about 665 million SCF/day (17.8 million N · m^3/day) of natural gas. New methanol projects are also likely to result in closures of some existing plants with higher production costs, but this provides limited opportunities and does not support higher market prices. Many of the alternative uses for methanol (i.e., fuel cells or conversion to olefins) require lower methanol prices to be competitive in their respective markets. In the economic comparisons below, an alternative methanol price of $85/MT is used to give an example of the economics of GTL/methanol integration with the methanol directed toward alternative markets such as MTO. In such a case the economics of the GTL and integrated GTL/methanol plants are essentially the same. See Table 10.2.3.

If methanol is converted to olefins and polyolefins, it further increases the value added for products derived from natural gas. Other papers have compared the economics of remote gas strategies including LNG, GTL, and GTP.[5,6] GTP offers attractive economics at about the same investment level as GTL because of the higher value of the polyolefin products compared to liquid fuels, even when those fuels command a price premium over conventional fuels. GTP offers gross profits equivalent to $5.70/1000 SCF of gas consumed. This is more than four times the corresponding gross profit offered by GTL. This helps GTP to be economical at a smaller scale than GTL, so it can be utilized in moderate- as well as large-sized gas fields.

TABLE 10.2.3 Economic Comparison of GTL Integrated with Methanol Production

	GTL	GTL/MeOH at $120/MT	GTL/MeOH at $85/MT
Investment, million $	1250	1264	1264
Gas consumed, million SCF/day	450	450	450
GTL products, BPSD	50,000	31,000	31,000
Methanol product, MT/D		5309	5309
Gas cost, million $/yr	80	80	80
Operating cost, million $/yr	78	78	78
Total cash cost, million $/yr	158	158	158
Product revenue, million $/yr	391	459	396
Transportation costs	−27	−38	−38
Net revenue, million $/yr	364	421	357
Gross profit, million $/yr	206	263	200
Gross profit, $/million SCF of gas	1.34	1.72	1.31
Gross profit, $/kN · m^3 of gas	50.0	64.2	48.9
Simple ROI	16.5%	20.7%	15.7%
IRR (pretax)	12.8%	16.6%	12.1%

TABLE 10.2.4 Economic Comparison of GTL Integrated with Polyolefin Production

	GTL	GTP	GTL/GTP	GTL/cracker/ polyolefin
Investment, million $	1250	1210	2030	2230
Gas consumed, million SCF/day	450	155	434	450
GTL products, BPSD	50,000	—	31,000	22,000
Polyethylene, MT/D	—	882	882	1300
Polypropylene, MT/D	—	882	882	490
Other by-products, MT/D	—	—	—	1491
Gas cost, million $/yr	80	27	77	80
Operating cost, million $/yr	78	99	150	176
Total cash cost, million $/yr	158	126	227	256
Product revenue, million $/yr	391	452	694	708
Transportation costs	−27	−24	−41	−42
Net revenue, million $/yr	364	428	653	666
Gross profit, million $/yr	206	301	426	410
Gross profit, $/million SCF of gas	1.34	5.72	2.89	2.68
Gross profit, $/kN · m^3 of gas	50.0	213.5	107.9	100.0
Simple ROI	16.5%	24.9%	21.0%	18.4%
IRR (pretax)	12.8%	20.0%	16.9%	14.6%

GTL offers huge potential for gas utilization because it links natural gas to markets historically supplied by products derived from crude oil. This is of strategic importance to many, because the world's gas reserve base is greater than the oil reserve base and gas discovery rates exceed oil discovery rates. When crude oil prices are high, GTL can offer attractive economics, but the potential for lower oil prices raises concerns about economic risks for GTL. One way to help mitigate such risks is to produce products with greater value margins. This is evidenced by the GTL/GTP example shown in Table 10.2.4.

In the example for GTL/GTP, 38 percent of the syngas was used for methanol production and subsequently converted to primarily ethylene and propylene and then converted to polyethylene and polypropylene. The remaining 62 percent of the syngas was converted to FT liquids. Although this requires a substantially greater investment, it doubles the gross profits per thousand standard cubic feet of natural gas consumed and increases the project IRR from about 13 percent for GTL to almost 17 percent for the integrated GTL/GTP project. In this example the MTO C_4+ by-products were used as fuel. This minimizes the amount of by-products from the complex but only provides the minimum value for this material. If these by-products are either shipped separately or blended into the FT liquid product streams, they can provide a significantly higher value and thereby further increase the IRR of the project.

Also shown in Table 10.2.4 are the economics of integrating a conventional steam cracker and polyolefin plants with GTL. There is less synergy in this integration because all the syngas must first be converted to FT liquids. In this example, 22,000 BPSD of FT liquids would remain for shipping in addition to more than 350,000 MTA of cracker liquid by-products. These by-products would consist of crude C_4's (~33 percent), pyrolysis gas (~59 percent), and fuel oil (~8 percent). With a substantially higher investment, additional facilities could be installed to recover butadiene, benzene, toluene, and xylenes from these streams. This would also require additional product storage for these extra products.

The GTL/cracker integration option offers slightly better economics compared to GTL, but it requires the largest investment, produces the largest number of products to be shipped from the remote location, and is less economical than the GTL/GTP integration.

The demand for additional ethylene and propylene capacity is expected to require about 60 million MTA of additional ethylene production and 30 million MTA of additional propylene production by the year 2015. If we assume that about 3 million BPSD of GTL capacity comes on-stream during this same period, then approximately 840,000 BPSD of FT-derived naphtha would be produced and 27 billion SCF/day (723 million N·m^3) of natural gas would be consumed. If this naphtha were cracked to produce ethylene and propylene, then about 14 million MTA of ethylene and 5 million MTA of propylene would be produced. This leaves over 75 percent of the additional ethylene production and over 80 percent of the additional propylene production to be supplied by other sources.

If this same amount of natural gas were consumed in an integrated GTL/GTP facility, about 1.86 million BPSD of GTL liquids or 521,000 MTA of FT-derived naphtha would be produced. In addition, about 18.4 million MTA of ethylene and 18.4 million MTA of propylene would be produced by the MTO process. Assuming the naphtha is shipped to other locations to be cracked to light olefins, this would bring the total ethylene production to 27.1 million MTA (18.4 + 8.7) and the total propylene production to 21.5 million MTA (3.1 + 18.4). This still leaves about 55 percent of additional ethylene and almost 30 percent of additional propylene demand remaining for supply by other routes.

ECONOMIC SENSITIVITY

The economic impact of various corresponding crude oil prices is shown in Fig. 10.2.6. In general, polyolefin prices tend to trend along with crude oil prices, but there can be a lot of scatter in the prices due to market conditions. However, each of these options achieves better economics as oil prices increase. GTL is slightly more sensitive to crude oil pricing, and it will start to approach GTP economics as crude oil prices reach close to $30/bbl.

Stand-alone GTL projects can look attractive when crude oil prices are about $20/bbl or higher. Integrated GTL/GTP projects can offer similar returns when market prices correspond to crude oil priced at $16/bbl or higher.

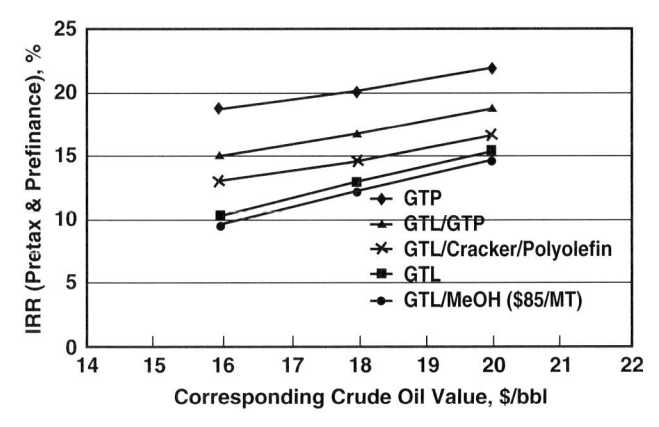

FIGURE 10.2.6 Economic sensitivity to corresponding crude oil prices.

CONCLUSIONS

- Producing products with higher value than fuel-grade materials requires greater investment but enhances GTL project economics.
- Olefin and polyolefin production offers larger market potential for remote gas utilization compared to conventional methanol markets.
- New methanol production technology combined with remote gas pricing can provide methanol delivered at less than $90/MT.
- The UOP/Hydro MTO process provides an ideal link between methanol and polyolefin production at world-scale capacities.
- GTP offers the highest returns for remote gas monetization.
- GTL/GTP integration offers large market potential for remote gas monetization and significantly better economics compared to stand-alone GTL projects, but at a higher investment cost.
- GTL/GTP integration offers lower cost and better economic returns compared to the integration of GTL with conventional cracking and polyolefin production facilities.
- GTL/GTP integration can offer attractive project economics at market prices corresponding to $16/bbl crude oil and higher.

REFERENCES

1. S. L. Wedden, "GTL Prospects," *Oil Gas J.,* 58–63, Mar. 12, 2001.
2. C. Eng, G. A. Peterson, T. Fuglerud, S. Kvisle, and H. Nilsen, "The UOP/Hydro MTO Process for Higher Natural Gas Profitability," Association Française de Techniciens et Professionels du Pétrole (AFTP) Seminar, Paris, France, Oct. 7, 1998.
3. *Proceedings of the Gas-to-Liquids Clean Fuels Strategy Conference,* London, November 1998.
4. Chem Systems, "Developments in Methanol Production Technology," Process Evaluation Research Planning (PERP), 96/97S14, August 1998.
5. B. V. Vora, C. N. Eng, E. C. Arnold, H. Nilsen, S. Kvisle, and T. Fuglerud, "Natural Gas Utilization at Its Best," presented at Petrotech 98, Bahrain, September 1998.
6. J-M. Jaubert, F. Bouvart, S. A. Gembicki, and J. M. Andersen, "Natural Gas to Polyolefins," Qatar, March 2001.

CHAPTER 10.3
UOP OLEFLEX™ PROCESS

Joseph Gregor and Daniel Wei
UOP LLC
Des Plaines, Illinois

INTRODUCTION

The UOP Oleflex™ process is catalytic dehydrogenation technology for the production of light olefins from their corresponding paraffin. An Oleflex unit can dehydrogenate propane, isobutane, normal butane, or isopentane feedstocks separately or as mixtures spanning two consecutive carbon numbers. This process was commercialized in 1990, and by 2004 more than 1,250,000 metric tons per year (MTA) of propylene and more than 2,800,000 MTA of isobutylene were produced from Oleflex units located throughout the world.

PROCESS DESCRIPTION

The UOP Oleflex process is best described by separating the technology into three different sections:

- Reactor section
- Product recovery section
- Catalyst regeneration section

Reactor Section

Hydrocarbon feed is mixed with hydrogen-rich recycle gas (Fig. 10.3.1). This combined feed is heated to the desired reactor inlet temperature and converted at high monoolefin selectivity in the reactors.

The reactor section consists of several radial-flow reactors, charge and interstage heaters, and a reactor feed-effluent heat exchanger. The diagram shows a unit with four reactors, which would be typical for a unit processing propane feed. Three reactors are

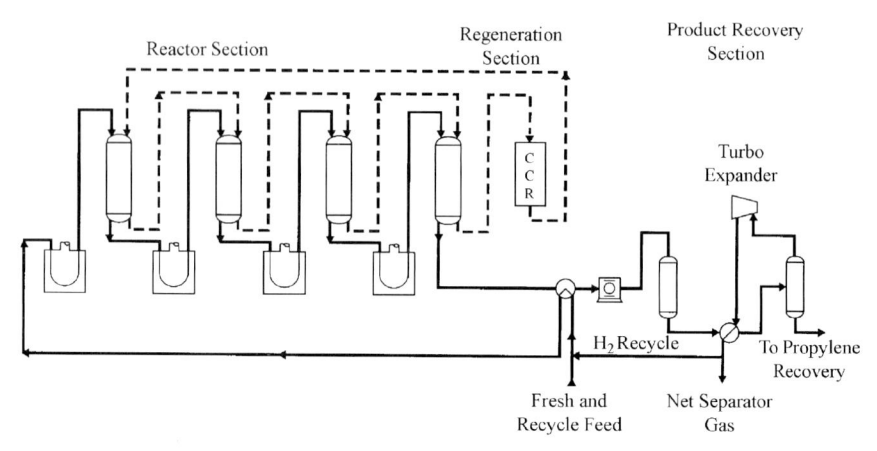

FIGURE 10.3.1 Oleflex process flow.

used for butane or isopentane dehydrogenation. Three reactors are also used for blends of C_3–C_4 or C_4–C_5 feeds.

Because the reaction is endothermic, conversion is maintained by supplying heat through interstage heaters. The effluent leaves the last reactor, exchanges heat with the combined feed, and is sent to the product recovery section.

Product Recovery Section

A simplified product recovery section is also shown in Fig. 10.3.1. The reactor effluent is cooled, compressed, dried, and sent to a cryogenic separation system. The dryers serve two functions: (1) to remove trace amounts of water formed from the catalyst regeneration and (2) to remove hydrogen sulfide. The treated effluent is partially condensed in the cold separation system and directed to a separator.

Two products come from the Oleflex product recovery section: separator gas and separator liquid. The gas from the cold high-pressure separator is expanded and divided into two streams: recycle gas and net gas. The net gas is recovered at 90 to 93 mol % hydrogen purity. The impurities in the hydrogen product consist primarily of methane and ethane. The separator liquid, which consists primarily of the olefin product and unconverted paraffin, is sent downstream for processing.

Catalyst Regeneration Section

The regeneration section, shown in Fig. 10.3.2, is similar to the CCR™ unit used in the UOP Platforming™ process. The CCR unit performs four functions:

- Burns the coke off the catalyst
- Redistributes the platinum
- Removes the excess moisture
- Reduces the catalyst prior to returning to the reactors

The slowly moving bed of catalyst circulates in a loop through the reactors and the regenerator. The cycle time around the loop can be adjusted within broad limits but is

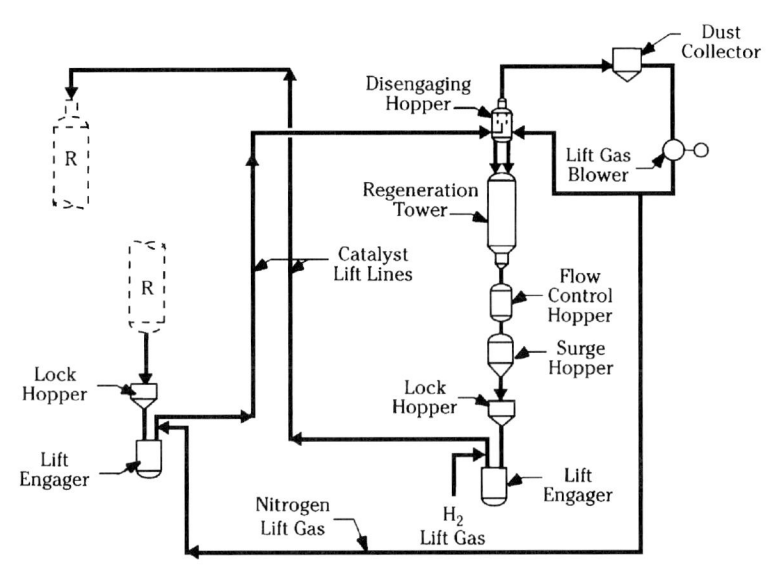

FIGURE 10.3.2 Oleflex regeneration section.

typically anywhere from 5 to 10 days, depending on the severity of the Oleflex operation and the need for regeneration. The regeneration section can be stored for a time without interrupting the catalytic dehydrogenation process in the reactor and recovery sections.

DEHYDROGENATION PLANTS

Propylene Plant

Oleflex process units typically operate in conjunction with fractionators and other process units within a production plant. In a propylene plant (Figure 10.3.3), a propane-rich liquefied petroleum gas (LPG) feedstock is sent to a depropanizer to reject butanes and heavier hydrocarbons. The depropanizer overhead is then directed to the Oleflex unit. The once-through conversion of propane is approximately 40 percent, which closely approaches the equilibrium value defined by the Oleflex process conditions. Approximately 90 percent of the propane conversion reactions are selective to propylene and hydrogen; the result is a propylene mass selectivity in excess of 85 wt %. Two product streams are created within the C_3 Oleflex unit: a hydrogen-rich vapor product and a liquid product rich in propane and propylene.

Trace levels of methyl acetylene and propadiene are removed from the Oleflex liquid product by selective hydrogenation. The selective diolefin and acetylene hydrogenation step is accomplished with the Hüls SHP process, which is available for license through UOP. The SHP process selectively saturates diolefins and acetylenes to monoolefins without saturating propylene. The process consists of a single liquid-phase reactor. The diolefins plus acetylene content of the propylene product is less than 5 wt ppm.

Ethane and lighter material enter the propylene plant in the fresh feed and are also created by nonselective reactions within the Oleflex unit. These light ends are rejected from

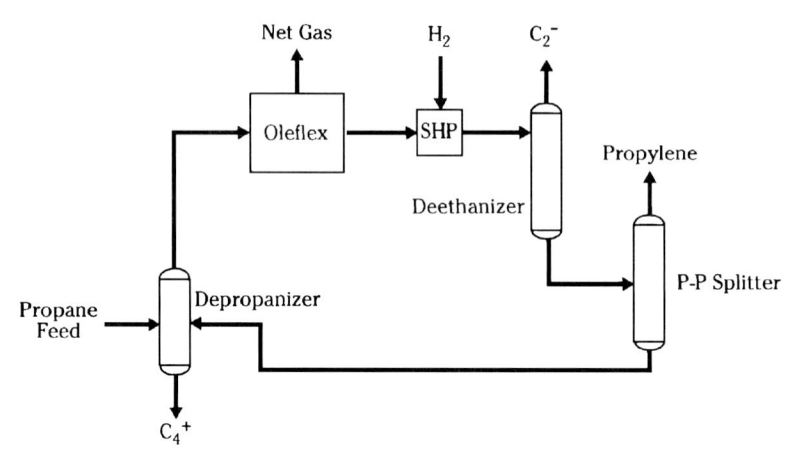

FIGURE 10.3.3 C_3 Oleflex plant.

the complex by a deethanizer column. The deethanizer bottoms are then directed to a propane-propylene (P-P) splitter. The splitter produces high-purity propylene as the overhead product. Typical propylene purity ranges between 99.5 and 99.8 wt %. Unconverted propane from the Oleflex unit concentrates in the splitter bottoms and is returned to the depropanizer for recycle to the Oleflex unit.

Ether Complex

A typical etherification complex configuration is shown in Fig. 10.3.4 for the production of methyl tertiary butyl ether (MTBE) from butanes and methanol. Ethanol can be substituted for methanol to make ethyl tertiary butyl ether (ETBE) with the same process configuration. Furthermore, isopentane may be used in addition to or instead of field butanes to make tertiary amyl methyl ether (TAME) or tertiary amyl ethyl ether (TAEE). The complex configuration for a C_5 dehydrogenation complex varies according to the feedstock composition and processing objectives.

Three primary catalytic processes are used in an MTBE complex:

- Paraffin isomerization to convert normal butane into isobutane
- Dehydrogenation to convert isobutane into isobutylene
- Etherification to react isobutylene with methanol to make MTBE

Field butanes, a mixture of normal butane and isobutane obtained from natural gas condensate, are fed to a deisobutanizer (DIB) column. The DIB column prepares an isobutane overhead product, rejects any pentane or heavier material in the DIB bottoms, and makes a normal butane sidecut for feed to the paraffin isomerization unit.

The DIB overhead is directed to the Oleflex unit. The once-through conversion of isobutane is approximately 50 percent. About 91 percent of the isobutane conversion reactions are selective to isobutylene and hydrogen. On a mass basis, the isobutylene selectivity is 88 wt %. Two product streams are created within the C_4 Oleflex unit: a hydrogen-rich vapor product and a liquid product rich in isobutane and isobutylene.

The C_4 Oleflex liquid product is sent to an etherification unit, where methanol reacts with isobutylene to make MTBE. Isobutylene conversion is greater than 99 percent, and

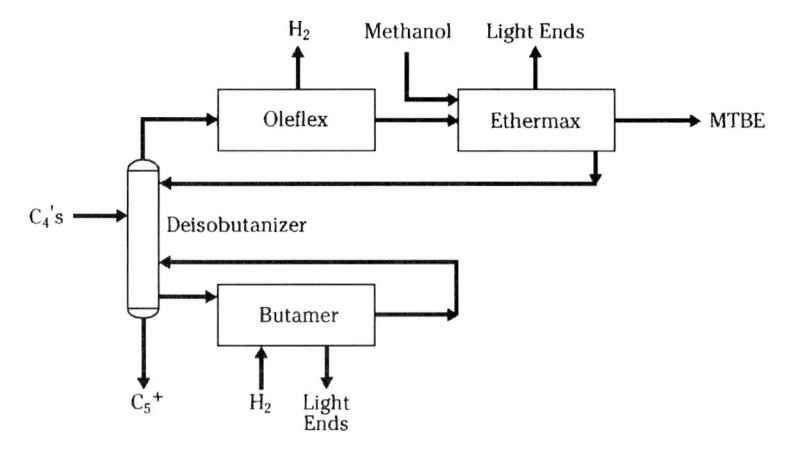

FIGURE 10.3.4 MTBE production facility.

the MTBE selectivity is greater than 99.5 percent. Raffinate from the etherification unit is depropanized to remove propane and lighter material. The depropanizer bottoms are then dried, saturated, and returned to the DIB column.

PROPYLENE PRODUCTION ECONOMICS

A plant producing 350,000 MTA of propylene is chosen to illustrate process economics. Given the more favorable C_4 and C_5 olefin equilibrium, butylene and amylene production costs are lower per unit of olefin when adjusted for any differential in feedstock value. The basis used for economic calculations is shown in Table 10.3.1. This basis is typical for U.S. Gulf Coast prices prevailing in mid-2004 and can be used to show that the pretax return on investment for such a plant is approximately 24 percent.

Material Balance

The LPG feedstock is the largest cost component of propylene production. The quantity of propane consumed per unit of propylene product is primarily determined by the selectivity of the Oleflex unit because fractionation losses throughout the propylene plant are small. The Oleflex selectivity to propylene is 90 mol % (85 wt %), and the production of 1.0 metric ton (MT) of propylene requires approximately 1.2 MT of propane.

An overall mass balance for the production of polymer-grade propylene from C_3 LPG is shown in Table 10.3.2 for a polymer-grade propylene plant producing 350,000 MTA, based on 8000 operating hours per year. The fresh LPG feedstock is assumed to be 94 LV % propane with 3 LV % ethane and 3 LV % butane. The native ethane in the feed is rejected in the deethanizer along with light ends produced in the Oleflex unit and used as process fuel. The butanes are rejected from the depropanizer bottoms. This small butane-rich stream could be used as either a by-product or as fuel. In this example, the depropanizer bottoms were used as fuel within the plant.

The Oleflex process coproduces high-quality hydrogen. Project economics benefit when a hydrogen consumer is available in the vicinity of the propylene plant. If chemical

10.32 PROPYLENE AND LIGHT OLEFINS

TABLE 10.3.1 Utility, Feed, and Product Valuations for Economic Calculations

Utility values		
Fuel gas	$2.80/million Btu	$11.10/million kcal
Boiler feed water	$0.45/klb	$1.00/MT
Cooling water	$0.12/kgal	$0.03/m^3
Electric power	$0.05/kWh	$0.05/kWh

Feed and product values		
C$_3$ LPG (94 LU % propane)	$0.35/gal	$180/MT
Propylene (99.5 wt %)	$0.19/lb	$420/MT

Note: MT = metric tons.

TABLE 10.3.2 Material Balance for a 350,000-MTA Propylene Plant

	Flow rate, MT/h	Flow rate, MTA
Feed:		
C$_3$ LPG (94 LV % propane)	55.00	440,000
Products:		
Propylene (99.5 wt %)	43.75	350,000
Fuel by-products	11.25	90,000
Total products	55.00	440,000

Note: MT/h = metric tons per hour; MTA = metric tons per annum.

hydrogen cannot be exported, then hydrogen is used as process fuel. This evaluation assumes that hydrogen is used as fuel within the plant.

Utility Requirements

Utility requirements for a plant producing 350,000 MTA of propylene are summarized in Table 10.3.3. These estimates are based on the use of an extracting steam turbine to drive the Oleflex reactor effluent compressor. A water-cooled surface condenser is used on the steam turbine exhaust. A condensing steam driver was chosen in this example for the propane-propylene splitter heat-pump compressor.

Propylene Production Costs

Representative costs for producing 350,000 MTA of polymer-grade propylene using the Oleflex process are shown in Table 10.3.4. These costs are based on feed and product values defined in Table 10.3.1. The fixed expenses in Table 10.3.4 consist of estimated labor costs and maintenance costs and include an allowance for local taxes, insurance, and interest on working capital.

TABLE 10.3.3 Net Utility Requirements for a 350,000-MTA Propylene Plant

		Utility cost	
Utility requirements	Consumption	$/h	$/MTA C_3
Electric power	6,500 kW	325	7.43
Boiler feed water	10 MT/h	10	0.23
Cooling water	6,000 m³/h	180	4.11
Fuel gas	(13.1 million kcal/h)	145	3.31
Net utilities			15.08

Note: MTA = metric tons per annum; MT/h = metric tons per hour.

TABLE 10.3.4 Cost for Producing 350,000 MTA of Polymer-Grade Propylene Using the Oleflex Process

	Revenues, million $/year	Costs Million $/year	$/MT C_3
Propylene product	147.0	—	—
Propane feedstock	—	79.2	226.3
Net utilities	—	5.3	15.1
Catalyst and chemicals	—	3.8	10.9
Fixed expenses	—	7.0	20.0
Total	147.0	95.3	272.3

Note: MTA = metric tons per annum; MT = metric tons.

Capital Requirements

The ISBL erected cost for an Oleflex unit producing 350,000 MTA of polymer-grade propylene is approximately $145 million (U.S. Gulf Coast, mid-2004 erected cost). This figure includes the reactor and product recovery sections, a modular CCR unit, a Hüls SHP unit, and a fractionation section consisting of a depropanizer, deethanizer, and heat-pumped P-P splitter. The costs are based on an extracting steam turbine driver for the reactor effluent compressor and a steam-driven heat pump. Capital costs are highly dependent on many factors, such as location, cost of labor, and the relative workload of equipment suppliers.

Total project costs include ISBL and OSBL erected costs and all owner's costs. This example assumes an inclusive mid-2004 total project cost of $215 million, including

- ISBL erected costs for all process units
- OSBL erected costs (off-site utilities, tankage, laboratory, warehouse, for example)
- Initial catalyst and absorbant loadings
- Technology fees
- Project development including site procurement and preparation

Overall Economics

Because the feedstock represents such a large portion of the total production cost, the economics for the Oleflex process are largely dependent on the price differential between propane and propylene. Assuming the values of $180/MT for propane and $420/MT for propylene, or a differential price of $240/MT, the pretax return on investment is approximately 24 percent for a plant producing 350,000 MTA of propylene.

CHAPTER 10.4
ABB LUMMUS GLOBAL PROPYLENE PRODUCTION VIA OLEFINS CONVERSION TECHNOLOGY

Catherine A. Berra and James T. C. Wu
ABB Lummus Global
Houston, Texas

INTRODUCTION

Propylene, a key petrochemical building block, is used to produce a wide range of polymers and intermediates. The major derivatives of propylene are polypropylene, acrylonitrile, propylene oxide, oxo chemicals, and cumene, which are used to make packaging, coatings, textiles, automobiles, medical products, fibers, and other consumer products. As the need for these products grows, so will the need for propylene. Global demand for propylene is forecasted to increase 6 to 8 percent per year and is expected to outgrow current global capacity. Alternative sources of propylene will have to be found to keep up with the growing demand. There is an increasing interest by petrochemical plant owners/refiners to find alternative means of propylene production that would not adversely affect the production of their primary product and at the same time would have low capital investment cost, be energy efficient, and improve gross margin.

As shown in Fig. 10.4.1, the majority of the world's propylene supply is produced in steam crackers and fluidized catalytic cracking (FCC) units. In these processes, propylene is produced as a by-product of ethylene production or transportation fuels. Steam crackers can increase propylene production by reducing cracking severity within the limits of existing equipment. However, many of the latest steam crackers use ethane feed, which produces less propylene than liquid (naphtha/gas oil) crackers. Historically, FCC units also balanced the propylene supply by varying severity (reducing gasoline production). Propane dehydrogenation—another source of propylene—is used when there is an ample supply of low-cost propane available as feedstock. Olefins conversion technology (OCT) from ABB Lummus Global (Lummus) is emerging as the low-capital and low-energy-consumption option for

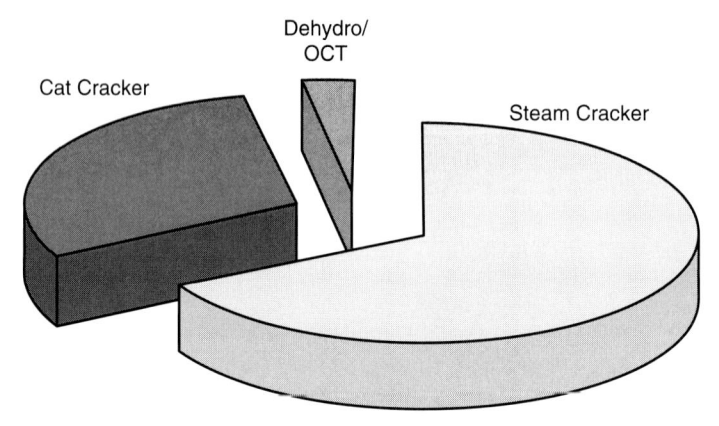

FIGURE 10.4.1 World propylene sources, which produce about 60 million MTA.

the production of propylene. OCT uses olefins metathesis, an equilibrium reaction between two olefins where the double bond of each is broken and new olefins are formed from the exchange of parts of the reactants. For maximum propylene production, OCT reacts butenes and ethylene to form propylene.

With the demand for propylene outpacing the demand for ethylene and C_4's, the metathesis process offers the potential for significant improvement in a steam cracker's or FCC unit's operating margins by increasing propylene production while reducing C_4 product. For the production of propylene, the metathesis reaction consumes 740 kg of 2-butene and 300 kg of ethylene to form 1000 kg of propylene (and 40 kg of C_5 and heavier olefins). Upgrading 740 kg of C_4 olefins, valued slightly above fuel value, to propylene value while reacting only 300 kg of ethylene, which has ranged in value between 90 and 125 percent of propylene over the last decade, increases operating margins. Most recently, propylene price has exceeded ethylene price in many areas of the world.

OCT provides superior economics to competing routes to propylene. In addition, it is an excellent way to vary product slate to cope with the fluctuating demands of downstream operations. It can be built as a stand-alone unit or be integrated with a steam cracker or an FCC unit for improved flexibility and performance.

DEVELOPMENT AND COMMERCIAL HISTORY

Chemists first recognized the metathesis reaction of olefins in the 1950s. The first significant application of metathesis for propylene production was designed by ABB Lummus Global and was commissioned by Lyondell Petrochemical in 1985.

Lummus acquired the technology in 1997 and initiated an intense development program to optimize and broaden the application of metathesis chemistry. Lummus continues to carry out comprehensive pilot plant studies and design developments for process improvements. Sabina, a BASF/Fina joint venture in the United States, will start up an OCT unit mid-2004 as part of the world's largest single-train olefins plant. Mitsui Chemicals, Japan, will start up an OCT unit in late 2004, and Shanghai SECCO Petrochemical Company, China, is scheduled to start up a unit in 2005. Two other units in the Far East, both using refinery C_4's as a feedstock and one using ethylene recovered from FCC offgases, are under design. Combined, these units will have a nameplate capacity over 1.5 million metric tons per annum (MTA) of propylene.

PROCESS CHEMISTRY

For the production of propylene from ethylene and butenes, two main equilibrium reactions take place: metathesis and isomerization. Propylene is formed by the metathesis of ethylene and 2-butene, and 1-butene is isomerized to 2-butene as 2-butene is consumed in the metathesis reaction. In addition to the main reaction, side reactions between olefins also occur.

Main reactions:

$$\underset{\text{Ethylene}}{C_2H_4} + \underset{\text{2-Butene}}{2\text{-}C_4H_8} = \underset{\text{Propylene}}{2\ C_3H_6} \quad \text{Metathesis}$$

$$\underset{\text{1-Butene}}{1\text{-}C_4H_8} = \underset{\text{2-Butene}}{2\text{-}C_4H_8} \quad \text{Isomerization}$$

Typical side reactions:

$$\underset{\text{Propylene}}{C_3H_6} + \underset{\text{1-Butene}}{1\text{-}C_4H_8} = \underset{\text{Ethylene}}{C_2H_4} + \underset{\text{2-Pentene}}{2\text{-}C_5H_{10}}$$

$$\underset{\text{2-Butene}}{2\text{-}C_4H_8} + \underset{\text{1-Butene}}{1\text{-}C_4H_8} = \underset{\text{Propylene}}{C_3H_6} + \underset{\text{2-Pentene}}{2\text{-}C_5H_{10}}$$

$$\underset{\text{1-Butene}}{2\ 1\text{-}C_4H_8} = \underset{\text{Ethylene}}{C_2H_4} + \underset{\text{3-Hexene}}{3\text{-}C_6H_{12}}$$

Based on the reaction stoichiometry, 3 t of propylene are produced from 2 t of butylene and 1 t of ethylene; however, the side reactions result in a slight reduction in feed ethylene consumption and a corresponding increase in 2-butene consumption.

PROCESS DESCRIPTION

Figure 10.4.2 is a simple flow schematic of the OCT process.

Feedstock Treatment

Polymer-grade or dilute ethylene feed streams can be used. Any saturated hydrocarbons, such as ethane and methane, do not react, so the technology can be used with a variety of C_4 streams (e.g., FCC mixed C_4's, steam cracking mixed C_4's, butadiene extraction C_4 raffinate, MTBE C_4 raffinate). The metathesis catalyst is sensitive to catalyst poisons such as arsine, mercury, oxygenates, mercaptans, nitriles, etc., so the feed is treated prior to entering the metathesis reactor. If butadiene is not recovered as a product from a C_4 stream, economics suggest that it be selectively hydrogenated to produce additional butylenes feed for the OCT metathesis reactor. The selective hydrogenation unit offers a highly selective catalyst for the hydrogenation of butadiene to butenes with minimal saturation losses.

Depending on the quantity of isobutanes and isobutenes in the C_4 feed, the unit design may include a deisobutanizer or deisobutenizer to extend reactor run length between regenerations and reduce OCT unit throughput, resulting in an overall lower capital cost plant. This is a catalytic distillation tower that isomerizes 1-butene to 2-butene (i.e., CD*Hydro*® technology) and fractionates isobutane/isobutene from normal butenes to maximize recovery of OCT feed.

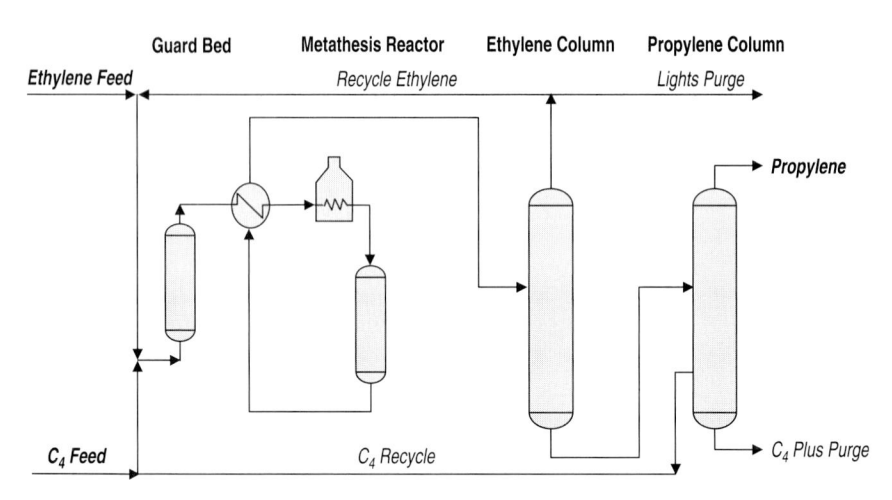

FIGURE 10.4.2 Olefins conversion technology flow diagram.

Reaction and Regeneration

Feed C_4's are mixed with ethylene and heated prior to entering the vapor-phase fixed-bed metathesis reactor where isomerization and equilibrium reactions take place. As mentioned previously, the catalyst promotes the reaction of ethylene and 2-butene to form propylene and simultaneously isomerizes 1-butene to 2-butene. The per-pass conversion of 2-butene is greater than 60 percent, with overall selectivity to propylene exceeding 92 percent. The reactions occurring are essentially isothermal. The OCT reactor effluent contains mainly propylene and unreacted feed. It is cooled and chilled prior to entering the product recovery section.

For regeneration, the coke deposited on the catalyst is burned in a controlled nitrogen-air atmosphere.

Recovery Section

The OCT reactor effluent contains a mixture of propylene, unconverted ethylene and butenes, and some C_5 plus components from side reactions. After cooling, the reactor effluent is sent to the recovery section, which consists primarily of two towers. The first tower separates unreacted ethylene for recycle to the OCT reactor. The second tower processes bottoms from the ethylene recovery tower to produce a polymer-grade propylene overhead product and a C_4 recycle stream. Unlike other routes to propylene, a propylene/propane superfractionator is not required to produce polymer-grade propylene product. Propane is not a reaction by-product, so the only propane in the system is that which is present in the feed. Purge streams containing nonreactive light material and butanes and heavier are removed and sent to OSBL.

PROCESS ECONOMICS

Whether combined with a steam cracker or an FCC unit or as a stand-alone project, the OCT process is a low-cost option for propylene production. It provides product flexibility

TABLE 10.4.1 Material Balance for a Naphtha Steam Cracker

Propylene-to-ethylene ratio	0.65
Naphtha feed consumption	3335 kMTA
Ethylene production	1000 kMTA
Propylene production	650 kMTA
Benzene production	195 kMTA

to increase propylene production and upgrade excess butylenes for higher total product value and improved margins. As an example, we will discuss steam cracker applications.

Typical steam crackers with liquid feedstocks produce ethylene, propylene, and mixed C_4's as a by-product. With C_4's and ethylene readily available, OCT provides an excellent vehicle to upgrade C_4's to high-value propylene. When integrated with a steam cracker, OCT optimizes propylene production and lowers energy consumption.

OCT typically can be applied to the ethylene plant material balance in two ways: to produce a propylene-to-ethylene product (P/E) ratio of 0.6 to 0.7 or to maximize the propylene production from steam cracking to produce a P/E ratio of greater than 1.0.

Table 10.4.1 shows a typical material balance for a naphtha steam cracker producing 1 million MTA ethylene. At a constant net ethylene and propylene production, the steam cracker integrated with the OCT unit considerably improves the overall plant material balance: The integrated case consumes 2 percent less fresh feedstock, produces 50 percent more benzene, and produces only 60 percent of lower-valued pyrolysis gasoline. Using the average U.S. Gulf Coast feedstock and by-product prices for the period from 1991 to 2000, the gross margin (value of products and by-products minus feedstock costs) of the integrated steam cracker/OCT unit is improved by US$16 million per year, a 3 percent increase. The historical prices for other regions of the world result in further improvement.

Another significant benefit is that the energy consumption of the integrated steam cracker/OCT unit is 13 percent lower. The reason for this reduction is that less olefins are produced by thermal cracking in the integrated case, thereby lowering the fired duty of the cracking heaters and the energy consumed in the recovery section. That is, in the steam-cracker-only case, thermal cracking produces 1.65 million MTA of ethylene and propylene. In the integrated case, only 1.48 million MTA of ethylene and propylene are produced by thermal cracking, with the remaining propylene being produced by the energy-neutral metathesis reaction. The 13 percent reduction in energy consumption results in a 13 percent reduction in greenhouse gas emissions. If we define net margin as "value of products and by products minus feedstock and energy costs," the lower energy costs result in an increase in net margin of US$26 million per year for the integrated steam cracker/OCT unit versus a steam cracker alone. This is a 6 percent improvement. This is illustrated in Fig. 10.4.3, which shows that the integrated steam cracker/OCT unit results in

- Reduced fuel consumption
- Additional benzene production
- Reduced low-value gasoline by-products
- Reduced energy consumption

Investment costs are also lower. Capital costs are reduced over US$30 million in total ISBL and OSBL costs. The investment costs associated with the ISBL ethylene plant are reduced as a result of lower plant throughput, lower fired duty, and a significant reduction in the size of the propylene fractionator system, which is the single most costly tower system in the ethylene plant. The investment cost of the pyrolysis gasoline hydrotreater is reduced due to lower throughput and lower diolefin/olefin content. The investment cost of

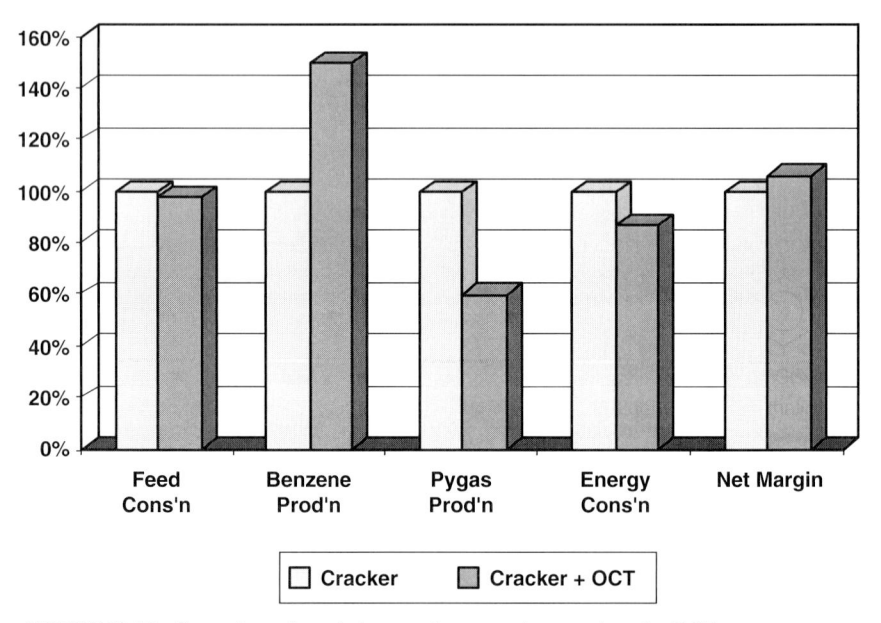

FIGURE 10.4.3 Comparison of stand-alone cracker versus integrated cracker/OCT.

the aromatics extraction unit is also reduced because the aromatic content of the pyrolysis gasoline is higher. Finally, OSBL costs are reduced as a result of the 13 percent reduction in energy consumption. The savings associated with these units more than offset the investment costs associated with the OCT unit.

Metathesis chemistry can be applied to the ethylene plant to push propylene production above conventional levels. The integrated steam cracker/OCT complex operating in maximum propylene mode can produce a P/E ratio exceeding 1.0. This far exceeds the steam-cracker-only conventional maximum P/E ratio of 0.65 to 0.7. The higher P/E ratio substantially improves operating margins compared with conventional crackers.

When applied to FCC units, OCT maximizes propylene and gasoline yield. In conclusion, OCT can be applied to both steam crackers and FCC units to increase propylene production and improve overall profitability.

SUMMARY OF PROCESS FEATURES

- The per-pass conversion of butylene is greater than 60 percent, with overall selectivity to propylene ranging between 92 and 96 percent depending on feedstock composition and reactor operating conditions.
- Ultra-high-purity propylene, exceeding polymer-grade specification, is produced without a propylene fractionation system because the only source of propane is that contained in the C_4 and ethylene feeds.
- There is greatly reduced energy consumption and corresponding greenhouse gas emission compared with other commercial routes to propylene (steam cracking, FCC, or propane dehydrogenation).

- There is no energy input to the reaction step. In fact, the reaction is mildly exothermic. Energy is only consumed in the separation of reaction by-products.
- No superfractionators are required and no special metallurgy, keeping capital cost low.
- The process can be integrated easily with an FCC unit or an ethylene plant for optimization of energy and products.

CONCLUSION

Demand forecasts for ethylene and propylene indicate that the growth rate of propylene will exceed the growth rate of ethylene. The olefins conversion technology (OCT) provides a low-cost solution to expand the propylene flexibility and profitability of the major propylene production processes, namely, FCC and steam crackers. In FCC applications, OCT converts by-product butylenes and ethylene from fuel gas to polymer-grade propylene. In steam cracker applications, OCT can increase steam cracker P/E ratio from the traditional 0.45 to 0.65 range to values greater than 1.0. It also improves the flexibility of the product slate to match varying market demands.

CHAPTER 10.5
PROPYLENE VIA CATOFIN® PROPANE DEHYDROGENATION TECHNOLOGY

V. K. Arora
ABB Lummus Global
Bloomfield, New Jersey

INTRODUCTION

The CATOFIN® dehydrogenation process is a catalytic, highly selective process for the on-purpose production of olefins. The CATOFIN propane dehydrogenation technology employs the same reactor system used in CATOFIN isobutane dehydrogenation and the CATADIENE® process for the production of butadiene. Table 10.5.1 lists all the CATOFIN/CATADIENE units currently in operation.

Approximately 3 percent of the worldwide propylene production of about 55 million metric tons per year (MTA) is produced in propane dehydrogenation (PDH) plants. Most propylene (over 65 percent) is produced from steam crackers as a coproduct with ethylene. The remainder is recovered from fluid catalytic cracking units. Since propylene demand is growing significantly faster than ethylene demand, the role of on-purpose propylene production technology is expected to grow.

The CATOFIN process is an extension of the CATADIENE process, which was employed in the 1960s and 1970s for the production of butadiene from *n*-butane feedstock. In the late 1980s, an oxygenate mandate for the U.S. gasoline pool created a huge demand for isobutylene as a raw material for methyl-tertiary-butyl-ether (MTBE) production. Many iso-C_4-based CATOFIN units were then built for isobutylene production to meet the increasing demand for MTBE. A total of eight CATOFIN iso-C_4 dehydrogenation units were built by the early 1990s, producing about 2.8 million MTA of isobutylene, which in turn produced more than 30 percent of the worldwide MTBE. Most of these CATOFIN units are still in operation.

Three CATOFIN PDH units have been designed with a total production capacity of about 1 million MTA of propylene, including the world's largest PDH unit (455,000 MTA).

TABLE 10.5.1 CATOFIN/CATADIENE Plants Currently in Operation

Client, location	On-stream date	Capacity, MTA	Products primary
Saudi Polyolefins Co., Al-Jubail, Saudi Arabia	Under construction	455,000	Propylene
Dubai Natural Gas	1995	315,000	Isobutylene for MTBE
SABIC/IBN SINA, Al-Jubail, Saudi Arabia	1994	452,000	Isobutylene for MTBE
SABIC/IBN ZAHR 2, Al-Jubail, Saudi Arabia	1993	452,000	Isobutylene for MTBE
North Sea Petrochemicals, Antwerp, Belgium	1991	250,000	Propylene
Super-Octanos, Jose, Venezuela	1991	315,000	Isobutylene for MTBE
Texas Petrochemicals, Houston, Texas	1990	235,000	Isobutylene for MTBE
SABIC/IBN ZAHR 1, Al-Jubail, Saudi Arabia	1988	310,000	Isobutylene for MTBE
TMI, Tobolsk, Russia	1988	180,000	Butadiene

Note: This list does not include several plants that were built and operated but are now shut down due to local economic reasons.

PROCESS CHEMISTRY

The main reaction in the CATOFIN PDH process is the dehydrogenation of propane in the presence of a proprietary catalyst that does not contain any noble metal:

$$C_3H_8 \leftrightarrow C_3H_6 + H_2$$

Side reactions occurring simultaneously with the main reaction cause the formation of some light and heavy hydrocarbons, as well as the deposition of coke on the catalyst.

The main dehydrogenation reaction is endothermic, with a heat of reaction of 720 kcal/kg. The propylene product yield increases with lower pressure and higher temperatures until the feed hydrocarbon temperature exceeds 600°C, where the decrease in selectivity outweighs the conversion improvement.

The CATOFIN PDH process differentiates itself by operating the reactors at subatmospheric pressure. This reduced pressure drives the reaction to high conversion at high selectivity while eliminating the need for hydrogen, which is used in other reaction systems.

PROCESS DESCRIPTION

Figure 10.5.1 is a simple process flow schematic of the CATOFIN PDH process.

Reaction Section

The reaction section of the CATOFIN PDH system consists of a single train of three or more parallel fixed-bed reactors and a regeneration air system. The reactors operate in a cyclic manner so that at any time some reactors are on-stream (for dehydrogenation reaction), some reactors are on reheat/regeneration, and some are on standby. In one

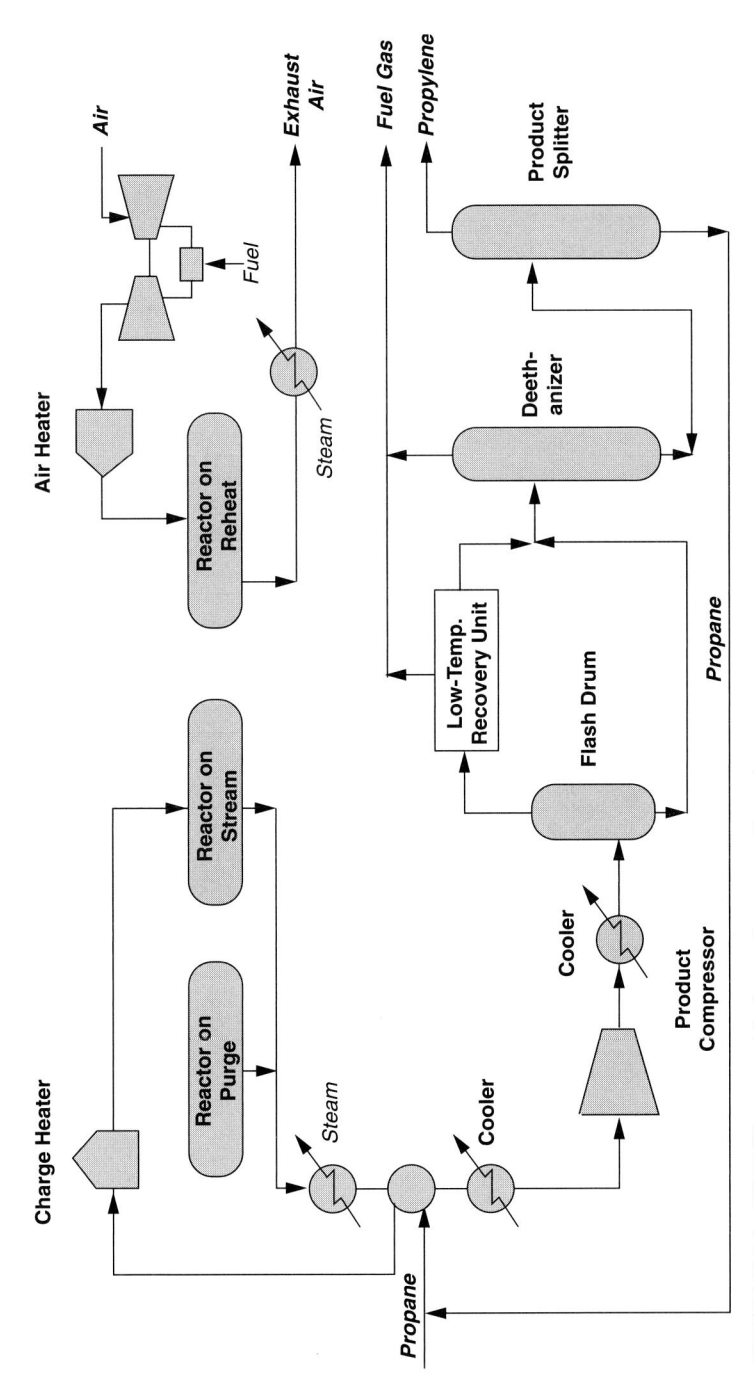

FIGURE 10.5.1 CATOFIN propane dehydrogenation process flow scheme.

10.45

complete cycle—which takes about 10 minutes—the reactor dehydrogenates the hydrocarbon feed, is purged with steam, is blown with air to reheat and decoke the catalyst, is evacuated, and undergoes reduction. Each reactor undergoes this same cycle in a defined automated sequence.

The sequence logistics result in continuous uninterrupted flow of hydrocarbon and air through the entire unit. Hydraulically operated isolation valves that control the flow of process streams to the individual reactors are actuated by centralized cycle timing instrumentation. Mixing of air and hydrocarbon streams is prevented by electrically interlocking the valve operators.

The fresh propane feed is combined with recycle feed from the bottom of the product splitter and vaporized by heat exchange with process streams in the compression section and by steam. The vaporized total feed is then heated by exchange with the reactor effluent in the reactor feed-effluent exchanger. The total feed is then raised to reaction temperature in the gas-fired charge heater and sent to the reactors. Hot effluent from the reactors is cooled by generating steam in the reactor effluent steam generator and by heat exchange with the reactor feed and flows to the compression section of the plant.

In the reactors, the hydrocarbon reactions take place at subatmospheric pressure. At the end of the reaction step, while the system is still under vacuum, the reactor is thoroughly purged with steam, thereby stripping residual hydrocarbons from the catalyst and reactor into the recovery system.

Reheat/regeneration air is supplied typically by a regeneration gas turbine or an air compressor and is heated in the regeneration air heater before passing through the reactors. The regeneration air serves to restore the temperature profile of the bed to its initial on-stream condition in addition to burning the coke off the catalyst. Additional heat is added during the regeneration period by the controlled injection of fuel gas, which is combusted within the catalyst bed.

When reheat/regeneration is completed, the reactor is reevacuated for the next on-stream period. Prior to the introduction of the next propane charge, hydrogen-rich fuel gas is introduced into the reactor for a short period of time to remove adsorbed oxygen from the catalyst bed and, in so doing, add heat. This reduction step also decreases the loss of feed via combustion during the hydrocarbon on-stream period.

The regeneration air stream leaving the reactors flows to a thermal incinerator. The air stream will contain small amounts of CO and hydrocarbons, which are converted to CO_2 and water by thermal or catalytic incineration. NO_x level can be reduced by adding an SCR (selective catalyst reduction) catalyst bed. The flue gas then flows to the waste heat boiler—which generates and superheats additional high-pressure steam—before it is discharged into a stack.

Compression Section

In this section the reactor effluent gas is cooled and then compressed to a suitable level for the operation of the recovery section. Any water that condenses after each stage of compression is separated in the interstage knockout drums and is further steam stripped of hydrocarbons before routing to offsites. The compressor discharge vapor is cooled, and the resulting vapor-liquid is separated in the low-temperature recovery flash drum. The vapor and liquid streams are sent separately to the recovery section.

Recovery Section

The recovery section removes inert gases, hydrogen, and light hydrocarbons from the compressed reactor effluent. The propylene, propane, and heavier components are sent to the product purification section.

The reactor effluent condensate from the compression section is dried and sent to the deethanizer for removal of light hydrocarbons (methane, ethylene, ethane, and inert gases). Uncondensed reactor effluent flows to the low-temperature recovery section, where it is further cooled to condense and recover the remaining C_3 and heavier hydrocarbons. The recovered C_3 stream is also sent to the deethanizer. The reject gas stream from the low-temperature recovery section can be sent to a PSA unit to recover hydrogen, if desired. The remaining gas is sent to the reactor section reduction gas surge drum and the facility fuel gas header. The deethanizer bottoms liquid flows forward to the product purification section.

Product Purification Section

The product purification section is designed to recover a high-purity polymer-grade propylene product from propane and heavier material. The deethanizer bottoms stream from the recovery section is pumped through a sulfur-removal unit before being sent to the product splitter. An open-loop "heat pump" system is used to provide reboiler heat. The product splitter produces 99.5 wt % or higher propylene purity overhead product. The bottoms stream from the product splitter is returned to the CATOFIN reactor section as recycle propane feed.

PROCESS ECONOMICS

The determining factor for economic viability of a PDH unit is the price of propane. The price differential between propane and propylene and the plant location relative to market demand are also important. These factors make regions such as the Middle East preferred locations for PDH plants.

A budgetary installed cost for the ISBL portion of a world-scale CATOFIN PDH unit (455,000 kMTA) is approximately US$180 million. The cost of production (including the capital and personnel elements) for a typical Middle East location is estimated to be in the range of US$260 to US$290 per ton of propylene. Figure 10.5.2 shows typical ranges and the relative contribution of different elements of cost of production.

FEEDSTOCK AND UTILITY CONSUMPTION

The CATOFIN PDH unit can be designed to process a wide variety of feedstocks. It can easily handle feedstocks that typically contain 95 to 97 mol % propane (Table 10.5.2), unlike other technologies that require higher-purity feedstocks. Lower-purity propane streams (containing as much as 3 mol % olefins and about 7 mol % C_4's) have been handled in the commercial CATOFIN PDH units without any feed pretreatment; however, any increase in the level of the main impurities would increase the amount of offgas and/or other by-products and would affect the utility consumption, etc.

Typical feedstock and utilities consumptions for a world-scale CATOFIN PDH unit producing 99.5 mol % purity propylene are shown in Tables 10.5.3 and 10.5.4, respectively. Feedstock purity will affect these values to some extent.

PRODUCT QUALITY AND BY-PRODUCTS

Typical propylene product specifications are shown in Table 10.5.5. Offgas from the low-temperature recovery section and the deethanizer, as well as small quantities of heavier

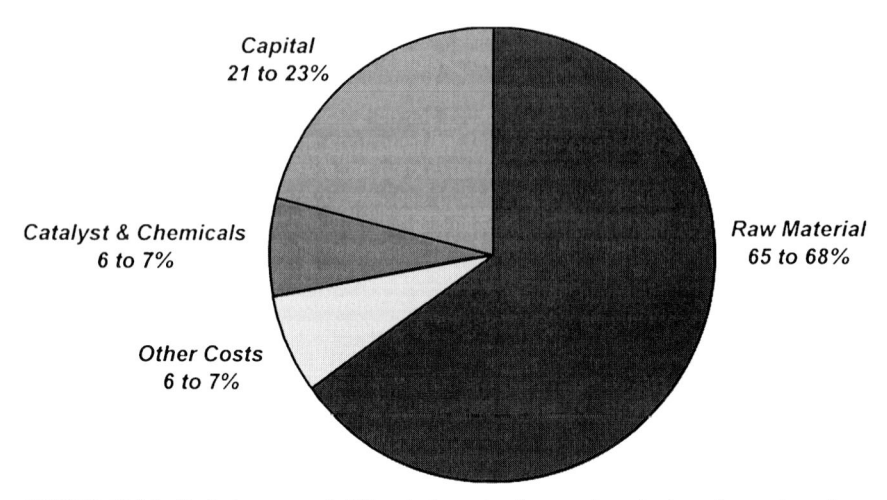

FIGURE 10.5.2 Typical ranges of different elements of cost of production of propylene from CATOFIN process.

TABLE 10.5.2 Typical Feedstock Composition

Propane	95–97 mol %
Ethane	1.0–3.0 mol %
Butanes	1.0–3.0 mol %
Pentanes and heavier	Nil
Water	10 ppmw max
Sulfur	30 ppmw max
Metals	5 ppmw max

TABLE 10.5.3 Raw Material per Metric Ton of Product

Propane feed	536,900 MTA (100% C_3 basis)
Polymer-grade propylene product	455,000 MTA

TABLE 10.5.4 Typical Utilities per Metric Ton of Product

Power	60 kWh
Boiler feed water	0.67 metric ton (net import)
Cooling water	78 m^3
Fuel	2.5–3.0 MW
Inert gas	15–25 N·m^3

TABLE 10.5.5 Typical Product Specifications

Propylene	99.5 wt % min
Ethane + propane	0.5 wt % max
Ethylene	100 wppm max
Diolefins + acetylenes	25 wppm max
Total sulfur as H_2S	1 wppm max
Carbon monoxide	30 wppb max
Carbon dioxide	2 wppm max
Oxygen	2 wppm max
Water	2 wppm max

hydrocarbon waste streams, are normally burned as fuel. Approximately 65 to 70 percent of the hydrogen in the offgas from the low-temperature section can be recovered as high-purity hydrogen (99.99 percent), if desired, in a pressure swing adsorption (PSA) unit.

CATALYST AND CHEMICAL CONSUMPTION

The dehydrogenation reactors contain a mixture of alumina-based proprietary catalyst and inerts. The expected catalyst life is 2 years. The inert materials are recovered and reused when the catalyst is changed. The waste heat boiler is also provided with SCR catalyst to minimize NO_x emissions.

Small quantities of sulfiding agent, neutralization agent, wash oil, and adsorbents are used in the process.

The annual cost for catalyst and chemicals/adsorbents based on producing 455,000 MTA of polymer-grade propylene is between US$8.0 and US$9.0 per metric ton of propylene product.

ENVIRONMENTAL EMISSIONS

The emissions from the CATOFIN PDH process are marginal. The liquid effluents include wastewater (approximately 0.1 m^3 per metric ton of propylene), which is easily treatable in the plant's wastewater facility. NO_x emissions are minimized by using low-NO_x burners in the fired sources in addition to an SCR system in the waste heat boiler. Spent catalyst and adsorbents typically are landfilled without any prior treatment.

SUMMARY OF TECHNOLOGY FEATURES

- Feedstock consumption is lower (averages less than 1.2 metric tons of propane per metric ton of propylene over the catalyst life) than other PDH technologies for the following reasons:
  - Lower-pressure operation favors higher conversion.
  - Lower residence time favors higher selectivity.
  - Lower hydrocarbon preheat temperature (not exceeding 600°C) favors higher selectivity (average ~ 89 percent).

- The process has a high tolerance for feed impurities. Olefins (~3 wt %) and total C_4's (~7 wt %) in the feed to the CATOFIN reactors have been demonstrated commercially for a long period of time.
- No hydrogen recycle is required for coke suppression due to the mild operating conditions. The addition of hydrogen reduces propylene yield and also increases energy demand due to higher compression duty.
- The small amount of coke make (less than 0.1 wt % of catalyst) is burned off completely in each regeneration cycle, and the associated heat is used for the dehydrogenation reaction.
- Sulfur injection for metal passivation is very low due to lower operating pressure.
- The catalyst cost is low because it uses nonprecious metal.
- There are no catalyst losses in the fixed-bed configuration of the reactor.
- The catalyst is regenerated by hot air instead of by chemicals.
- Mild operating conditions pose no significant fouling problems.
- The high-efficiency cold box design minimizes equipment count and refrigerant compressor power demand.
- The low-pressure product splitter eliminates the need for feed pumps.
- The low-pressure product splitter, integrated with the propylene refrigeration system as an open-loop heat-pump system, eliminates the need of a separate propylene compressor for the product splitter.
- The process generates sufficient steam to meet the demand of its process users and the compressor drivers (i.e., no steam import).

PART · 11

STYRENE

CHAPTER 11.1
LUMMUS/UOP "CLASSIC" STYRENE TECHNOLOGY AND LUMMUS/UOP SMART[SM] STYRENE TECHNOLOGY

Stephen Pohl and Sanjeev Ram
ABB Lummus Global
Bloomfield, New Jersey

INTRODUCTION

Styrene monomer (SM), or vinylbenzene, is an important industrial chemical used in the manufacture of a variety of polymers. The activity of the vinyl group makes SM easy to polymerize on its own or with many other monomers. Commercial processes for the production of styrene by the dehydrogenation of ethylbenzene (EB) were developed in Germany and the United States in the 1930s. The demand for synthetic rubber during World War II resulted in a large boost in production, and growth in the use of styrenic polymers has continued to fuel demand. Production capacity for styrene in 2003 exceeded 22 million metric tons per annum (MTA).

Over 60 percent of all styrene currently produced is used in the manufacture of polystyrene. The remainder is used for the production of a wide range of liquid and solid copolymers, such as styrene acrylonitrile (SAN), acrylonitrile butadiene styrene (ABS), styrene butadiene (SB) latex, and styrene butadiene rubber (SBR).

Early plants for the dehydrogenation of EB used either an adiabatic reactor with a large cofeed of steam or a tubular reactor with a more isothermal temperature profile. As plants became larger, the adiabatic reactor system proved to be more economical and eventually evolved into a two-reactor system with interstage reheat. This configuration continues to be the most widely employed. Improvements in the iron oxide catalysts and in plant designs have allowed a reduction in the amount of cofeed steam. Modern styrene processes use less than one-half the steam of the early designs and have around one-third the losses to byproducts.

11.4 STYRENE

PROCESS PERSPECTIVE

The Lummus/UOP "classic" SM dehydrogenation technology was first put into commercial practice in 1972. The unique dehydrogenation reactor system is designed to operate at the minimum economical operating pressure ("deep" vacuum) to achieve the highest SM selectivities at high conversions. Since its introduction, the technology has been selected for more than 50 major projects; 42 plants are currently in operation. The complete list of "classic" SM units licensed by Lummus/UOP is shown in Table 11.1.1.

The Lummus/UOP "classic" SM dehydrogenation technology uses commercial iron oxide–based dehydrogenation catalysts in two high-temperature (>600°C) adiabatic reaction stages. Interstage reheating is required because of the endothermic nature of the dehydrogenation reaction. Reactor temperatures are maintained by heat exchange with superheated (up to 900°C) steam, which is used subsequently as cofeed to the reactor.

TABLE 11.1.1 "Classic" SM Project Awards

Client	Location	MTA	Awarded	Status
Chevron Phillips Chemical	Saudi Arabia	715,000	2002	Engineering
BP-SPC (SECCO)	China	500,000	2002	Engineering
Asahi Kasei	Japan	330,000	2001	Construction
BYPC	China	84,000	2001	Operating
Kaucuk	Czech Republic	170,000	2000	Operating
P.T. SMI II	Indonesia	200,000	1996	Operating
SADAF	Saudi Arabia	500,000	1996	Operating
Kaucuk	Czech Republic	130,000	1996	Operating
Innova	Brazil	180,000	1996	Operating
Nova (3 awards)	Canada	427,000	1996	Operating
PT TPPI	Indonesia	500,000	1996	On hold
Jilin	China	100,000	1996	Operating
Chemical Company Dwory	Poland	100,000	1995	Operating
BASF	Belgium	245,000	1995	Operating
TPI	Thailand	200,000	1995	Operating
Yangzi/BASF	China	120,000	1994	Operating
Angarsk	Russia	120,000	1993	On hold
Maoming	China	100,000	1992	Operating
AKPO	Former CIS	370,000	1992	On hold
Mitsubishi Chemicals	Japan	350,000	1992	Operating
Grand Pacific (2 awards)	Taiwan	130,000	1992	Operating
Supreme	India	100,000	1992	On hold
Supreme (Polychem)	India	60,000	1992	On hold
Idemitsu IV	Malaysia	200,000	1992	Operating
Pemex II	North America	250,000	1992	On hold
Chiba Styrene Monomer	Japan	250,000	1992	Operating
Mitsui Chemical	Japan	240,000	1991	Operating
TPC	Iran	95,000	1991	Operating
P.T. GSP	Indonesia	120,000	1990	On hold
Petroflex	Brazil	80,000	1990	Canceled
Confidential	Asia	80,000	1990	Canceled
P.T. SMI I	Indonesia	100,000	1990	Operating
NSMC	Japan	200,000	1988	Operating
LG II	Korea	180,000	1988	Operating
Idemitsu III	Japan	200,000	1988	Operating

TABLE 11.1.2 SMART Styrene Project Awards

Client	Location	MTA	Awarded	Startup	Status
BYPC	China	84,000	2001	2001	Operating
Kaucuk	Czech Republic	170,000	2000	2001	Operating
Nova	Canada	427,000	1996	1998	Operating
Angarsk	Russia	120,000	1993	—	On hold*
Mitsubishi	Japan	350,000	1992	1995	Operating
TPC	Iran	95,000	1991	1998	Operating
Petroflex	Brazil	80,000	1990	—	Canceled*
Confidential	Asia	80,000	1990	—	Canceled*

*Basic engineering designs have been completed.

The Lummus/UOP SMART[SM] SM process uses in situ oxidative reheat technology to eliminate the need for some of the heat-exchange equipment. The oxidative reheat also eliminates a portion of the hydrogen produced in the dehydrogenation reaction, releasing the constraint of thermodynamic equilibrium. In conventional dehydrogenation units, selectivity and conversion are linked so that high conversion and high selectivity cannot be achieved simultaneously. To achieve high selectivity, conversion is limited to less than 70 percent. However, the SMART SM technology permits both high selectivity and high conversion and also results in considerably less flow through the plant for a given SM capacity, thereby increasing efficiency and reducing capital cost.

In this respect, the SMART SM process is uniquely qualified for increasing the capacity of existing plants at the lowest investment. The combination of higher EB conversion and interstage reheat provided by the oxidative reheat step results in debottlenecking the usual constraining equipment—EB/SM fractionation and steam superheater systems—to provide more than 50 percent additional capacity.

The introduction of oxidative reheat eliminates the indirect reactor reheat equipment, simplifies the steam superheat system, and significantly reduces EB recycle because of high EB conversion per pass. This permits a significant expansion of plant capacity without the need to replace major and expensive equipment, such as the steam superheater, the waste heat exchangers on the reactor effluent stream, the offgas compressor, and the EB/SM splitter. The lower EB recycle flow rate reduces utility consumption.

The first commercial SMART SM unit to come on stream was in 1995 when Mitsubishi started its plant in Kashima, Japan. Table 11.1.2 lists the SMART SM projects awarded and identifies the five plants currently in operation.

The 130,000-MTA SM plant of Kaucuk a.s. in the Czech Republic was expanded to 170,000 MTA with the addition of a SMART reactor. The capacity of the 60,000-MTA plant of Beijing Yanshan Petrochemical Corporation (BYPC) similarly was increased to 84,000 MTA. Due to the simplicity of the plant modifications, both projects were completed within 12 months from project award.

PROCESS CHEMISTRY

The major reaction in the "classic" styrene (SM) process is the dehydrogenation of EB to SM and hydrogen (H_2). Side reactions produce benzene and toluene and some light compounds. The reactions are represented in Fig. 11.1.1.

The major reactions in the SMART process are the dehydrogenation of EB and the subsequent oxidation of the hydrogen (H_2) produced in the first reaction. Dehydrogenation side reactions produce benzene and toluene and some light compounds. The reactions are shown in Fig. 11.1.2.

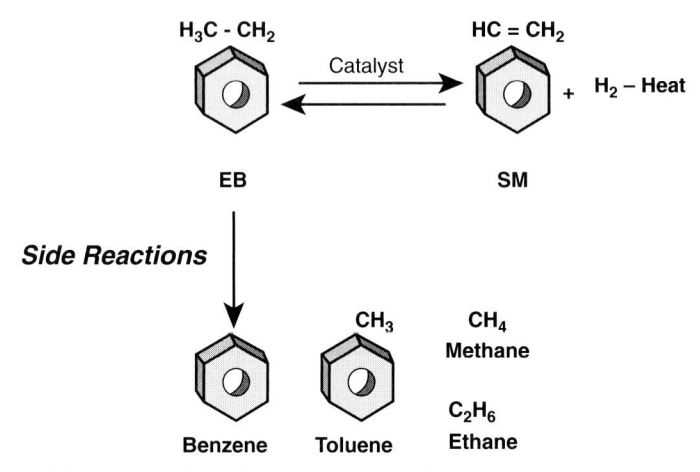

FIGURE 11.1.1 "Classic" styrene process chemistry.

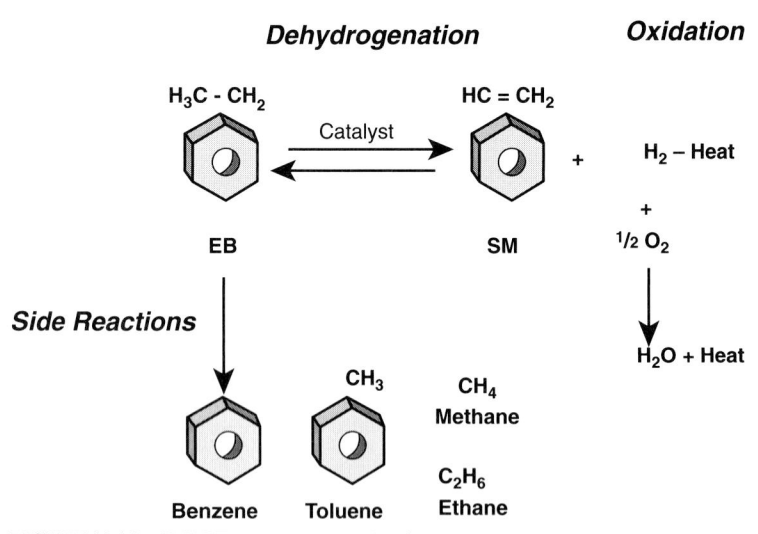

FIGURE 11.1.2 SMART styrene process chemistry.

PROCESS DESCRIPTIONS

"Classic" Styrene

Figure 11.1.3 is a simple process flow diagram of the "classic" SM process. EB (fresh and recycle) and a portion of steam are combined and superheated. The EB-steam mixture is

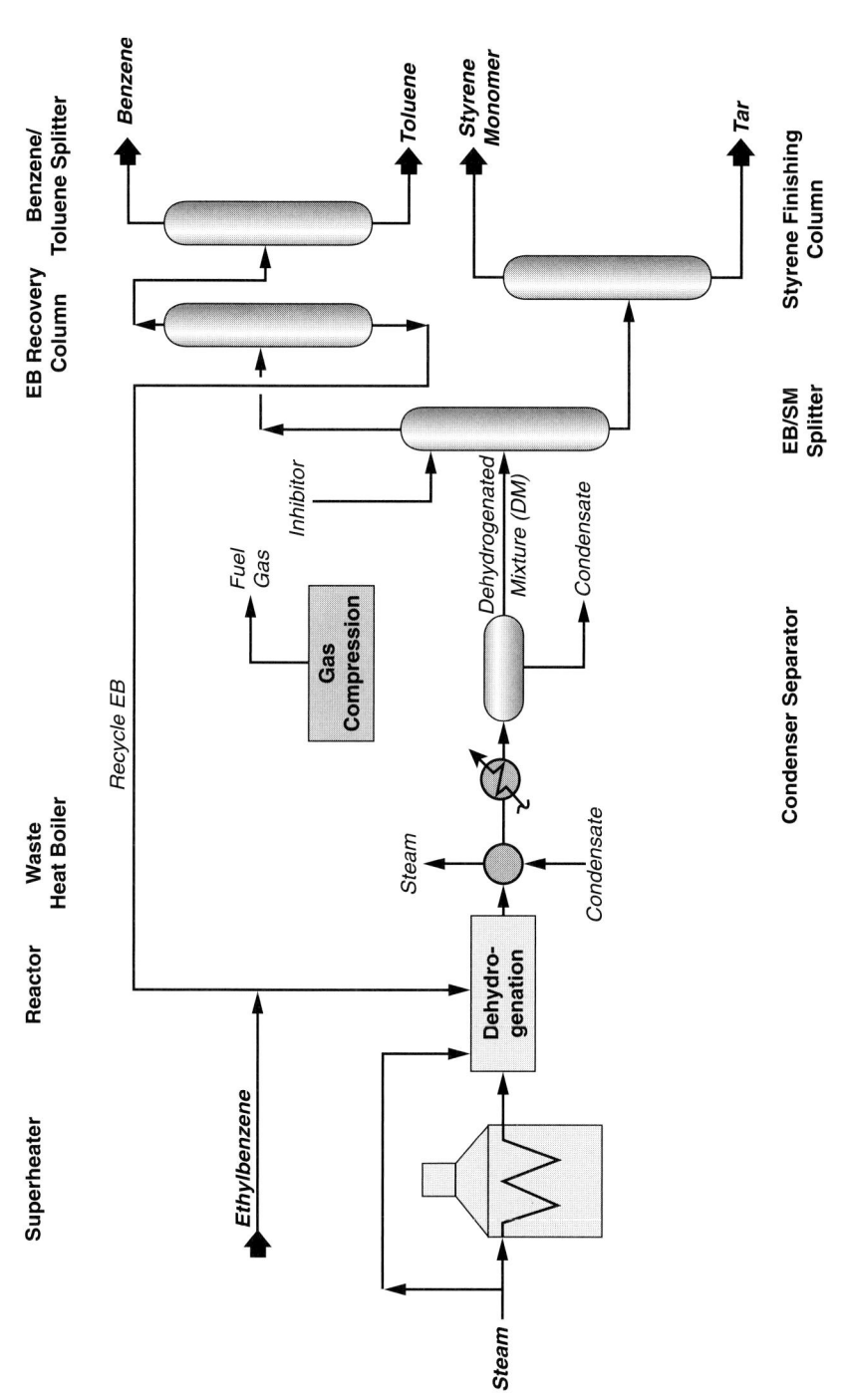

FIGURE 11.1.3 "Classic" styrene process flow diagram.

11.7

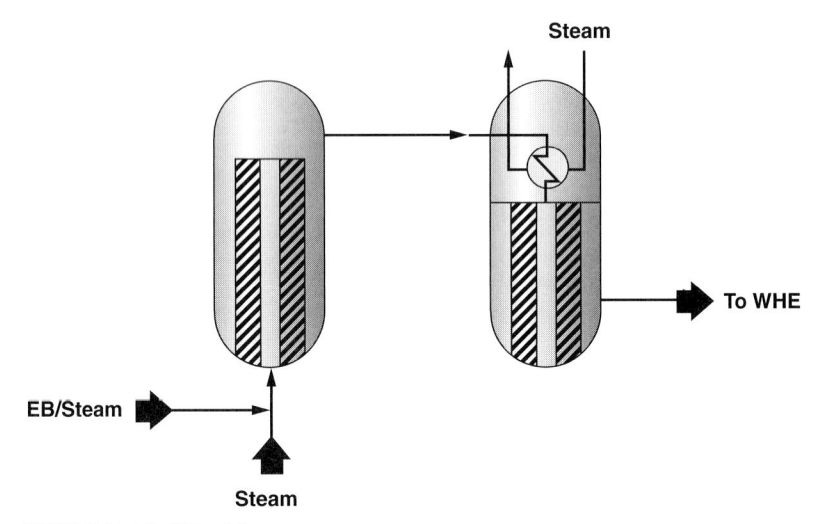

FIGURE 11.1.4 "Classic" styrene reactor system.

combined with additional superheated steam and dehydrogenated in a multistage radial-flow catalytic reactor system (Fig. 11.1.4).

The hot reactor effluent is cooled and condensed in a waste heat exchanger to recover heat. The noncondensable gases are compressed and then treated to recover hydrocarbon prior to being fed to the fuel gas system. The liquid is separated into an aqueous condensate layer and the organic dehydrogenated mixture (DM).

Product SM is separated from DM (styrene, unreacted EB, benzene, toluene, and higher boilers) by fractionation. The DM is first distilled in the EB/SM splitter to take EB and lighter components overhead. The EB/SM splitter overhead is fed to the EB recovery column, where by-products benzene and toluene are separated from EB. The EB stream is recycled to the dehydrogenation reactor. The EB recovery column overhead stream is further split into useful high-purity benzene and toluene by-products or sent to an adjacent refinery as a benzene-toluene mixture.

EB/SM splitter bottoms (SM and heavier) are fractionated in the SM finishing column, which produces SM product as overhead and a very small amount of tar as bottoms. The tar normally is used as fuel in the steam superheater.

SMART Styrene

Except for a small difference in the reactor design, the process flow diagram for the SMART SM process is identical to the "classic" SM scheme. As shown in Fig. 11.1.5, the multistage radial-flow catalytic reactor system contains two catalysts, a conventional EB dehydrogenation catalyst and an oxidation catalyst. In the oxidative reheat step, hydrogen formed from the EB-to-SM endothermic dehydrogenation reaction is reacted with oxygen over a highly selective catalyst. The removal of hydrogen from the process substantially increases single-pass EB conversion while maintaining high SM selectivity. The heat generated is used to reheat the reaction mixture, thereby avoiding expensive interstage heat-exchange equipment.

The hot reactor effluent is cooled and condensed in a waste heat exchanger to recover heat. The noncondensable gases are compressed and treated for hydrocarbon recovery prior to sending to the fuel gas system. The liquid is separated into an aqueous condensate

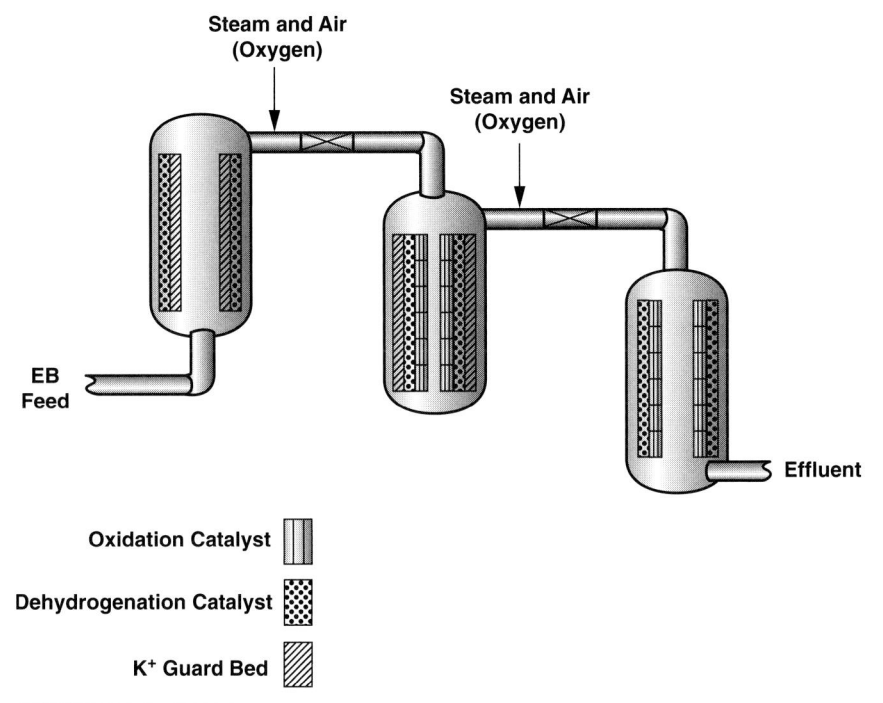

FIGURE 11.1.5 SMART styrene reactor system.

TABLE 11.1.3 Typical Range of Operating Parameters

	"Classic" SM	SMART SM
Expected consumption	1.055–1.060	1.065–1.075
Steam-to-oil ratio	1.0–1.15	1.15–1.50
Catalyst run length	2.5 yr	2.5 yr
Liquid hourly space velocity	0.45 h^{-1}	0.45 h^{-1}
Conversion per pass	63%–65%	65%–75%
Reactor temperatures, SOR/EOR	622/627°C	620/620°C
Last reactor outlet pressure, psia	5.8	6.0–8.0

layer and the organic dehydrogenated mixture (DM). The separation of product SM from DM and all steps thereafter are the same as in the "classic" SM process.

The typical range of operating parameters for the two processes is shown in Table 11.1.3.

ECONOMICS

Investment Costs

A typical "classic" styrene unit of 500,000-MTA capacity has an ISBL U.S. Gulf Coast third-quarter 2003 investment cost of approximately US$75 million.

TABLE 11.1.4 Typical Ethylbenzene Feed Specifications

Ethylbenzene	99.8–99.95 wt %
Color, APHA	15
Specific gravity (15.5°C/15.5°C)	0.869–0.872
Acidity	Neutral
Benzene	Balance
Toluene	500 wt ppm
Nonaromatics	500 wt ppm max
Diethylbenzenes	10 wt ppm max
C_{10} aromatics	300 wt ppm max

TABLE 11.1.5 Typical Styrene Product Quality

Styrene	99.9 wt % min*
Color, APHA	10 max
Polymer	10 wt ppm max
Sulfur (as S)	1 wt ppm max
Aldehydes (as benzaldehyde)	30 wt ppm max
Peroxides (as hydrogen peroxide)	20 wt ppm max
Chlorides (as chlorine)	2 wt ppm max

*Purity of 99.95+ wt % can be attained.

TABLE 11.1.6 Raw Materials/Utilities per Metric Ton Product

Ethylbenzene, kg	1055
Utilities, $US	29

Feedstocks

EB feedstock for "classic" and SMART SM plants typically has the specifications shown in Table 11.1.4.

Product Quality and Yields

The typical SM product quality is shown in Table 11.1.5. SM product purity ranges from 99.85 to 99.95 percent. The process provides high product yield due to a unique combination of catalyst and operating conditions used in the reactors and the use of a highly effective polymerization inhibitor in the fractionation columns.

Raw Materials and Utilities Consumption

Table 11.1.6 summarizes raw materials and utilities consumptions for the process.

SUMMARY OF PROCESS FEATURES

- The designs use a unique reaction system to achieve high single-pass conversion and high selectivity while providing safe, reliable operation. Consisting of the steam superheater, superheated steam transfer piping, reactors, and reactor effluent exchanger, the reaction system is designed as a mechanically and thermally integrated unit. This results in a safe, compact, and reliable design.
- Most of the fuel requirement for the steam superheater is supplied by the dehydrogenation offgas and styrene tar.
- Based on operating experience, the useful life of the dehydrogenation catalyst and the oxidation catalyst is about 24 to 30 months.
- A proprietary combination styrene polymerization inhibitor is used in the styrene distillation system. A true inhibitor is combined with a retardant inhibitor to provide for the maximum cost benefit.
- A patented azeotropic heat-recovery system recovers about 500 kcal/kg SM product of low-level energy for use in the SM plant. The system does not require any compression equipment.
- Pressure drop throughout the dehydrogenation system, including the dehydrogenation reactors, waste heat exchangers, and the reactor effluent condensing system, is minimized to provide the highest possible pressure at the compressor suction while at the same time providing the lowest practical pressure in the dehydrogenation reactors. Thus low pressure, which favors selectivity, is achieved in the reactors without incurring penalties in compression energy or capital cost.

The Lummus/UOP "classic" SM technology offers the following advantages based on the integrated dehydrogenation reactor system design and "deep" vacuum/adiabatic conditions:

- Lowest requirement for steam cofeed to the reactors (steam-to-oil ratio) for an adiabatic technology. With the azeotropic heat-recovery system, total steam requirements are lower than any competing process.
- Long reactor run lengths while using advanced high-selectivity dehydrogenation catalyst.

The Lummus/UOP SMART SM technology offers the following process advantages over conventional EB dehydrogenation-only SM technologies:

- Removal of hydrogen shifts the dehydrogenation reaction equilibrium to obtain up to 80 percent single-pass EB conversion, thereby providing low-cost plant expansion.
- Oxidative exothermic heat of reaction provides reactor interstage heating.
- The steam superheater system is simpler.
- Competitive SM selectivities are obtained at high EB conversion.
- Oxidation catalyst life matches dehydrogenation catalyst life of about 24 to 30 months.

CHAPTER 11.2
STONE & WEBSTER (BADGER) STYRENE TECHNOLOGY

Vincent Welch
Badger Licensing LLC
Cambridge, Massachusetts

INTRODUCTION

Badger Licensing LLC, now part of Stone & Webster, jointly licenses technology for the manufacture of styrene monomer (SM) with ATOFINA Petrochemical. Since the partnership was formed in the early 1960s, the Badger-ATOFINA partnership has developed six unique generations of technology and has licensed 47 plants having a cumulative annual operating capacity in excess of 9 million metric tons. To provide the styrene plant with feedstock ethylbenzene (EB), Badger jointly licenses the complementary EB technology in partnership Exxon/Mobil. Given that there is little or no use for ethylbenzene except as feedstock for styrene production, the two technologies are highly integrated and usually are licensed together. Moreover, Badger licenses two styrene-related ancillary technologies: phenylacetylene reduction technology, which greatly reduces the concentration of trace levels of phenylacetylene impurity contained in the finish styrene product, and catalyst stabilization technology, an innovative technique that replaces the key activity promoter contained on the dehydrogenation catalyst, thus greatly extending catalyst life and/or increasing EB conversion.

STYRENE INDUSTRY

Styrene is one of the most important monomers produced by the petrochemical industry today. It is a vital commodity chemical traded in large volumes across the industrialized world. In 2005, the demand for styrene monomer will approach 24 million metric tons per annum (MTA). For the next 4 to 5 years, the projected worldwide annual market growth rate will average about 4.5 percent. In developing regions, such as Northeast Asia, most notably in China and India, the market growth is increasing more rapidly. Styrene demand in this region is currently much greater than manufacturing capabilities. As a result, styrene is imported to these regions from the Middle East, Korea, Singapore, and Japan. Given the relatively low cost of ethylene feedstock and energy, producers located in countries such as Saudi Arabia have a distinct advantage in variable operating cost as compared with others outside the region. In the long term, styrene producers in North America, faced with escalating energy prices and more stringent environmental regulations, will need to improve efficiencies to remain competitive. Not unexpectedly, several U.S. producers are currently revamping their older plants to achieve significantly reduced energy consumption.

Styrene is used widely in polymers for several reasons. Unlike some other monomers, styrene is a liquid that has a low vapor pressure at ambient conditions, making it easy to ship, handle, and store. In relation to other common polymer building blocks, the consumption of styrene ranks fourth behind polyethylene, vinyl chloride, and polypropylene. Styrene can be polymerized or copolymerized by many common methods to a large number of polymers of different properties.

The raw materials for SM production are benzene and ethylene, both of which are produced in very large quantities and are readily available from most petrochemical or refinery complexes. Frequently, EB/SM units are located in close proximity to an ethylene cracker, the exception being on the U.S. Gulf Coast, where ethylene is usually supplied by pipeline. Conversely, benzene can be supplied from within the refinery-petrochemical complex or shipped to more remote locations either by pipeline or barge. To be most competitive, many SM producers are back-integrated; that is, they have either an ethylene cracker or benzene production facilities, and some producers have both.

Industry-wide, there are about 50 companies that produce SM, some of the largest being Dow, BASF, ATOFINA, SABIC, and Shell. Many companies have multiple production facilities often located at the same site. While the number of plants operating within the industry fluctuates year to year, presently there are approximately 85 individual plants producing styrene. The nameplate capacity of styrene plants varies greatly, ranging from 35 kMTA to over 700 kMTA, with the average-sized plant being roughly 250 kMTA. For reasons of investment economies of scale and lower overhead costs, the nameplate capacity of new grassroots facilities is typically in the range of 500 to 700 kMTA.

The marketplace and the chemistry of downstream polymer application largely dictate product quality specifications for SM. In recent years, there has been a trend toward higher and higher purity. While for decades the typical market-grade SM purity had been 99.80 wt %, most newer plants now manufacture 99.90 wt % or even as high as 99.95 wt %. In stark contrast, a few SM plants that are located in close proximity to polystyrene units produce low-purity styrene, 99.7 wt % or less. In these cases, the SM product impurities (nonpolymerizables) are recycled to the styrene unit, thus avoiding the associated yield loss. Impurities of most interest to polymer manufactures are ethylbenzene, α-methylstyrene, phenylacetylene, cumene, xylenes, aldehydes, peroxides, and polymer.

USE OF STYRENE MONOMER

SM is a basic building block for the manufacture of a broad range of plastics materials. Styrene undergoes polymerization by all the common methods used in plastics technology to produce a wide variety of polymers and copolymers. The most important products are solid polystyrene (PS), general-purpose, crystal, and high-impact grades; expandable polystyrene (EPS); styrene butadiene latex (SBL); acrylonitrile-butadiene-styrene/terpolymer (ABS); unsaturated polyester resins (UPR); and styrene-butadiene rubber (SBR). Styrene and its derivates are used in packaging, insulation, construction materials, automobiles, household goods, toys, paper coatings, eating utensils, and electric appliances.

TABLE 11.2.1 Boiling Point of Pure Styrene as a Function of Pressure

Pressure (mmHg)	Temperature (°C)
5	21.7
50	65.6
100	82.2
200	101.1
760	145.5

The polystyrene and expandable polystyrene applications account for about 65 percent of the market, whereas ABS consumes about 12 to 15 percent of the market.

PROPERTIES

Styrene is moderately toxic, highly flammable, and highly reactive. It is a colorless oily liquid with a distinct sweet, aromatic odor. It can be detected by smell at very low concentrations, having an odor threshold of less than 0.5 ppm in air. It is slightly soluble in water and infinitely soluble in ethanol, acetone, carbon tetrachloride, and *n*-heptane. Styrene has a molecular weight of 104.15, boils at 145.5°C at atmospheric pressure, and freezes at −30.6°C. The density of liquid styrene at 20°C is 0.906 g/cm^3. The boiling point of pure styrene as a function of pressure is shown in Table 11.2.1.

Nearly all styrene produced commercially is consumed in polymerization or copolymerization processes. The polymerization reaction is exothermic and, if contained, may become violent. If the heat of reaction is not removed, the bulk temperature may rise to a level at which the polymerization is self-sustaining and will very rapidly evolve large quantities of heat. At its atmospheric boiling point of 145.5°C, the rate of polymerization is nearly 38 percent per hour. On exposure to air and light, it slowly polymerizes and reacts with oxygen to form peroxides and aldehydes. Styrene polymerization generally takes place by free-radical reactions initiated thermally or catalytically. Crucial to the manufacturing of the monomer, particularly in the purification stage, is the fact that the rate of styrene polymerization can be retarded and/or inhibited by the addition of a variety of additives collectively called *inhibitor chemicals*.

STYRENE MANUFACTURING

In commercial production, with one inconsequential exception, styrene is produced solely using ethylbenzene (EB) as the feedstock. As a general rule, styrene production facilities are integrated with the upstream EB plant. From an energy standpoint, integration is important because the styrene plant is a large consumer of steam, whereas the EB plant, given the exothermic nature of the alkylation reaction, is a net exporter of energy. For production of EB, Badger licenses the highly efficient EBMax process in partnership with Exxon/Mobil.

The manufacture of styrene via ethylbenzene consumes more than 50 percent of the commercial benzene worldwide. For nearly 60 years, the most economical route for the production of SM has been the dehydrogenation of EB. Modern dehydrogenation processes perform the endothermic reaction at low pressure, usually vacuum conditions with an excess of steam in the presence of potassium-promoted iron oxide catalyst.

At present, about 70 percent of the SM produced is via the dehydrogenation route, which is more often than not accomplished adiabatically in the fixed-bed radial-flow reactors. Of the more than 70 EB dehydrogenation units operating worldwide, most use two reactors in series. Presently, about a half dozen units make use of three reactors in series. With one exception, the third reactor in these plants was added to an original two-reactor system as part of a revamp or capacity expansion. For grassroots plants, the two-reactor system is inherently more economical both from a capital investment and operating cost standpoint.

EB dehydrogenation units not using the adiabatic approach employ various forms of a tubular isothermal reactor. Given its relatively high equipment cost, limited capacity, and low selectivity, the economics of tubular isothermal technology have fallen short of those of conventional adiabatic technology. Consequently, the number of commercial isothermal units has declined in recent years.

The remaining non-dehydro-based styrene production is obtained from the coproduction of propylene oxide (PO) and styrene (SM). In comparison with the dehydrogenation route, the coproduct route is vastly more complex and requires significantly higher capital investment. Furthermore, the smaller PO market typically drives the operating economics of these plants. As such, in recent years, several PO-only processes have been developed, thus potentially reducing the importance of the PO-SM route.

PROCESS CHEMISTRY

The dehydrogenation reaction of EB to styrene is endothermic, approximately 300 to 320 kcal/kg, and equilibrium-limited. Owing to the endothermic nature of the dehydrogenation, the reaction rate drops with the temperature as the steam EB mixture flows through the catalyst bed. Even with high steam-to-EB ratios, if no heat is added, EB conversion in a single reactor is limited to about 40 percent. To achieve the most economical EB conversions, the process usually is carried out in at least two stages with intermediate reheat between stages. The dehydrogenation of EB is a reversible reaction and takes place over potassium-promoted iron oxide catalyst in the presence of steam.

$$\underset{\text{Ethylbenzene}}{C_6H_5\text{-}CH_2CH_3} \underset{\text{Steam}}{\overset{\text{Catalyst}}{\rightleftarrows}} \underset{\text{Styrene}}{C_6H_5\text{-}CH=CH_2} + \underset{\text{Hydrogen}}{H_2}$$

The reaction is highly temperature-dependent. Reducing operating pressure and/or adding steam to reduce the partial pressure of the reactants favors the desired forward reaction. Commercially, the reaction is carried out at temperature ranging from 590 to 650°C at pressures from 200 mmHg abs to slightly above atmospheric pressure. Grassroots plants designed by Badger use the deepest vacuum possible, only limited by the ability to condense the effluent and the capacity of vent gas compression equipment. Conversion and selectivity to styrene are affected by a number of interdependent variables: temperature, pressure, steam-to-EB ratio, space velocity, and catalyst type. In the Badger process, the combination of steam-to-EB ratio and operating pressure is adjusted such that selectivity to styrene is approximately 97 mol % or greater.

High temperature favors dehydrogenation (EB conversion) both kinetically and thermodynamically but also increases by-product formation and decreases selectivity to styrene. Both conversion and selectivity increase with decreasing pressure and higher steam ratios. Badger plants with the latest-generation technology typically operate at a molar steam-to-EB ratio of 7:1 or less (1.2 wt/wt).

Dehydrogenation catalyst is supplied to the industry by several manufactures, including Criterion (CRI), Süd-Chemie, BASF, and Dow. EB dehydrogenation catalyst consists primarily of ferric oxide with potassium oxide as promoter. Other metals oxides have been used as supplemental promoters, including chromium, cerium, tin, calcium, platinum, and titanium. By and large, the catalyst supply market is dominated by two companies, CRI and Süd-Chemie. For low steam-to-EB applications, CRI's Hypercat catalyst is the most stable and selective catalyst on the market. The Hypercat catalyst is also known for its high mechanical strength and low attrition rate. Over the last decade, most Badger plants have used catalyst supplied by CRI.

Space velocity primarily affects the temperature needed to achieve a given conversion. Reactor inlet temperatures must be increased as space velocity is increased, and vice versa. Because the catalyst gradually loses activity, principally as a result of loss of potassium promoter, it is common practice to raise reactor inlet temperatures to compensate for the lost activity and maintain constant conversion. When the reactor inlet temperatures finally reach the mechanical design limit of the equipment, the plant may extend catalyst life by reducing throughput and thus the space velocity through the reactors. To minimize costs associated with frequent turnarounds for catalyst replacement, the space velocity typically is set to achieve between a 24- and 36-month run length.

As noted earlier, as a catalyst run progresses and the reaction section has reached the mechanical design limits (temperatures), the producers must either shut down to replace catalyst or reduce throughput, both of which have severe economic penalties. To counteract the catalyst aging process, Badger developed a novel technique to stabilize or reactivate dehydrogenation catalyst called *catalyst stabilization technology* (CST). Now proven in two commercial facilities, the basic concept of CST is to replenish to volatile activity promoters, usually potassium compounds, while the plant is in operation. In so doing, catalyst activity loss is greatly slowed, thus allowing catalyst replacement to be deferred. Data derived from both commercial plant and pilot plant operations indicate that CST can extend catalyst life by as much as 50 percent.

The main by-products of the dehydrogenation reaction are toluene and benzene. Although technically a yield loss, both are high-value products. Benzene is many times recycled to the EB unit, whereas toluene is either returned to an aromatics unit or sold to refiners. Toluene is produced via both EB and styrene, with the greater portion coming from the styrene route.

$$C_6H_5\text{-}CH_2CH_3 \rightarrow C_6H_5\text{-}CH_3 + H_2 + C$$
Ethylbenzene → Toluene + Hydrogen + Carbon

$$C_6H_5\text{-}CH=CH_2 \rightarrow C_6H_5\text{-}CH_3 + C$$
Styrene → Toluene + Carbon

Similarly, benzene is also formed via EB and styrene by the following:

$$C_6H_5\text{-}CH_2CH_3 \rightarrow C_6H_6 + CH_3 + C$$
Ethylbenzene → Benzene + Methane + Carbon

$$C_6H_5\text{-}CH=CH_2 + H_2 \rightarrow C_6H_6 + CH_3 + C$$
Styrene + Hydrogen → Benzene + Methane + Carbon

Impurities that affect product purity include α-methylstyrene and cumene. These heavier C_9 compounds are produced in the dehydro reactors, Cumene is formed mainly in the reactor void areas (i.e., noncatalytically), with EB being the primary precursor.

$$2C_6H_5\text{-}CH_2CH_3 \rightarrow C_6H_5\text{-}C_3H_7 + C_6H_5\text{-}CH_3$$
Ethylbenzene → Cumene + Toluene

A large portion of the cumene formed by this reaction, as well as any cumene that enters the styrene plant with fresh EB, is dehydrogenated catalytically to form α-methylstyrene.

$$C_6H_5\text{-}C_3H_7 \rightarrow C_6H_5\text{-}C_3H_5 + H_2$$
Cumene → α-Methylstyrene + Hydrogen

In general, however, α-methylstyrene is made mostly on the catalyst, with styrene being the primary precursor.

$$2C_6H_5\text{-}CH=CH_2 + CH_4 \rightarrow C_6H_5\text{-}C_3H_5 + C_6H_5\text{-}CH_3$$
Styrene + Methane → α-Methylstyrene + Toluene

Steam plays a very important role and is added for three reasons. First, it serves as a heat source for the reaction. Second, the dilution effect of the steam lowers the partial pressures of styrene and hydrogen to shift the thermodynamic equilibrium favoring the conversion to EB. Finally, the steam reacts with coke deposited on the catalyst surface through the water-gas shift reaction and thus maintains catalyst activity. Carbon deposits on the catalyst as a by-product of benzene and toluene formation are readily removed as carbon dioxide by reaction with steam. Given the high temperatures of the dehydrogenation reactors, the production of carbon dioxide is an order magnitude greater than that of carbon monoxide.

$$C + H_2O \rightarrow CO + H_2$$
Carbon + Water → Carbon monoxide + Hydrogen

$$CO + H_2O \rightarrow CO_2 + H_2$$
Carbon monoxide + Water → Carbon dioxide + Hydrogen

PROCESS DESCRIPTION

The Badger/ATOFINA styrene process consists of two main sections: reaction and distillation. The reaction section includes a steam superheater, two radial-flow reactors, an interreactor reheat exchanger, a feed/effluent heat exchanger, an effluent condenser, a decanter/vapor/liquid separator, and a vent gas compression system. As noted earlier, the Badger/ATOFINA dehydrogenation reaction system uses two fixed-bed radial-flow reactors in series (Fig. 11.2.1).

Prior to the reactors, fresh and recycle EB (recovered in the distillation section) is mixed with low-pressure steam upstream of the feed/effluent exchanger. Steam is added in this step to prevent coke formation and subsequent metal carburization at high temperatures. In the feed/effluent exchanger, the two-phase feed is first vaporized and then superheated. Next, this EB-steam stream is combined with additional superheated steam, bringing the entire feed mixture up to the required reaction temperature. Prior to entering the primary reactor, the steam from the superheater is thoroughly mixed with the EB-steam feed in a series of proprietary mixing devices to ensure a homogeneous feed mixture and thus a uniform steam-to-EB distribution. High fluid velocities and short residence time are desirable for minimizing thermal noncatalytic reactions upstream of the catalyst bed. The Badger/ATOFINA process uses proprietary internals in the core of the reactor to both reduce residence time and uniformly distribute the feed across the length of the catalyst bed. The inlet temperature of the primary reactor is set such that EB conversion is approximately 35 to 40 percent.

The effluent from the primary reactor is reheated by exchange with superheated steam in an exchanger located directly upstream of the secondary reactor. To minimize thermal noncatalytic reactions, the reheater is close-coupled with the secondary reactor. Once reheated, the effluent flows to the secondary reactor, where additional conversion takes place. Depending on raw materials and energy prices, the optimal overall EB conversion is typically 60 to 70 percent per pass. For a specific catalyst, higher conversion requires higher reactor operating temperatures, which accordingly reduce selectivity and catalyst life. As outlined in the chemistry discussion, in addition to styrene, small amounts of ben-

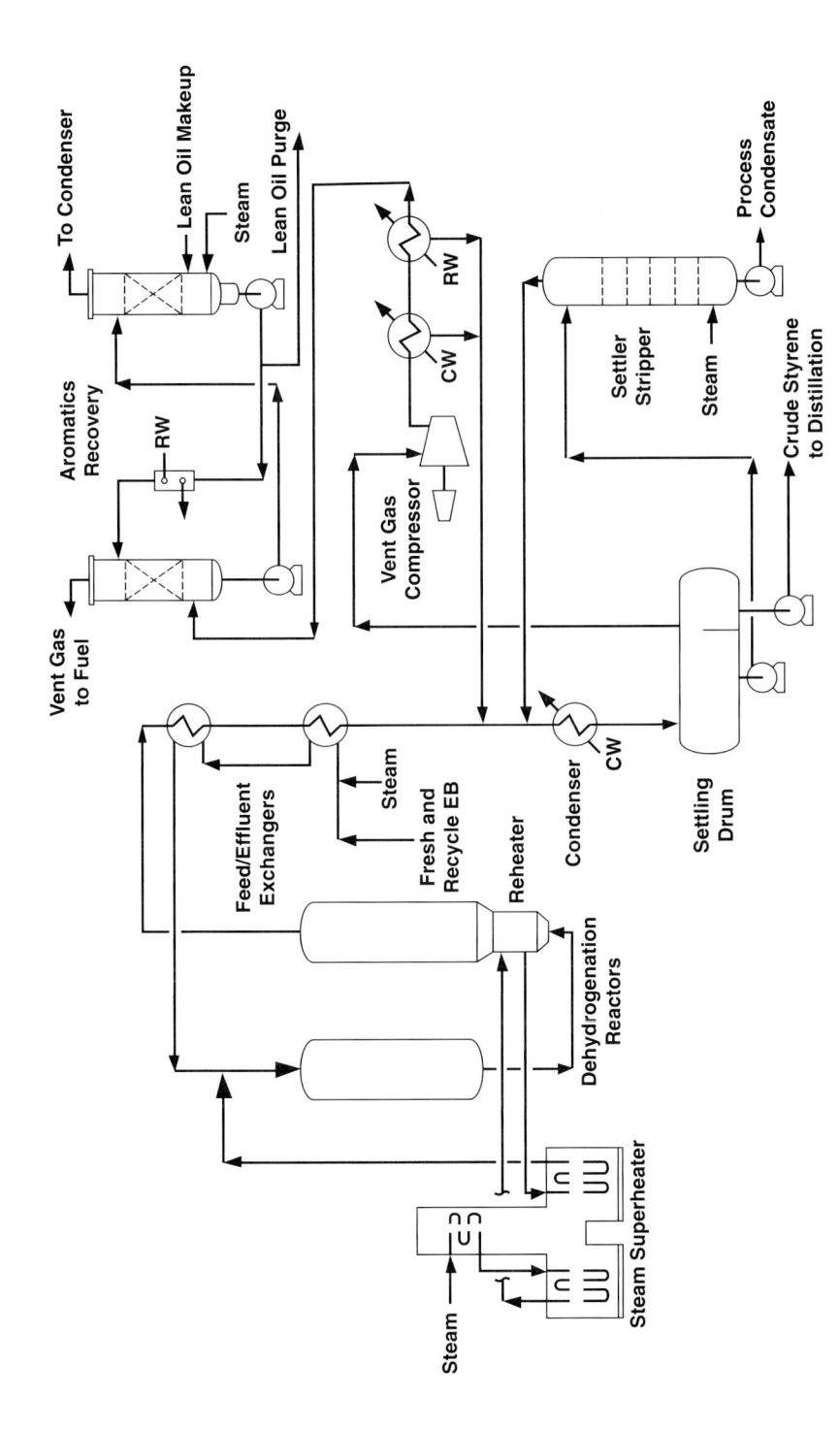

FIGURE 11.2.1 Typical ATOFINA/Badger dehydro flowscheme.

11.19

zene and toluene (B/T) and other trace impurities are produced in the reactors. The recovery of B/T and the removal of most of the trace components are effectively accomplished in the SM distillation section.

From the secondary reactor, the effluent is first cooled against reactor feed in a stacked feed/effluent exchanger. The cooled reactor effluent from the feed/effluent exchanger is then partially condensed in the crude styrene condenser. Depending on site-specific economics, this condenser is either a shell-and-tube exchanger using cooling water or an air-fin type of exchanger. Furthermore, in some plants where the maximum ambient air temperature is high (i.e., 40 to 50°C), the air-fin exchanger is followed by a cooling water trim condenser.

Badger/ATOFINA also offer an optional heat-recovery scheme that recovers energy from the condensing reactor effluent stream. Following the recovery of relatively high-level energy from the effluent in the feed/effluent exchanger, the effluent is further cooled and partially condensed by vaporizing an EB-water azeotrope. In a prescribed ratio, EB and water form a minimum-boiling-point azeotrope in which the boiling point of the mixture is significant lower than that of the two pure components. When vaporizing the azeotrope, for every mole of EB vaporized, between 2.5 and 3.2 mol of steam are vaporized as well. Once vaporized and compressed, this stream is then used as reactor feed, thus reducing the net steam consumption to the reactor system. From the azeotropic vaporizer, the reactor effluent is partially condensed, and thus cooling water requirements are also significantly reduced. With the ever-escalating natural gas prices in the United States and the relatively high cost of steam and fuel in the Far East, this heat recovery scheme has a reasonably attractive payout time. Currently, this scheme is used in five Badger plants.

From the condenser, the effluent flows to a decanter-type settling drum. The liquid organic and aqueous phases are separated, and the vapor stream (vent gas) is disengaged. The aqueous phase (process condensate), containing dissolved trace hydrocarbons, is directed to a stripping column where hydrocarbons and light gases, mainly carbon dioxide, are removed by direct steam stripping. From the stripper, process condensate is filtered and reused as boiler feedwater either to process steam generators or to offsite boilers.

Vent gas from the settling drum, consisting of noncondensables (primarily hydrogen, carbon dioxide, and methane), aromatics, and water, flows to the suction of the vent gas compressor. From the compressor, the vent gas is cooled in a series of exchangers, thereby condensing much of the water and aromatics. Condensed liquids are then recycled to the crude settling drum. Following the cooling step, the vent gas is directed to an absorber/stripper system for recovery of residual aromatics. In the absorber column, the vent gas stream is contacted with a chilled lean-oil stream, usually a heavier aromatic stream such as EB unit residue. In the absorber, nearly all aromatics are removed from the vent gas. Rich oil from the bottoms of the absorber is then directed to a stripping column, where lean oil is regenerated by steam stripping. From the overhead of the stripping column, the recovered aromatics and stripping steam are recycled to the settling drum via the crude styrene condenser. The hydrogen-rich vent gas from the styrene unit frequently is burned as fuel either in the process steam superheater or in an offsite boiler. A small number of plants choose to recover high-purity hydrogen from the vent gas using PSA technology. When compared with using it as fuel, hydrogen recovery (PSA) requires that the vent gas be compressed to significantly higher pressures.

The decanted hydrocarbon phase (crude styrene) from the settling drum is pumped to the distillation section for purification. The styrene purification section consists of three low-temperature distillation column systems all operating under vacuum. The function of these columns is to produce a high-purity styrene monomer product, recover the valuable benzene/toluene by-product, and recycle unconverted EB back to the reactor section.

As discussed, the rate of styrene polymerization is significant even at moderately low temperatures. To minimize polymer formation (yield loss), styrene distillations are carried out at relatively low temperature, usually 100°C or less, in the presence of retarder/inhibitor chemicals. Depending on separation requirements and the system's liquid holdup volume, the overhead pressure of the styrene distillation columns is typically between 30 and 200 mmHg absolute.

As noted, to lessen polymer formation throughout the distillation section, it is standard practice to add retarder and/or inhibitor chemicals to the crude styrene feed upstream of the distillation columns. Most inhibitors applied today are aromatic compounds with hydroxyl, nitro, or amine functional groups, such as dinitrophenol, dinitrocresol, and phenylenediamine. Various inhibitors are supplied to the industry by companies such as Crompton, GE Betz, Ondeo Nalco, and AH Marks. Given the highly competitive nature of inhibitor supply, many of the additive chemicals are proprietary and/or covered by patent protection.

These chemical additives act as radical terminators that either hinder (retard) or inhibit the polymerization reaction. In addition to inhibition efficiency, a number of other factors must be considered in selection of inhibitors: toxicity, thermal stability, volatility, viscosity reduction, and the requirement for the presence of molecular oxygen. Likewise, most inhibitor chemicals are heavy nonvolatile compounds that exit the process with the residue stream, which is eventually burned as fuel. As a consequence, nitrogen contained in the inhibitor forms NO_x during the combustion process. As a result, one of the more important factors for selecting an inhibitor today is the nitrogen content relative to the chemical's inhibition effectiveness. As regulatory laws become more stringent with regard to NO_x emissions, in particular in the United States and Europe, the industry is gradually moving toward additives that contain less or little nitrogen.

The basic configuration of the distillation section is shown in Fig. 11.2.2. In the first column, the B/T column, benzene and toluene produced in the dehydrogenation reactors are separated from ethylbenzene and styrene. The overhead distillate from the B/T column is often sent to a benzene-toluene splitter to recover benzene for recycle to the EB unit. Toluene by-product is returned to an aromatics unit, sold as a by-product, or converted to benzene in a dealkylation reactor. With the high styrene selectivity of most modern plants, installation of a dedicated toluene dealkylation unit is difficult to justify.

The bottoms stream from the benzene-toluene column is sent to the EB recycle column, where C_8's, mainly ethylbenzene, are separated from styrene and recycled back to the reactors. This separation is of particular importance because any C_8's (ethylbenzene and xylenes) that exit the column bottoms becomes a styrene product impurity. Because of their relatively close boiling points, the separation of EB and styrene is difficult. At 100 mmHg, the boiling point of EB is 74°C as compared with 82°C for styrene. On average for a grassroots design, the EB (C_8's) specification for the SM product is between 100 and 1000 ppm wt. Nevertheless, to differentiate their product from that of other SM producers, a relatively large number of plants produce less than 100 ppm, and a few produce less that 10 ppm. Despite the fact that significantly more energy is required to manufacture the higher-quality product, SM producers rarely can capture a higher price.

The combination of close boiling point and low EB content in the bottoms results a large number of theoretical distillation stages to accomplish the required EB-styrene separation. Furthermore, the operating pressure of the column must be kept low to reduce operating temperatures and thus polymer formation. Until the mid-1980s, specialized sieve trays were used to accomplish this separation. Even with their relatively low pressure drop compared with standard sieve trays, the use of trays resulted in bottoms temperatures in exccss of 120°C. The other major drawback to trayed column design was the relatively large liquid holdup volume on the trays. Unavoidably, both the higher bottoms temperature and holdup volume resulted in high polymer formation rates.

Standard practice today is to use high-efficiency low-pressure drop-structure packings such as those provided by Koch-Glitsch and Sulzer Chemtech. Compared with the trays, structure packing has two important advantages: First, the pressure drop per theoretical stage is much lower, and second, the liquid holdup is significantly reduced. Depending on the density of packing used (surface area per unit volume), the HETP (height per theoretical plate) is on the order of 250 to 500 mm with a pressure drop per stage of less 1 mmHg.

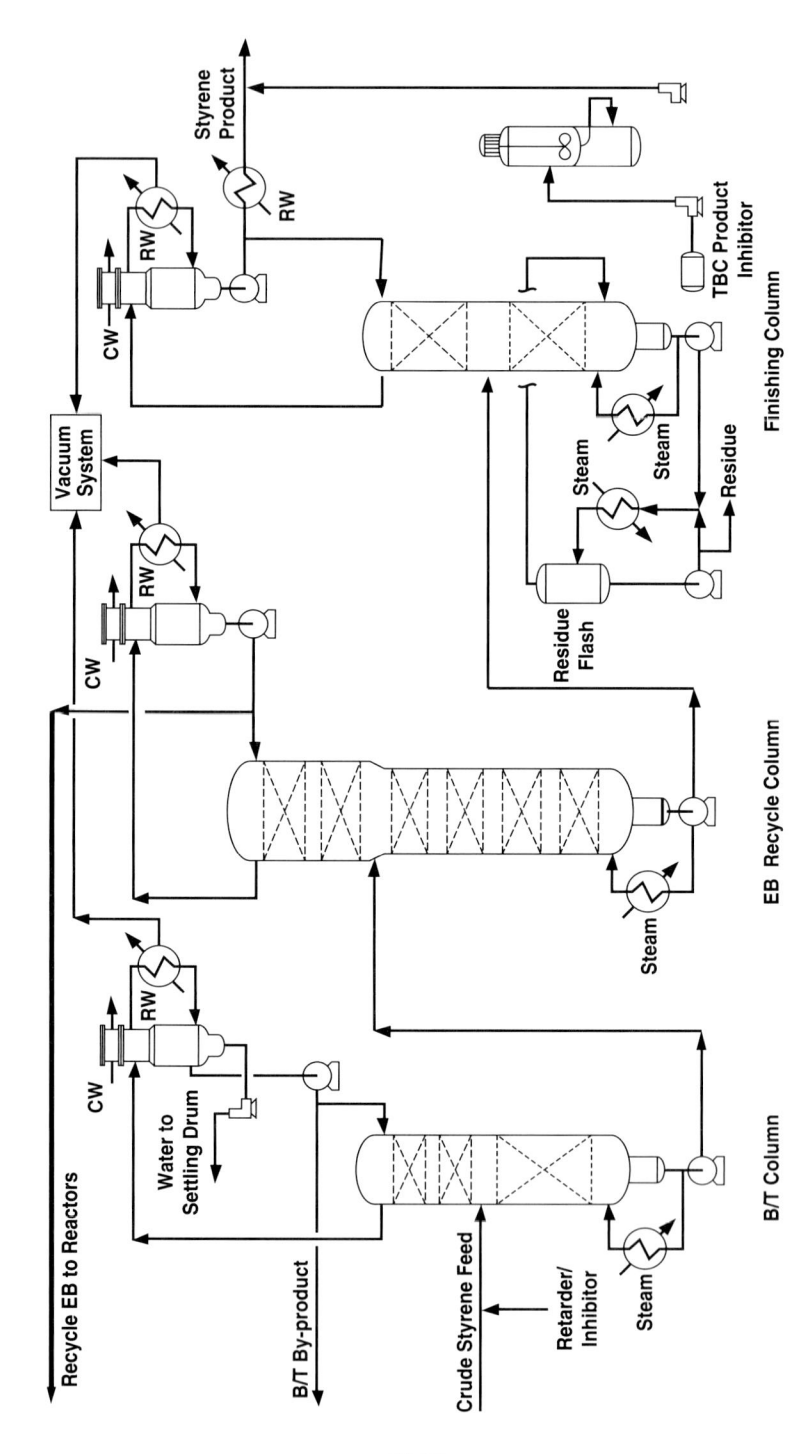

FIGURE 11.2.2 Basic configuration of the distillation section.

Having these beneficial characteristics, the use of packing permits optimization of the column's operating pressure with respect to capital investment and operating costs (steam, cooling water, polymer make, and inhibitor consumption). By and large, the optimal design results in a bottoms temperature of less than 100°C and in extreme cases around 85°C. Overall, the combination of lower pressure drop (lower bottoms temperatures) and reduced liquid holdup has significantly reduced the amount of polymer formed compared with the trayed column design. Correspondingly, inhibitor consumption is also reduced to a great extent with the use of structured packing.

The bottoms stream from the EB recycle column is fed to the styrene finishing column. The feed to the column contains styrene, small amounts of ethylbenzene, C_9 aromatics, heavy polycyclic compounds, polymer, and inhibitor. The key separation of the finishing column is to remove C_9 aromatic compounds, primarily α-methylstyrene, from the finished styrene product. Purified styrene recovered overhead as a distillate product is first chilled prior to being sent to offsite storage. The heavier materials from the column bottoms are pumped to a flash drum system for recovery of styrene from the residue. The temperature of this pumparound flash system is set such that the residue contains between 5 and 10 wt % styrene. Additionally, some plants control the total amount of volatile components in this residue stream to meet either a viscosity or flash-point target. As mentioned earlier, styrene residue is burned as fuel either in process-fired heaters or in offsite boilers.

Styrene product normally is stored in refrigerated or chilled tanks. If the styrene is not consumed immediately, to prevent polymer formation, it must be inhibited for storage and transportation with 4-*tert*-butyl-catechol (TBC). The standard industry addition rate of TBC is 10 to 15 ppm by weight. This level of inhibitor permits its use by many polymerization processes without its removal, while still providing adequate shelf life. For TBC to be effective, the storage temperature must carefully controlled, and dissolved oxygen must be present. For example, at a 10-ppm level and a storage temperature of 24°C, the maximum recommended storage time is about 4 to 5 weeks, whereas at 15°C it can be as long as 2 to 3 months.

PRODUCT SPECIFICATION

Both the marketplace and the specific polymer application largely dictate SM specifications. By and large, however, the industry is gravitating toward higher and higher purities. While 10 years ago the typical design purity was 99.80 wt %, today plants are designed for at least 99.90 wt %, with some producers requesting as high as 99.95 wt %. Generally, 99.9 percent purity reduces all impurities to sufficiently low levels for virtually all polymer applications. Reasons why consumers may be wanting higher purity include avoiding taste detection in foamed food packaging, reducing water treatment from expandable polystyrene plants, and reducing emissions from finished products such as latex carpet backing. Specifications of typical polymer-grade SM product are shown in Table 11.2.2.

OPERATING ECONOMICS

As a general rule, styrene plants are highly integrated from a steam balance standpoint with the upstream EB unit. Overall, the EB unit is a net exporter of steam, whereas the styrene unit is a large consumer of steam. Most steam consumers in the styrene unit require steam at low pressure levels, typically in the range of 1.0 to 5.0 kg/cm^2 · g. Over 80 percent of the variable operating cost of a styrene unit is attributable to raw materials, benzene and ethylene via ethylbenzene. As a result of the counterbalancing effect of steam-to-EB ratio versus styrene selectivity, the operating conditions of each styrene plant

TABLE 11.2.2 Specifications for Typical Polymer-Grade SM Product

Styrene	99.8–99.95 wt %
C_8's (EB + xylenes)	100–1000 ppm wt
α-Methylstyrene	150–500 ppm wt
Phenylacetylene	100 ppm wt
Polymer	5–10 ppm wt
Aldehydes (benzaldehyde)	30–50 ppm wt
Peroxides (hydrogen peroxide)	30 ppm wt
Inhibitor (TBC)	10–15 ppm wt
Chlorides	2–5 ppm wt
Sulfur	2–5 ppm wt
Color, APHA	10–15 max

TABLE 11.2.3 Typical Raw Material and Energy Consumptions

Item, units	Unit/kg SM
Ethylbenzene, kg	1.057
B/T by-product, kg	0.029
Steam, kg	2.550
Fuel, 1000 kcal	0.090
Cooling water, m^3	0.215

are optimized for local pricing of raw materials and energy (primarily fuel and steam). The unit consumptions stated in Table 11.2.3 are typical for a stand-alone SM plant located in the U.S. Gulf Coast area.

Note that the consumption figures presented in the table include only styrene unit onsite equipment items. The fuel consumption figure assumes that both the vent gas and styrene unit residue are burned as fuel.

CHAPTER 11.3
POLIMERI EUROPA STYRENE PROCESS TECHNOLOGY

Leonardo Trentini and Armando Galeotti
"C. Buonerba" Research Centre
Mantova, Italy

INTRODUCTION

Styrene is one of the most important basic chemicals. Its extensive production started in United States during the World War II, when the supply of natural rubber from South Asia was cut off from access by allied countries. After the war, the increasing use of styrene for synthetic rubber and for polystyrene forced a worldwide expansion of plants. In the meantime, technology and catalyst improvement increased enormously plant capacity, efficiency, and environmental safety. In 2003, the world styrene capacity is close to 25 million metric tons per year (MTA).

In the early 1960s, Polimeri Europa (at that time Montedison and then EniChem) started the production of styrene in Mantova, Italy. Driven by its research center in Mantova, in the second half of the 1970s the company began the development of a proprietary technology by improving the process and key equipment of existing units. The final result of this effort was the design of a new styrene production unit that went on-stream in the early 1990s in Mantova. The opportunity to run proprietary styrene plants has allowed Polimeri Europa to acquire great expertise in all different aspects of the styrene technology and any event that can occur during industrial runs.

In the present competitive styrene market, small differences in raw materials and utilities-specific consumption can heavily influence profit. For this reason, it is very important to take into account anything that can affect the choice of process parameters for a new unit. The aim is to find the most optimized design, specifically tailored for site conditions and customer needs. Taking advantage of its long experience in both design and manufacturing of styrene units, Polimeri Europa is now in a position to offer a competitive styrene technology.

PROCESS CHEMISTRY

Nearly 90 percent of the world styrene production comes from gas-phase ethylbenzene dehydrogenation. Ethylbenzene dehydrogenation may be carried out by a simple thermal reaction. The use of catalysts, however, allows a much higher selectivity to styrene. The reaction is highly endothermic ($\Delta H°_{610°C}$ = 29.80 kcal/mol) and occurs with an increased number of moles, as shown below:

$$C_6H_5CH_2CH_3 \leftrightarrow C_6H_5CHCH_2 + H_2$$

It is an equilibrium reaction and for this reason is thermodynamically limited. The equilibrium constant K_p is calculated as follows:

$$K_p = \frac{p_{sty}\, p_{H_2}}{p_{eb}}$$

where p_i is the partial pressure of the component i, and it can be written also as

$$K_p = \frac{n_{sty} n_{H_2}}{n_{eb}} \frac{P}{\Sigma n_i}$$

where n_i is the number of moles of the component i and P is the total pressure of reactor system.

K_p is constant at constant temperature. According to the second equation, it is possible to increase the products (n_{sty} and n_{H_2}) by decreasing the total operating pressure, the other numerator term, in the reaction system. In such a way, not only will ethylbenzene conversion increase, but the selectivity to styrene also will be improved.

Before entering reaction, steam is added to the ethylbenzene. It has the essential role to supply the heat needed by the reaction, above all in an adiabatic condition. Looking at the second equation above, it is possible to increase ethylbenzene conversion also by increasing the total number of moles in the system (Σn_i), reducing the ethylbenzene partial pressure by dilution. Thus it is clear that steam also has a direct thermodynamic effect on the reaction equilibrium: The ethylbenzene conversion increases if the steam-to-ethylbenzene ratio increases.

Steam has other two positive effects on the catalyst:

- It decreases the amount of coke or coke precursors formed by parallel cracking reactions of ethylbenzene by steam re-forming reactions.
- It avoids catalyst overreduction and deactivation by controlling the oxidation state of iron; steam that has oxidant proprieties balances the high reduction capacity of hydrogen at high concentration and temperature.

The main reaction by-products are toluene and benzene. It is believed that they come from the following reactions:

$$C_6H_5CH_2CH_3 + H_2 \rightarrow C_6H_5CH_3 + CH_4$$
($\Delta H°_{610°C}$ = 24.26 kcal/mol)

$$C_6H_5CH_2CH_3 + H_2 \rightarrow C_6H_6 + C_2H_4$$
($\Delta H°_{610°C}$ = −15.56 kcal/mol)

Carbon dioxide is also an important by-product. It comes from the steam re-forming reaction with the coke settled on the catalyst. Then, owing to the high operating temperature, a small amount of heavy components is formed. Nevertheless, their quantity is important because they cause material consumption, and as a result, they affect the overall process selectivity.

The reaction rate for the main reaction can be written as

$$r_{sty} = k_{sty} \frac{b_{eb}\left(p_{eb} - \dfrac{p_{sty}\,p_{H_2}}{K_p}\right)}{1 + b_{eb}p_{eb} + b_{sty}p_{sty}}$$

where k_i is the kinetic constant and b_i is the adsorption coefficient on catalyst active sites of i component.

The partial pressures of ethylbenzene (p_{eb}), styrene (p_{sty}), and hydrogen (p_{H_2}) affect the reaction rate. The partial pressures of styrene and hydrogen inhibit the reaction rate because they appear in the numerator of the subtractive factor. The result will be that the more ethylbenzene conversion increases and the styrene concentration becomes closer to the thermodynamic equilibrium, the more the reaction rate will decrease.

The following rate equations are developed for the main by-products:

$$r_{tol} = k_{tol} \frac{b_{eb}p_{eb}}{1 + b_{eb}p_{eb} + b_{sty}p_{sty}}$$

$$r_{bz} = k_{bz} \frac{b_{eb}p_{eb}}{1 + b_{eb}p_{eb} + b_{sty}p_{sty}}$$

Contrary to the preceding styrene equation, in this case an inhibiting factor of styrene and hydrogen is not present. The result is that at high conversion (approaching the equilibrium), the styrene reaction rate becomes slower and slower than the toluene and benzene reaction rates, thus decreasing selectivity to styrene.

DESCRIPTION OF THE PROCESS FLOW

Dehydrogenation Reaction

Figure 11.3.1 shows a typical dehydrogenation unit. Fresh and recycled ethylbenzene is mixed with superheated steam just before the inlet of the first reactor. Then the reaction mixture flows through the catalyst bed between two concentric screens, diffusing radially from the center toward the external shell. Radial design provides a large flow area and minimizes the pressure drop through the catalyst.

Generally, styrene processes have adiabatic reactors. Since the dehydrogenation reaction is endothermic, the temperature of the reaction mixture and catalyst bed decreases from inlet to outlet of the reactor. The higher the conversion in the reactor, the more the temperature decreases. A typical inlet temperature of an industrial adiabatic reactor is about 600°C, and the temperature decrease in the reactor is usually more than 50°C. The reaction rate at the reactor outlet temperature is very slow. For this reason, it is necessary to reheat the reaction mixture before it enters in another stage to get conversions closer to the thermodynamic equilibrium of the reaction. Two or three reactors are used in series with an interstage reheater. The heat is provided by

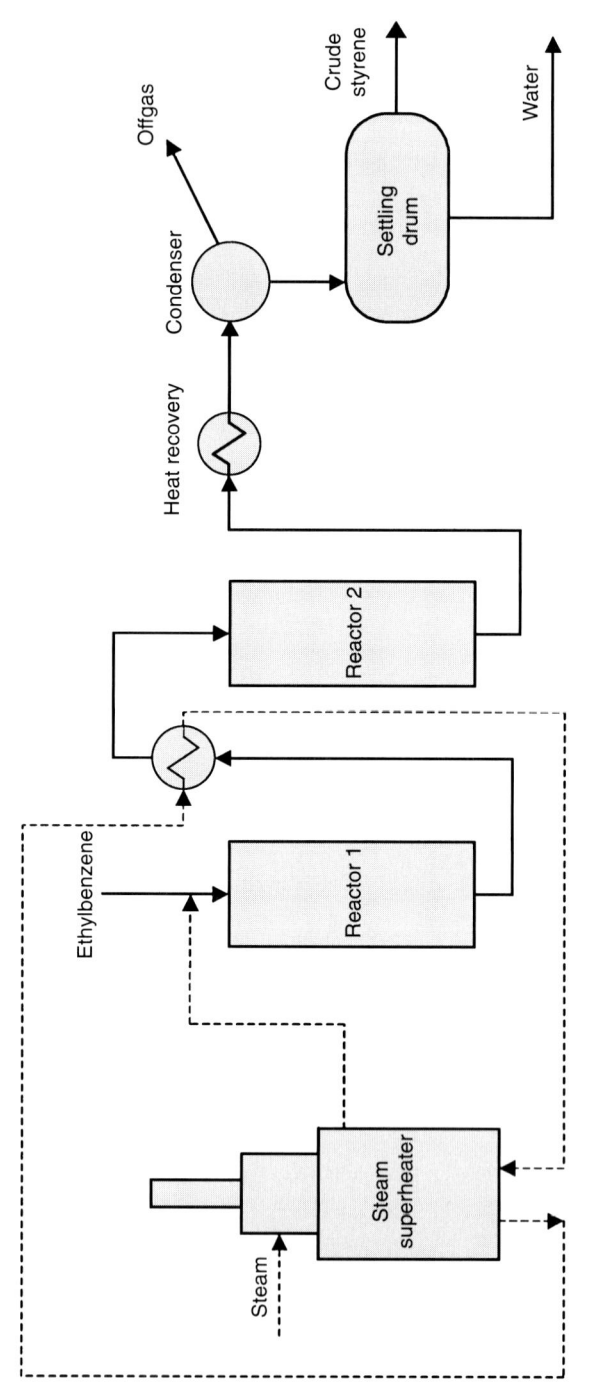

FIGURE 11.3.1 Dehydrogenation section (two reaction stages).

superheated steam that, at the end of the cycle, will enter the first reactor together with ethylbenzene.

The final dehydrogenated mixture, after heat recovery, is sent to the following section, where organics and steam are condensed and separated from offgases (hydrogen, carbon dioxide, methane, ethylene, etc.). This section is generally known as the *cold zone,* whereas the reaction section, including heat superheating and effluent heat recovery, is called the *hot zone.*

Offgas is removed by a compressor that ensures subatmospheric pressure in the reactors. Water and organic phase are separated in the settling drum. Water is then purified and reused normally, whereas the organic phase is sent to the distillation section.

A simple basic process scheme is presented in Fig. 11.3.1. With reference to this scheme, Polimeri Europa can offer different process solutions in order to best fit a customer's need.

It is the case, for example, of the hot zone where the following main alternatives are available:

1. *Reaction.* Depending on the overall balance of the different economic aspects (related to energy, raw materials, catalyst, and capacity), one of the following two options can be provided:
   a. Two reaction stages
   b. Three reaction stages
2. *Heat recovery on reaction effluent.* According to steam-EE cost ratio, one of the following two options is available:
   a. In a first step, preheat ethylbenzene plus primary steam vapor mixture, and then, in a second one, boil water and superheat steam before being sent to furnace.
   b. Generate only medium-pressure superheated steam to be fed to a turbine as driver of the vacuum offgas compressor; the ethylbenzene plus primary steam vapor mixture will be preheated in the convective zone of the steam superheater.

Distillation

Figure 11.3.2 shows a typical distillation train in a Polimeri Europa styrene unit. Crude styrene, containing styrene (about 70 percent), ethylbenzene, benzene, toluene, and some heavy components, is fed to the first column, where benzene and toluene are recovered in the overhead. The mixture then passes to another column, generally operating at atmospheric pressure, where benzene and toluene are separated.

Ethylbenzene, styrene, and tar are sent to the next column of the train, where ethylbenzene is separated from the styrene. This is a quite difficult separation because the normal boiling points of ethylbenzene and styrene differ by only 9°C. The column, equipped with structured packing, is operated under vacuum; to get the required styrene purity (usually not less then 99.9 percent), the residual ethylbenzene content in the bottom stream must be very low (generally no more than 100 ppm). To achieve this target, many theoretical stages are required, usually at least 80. Ethylbenzene, with some residual styrene (normally no more than 2 percent), is recovered overhead and recycled to the dehydrogenation section.

The bottoms are pumped to the next column, where pure styrene is recovered from the top. The bottoms are further processed in finishing equipment to recover additional styrene from the residue. In the overhead vapor of styrene column *p-tert*-butyl-catechol (TBC) is added to prevent the polymerization of styrene during storage.

Although distillation columns work in vacuum, bottoms temperature can vary from 85 to 110°C. For this reason, it is necessary to add an inhibitor to avoid loss of styrene due to polymerization. This is a very critical design feature because the formation of polymer within the distillation train can have a negative impact not only on raw materials con-

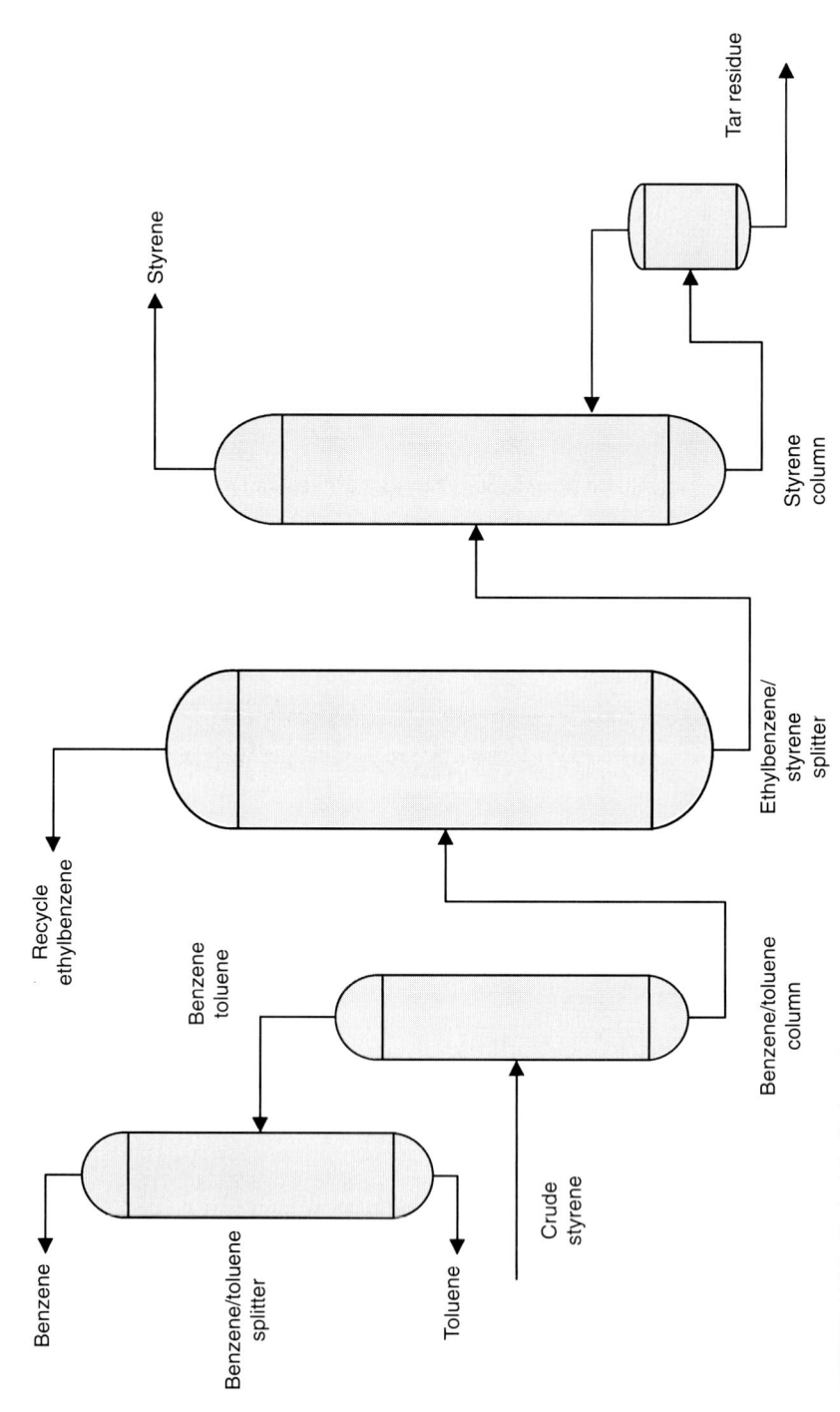

FIGURE 11.3.2 Typical distillation train.

sumption but also on the maintenance cost and, more than this, on plant reliability. Different substances can be used for this purpose. Most common families are nitroderivatives, free radicals, etc.; they can be used alone or in combination with each other to have a possible synergetic effect. To prevent polymer formation, all the systems used generally are called *inhibitors,* but these compounds can be divided really into two types: true inhibitors and retarders. A true inhibitor reduces the polymerization rate to near zero until it is used up, after which the rate increases to what it would be if no inhibitor were present. A retarder reduces the polymerization rate to a very low level, which gradually increases over time. Each one has advantages and disadvantages. Inhibitors generally give less polymer formation during normal processing but are used up relatively quickly. Retarders, on the other hand, while not quite as efficient in preventing polymerization, retain some effectiveness for an extended period. This is very important in case of a system shutdown, where the retarder will provide protection during the time required to restore normal operation or for column temperatures to drop to a point where polymerization is negligible. For this reason, it is not advisable to use a true inhibitor as the sole polymerization preventer. A combination of inhibitor and retarder can result in an excellent performance (case D, Fig. 11.3.3).

A careful monitoring of the inhibitor content in the distillation section is particularly important to prevent styrene-divinylbenzene polymer formation, as well as polystyrene (divinylbenzene is a by-product present in crude styrene below 100 ppm). Styrene-divinylbenzene polymer is a cross-linked and insoluble material called "popcorn polymer" because of its appearance. Popcorn polymer contains stuck radicals and can initiate the polymerization of styrene. When styrene monomer diffuses into popcorn polymer, it adds to the stuck radicals and the popcorn swells irreversibly. The swelling also fragments the popcorn, breaking bonds and generating new radicals.

An extensive study of the distillation section has been carried out by Polimeri Europa that results in an up-to-date design, including special arrangements suitable to achieve low material losses and energy consumption with very low inhibitor consumption. A wide

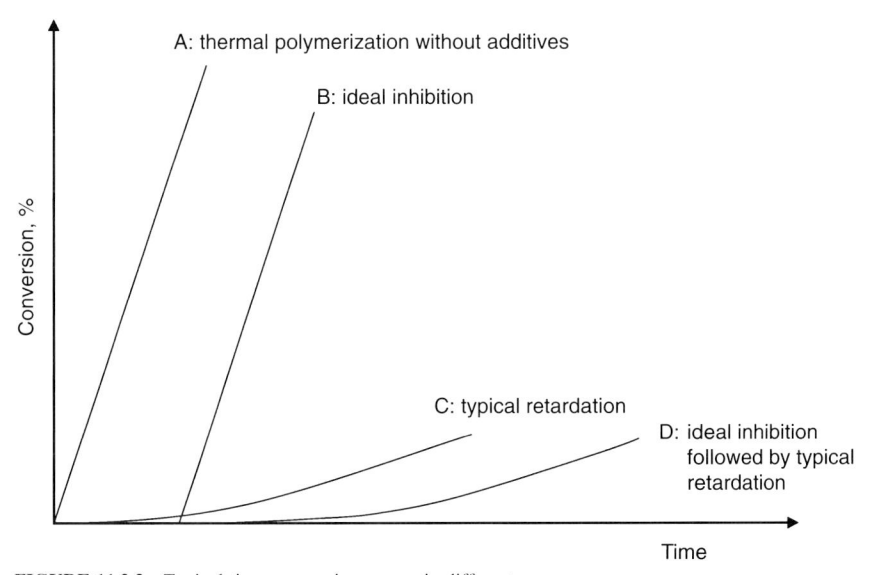

FIGURE 11.3.3 Typical time-conversion curves in different cases.

range of inhibitors and retarders has been tested and can be proposed. Polimeri Europa also has developed a simulation program suitable to optimize inhibitor/retarder dosage against polymer formation prediction.

PROCESS AND MECHANICAL DESIGN ADVANCED FEATURES

When a new styrene unit is designed, many different targets should be achieved. These include:

- Low ethylbenzene consumption
- High conversion/yield
- Low energy consumption
- Long catalyst life
- Plant reliability
- Low maintenance cost

Because some of these targets conflict with each other at least in part, a compromise has to be found, case by case, in relation with site conditions and customer needs. With this aim, Polimeri Europa can provide appropriate solutions, properly making the most of different design features.

Research and Operation

Polimeri Europa's background and expertise in styrene technology come from

- Direct manufacturing experience, such as startup and shutdown, maintenance and safety, and environmental issues
- Laboratory and pilot plant testing on dehydrogenation catalysts and polymerization inhibitors/retardarders.

The continuous monitoring of production facilities and the results of laboratory tests feed continuous improvement of the technology.

Process Design

Polimeri Europa technology is flexible; as mentioned earlier, different solutions can be adopted affecting energy balance, process performances, and catalyst life. An important additional tool is the thermal and fluodynamic analysis (by CFD) applied to hot area reactors/exchangers and interconnecting piping. Main targets that can be achieved with such a system are increased heat recovery, minimal pressure drop, and reduced lack of homogeneity in distribution.

Specific studies have been carried out in following cases:

- EB-steam mixer in order to reduce ethylbenzene residence time at high temperature
- Reactor shape (inner cone, outer vessel) in order to operate at a low pressure drop without badly affecting the flow profile across catalyst beds
- Heat exchangers in order to balance the internal pressure drop and to equalize the flow profile at the shell inlet

The great attention to minimizing the hot void volume, where low-selectivity reactions take place, and the precise choice to run the reaction at very low pressure have led to an improved design that maximizes catalyst performance and overall process selectivity.

Mechanical Design

All aspects of mechanical design are of great concern to Polimeri Europa. Accuracy in this field is really important because it directly affects unit reliability. Thermal and mechanical analyses (including FEA) are the most appropriate tools. The main targets are to minimize temperature gradients, smooth discontinuity in wall thickness, and improve stress distribution.

Specific studies have been performed in following cases:

- Reactor baskets
- Large inlet/outlet nozzles
- Tube sheet/vessel attachment
- Guides and supports for pipe and equipment
- Earthquake effects

Owing to the great care in details and choice of the best materials and generally of the safest technical solutions, even if they are slightly more expensive in terms of starting investment cost, all plant components, such as hot zone vessels and interconnecting pipelines, will reach a high level of mechanical reliability.

The preceding accomplishments should not be conceived as separate ones. Often it is necessary to be able to make good use of them all together so that a good compromise can be reached.

It is the case, for example, of hot zone layout, where the mutual relevant positions of reactors and exchangers are selected in order to minimize residence time at high temperature in interconnecting pipelines, taking account of both pressure drop and piping elasticity.

With all the above-mentioned guidelines in mind, the new unit will be optimized according to the requirements of the licensee, balanced from all process points of view, and mechanically reliable.

PROCESS PERFORMANCE

For a styrene unit operating with three reactor stages, a typical set of process parameters is reported in Table 11.3.1. The consumption of utilities (electric power, steam, and fuel gas) is typically related to the scheme adopted and depends on customer needs.

TABLE 11.3.1 Specific Consumption for a Typical Styrene Unit

EB	1055 kg/t styrene
Steam/oil	1.4
Run length	3 yr
SM purity	99.9%
Conversion	72%

COMMERCIAL EXPERIENCE

Two styrene plants based on proprietary technology are on-stream in Italy:

- 1972, 430 KTA, two reaction stages, with feed/effluent heat recovery
- 1992, 190 KTA, three reaction stages, with medium-pressure superheated steam generation, to drive offgas compressor

A third unit (60 KTA) is on-stream in the United Kingdom. A styrene unit (600 KTA) licensed by Polimeri Europa will be started up in 2005 in Iran.

PART 12

TEREPHTHALIC ACID

CHAPTER 12.1
E PTA: THE LURGI/ EASTMAN/SK PROCESS

Frank Castillo-Welter
Lurgi Oel Gas Chemie GmbH
Frankfurt am Main, Germany

INTRODUCTION

E PTA (Eastman polymer-grade terephthalic acid) is an excellent raw material to produce polyethylene terephthalate resin (PET), which is used for engineering plastics and packaging materials, such as bottles and other food containers, as well as films. E PTA is also the appropriate raw material to produce polyester fibers in the textile industry. The process is proven to be suitable for the production of all kinds of fibers and containers without limitation, including hot fill, at international first-grade quality.

In 2000, Eastman Chemical Co. and Lurgi Oel Gas Chemie GmbH formed an alliance to license the E PTA technology. Based on the design of Eastman's Rotterdam plant with a nameplate capacity of 290 kMTA, which was started in 1997, Eastman and Lurgi executed an extensive value engineering program for investment and operating cost optimization. To provide assurance, know-how, and the best assistance to the licensees regarding fiber production in all qualities from E PTA, a cooperative agreement with SK Chemicals was concluded in 2002. From that time, the technology has been offered as the Lurgi/Eastman/SK E PTA process.

Thus E PTA technology is based on the combined resources of Eastman, SK Chemicals, and Lurgi. With more breadth, experience, and capability than any single company, this partnership can deliver the terephthalic acid technology needed for tomorrow's market.

Eastman, with a production of more than 1.5 million MTA of polyester, is the world's largest PET producer for containers and is a guarantor of the top-quality bottles produced from E PTA. The production includes PET for water, soft drink, and beer bottles, as well as PET for food and custom containers. The technology has been proven by Eastman's installations in the United States and Europe for more than 40 years.

SK Chemicals operates its E PTA facility with a capacity of more than 420 kMTA and uses E PTA to produce all kinds of polyester fibers, including fully drawn yarn, microdenier, and bicomponent fibers. Through a joint venture with Huvis, SK Chemicals is the largest staple fiber producer in the world.

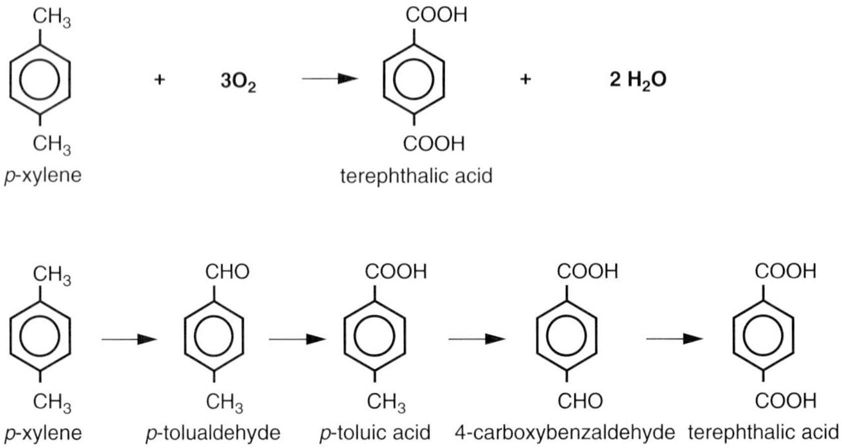

FIGURE 12.1.1 Main process chemistry.

Lurgi is responsible for licensing and engineering. With its vast experience in technology development, research and development (R&D), value engineering, plant design, and worldwide contracting and execution of five terephthalic acid projects, Lurgi is ideally qualified to assist in implementing a world-scale E PTA plant safely, on time, in quality, and within budget.

CHEMISTRY OVERVIEW AND PRODUCT SPECIFICATION

The chemistry for the formation of terephthalic acid (TPA) from *para*-xylene and oxygen is shown in Fig. 12.1.1.

A typical specification for E PTA product is given in Table 12.1.1.

PROCESS DESCRIPTION

A simplified flow diagram for the E PTA process is shown in Fig. 12.1.2.

Process Tank Farm

The process tank farm consists of a wash acid tank, an oxidizer reflux tank, and a filtrate tank. The wash acid tank serves as the acetic acid reservoir to supply acid deficiency to the filtrate tank and to the oxidizer reflux tank. The wash acid tank also supplies the high- and low-pressure wash acid system for purge and flush connections in slurry service.

Weak acid from the oxidizer reflux tank is sprayed into the top of the oxidizers in order to prevent solid carryover to the water removal column.

The filtrate tank collects the prepared catalyst solution and hydrogen bromide, as well as filtrate from the filtrate treatment, the E PTA rotary vacuum filters, and various other points in the process. Filtrate with a controlled catalyst and acid concentration is fed together with *p*-xylene into the oxidizers.

TABLE 12.1.1 E PTA Product Specification

Item	Typical
Appearance	White dry crystals, free flowing
Acid value, mg KOH/g	675
Ash content (total metal), wt ppm	<5.0
Iron (Fe), wt ppm	<0.8
Cobalt, wt ppm	<3
4-Carboxybenzaldehyde, wt ppm	<200
para-Toluic acid, wt ppm	<10
Volatiles (acetic acid), wt %	<0.04
Transmittance, % @ 340 nm	86 ± 3
b^*	3.2
L^*, %	98.5–99
ΔY	<2
Average particle size (wet analysis), μm	85

Oxidation and Water Removal

The oxidizers serve as the primary reactor vessels for converting *p*-xylene to terephthalic acid. Liquid feed for the oxidizers is *p*-xylene and filtrate from the filtrate tank. Catalyst concentrations in the oxidizers are controlled by maintaining a target concentration in the filtrate tank. Compressed air from the main air compressor is sparged via an air ring at the base of the vessels to provide agitation and proper mixing. The air rate is adjusted to maintain a defined oxygen concentration in the noncondensable portion of the oxidizer offgas. Proper operation of the oxidizer requires intimate mixing of the catalyst, air, and *p*-xylene. The aerated reaction mixture containing the terephthalic acid particles is essentially homogeneous. The combination of temperature, pressure, and catalyst levels produces a nearly complete conversion of *p*-xylene to terephthalic acid. Oxidizer reflux (a recycled solution of relatively weak acetic acid) is fed through spray nozzles in the vapor head space to reduce solid carryover and provide about one stage of acetic acid and water separation.

Because of the high level of impurities, especially 4-CBA, the product from these oxidizers is shown as crude terephthalic acid (CTA). This slurry is pumped to the centrifuges.

The oxidizers are connected to the water removal column (WRC) via large vapor lines. The water removal column is the primary means of water removal from the process. An inability to remove the water would result in catalyst deactivation of the oxidizers. The oxidation reactions in the oxidizers are exothermic, and the heat evolved vaporizes acetic acid, water, and low-boiling compounds. This vapor, along with nitrogen, unreacted oxygen, and smaller amounts of carbon monoxide and carbon dioxide, is fed to the base of the water removal column. This vapor has enough enthalpy to negate the need for a reboiler.

The underflow from the water removal column, containing mainly acetic acid, is used as the liquor interchange in the E PTA centrifuges and as the feed to the acetic acid vaporizer. The excess underflow is cooled down and returned to the wash acid system. The water removal column overhead is fed to the WRC condensers. The condensate is further cooled by the WRC reflux cooler and either is pumped back to the top of the column as reflux or is sent to a wastewater treatment system. The noncondensable overhead vapor from the water removal column is composed primarily of reaction offgases, low-boiling organics, and trace amounts of bromine-containing species such as methyl bromide. This stream is sent to offgas treatment.

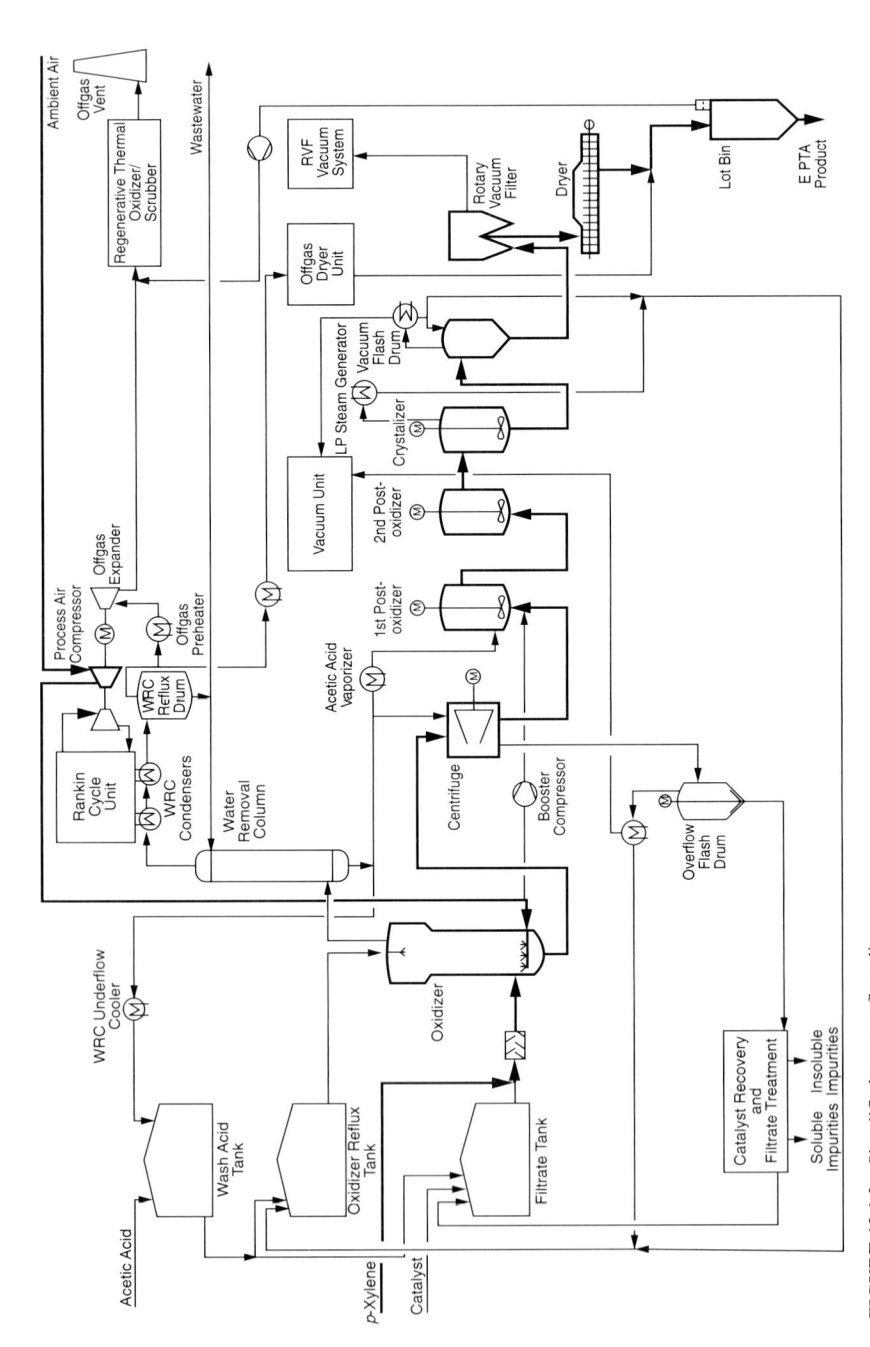

FIGURE 12.1.2 Simplified process flow diagram.

The WRC condensers transfer their heat to a cycle process for power recovery. The heat is used to heat up and evaporate the low-boiling circulating media of the cycle, which is sent to the energy recovery gas expander.

Offgas Treatment

The offgas treatment process can be broken down in two sections: air compression and power recovery system and the regenerative thermal oxidation and scrubber system.

The main purpose of regenerative thermal oxidation is to destroy organics and remove bromine-containing compounds from the E PTA offgas before the offgas is vented to the atmosphere. Because the offgas was in contact with copious amounts of liquid in the water removal column, the offgas is near or at its dew point. In order to ensure that the offgas is liquid-free before entering the offgas expander, the overhead vapor from the water removal column reflux drum is superheated.

The offgas is then routed through the offgas expander for energy recovery. The offgas expander drives together with an electrical motor and the energy recovery gas expander the main process air compressor of the plant. The offgas stream from the offgas expander is combined with the recycled offgas from the convey gas system and fed to the regenerative thermal oxidizer (RTO), where the organic volatile components such as acetic acid are oxidized to carbon dioxide and water. Methyl bromide is converted into bromine (Br_2) and hydrogen bromide (HBr). The typical oxidation efficiencies of the RTO are 99 percent oxidation of volatile organics, methyl bromide, and carbon monoxide.

The oxidized offgas is ready for further treatment in the caustic scrubber system for removing elemental bromine and hydrogen bromide.

Air for the (two) postoxidizers is provided by an electrical-driven booster air compressor, which is fed from the main process air compressor.

CTA Separation

As stated earlier, the slurry underflow of the oxidizers contains high levels of impurities, which need to be either removed or reacted to produce terephthalic acid. The E PTA process is designed to accomplish this goal.

The solid phase in the oxidizer underflow is separated from the contaminated acetic acid solution using decanter centrifuges. The CTA cake from the centrifuges is diluted with fresh wash acid from the water removal column. The resulting centrifuge underflow slurry is fed to the agitated postoxidizers. The centrifuge overflow, which consists of the original acid-catalyst carrier, CTA solids, and impurities, is flashed and then sent to the filtrate treatment unit for removal of insoluble and soluble impurities, as well as for recovery of catalyst and acetic acid.

The liquor exchange in the centrifuges is important for several reasons: First, many of the impurities such as *p*-toluic acid, isophthalic acid, etc., are fairly soluble in acetic acid. By exchanging the solvent liquor, these impurities are effectively removed from the final product. Additionally, the catalysts (Co, Mn, and Br) form soluble complexes in water and acetic acid. By exchanging the liquor, a large percentage of these catalysts can be recycled.

Postoxidation and Crystallization

The postoxidizers serve as reactors for completing the reactions of partially oxidized products of *p*-xylene (*p*-toluic acid and 4-CBA) to terephthalic acid (TPA), resulting in a more complete conversion of *p*-xylene to TPA. The primary obstacle to achieving high conversion of *p*-xylene to TPA is mass-transfer limitations associated with oxygen diffusion to embedded 4-CBA and *p*-toluic acid in the terephthalic acid. To achieve very low *p*-toluic acid and 4-CBA levels, some form of postoxidation or long-residence-time oxidation is necessary. This is accomplished by operating the postoxidizers at relatively high temperatures and pressures. A small amount of compressed air is fed to the first postoxidizer as an oxygen source for the oxidation reaction. Hot acetic acid vapor is also fed to the first postoxidizer to maintain the temperature and pressure. A series configuration for postoxidation leads to lower *p*-toluic acid and 4-CBA levels in the final product (E PTA).

Support equipment for the postoxidizers includes an acetic acid vaporizer. The acid vaporizer receives acid from the water removal column underflow and is heated by high-pressure (HP) steam to achieve the required temperatures and vaporization.

The slurry effluent from the second postoxidizer is fed to the crystallizer, which serves as an adiabatic flash vessel.

From the crystallizer, vent gas energy is recovered by producing low-pressure (LP) steam by means of an LP steam generator. The generated steam is used internally for LP steam consumers. The partially condensed vent is subsequently routed to the crystallizer condenser and finally transported to the oxidizer reflux tank.

The slurry stream from the crystallizer is fed to the vacuum flash drum and postoxidizer product drum, which is another adiabatic flash tank operated under vacuum conditions. Here is where the final crystallization occurs. The vapor from the vacuum flash drum is routed to the vacuum flash drum condenser, where the bulk of the vapor is condensed. Support equipment for the crystallization and flash cooling system includes a vacuum unit.

Product Separation and Drying

Once the postoxidizer product has been flash-cooled, it is sent to the continuous filter/dryers. The purpose of the rotary vacuum filters is to separate the E PTA from the acetic acid–catalyst liquid carrier, wash the cake, and discharge the wet cake into the respective dryers. The filter drum, covered with a cloth medium and operated under vacuum, draws E PTA slurry from a reservoir. Acetic acid–catalyst solution is pulled through the cloth and collected in a filtrate receiver drum, and the E PTA is collected on the cloth. Next, the cake is washed with cooled acetic acid from the wash acid tank. The wet cake falls through a chute into the top tier of the dryer system.

A rotating hollow screw heated by steam evaporates acetic acid from the cake and moves the cake through the tiered dryer system. The dried E PTA powder falls through a rotary valve located beneath the dryer. Dried E PTA powder is gas-conveyed to an E PTA silo.

Convey gas is taken from the water removal column reflux drum. This pressurized gas is cooled and dried by an offgas dryer unit before being distributed into the convey systems. The convey gas from the silos is collected from the silos' dust collectors, filtered, cooled, and transported by a blower to the RTO unit for destruction of contaminants.

Filtrate Treatment and Catalyst Recovery

The overall goal of the filtrate treatment unit is to remove impurities that are present in the filtrate either due to the purity of *p*-xylene or due to unwanted side reactions occurring in the oxidizers. These impurities need to be removed at the lowest catalyst and acetic acid loss while also keeping operational and capital costs as low as possible.

The filtrate treatment unit is build up out of several process units consisting of evaporation, (azeotropic) distillation, extraction, filtration, and a process vent scrubber for gas cleaning.

HIGHLIGHTS AND BENEFITS OF E PTA TECHNOLOGY

The E PTA technology provides remarkable advantages compared with conventional PTA processes.

Unique Mild Oxidation Process

- Minimum *p*-xylene consumption (646 kg/t E PTA)
- Maximum yield (98.3 percent)
- Lowest acetic acid consumption (36 kg/t E PTA)
- Lowest catalyst concentration

Resulting benefit: Variable cost savings

- Lowest temperature/pressure equipment
- Smallest air compressor

Resulting benefit: Investment/operating cost reduction

Unique Oxidation Reactor

- No internals/agitator

Resulting benefit: Investment/maintenance cost reduction

Unique Catalyst Recovery

- Maximum recovery (>95 percent)

Resulting benefit: Variable cost savings

Unique Energy Integration

- Simple setup
- Power generation of 10,000 kW

Resulting benefit: Significant variable cost savings

Unique Offgas Treatment

- No fired incinerator
- No precious metal catalyst
- Easy operation
- In compliance with German "TA Luft" (clean air act) regulation

Resulting benefit: Investment/operating cost reduction

No Purification by Hydrogenation, Therefore:

- No high-pressure/downstream stationary-rotating equipment
- No precious metal catalyst
- No extremely high-temperature heat source (super-high-pressure steam/hot oil)
- Less waste water output to biologic wastewater treatment
- Fewer operating personnel
- Independent routine plant shutdown (no annual catalyst change)

Resulting benefit: Significant investment/operation/maintenance cost savings

Minimum Environmental Impact

- One-third wastewater flow
- One-fifth wastewater load
- Cleanest reactor offgas
- Easy-to-treat solid waste

Resulting benefit: Significant investment/operation cost savings

Outstanding Plant Reliability

- Simple and easy-to-operate equipment
- No solids plugging/scaling issues
- Rotterdam operation records prove 8300 h/yr on-stream time

Resulting benefit: Significant profit gains

Additional Earnings

These benefits result in a reduced production cost of about US$50 per ton of E PTA product depending on applicable local conditions compared with the conventional PTA processes.

Figure 12.1.3 presents an aerial view of Eastman's E PTA Plant in Rotterdam. The oxidizers are shown on the right side of the picture.

ECONOMICS OF E PTA TECHNOLOGY

Specific Consumption Figures

The expected consumption of catalyst and chemicals to be used in the E PTA process is given in Table 12.1.2. The figures, in tons per ton of E PTA, are the ratio of the material/chemical used in tons per ton of E PTA product. The expected consumption of utilities in the E PTA process is presented in Table 12.1.3.

FIGURE 12.1.3 Eastman's E PTA plant in Rotterdam, Holland.

TABLE 12.1.2 Raw Materials, Catalyst, and Chemical Consumption

Item	Consumption
p-Xylene, t/$t_{E\ PTA}$	0.646
Acetic acid, t/$t_{E\ PTA}$	0.036
Cobalt (as Co metal), t/$t_{E\ PTA}$	0.012×10^{-3}
Manganese (as Mn metal), t/$t_{E\ PTA}$	0.001×10^{-3}
Bromide (as HBr 48 wt % solution), t/$t_{E\ PTA}$	0.0012
n-Propyl acetate (NPA), t/$t_{E\ PTA}$	0.0002
Caustic soda (50 wt % solution), t/$t_{E\ PTA}$	0.95×10^{-3}
Sodium hydrogen sulfite (40 wt % solution), t/$t_{E\ PTA}$	0.88×10^{-3}

TABLE 12.1.3 Utility Consumption

Item	Consumption
Electricity, kWh/$t_{E\ PTA}$	96
High-pressure steam, t/$t_{E\ PTA}$	0.355
Medium-pressure steam, t/$t_{E\ PTA}$	0.453
Demineralized water, t/$t_{E\ PTA}$	0.4
Cooling water ($\Delta t = 10°C$), t/$t_{E\ PTA}$	251
Instrument air, N · m^3/$t_{E\ PTA}$	7.9

E PTA Production Cost

Information about investment and production cost for a typical E PTA plant is given in Table 12.1.4 on p. 12.12.

COMMERCIAL EXPERIENCE

Lurgi's experience regarding terephthalic acid projects and Eastman's references for production facilities of E PTA are listed in Table 12.1.5 and Table 12.1.6, respectively, on p. 12.12.

TABLE 12.1.4 Cost Estimate and Production Cost for a Typical Grassroots Plant

No.	Cost element	Cost
1.	Investment ISBL (without license fee)	US$136 million
2.	Production cost	US$458/t $_{E\ PTA}$
2.1	Variable cost	
	a. Raw materials, catalyst, chemicals	US$386/t $_{E\ PTA}$
	b. Utilities and waste water treatment	US$23/t $_{E\ PTA}$
2.2	Fixed cost (labor—14 per shift, plant management, maintenance, etc.)	US$18/t $_{E\ PTA}$
2.3	Capital service, including depreciation (10% depreciation + 5% interest rate of fixed capital without working capital)	US$31/t $_{E\ PTA}$

Note: Capacity, 660 kMTA; on-stream time, 333 days/yr; time of estimate, Q4, 2003.

TABLE 12.1.5 Lurgi's Involvement in E PTA/PTA Technology

Year	Country	Technology	Capacity	Scope
1982	China	Amoco	36 kMTA PTA	EPCM
1989	China	Amoco	450 kMTA PTA	EPCM
1991	Spain	Amoco	250 kMTA PTA	EP
1997	India	Mitsui	250 kMTA PTA	EP
2003	China	Lurgi/Eastman/SK	600 kMTA E PTA	EP

TABLE 12.1.6 Eastman's Reference List for E PTA Production

E PTA producers	Site	Country	Capacity
Eastman	Kingsport, Tenn.	United States	215 kMTA
Eastman	Columbia, N.C.	United States	240 kMTA
Eastman	Rotterdam	Netherlands	290 kMTA
SK Chemicals*	Ulsan	South Korea	420 kMTA
PT Polysindo*	Karawang	Indonesia	350 kMTA

*With full fiber application.

PART 13

XYLENES

CHAPTER 13.1
EXXONMOBIL PXMAX[SM] P-XYLENE FROM TOLUENE

Terry W. Bradley
ExxonMobil Chemical Company
Baytown, Texas

INTRODUCTION

The traditional method of producing *p*-xylene (PX) has been catalytic reforming of naphtha with recovery of the contained PX and isomerization of the *o*-xylene and *m*-xylene to increase the PX yield. As polyester, the predominant end-use application for PX, has displaced other materials in many widely used fiber, film, and bottle applications, PX has enjoyed one of the highest growth rates among petrochemical products. With this rapid growth in PX demand, the industry has sought processes using alternative feedstocks for PX.

The process that has emerged to help meet the demand growth for PX is toluene disproportionation. Toluene disproportionation technology arose from the discovery of shape-selective zeolite catalysts in the early 1970s and the ability to apply these unique materials to petrochemical processes. The first generation of toluene disproportionation technology was introduced in 1975 and was known as the Mobil Toluene Disproportionation (MTDP[SM]) process. MTDP produces a xylene fraction with a PX concentration of approximately 24 percent, which is limited by the chemical equilibrium between PX and the other xylene isomers. The PX in the xylene fraction is recovered via crystallization or adsorption. The remaining xylene isomers, *m*-xylene and *o*-xylene, typically are processed through a xylene isomerization unit to maximize the yield of PX. This results in a large recycle loop to completely convert the isomers to PX.

In the 1990s, the Mobil Selective Toluene Disproportionation (MSTDP[SM]) process was introduced, representing a major advancement of PX technology. The process relies on a novel in situ pretreatment of the catalyst that for the first time made it possible to exceed the equilibrium PX concentration, producing over 80 percent PX in the xylene fraction. Selectivation of disproportionation catalyst was improved and extended later in the decade with the introduction of the PxMax[SM] process. In PxMax, the catalyst is selectivated during manufacture and produces over 90 percent PX in the xylene product. The dramatically higher PX concen-

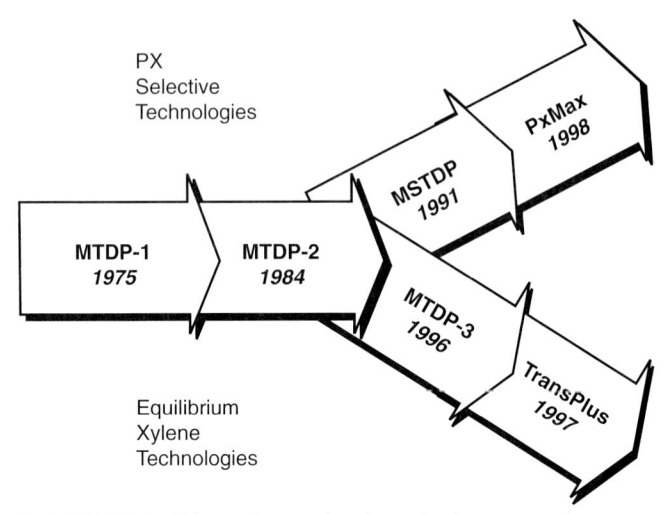

FIGURE 13.1.1 Toluene disproportionation technology progression.

tration greatly reduces the cost of the PX purification facilities and eliminates the need for xylene isomerization to maximize the PX yield. The relative progression of ExxonMobil toluene disproportion technology is shown in Fig. 13.1.1.

PROCESS CHEMISTRY

The key to the dramatic selectivity of the PxMax catalyst in comparison with conventional toluene disproportionation technology is the shape selectivity of the zeolite catalyst, combined with a novel selectivation agent. The properly selectivated catalyst takes advantage of the relative diffusivity of PX as compared with the larger o-xylene and m-xylene isomers. When toluene enters the zeolite, it catalytically disproportionates to benzene and an equilibrium mixture of the three xylene isomers. Because of its greater relative diffusivity, the PX exits the pore over 100 times faster than the other two isomers. The o-xylene and m-xylene remain in the catalyst exposed to its isomerization function and reequilibrate to form more PX. The selectivation procedure also deactivates sites on the exterior surface of the catalyst particle, reducing the potential for the PX to reisomerize when it exits the catalyst pore. Hydrogen is circulated through the reactor to suppress coke formation and to maintain a desirable hydrocarbon partial pressure. Figure 13.1.2 demonstrates the performance of the zeolite catalyst in the PxMax process.

Owing to the increased activity of the catalyst used in the PxMax process, start-of-run temperatures are as much as 50°C lower than those of MSTDP. The lower temperature reduces energy consumption and suppresses light gas formation, reducing loss of feedstock to low-value by-products.

Without the need for high-temperature in situ selectivation capability, the plant will have less equipment and lower-temperature metallurgy, resulting in lower capital costs for grassroots units or the ability to retrofit the technology into a broader range of conventional hydroprocessing units that may be idle or spare. Most important, the selectivation is permanent, and no online selectivation is required following either the initial catalyst fill or regeneration.

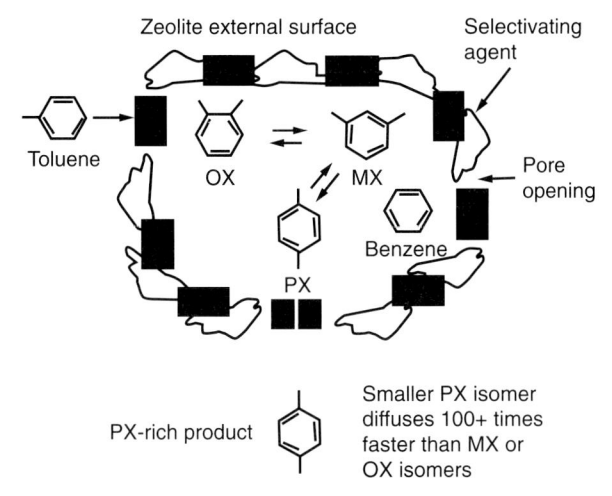

FIGURE 13.1.2 Model for *p*-xylene selectivity in zeolite.

PROCESS DESCRIPTION

The process flow for toluene disproportionation consists of three main parts: reactor section, fractionation, and PX recovery. Incoming toluene feedstock and unreacted toluene recycled from the fractionation section are mixed with hydrogen, preheated against reactor effluent, and introduced to the reactor. Hydrogen typically is controlled to 1.0 to 1.5 hydrogen-to-hydrocarbon molar ratio in the reactor feed stream. The hydrogen helps maintain a desirable hydrocarbon partial pressure and minimizes catalyst coking. This hydrogen addition rate is significantly below the 1.0 to 2.5 ratio used by other toluene disproportionation technologies. The reactions are carried out in the vapor phase and are mildly exothermic. The reaction section is shown in Fig. 13.1.3.

The reaction products are cooled and introduced to a stabilizer tower, where unreacted hydrogen is separated from the liquid aromatics. A portion of the overhead gas is purged to fuel to control circulating hydrogen purity, with the remainder compressed, mixed with incoming makeup hydrogen, and returned to the reactor. The purge of the hydrogen loop is necessary to control hydrogen purity because a small volume of light hydrocarbon gas is produced in the reactor through cracking side reactions.

The liquid aromatics from the stabilizer bottoms proceed to subsequent conventional fractionation steps in which the benzene product, unreacted toluene, mixed xylenes, and a small quantity of C_9+ aromatics are separated. The benzene product recovered from PxMax is produced at industry standard quality without further processing. Unreacted toluene is recycled to the reactor inlet. Mixed xylenes are taken to the PX recovery section for purification of the PX product.

Figure 13.1.4 shows ExxonMobil Chemical's PxMax plant in Beaumont, Texas. The Beaumont facility was the first grassroots application of PxMax.

OPERATING PERFORMANCE

The principal benefit of PxMax versus other selective toluene disproportionation technologies is the 90+ percent purity of PX in the mixed-xylene stream. In addition to maximizing the

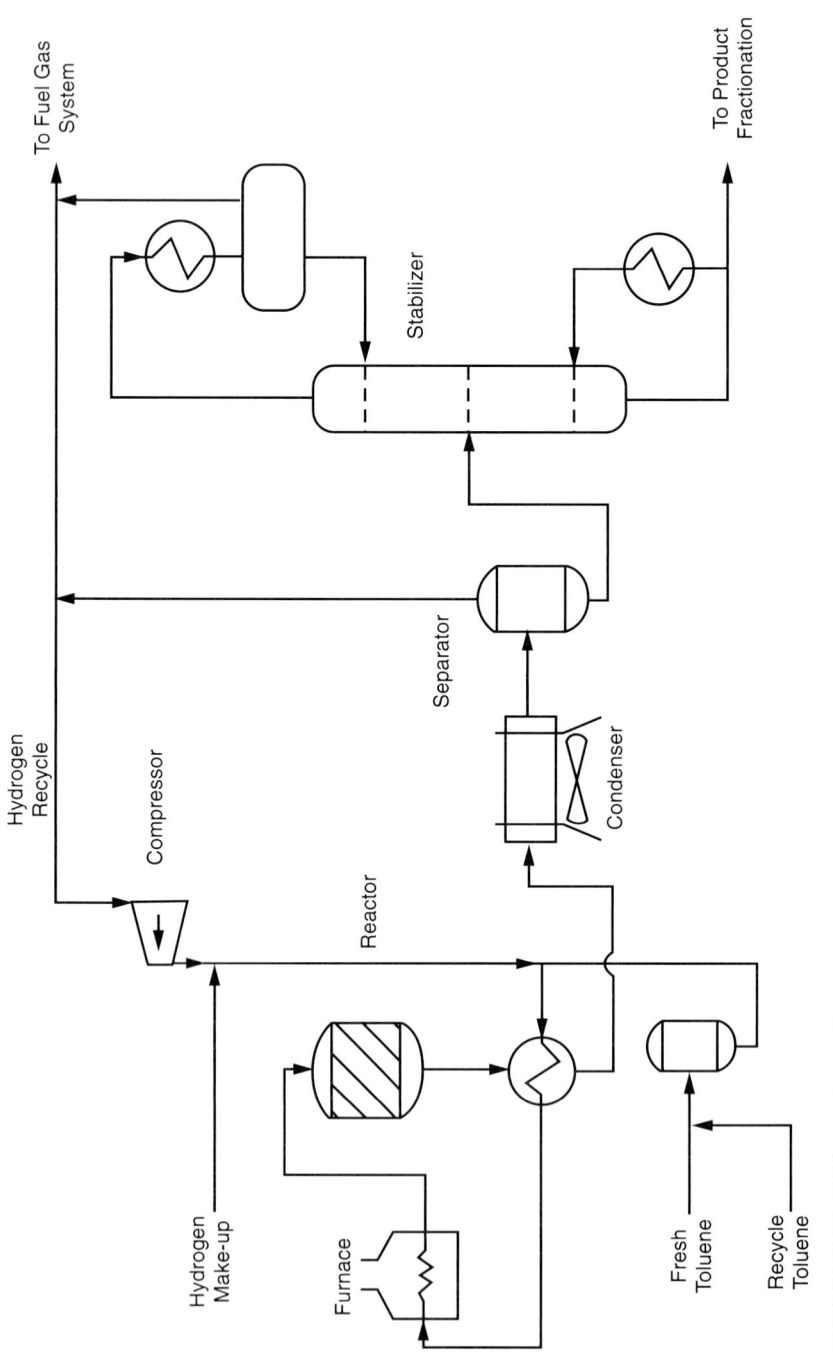

FIGURE 13.1.3 PxMax reaction section.

FIGURE 13.1.4 ExxonMobil Beaumont PxMax plant. Shown in the photograph are the five distillation towers that separate the products from the PxMax process reactor (toluene, benzene, unpurified *p*-xylene, heavy aromatics, and light gas). The reactor vessel is the tallest of the three vessels near the road, at the right of the photograph. In this unit, *p*-xylene is recovered by crystallization. The crystallizers are the row of vertical vessels behind the distillation towers.

PX production from a given volume of toluene feedstock, this purity level minimizes the size and cost of the PX recovery facilities. Additionally, at this purity level, much less expensive two-stage crystallization technology for PX recovery becomes an option in grassroots plants, offering further opportunity for reducing costs. Lower start-of-cycle temperature contributes to the improved selectivity but also provides utility savings and longer cycle lengths and minimizes downgrade of feedstock to low-value fuel and LPG by-products. The PxMax units currently in operation have demonstrated cycle lengths of over 5 years, in contrast to less than 2 years with other disproportionation technologies.

With regard to benzene production, PxMax provides lower benzene production than other PX-selective toluene disproportionation technologies during normal operation, but the process and catalyst are capable of operating at higher severity and producing more benzene when market forces provide incentive for the extra production. The higher-severity operation does not add significantly to catalyst aging. The benzene product meets industry standard quality without being further processed through an extraction unit.

PXMAX RETROFIT AND DEBOTTLENECK APPLICATIONS

One of the key benefits of the PxMax process is its ability to debottleneck or expand the capacity of existing PX recovery units, which recover PX from mixed-xylene feeds, such as from xylene isomerization or nonselective TDP processes. Three case studies were performed to demonstrate the debottlenecking potential of PxMax. The scenarios considered include

1. Add a grassroots PxMax unit to an existing mixed-xylenes PX recovery train.
2. Retrofit PxMax into a conventional STDP unit.
3. Retrofit PxMax into a nonselective TDP unit.

There are two techniques commonly used for recovering PX: crystallization and adsorption-based technologies. Studies have shown that implementation of PxMax can expand the PX production capacity of a crystallizer recovery unit by 20 to 40 percent. Since the most prevalent technique currently used in industry for PX recovery is adsorption, this chapter will focus on how PxMax can increase the PX production capacity of an adsorption-based recovery unit such as Axens' Eluxyl™ process.

AROMATICS COMPLEX AND PXMAX UNIT DESCRIPTION

The typical aromatics complex, displayed in Fig. 13.1.5, contains a reformer, an extraction unit with downstream recovery fractionation, and a xylenes loop, which consists of the xylene column, the PX recovery unit, and the xylene isomerization unit.

In most cases, the ability to expand the PX complex is limited by the recovery unit capability. The two main causes of an adsorption-based PX recovery bottleneck are (1) the maximum C_8 aromatic feed rate to the recovery unit, which is dictated by the capacity of the adsorption chambers or a hydraulic limitation, and (2) the maximum PX-depleted C_8 raffinate rate, which is dictated by the fractionation capacity of the raffinate column. Another potential bottleneck in the PX recovery unit is the fractionation capacity limit of the extract or finishing columns. The actual constraint depends on the specific site configuration.

To increase the PX capacity of the complex without requiring additional capital investment for PX recovery or xylene isomerization, the toluene derived from reformate, along with any toluene produced by xylene isomerization, can be fed to a PxMax unit (inside the dashed box in Fig. 13.1.5).

Examples of various scenarios where PxMax can be used to overcome the PX recovery unit bottlenecks follow.

CASE I: GRASSROOTS PXMAX UNIT

This case demonstrates the benefit of adding a grassroots PxMax unit to an existing xylenes loop. For the base case (case I, base), 30 KBD of CCR-based reformate with a typical composition feeds an aromatics complex having a design capacity of 350,000 metric tons per annum (MTA) PX. To achieve this PX capacity, purchased mixed xylenes must be brought into the complex. This scenario, in which the PX plant is reformate feed limited, is typical for the Asia-Pacific region, where the predominant growth in the PX market is taking place. Typical PX recovery and product purity were assumed for the PX recovery unit, and the xylene isomerization unit has been assumed to be operating at 75 percent ethylbenzene conversion with yields similar to conventional ExxonMobil ethylbenzene dealkylation–based

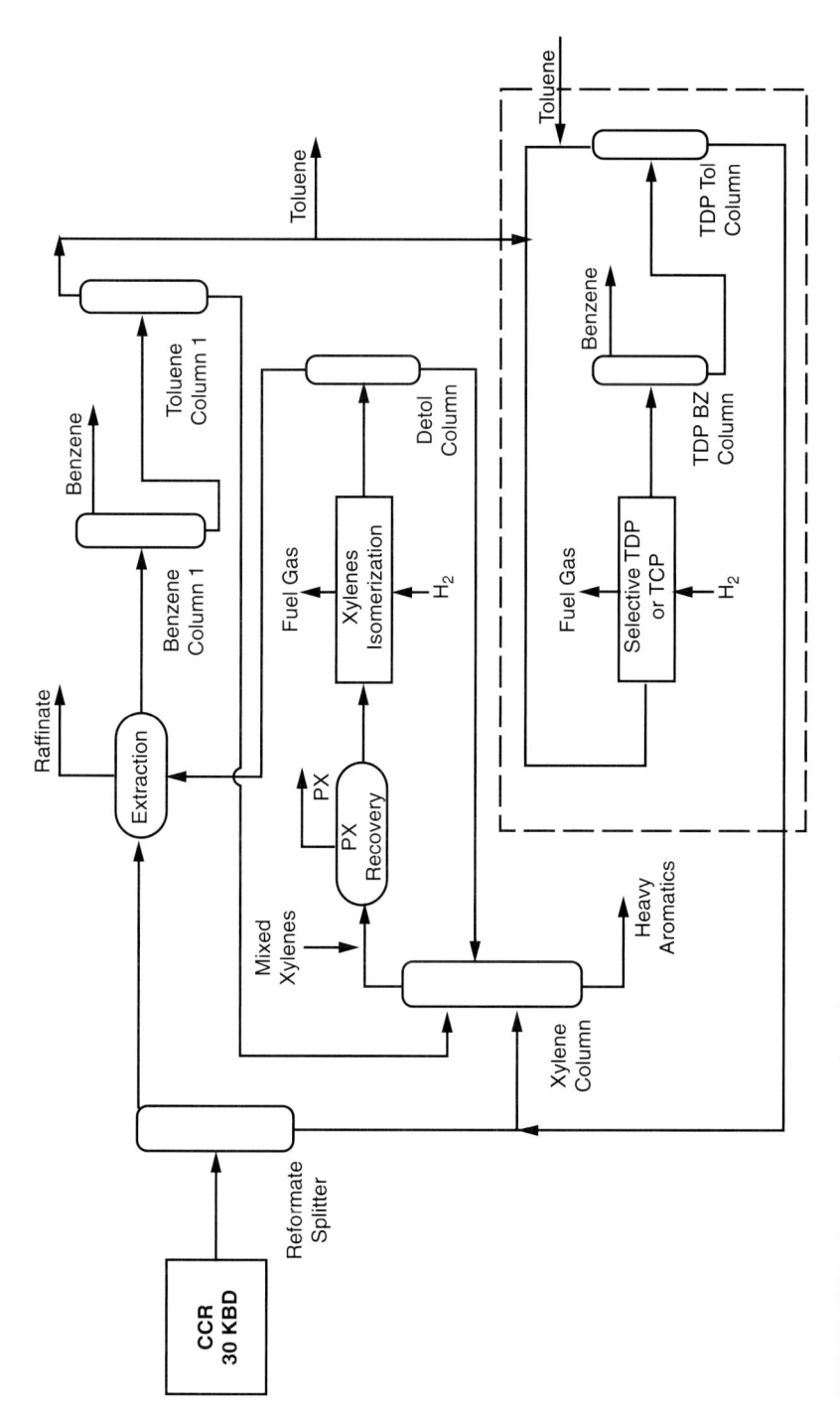

FIGURE 13.1.5 Aromatics complex schematic.

TABLE 13.1.1 Incentives for Grassroots PxMax Addition

Case description	Case I, base: Base-case xylenes loop	Case IA: Grassroots PxMax	Case IB: Grassroots PxMax
Recovery unit constraint	C_8 feed or C_8 raffinate	C_8 feed	C_8 raffinate
Feeds, kMTA			
Reformate	Base	0	0
Purchased mixed xylenes	40	−40	−15
Hydrogen	Base	+1	+2
Products, kMTA			
Fuel gas	Base	+6	+9
Extraction raffinate	Base	0	0
Benzene	Base	+94	+121
Toluene	273	221	273
p-xylene	350	427	472
Heavy aromatics	Base	+5	+8
Net income,*,† M$/yr	Base	+11	+17
Capital investment for grassroots PxMax unit,‡ M$	—	35	40

*Incremental net income (revenue minus operating cost) defined as difference between adding PxMax unit and base case. Operating cost accounts for utilities, labor, catalyst, freight, return on working capital, and other miscellaneous costs.

†Pricing based on CMAI historical average from 1987 to 2002. Past performance is no guarantee of future market performance.

‡Grassroots PxMax capital estimate includes ISBL (feed pretreatment, reactor circuit, fractionation), OSBL, engineering design, and contractor's fee.

xylene isomerization technologies. The base case assumes that the PX recovery unit is the limiting constraint.

The benefits of PxMax have been quantified for two scenarios. In one case, case IA, the PX recovery unit is operating at the C_8 feed limit. In the second case (case IB), the PX recovery unit is operating at the C_8 raffinate limit. Only toluene derived internally from reformate is used for the PxMax feed. Table 13.1.1 shows the results for case I.

As shown by case IA, adding a PxMax unit while operating at the same PX recovery unit C_8 feed rate as the base case results in increasing the PX production capacity of the complex by 77,000 MTA and increasing the benzene production capacity by 94,000 MTA. This can be accomplished while eliminating the purchased mixed xylenes (40,000 MTA) and using only 221,000 MTA of toluene as feed to the PxMax unit. As shown by case IB, if one operates at the same C_8 raffinate rate as in the base case, adding the PxMax process, which converts all the toluene (273,000 MTA), and purchasing only 25,000 MTA of mixed xylenes enable the PX production capacity to be increased by 122,000 MTA with +121,000 MTA benzene coproduction. If the recovery unit limits the PX production capacity, implementing a grassroots PxMax unit can lead to $11 million to $17 million increase in annual net income.

In these two scenarios, the ability to debottleneck the recovery unit results from the ultrahigh concentration of PX in the PxMax product. Also, for a constant rate to the PX recovery unit, the lower concentration of *m*-xylene, *o*-xylene, and ethylbenzene in the PxMax product results in a significant rate reduction in the C_8 raffinate stream. This allows one to increase the feed rate to the PX recovery unit, assuming that there is sufficient capacity, while not exceeding the C_8 raffinate limit.

CASE II: RETROFIT OF SELECTIVE TDP TO PXMAX

This case study demonstrates the benefit of retrofitting an existing conventional STDP technology, e.g., MSTDP, to the PxMax process. For the base case (case II, base), the aromatics complex, which was shown in Fig. 13.1.5, processes 30 KBD reformate, all the toluene production (273,000 MTA) is converted across the MSTDP unit, and there are no purchased mixed xylenes. The resulting PX production rate is 446,000 MTA. In this scenario, the PX recovery unit is operating at its C_8 raffinate limit.

The benefits of retrofitting the MSTDP unit with PxMax stem from the improved PX purity, the improved product yields, and the more favorable process conditions. These benefits have been demonstrated in cases IIA and IIB, which are shown in Table 13.1.2.

In case IIA, which represents a simple drop-in catalyst replacement, the MSTDP catalyst has been replaced with a catalyst used in the PxMax process. To maintain the C_8 raffinate at the same rate as the base case and thus maximize PX capacity, 18,000 MTA of additional mixed xylenes have been brought into the complex. This simple PxMax retrofit results in increasing the PX capacity by 20,000 MTA (5 percent) while reducing both benzene and fuel gas production. These combined benefits result in a $3 million annual increase in net income. Minimal capital investment is required for this scenario because the retrofit essentially would be a drop-in catalyst replacement.

The ability of the PxMax process to operate at a potentially higher weight hourly space velocity (WHSV) and a lower hydrogen-to-hydrocarbon molar ratio compared with MSTDP enables expansion of the STDP unit by 20 to 100 percent, assuming that there are no other unit constraints. In case IIB, which represents a STDP reactor loop limit, the fresh toluene feed rate to the STDP unit has been doubled (546,000 MTA fresh feed rate to STDP) by purchasing 273,000 MTA of toluene. In this scenario, the PX and benzene capacities increase by 144,000 MTA (+30 percent) and 120,000 MTA, respectively, while backing out all (18,000 MTA) of the purchased mixed xylenes and still operating at the C_8 raffinate con-

TABLE 13.1.2 Incentives for Retrofitting an MSTDP Unit to PxMax

Case description	Case II, base: Base-case STDP unit	Case IIA: PxMax, no TDP expansion	Case IIB: PxMax, reactor loop limit
Recovery unit constraint	C_8 raffinate	C_8 raffinate	C_8 raffinate
Feeds, kMTA			
Reformate	Base	0	0
Purchased toluene	0	0	+273
Purchased mixed xylenes	0	+18	0
Hydrogen	Base	+1	+2
Products, kMTA			
Fuel gas	Base	−1	+8
Raffinate	Base	0	0
Benzene	Base	−3	+117
Toluene	Base	0	0
p-xylene	446	466	585
Heavy aromatics	Base	+3	+11
Net income,*,† M$/yr	Base	+3	+20

*Incremental net income (revenue minus operating cost) defined as difference between adding PxMax unit and base case. Operating cost accounts for utilities, labor, catalyst, freight, return on working capital, and other miscellaneous costs.
†Pricing based on CMAI historical average from 1987 to 2002. Past performance is no guarantee of future market performance.

straint. If MSTDP is replaced with PxMax and the reactor loop is expanded by 100 percent, the annual net income can be increased by $20 million. Some equipment modifications may be required to take advantage of the entire expansion capability offered by PxMax; however, given the high increase in annual net income, high return on capital is likely.

CASE III: RETROFIT OF NONSELECTIVE TDP TO PXMAX

This case study, shown in Table 13.1.3, illustrates the benefits of retrofitting a nonselective TDP unit, such as MTDP-3, to PxMax. The base case (case III, base) is similar to case II, base. The aromatics complex processes 30 KBD of reformate, all the in-house toluene is converted across the TDP unit, and no purchased mixed xylenes are brought into the complex. ExxonMobil's MTDP-3 technology, which converts toluene to benzene and mixed xylenes at high yield, is used as the basis for this case. The base-case PX production capacity is 446,000 MTA. Similar to case II, we have assumed that the PX recovery unit is operating at its raffinate limit.

By substituting the high-purity PX PxMax product for the mixed-xylene TDP product, the xylenes loop circulation rate is reduced substantially. This enables supplemental mixed xylenes to be brought into the complex to use all the PX recovery unit capacity. Also,

TABLE 13.1.3 Incentives for Retrofitting a TDP Unit to PxMax

Case description	Case III, base: Base-case TDP unit	Case IIIA: PxMax, reactor feed limit	Case IIIB: PxMax, use all in-house toluene	Case IIIC: PxMax, purchased toluene/no purchased xylenes
Recovery unit constraint	C_8 raffinate	C_8 raffinate	C_8 raffinate	None
Feeds, kMTA				
Reformate	Base	0	0	0
Purchased toluene	0	0	0	+226
Purchased mixed xylenes	0	+148	+141	0
Hydrogen	Base	+3	+4	+3
Products, kMTA				
Fuel gas	Base	+8	+12	+10
Raffinate	Base	0	0	0
Benzene	Base	−29	+18	+103
Toluene	Base	+107	0	0
p-xylene	446	516	563	563
Heavy aromatics	Base	−5	−2	−1
Net income,*,† M$/yr	Base	+5	+12	+17
Capital investment for grassroots PxMax Unit,‡ M$	—	—	40	60

*Net income (revenue minus operating cost) defined as difference between adding PxMax unit and base case. Operating cost accounts for utilities, labor, catalyst, freight, return on working capital, and other miscellaneous costs.

†Pricing based on CMAI historical average from 1987 to 2002. Past performance is no guarantee of future market performance.

‡Grassroots PxMax capital estimate includes ISBL (feed pretreatment, reactor circuit, fractionation), OSBL, engineering design, and contractor's fee.

because TDP and PxMax operate at considerably different toluene conversions (48 percent for TDP versus 30 percent for PxMax), implementing PxMax using the existing TDP train results in less toluene processing capability due to bottlenecks encountered in the reactor section and downstream fractionation. As a result, a portion of the internally produced toluene must be removed from the PxMax unit feed and sold as product. As shown by case IIIA, assuming that the PxMax reactor feed rate is the same as in the base case, retrofitting the TDP unit to PxMax results in +70,000 MTA (+15 percent) PX production while using 107,000 MTA less toluene and producing 29,000 MTA less benzene. This retrofit can be accomplished with minimal equipment modifications, resulting in an additional annual net income of $5 million for the aromatics complex.

To retrofit the TDP unit to PxMax using all the in-house toluene, the TDP reactor section and fractionation train can be modified as required. As shown in case IIIB, retrofitting the TDP unit to PxMax while using all the in-house toluene and feeding 141,000 MTA purchased mixed xylenes expands the PX production capacity by 111,000 MTA (25 percent) and increases the annual net income by $12 million. If insufficient mixed xylenes are available, an alternative approach for increasing the PX production capacity is to further expand the capacity of the PxMax unit. As shown in case IIIC, using all the internal toluene as well as purchasing 226,000 MTA of toluene also will increase the PX production by 111,000 MTA (25 percent), resulting in a $17 million increase in the annual net income. If grassroots TDP reactor circuit and fractionation equipment were required for these two scenarios, the capital investment requirement would be $40 million and $60 million for case IIIB and case IIIC, respectively.

CONCLUSION

PxMax, which has been developed through major advancements in zeolite and selectivation technology at ExxonMobil, offers substantial debottlenecking potential to existing PX recovery units. Three case studies have been evaluated including (1) the addition of a grassroots PxMax unit to an existing xylenes isomerization recovery loop, (2) the retrofit of a selective toluene disproportionation unit to PxMax, and (3) the retrofit of a nonselective toluene disproportionation unit to PxMax. It has been estimated that for these three scenarios, the PX production capacity can be expanded by 15 to 30 percent. The expansion capability of the PxMax process stems from numerous advantages, including an ultra-high PX selective product, an excellent yield slate, the simplicity of its operation, and very long operating cycles. Of critical importance is that PxMax can be implemented with minimal technical risk because the process is a simple hydroprocessing technology. Since its commercial introduction in 1996, PxMax has been deployed in seven applications around the world, including both grassroots facilities and retrofit of existing plants. Table 13.1.4 shows the current PxMax applications.

TABLE 13.1.4 PxMax Commercial Experience

Location	Application type	Startup
Chalmette, La.	Retrofit	1996
Beaumont, Tex.	Grassroots	1997
Baytown, Tex.	Retrofit	1998
Japan	Retrofit	2001
Undisclosed	Retrofit	2002
Korea	Grassroots	2003
Undisclosed	Retrofit	2003

CHAPTER 13.2
EXXONMOBIL XYMAX^{SM} XYLENE ISOMERIZATION

Terry W. Bradley
ExxonMobil Chemical Company
Baytown, Texas

INTRODUCTION

Advances in ExxonMobil zeolite catalysts have led to significant improvements in xylene isomerization technology available to the petrochemical industry. ExxonMobil's most advanced isomerization process is known as XyMax^{SM}. XyMax is tailored for *p*-xylene (PX)–depleted streams such as those obtained when adsorptive-type PX recovery units are used. Compared with other conventional technologies, XyMax offers enhanced PX yield, lower xylene loss, higher ethylbenzene (EB) conversion, and debottleneck potential in existing plants.

Since the discovery of ZSM-5, a zeolite catalyst used initially as a fluid catalytic cracking catalyst, ExxonMobil has been a leader in zeolite technology for over 30 years. This expertise was applied quickly to developing catalysts and processes for petrochemicals with the introduction of the Mobil Vapor Phase Isomerization (MVPI^{SM} process in 1975). MVPI used the first high-activity zeolite-based xylene isomerization catalyst. In 1978, the Mobil Low Pressure Isomerization (MLPI^{SM}) process was introduced and was capable of operating without hydrogen circulation while achieving low xylene losses and long catalyst cycle lengths.

In 1981, the Mobil High Temperature Isomerization (MHTI^{SM}) process was introduced, and it revolutionized aromatics production. MHTI was capable of operating at higher EB conversions with lower xylene losses than its predecessors. As a result, many isomerization units of all types replaced their catalyst with MHTI to expand PX production capacity. MHTI and the ExxonMobil isomerization technologies that followed it remove EB by dealkylating it to produce benzene and ethane.

In 1990, the Mobil High Activity Isomerization (MHAI^{SM}) process pushed the capability of zeolite-based processes even further by enabling operations at even higher EB conversions with lower temperatures and lower xylene losses. MHAI operated at EB conversions in the range of 60 to 75 percent, high weight hourly space velocity, and low

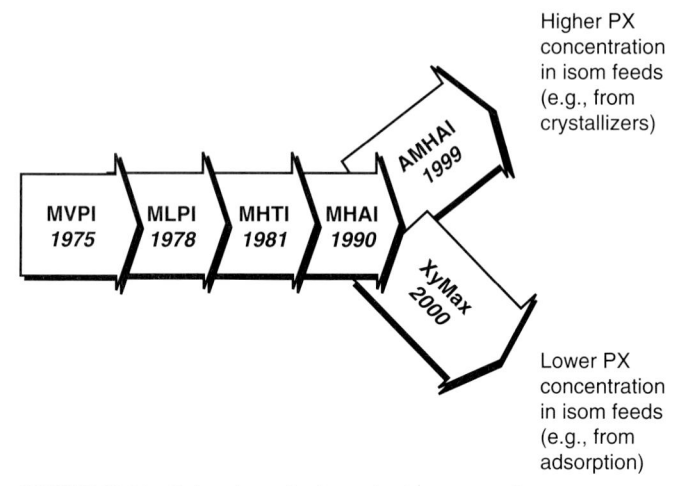

FIGURE 13.2.1 Xylene isomerization technology progression.

hydrogen-to-hydrocarbon molar ratios. These features allowed producers to further debottleneck units with minimal or no hardware modifications.

In 1999, Advanced MHAI (or AMHAI^{SM}) further reduced xylene losses, offering a wide range of EB conversion and operating at significantly lower temperatures than its predecessors. AMHAI is particularly suited for xylene feedstocks from plants using crystallization for PX recovery containing a higher level of PX than is seen in xylenes from an adsorptive-type recovery unit.

The most recent addition to the ExxonMobil xylenes isomerization technology portfolio is XyMax, the primary subject of this chapter. The catalyst developed for the XyMax process was specifically formulated to take advantage of low-PX-content feeds, such as those found at sites employing adsorptive-type PX recovery units. XyMax dramatically reduces xylene losses at high EB conversions. The relative progression of ExxonMobil xylene isomerization technology is shown in Fig. 13.2.1.

PROCESS CHEMISTRY

The primary chemical reactions in XyMax are the conversion of EB to benzene and ethylene, cracking of nonaromatics, and isomerization of the PX-depleted feedstock to an equilibrium mixture of xylenes. These reactions take place in a fixed-bed reactor with two distinct zeolite catalysts. In the top bed, the catalyst is designed to convert EB to benzene and ethylene, via dealkylation, and to crack nonaromatics. The ethylene produced is largely hydrogenated to ethane in the top bed also, reducing the likelihood of xylene loss through alkylation to heavy aromatics. The catalyst in the bottom bed is optimized for complete xylene isomerization to near-equilibrium levels of PX. The relative volumes of the two catalyst beds are tailored to fit the specific feed composition of each application.

Isomerization technologies are designed to manage the EB contaminant using one of two reactions: (1) EB dealkylation (to benzene and ethylene) or (2) EB isomerization (to xylenes). ExxonMobil's dual-bed isomerization technologies all use the EB-dealkylation reaction to remove EB. Dealkylation has the inherent advantage that it is not equilibrium-constrained. With the high-activity XyMax catalyst, high levels of EB conversion are attained. By contrast, EB conversion by isomerization is limited by the chemical equilibrium, resulting

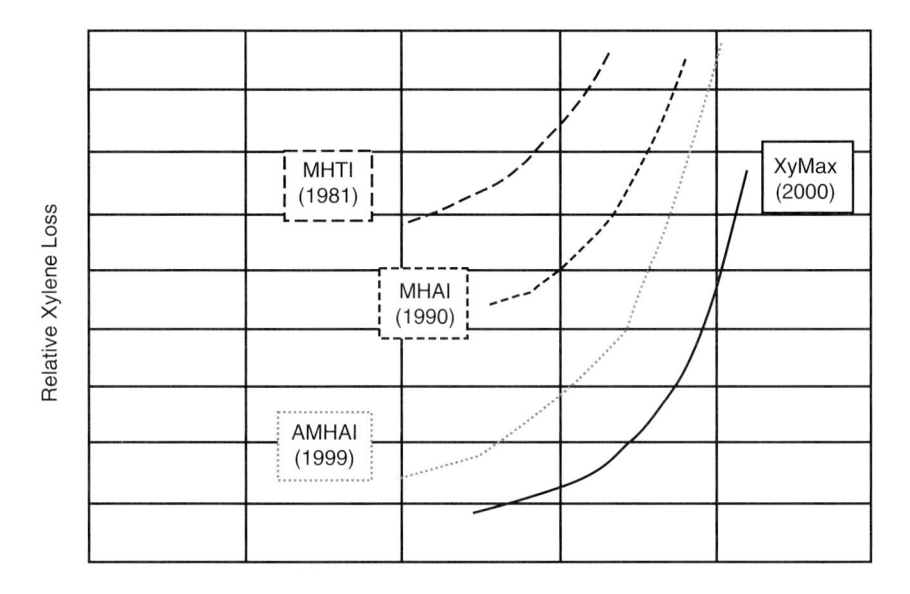

FIGURE 13.2.2 Ethylbenzene conversion in xylene isomerization.

in much more modest EB conversions, higher EB recycle volume, and higher costs. In situations where ExxonMobil has retrofitted EB-dealkylation-type technologies into units previously employing EB-isomerization-type technology, PX production increases of up to 40 percent have been observed. The primary modification typically required for conversion to an EB-dealkylation technology is the addition of a benzene fractionation tower.

The EB-dealkylation feature of XyMax is important in understanding XyMax's advantages versus more conventional technologies. EB is an undesirable component in C_8 aromatics recovered from reformate. High EB conversion is critical to the efficient operation of the xylene isomerization loop. Each successive generation of ExxonMobil xylene isomerization technology has pushed EB conversion to a new high. EB not converted in the reactor is recycled through the process, making PX recovery more difficult, consuming energy, and occupying capacity that otherwise could be used to produce PX. Therefore, higher EB conversions often provide an opportunity to debottleneck existing PX recovery units. The relative improvement in EB conversion overtime is shown in Fig. 13.2.2.

Another group of unwanted constituents of reformate with potential to interfere with efficient product recovery is nonaromatics. The top catalyst bed also serves to hydrocrack nonaromatics to light gas that is removed easily by a stabilizer tower downstream of the reactor. Like EB, nonaromatic conversion is important to avoid concentration in the isomerate recycle stream, occupying capacity and increasing costs.

PROCESS DESCRIPTION

The primary route to PX is through catalytic reforming of naphtha to produce a mixed-xylene feedstock from which PX is recovered. Following recovery of the PX in the incoming feedstock, the PX-depleted xylene stream is isomerized to convert *o*-xylene (OX) and *m*-xylene (MX) to an equilibrium mixture including PX.

The incoming xylene feedstock is combined with recycled xylenes and fractionated to remove C_9+ aromatics because heavy aromatics can promote coking on the XyMax catalyst, potentially reducing activity and ultimately cycle length. The fractionated xylenes are then fed to a PX recovery unit. PX recovery typically is accomplished either by fractional crystallization or through adsorptive-type processes. In crystallization, the differences in freezing points of the three isomers are used to accomplish the separation. In absorptive processes, the isomers are separated on the basis of differing rates of adsorption and desorption on a specialized material. Both processes are used widely in the petrochemical industry.

The OX and MX exiting the recovery unit are mixed with hydrogen, preheated against reactor effluent, and then further heated in a fired heater to approximately 405°C. The heated mixture is introduced to the isomerization reactor. Hydrogen is consumed in the reactor to saturate the ethylene produced from EB dealkylation and trace amounts of olefins in the feed. The hydrogen also inhibits coking of the catalyst and controls hydrocarbon partial pressure to a desirable range. A minimum 80 vol % purity of the circulating hydrogen is maintained by addition of makeup hydrogen. The hydrogen addition rate to the reactor typically is controlled to a 1.0 to 1.5 hydrogen-to-hydrocarbon molar ratio, depending on the specifics of the application.

In the reactor, EB and nonaromatics are converted in the top bed, and xylene isomerization is performed in the bottom bed, with an equilibrium mixture of xylenes produced from the bottom of the reactor. The reactor effluent is cooled against reactor feed and further cooled with cooling water before entering a separator drum to remove unreacted hydrogen, which is then compressed for recycle to the reactor. The liquid product from the separator is fractionated in a stabilizer tower to further separate light gases overhead, benzene/toluene, and the isomerized xylenes. The isomerized xylene stream is recycled to the start of the process, where it is blended with incoming fresh feed.

A key feature of XyMax and AMHAI technologies, as compared with other isomerization technologies, is use of the unique stacked dual-bed catalyst system. Utilizing ExxonMobil's expertise in shape-selective catalysis, the dual catalysts are tailored to optimize EB dealkylation and cracking of nonaromatics in the top bed and achieve isomerization in the bottom bed. The bottom bed promotes the isomerization while minimizing unwanted disproportionation and transalkylation reactions that destroy xylenes.

ExxonMobil dual-bed technologies are drop-in replacements for many EB-dealkylation- or EB-isomerization-type technologies. While the dual-bed system is unique, the two catalysts can be loaded into any down- or upflow reactor because the two beds are separated only by a screen. Radial-flow reactors also have been retrofitted successfully and inexpensively to enable the use of ExxonMobil dual-bed technologies. The ExxonMobil XyMax process is shown in Fig. 13.2.3.

OPERATING PERFORMANCE

XyMax's enhanced yield performance is achieved primarily by an approximate 65 percent reduction in xylene losses compared with MHAI when operated with low PX feeds and at constant EB conversion. The lower xylene losses are a result of high selectivity of the proprietary EB-dealkylation catalyst employed. In addition, undesirable reactions are suppressed. The net result is higher PX production and reduced production of toluene and heavy aromatics.

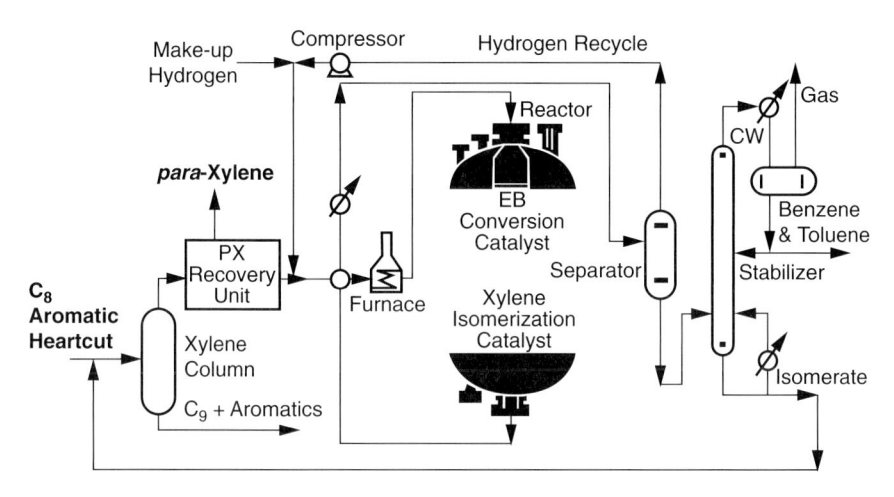

FIGURE 13.2.3 ExxonMobil dual-bed isomerization process.

XyMax yields are compared with MHAI and AMHAI processes, as shown in Table 13.2.1. Absolute feed and product shifts are in thousands of metric tons per year (kMTA). At constant PX production of 300 kMTA, an economic benefit is realized by reducing the fresh feed requirement by 19,100 MTA compared with MHAI. At constant fresh feed, PX capacity can be increased by up to 15,600 MTA. Operating conditions are equivalent for each set of yields.

This example was prepared for a site using an adsorptive-type PX recovery unit. XyMax is specifically designed to process the low-PX-containing xylene streams produced by adsorptive recovery units. In situations where xylene feeds contain more PX, such as from plants using crystallization, AMHAI is usually recommended.

In addition to the flexibility to operate the unit for either maximum PX or conservation of raw materials, other options are open to the producer using XyMax. For example, improved flexibility and debottlenecking potential are also the result of the higher EB conversion and lower hydrogen-to-hydrocarbon molar ratio. XyMax can be optimized to operate at EB conversions as high as 85 percent, higher than any other commercialized isomerization technology. Higher EB conversion increases the fresh feed processing capacity of the xylenes isomerization unit by reducing recycle loop volume and by concentrating xylenes in the feed.

Because of the ultrastable nature of the catalysts used in the XyMax process, the hydrogen-to-hydrocarbon ratio can be reduced to as low as 1.0. This ratio is lower than any other commercially available EB-dealkylation-based xylene isomerization technology and substantially lower than EB-isomerization-based technologies. The lower hydrogen-to-hydrocarbon ratio has a major impact on plant economics in both grassroots and retrofit applications. In grassroots applications at a fixed mixed xylene feed rate, the size of the recycle gas compressor and associated gas-handling facilities can be reduced, leading to significant investment cost savings. In retrofit applications, utility costs can be reduced at constant PX production. A lower hydrogen-to-hydrocarbon ratio also benefits retrofit applications by enabling a debottleneck of the xylenes isomerization unit in cases where the mixed-xylene feed is limited by recycle compressor size or hydraulic constraints.

In addition to its more favorable yields, the catalyst employed in XyMax converts more nonaromatics in the feed than do the catalysts used in the MHAI process. Hence XyMax

TABLE 13.2.1 XyMax Yields

| | Conv. MHAI | Absolute shift vs. conventional MHAI ||||
| | | Advanced MHAI || XyMax ||
		Const. PX	Const. FF	Const. PX	Const. FF
Mixed xylenes, kMTA	Base	(6.5)	—	(19.1)	—
SOC temp, °C	Base	−20	−20	Base	Base
Products, kMTA					
Light gas	Base	(0.1)	0.2	(0.4)	0.4
Benzene	Base	1.5	20	1.6	3.5
Toluene	Base	(6.4)	(6.0)	(14.8)	(14.3)
p-xylene	300	—	5.1	—	15.6
Heavies	Base	(1.5)	(1.3)	(5.5)	(5.1)

Bases: (1) Fresh feed, wt %: EB-15, PX-21, MX-43, OX-21; (2) constant EB conversion; (3) adsorptive-type p-xylene recovery.

is better able to process unextracted feeds, allowing producers to use lower-cost xylene sources and make extraction unit capacity available to process other feeds.

XYMAX CYCLE LENGTH

Cycle length is a typical concern for prospective licensors of a process using a fixed-bed catalyst. Long operating cycles between regenerations of 3 to 5 years are expected because of the following three attributes:

1. *Dual-bed catalyst system.* In XyMax, the EB conversion occurs in the top-bed catalyst, which is both highly selective and highly resistant to coke formation. With the EB dealkylation essentially complete, the bottom-bed catalyst performs the xylene isomerization function in a low-severity, low-aging-rate environment.
2. *Highly refined catalyst metal function.* Highly selective metal functionality leads to minimal product downgrade and catalyst aging.
3. *Tailored catalyst acidity function that produces fewer coke precursors.* The catalyst acid function is designed so that the highest acid activity is located where the catalyst is most selective and the milder acid activity is located where the catalyst is less selective. This distribution of the acid function minimizes undesirable chemistry and associated catalyst aging.

The MHAI process also employs a dual-bed catalyst system that has experienced first cycles of over 4 years, with one licensee that operated for 10 years, without regeneration. The XyMax process catalyst is expected to be at least as stable as those used in MHAI.

Another indication of catalyst stability is the first commercial application of XyMax at ExxonMobil's Singapore aromatics complex. This catalyst has been in operation for over 4 years with a target EB conversion during that time of 80 percent. Based on observed aging rate, cycle length of over 5 years is expected.

COMMERCIAL EXPERIENCE

As outlined in the introduction, ExxonMobil xylene isomerization technologies have a presence worldwide. There are currently 21 units worldwide using ExxonMobil xylene isomerization technologies. These units represent over 30 percent of the world's PX production capacity from xylene isomerization. ExxonMobil's most advanced isomerization technology, XyMax, was introduced at ExxonMobil's Singapore aromatics complex in 2000. Other applications of AMHAI and XyMax are shown in Table 13.2.2.

TABLE 13.2.2 XyMax and AMHAI Commercial Experience

	AMHAI	XyMax
1999		
Chalmette, La.	X	
Japan	X	
2000		
Singapore		X
Japan	X	
North America	X	
Japan	X	
2002–2003		
Undisclosed		X
Rotterdam		X
Baytown, Tex.		X
Undisclosed	X	
Asia	X	

CHAPTER 13.3
UOP PAREX™ PROCESS FOR *P*-XYLENE PRODUCTION

Scott E. Commissaris
UOP LLC
Des Plaines, Illinois

INTRODUCTION

The UOP Parex™ process is an innovative adsorptive separation method for the recovery of *para*-xylene from mixed xylenes. The term *mixed xylenes* refers to a mixture of C_8 aromatic isomers that includes ethylbenzene, *p*-xylene, *m*-xylene, and *o*-xylene. These isomers boil so closely together that separating them by conventional distillation is not practical. The Parex process provides an efficient means of recovering *p*-xylene by using a solid zeolitic adsorbent that is selective for *p*-xylene. Unlike conventional chromatography, the Parex process simulates the countercurrent flow of a liquid feed over a solid bed of adsorbent. Feed and products enter and leave the adsorbent bed continuously at nearly constant compositions. This technique is sometimes referred to as *simulated-moving-bed* (SMB) separation.

In a modern aromatics complex (Fig. 13.3.1), the Parex unit is located downstream of the xylene column and is integrated with a UOP Isomar™ unit. The feed to the xylene column consists of the C_8+ aromatics product from the CCR™ Platforming™ unit together with the xylenes produced in the Tatoray unit. The C_8 fraction from the overhead of the xylene column is fed to the Parex unit, where high-purity p-xylene is recovered in the extract. The Parex raffinate is then sent to the Isomar unit, where the other C_8 aromatic isomers are converted to additional *p*-xylene and recycled to the xylene column. A complete description of the entire aromatics complex may be found in Chapter 2.1 of the *Handbook of Petroleum Refining Processes*, 3d ed., edited by Robert A. Meyers (McGraw-HIll, 2004).

UOP Parex units are designed to recover more than 97 wt % of the *p*-xylene from the feed in a single pass at a product purity of 99.9 wt % or better. The Parex design is energy-efficient, mechanically simple, and highly reliable. On-stream factors for Parex units typically exceed 95 percent.

PAREX VERSUS CRYSTALLIZATION

Before the introduction of the Parex process, *p*-xylene was produced exclusively by fractional crystallization. In crystallization, the mixed-xylenes feed is refrigerated to $-75°C$

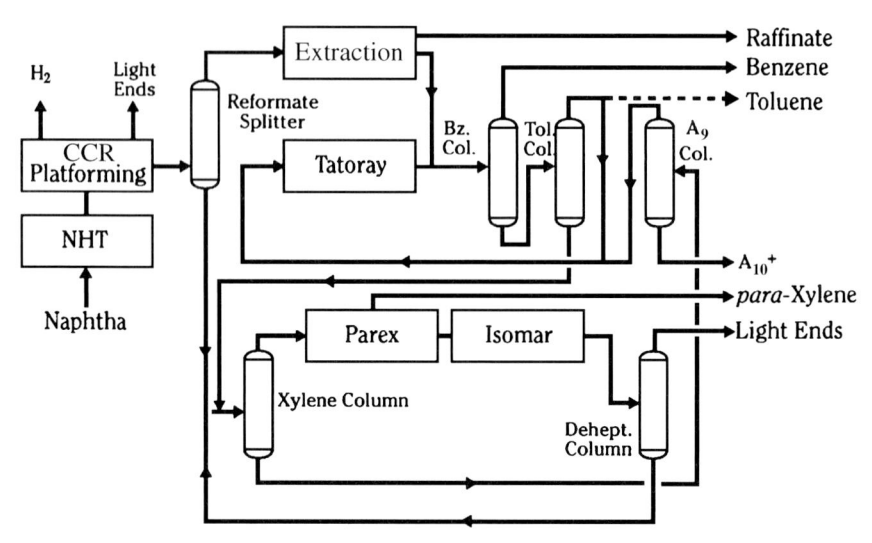

FIGURE 13.3.1 UOP aromatics complex—maximum p-xylene.

($-100°F$), at which point the p-xylene isomer precipitates as a crystalline solid. The solid is then separated from the mother liquor by centrifugation or filtration. Final purification is achieved by washing the p-xylene crystals with either toluene or a portion of the p-xylene product.

Soon after it was introduced in 1971, the UOP Parex process quickly became the world's preferred technology for p-xylene recovery. Since that time, virtually all new p-xylene production capacity has been based on the UOP Parex process (Fig. 13.3.2).

The principal advantage of the Parex adsorptive separation process over crystallization technology is the ability of the Parex process to recover more than 97 percent of the p-xylene in the feed per pass. Crystallizers must contend with a eutectic composition limit that restricts p-xylene recovery to about 65 percent per pass. The implication of this difference is clearly illustrated in Fig. 13.3.3: A Parex complex producing 250,000 metric tons per annum (MTA) of p-xylene is compared with a crystallizer complex producing 168,000 MTA. The upper numbers in the figure indicate the flow rates through the Parex complex; the lower numbers indicate the flow rates through a comparable crystallizer complex. A Parex complex can produce about 50 percent more p-xylene from a given-size xylene column and isomerization unit than a complex using crystallization. In addition, the yield of p-xylene per unit of fresh feed is improved because a relatively smaller recycle flow means lower losses in the isomerization unit. The technologies also could be compared by keeping the p-xylene product rate constant. In this case, a larger xylene column and a larger isomerization unit would be required to produce the same amount of p-xylene, thus increasing both the investment cost and the utility consumption of the complex.

A higher p-xylene recycle rate in the crystallizer complex not only increases the size of the equipment in the recycle loop and the utility consumption within the loop, but also makes inefficient use of the xylene isomerization capacity. Raffinate from a Parex unit is almost completely depleted of p-xylene (less than 1 wt %), whereas mother liquor from a typical crystallizer contains about 9.5 wt % p-xylene. Because the isomerization unit cannot exceed an equilibrium concentration of p-xylene (23 to 24 wt %), any p-xylene in the feed to the isomerization unit reduces the amount of p-xylene produced in that unit per pass. Thus, the same isomerization unit produces about 60 percent more p-xylene per pass when processing Parex raffinate than it does when processing crystallizer mother liquor.

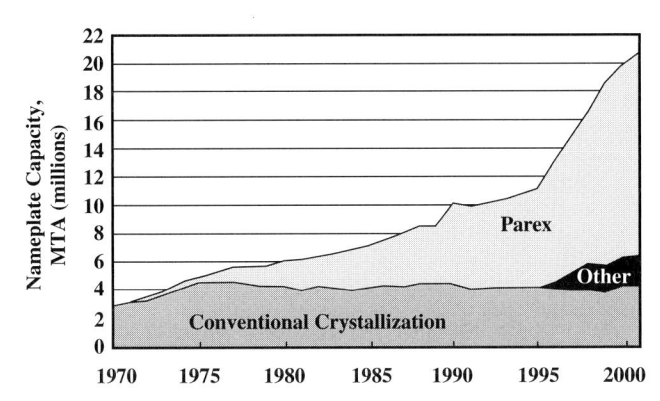

FIGURE 13.3.2 Total installed Parex capacity. In 2002, Parex represented 69 percent of the total capacity.

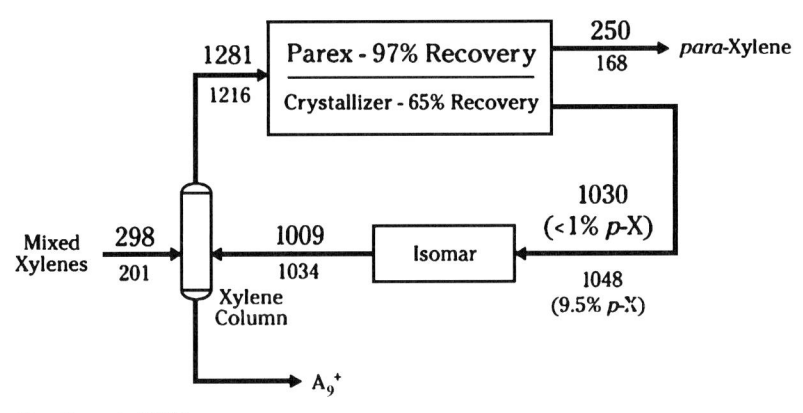

FIGURE 13.3.3 Comparison of Parex with crystallization.

In 1997, UOP, Washington Group International, and Niro Process Technology recognized the value that the three companies could bring to the marketplace by consolidating their combined 80+ years of process design and know-how to reevaluate p-xylene production from a multidiscipline perspective. In 1998, this alliance introduced the HySorb XP process, a simplified, single-chamber, light desorbent adsorption process coupled with single-stage crystallization and Niro wash column technology. This combination of technologies, when integrated into existing multistage crystallization facilities, can increase p-xylene production by as much as 500 percent. The HySorb process produces a 95 wt % p-xylene concentrate, eliminating eutectic constraints and enabling single-stage crystallization recoveries above 90 percent at much improved utilities' consumption. p-Xylene production using single-stage crystallization is categorized as "Other" in Fig. 13.3.2. Economic studies indicate that the HySorb XP configuration does not provide any cost or performance advantages relative to the Parex process for new grassroots designs or for expanding the production capacity of an existing Parex unit.

PROCESS PERFORMANCE

The quality of *p*-xylene demanded by the market has increased significantly over the last 20 years. When the Parex process was introduced in 1971, the standard purity for *p*-xylene sold in the market was 99.2 wt %. By 1992, the purity standard had become 99.7 wt %, and the trend toward higher purity continues. All Parex units built after 1991 are designed to produce 99.9 wt % pure *p*-xylene at 97 wt % recovery per pass. Most older Parex units can also be modified to produce 99.9 wt % purity.

FEEDSTOCK CONSIDERATIONS

Most of the mixed xylenes used for *p*-xylene production are produced from petroleum naphtha by catalytic re-forming. Modern UOP CCR Platforming units operate at such high severity that the C_8+ fraction of the re-formate contains virtually no nonaromatic impurities. As a result, these C_8 aromatics can be fed directly to the xylene recovery section of the complex. In many integrated aromatics complexes, up to one-half the total mixed xylenes are produced from the conversion of toluene and C_9 aromatics in a UOP Tatoray unit.

Nonaromatic impurities in the feed to a Parex unit increase utility consumption and take up space in the Parex unit, but they do not affect the purity of the *p*-xylene product or the recovery performance of the Parex unit.

Feedstocks for Parex must be prefractionated to isolate the C_8 aromatic fraction and clay treated to protect the adsorbent. If the Parex unit is integrated with an upstream refinery or ethylene plant, prefractionation and clay treating are designed into the complex. If additional mixed xylenes are purchased and transported to the site, they first must be stripped, clay treated, and rerun before being charged to the Parex unit. In general, feed to a Parex unit should meet the specifications outlined in Table 13.3.1.

DESCRIPTION OF THE PROCESS FLOW

The flow diagram for a typical Parex unit is shown in Fig. 13.3.4. The separation takes place in the adsorbent chambers. Each adsorbent chamber is divided into a number of adsorbent beds. Each bed of adsorbent is supported from below by specialized internals that are designed to produce highly efficient flow distribution. Each internals assembly is connected to the rotary valve by a "bed line." The internals between each adsorbent bed are used to inject or withdraw liquid from the chamber and simultaneously collect liquid from the bed above and redistribute the liquid over the bed below.

The Parex process is one member of UOP's family of Sorbex™ adsorptive separation processes. The basic principles of Sorbex technology are the same regardless of the type of separation being conducted and are discussed in Chapter 10.3 of the *Handbook of Petroleum Refining Processes*, 3d ed., edited by Robert A. Meyers (McGraw-Hill, 2004). The number of adsorbent beds and bed lines varies with each Sorbex application. A typical Parex unit has 24 adsorbent beds and 24 bed lines connecting the beds to the rotary valve. Because of practical construction considerations, most Parex units consist of two adsorption chambers in series with 12 beds in each chamber.

The Parex process has four major streams that are distributed to the adsorbent chambers by the rotary valve. These *net* streams include

- *Feed in:* Mixed-xylenes feed
- *Dilute extract out:* *p*-Xylene product diluted with desorbent

TABLE 13.3.1 Parex Feedstock Specifications

Property	Specification
p-Xylene, min, wt %	18
Ethylbenzene, max, wt %	20
Toluene, max, wt %	0.5
C_9 and higher-boiling aromatic hydrocarbons, max, wt %	1.5
Nonaromatic hydrocarbons, max, wt %	0.3
Nitrogen, max, mg/kg	1.0
Sulfur, max, mg/kg	1.0
Acidity	No free acid
Appearance	*
Relative density, 15.56/15.56°C or	0.865–0.875
Density, 20°C, g/cm^3	0.862–0.872
Color, max Pt/Co scale	20
Distillation range, at 101.3 kPa (760 mmHg) pressure, max, °C	5
Initial distillation temperature, min, °C	137
Dry point, max, °C	143

*Clear liquid free of sediment and haze when observed at 18.3 to 25.6°C (65 to 78°F).

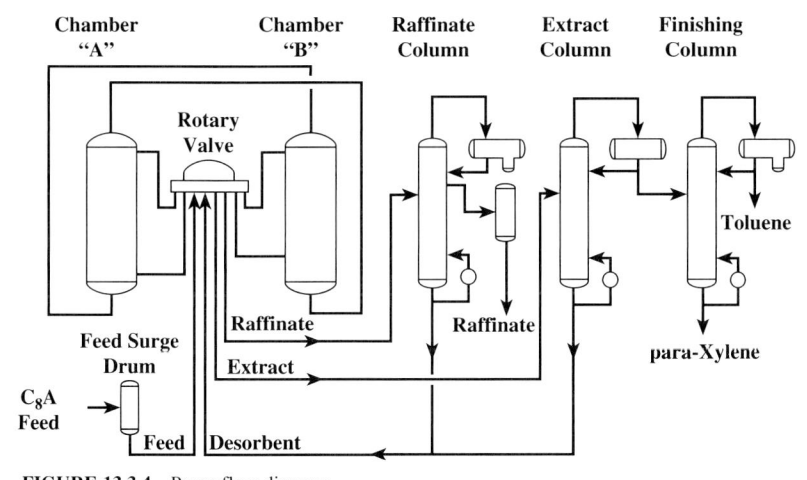

FIGURE 13.3.4 Parex flow diagram.

- *Dilute raffinate out:* Ethylbenzene, *m*-xylene, and *o*-xylene diluted with desorbent
- *Desorbent in:* Recycle desorbent from the fractionation section

At any given time, only four of the bed lines actively carry the net streams into and out of the adsorbent chamber. The rotary valve is used to periodically switch the positions of the liquid feed and withdrawal points as the composition profile moves down the chamber. A pump provides the liquid circulation from the bottom of the first adsorbent chamber to the top of the second. A second pump provides circulation from the bottom of the second adsorbent chamber to the top of the first. In this way, the two adsorbent chambers function as a single, continuous loop of adsorbent beds.

The dilute extract from the rotary valve is sent to the extract column for separation of the extract from the desorbent. The overhead from the extract column is sent to a finishing column, where the highly pure p-xylene product is separated from any toluene that may have been present in the feed.

The dilute raffinate from the rotary valve is sent to the raffinate column for separation of the raffinate from the desorbent. The overhead from the raffinate column contains the unextracted C_8 aromatic components: ethylbenzene, m-xylene, and o-xylene, together with any nonaromatics that may have been present in the feed. The raffinate product is then sent to an isomerization unit, where additional p-xylene is formed, and then recycled to the Parex unit.

The desorbent from the bottom of both the extract and raffinate columns is recycled to the adsorbent chambers through the rotary valve. Any heavy contaminants in the feed accumulate in the desorbent. To prevent this accumulation, provision is made to take a slipstream of the recycle desorbent to a small desorbent rerun column, where any heavy contaminants are rejected. During normal operation, mixed xylenes are stripped, clay treated, and rerun prior to being sent to the Parex unit. Thus, few heavy contaminants need to be removed from the bottom of the desorbent rerun column.

EQUIPMENT CONSIDERATIONS

UOP supplies a package of specialized equipment that is considered critical for the successful performance of the Parex process. This package includes the rotary valve; the adsorbent chamber internals; and the control system for the rotary valve, pumparound pump, and net flows. The erected cost estimates that UOP provides for the Parex process include the cost of this equipment package.

The rotary valve is a sophisticated, highly engineered piece of process equipment developed by UOP specifically for the Sorbex family of processes. The UOP rotary valve is critical for the purity of the p-xylene product and for the unsurpassed reliability of the Parex process. The design of the UOP rotary valve has evolved over 40 years of commercial Sorbex operating experience.

The adsorbent chamber internals are also critical to the performance of the Parex process. These specialized internals are used to support each bed of adsorbent and to prevent leakage of the solid adsorbent into the process streams. Each internals assembly also acts as a flow collector and distributor and is used to inject or withdraw the net flows from the adsorbent chamber or redistribute the internal liquid flow from one adsorbent bed to the next. As the size of Parex units has increased over the years, the design of adsorbent chamber internals has evolved to ensure proper flow distribution over increasingly larger-diameter vessels.

The Parex control system supplied by UOP is a specialized system that monitors and controls the flow rates of the net streams and adsorbent chamber circulation and ensures proper operation of the rotary valve.

Because of the mild operating conditions used in the Parex process, the entire plant may be constructed of carbon steel.

The Parex process is normally heat-integrated with the upstream xylene column. The xylene column is used to rerun the feed to the Parex unit. The mixed xylenes are taken overhead, and the heavy aromatics are removed from the bottom of the column. Before the overhead vapor from the xylene column is fed to the adsorption section of the Parex unit, it is used to reboil the extract and raffinate columns of the Parex unit.

UOP offers High Flux™ high-performance heat-exchanger tubing for improved heat-exchange efficiency. High Flux tubing is made with a special coating that promotes nucleate

TABLE 13.3.2 Investment Cost and Utility Consumption*

Estimated erected cost, million $US:	
Xylene column	32
Parex unit	98
Utility consumption:	
Electric power, kW	5300
Medium-pressure steam, MT/h (klb/h)	20 (credit 44.05)
Cooling water, m^3/h (gal/min)	174 (766)
Fuel fired, million kcal/h (million Btu/h)	125 (497)

*Basis: 700,000 MTA of *p*-xylene product.

boiling and increases the heat-transfer coefficient of conventional tubing by a factor of 10. Specifying UOP High Flux tubing for the reboilers of the Parex fractionators reduces the size of the reboilers and may also allow the xylene column to be designed for lower-pressure operation. Designing the xylene column for lower pressure reduces the erected cost of the column and lowers the utility consumption in that column.

UOP also offers MD™ distillation trays for improved fractionation performance. The MD trays are used for large liquid loads and are especially effective when the volumetric ratio between vapor and liquid rates is low. The use of MD trays provides a large total weir length and reduces froth height on the tray. Because the froth height is lower, MD trays can be installed at a smaller tray spacing than conventional distillation trays. The use of MD trays in new column designs results in a smaller required diameter and lower column height. Consequently, MD trays are often specified for large xylene columns, especially when the use of MD trays can keep the design of the xylene column in a single shell.

CASE STUDY

A summary of the investment cost and utility consumption for a typical Parex unit is shown in Table 13.3.2. The basis for this case is a Parex unit producing 700,000 MTA of 99.9 wt % pure *p*-xylene product. This case corresponds to the case study for an integrated UOP aromatics complex presented in Chapter 2.1 of the *Handbook of Petroleum Refining Processes*, 3d ed., edited by Robert A. Meyers (McGraw-Hill, 2004). Because the Parex unit is tightly heat-integrated with the upstream xylene column, the investment cost and utility consumption estimates include both. The estimated erected costs for these units assume construction on a U.S. Gulf Coast site in 2002. The scope of the estimate includes engineering, procurement, erection of equipment on the site, and the initial inventory of Parex adsorbent and desorbent.

COMMERCIAL EXPERIENCE

UOP's experience with adsorptive separations is extensive. Sorbex technology, which was invented by UOP in the 1960s, was the first large-scale commercial application of continuous adsorptive separation. The first commercial Sorbex unit, a Molex™ unit for the separation of linear paraffins, came on-stream in 1964. The first commercial Parex unit came on-stream in 1971. UOP has licensed more than 100 Sorbex units throughout the world. This total includes 73 Parex units, of which 71 units are in operation and 2 others are in various stages of design and construction. UOP Parex units range in size from 24,000 MTA of *p*-xylene product to more than 700,000 MTA.

BIBLIOGRAPHY

Jeanneret, J. J.: "Developments in *p*-Xylene Technology," *DeWitt Petrochemical Review,* Houston, March 1993.

Jeanneret, J. J.: "*p*-xylene Production in the 1990s," UOP Technology Conferences, various locations, May 1995.

Jeanneret, J. J., C. D. Low, and V. Zukauskas: "New Strategies Maximize *p*-xylene Production," *Hydrocarbon Processing,* June 1994.

Prada, R. E., et al.: "Parex Developments for Increased Efficiency," UOP Technology Conferences, various locations, September 1992.

PART 14

POLYETHYLENE

CHAPTER 14.1
BASELL *SPHERILENE* TECHNOLOGY FOR LLDPE AND HDPE PRODUCTION

Maurizio Dorini and Gijs ten Berge
Basell Polyolefine GmbH
Frankfurt, Germany

GENERAL PROCESS DESCRIPTION

The *Spherilene* technology is a swing process that has been designed to allow the production of a flexible, high-quality product-grade slate including high-density polyethylene (HDPE), medium-density polyethylene (MDPE), linear low-density polyethylene (LLDPE), and very-low-density polyethylene (VLDPE). The process consists of a polymerization unit with two gas-phase fluidized-bed polymerization reactors in series, in which ethylene and comonomers such as butene-1 are converted to various types of polymers. Downstream of the reaction section is a product finishing section, additivation, and extrusion.

PROCESS CHEMISTRY AND THERMODYNAMICS

Conventional polyethylenes are thermoplastic materials with property characteristics usually identified in terms of the following main parameters:

- Molecular weight (MW), characterized by the MIE (g/10 min)
- Molecular weight distribution (MWD), expressed by the ratio MIF/MIE
- Density (g/cm^3), determined by the polymer chain branching and structure

Nowadays, complex polymer design and significant process developments can be obtained using gas-phase processes.

Linear Polyethylene

The family of linear polyethylenes includes

- High-density PE (HDPE)
- High-density high-molecular-weight PE (HMWPE)
- Ultrahigh-molecular-weight PE (UHMWPE)
- Low-density PE (LLDPE)
- Very-low-density PE (VLDPE)

All of these are manufactured via low-pressure technologies, and—unlike the conventional high-pressure LDPE—the crystallinity level is controlled only by the comonomer content.

The comonomer is grafted along the molecular backbone, forming short side-chain branches that "disturb" the ability of the macromolecules to "pack" and so to form crystals. Consequently, the higher the comonomer content, the lower is the polymer density. In LDPE from the high-pressure process, both short- and long-chain branches are present; the polymer is constituted only by ethylene monomer units (homopolymer), and the density level is affected by variations in the process parameters, such as pressure and temperature. In linear PEs, the copolymerization takes place by introducing α-olefins that go from propylene to octene.

The polymer properties vary based on comonomer type and amount. Generally, some properties improve with an increase in the short branch lengths passing from propylene → butene → hexene/4-methyl-1-pentene → octene.

The absence of comonomer produces homopolymer, which obviously shows the highest density among the components of linear PEs.

High-Density PE (HDPE). HDPE is the oldest member among the linear polyethylenes, with commercial production starting in 1956. Today, HDPE is manufactured using three process technologies at low to medium pressure: slurry, gas phase, and solution. Two major classes of catalysts are used: those based on chromium oxide and those based on Ziegler-Natta systems. The former generally is able to produce HDPE with medium-broad MWD, whereas the latter produces narrower MWD polymers.

Plant assets where two or more polymerization reactors are available can produce "bimodal" HDPE with broad MWD, even if Z-N catalysts are used.

By convention, HDPE density ranges from 0.940 to 0.970 g/cc. The correspondent crystallinity level is between 60 and 80 percent, and the melting temperature is from 130 to 135°C. Polymers with density in the range 0.926 to 0.940 g/cc are classified as medium-density PEs.

The molecular weight range of the HDPEs is very broad; the fluidity index, inversely proportional to the MW, is by convention between 0.4 and more than 25 g/10 min (melt index test condition E).

Unmodified HDPE (homopolymer) shows the highest crystallinity value and consequently high stiffness, heat resistance, and barrier properties. Modification, normally with 1 to 2 percent of comonomer, reduces the crystallinity and the stiffness level of the polymer but strongly improves the stress-cracking resistance.

High-Density High-Molecular-Weight PE (HDWPE). HMWPEs are characterized by a higher MW than the traditional HDPEs. Their weight-average MW is between 200,000 and 500,000, and the typical fluidity index (melt index test condition F) is between 1 and 15 g/10 min. HMWPEs can be manufactured either with slurry or gas-phase technologies, and both the chromium oxide and Z-N catalyst systems can be

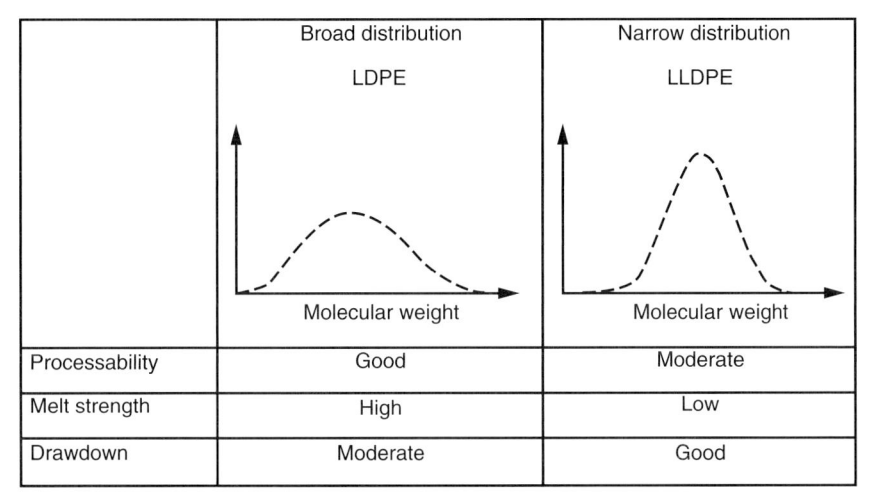

	Broad distribution LDPE	Narrow distribution LLDPE
Processability	Good	Moderate
Melt strength	High	Low
Drawdown	Moderate	Good

FIGURE 14.1.1 Molecular weight distribution of polyethylenes.

used. Commercial grades are mainly copolymers containing small quantities of comonomer, having a density range between 0.944 and 0.955 g/cc.

A very important property to be controlled during HMWPE polymerization is the MWD. The high MW of the resin makes it difficult to be processed with conventional equipment because of its high melt viscosity. A too narrow MWD makes the polymer almost impossible to be converted economically, whereas an MWD that is too broad will result in low melt strength and flow instability.

Therefore, the resin must have a relatively broad MWD in order to get the best balance between processing demand and end-use properties. Today, bimodal resins provide the opportunity to better control this balance, producing the required amount of low- and high-molecular-weight fractions (Figs. 14.1.1 and 14.1.2).

HMWPE is mainly transformed by extrusion technology. Thanks to its excellent balance among properties, such as rigidity, stress-cracking resistance, and impact and abrasion resistance, the polymer is used for special applications such as high-strength thin film, pressure pipe, drums, and sheet for industrial applications.

Ultrahigh-Molecular-Weight PE (UHMWPE). UHMWPE is a high-density PE with a molecular weight of over 3 million; polymer chains are 10 to 20 times longer than HDPE. Chemically, HDPE and UHMWPE are identical; both are straight-chain linear polymers, but the MW of conventional HDPE is rarely above 50,000. Crystallinity of UHMWPE is typically about 45 percent.

The polymerization process for UHMWPE employs slurry technology using the Z-N catalytic system.

Grades of UHMWPE with MWs between 3 and 6 million are available on the market, corresponding roughly to intrinsic viscosities of 20 and 30 dL/g, respectively.

The long molecular chains are responsible for the difficulty in processing the material on conventional molding and extrusion equipment. UHMWPE normally is transformed by compression molding, and then with the use of sharp tools, it is machined easily.

Using sophisticated spinning and stretching techniques, high-strength UHMWPE fibers can be produced. These fibers, thanks to their very high modulus and low density, are replacing aramidic fibers in application sectors such as reinforcement of composite structures and racing sails.

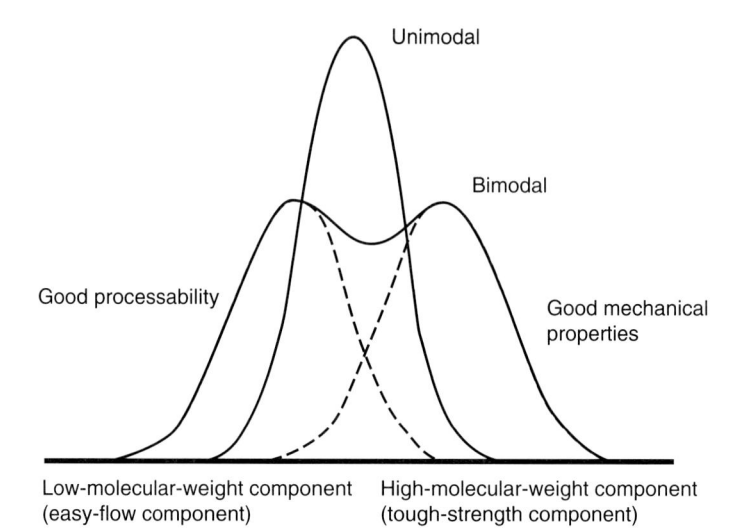

FIGURE 14.1.2 Unimodal and bimodal molecular weight distribution for polyethylene.

UHMWPE is a real engineering plastic, and thanks to its superior impact, abrasion, chemical, and corrosion resistance, it has found applications in very specialized market segments. Gears, rollers, and sliding surfaces and mechanisms made of UHMWPE are used successfully in many industrial sectors.

Linear Low-Density PE (LLDPE). LLDPE is a linear polyethylene with short side chains, which are able to reduce its density to 0.925 g/cc or less. Among linear PEs, LLDPE is the youngest member with consistent commercial presence since 1980. Today it represents about the 23 percent of total PE consumption.

The major polymerization processes for the production of LLDPE are the gas-phase and solution low-pressure technologies. Either the Ziegler (titanium) or Phillips (chromium) catalyst systems can be used.

Recently, small quantities of LLDPE grades have been produced using Kaminsky-type (metallocene) catalysts. Compared with the traditional catalyst systems, these "single site" catalysts result in a narrower MWD and a more uniform distribution of the comonomer. Consequently, these resins show better mechanical properties but lower processability than the corresponding traditional LLDPE grades.

The copolymerization of ethylene is done with α-olefins such as butene, hexene, or octene, the amount of which determines the density level. The length of the polymer side chains affects its mechanical properties.

As noted previously, LLDPE differs structurally from conventional LDPE owing to the absence of long-chain branches and a narrow MWD. Consequently, the two polymers have different rheologic and physicomechanical properties. During the extrusion process, owing to its narrow MWD and absence of long-chain branching, LLDPE is less shear-sensitive, and so the melt is more viscous than the corresponding LDPE grades.

Therefore, the film extrusion parameters are different from those used to transform conventional LDPE. The higher viscosity of LLDPE requires the use of more power from the extruder and gives rise to higher melt temperature and pressure. Die gaps have to be widened to avoid lower outputs because of a high backpressure and possibility of melt fracture. A more intense and sophisticated cooling system of the bubble also is required to avoid bubble instability in the blown-film production.

Many converters who use conventional LDPE extruders blend LLDPE with LDPE up to 50 percent to maximize the advantage of both the polymers: the good processability and optical properties of LDPE and the good mechanical properties of LLDPE.

The density level of LLDPE ranges between 0.910 and 0.925 g/cc (the same as LDPE), but its crystallinity, owing to the absence of long-chain branches, is higher than conventional LDPE, as is its melting point, which is 10 to 15°C higher. Because of its higher crystallinity, LLDPE shows inferior optical properties and better mechanical properties than conventional LDPE.

The major market for LLDPE is in the film sector, with more than 70 percent of its consumption. LLDPE is replacing LDPE in many applications. Areas where LDPE will continue to be dominant are the extrusion-coating field (where good processability is required) and film at very high transparency.

Very-Low-Density PE (VLDPE/ULDPE). These linear copolymers of ethylene show a density that is lower than 0.915 g/cc. Commercially available VLDPE grades typically range from 0.880 to 0.910 g/cc and include ultralow-density PE.

These PEs are manufactured by copolymerization of ethylene with the already mentioned α-olefins using either gas-phase or solution-process technology. The catalyst system used is the Z-N-Ti-based, because the chromium oxide system is not able to reach such low crystallinity levels.

Owing to their very low density, these polymers show high flexibility, puncture resistance, tear strength, and sealability. The VLD polymers have similar properties to 4 to 18 percent EVA copolymers in terms of flexibility and sealability while retaining the physical properties of LLDPE. Rheologic properties and thus the behavior during conversion processes are similar to those of LLDPE.

Major markets for VLDPEs are those where toughness, puncture resistance, and impact strength in combination with softness and flexibility are required, such as film for meat packaging and medical packaging, stretch and shrink films, soft bottles, tubing, and soft toys.

SPHERILENE PROCESS PERSPECTIVE

An Overview of *Spherilene* Development

In 1968, one of Basell's predecessor companies (Montedison) discovered magnesium chloride–supported (MgCl) high-yield catalysts. This revolutionized the chemical industry and boosted polyethylene activities. The refinement of gas-phase and bulk polymerization reactors (the revolutionary reactor granule technology, RGT) led to development of the *Spheripol* process for the production of polypropylene in 1982 and extended the properties of the polyethylene product families.

From 1990 to 1993, an intense research program at three pilot plants led to the development of a new, innovative gas-phase PE "swing" technology: the *Spherilene* process. This was developed to fully exploit the combination of gas-phase technology and product-mix flexibility: from VLDPE with a density below 0.900 g/cc to HDPE with a density of 0.962 g/cc using a single family of Ziegler-Natta titanium-based catalysts.

Spherilene technology for polyethylene is based on the well-established RGT. Ziegler spherical-form-supported catalyst particles act as "microreactors" throughout the process. This occurs in the first and second gas-phase reactors, where the "polymerizing particle" gradually "grows on itself," retaining its initial spherical form and allowing for perfect control of the final polymer morphology.

Today's generation of catalysts fully exploits the potential of the *Spherilene* process by their ability to produce new families of reactor-based products displaying highly controlled, improved properties.

Spherilene Process Key Characteristics

Safety. The *Spherilene* process technology uses medium-pressure gas-phase reactors that operate with low residence time, indicating that the monomer holdup is minimized and the plant can be operated safely. In order to increase the already high safety level of the technology, the design of each *Spherilene* process plant includes a number of safety features, such as

- A proprietary catalyst deactivation system that immediately stops all reaction
- Interlocks that help to prevent operator error
- Computer-controlled emergency shutdown systems
- An uninterruptible power supply (UPS) for computer control and critical instrumentation control
- Gas detectors that instantly determine and highlight (on a graphic easy-to-read board) the source of any hydrocarbons in the event of leakage into the atmosphere
- Automatic fire protection systems

Depending on the severity of the situation, a plant can be shut down manually in a step-by-step, controlled fashion more rapidly by using both manual and computer controls or by instant automatic shutdown.

Environment. *Spherilene* process units have minimal environment impact. Use of a light hydrocarbon as catalyst carrier instead of inert gas allows full recovery of the unreacted monomers, which are recycled back to the reactors. If necessary, other discontinuous hydrocarbon purges can be sent to "offgas recovery" (purge of ethane coming with the ethylene feed) for use as a fuel supply or to a flare system.

The *Spherilene* process does not use heavy hydrocarbon diluent nor contaminant chemicals, and the only wastewater (steam condensate quality) is released from the steaming/drying section of the plant.

PROCESS DESCRIPTION

Spherilene technology using catalysts supplied by Basell has the unique capability to produce uniform polymer spheres directly in the reactor, which differs considerably from the small, irregularly shaped granular particles produced with other technologies.

The *Spherilene* process involves the following stages (Fig. 14.1.3):

- Catalyst activation and feeding
- Polymerization: two-stage gas-phase polymerization
- Finishing

Properly proportioned catalyst components are fed to the polymerization section, together with a certain amount of a light, inert hydrocarbon used as a carrier.

Each reaction system consists of a polymerization reactor. A centrifugal compressor circulates gas through the distribution grid to maintain the proper fluidization velocity. Polymerization heat removal is obtained by a cooler in the circulation gas loop. Polymer quality is controlled through gas composition. The production rate is controlled by the monomer feed rate and by polymer residence time.

Polymer is discharged from the first reactor system, via a proprietary device, to a low-pressure vessel. Degassed polymer from the low-pressure vessel is fed by a proprietary

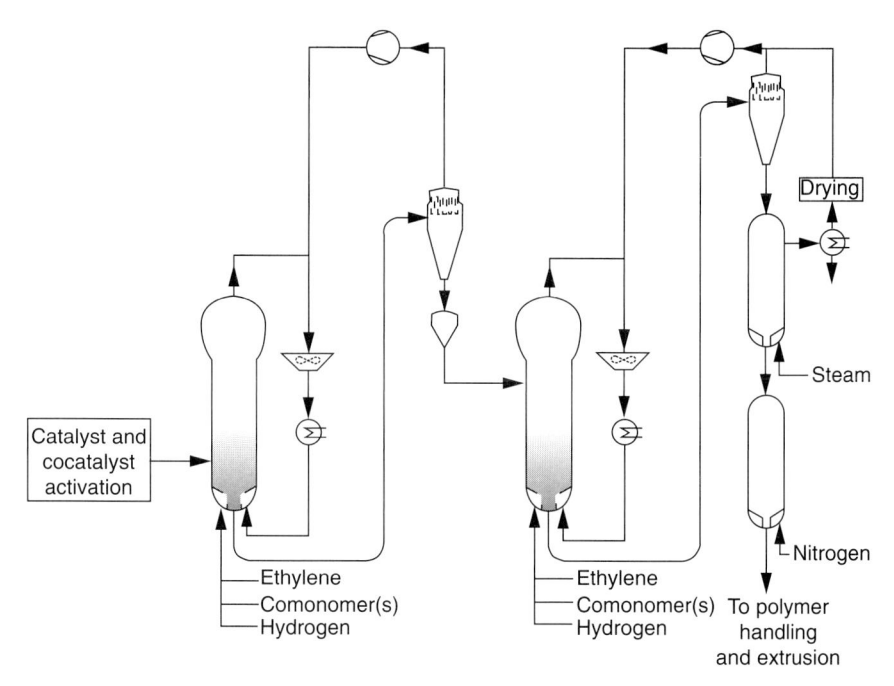

FIGURE 14.1.3 Simplified process flow scheme for the *Spherilene* process.

"lock hopper" system to the second gas-phase reaction stage while the gas is recycled to the first reactor.

No substantial quantity of gas from the first reactor enters the second reactor. Therefore, a completely independent gas composition can be created in the second gas-phase reactor, allowing the growth of a new, different (when required) polymeric fraction in the polymer particle's matrix that had been developed in the first stage.

Pressure and temperature conditions can be selected independently for each reaction stage; no additional catalytic components are required. The configuration of the second reactor system is similar to that of the first one. Polymer is discharged continuously to a low-pressure vessel, from which the gas is recompressed and recycled back to the second reactor.

Polymer is then discharged to a proprietary unit for final monomer stripping and catalyst deactivation in the spherical product particles. A countercurrent steam flow completely strips off hydrocarbons from the polymer, which are recovered, dried, and recycled to the reaction.

The wet polymer coming from the steaming section is finally dried by nitrogen in a closed loop system. The polymer is transported to the additivation and extrusion section, where stabilizers are added, and the polymer can be extruded to pellets.

Spherilene process plants are designed to meet the particular requirements of individual licensees. *Spherilene* process facilities (see Fig. 14.1.4) can be designed with capacities ranging from 100,000 to 350,000 metric tons per annum (MTA). With *Spherilene* technology, switching from one grade to another is particularly easy and inexpensive because all grades are made with a single family of fully compatible Ziegler-Natta titanium-based catalysts. The high catalyst reactivity allows low polymer holdup in the reactors and minimizes the volume of transition material.

FIGURE 14.1.4 *Spherilene* process unit.

PRODUCTS AND APPLICATIONS

The *Spherilene* technology is capable of covering vast product areas (HDPE, LLDPE, and VLDPE) with a single family of titanium-based Ziegler-Natta catalysts.

LLDPE

Linear low-density polyethylene (LLDPE) is obtained by polymerization of ethylene with a dedicated Ziegler-Natta catalyst system in the presence of small amounts of comonomers such as butene or hexene. The resulting polymer chain is mainly a linear molecule, while the short side branches give the polymer its low density. The range of density produced in the *Spherilene* technology varies from 910 to 930 g/L.

Film is by far the largest segment of the LLDPE market:

- Agricultural film forms the bulk of the film market; main applications are greenhouse film, mulch film, field and row covers, pond liners, and wind and sun screens.
- The second largest end-use film application is for bags of two types: retail bags and bags for soft goods (such as textile, clothing, etc.). The demand for heavy-duty sacks for bulk food and fertilizer packaging is also growing.
- The third sector is for stretch film to wrap pallets and shipping units, and automated wrapping.

Smaller segments include injection molding, with housewares being the main end-use application. VLDPE is also used to make geomembranes.

LDPE Replacement by LLDPE

The high cost and lower mechanical performance of LDPE in most markets are growing incentives to modify and upgrade equipment to increase substitution of LDPE by LLDPE (traditional and easy-processing and/or second-generation LLDPE). New equipment and modifications to the existing processing machines already make higher LLDPE/LDPE content ratios possible.

At the same time, there is a drive toward easy processing of second-generation grades capable of being used in the old units. New polyethylene projects will pay more and more attention to these second-generation easy-processing LLDPE technologies.

HDPE

High-density polyethylene (HDPE) is obtained by polymerization of ethylene with a dedicated Ziegler-Natta catalyst system. It is homopolymer polyethylene, or it can polymerize in the presence of very small amounts of comonomer, producing an HDPE copolymer with a slightly lower density. The range of HDPE density produced in the *Spherilene* technology varies from 940 to 965 g/L.

HDPE can be processed with the major conversion technologies; blow molding, extrusion, and injection molding are the most used. High-MW, broad-MWD grades are more suitable for the blow molding and extrusion technologies, whereas products with high fluidity (low MW) and narrow MWD normally are used for the injection molding process.

More than 55 percent of the HDPE consumption is in the packaging industry (bottles, drums, film, crates, etc.). The rest of the applications are equally distributed in the industrial and building (pipe, profiles, sheets), consumer (housewares, toys, etc.), and agricultural (pipe, nets, sheets, etc.) sectors.

- Blow molding is the largest end-use market. The use of HDPE for small containers is growing rapidly, but also the drum market for chemicals, oil, agriculture, and bulk food products is being targeted
- Film is the second-largest end-use market, most of which is for bags.
- Injection molding is another large end-use market with three major applications: toys, housewares, and crates.
- The pipe extrusion market consists primarily of corrugated drainage and irrigation pipe, potable water pipe, and gas distribution pipe.
- Fiber and filament constitute another relevant end-use segment, which consists of the production of woven sacks, sandbags, netting, and nonwoven fabric used for clothing/medical gowns. The woven sacks (by raffia texturization) market is the largest application.
- MDPE for rotomolding (used for tanks) is also available.

For monomodal HDPE with increasing density, impact strength, transparency, and stress cracking resistance will decrease.

Single gas-phase polymerization technologies use different families of catalyst (Cr-based) to broaden MWD. The *Spherilene* process—owing to its cascade reactor configuration—is capable of combining the advantages of using a single family of Ziegler-Natta catalysts with the benefit of bimodal operating mode for the production of improved BMWD HDPE structure.

HP Products

The HP (high performance) group of PE products produced with *Spherilene* technology is a proprietary new family of grades that combine ease of processability comparable with LDPE with enhanced mechanical properties in a wide range of densities:

- HP-VLDPE
- HP-LLDPE
- HP-MDPE
- Quattropolymers

  It has been demonstrated that

- The processability of the HP-LLDPE types is similar to that of the equivalent LLDPE/LDPE blend (LDPE 15 ÷ 25 percent) applications.
- General mechanical properties, and impact resistance in particular, of terpolymer HP grades are better than those of conventional butene grades, even using the cheaper propylene as component.
- General mechanical properties, and impact resistance in particular, of quattropolymer HP grades (based on a particular combination of butene-1, propylene, and hexene-1) are similar to or better than those conventional HAO (hexene/octene) grades, even using cheaper propylene and reduced amounts of hexene as components.
- The blown-film bubble stability of HP-LLDPE during extrusion is superior to that of conventional grades.
- The HP-LLDPE film stretch applications demonstrate superior stretchability in commercial application.
- In general, during the production of HP specialities, plant stream factor and raw materials cost are equivalent to butene grades and about 8 to 10 percent less than conventional HAO grades for the same applications.

PROCESS ECONOMICS

Reliability

Good morphology (polymer in spherical form, high bulk density, and size distribution in the range 0.5 to 3 mm) is the first guarantee for a trouble-free process: no fines buildup, no fines agglomeration, no fouling. Owing to the absence of chunk formation, the unique morphology also allows for a continuous, free-flowing polymer discharge from the reactor through a simple, reliable control valve.

Spherilene technology requires no startup seed resins. In other processes, the catalyst direct injection technique requires accurate dispersion inside a polymer mass, and as a consequence, the reactor cannot be started while empty. As such, additional equipment is needed to supply "seed resins" for the startup. The *Spherilene* process enables the plant to start up with empty reactors or to empty or refill reactors during grade transition.

In addition to high operating rates, products manufactured using the *Spherilene* process maintain a consistent level of high quality from one run to the next and are identical between plants, wherever located.

TABLE 14.1.1 Typical Specific Consumption (per 1000 kg of PE Produced)

	HP (butene-1 film grades)	LLDPE (butene-1 film grades)	HDPE (homopolymer injection molding)
Monomers (kg)*	1002	1002	1002
Hydrogen (kg)†	0.14	0.17	0.60
Catalysts and chemicals ($)	28–32	26–30	26–30
Electric power (kWh)‡	420–430	410–420	410–420
Steam (kg)§	235	200	200
Cooling water (m^3)¶	170	150	150
Nitrogen (N · m^3)	15–24	13–20	13–20

*Net consumption of 100 percent monomers/comonomers included.
†Depending on reactor gas composition, according to polymer MI.
‡Extrusion included, consumption depending on product MI.
§Includes low- and medium-pressure steam.
¶At $\Delta T = 10°C$.

Consumptions

Spherilene process plant capital and operating costs are competitive (Table 14.1.1) as a result of

- Small reactor volumes
- Very efficient transition time
- Low monomer consumption
- Optimal utilization of low-cost monomers
- Superior process reliability

CHAPTER 14.2
BORSTAR LLDPE AND HDPE TECHNOLOGY

**Tarja Korvenoja, Henrik Andtsjö,
Klaus Nyfors, and Gunnar Berggren**
*Borealis Polymers O/Y
Porvoo, Finland*

PROCESS DESCRIPTION

General

The Borstar PE process was developed primarily to produce high-strength, bimodal-type polyethylene. The reactor part of the process consists of a cascade of a slurry-loop reactor followed by a gas-phase fluidized-bed reactor. The process is suitable for producing a wide range of low- to high-density bimodal polyethylene grades utilizing Ziegler-Natta-type catalysts. The low-molecular-weight part is made in the loop reactor, and the high-molecular-weight part is made in the gas-phase reactor (GPR).

The first commercial plant was started up in 1995 in Finland. A photograph of a typical plant is shown in Fig. 14.2.1.

The process consists of the following areas (Fig. 14.2.2):

Wet end:

- Feed preparation
- Catalyst handling
- Loop reactor
- Diluent separation and recovery
- Gas-phase reactor and polymer purging

Dry end:

- Pelletizing
- Blending
- Bagging and storage

FIGURE 14.2.1 Borstar PE plant in Finland.

Wet End

The Borstar PE process consists of a prepolymerizer, loop reactor, and gas-phase reactor in series. The main raw material is ethylene. 1-Butene normally is used as a comonomer to control the density, but also higher α-olefins can be used. The design has been made for operation with Borealis' proprietary Ziegler-Natta-type catalyst, although other types of catalysts also can be used, e.g., a single-site catalyst. TEAL is used as a cocatalyst. Hydrogen is used for control of molecular weight. Propane is used as an inert diluent in the prepolymerizer and in the loop reactor. In the gas-phase reactor, propane is used solely or in a mixture with nitrogen as an inert gas. Propane also is used for catalyst preparation and feeding.

Feed Preparation. Feed preparation includes purification and handling of raw materials for proper reactor conditions. The reactor feeds have to be purified to remove catalyst poisons before being fed to the reactors. Typical catalyst poisons are water, oxygen, carbon monoxide, carbon dioxide, and sulfur components. The purification is carried out in beds filled with catalysts or absorbents dedicated to each poisonous component using separate purification trains for each feed stream. The types of catalysts used in the beds are available from catalyst and absorbent

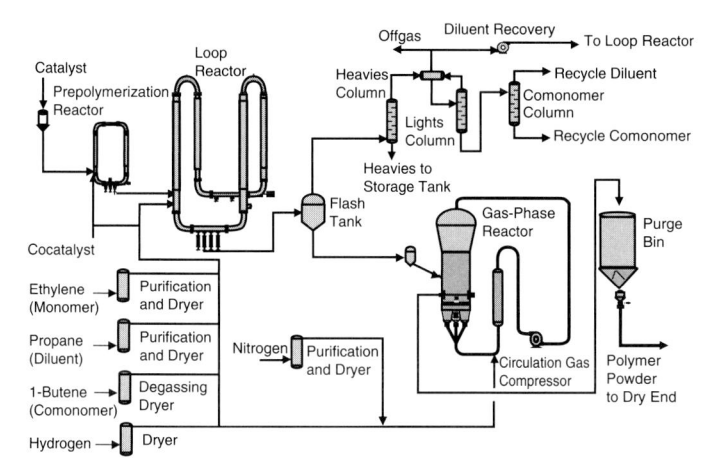

FIGURE 14.2.2 Borstar PE process flow diagram.

vendors. Each bed is designed for a certain lifetime, e.g., 1 month or 1 year, depending on the residence time requirements, after which the bed is normally regenerated for reuse.

Typical purification and handling of each raw material used in the Borstar process are as follows:

Ethylene. Ethylene typically is purified in three beds, CO treater, O_2 treater, and dryer, for removal of water and other polar components before being fed into the reactors. Ethylene is compressed before being fed to the prepolymerizer and the loop reactor.

Comonomer. The comonomer handling consists of a comonomer holdup tank, feed pumps, and a dryer.

Hydrogen. Hydrogen purification normally is not required owing to high purity level of the available hydrogen. Hydrogen handling includes compressing to the prepolymerizer and loop reactor pressure level.

Diluent. Propane purification typically consists of dryers, sulfur-removal treaters, an arsine treater, a hydrotreater, and lights removal (degassing). Purified diluent is pumped into the process from the diluent degasing tank.

Nitrogen. Purified nitrogen is required for catalyst handling, so part of the nitrogen is purified in a deoxo treater (O_2 removal) and a dryer.

Catalyst and Cocatalyst Handling. The solid catalyst is delivered in dry form under nitrogen blanket in catalyst transport drums. The transport drums are emptied into the catalyst feed tanks, and from there the catalyst is fed with a special feeder valve into the prepolymerization reactor. The cocatalyst, triethyl aluminium (TEAL), is stored in a bunkered area under nitrogen pressure.

Catalyst features include

- High activity under all process conditions
- Long polymerization lifetime
- Good particle morphology
- High comonomer and hydrogen sensitivity
- Well-reproducible preparation

Prepolymerizer. The prepolymerizer is a small loop reactor where the catalyst is coated with polymer to prevent the catalyst from breaking later in the process. The catalyst is fed into the prepolymerizer, where specified amounts of cocatalyst, ethylene, hydrogen, and 1-butene (depending on the recipe) also are fed. The residence time of the prepolymerized catalyst is controlled by the feed of diluent. The prepolymerized catalyst is fed continuously from the prepolymerizer to the loop reactor.

Loop Reactor. The low-molecular-weight part of bimodal polyethylene is produced in the loop reactor. The polymerization takes place at supercritical conditions in a propane-diluent slurry. All raw materials (ethylene, 1-butene, hydrogen, and recycled diluent) are fed continuously into the reactor together with partly polymerized catalyst from the prepolymerizer.

The reactor temperature is kept at 85 to 100°C depending on the polymer being produced. The vapor pressures of the diluent, dissolved monomer, and hydrogen determine the minimum reactor pressure. Typically, about 65 bar(g) is used; i.e., the reactor is operating under supercritical conditions for propane.

In the loop reactor, the polyethylene forms discrete particles on each grain of catalyst. The reaction mixture is circulated at high velocity by an inline axial-flow pump. The polymer concentration in the slurry is 30 to 50 wt %. It is controlled by the diluent feed rate into the reactor and by solids withdrawal from the reactor. The rate of polymerization is controlled by ethylene concentration and catalyst feed rate.

The loop reactor consists of jacketed vertical pipes and connecting top and bottom bends forming a loop. The heat of reaction is removed into a closed-loop cooling water system. The reaction temperature is controlled by the cooling water temperature. The cooling water system consists of a circulation pump, cooler, heater, and expansion tank. The polymerization is a highly exothermic reaction (3700 kJ/kg). Production of 1 t/h polyethylene generates about 1 MW heat.

The polymer slurry is withdrawn from the loop reactor through a flash pipe to a flash tank. The reaction in the loop reactor can be terminated with CO injection, or the reactor contents can be emptied into the dump tank in process-upset conditions.

Diluent Separation. Polymer slurry from the loop reactor is discharged to the flash tank, where diluent and unreacted monomer and comonomer are vaporized and separated from the polymer powder. The vapor stream is then passed through a recovery and purification train and circulated back to the reactor. As a result of a very efficient recovery system, the monomer and comonomer consumption can be kept low, and the diluent losses are very small.

Polymer separated in the flash tank is discharged continuously from the bottom through a screen to the GPR.

Diluent Recovery. The diluent recovery area is designed to recover diluent (propane), monomer (ethylene), and comonomer (1-butene) and to recycle the components back to the loop reactor. Another target is to remove impurities such as catalyst poisons (CO) and to prevent accumulation of inert components (ethane, methane, nitrogen, etc.), as well as heavy hydrocarbons and oligomers.

Typically, this is done as follows: Flash gas from the flash tank is routed to a heavies column, where heavy components are separated from the stream and discharged at the bottom of the column. The overhead of the heavies column is partly condensed, and part of the liquid distillate is pumped to the lights column. The noncondensable fraction is routed to the light gas compressor.

The bottom product of the lights column contains mainly propane but also some 1-butene. The vapor stream from the top of the column is combined with the noncondensable overhead from the heavies column. Most of the combined gas is compressed with the light gas compressor and returned to the loop reactor. A fraction of the gas is taken out from the compressor to remove light inerts and catalyst poisons. This stream is sent to a cracker for recovery.

The bottom product from the lights column can be routed to the comonomer column. Purified recycle comonomer is taken from the bottom and recycle diluent from the top of the comonomer column. Recycle diluent is collected in the recycle diluent buffer tank and is further pumped back to the prepolymerizer and the loop reactor.

When producing high-density grades, the comonomer column can be bypassed, and recycle diluent is taken directly as bottom product of the lights column.

Gas-Phase Reactor and Polymer Purging. Polymer is conveyed into the GPR for polymerization without addition of new catalyst. The GPR is a fluidized-bed reactor. High-molecular-weight copolymers are produced in this reactor, thus broadening the molecular-weight distribution (MWD) of the final polymer product. The production ratios between the reactors can be regulated according to the targeted product properties. The reactor normally is operated at 80 to 90°C and 20 bar.

Polymer from the loop reactor is still active and continues to polymerize in the GPR. Fluidizing gas, circulated with the circulation gas compressor, is cooled in the circulation gas cooler to remove the heat of reaction and compression. The required amount of cooling depends on the current production rate and product grade.

1-Butene for product density control, hydrogen for MFR control, and ethylene for production rate control are injected into the circulation gas line. Propane and nitrogen are used as inert components in the fluidizing gas. Carbon monoxide (CO) is used to stop the reaction in the GPR in case of process upset or emergency.

The reactor temperature is controlled by adjusting the inlet temperature of circulation gas. The reactor pressure is controlled by adjusting the inerts. The reactor level is controlled by product takeoff. Polymer is withdrawn from the reactor continuously under level control. There is also a batch outlet system.

Gases released from the polymer feed system and product outlet system are recovered, compressed in the gas recovery compressor, and recycled back to the GPR.

After the GPR, polymer powder enters the purge bin. Powder is treated with hot nitrogen and steam to reduce hydrocarbon content and to deactivate catalyst residues. From the purge bin, polymer is transferred with a closed-loop nitrogen-conveying system to the powder feed bins in the dry end section.

Dry End (Fig. 14.2.3)

Pelletizing Building. From the purge bin, the polymer powder is pneumatically transferred to the powder feed bins. These powder silos serve as a buffer between the reaction section and the pelletizing. Polymer powder from the powder feed bins is melted, homogenized, and formed into suitable pellets in the extruder. At the same time, the required additives, including Carbon Black Master Batch (CBMB) for black products, are mixed in. Borstar technology uses a twin-screw extruder suitable for both natural and black grades.

The system upstream of the extruder is slightly pressurized by nitrogen to ensure an oxygen-free material to the extruder. Downstream of the extruder, the process consists of the pellet water system and the air-conveying system transferring the pellets into pellet blenders or transition silos.

The pellet water system, consisting of pellet water circulation equipment, a centrifugal dryer, and a product screen, is used for cooling, drying, and transferring the polymer pellets.

Blender Area. Dried pellets are pneumatically conveyed from pelletizing to pellet blenders. Homogenization is done by circulating the pellets with a pneumatic conveying system.

14.20 POLYETHYLENE

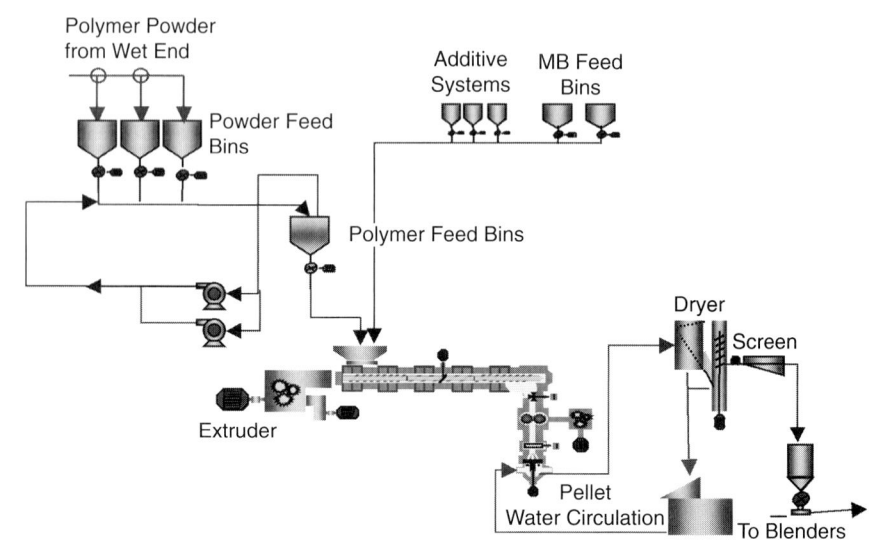

FIGURE 14.2.3 Borstar PE process, dry end.

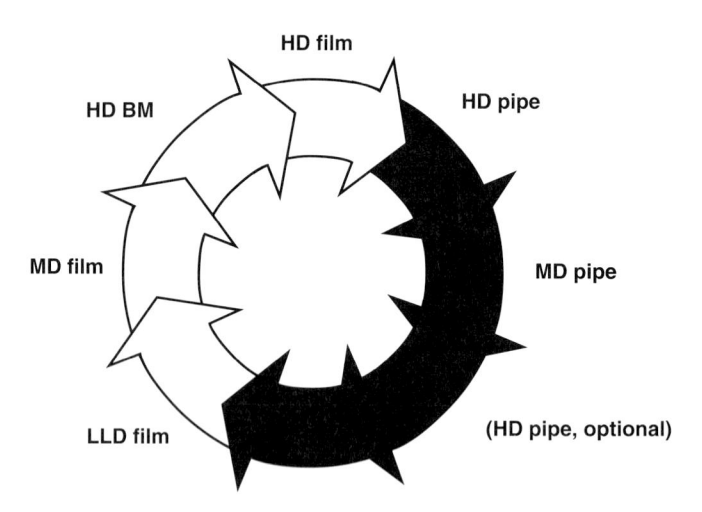

FIGURE 14.2.4 Typical Borstar PE production cycle (5 to 6 weeks in length).

From the blenders, the polymer pellets are pneumatically transferred to silos through elutriators. The elutriators are used to remove dust and streamers. The product is conveyed into the elutriator at the top, and clean air flows countercurrently from the bottom. Cleaned product is discharged from the bottom and conveyed to the storage silos.

Bagging and Storage. Pellets from silos are pneumatically conveyed to the bagging lines. Each bagging line also includes a palletizing and a shrinking line. Alternatively, the pellets can be delivered as bulk transport in trucks (Fig. 14.2.4).

ADVANCED PROCESS CONTROL

Introduction

BorAPC is the proprietary advanced process control system developed by Borealis for the control of polyolefin plants. The purpose of BorAPC is to

- Stabilize the production in the reactors
- Improve first-time-right yields of prime-grade polymer product through online closed-loop control of the polymerization reaction
- Maximize use of production capacity by operating with full and stable control near the maximum capability while still respecting the constraints of the plant
- Improve plant performance during grade transitions

BorAPC helps the process operator to monitor and control the polymerization reactors and the key quality properties of the polymers produced and to optimize the performance of the plant.

There are two main functions:

- *Supervisory control* in the form of nonlinear multivariable predictive controllers that will stabilize the conditions in the reactors and the polymer quality properties. This is closed-loop real-time process control.
- *Basic calculations* are online process calculations that quantify and predict important variables that cannot be measured directly. This information is used both for the online process control to assist the operator and for offline use by plant engineers in reporting and follow-up.

BorAPC is developed based on Borealis' operational experience and has already been installed in several polymerization plants. It can be installed together with any modern DCS control system.

Technology

Multivariable. BorAPC contains the multivariable controller. Polymerization processes have strong interactions, which means that changes in one manipulated variable (MV) will influence several of the controlled variables (CVs). Multivariable control is a tool that systematically compensates for these interactions and calculates an optimized set of moves for the MVs that will bring all the CVs to their setpoints. This brings a significant advantage over common single-input, single-output controllers, which cannot counteract the interactions.

Nonlinear. BorAPC differs from almost all other multivariable controllers by the very important advantage that BorAPC has nonlinear fundamental models that are valid over the whole operating window of the plant, so the controller also can be used during most of the grade transitions. This is a strong advantage compared with conventional linear controllers that have to switch off and change model parameters at grade transition.

Constraint Handling. Another important advantage of BorAPC compared with DCS controllers is the ability to observe and respect multiple constraints. For example, the plant can be operated at maximum capacity and still observe and respect constraints such as reactor circulation pump load, loop reactor solids content, and cooling water system capacity.

Advanced Process Control System

BorAPC has multivariable model-based predictive controls that are especially suitable to handle control of polymerization processes. Examples of controlled variables include

- Concentrations of catalyst, ethylene, and comonomer in the reactors
- Amount of polymer solids in the reactors
- Reaction rates in each reactor
- The balance ("split") between reaction rates in the prepolymerization, loop, and gas-phase reactors

BorAPC has a fundamental physical nonlinear process model that uses measurement information from the plant to estimate key process variables and predict their future behavior. The predictions are used to assist the operator with information and for closed-loop multivariable control. The BorAPC control application is implemented in a separate computer with an interface to the DCS control system. An estimator uses online information from process measurements and analysis data from the laboratory to update selected parameters in the prediction models to ensure that the prediction models reflect the real plant as closely as possible.

Basic Calculations

The basic calculation package helps the operator by providing online process calculations for a large number of important variables, such as

- Production rate based on energy balances
- Solids concentrations in the slurry

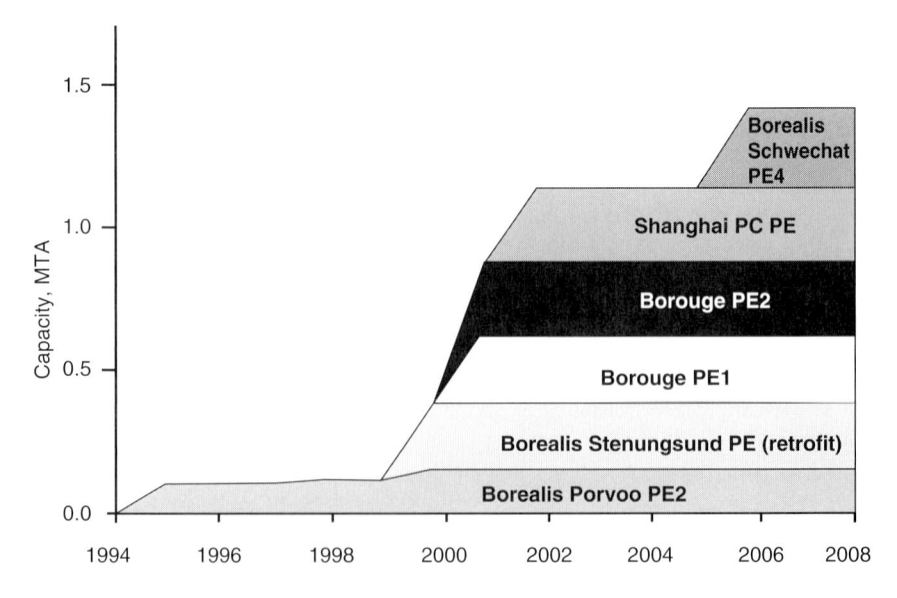

FIGURE 14.2.5 Borstar PE capacities.

- Critical temperature and pressure
- Liquid and gas densities
- Physical and thermodynamic properties
- MFR and density of polymers produced

The basic calculation package provides real-time property calculations with accuracy at least as good as conventional steady-state process-simulation tools. It can be implemented in a separate computer or be configured as a set of equations in the DCS system.

CAPACITIES AND LOCATIONS OF BORSTAR PE PLANTS

The spread of Borstar PE technology can be seen in Fig. 14.2.5. All other capacities are realized, and the plants are running already except the PE4 Schwechat, which is to be started in October 2005.

BORSTAR PE PRODUCTS

Using the Borstar PE process, a full range of products can be made, from LLDPE to HDPE, from low-molecular-weight to high-molecular-weight, and with variations in MWD. In addition, replacement grades for high-pressure LDPE can be produced. The product range will be further expanded when the Borealis proprietary single-site catalyst is introduced to the process.

The products display excellent properties in various applications. Even with economical 1-butene as comonomer, superior properties can be achieved compared with any unimodal products. The product properties can be further enhanced with other comonomers. Comonomer combinations, such as different comonomers in two reactors, are also possible, thus providing a new approach to tailor-made product properties.

One of the main advantages with Borstar PE products is their combination of superior mechanical properties with excellent processability. The key to achieve this combination is the tailor-made MWD, thanks to the bimodal process of two independent reactors and the specially designed catalyst. This maximizes the number of polymer chains improving processability and stiffness simultaneously with other chains giving mechanical strength. At the same time, the number of very short polymer chains typically producing smoke and odor problems during extrusion can be minimized. Extremely high-molecular-weight material producing homogeneity problems in extrusion also can be minimized. This kind of tailor-made MWD is impossible to achieve in single-reactor lines. Furthermore, comonomer distribution in the polymer chains can be tailored, thanks to two independent reactors running in series. For example, comonomer can be introduced only to certain chain lengths, contributing to a maximum amount of tie molecules between polymer lamellae and thus giving strength to the product.

Excellent mechanical properties provide clear benefits to a plastic converter. Products can be introduced into totally new areas requiring high strength, e.g., higher pressure classifications for pipes or increased pipe diameters. Up to 30 percent of material can be saved due to this higher strength. In the case of LLD film, material savings of 31 to 47 percent have been experienced in diaper bags and lamination films compared with typical LD/HD or LD/LLD blends. In household and industrial chemicals

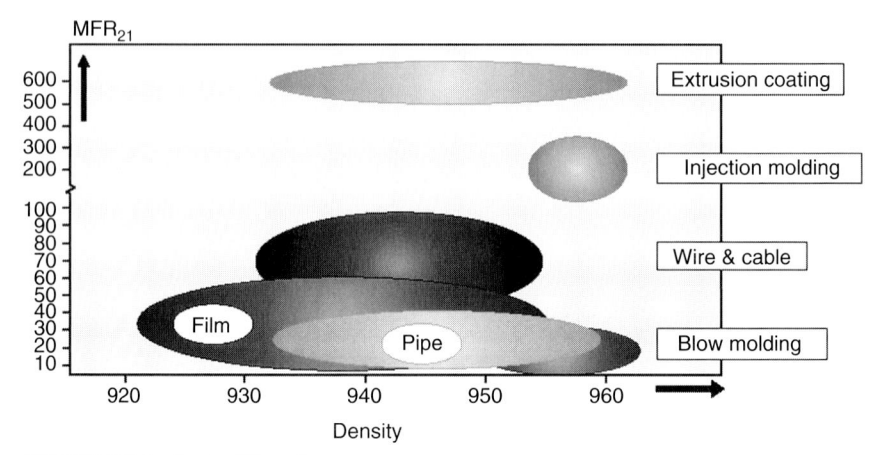

FIGURE 14.2.6 Borstar PE product range.

(HIC) blow-molding applications, 20 to 30 percent material saving has been experienced compared with typical unimodal grades. An additional benefit is that owing to improved mechanical properties in a virgin resin, a higher degree of recycling of plastic products can be achieved (Fig. 14.2.6).

Film

The HD, MD, and LLD film grades of the Borstar PE technology cover a wide range of applications and technologies in the film packaging industry. The main advantage of Borealis bimodal film products is good mechanical properties combined with excellent processability, which enables significant materials savings in several applications.

The typical melt/flow rate (MFR) and density values of Borstar PE film products are as follows:

	MFR_{21}	Density, kg/m^3
Borstar PE HD film	6	946
Borstar PE MD film	20	931
Borstar PE LLD film	22	923
Borstar PE new LLD film	$MFR_2 = 1$	923

LLD Film. Normal LLD produced in one step in a low-pressure reactor has better mechanical properties than high-pressure LDPE material. On the other hand, high-pressure material is easier to extrude owing to long chain branching. Since LLD is more difficult to extrude, it has not replaced LD as fast as one might expect despite its good mechanical properties. However, environmental pressures are leading to a trend toward downgauging of film thickness. Therefore, LLDPE is continuing to replace LDPE in many applications.

The idea with bimodal LLD is to make a product that is easy to process and has excellent mechanical properties. Owing to enhanced processability compared with unimodal LLD, totally new types of materials can be made with the Borstar PE process.

The key application properties of LLD are excellent processability, including high drawdown and bubble stability, combined with very good mechanical properties, such as high impact resistance, high tensile and yield strength, and hot tack properties.

Typical applications for superstrong Borstar LLDPE film are industrial liners, heavy-duty sacks, compression packaging for diapers and insulation materials, deep freeze packaging, and agricultural applications.

A high-slip version of Borstar LLDPE film resin is also available. Also, a new LLDPE film for LDPE coextrusion has been developed.

MD Film. Borstar MDPE film resin is flexible but stiffer than LLDPE. Typical applications for MD film resin are carrier bags, thin pouches, and compression packaging.

HD Film. Borstar HDPE film resins typically are used in carrier bags, industrial liners, heavy-duty bags, and other applications where high strength is needed. HD films are more hazy compared with lower-density LLDPE and LDPE materials. Among the HD film resins, bimodal materials display superior mechanical properties.

The most important properties of HD film materials are high dart drop, balanced tear strength, good film microstructure, and low gel level. The demanding combination of these properties has been achieved by a proprietary catalyst combined with process conditions in which a relatively broad MWD material is produced. The low-molecular-weight part produced in the loop reactor ensures high shear thinning and thus easy processability. The high-molecular-weight part produced in the gas-phase reactor attains good bubble stability and superior mechanical properties.

Blow Molding

The Borstar PE technology offers blow-molding grades for bottles and small containers. These products have high strength, allowing thinner wall-molded articles to be blown. Blow-molding grades for large containers with a very good combination of extrudability and mechanical properties are also being developed.

HIC Blow Molding. The typical MFR and density values of Borstar PE HIC blow-molding product are an MFR21 of 28 and a density of 958 kg/m3. The main benefit of Borstar PE blow-molding grades for bottles and small containers is light weight based on excellent environmental stress cracking resistance (ESCR), good processability, and low swell. Easy extrudability, thanks to bimodal MWD, gives a high-quality surface and increased throughput compared with conventional unimodal grades. Low swell provides the possibility of using a wider die gap, which allows more stable production with lower wall thickness and thus allows downgauging possibilities. With controlled comonomer distribution (which means as high a density as possible in the loop reactor and a low density in the gas-phase reactor), an ESCR 5 to 10 times higher can be achieved. This means one grade for all applications. A high ESCR also enables use of higher density (958 kg/m^3), which clearly improves the top load strength despite a lower bottle weight being used. Owing to the lower ESCR of conventional blow-molding grades, the use of high densities such as 958 kg/m^3 is not possible.

The main applications for the blow-molding grade for bottles are different containers for household and industrial chemicals (HIC), such as detergents, shampoo, oil, and cleaners. Containers up to 10 L can be made with this grade.

The benefits of Borstar bimodal blow-molding grade can be seen in Fig. 14.2.7.

Injection Molding

Injection molding is the most important method in plastic processing technology. It is especially well suited for mass-produced goods in an almost unlimited number of applications,

- Bimodal HDPE has 3-4 times higher ESCR at the same top load
- Bimodal HDPE has 15-20% lower swell compared to unimodal
- Borstar HDPE gives lower melt temperature and lower melt pressure
- Borstar HDPE is ideal for producing lightweight/thin-wall bottles at high production rate and lower energy consumption

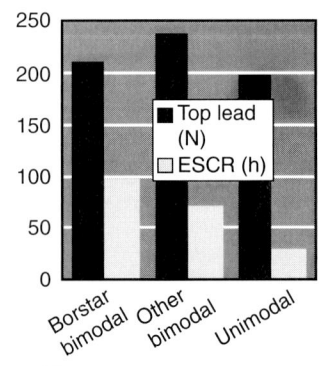

FIGURE 14.2.7 Borstar HDPE versus unimodal HDPE in blow molding.

such as crates, caps and closures, bumpers, household articles, compact discs, toys, toothbrushes, garden furniture, lids, etc. Conventional injection molding is used mostly with one mold, or many parts can be molded simultaneously.

Important properties for PE to be used in injection molding are shrinkage during processing and flow behavior. Important properties for final products are impact resistance, stiffness, creep, warpage, and shrinkage. Typically, injection-molding materials have a relatively high MFR to allow a high production speed.

In the Borstar PE process, although it is a bimodal process, the most typical unimodal injection-molding grades are also produced with the following melt/flow rates and densities:

	MFR_2	Density, kg/m^3
Injection-molding grade 1	8	964
Injection-molding grade 2	20	958
Injection-molding grade 3	4	954

Grade 1 is a high-density PE with an excellent impact strength and toughness. This grade is ultraviolet (UV)–stabilized, which makes it suitable for outdoor applications. It is intended for products requiring extremely high impact strength and good flow properties. Typical applications are waste bins, fish crates, meat crates, and packaging and storage boxes for the mechanical industry, as well as various technical products.

Grade 2 is a high-density PE with good stiffness properties. The grade is designed for injection molding of articles where good rigidity and flow properties are required. Typical applications are crates, pails, and housewares.

Grade 3 is a bimodal high-density PE with superior ESCR, excellent flow properties, and good impact/stiffness balance. This is the very first bimodal injection-molding grade to the IM market. Typical applications are pails for solvent-based liquids, caps and closures, wastebins, and storing cases/boxes for the mechanical industry.

Pipe

Pipe materials can be divided into pressure-pipe materials and materials without pressure ratings. The two highest PE pressure classifications are PE80 and PE100. Both these classifications can be produced with the Borstar PE process.

The typical MFR and density values of Borstar PE pressure pipe products are as follows:

	MFR$_5$	Density, kg/m^3
Borstar PE100 HD pipe	0.4	959
Borstar PE80 HD pipe	0.5	956
Borstar PE80 MD pipe	0.9	950

Pressure pipes are used mainly for water and gas applications. Water pipes typically are black or blue, whereas gas pipes are yellow or orange. It is also possible that the pipe itself is black but with blue or yellow/orange stripes to identify whether it is a water or gas pipe.

The most important product properties are good processability, high resistance to hydrostatic pressure and to fracture propagation, high resistance to creep, excellent ESCR, and good impact strength, also at low temperatures.

Within the HD segment, development is heading toward higher pressure class pipes, the so-called MRS 10 resins (PE100) being the required standard in many markets already today. Minimum required strength (MRS) is the extrapolated design stress excluding a safety factor. For PE100, MRS at 20°C and 50 years is 10 MPa. This makes it possible to produce pipes with 30 percent less material consumption and to use pipes at higher pressures. It also opens up new segments within the pipe application field (e.g., the possibility to make plastic pipes with larger diameters).

However, even if the clear development trend in HD pipes is toward higher than PE100 pipes, the second-generation HD pipe, the so-called MRS 8 resin (PE80), is still used widely. More flexible Borstar MD pipe resins also meet the PE80 classification. In PE80 pipes, the extrapolated design stress excluding a safety factor at 20°C and 50 years is 8.0 MPa.

Slow crack growth and rapid crack propagation have been difficult to meet at the same time. However, Borstar products meet these requirements owing to tailored molecular weight and comonomer distribution.

ESCR is a very important property in pipe resins. In the case of bimodal resins, this property is excellent, provided that enough comonomer is incorporated in the high-molecular-weight fraction.

Borstar pipe products have pronounced shear thinning, which leads to good processability of these resins on pipe extrusion lines. Low taste and odor levels, which are important properties for water pipe material, are achieved by using a proprietary catalyst, effective powder degassing, and catalyst/cocatalyst deactivation in purge and surge tanks. Borstar pipes are also suitable for very large diameters owing to low sag properties.

Figure 14.2.8 shows why bimodality is so important in creating a material for high-pressure pipes.

Wire and Cable

The Borstar HDPE wire and cable (W&C) resins are targeted at standard and special jacketing applications. However, several new grades are in the development phase. The Borstar PE process is ideal for producing a wide range of W&C resins from low-density LLD to high-density HD products.

Commercial Borstar HDPE products have the following MFR and density values:

	MFR$_2$	Density, kg/m^3
Borstar PE HD for special applications	1.7	954
Borstar PE HD for standard jacketing	0.5	954

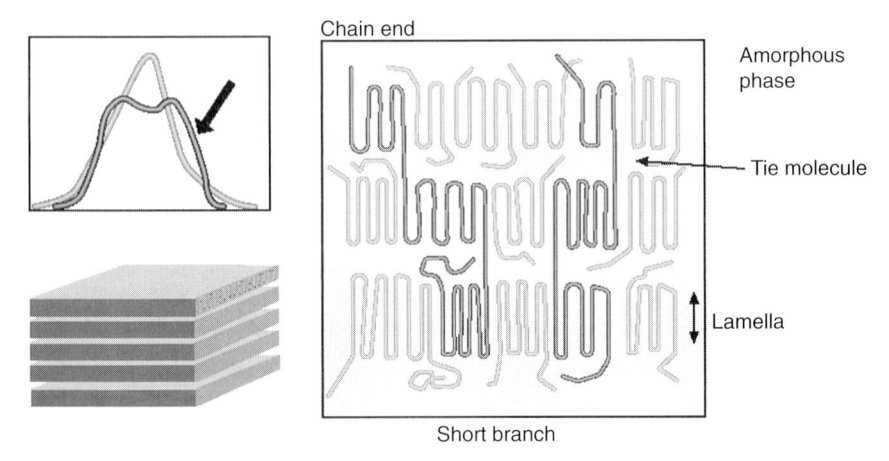

FIGURE 14.2.8 Why bimodal is so good in PE pipes.

The key application properties for these materials are good processability, low shrinkage, good ESCR, good mechanical strength, and high surface hardness.

Several trends for jacketing can be distinguished. In the low-density area there is a trend from high pressure (LD) toward linear material (LLD) mainly because of improved mechanical properties. Furthermore, there is, in general, an increasing demand for colorable grades, which require improved UV stabilization. One reason is the increasing optical fiber market, which often requires colorable jacketing grades. Colorable jacketing grades also can be used for polyvinyl chloride (PVC) replacements.

There is a general trend in jacketing for PE to replace PVC. One problem is the higher shrinkage of PE compared with PVC.

The bimodal Borstar PE process allows a good combination of mechanical properties and low shrinkage to be achieved. A critical factor is the possibility to tailor MWD. This is important in being able to achieve the right balance between processability and shrinkage of the material.

Extrusion Coating

Extrusion coating is run at a very high line speed. This makes the application very demanding. The viscosity of the material must be low at the processing conditions to ensure a high throughput.

Bimodal Borstar PE technology offers a tailor-made PE with balanced draw down (low-molecular-weight component) and neck-in (high-molecular-weight component). In certain paper-paperboard-PE combinations, improved mechanical properties such as bursting strength, puncture resistance, tensile strength, and elongation are obtained without compromising processability. Alternatively, equal performance can be achieved with lower layer thickness (source reduction). In addition, the organoleptic (sense perception) properties of Borstar-coated packaging material are superior to those of conventional materials. Borstar Technology has extrusion-coating grades for both LLD and HD applications.

TABLE 14.2.1 Typical PE Consumption of Raw Materials and Utilities per Ton PE

	HDPE	LLDPE
Total monomers, kg/t	1010	1010
Ethylene, kg/t	993	920
1-Butene, kg/t	17	90
Propane diluent, kg/t	10	10
Hydrogen, kg/t	0.3	0.3
Catalyst, kg/t	0.08	0.04
Power, kWh/t	500	500
LP steam, kg/t	200	300
Nitrogen, N·m^3/t	80	80

PROCESS ECONOMICS

The production cost of polyethylene is highly dependent on the cost of the monomer and comonomer. The investment cost for a dual-reactor process for production of bimodal products is slightly higher than that for a single-reactor plant of the same capacity. On the other hand, a bimodal product has a higher sales price.

At the Polyolefins XI conference held in 1999, Kenneth B. Sinclair presented a comparison study of the new generation of PE technologies. Some of his findings were as follows:

- In investment costs, all technologies fall within ±15 percent of the mean.
- The production cost difference among technologies is small, less than $50/t.
- The product sales price spread (PE − C_2) is three times higher than variations in costs.

He concluded that the profitability of a plant is more dependent on product price than on investment or operation costs. The investment costs for a plant are highly dependent on the scope and also differ vastly in Europe, the Middle East, and the United States.

See Table 14.2.1.

Wastes and Emissions

The emissions and wastes are minimal in this kind of plant. The only by-product formed in the process is a small amount of oligomers that can be used as a fuel in a boiler. Since hydrocarbons are used in the plant, fugitive emissions cannot totally be avoided.

Some wastewater is generated in the pellet water system because the content needs to be exchanged gradually. Some polymer spills are washed away with water and end up in the skimmer pit, where it is recovered. The purification beds are replaced at intervals of some years and are sent to recycling or waste destruction.

CHAPTER 14.3
CHEVRON PHILLIPS SLURRY-LOOP-REACTOR PROCESS FOR POLYMERIZING LINEAR POLYETHYLENE

Mike Smith
Chevron Phillips Chemical Company LP
Kingwood, Texas

HISTORY

The original design for the slurry-loop reactor was invented by Phillips Petroleum Company in its research center at Bartlesville, Oklahoma, in the late 1950s. The first commercial polyethylene (PE) slurry-loop reactors were commissioned in 1961 near Houston, Texas. These first commercial reactors were just 2700 gal in size compared with the largest to date, at 35,000 gal. Since production rate is proportional to reactor size, reactor nameplate production capability has increased by more than tenfold in the last 40 years.

This slurry process was a momentous improvement over older processes for polymerizing PE, such as solution reactors, continuous-stirred tank reactors in series (CSTR), and high-pressure PE reactors (e.g., autoclave and tubular). It was so momentous that five licenses were signed within the first year of operation, with many more to follow. Phillips Petroleum commercialized a solution process in 1956 to produce PE. Although this solution process produced quality high-density PE (HDPE), it had high operating costs owing to higher energy consumption and was unable to produce higher-molecular-weight PE. Therefore, solution-reactor technology was not as profitable as slurry-loop-reactor technology for producing PE and subsequently was shut down at a later time.

In the early 1950s, Phillips Petroleum research chemists Paul Hogan and Robert Banks invented a new family of polymerization catalysts that had chrome supported on silica that lead to the monumental discoveries of linear HDPE and polypropylene (PP).

During the 1950s, other chemical companies also discovered catalysts that would produce linear PE, but the resulting polymer was narrow in molecular weight distribution (MWD). In order to produce a broader-MWD polyethylene that would have better physical properties and easier processability, these companies found it necessary to link two or more reactors together in series. One reactor would make low-molecular-weight PE, and the other reactor would make high-molecular-weight PE, such that the resulting final PE would be broad in molecular weight distribution. Of course, two reactors were more expensive than just building a single reactor. The Chemicals Division of Phillips Petroleum, now Chevron Phillips Chemical Company, was able to produce broad-MWD polyethylene in a *single* reactor because of its novel, proprietary chrome silica catalysts, which has allowed the slurry-loop process to become the most popular for producing HDPE in the world.

PROCESS DESCRIPTION

Overview

At the heart of Chevron Phillips Chemical Company's (CPChem) continuous PE process is the slurry-pipe-loop reactor. It is simply sections of carbon steel pipe joined together in the form of a continuous loop (Fig. 14.3.1). It contains a low-boiling hydrocarbon (isobutane), which is a poor solvent for PE. Isobutane is used as a diluent for suspending the catalyst, dissolving the monomer (ethylene gas) and comonomer (1-hexene), and transferring the exothermic heat of the polymerization reaction to the cooling surfaces of the reactor. This loop reactor is operated liquid-full at about 600 psig (40 kg/cm^{2}g) pressure, and an axial-flow pump is used to circulate the reactants and resulting polymer particles. The PE polymerization reaction generates 1460 Btu/lb of PE produced, of which a net of 1400 Btu/lb of PE must be removed through the reactor wall.

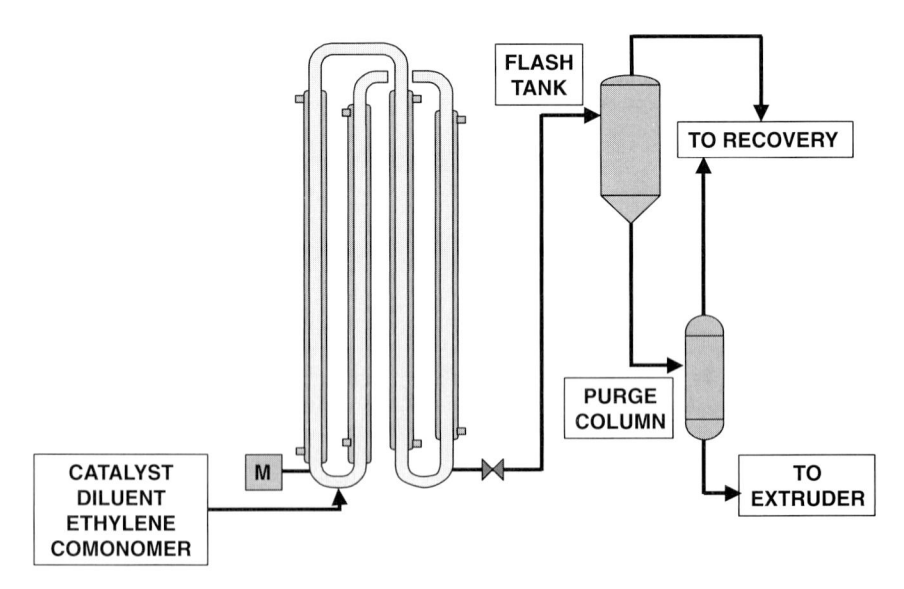

FIGURE 14.3.1 Chevron Phillips Chemical Company polyethylene slurry-pipe-loop reactor.

If heat is not removed efficiently from the reactor, the polymer could melt and go into solution. The loop reactor is ideal for removing this heat and does so by a countercurrent flow of cooling water in a pipe jacket that extends almost the entire length of the reactor.

Since the PE particles do not reach temperatures high enough to go into solution, they form discrete white particles the size of sand that form a slurry in the liquid isobutane and can be easily separated physically from the isobutane downstream of the reactor with minimal energy input. This separation is done by allowing a slipstream of reactor slurry to leave the process and drop from 40 kg/cm^{2}g pressure to almost atmospheric pressure. This pressure drop vaporizes the isobutane diluent, whereas the remaining polymer particles drop by gravity into a fluidized bed of polymer particles, where most of the remaining hydrocarbons are removed by a countercurrent flow of nitrogen. Unreacted ethylene, 1-hexene, and isobutane normally are recycled to the reactor.

Once the PE particles are relatively dry, they can be pneumatically transported to a commercial extruder, where they are combined with small amounts of special chemicals to protect or enhance the polymer. These additives are further mixed into the polymer as it is melted in the extruder and subsequently pelletized.

Depending on the producer, the polyethylene pellets can be packaged in bags, boxes, sea containers, hopper trucks, or hopper train cars. It is even possible to package the unpelletized PE particulate, but this is not common because its shelf life is shorter and it is less cost-effective to ship.

Catalyst Families Used in the CPChem Slurry-Loop Process

There are three major catalyst families that can be used in the Chevron Phillips Chemical slurry-loop polymerization process to produce linear polyethylene, as listed in Fig. 14.3.2.

By far the most popular is the chrome silica catalyst family. These catalysts can produce medium- to broad-MWD PE with densities from 0.920-g/cm^3 copolymers to homopolymers with densities greater than 0.960 g/cm^3.

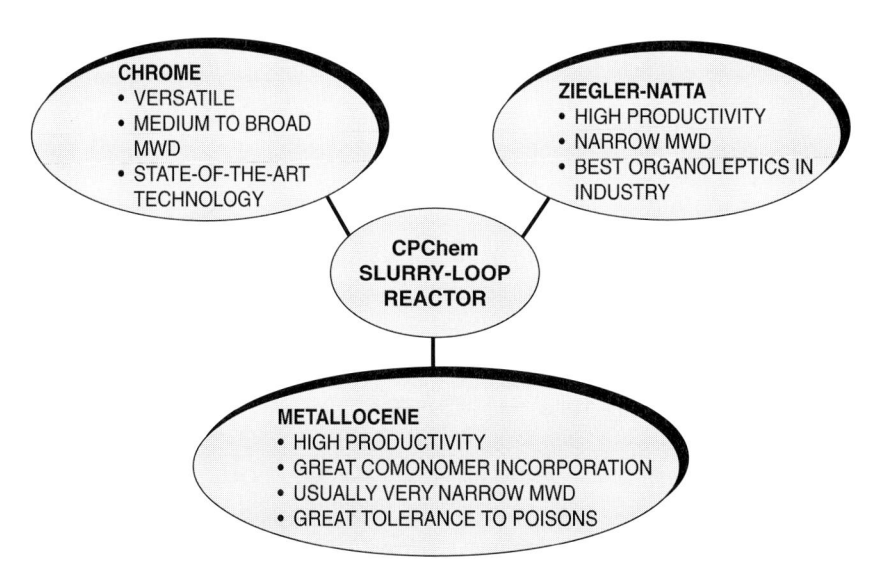

FIGURE 14.3.2 CPChem proprietary catalysts.

Chrome Silica Catalysts

Chevron Phillips Chemical Company's proprietary chrome silica catalysts are available from third-party manufacturers. This type of catalyst comes as a dry powder in either cardboard-fiber or metal drums.

The chrome is supported on highly porous silica (Fig. 14.3.3) and must be activated at elevated temperatures in a fluidized-bed activator in order to convert it to a valence state that enables polymerization in the reactor.

This conversion is done in a catalyst activator unit using prescribed recipes that control temperature, activation time, air flow, and different adjuvants that can be added to control the molecular parameters of the resulting polymer.

The unactivated catalyst can be activated using any one of several different recipes (Fig. 14.3.4), which enables the producer the flexibility of tailor-making a specialized chrome silica activation just hours before it will be used in the reactor. Just as in purchasing paint in a store, where only a few base colors are kept on hand and then pigments are added to obtain the custom color needed, so too can the slurry-loop producer keep only a few base chromium catalysts in inventory before customizing them in the activation process. This provides the producer maximum flexibility in controlling the polymer's molecular parameters using this just-in-time catalyst modification system.

The catalyst activation unit consists of an internal vessel that contains the catalyst bed. This internal vessel and the catalyst are heated indirectly by passing exhaust gases from a gas-fired heater over the outside of this vessel while the catalyst is being fluidized inside. The fluidizing gas discharges through a filter that returns catalyst particles back to the catalyst bed. An outer vessel contains the hot flue gas and directs it upward through the annular space, where it is cooled and filtered in a HEPA filter before being discharged to the atmosphere.

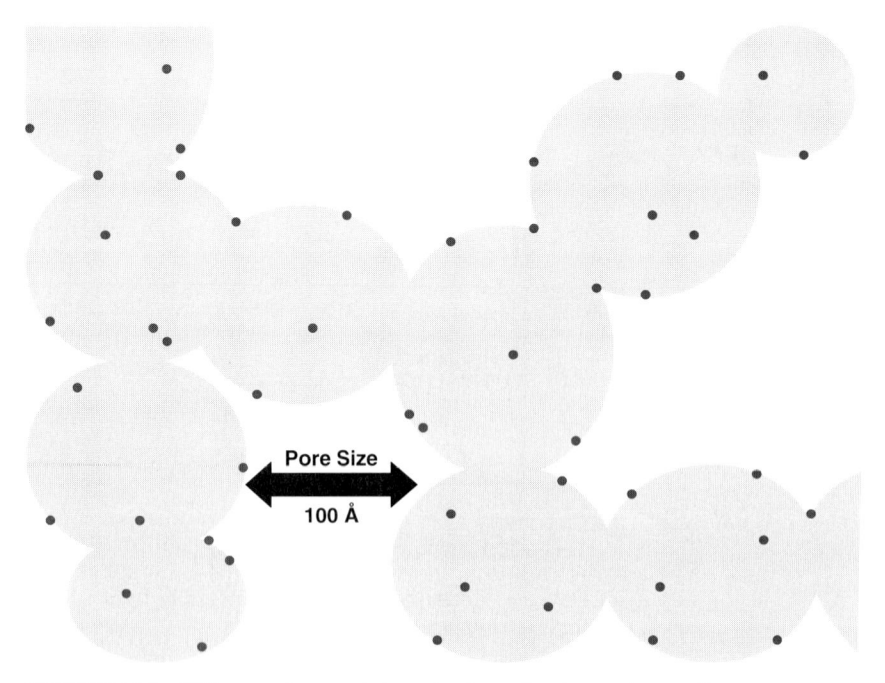

FIGURE 14.3.3 CPChem catalyst morphology showing active chrome sites on silica.

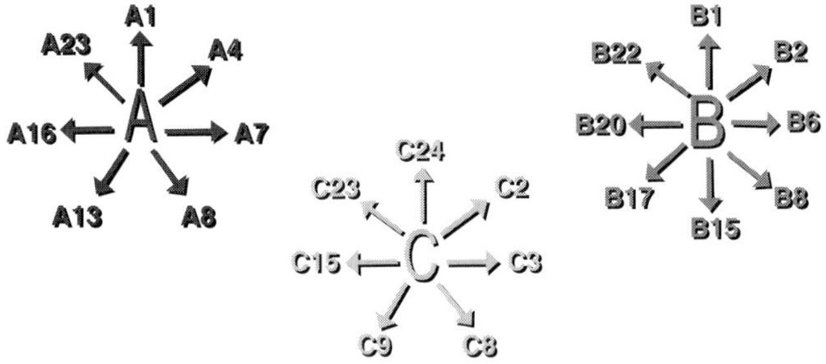

FIGURE 14.3.4 Minimal chromium catalyst inventory, just-in-time activated chromium catalyst (where A, B, C, ... = unactivated chromium catalysts, and 1, 2, 3, 4, ... = activation recipes).

The conversion of the valence state of the chromium on the catalyst is controlled by the temperature of the heated exhaust gases and the fluidization rate of the catalyst.

After completion of the activation step, the activated catalyst is discharged under inert conditions into a storage vessel. This vessel is then taken to the reactor area for loading into the catalyst feed system. A typical catalyst batch will sustain 1.5 to 2 days of reactor operation for a world-scale plant. Catalyst productivities for chrome silica catalysts typically range from 2500 to 3500 kg/kg. Catalyst productivity is affected by the specific grade of catalyst and the purity of the plant feedstocks. Reactor control parameters can be adjusted to make modest improvements in catalyst productivity. Catalyst productivity is sufficient to not require the removal of catalyst residue from the polymer product.

Titanium Magnesium Catalysts

A second family of proprietary catalysts that have been developed by Chevron Phillips Chemical Company is composed of titanium and magnesium. These are also known generically as *Zeigler-Natta catalysts.* They are sold by a third- party vendor. These catalysts are received as a hydrocarbon slurry in drums. Since the catalyst is already activated, it can be charged directly to the catalyst feed system. PE resins made with these catalysts are narrow in MWD and have some of the lowest organoleptics (least amount of taste and odor) in the industry.

Metallocene Catalysts

Metallocenes comprise the third family of proprietary catalysts being used in the Chevron Phillips Chemical PE process. These are usually single-site catalysts whose constrained geometry produces unique PE—usually very narrow in MWD. These metallocenes have even better productivity than titanium magnesium catalysts (see Fig. 14.3.8).

SLURRY-LOOP REACTOR

Feedstock Treatment

All feeds to the reactor are passed through solid-bed desiccants to remove catalyst poisons. Typical catalyst poisons include water, alcohol, CO_2, CO, acetylene, carbonyls, and sulfur compounds. Normally, this is a single treatment step per feed, not necessarily requiring additional treatment steps. Typical service requires two vessels, one in operation and the other in regeneration or on standby.

Depending on the quality of the comonomer feedstock, a separate degassing step may be required to ensure polymerization quality.

Reaction Area

A continuous pipe loop is used for the reactor. Isobutane diluent is fed to the reactor to dissolve the olefin feeds and provide a transport medium for the polymer particles that form a slurry in the loop reactor. The polymer particles formed do not go into solution but remain as discrete particles in the slurry. An axial-flow pump is used to circulate the slurry around the pipe loop at a velocity sufficiently high to avoid polymer saltation.

The reactor pipe loop is jacketed by an outer pipe along its vertical legs. Coolant water is circulated in a countercurrent flow through the annular area between the reactor pipe and the outer pipe (Fig. 14.3.5). This coolant water removes the heat of polymerization very efficiently. This reactor coolant circulates in a closed system and uses clean water such as condensate or demineralized water to minimize any fouling on the coolant side of the reactor. A portion of the circulated coolant is cooled by heat exchange with cooling tower water. Reactor temperature is controlled by precise control of the temperature of the coolant returning to the reactor jackets.

The properties of the resulting PE are affected by variables such as reactor temperature, monomer and comonomer concentrations, and residence time, to name a few. Control systems are provided to control these variables within required guidelines to produce on-specification product at design rates.

If chrome catalyst is used, it is brought from the catalyst activation area and charged to a mix vessel, where it is mixed with diluent and fed to the reactor. Catalyst feed is regulated to control the reactor production rate.

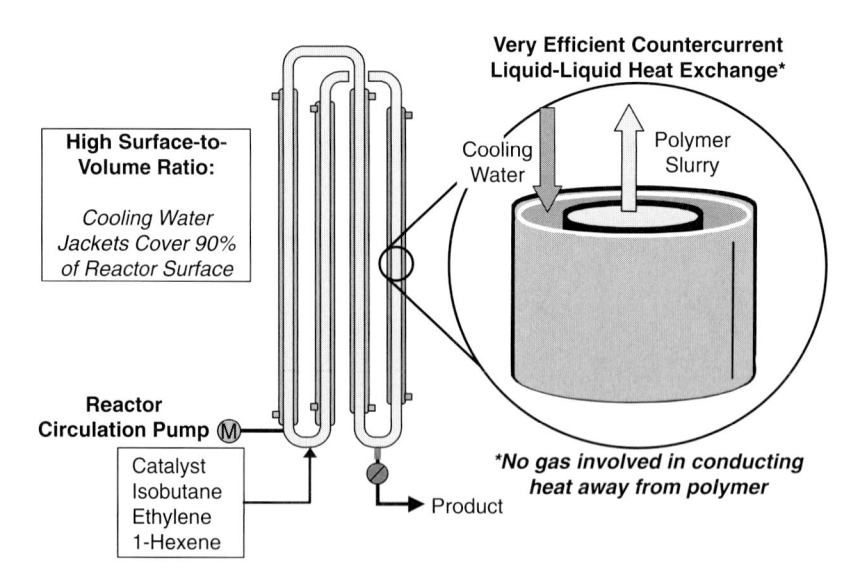

FIGURE 14.3.5 CPChem slurry-pipe-loop reactor has high heat-removal capacity.

Polymer Recovery

The reactor product is removed continuously in a concentrated slurry from the reactor. This slurry is heated to a controlled temperature and then flashed to separate the hydrocarbon diluent and unreacted monomer and comonomer from the polymer product.

The recovered polymer falls by gravity into a purge vessel, where any remaining hydrocarbon is stripped from the polymer with the use of nitrogen. The nitrogen and stripped hydrocarbons are removed from the purge vessel and sent to a recovery unit. The hydrocarbons are returned to the diluent recovery system, and the purified nitrogen is returned to the purge vessel.

The purged polymer is then transferred to the extruder feed system or to intermediate storage. A closed-loop powder transport system uses inert gas to transfer the polymer.

Diluent Recovery

The flashed hydrocarbons exit the top of the flash vessel. They are filtered and condensed and sent to an intermediate storage vessel. The liquids from this storage vessel are pumped back to the reactor. A portion of these liquids is further purified to remove inerts received with the ethylene. A vent stream from this purification process is primarily ethylene and can be returned to the ethylene plant for recovery.

There is also a very small, intermittent stream of heavier hydrocarbons sent to the flare system in the purification step. This small stream is primarily unreacted comonomer and any impurities received with the comonomer feed. This purge stream also can be used during transitions to higher-density products.

Control System

Typically, a distributive control system (DCS) is provided for plant control. Normal regulatory control is all that is required to operate the plant safely and to produce high-quality product. An advanced process control system (APCS) can be added to enhance plant operations but is not necessary.

POLYMER FINISHING AND PACKAGING

The dry polymer powder is mixed with small amounts of various chemical additives before being fed to an extruder. The extruder melts, mixes, and pelletizes the polymer and additives. The polymer pellets are screened to ensure size uniformity and are transferred to product blending and storage. Tanks with once-through blending capability are available to achieve product uniformity within product lots.

The PE pellets are then transferred to product load-out areas, where they can be packaged in bags, boxes, hopper trucks, or hopper train cars. In some cases, unpelletized polymer powder can be shipped, but this is less common.

UTILITIES

Utility systems are required for electricity, steam, condensate, nitrogen, cooling water, fuel gas, instrument air, and flare gas.

Approximately 50 percent of the electrical consumption is due to the extruder. The reactor circulation pump is the second largest user. Other major users are the reactor coolant pump, the reactor diluent feed pump, a purge-gas compressor, and the solids-transfer blowers. These users account for over 80 percent of the power consumption.

The largest steam usage is reactor product heating. A lesser amount of steam is consumed in the diluent purification area. There are two intermittent steam users—one for reactor startup and the other for extruder startup. A condensate system typically is provided. Recovered condensate can make up to the pellet water system or be returned to the steam provider.

Nitrogen is used throughout the plant for purging. The single largest consumer of nitrogen is regeneration of the feedstock treatment media. A small compression system is used to supply high-pressure nitrogen to a few users in the reaction area.

Most of the cooling water is used to remove the heat of polymerization from the reactor coolant system. Another large user is the pellet water system. Other users are in the diluent purification area and in the solids-transfer systems.

Fuel gas is consumed primarily by the flare system to maintain sufficient hydrocarbon content for combustion. A lesser amount of fuel gas is consumed by the catalyst preparation area.

Instrument air is used primarily for the instrumentation, with a smaller amount sent to the catalyst preparation area.

Loss of power or cooling water requires sending the reactor contents to a recovery vessel. A portion of the hydrocarbon content is flashed and is directed to the flare system. This is typically the largest relief scenario for sizing the flare system.

TECHNICAL ADVANTAGES OF THE CHEVRON PHILLIPS SLURRY-LOOP PROCESS FOR PE

The CPChem PE process offers many advantages compared with competing PE processes. Some of the major advantages are summarized as follows.

Efficient Monomer Use

The ethylene conversion per reactor pass is in excess of 96 percent because the loop reactor does not rely on ethylene circulation for reaction heat removal. The small amount of unconverted ethylene normally is recycled back to the ethylene unit for complete recovery. Hexene-1 is the comonomer selected for the CPChem loop reactor because of its cost-benefit ratio. Hexene-1 provides a significant improvement in product properties as compared with butene-1 when used as a comonomer in polyethylene. Octene-1 provides an additional benefit but at a much higher cost. All three families of catalysts used to copolymerize ethylene and 1-hexene in the slurry-loop reactor do an excellent job of comonomer incorporation.

Efficient Heat Removal

The polymerization of ethylene is a highly exothermic reaction. Efficient heat removal from the reaction mass is very important for the production of quality products. The loop reactor is essentially a double-pipe heat exchanger that is a very efficient heat-transfer device, especially since the contents of the reactor are being circulated at a high flow rate that results in turbulent flow. Since the circulation rate of the slurry inside the loop reactor is high, the heat-transfer rate is also high. The overall heat-transfer rate during regular production is monitored, and no significant change occurs over time. This turbulence mixes the reactor contents and keeps the reactor walls clean and free of resin deposits. The heat generated by the reaction is transferred from the growing polymer particles into the liquid diluent and then is passed into the cooling water through the reactor walls. All heat transfer is efficient liquid-solid transfer, not inefficient gas-solid transfer, as in some other processes.

Uniform Reactor Contents

The CPChem loop reactor produces the most uniform PE owing to the turbulent state of the reactor contents and the reactor's very precise cooling system. Because of this turbulence, there is no eddying or dead spots inside the loop reactor. This ensures a uniform distribution of both reactor contents and reactor temperature, virtually eliminating hot spots within the reactor that could produce off-specification products. Temperature typically is controlled within 0.1°C both across the cross section of the reactor and longitudinally along the axis of the reactor. Temperature uniformity is important because temperature controls molecular parameters that affect the processability and physical properties of the resin. Compared with PE made by any other process, PE made in the slurry-loop process is considered the most uniform from lot to lot and run to run. Internal ethylene gas flow for gas-phase reactors often has nonuniform flow, which leads to nonuniform reactor temperature, resulting in localized hot spots. These hot spots subsequently can lead to the formation of chunks of melted polymer and variability in product quality.

Reliability

The CPChem loop reactor has a very high on-stream service factor. On-stream factors of 97 to 98 percent plus are not uncommon among the over 80 reactors in operation. One reason for the very high loop-reactor on-stream factor is that there is no reactor fouling or external heat exchanger fouling during steady-state operations that would require reactor shutdown for cleanup. Another reason for CPChem's high on-stream factor is the use of installed spare equipment where appropriate.

Quick Product Transitions

If it is desirable to produce several PE grades to satisfy a market, the CPChem loop-reactor process can do so easily. Because of the reactor's short residence time, it can make very quick grade transitions from one type to another. This includes more radical grade changes, such as switching from linear low-density polyethylene (LLDPE) to high-density polyethylene (HDPE), which only takes a few hours with this process. The loop reactor typically operates around 1 hour of residence time, but it is not unusual to operate at residence times as short as 35 to 45 minutes. Reductions in residence time can be achieved by use of higher-activity catalysts.

The amount of catalyst residue left in the polymer is so low, even at the shorter residence times, that the quality of the resin as judged by customers is not affected. Competing gas-phase technologies can have transitions that take as long as several days (Fig. 14.3.6). The longer the transition time, the more off-specification resin one produces.

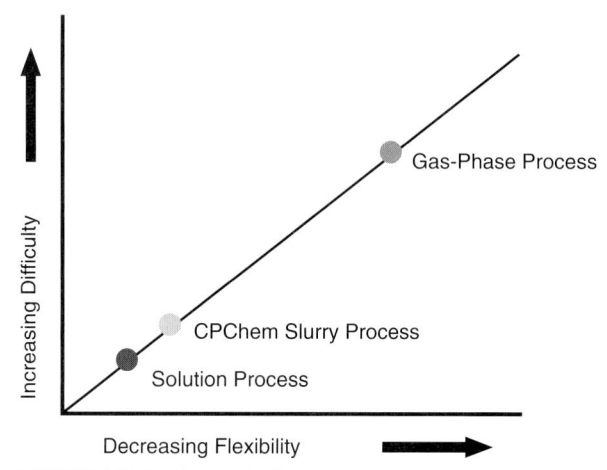

FIGURE 14.3.6 True "swing" process between LLDPE and HDPE because of quick product transitions.

No Limitation on Molecular Weight or Density

With a slurry-reactor process, the viscosity of the fluid inside the loop reactor is independent of resin molecular weight, unlike a solution process. This independence allows the production of very-high-molecular-weight resins. Processes that dissolve the polymer in a solution are limited by viscosity as the molecular weight increases. Therefore, they cannot produce the more demanding very-high-molecular-weight resins for applications such as 200-L drums or automotive fuel tanks. The Chevron Phillips slurry PE process can produce extra-high-molecular-weight PE (Fig. 14.3.7) using almost any of its catalysts.

Prior to 1990s, the slurry-loop reactor did not produce PE much below a density of 0.935 g/cm^3. That limitation was lifted when new catalysts and process improvements allowed the Chevron Phillips slurry loop to make resins below 0.920 g/cm^3 (see Fig. 14.3.7) commercially. These lower-density resins can be made with any of the catalyst families, but the most common lower-density resins being produced commercially are mLLDPE and LDLPE, an easier-processing version of LLDPE.

No Unwanted By-Products

Because the catalysts used in the CPChem process are so active, no waxes or greases are formed. In other processes, waxes are formed and must be removed from the recycle streams, adding capital and operating costs to the process.

The loop reactor uses isobutane as a reactor diluent. Since isobutane has a low boiling point, separating the polymer from the hydrocarbon is very easy and energy-efficient. Use of heavier diluents leads to increased costs in separating the polymer from its hydrocarbon diluent and then recovering the hydrocarbon diluent by extraction or washing steps.

High-Activity Catalyst Capability

Because the loop reactor forms polymer in a liquid diluent, the heat of polymerization can be removed very quickly from the polymer particle. The pipe-loop reactor itself is an efficient heat-transfer device that is ideally suited to the use of very reactive catalysts such as

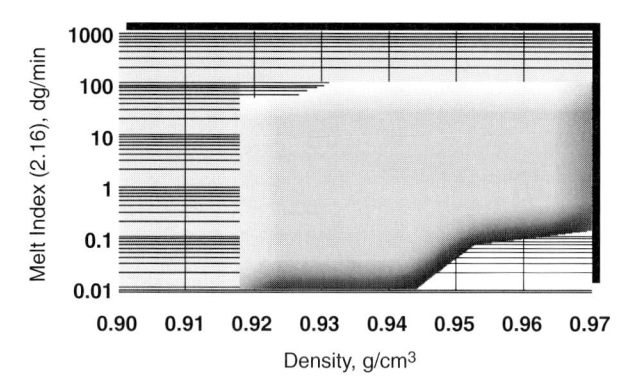

FIGURE 14.3.7 Chevron Phillips slurry-loop reactor produces full density spectrum.

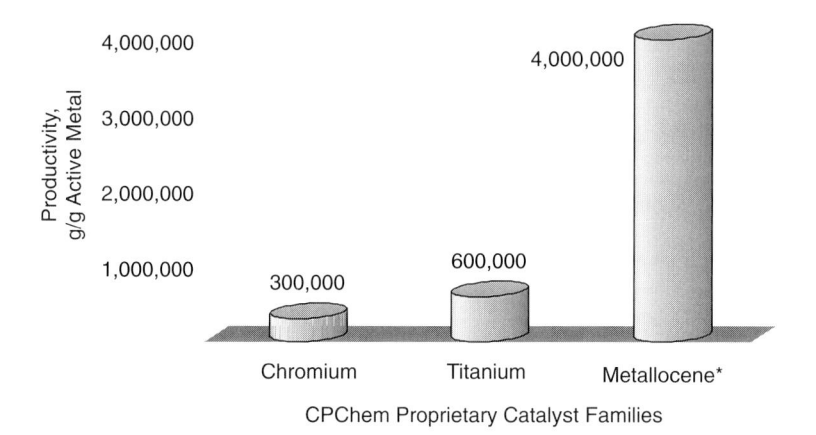

FIGURE 14.3.8 Chevron Phillips process operates well even with very-high-productivity catalysts.

*Laboratory experiments have demonstrated some variations with productivities 10 times greater than this

metallocenes (Fig. 14.3.8). Processes that rely on inefficient gas-solids heat transfer have difficulty with high-activity catalysts owing to poor heat removal leading to molten polymer lumps and chunks forming. A prereactor often must be added upstream of these gas-phase reactors to slow down the catalyst before polymerization can take place in a controlled but less efficient fashion.

Constructability

The loop reactor is constructed of rolled carbon steel pipe that is relatively easy to design and inexpensive to construct. Other processes may have complex reactor vessels with elaborate geometry, surface finishes, high-pressure seals, or multiple vessels (Fig. 14.3.9).

14.42 POLYETHYLENE

- Simple, modular design allows for lower on-site construction cost

- Reactor is sections of pipe, not an unusually shaped vessel requiring expensive casting

- Material of construction is inexpensive carbon steel

- Compact footprint

FIGURE 14.3.9 Chevron Phillips chemical pipe-loop reactor advantages.

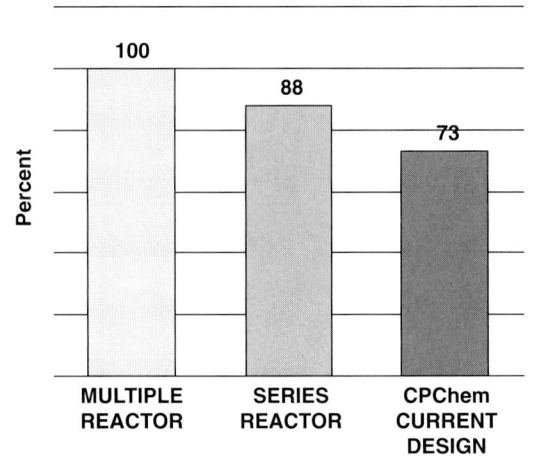

FIGURE 14.3.10 CPChem loop capital cost (estimate) comparison with multiple reactor systems.

The Chevron Phillips slurry-loop reactor pipes can be preassembled off-site into clusters of four or two vertical legs, if desired; shipped in by barge; and then hoisted into place. Preassembly of the reactor saves several days of expensive crane time. The slurry-loop reactor vertical legs also can be assembled onsite, if desired, before being hoisted into place.

Of course, a multiple-reactor system with the same capacity as a single-reactor system would have a higher capital cost. Any process that uses two reactors in series, for example, would cost approximately 20 percent more than a single reactor of the same capacity

(Fig. 14.3.10). If more than two reactors were required to polymerize ethylene to make PE, the capital cost would be 30 percent or greater than a single reactor of the same capacity.

Expandability

The loop-reactor output is very easily expanded by extending the reactor length (increasing its volume). This is done by adding more pipe in the form of additional vertical legs. Additional heat-transfer surface is automatically included with the addition of each leg in the form of coolant jacketing (Fig. 14.3.11). Most other technologies cannot expand the size of their reactors and must instead build and install an additional reactor with additional infrastructure.

Loop Not Limited in Capacity

The largest single Chevron Phillips loop reactor has a nameplate capacity of 320,000 MTA. A feasibility study has been completed on a 400,000-MTA reactor (Fig. 14.3.12). Even this capacity is not the true limit for a single reactor using the Chevron Phillips slurry-loop process.

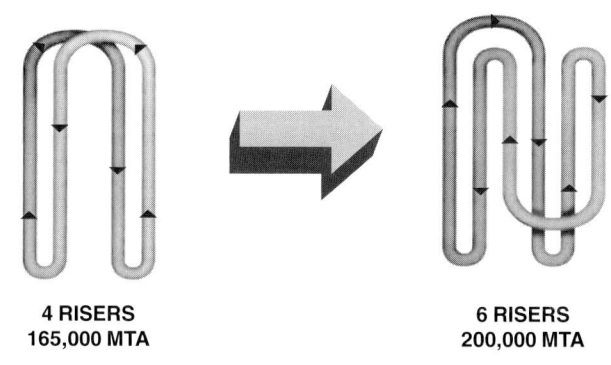

4 RISERS
165,000 MTA

6 RISERS
200,000 MTA

FIGURE 14.3.11 Chevron Phillips loop-reactor capacity is easy to expand.

FIGURE 14.3.12 World-scale CPChem reactors provide excellent size flexibility.

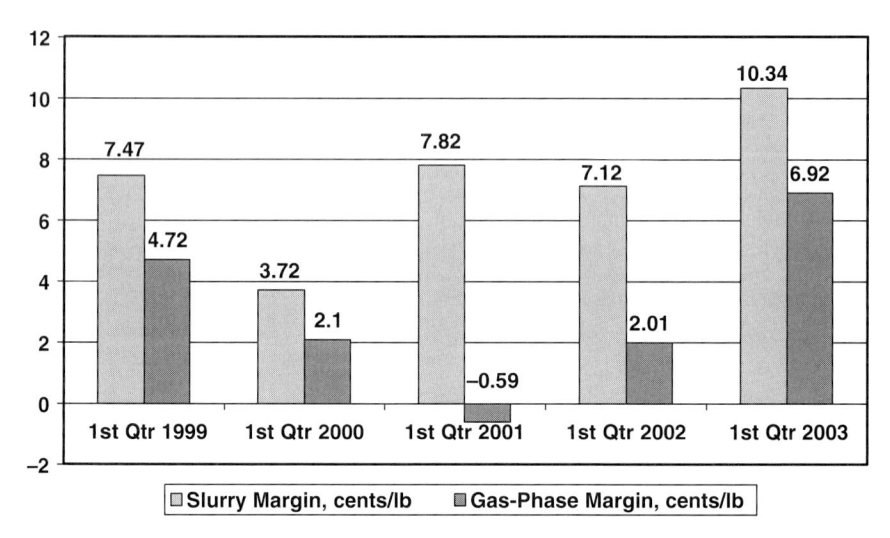

FIGURE 14.3.13 Chevron Phillips slurry polyethylene process profit margin is consistently higher than the nearest competitor.

Profitability

Over the years, the Chevron Phillips slurry process has consistently outperformed HDPE made via all other processes in terms of profit margin (Fig. 14.3.13). Although operating costs may be somewhat similar between the Chevron Phillips process and its nearest competitor, the gas-phase process, the HDPE made with the CPChem process commands a higher premium on the open market because of superior properties and processability.

SUMMARY

Because the Chevron Phillips slurry-loop reactor has a long history of producing top-quality HDPE, it has become the number one process in both North America and the world, with over 86 reactors in service. This accounts for over one-half of North America's HDPE production and over one-third of the world's production.

Recent process improvements to the Chevron Phillips slurry-loop reactor have enabled producers to significantly reduce both their capital cost and operating cost to manufacture these resins, making it the most competitive in the industry.

Innovations in polymer physical and processing properties have enabled the users of this process to command a premium price for resins produced with the slurry-loop-reactor process. These innovations include the ability to produce premium LLDPE and mLLDPE resins, so the process can now produce the full spectrum of linear polyethylene.

Low operating costs and premium prices for product have continuously yielded the best profit margin for the slurry-loop-reactor process compared with any other process.

There is currently no known capacity limitation for this process, which positions it to do well in the future, when larger reactors will be desired to maximize the economy of scale.

CHAPTER 14.4
EXXONMOBIL HIGH-PRESSURE PROCESS TECHNOLOGY FOR LDPE

Charles E. Schuster
ExxonMobil Chemical Company
Baytown, Texas

INTRODUCTION

Polyethylene's discovery was the unexpected result of an experiment undertaken in the course of fundamental research. In the early 1930s, without a direct commercial target in view, Imperial Chemical Industries (ICI) in England decided to evaluate the effect of ultra-high pressures on some 50 chemical reactions.

In 1933, an experiment was carried out compressing ethylene gas to 1400 bar. A white solid was formed in the heavy steel vessel, which proved to be low-density polyethylene (LDPE). Subsequent work showed that minute traces of oxygen had caused the polymerization but that larger amounts of oxygen would result in violent explosions. ICI's Fawcett and Gibson were credited with the discovery and acquired a patent in 1936. By the end of 1935, a pilot plant with an 80-mL reactor was producing small quantities of LDPE with some very interesting product properties.

Outstanding electrical insulation properties were combined with toughness, flexibility, lightness, and inertness. When the first commercial plant started at the end of 1939, the LDPE was used in communication cables and radar, one of the most important scientific weapons of World War II. By the end of the war, England was turning out 20 times as much polyethylene (PE) as at the start of the war.

The original reactors were based on an autoclave process, relying on backmixing of the hot reactants with cold incoming ethylene to keep the reactants stable. Later, a tubular reactor process was developed with a plug-flow system resulting in a very consistent molecular weight. Currently, both types of reactor systems are still in use commercially, and although they operate at very high pressures (up to 3000 bar), some of these reactors have been in continuous service for over 30 years.

FIGURE 14.4.1 ExxonMobil LDPE facility in Meerhout, Belgium.

ExxonMobil Chemical Company (ExxonMobil) entered the LDPE business in 1968 with the startup of the Baton Rouge Plastics Plant in Louisiana. ExxonMobil and its affiliates currently operate LDPE facilities not only in Louisiana but also in Texas, Belgium, Australia, and Saudi Arabia. The ExxonMobil LDPE facility in Meerhout, Belgium, is shown in Fig. 14.4.1.

Through the years, ExxonMobil has continued to develop and enhance the basic process technology because it remains one of the acknowledged industry leaders in the production of LDPE. Over 35 years of operating experience while developing LDPE products and processes form the backbone of the ExxonMobil high-pressure (HP) process technology for LDPE. This technology provides licensees with a flexible process (based on either a tubular or autoclave reactor) yielding a broad product portfolio using a high-quality and cost-effective process design.

REACTION MECHANISM

In the ExxonMobil high-pressure process technology for LDPE, the polymerization reaction is a free-radical polymerization, with organic peroxides providing the source of the free radicals. Early LDPE process technology used air (oxygen) as the source of the free radicals, but this technology essentially has been replaced by organic peroxides.

A free radical is a short-lived reactive intermediate with an unpaired electron. The reaction starts when a free radical reacts with an ethylene molecule, forming a new radical that propagates the reaction with ethylene molecules until the growth of long chain molecules ends. The overall polymerization reaction is

$$n(CH_2=CH_2) \rightarrow (-CH_2-CH_2-)_n$$

At the high pressures involved, the polymerization step is very rapid. Although the commercial PEs have a very complex structure at the molecular level, the polymerization process can be described by the classic kinetic description of the free-radical (addition) polymerization.

When simplifying the free-radical mechanism, the following four steps are distinguished:

1. *Initiation.* Free-radical sites for polymerization are formed by reaction between primary initiator free-radical fragments and vinyl monomer molecules.
2. *Propagation.* Polymerization then proceeds through a series of additions of monomer molecules to the growing polymer chain, with the free-radical site jumping to the end of the growing chain after each addition.
3. *Termination.* Active free radicals are destroyed by two free-radical sites coming together and reacting to form either one or two dead polymer chains.

These three reaction steps (initiation, propagation, and termination) are the part of the mechanism that determines the polymerization rate. There is another class of reactions that determines the molecular weight and molecular weight distribution of the polymer. These are

4. *Chain-transfer reactions.* Active free-radical sites at the ends of growing chains jump to another site on the same polymer molecule, another polymer molecule, or a solvent, monomer, or modifier molecule. Chain transfer affects the size, structure, and end groups on the polymers. Chain-transfer agents (modifiers) are able to donate hydrogen atoms to this reaction.

Peroxide type and concentration, as well as reactor temperature, control the initiation rate. The propagation rate increases with pressure and temperature. The rate of chain termination is controlled by free-radical concentration and by the amount of chain-transfer agent present in the reaction mixture.

During propagation, significant heat is liberated as ethylene double bonds are converted to single bonds. The heat of polymerization is roughly 22 kcal/mol. However, since the molecular weight of ethylene is low, this heat of polymerization is high on a weight basis, thus making removal of this heat of reaction critical on a commercial scale.

Physical properties of LDPE are controlled primarily by average molecular weight, molecular weight distribution, number of short-chain branches, and number of long-chain branches.

The degree of crystallinity (polymer density) is controlled primarily by the number of short branches (two to four carbons per branch). Short-chain branches are formed as the result of intramolecular chain transfer, which occurs simultaneously with chain growth. In this intramolecular chain transfer or back-biting reaction, the formation of a transient five-carbon ring with the resulting extraction of a hydrogen from the fifth carbon transfers the free radical from the chain end to the fifth carbon atom. The active site continues to propagate, leaving a butyl branch. A possible second back-biting step can convert the butyl branch to two ethyl branches. A temperature decrease or a pressure increase decreases this branch formation.

Rheologic properties of the molten polymer are controlled by the molecular weight, the molecular weight distribution, and the number of long-chain branches. The molecular weight of the polymer is characterized by a measurement called the *melt index.* The melt index measurement is the output of a standardized polymer test and has an inverse relationship to the molecular weight of the polymer. A low melt index represents an LDPE with a high molecular weight, and a higher melt index describes an LDPE with a lower molecular weight.

In the high-pressure LDPE process, the molecular weight of the individual polymer chains is not uniform. The breadth of this molecular weight distribution depends on the reactor process (tubular or autoclave) and can be further influenced by the reaction conditions.

Intermolecular chain transfer between a completed polymer molecule and a growing polymer radical or between two growing polymer radicals forms long-chain branches. The active radical attacks the dead chain at an internal carbon, transferring its activity to the dead chain, and terminates itself. The new species then continues to propagate from the internal free-radical site, forming a long-chain branch. This reaction increases with increased reaction temperature, decreased pressure, increased conversion, and of course, increased polymer residence time.

PROCESS OVERVIEW/DESCRIPTION

The overall processes for the tubular reactor and the autoclave reactor systems are very similar except for the design of the reactor system itself. The two reactor technologies have the capabilities of producing different products for different end-use markets. A simplified feed/product overview can be found in Fig. 14.4.2.

In the ExxonMobil high-pressure LDPE process, polymerization-grade ethylene is supplied to the battery limit and is boosted to 300 bar(g) by the primary compressor. This makeup ethylene is cooled and divided into two streams. One stream is fed to the secondary compressor suction, whereas the other is added to the ethylene-polymer mixture downstream of the reactor pressure control valve (also known as the *high-pressure letdown valve*) at the end of the reactor as "low-pressure cold quench."

In the suction manifold of the secondary compressor, the gas from the high-pressure recycle system is combined with the ethylene from the primary compressor. Furthermore, chain-transfer agent(s) and/or comonomers are injected. The secondary compressor will then boost the pressure from the primary compressor discharge pressure to the reactor operating pressure. The reactor operating pressure is typically 2500 to 3100 bar(g) for the tubular reactor and 1200 to 2000 bar(g) for the autoclave reactor.

The ethylene mixture then flows to the reactor system (either tubular or autoclave), where mixtures of organic peroxides diluted in hydrocarbon solvent are injected to initiate the exothermic polymerization reaction. The reaction mixture (polyethylene and unreacted ethylene) is then decompressed through the reactor pressure control valve to about 300 bar(g).

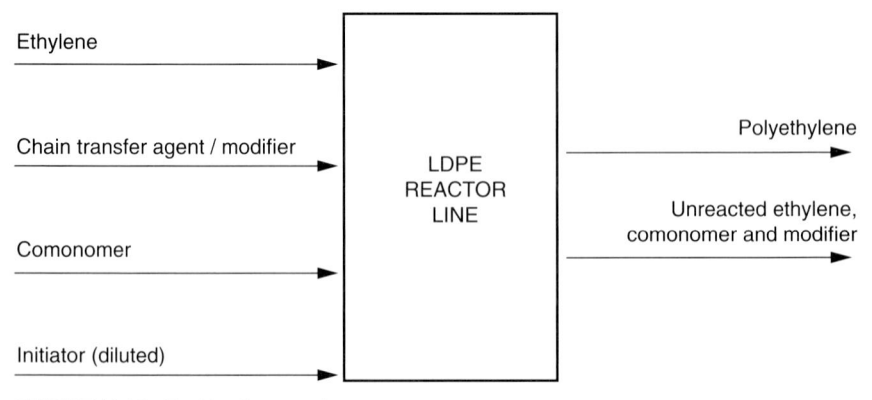

FIGURE 14.4.2 Feed/product overview.

Tubular Reactor

For tubular reactors, the discharge flow of the secondary compressor is divided in the number of streams required for the front and side streams of the reactor. The flow to the front of the reactor is heated, whereas the flows to the side-stream entry points are cooled. The side-stream flows are then fed to injection points along the length of the tubular reactor.

In essence, the tubular reactor is a plug-flow reactor. Multiple peroxide injection points are used along the length of the reactor to maximize conversion of ethylene to LDPE and optimize product properties. Three or more injection points are used in most ExxonMobil tubular reactors. Reactor conversion rates of approximately 40 percent have been achieved.

For illustrative purposes, a four-reaction-point reactor is shown in Figs. 14.4.3 and 14.4.4. In a four-reaction-point reactor, a mixture of peroxides in an organic solvent is injected at the reactor inlet and three downstream points. The peroxides initiate the free-radical exothermic polymerization reaction. The reaction heat is removed by an increase in the temperature of the ethylene, by injection of the cold side-stream ethylene, and by heat transfer through the reactor walls to a closed-loop jacket water cooling system. The conversion of ethylene to polymer is maximized by the use of reactor jacket water at cooling water temperatures. This cool jacket water maximizes the heat transfer through the thick-walled reactor pipes. The pressure drop along the length of the reactor is minimized through the use of a stepped profile on reactor tube diameters. In addition, this stepped profile optimizes the gas velocity along the length of the reactor, which virtually eliminates fouling on the interior wall of the tubular reactor.

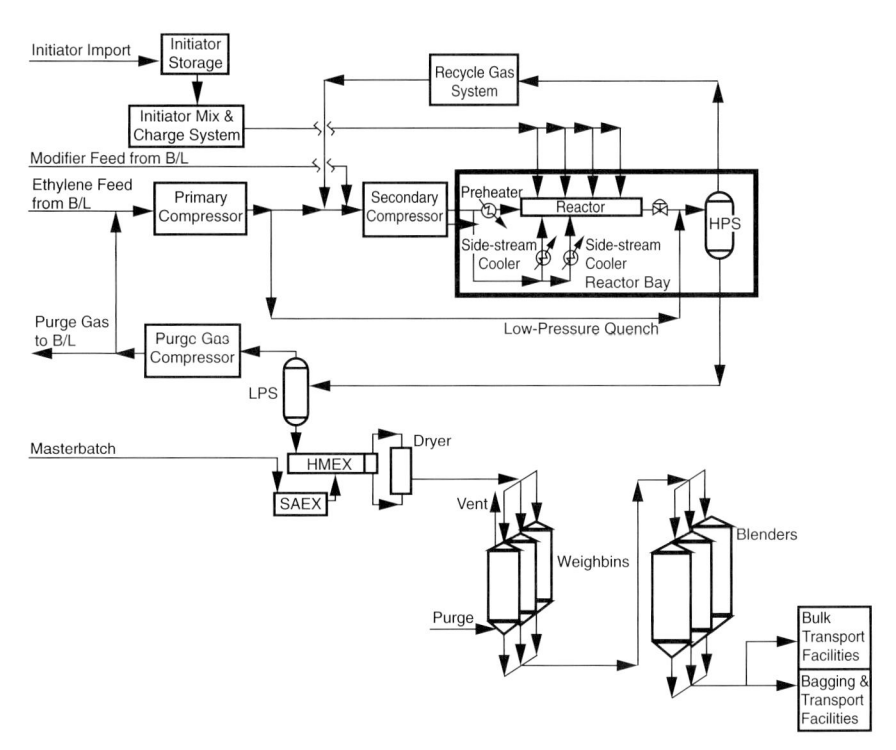

FIGURE 14.4.3 Process flow diagram of the ExxonMobil LDPE tubular process.

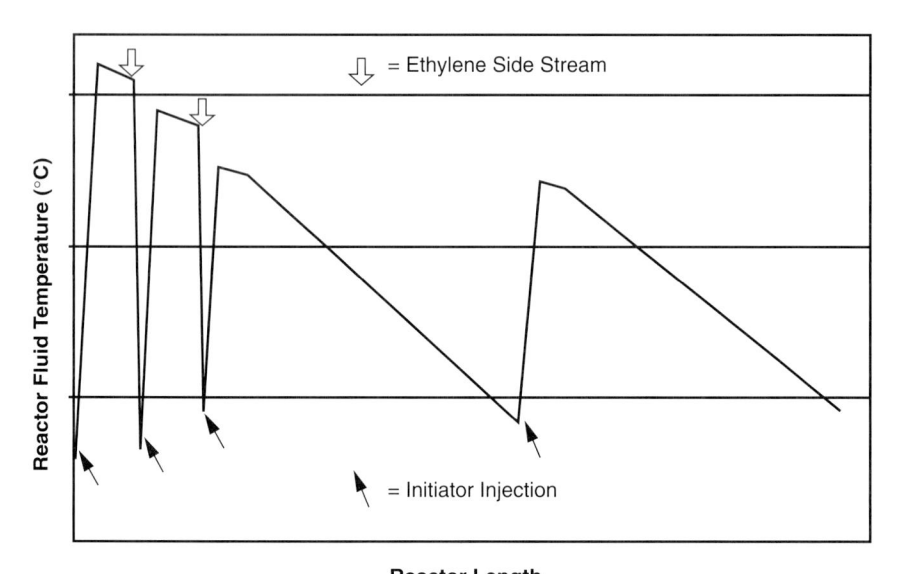

FIGURE 14.4.4 Tubular reactor temperature profile.

An extensive distributed computer control system is used to control the reactor temperature profile and other key process variables. In addition, multiple proprietary applications have been developed to maximize the use of online quality control for the key product properties, which results in outstanding product quality performance.

A typical profile of the reactor temperatures in a tubular reactor is shown in Fig. 14.4.4.

After polymerization, the reaction fluid is decompressed through the reactor pressure-control valve to about 300 bar(g) and cooled with the low-pressure cold quench from the primary compressor discharge. The mixture is then fed to the high-pressure separator, where polymer is separated from unreacted ethylene. The use of the low-pressure quench minimizes the formation of gels and improves the quality of those polymers destined for applications in high-clarity film packaging.

The molten polymer and a small amount of entrained ethylene are again decompressed to approximately 0.7 bar(g) through the low-pressure letdown valve and fed to the low-pressure separator. At this point, most of the remaining entrained ethylene is separated from the LDPE and is sent to the purge-gas compression system.

The purge-gas compressor boosts the low-pressure separator offgas to the suction pressure of the primary compressor (operated at ethylene pipeline pressure). At this point, part of the gas is purged from the LDPE line to the ethylene recovery system, whereas most of this gas is recycled via the primary compressor back to the reactor system.

The polymer melt from the low-pressure separator is fed to the hot-melt extruder. In the hot-melt extruder, additives are mixed in. This is typically done via a sidearm extruder using the masterbatch technique. Note that the masterbatch can contain slip, antiblock, and antioxidant. Alternatively, liquid additive injection can be considered.

The polymer strands leaving the die of the extruder are cut into pellets underwater. The pellets are cooled, dried, and pneumatically transferred to weighbins or blenders, in which the pellets are purged to reduce the residual ethylene concentration to a safe-to-store concentration (<50 wppm). After blending, the pellets are conveyed to bulk loading or bagging facilities.

Autoclave Reactor

The discharge flow of the secondary compressor is divided into a number of reactor feed streams. The reactor feed streams are cooled and then fed to the different injection points in the autoclave reactor. The autoclave reactor is a continuous-stirred-tank reactor (CSTR) with an agitator to promote good mixing. The multiple zones in the reactor allow for manipulation of the temperature profile for tailoring of product properties. Figure 14.4.5 illustrates the process flow diagram for the autoclave process.

Organic peroxide solutions are injected at multiple points into the reactor to initiate the free-radical exothermic polymerization reaction. Since the autoclave is an adiabatic CSTR reactor, addition of the cooler, fresh feeds of ethylene balances the heat of polymerization.

As noted earlier, an extensive distributed computer control system is used to control the reactor temperature profile and other key process variables. In addition, multiple proprietary applications have been developed to maximize the use of online quality control for the key product properties, which results in outstanding product quality performance.

After polymerization, the reaction fluid is decompressed through the high-pressure letdown valve to about 800 bar(g) and cooled with the product cooler. The mixture is then fed to the high-pressure separator, where polymer is separated from unreacted ethylene. From this point forward in the process, the process flow is identical to the process steps described earlier in the tubular reactor section.

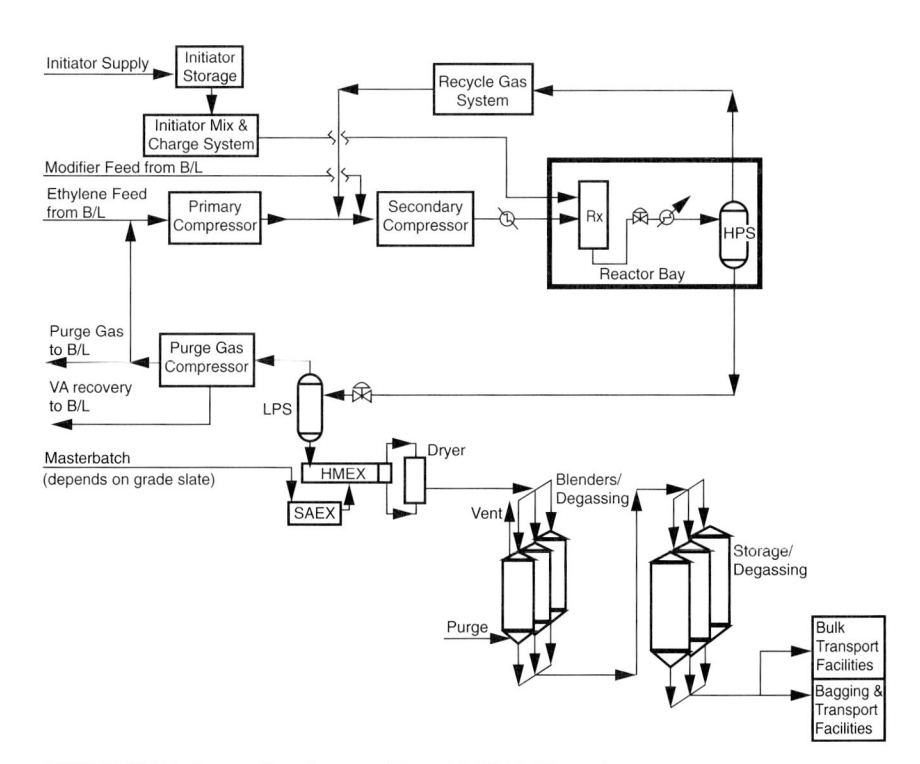

FIGURE 14.4.5 Process flow diagram of ExxonMobil LDPE autoclave process.

TABLE 14.4.1 Key LDPE Process Variables

Product property	Primary control process variable
Melt index	Chain-transfer agent concentration
Density	Comonomer type
	Reactor peak temperature
	Reactor density
Haze	Reactor pressure
	Reactor temperature

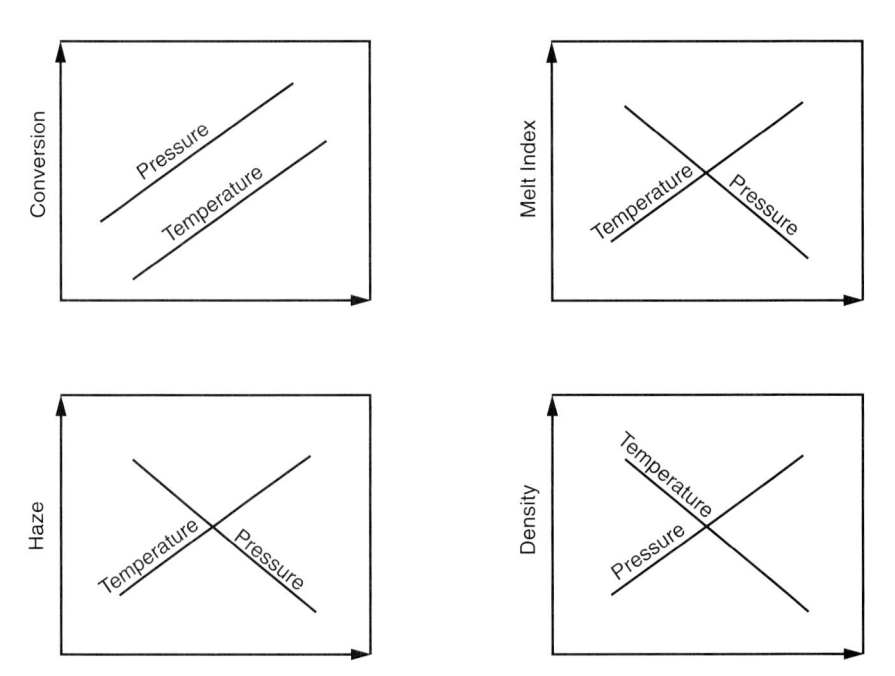

FIGURE 14.4.6 Process variable/product property relationships.

Key Polymer Properties and Process Variables

Table 14.4.1 lists the relationship between the important LDPE polymer properties and the key process variables that affect those properties.

The charts in Fig. 14.4.6 show the direct impact of reactor temperature and reactor pressure on these same properties.

Safety

The ExxonMobil high-pressure LDPE process has a history of delivering outstanding safety by design and in commercial operation. The reactor system (either tubular or autoclave) is protected by pressure-relief devices and a reactor interlock logic system that guarantee an immediate release of the reactor contents in case a runaway reaction occurs. A runaway reaction is often referred to as a *decomposition reaction*. Under specific conditions outside the normal operating range in the reactor, ethylene can undergo a highly exothermic decomposition. The reactions that occur during the decomposition of ethylene are the formation of carbon, hydrogen, and methane. This decomposition reaction causes a sharp increase in reactor pressure and reactor temperatures.

The heart of the emergency relief system is the reactor interlock logic system. This logic system is designed to take corrective action to slow down or stop the polymerization reaction by partially or totally depressurizing the reactor and surrounding process systems.

As a result of ExxonMobil's process design, computer control systems, and operating practices and procedures, reactor decompositions are very infrequent.

LDPE VERSUS LLDPE

When a manufacturer decides to produce low-density polyethylene, a choice often must be made between LDPE produced by the high-pressure process and LLDPE produced by the lower-pressure processes such as gas phase or solution. LDPE is a very versatile polymer that is used in a wide range of product applications where its balance of strength, stretch, clarity, sealing, low temperature impact, and adhesion characteristics brings value. LLDPE is used widely in the film market owing to its balance of toughness, puncture resistance, and stiffness. Table 14.4.2 summarizes the key differences between the high-pressure process for LDPE and the gas-phase process for LLDPE.

In a 2001 study, Nexant, Ltd.,[1] compared the cost to produce PE in the high-pressure process with the cost in a gas-phase process. LDPE requires slightly higher investment costs and uses more energy than some low-pressure gas-phase (GP) processes. However, in this study, these costs were offset by the higher comonomer costs required by the low-pressure process and the high operating efficiency (reactor operating percentage and product quality) of the high-pressure process. The results of this study are shown in Fig. 14.4.7. The cash costs to produce 1 t of PE were less for the high-pressure LDPE process.

The ExxonMobil high-pressure process technology delivers world-scale performance in the key areas of reactor service factor (online operating percentage), product quality, and reactor flexibility. The two world-scale tubular reactors at the ExxonMobil LDPE facility in Meerhout, Belgium, operate with greater than 98 percent reactor service factor. This high level of online operating hours is due to a very robust mechanical design, state-of-the-art maintenance procedures, rigorous inspection procedures, and advanced process control applications. These same two reactors produced more than 98 percent prime product. These impressive performance metrics are obtained while supporting a robust polymer grade slate and the resulting high number of reactor transitions in a given year.

TABLE 14.4.2 High-Pressure Process versus Low-Pressure Process (Gas Phase)

	High-pressure process	Gas-phase process
Reactor residence time	2–3 min (tubular reactor)	1–4 h
	20–40 s (autoclave reactor)	
Grade-change flexibility	Higher	Lower
Uses polar comonomer	Yes	No
Monomer cost	Lower	Higher
Energy cost	Higher	Lower
Investment cost	Higher	lower

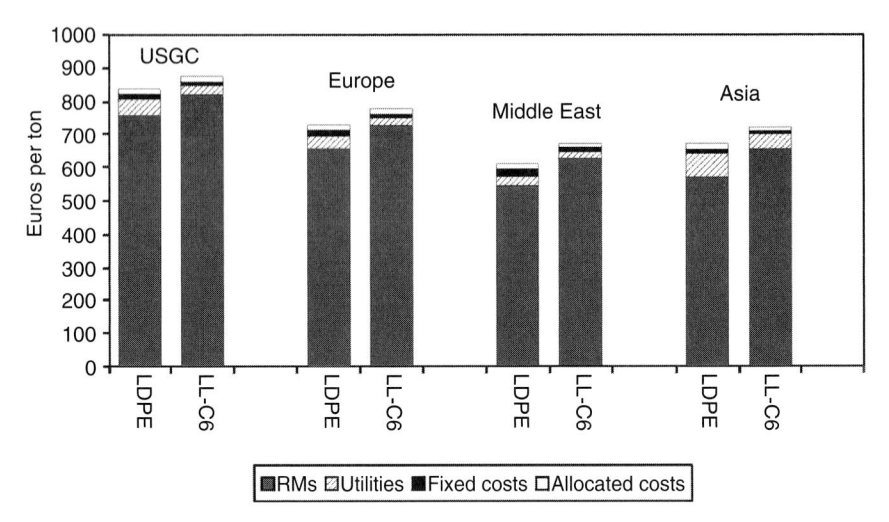

FIGURE 14.4.7 Regional cash cost comparison of HP/GP processes, Q2 2001 (350-kMTA capacity).

PRODUCT CAPABILITY/GRADE SLATE

With the ExxonMobil high-pressure process technology for LDPE, a broad and varied product grade slate is possible. With ExxonMobil high-pressure autoclave technology, the grade-slate capability is

- Homopolymer LDPE ranging from 0.910 to 0.935 g/cm^3 density
- Ethylene vinyl acetate copolymers up to 40 wt % vinyl acetate content
- Specialty copolymers of ethylene methyl acrylate, ethylene acrylic acid, and ethylene *n*-butyl acrylate
- Extrusion-coating grades
- High-clarity grades for film applications

With ExxonMobil high-pressure tubular technology, the grade-slate capability is

- Homopolymer LDPE ranging from 0.915 to 0.935 g/cm^3 density
- Ethylene vinyl acetate copolymers up to greater than 30 wt % vinyl acetate content
- Very-high-clarity grades for speciality film application

LDPE MARKETS

The applications (as shown in Table 14.4.3) and markets for LDPE are strong and continue to grow, varying from flexible packaging and agricultural film to electrical cable. After 60 years, the demand for LDPE produced via the high-pressure process continues to increase globally at an annual 2 percent growth rate (Fig. 14.4.8).

The high-clarity LDPE products are valued for their excellent optical properties in film applications. Medium-density grades offer increased stiffness, higher barrier, and higher

TABLE 14.4.3 End-Use Applications for LDPE

Application	Use
Film	65%
Extrusion coating	10%
Other extrusion	8%
Injection molding	7%
Blow molding	4%
Others	6%

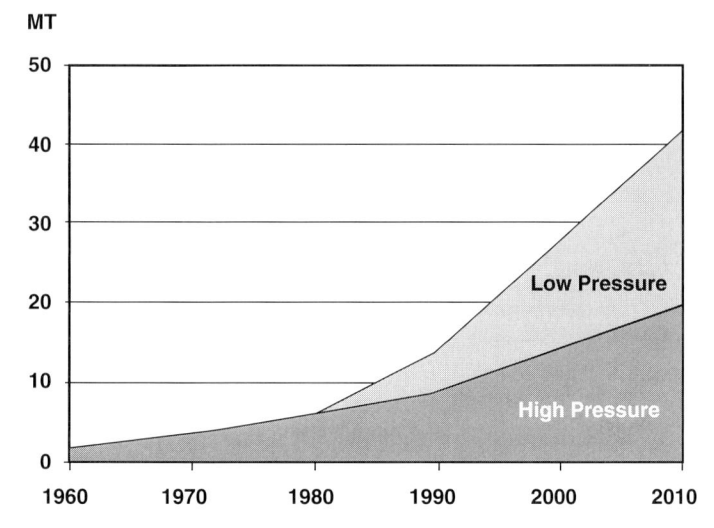

FIGURE 14.4.8 World market for polyethylene.

melting points for speciality film applications. EVA copolymers are used widely in elastic films. In addition, owing to their low sealing temperatures, high polarity, and superior optical properties, EVA copolymers also are used widely in specialty film, foam, adhesive, and extrusion applications.

Consistency of product properties and purity of product make LDPE an ideal polymer for critical applications for coatings and laminations. This is important in food packaging, where shelf life and appearance are essential marketing qualities. In addition, the easy sealing properties make LDPE and EVA copolymers the preferred materials for interior layers of multilayered composite films.

LDPE is the easiest PE polymer to process by the converter. LDPE's balance of properties and inherent high melt strength make it the material of choice for several applications:

- Wide film, where bubble stability is critical
- Foamed applications, where collapse resistance in the molten phase is essential
- Extrusion coating, where low neck-in properties are required

It also makes an ideal material for blending in any ratio with other polymers for improved processability.

Because of the diversified applications just described that use LDPE, it is easy to see why the demand for this versatile polymer continues to grow.

STRENGTHS OF EXXONMOBIL TECHNOLOGY

Polyethylene (LDPE, LLDPE, and HDPE) is a core business for ExxonMobil Chemical Company. Through the years, ExxonMobil has continued to invest in commercial LDPE facilities. Table 14.4.4 lists ExxonMobil LDPE production facilities. In addition, investments also have continued in the process and product technology required to develop and support these world-scale production facilities. Years of operating experience with over 25 reactors (ExxonMobil and licensees) has further enhanced commercial operating practices and procedures. As a result, the licensed process technology is a direct reflection of the commercial success of ExxonMobil and its position as an acknowledged leader in the LDPE industry. Figure 14.4.9 depicts a tubular reactor at an ExxonMobil licensee's facility.

The strengths of the ExxonMobil high-pressure process technology for LDPE are

- A robust high-pressure mechanical design supported by maintenance procedures that result in high operating efficiencies
- The ability to control the reactor temperature profile and reactor gas distribution to tailor the LDPE product properties
- Outstanding product quality based on well-defined operating procedures, clean product technology, and online quality control
- World-class reactor conversion rates—up to 40 percent
- Reactor designs that are optimized to achieve the best balance between highest reactor conversion rates and product properties
- Process and mechanical designs with capacities up to 400 kMTA

ExxonMobil began licensing its high-pressure LDPE process technology in the mid-1990s. A listing of the current ExxonMobil licensees can be found in Table 14.4.5. Licensees benefited from the strengths noted earlier and access to world-class ExxonMobil technology centers located in Baytown, Texas, and Brussels, Belgium. Today, all licensees are operating at greater than licensed capacity.

TABLE 14.4.4 ExxonMobil LDPE Facilities (2003)

Plant site	Country	Number of lines
Baton Rouge	United States	4 tubular
		2 autoclave
Meerhout	Belgium	2 tubular
Beaumont	United States	3 tubular
Antwerp	Belgium	6 autoclave
Al-Jubail*	Saudi Arabia	1 tubular
Sydney*	Australia	4 autoclave

*ExxonMobil joint ventures.

FIGURE 14.4.9 Tubular reactor at licensee.

TABLE 14.4.5 Licensees of ExxonMobil LDPE Process

Plant site	Country	Reactor type	Licensed capacity, kMTA	Startup year
Daesan	South Korea	Autoclave	85	1998
Johor	Malaysia	Tubular	200	1999
Al-Jubail	Saudi Arabia	Tubular	200	2000
Yanshan	China	Tubular	200	2002
Map Ta Phut	Thailand	Autoclave retrofit	90	2003
Sasolburg	South Africa	Tubular	220	2005

SUMMARY

The ExxonMobil high-pressure process technology for LDPE is a versatile technology characterized by a cost-effective plant design and the flexibility to produce a diversified grade slate with superior product quality. In addition, operating experience and state-of-the-art procedures result in an outstanding reactor service factor while meeting world-class safety and environmental standards. As a result, the ExxonMobil process is ideally suited to allow a licensee to succeed in the highly competitive LDPE marketplace.

REFERENCE

1. "High Pressure and Gas Phase Polyethylene Process Cost Comparison," Nexant ChemSystems, October 2001. (Nexant's Petroleum and Chemical Division acquired the ChemSystems business in 2001.)

DISCLAIMER

Copyright © 2003 ExxonMobil. While the information is accurate to the best of our knowledge and belief as of the date compiled, it is limited to the information as specified. No representation or warranty, expressed or implied, is made regarding the information or its completeness, merchantability, or fitness for a particular use. The user is solely responsible for all determinations regarding use, and we disclaim liability for any loss or damage that may occur from the use of this information. The terms "we," "our," "ExxonMobil Chemical," or "ExxonMobil" are used for convenience and may include any one or more of ExxonMobil Chemical Company, Exxon Mobil Corporation, or any affiliates they directly or indirectly steward. The ExxonMobil Chemical emblem and the "Interlocking X" device are trademarks of Exxon Mobil Corporation.

CHAPTER 14.5
POLIMERI EUROPA POLYETHYLENE HIGH-PRESSURE TECHNOLOGIES

Mauro Mirra
Polimeri Europa
Milan, Italy

INTRODUCTION

Low-density polyethylene (LDPE), the oldest polyethylene resin, is produced using high-pressure processes. The chemistry was discovered in Imperial Chemical Industries' laboratories, and the first commercial production started in the early 1940s. Since the late 1940s, two different high-pressure LDPE technologies have evolved, based on an autoclave reactor and a tubular one. These two processes were developed contemporaneously based on radical polymerization by organic peroxides, and both are operated widely in industry, chosen by the polyethylene (PE) producers mainly depending on the product-mix requirement.

Polyethylene resins from high-pressure technology, because of their intrinsic characteristics, are used today in different forms, covering most of the popular applications in the field of films such as carrier bags, packaging materials and agricultural film covers, surface coatings, pipes, and insulating materials.The density of LDPE lies between 915 and 935 g/L.

Polimeri Europa, over the years, has built a very valuable operation, developing and producing a large mix of ethylene vinyl acetate (EVA) copolymers with vinyl acetate monomer (VAM). New proprietary Ziegler-Natta-based catalytic systems have been set up recently for high-pressure PE processes that today allow production of most LLDPE grades and have widened the opportunities up to VLDPE specialties under 915 g/L density for film niches and to ULDPE elastomeric polymers up to 890 g/L.

It is a very simple job to retrofit an LDPE plant to a new process with Ziegler-Natta catalysts, achieving a wide product mix in linear low-density PE (LLDPE) and very-low-density PE (VLDPE) grades with a reasonable capital investment and competitive economics. Today, developments in mechanical manufacturing and the polymerization operation have increased the capabilities of new reactors, allowing large single-train designs to 400,000 metric tons per annum (MTA) and very efficient processes that can compete with LLDPE for capital investment and economics, making new LDPE projects even more attractive.

Polimeri Europa started its activity in the PE industry in the 1960s; today, more than 1.5 MTA of PE high-pressure technology capacity is in operation, including 800 kMTA of licensed capacities in Europe and elsewhere.

Polimeri Europa, the European coleader in PE production, is producing more than 1.5 MTA of resins by high-pressure technology, slurry, solution, and gas-phase processes, covering the LDPE, LLDPE, and HDPE areas.

POLIMERI EUROPA TRADEMARKS

Polimeri Europa started to license polyethylene high-pressure technology in the 1970s (two tubular reactors) in Algeria and Romania. In late 1980s, the activity really improved following the incorporation of CdF Chemie, the old autoclave reactor licensor, supporting licensees such as QAPCO having plants in Qatar, LG Chemicals and Toso in the Far East, IPCL in India, and Polilago in South America. During the 1990s, new contracts have been awarded in relevant autoclave and tubular reactor technologies, such as LDPE in the Middle East, QAPCO and EVA copolymers such as BOCW in China, and Formosa Plastics in Taiwan for a total 450 kMTA of production capacity.

Polimeri Europa technology offers a wide range of resins that are available on the market with different trademarks (Fig. 14.5.1).

Riblene®

These are highly competitive LDPE resins that are tailor-made for European market requirements, and cover most applications, general purpose and specialty. These are tubular reactor resins offering a complete range of densities and high molecular weight for the best result in processability and optical film properties; they are also highly downgaugable, making them suitable for high-clarity and lamination films. Cables grades are incuded because of resin purity. Autoclave reactor products are recommended for a very wide mix having high mechanical properties ranging from agricultural films, heavy-duty and shrink films, injection grades, and surface-coating materials.

GREENFLEX®

These are ethylene copolymers with VA comonomers, full range from 2 to more than 30 percent VA, that are suitable to satisfy more specialized market sectors such as high-clarity film and special films for greenhouses, tubing, and injection resins for shoes and foam grades, together with a wide range of base resins for hot-melt products.

FLEXIRENE®

This mix enables high-pressure technology to supply resins for most of the large LLDPE market, such as cast and stretch films, injection resins, and rotomolding, with densities ranging from 915 to 930 g/L.

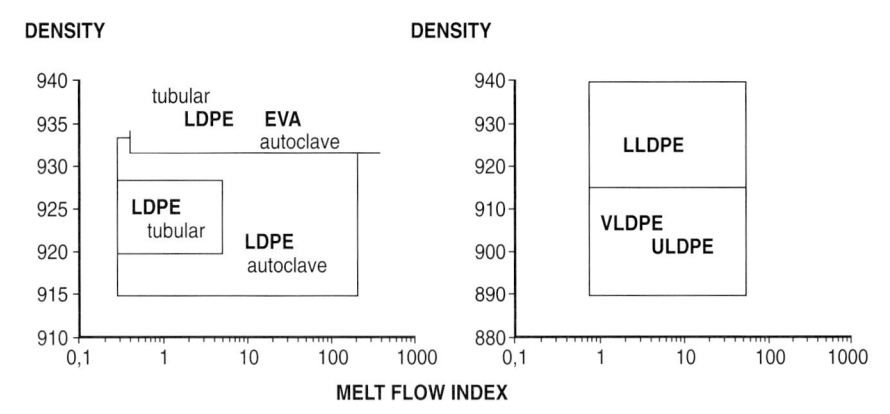

FIGURE 14.5.1 Resin mix ranges.

CLEARFLEX®

These are high-quality resins with a density lower than 915 g/L, suitable for the most valuable sectors of LLDPE films, resins for blowing and casting, and specialties for typical VLDPE applications such as sticky cast films and PVC substitution. The ULDPE area represents polymer families with densities ranging from 900 to 890 g/L and results in specialties for the elastomeric polymer markets and polymer modification such as shock-resistant thermoplastic olefins.

CHEMISTRY AND THERMODYNAMICS

The high-pressure polymerization LDPE is a highly exothermic free-radical-promoted reaction carried on in bulk by the action of organic peroxides and oxygen (Fig. 14.5.2). The use of hypercritical thermodynamic conditions enables the ethylene to act as the solvent for the polymer. Polymerization heat is about 920 kcal/kg, and it is removed mainly adiabatically, so for better reactor performance, the reaction should be initiated at the lowest temperature possible that allows building of the required molecular structure. The high-pressure process makes its own side chains for lowering resin density and to create long-chain branching properly to make LDPE easy to process. As far as reactor selectivity concerns, the characteristics of the molecular structure are correlated with the residence time distributions of the free radicals and the polymer. Free radicals in growing polymers mainly contribute to branching mechanism, with evidence of more branched molecular structures having better mechanical behavior and poorer optical properties. The tubular reactor produces more crystalline structures, higher polymer density at a lower molecular weight, and narrower molecular weight distributions, whereas the autoclave reactor gives better results in resin melt strength and swelling characteristics.

With ethylene copolymerization with VAM comonomer, a wide mix of resin can be produced at different VA contents. Copolymers with acrylic acid (EAA) and/or methyl acrylate (EMA) can be produced for downgauging similar to LLDPE.

Alternative specific Ziegler-Natta catalyst systems exploit an LLDPE coordination copolymerization mechanism, and α-olefins, C_4 (C_6), can be inserted in the chains based on their reactivity ratio, creating ordinate and regular branches that contribute to lowering

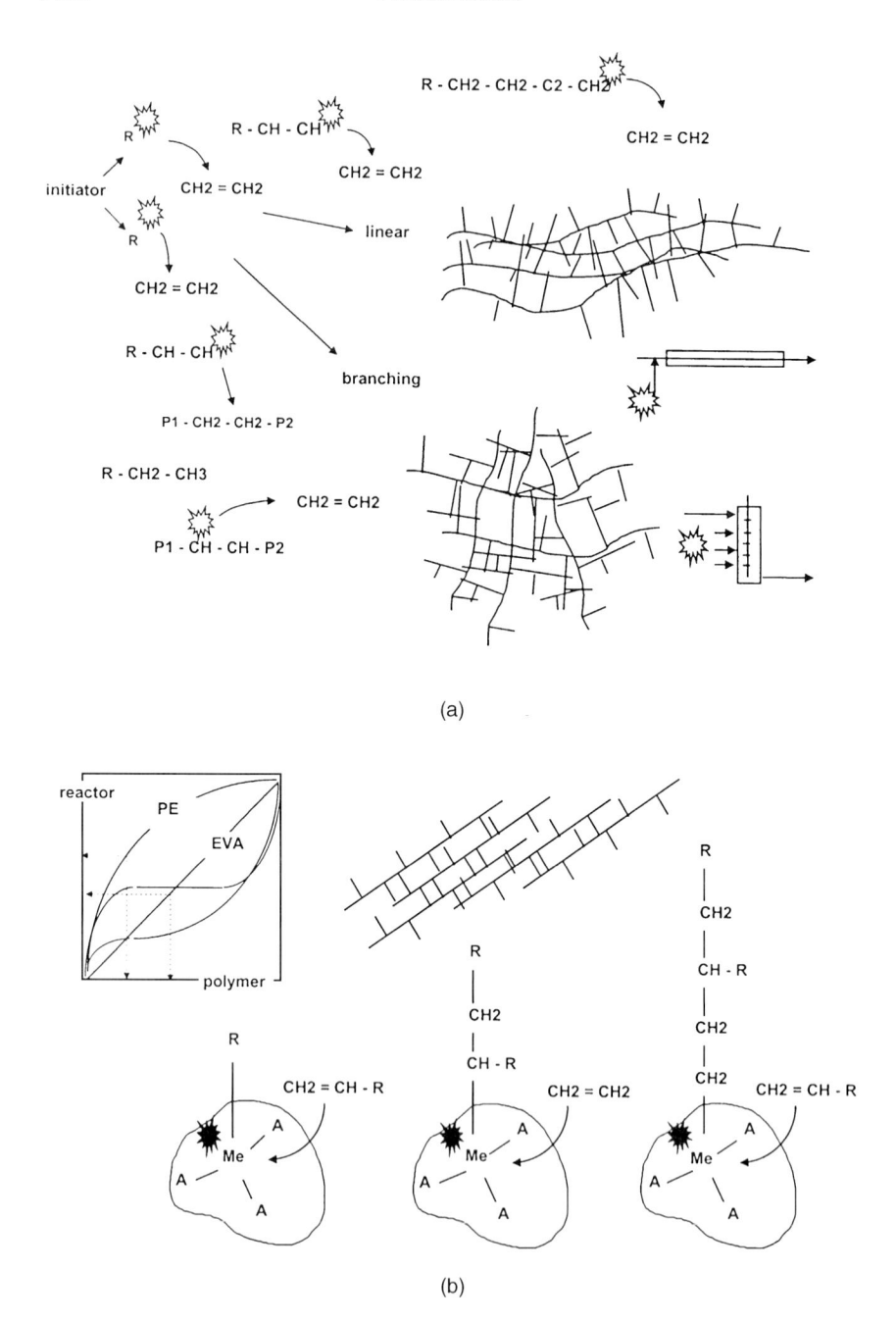

FIGURE 14.5.2 Polymerization mechanisms: (*a*) free-radical polymerization; (*b*) coordinated mechanism for copolymerization by Ziegler-Natta catalyst.

density proportionally and to increasing polymer melt strength, exalting the mechanical characteristics of the product.

While leading companies that offer a wide range of peroxides support the development of the peroxide catalysts for their specific applications, the proprietary Ziegler-Natta catalyst is the result of Polimeri Europa know-how.

There is an upper temperature operating limit at which ethylene starts to become thermodynamically unstable; thus the reaction is controlled strictly by injection of fresh cold ethylene into the reactor at several points and, in tubular reactors, also by refrigeration across the tube wall. In any case, a runaway reaction is considered possible, and an emergency letdown system is necessary for the offsets to vent the contents of the reactor to the atmosphere. Because of the quick response required, Polimeri Europa designed and set up a very reliable emergency reactor safety discharge system. It is of proprietary design and is recommended for dilution and quenching of the gas in order to avoid hydrocarbon vapor cloud ignition.

HIGH-PRESSURE REACTOR TECHNOLOGIES

Two reactor technologies are operated: an adiabatic stirred autoclave and a tubular reactor jacketed with a cooling medium (Fig. 14.5.3). High-pressure reactors have very short residence times, resulting in a very fast transition between different reactor grades. The process therefore is very flexible, and it allows a large number of product types to be made in a single reactor.

The polymer is obtained in molten form. Both reactor processes are operated cross-flow-wise by injections of fresh feed streams together with catalyst. The polymer concentration growing along the length of the reactor and the resin quality are maintained at the proper operating temperature by means of the peroxide types based on their specific decomposition temperature and their reaction rates. To control molecular weight, polar modifiers or aliphatic hydrocarbons are fed into the monomer stream.

The autoclave reactor process was the first in commercial operation, and the reaction is carried on in a stirred vessel that operates adiabatically without significant heat removal across its wall. The reaction mixture is precooled in a double-pipe heat exchanger to exploit the maximum adiabatic potential before the reactor. Temperature control is achieved by injection of fresh cold ethylene into the different zones, each running at different temperatures, setting up a designed temperature profile. Different peroxides are injected in the zones to fit the temperature profile for the running grade; they are dissolved in a hydrocarbon solvent, like a carrier. It is important that the initiators are consumed completely in the reactor because otherwise the reaction could continue producing upsets in the process and potentially dropping the quality of the products. The autoclave reactor has an agitator for good mixing, and it performs as an adiabatic series of continuous stirred-tank reactors (CSTRs). The electric motor driving the agitator is built inside the top zone of the reactor and is refrigerated by part of the ethylene feed stream. The elongated form of the reactor combines manufacturing requirements with a functional design. It is a long barrel with a length-diameter ratio close to 12; there are also internal sets and baffles that divide the reaction path length into zones at proper temperatures.

The volumes in the autoclave reactors can vary between 1000 and 2000 liters for the more powerful reactors, with a residence time in the range of some tenth of seconds. Thermoelements and pressure-relief devices installed in trough holes in the reactor walls are designed for monitoring and control of operating parameters: temperature and pressure ranges between 180 and 300°C and 180 and 220 MPa, respectively, and a pressure under 100 MPa for LLDPE copolymerization.

An excess of fresh ethylene is used to remove the heat generated by the exothermic polymerization reaction. The conversion of ethylene to polymer can be roughly calculated by

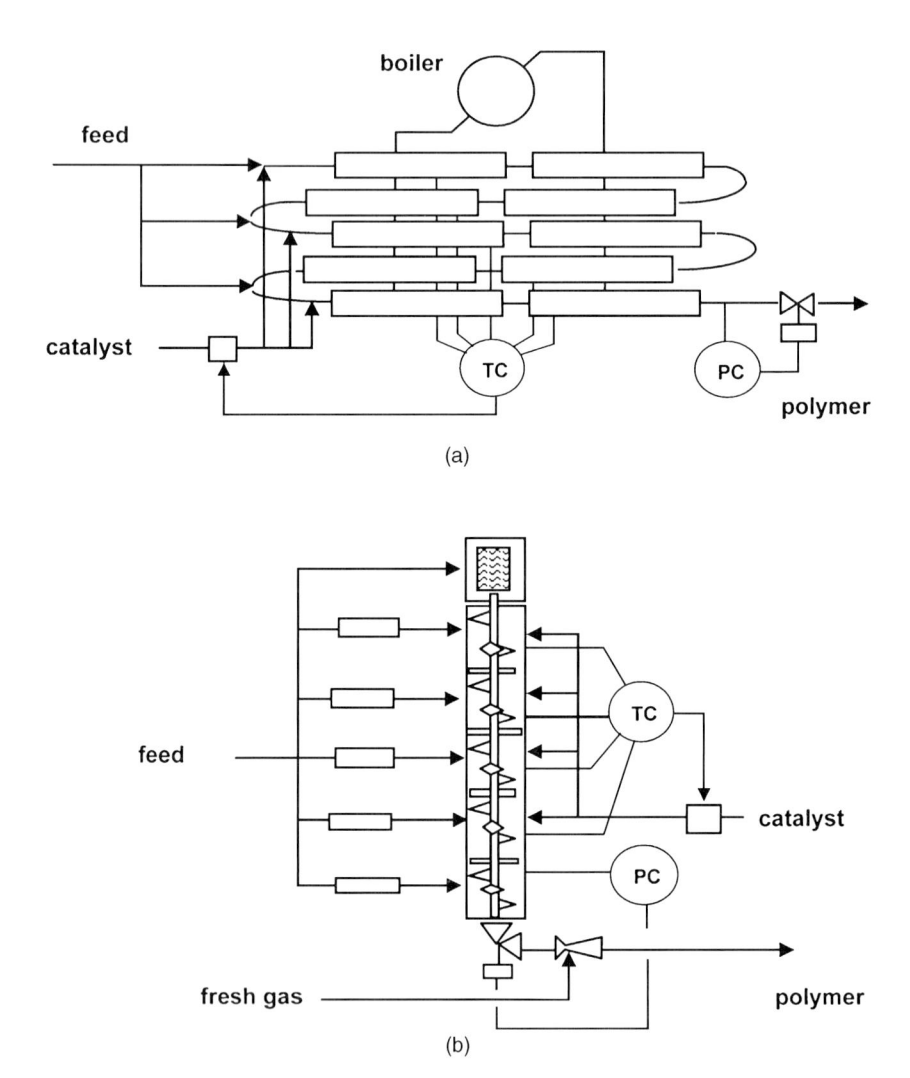

FIGURE 14.5.3 Polymerization reactors: (*a*) tubular reactor; (*b*) autoclave reactor.

the adiabatic temperature difference across the reactor, and it is close to 20 percent per pass. Because of the sensitive ethylene reverse Joule-Thomson effect, the polymer mixture heats up, but a jet pump, an ejector of proprietary design, provides to the polymer stream's cooling, contributing at the same time to energy savings in the process. Part of the gas, in fact, represents part of the fresh gas that has to be compressed.

In the tubular reactor process, the reactants are cooled in a long pipe by water in a jacket. Once the reactor is in operation, there is a net generation of steam; a boiler provides the steam production by boiling the hot water circulating in the reactor jackets from the polymerization heat removal equipment.

The mechanical design of the process side of the tubular reactor consists of pressure pipe with internal diameters mainly of 3 to 5 in having TSDs for more than 350 MPa pressure and thickness close to $1/2$ in. The reactors are built of tubes that are each 10 to 15 m long in a serpentine-like structure within a concrete bay.

The operating level ranges between 250 and 300 MPa at a peak temperature of 320°C. Part of the tube is devoted to preheating the gas to allow the peroxide to lift off and start the reaction. Several temperature peaks are designed into the temperature profile as the reaction proceeds, each one with fresh ethylene injection and a proper peroxide and oxygen balance. The end of the reactor acts like a cooler, controlling the temperature in the high-pressure separator (HPS).

Thermoelements are installed along the length of the reactor to follow the progress of the polymerization reaction. The temperature control of the polymer chain length does not provide enough freedom to tailor polymer properties. Therefore, a chain-transfer agent is necessary. Typically, polar modifiers are used, whereas at high polymerization temperature, even normally inert aliphatic hydrocarbons can be used. The conversion basically results from the temperature difference between the inlet and the outlet of the reactor and from the temperature peaks and cooling cycles, which provide benefits in temperature difference at the walls of the tube; thus the trend toward ever more reactive and multifunctional peroxides can contribute to revamping the capabilities of existing reactors.

Because of the heat transfer through the walls of the reactor, the tubular reactor has a higher conversion to polymer than the autoclave reactor, and conversions of up to 30 percent are achieved easily. The conversion to polymer influences the properties of the product. At higher conversion, the degree of branching increases despite resin clarity. The operating pressure is controlled by a valve at the reactor outlet, resulting in a wide pressure drop of hundreds of bars along the almost 2500 m of process tubing.

DETAILED PROCESS DESCRIPTION

The essential process design is summarized in Fig. 14.5.4 for both reactor types. There are no real differences in the schemes between the two reactor models apart from the inherent requirements of the different polymerization situations and the specific facilitites necessary for their operation. Mainly, some differences occur for historical reasons, resulting from the backrounds and experiences of the different licensors.

Because of the high operating pressure used, special equipment and technology are required. Key operating characteristics and design details generally are treated as proprietary information. The design standards for the reactors are related to thick-walled vessels and tubes. The high pressure requires the use of reciprocating compressors and pumps. The most typical and important compressor used in the high-pressure process is the two-stage hypercompressor with interstage gas cooling. It provides the reaction gas to the reactor compressed to a very high pressure (200 to 300 MPa for LDPE and 50 to 80 MPa for LLDPE), receiving the fresh ethylene makeup from the plant battery limit and collecting the unreacted gas recycle stream.

Booster and primary compressors, typical alternative multistage compressors with interstage gas cooling, raise the ethylene pressure to 30 MPa. The number of compression steps depends on ethylene supply conditions. Because of the need to remove the heat generated by the exothermic reaction, there is an exess of ethylene in the reactor. Unreacted gas is recycled back to the process, combined with the fresh ethylene at the discharge of the primary compressor, and fed to the suction of the hypercompressor at the 30 MPa.

Peroxide initiators or catalyst systems are adjusted to reaction levels; the amount added is controlled by adjusting the speed of the high-pressure pumps. The reactor pressure is reduced by a high-pressure letdown valve to 25 to 30 MPa. The temperature rises as the

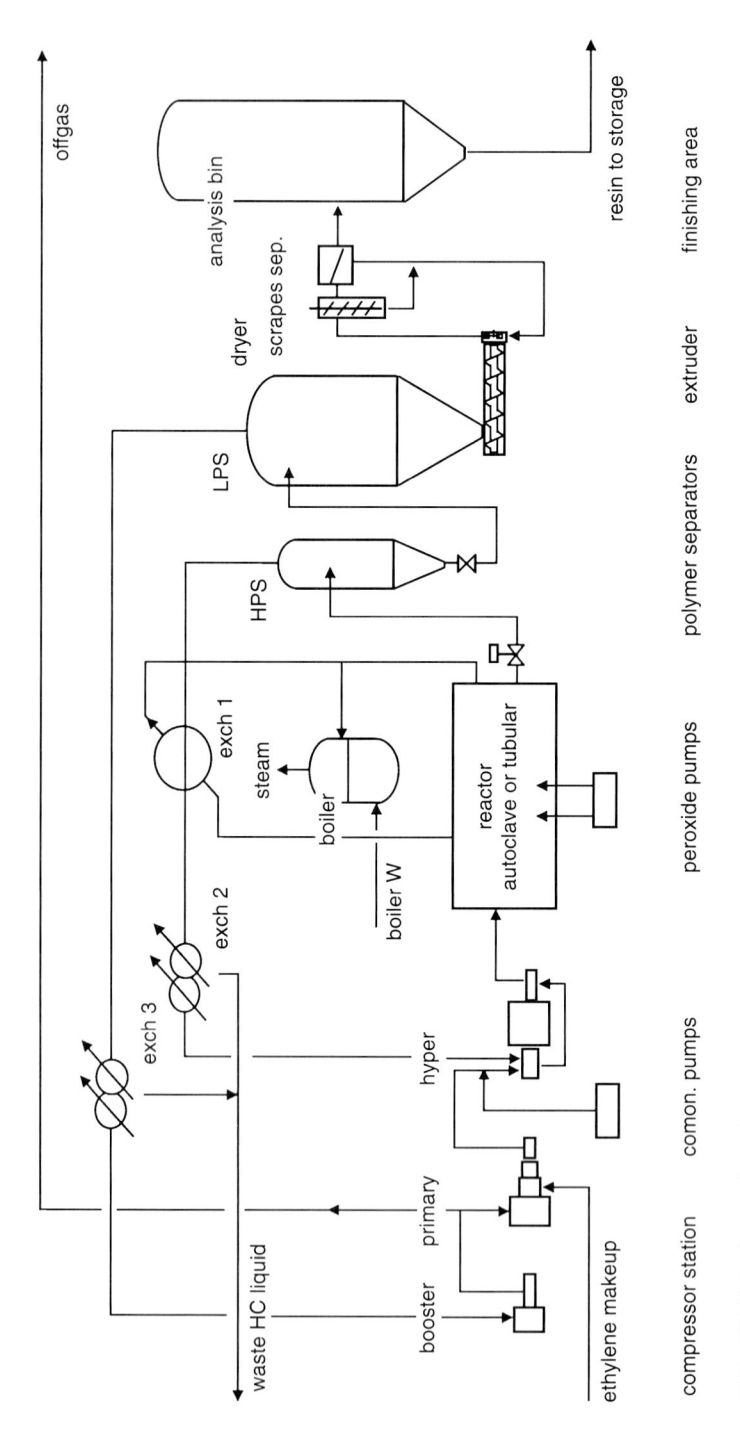

FIGURE 14.5.4 Process flow diagram.

pressure is reduced (Joule-Thomson effect). However, ethylene decomposition is the major limit in the exit-temperature monomer and polymer reaction mixture and thus controls the chemical conversion of the reactor.

The polymer and nonreacted gas are separated by adiabatic flash in a high-pressure separator (HPS) operating at 25 to 30 MPa. The gas stream from the HPS is cooled in a series of cooling water coolers. Part of the heat of reaction can be recovered in this section to generate low-pressure steam, significantly improving the energy efficiency of the process. A smaller separator that removes the waxy olygomers from the recycled gas follows each cooler.

An additional flash step is designed to remove the dissolved gas nearly completely from the molten polymer (the LPS) at 0.15 MPa and is operated as an extrusion hopper. This final pressure step is a compromise between a low level of residual volatile content in the product pellets and compression energy costs. The gas separated from the polymer in these two separation steps is also recycled back to the process, compressed up to the supply pressure level of fresh ethylene.

The low-pressure separator (LPS) is directly above the hot melt extruder. The melted polymer flows directly into the extruder screw and is fed to the underwater pelletizer.

A short single-screw extruder with an underwater pelletizer is recommended; it allows the operator to maintain the resin quality obtained in the reactor, combining lower investment cost with energy savings. Additives can be combined with the molten polymer in the extruder. After pelletizing, the product is dried, classified for scrapes separation, stored temporarily, and tested for quality in analysis bins. The resins can be blended in specially designed silos for lots homogenization to smooth the effects of small operating conditions. Because of the solubility of the ethylene in the melt polymer, to prevent the risk of fire in the silos, degassing operations are necessary after extrusion by venting the resin in air in specific silos before storage, bagging, or bulk loading.

REACTOR SAFETY DISCHARGE SYSTEM

In the high-pressure PE reactor, sometimes runaway reactions can occur with very fast increases in temperature and pressure; thus ethylene decomposition could take place. In the case of decomposition, or in the case of uncontrolled overpressure in the reactor, rupture disks and safety valves break or open, and gases are vented directly to the atmosphere through stacks. The hot gas mixture is dispersed into the atmosphere, and explosive conditions could be met.

To avoid such a situation, a protective system has been designed with the objectives of blanketing each stack of the reactor with nitrogen and injecting cold water for diluting and quenching the hot gases released from the high-pressure vessel in the stack. This excellent system, a proprietary design, has been used for a long time and is shown in Fig. 14.5.5.

Both stacks are connected to the reactor and are occluded by a rubber balloon inflated with nitrogen. The water vessel is under nitrogen pressure. When a rupture disk bursts, the pressure detector in the related stack notes the information and sends a signal to the controllig electric module. The module orders a valve to open very quickly, and water rushes through the injection devices to cool the gases. This happens very fast, in about 20 ms, fast enough to allow dilution and cooling of the hot hydrocarbon cloud before it reaches the outlet of the stack. The failure loop is managed by the controlling electronic module. A photograph of a typical plant is presented in Fig. 14.5.6.

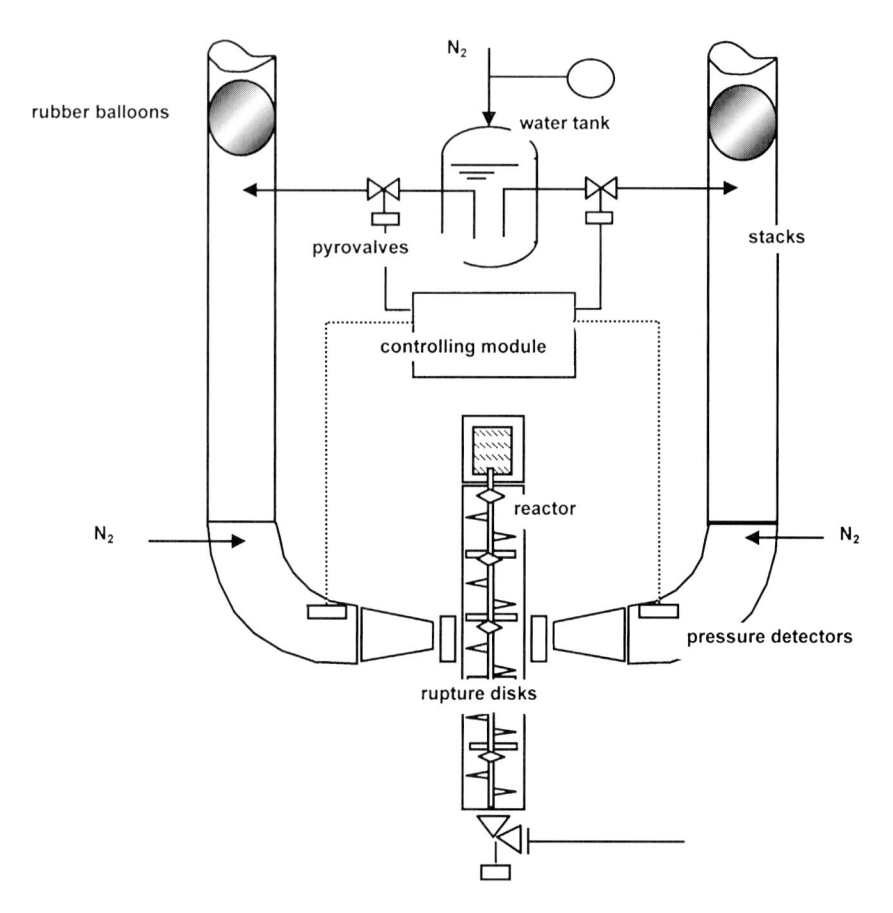

FIGURE 14.5.5 Emergency SD system.

FIGURE 14.5.6 Ferrara plant.

TABLE 14.5.1 Plant Consumption Figures (Average per Ton of Resin Mix)

	LDPE (tubular)	LDPE (autoclave)	LLDPE
Net raw materials, kg	1009	1009	1014
Utilities			
Electrical energy, kWh	900	970	640
Thermal energy, kg/FOE*	7.5	6	5
Steam production, t	1	0	0
Cooling water (circulating), cm	200	200	200
Nitrogen, N·cm	10	10	10
Catalyst, €	3	5.2	2
Chemicals and lubricants, €	4	4.7	6.5
Additive package, €	4.5	1.3	0.3

*kg/fuel oil equivalent, at 104 kcal/kg.

PROCESS PERFORMANCE

High conversion rates and large-scale reactors, specific equipment and machinery design, and proven operating procedures allow a minimization of the net monomer consumption and energy requirements. Assuming the reference ethylene specification, typical plant consumption figures are reported in Table 14.5.1.

The selectivity of the Ziegler-Natta catalyst and the massive use of a comonomer that increases the process compressor output and drops the operating reaction pressure to 80 MPa have a major impact on reactor capabilities, increasing resin production and cutting drastically plant energy consumption to half that of LDPE production. The production of all these copolymers requires some adaptations to be made to the LDPE base process, and this is a chance to restructure and revamp an existing LDPE plant.

PLANT BATTERY LIMITS

Modern crackers produce ethylene at sufficient purity to be used in the high-pressure process without the need for additional purification. The fresh ethylene normally can be delivered to the plant by a pipeline grid or directly from the cracker as gas or liquid at 99.9 percent purity and at between 1 and 10 MPa.

Chain-transfer agents or comonomers in general are supplied at chemical grade after control for approval. A small offgas stream is sent back to the cracker or to a dedicated purification unit to limit the buildup of impurities in the process.

Process offstreams are limited to waste and water. High-pressure PE plants produce only small quantities of waste. Scrap polymer formed during startup of the extruders, fines, and clusters are rejected during the classification of pellets; polymer pellets purged from conveying lines and packaging materials from incoming raw materials are the major non-hazardous wastes.

Potential hazardous wastes may contain only waste solvent, waste oil, and polymer waxes produced in the reactor and separated in the gas recycle stream after cooling. It is recommended that the waste solvent and oils be used as cracker feedstock or fuel. Sometimes concentrated polymer waxes can be sold as a by-product.

The plant is a small process water consumer. Process water and wastewater are restricted to the amount needed for steam production, cooling water towers, if any, and pellet cooling water systems.

Dust and particles in the air for pneumatically conveying pellets, which mainly contribute to the dust emission, can be minimized by applying such techniques as dense phase conveyors and cyclone and/or filters together with good operating practices.

Fugitive emissions are minimized by using specific equipment design and proper engineering STDs, providing fugitive loss assessment and equipment monitoring techniques.

Resin pelletized at a relative high temperature still contains residual monomer, comonomer, and solvent. Therefore, continuous organic emissions can occur during the finishing operations and in the blending silos and the first stage of pellet storage. This potential emission source is reduced by different techniques that attempt to minimize the hydrocarbon level in the polymer during processing. Thermal and catalytic incineration techniques or the option to send this stream to a furnace should be considered for treating VOCs in air purges coming from the silos.

CHAPTER 14.6
BASELL *HOSTALEN* TECHNOLOGY FOR BIMODAL HDPE PRODUCTION

Dr. Reinhard Kuehl and Gijs ten Berge
Basell Polyolefine GmbH
Frankfurt, Germany

GENERAL PROCESS DESCRIPTION

The *Hostalen* process technology is designed to produce from ethylene monomer (and butene comonomer) monomodal or bimodal high-density polyethylene (HDPE) of excellent quality. The process consists of two polymerization reactors that can be operated in parallel (unimodal product) or series (bimodal product); switching between the two modes requires only a short changeover time. Catalyst is injected in the stirred-slurry reactors, where the liquid phase consists of hexane as the suspending agent. After the reaction, the polymer is separated from the slurry and dried. The polymer powder is then transported to the additivation and extrusion section.

PROCESS CHEMISTRY

Reaction Mechanism

The *Hostalen* process is based on a Ziegler reaction mechanism with the so-called Ziegler catalysts. These catalysts are produced with $TiCl_4$ and $Al(C_2H_5)_3$ according to the following formula:

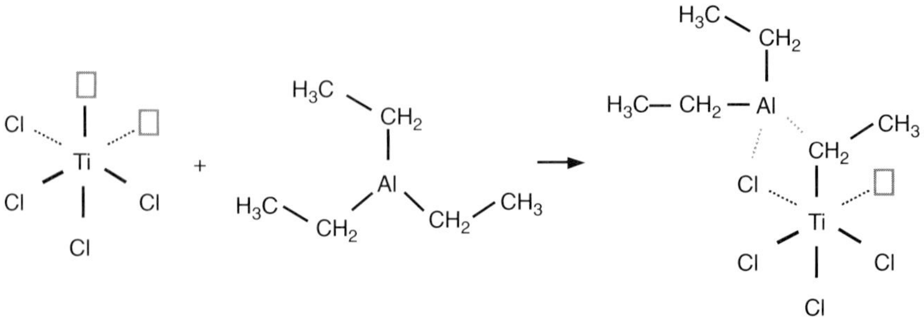

14.72 POLYETHYLENE

The residual compound is an octahedron Ti complex with a coordination gap that acts as an active center in which the polymer chain is initiated. This coordination gap is a vacant ligand position able to coordinate with a monomer molecule (shown in the next formula). This is the starting point for a polyethylene chain.

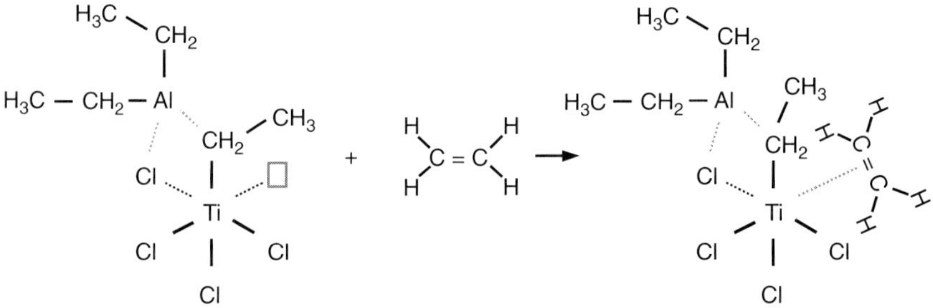

The chain propagation is an insertion process where the monomer molecule is bonded to the Ti complex and then inserted between the polymer chain and the metallic active center. This step is repeated many times. For simplification, the cocatalyst with the Al atom is not shown in the next formula.

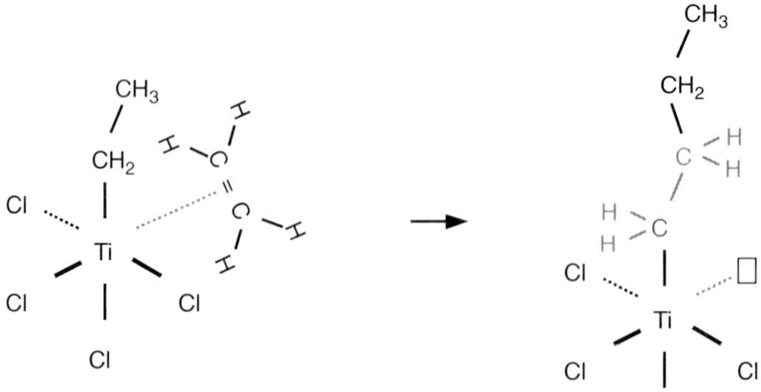

During the chain propagation step, there is also a chain transfer step that takes place in which the polymer chain is removed from the Ti complex and bonded to a monomer molecule or to the Al-alkyl. Both of the following reactions are possible:

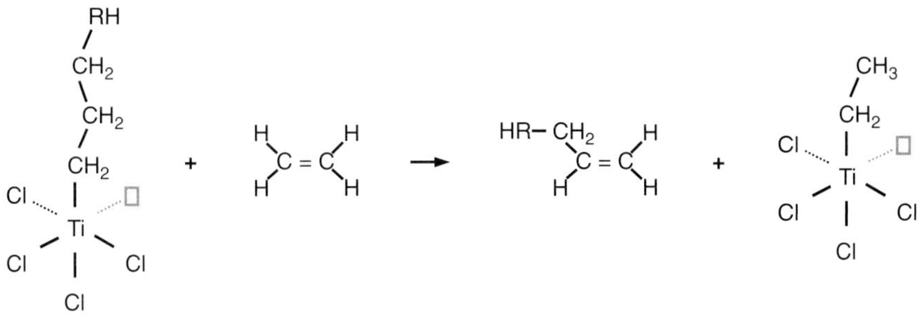

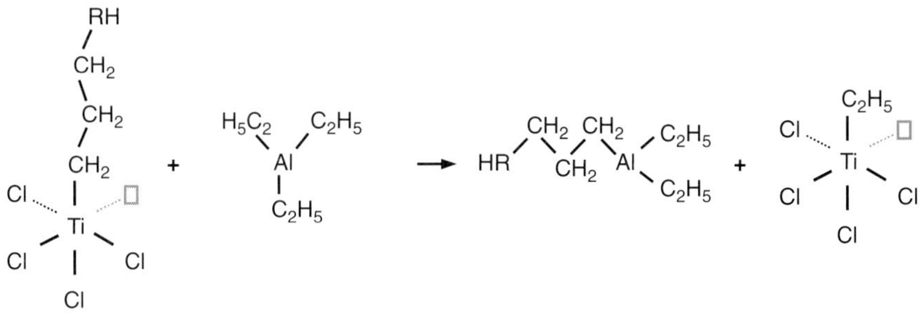

There also can be a deactivation step in which the polymer chain is terminated. The normal termination step is forced by adding hydrogen, which is bonded to the Ti complex and the polymer chain. The polymer chain is terminated and has reached the final molecular weight:

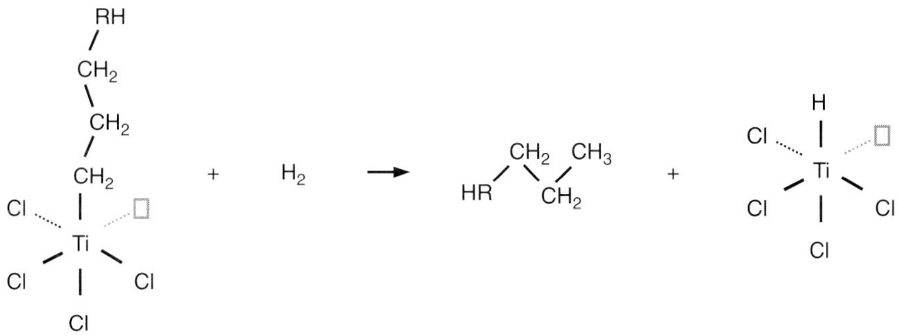

It is now apparent that hydrogen is used as a molecular weight controller, but in parallel, hydrogen is deactivating the Ti complex. To produce a low molecular weight, the hydrogen concentration has to be increased, and vice versa.

The main characteristics of a Ziegler-type catalyst are the productivity and the hydrogen response. The third generation of the catalyst displays an excellent hydrogen response combined with high productivity. This leads to a wide product range with custom-tailored products.

HOSTALEN PROCESS PERSPECTIVE

Basell's *Hostalen* process, the leading slurry-cascade technology, delivers products with outstanding HDPE bimodal molecular weight distribution. With total worldwide installed capacity now over 3 million metric tons per annum (MTA), customized HDPE products can be developed jointly with Basell licensees according to local market demands.

The *Hostalen* process has been scaled up continuously from the original 16-m^3 reactors to a current reactor design volume of around 200 m^3, facilitating a capacity of up to 350 kMTA. The extrusion technology also has developed accordingly, optimizing homogenization performance for bimodal product requirements.

Today's *Hostalen* process is the result of 40 years of continuous research and development that has resulted in ongoing catalyst, process, and product improvements.

An Overview of *Hostalen* Process Development

In 1953, Professor Karl Ziegler and his colleagues discovered and produced the first linear polyethylene chain in a low-temperature and low-pressure process by using a transition-metal catalyst in conjunction with an organoaluminum as cocatalyst. The first discontinuous production plant was constructed in 1955 using BC catalyst, the first Ziegler catalyst for the industrial production of polyethylene.

Since then, process and catalyst technologies have evolved considerably. Following development of the first high-mileage catalyst for the *Hostalen* process in 1969, Hoechst, one of the Basell predecessor companies, produced in 1971 the first bimodal HDPE in a discontinuous process, the so-called *Hostalen* K0 process.

Startup of the first *Hostalen* technology production plant using hexane as diluent and large reactor technology took place in 1975.

In the 1980s, the discovery of a second-generation high-yield catalyst facilitated continuous cascade production of unimodal and bimodal HDPE in a single plant. To improve product quality and increase the mechanical properties of the film, blow-molding, and pipe HDPE resin, the extrusion process was developed continuously and resulted in 1989 in the installation the first modern extrusion line with gear pumps.

From 1990 to 1995, an intensive research program developed a third-generation catalyst, *Avant* Z 501, and also led to a new dimension of pipe application, the PE 100 class. The start of the catalyst production plant in 1995 was combined with the startup of the new continuous polymerization and extrusion plant to produce best-in-class PE 80 and PE 100 pipe resins.

Today's generation of catalysts further enhances the capabilities of the *Hostalen* process, and ongoing developments are expected to deliver a single catalyst-cocatalyst system that can continuously enhance the range of HDPE applications.

Key Characteristics

The *Hostalen* process is a high-tech slurry-cascade process delivering high-quality multimodal HDPE products. A key feature is an excellent combination of processability with final product performance.

Depending on product needs, the *Hostalen* process can be operated using different setups of the two reactors. For unimodal HDPE, the reactors can be operated in parallel or in cascade under equal conditions; for bimodal HDPE, the reactors are operated in series under different reaction conditions (cascade). Both modes can achieve the same capacity, and the modes can be switched in a short transition time.

Safety and Loss Prevention. Plants using the *Hostalen* process can meet all environmental and safety requirements. The polymerization reaction takes place in hexane diluent under nitrogen blanketing at low pressure and low temperature.

Reducing Resource Intensity. High-mileage catalysts, the absence of undesired by-products from the reaction, and a closed-loop hexane system during polymerization help to reduce resource consumption and lower emissions from the *Hostalen* process.

Favorable Production Costs. High conversion rate up to 99.5 percent for ethylene; butene recovery and hexane recycle lead to minor losses on comonomer and diluent. Low catalyst cost because the catalyst preparation unit for second-generation catalyst THE/THB and THT may be included.

Design Flexibility. Single-line capacities are from 40 to 350 KTA. The maximum capacity of 350 KTA single line is restricted only by the extrusion section.

Process Versatility. Easy operability with a broad range of HDPE resins for the all HDPE applications. The *Hostalen* process allows the selective incorporation of comonomers into the high-molecular-weight phase and an extreme splitting of the two reaction stages, which enables the flow rate ratio (FRR) to be tailored to the needs of the customer.

Product Quality. Unmatched HDPE quality and product consistency due to excellent process stability.

Economical. The *Hostalen* process has cash cost advantages over other processes for bimodal HDPE production. There is no need for a monomer purification unit if the monomer is delivered from a state-of-the-art cracker.

PROCESS DESCRIPTION

The *Hostalen* process involves the following stages (Fig. 14.6.1):

- Catalyst preparation and feeding
- Polymerization
- Powder drying
- Extrusion and pellet handling
- Hexane recycling
- Butene recycling
- Tank farm

Catalyst Preparation

Production of THT/THE/THB catalyst is performed batchwise from four commercially available catalyst components in one catalyst preparation vessel under precisely defined conditions. Finished catalyst batches are transferred into catalyst dilution vessels and further diluted to the correct concentration.

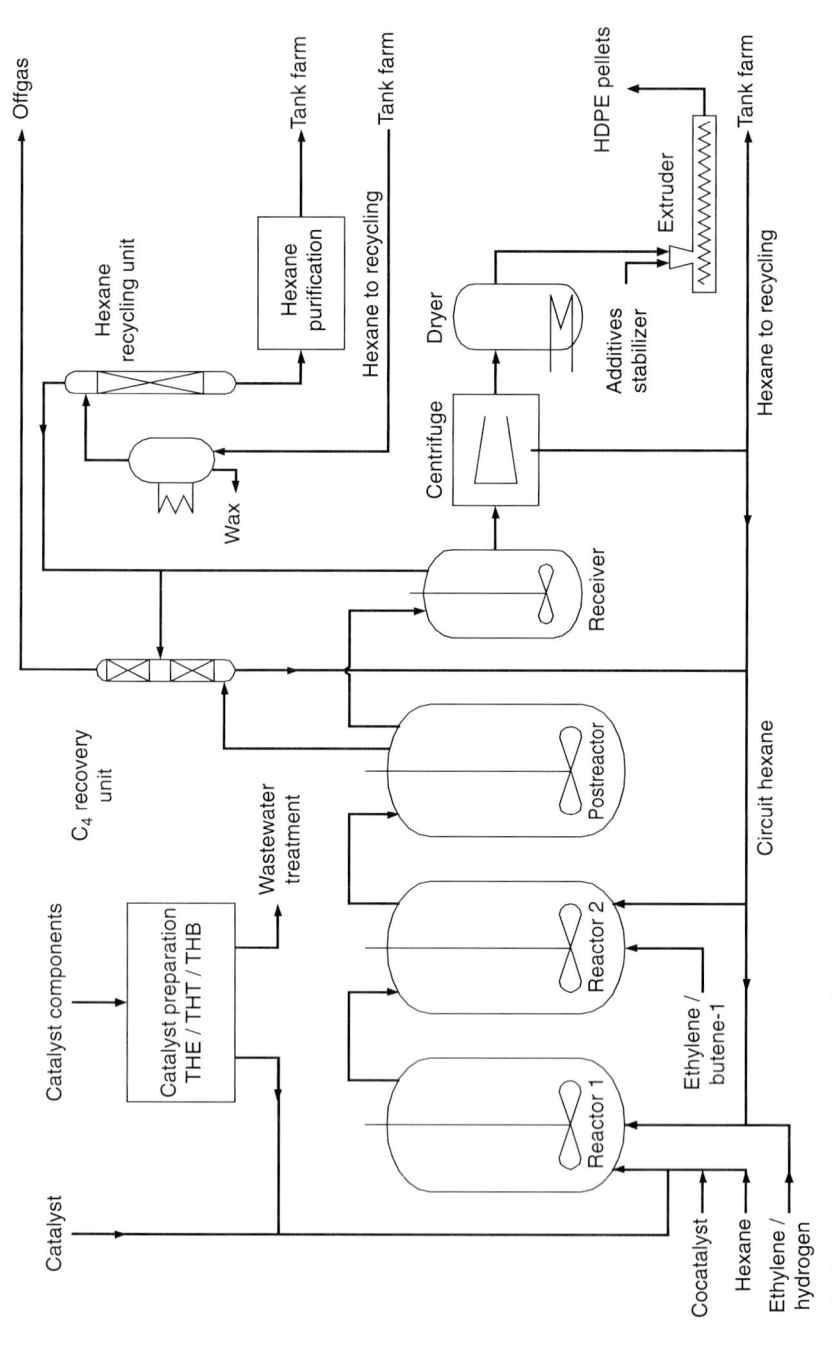

FIGURE 14.6.1 Simplified process flow diagram.

14.76

The dry *Avant* Z 501 catalyst, supplied by Basell, is filled directly into the catalyst dilution vessel and resuspended with hexane. From the dilution vessels, the catalyst suspension is fed via catalyst-dosing devices directly into the polymerization reactors.

Polymerization

The reactors are continuous-stirred-tank reactors (CSTRs) and are operated under different conditions and residence times. The main variable to determine reactor size is the space-time yield of the catalyst system. The reactor slurry is well homogenized by an agitator to guarantee the same reaction conditions all over the vessel. It is fed continuously with a mixture of monomers, hydrogen, catalyst/cocatalyst, and hexane recycled from the process. In the reactors, the polymerization of ethylene occurs as a strongly exothermic reaction at pressures of 5 to 10 bar and at temperatures ranging from 75 to 85°C. Polymerization heat is removed in a so-called outer-cooler system by means of cooling water. The reactive slurry is pumped through this outer cooler via a centrifugal pump. The polymerization, product properties, and yield are controlled by measuring the concentration of the monomers, hydrogen, and aluminium-alkyl inside the reactor and feeding them according to the recipe values. Average molar mass, molecular weight distribution, and polymer density in the reactors are determined by the catalyst system and by the concentrations of catalyst, comonomer, and hydrogen for the different PE grades.

BM Process Mode. The reactors are set up in cascade, with different conditions in each reactor, including low hydrogen and high comonomer content in the second. This allows production of HDPE with a bimodal molecular mass distribution and a defined comonomer content in the high-molecular-weight HDPE chains.

K2 Process Mode. The reactors are set up in a cascade with different reaction conditions in each reactor. This allows the production of HDPE with a broad molecular mass distribution and a defined comonomer content in the high-molecular-weight chains.

K1 Process Mode. Reactors run in parallel mode, i.e., under identical polymerization conditions. This process is used if a fairly narrow molecular weight distribution is required, e.g., for injection-molding products.

Polymer suspension or slurry flows from the reactors into a common postreactor where a final polymerization takes place, resulting in a conversion rate of more than 99 percent of the monomers used, simultaneously reducing the monomer loss. The suspension is then pumped to a suspension receiver vessel.

HDPE Powder Drying/Diluent Separation

Suspension leaving the receiver is separated in a decanter centrifuge into a liquid and a solid fraction. The solid part, which includes some residual moisture, is fed to a fluidized-bed dryer operated with nitrogen. The liquid part (recycle hexane) goes back to the reactors. In the dryer, the residual hexane is evaporated and then recovered in a condenser system. After condensing, the nitrogen is reused in the fluidized-bed dryer by a closed-loop system.

Dried HDPE powder passes through a sieve and is pneumatically conveyed by nitrogen directly to the extrusion section. For reliability reasons, usually powder silos are installed upstream the extruder.

Extrusion and Pelletization

From an intermediate storage bin, the powder is dosed into a mixing screw together with proportionally controlled stabilizers and optional fillers, pigments, etc. From here the mixture is fed to the extruder. Energy for processing the material in the extruder is introduced by the main motor through screw speed and the design of the screws. This extrusion know-how—especially for the bimodal products—is very important. In the mixing section of the extruder, the powder mixture is compressed, melted, and homogenized.

The melt is fed to an underwater pelletizer by a gear pump and then pressed through a die plate and cut into small pellets by rotating knives. The pellets are cooled in a bypassing water stream. Subsequently, the pellets are dried and pneumatically conveyed to the homogenization silos. This homogenization step is necessary owing to the request for high-quality consistency for the product application. After the final quality check, the lot of product is released and forwarded to the logistics center.

Hexane Recycling Unit

In the reactors, only part of the recycle hexane is used again. The other part has to be cleaned to prevent the accumulation of impurities, which are produced in side reactions or introduced with the feed streams. This hexane cleaning system consists of a two-step distillation process. In the first step, the hexane is evaporated, and the remaining heavy ends are removed and fed to a further workup section. The overhead stream of evaporated hexane is further treated in a distillation column. The light ends are separated from the hexane and fed to the butene recovery unit. Finally, the hexane is dried in adsorbers and fed back into the *Hostalen* process.

Butene Recovery Unit

Tail gases from the postreactor and light ends from the hexane distillation unit are fed to a scrubber operated with purified hexane. The unreacted comonomer is recovered and recycled to the reactors. Remaining offgases can be guided to a boiler station for energy recovery or burning in a flare.

Tank Farm

The tank farm consists normally of tanks for recycle hexane, for purified hexane, for comonomers, and for cocatalysts. The size of the hexane tanks depends on the availability of fresh hexane on the market and on the total hexane holdup of the polymerization area. The size of the recycle hexane tank should be big enough to buffer the holdup of the plant in case of a shutdown. The tanks for cocatalysts are equipped with a special unloading station for the Al-alkyls. After the unloading, the cocatalysts are diluted to a concentration that can be used inside the plant. There has to be a safety distance from the tank farm to the process equipment and a special safety distance from the Al-alkyl tank and unloading station to the process and the tank farm.

Safety and Environment

Hostalen process plants (see Fig. 14.6.2) are built according to Basell safety design criteria and must undergo safety audits prior to commissioning and startup. All licensed sites receive visits from a team of specialists with experience in diverse fields, including

FIGURE 14.6.2 Polymerization and hexane recycling section.

- Safety and loss prevention
- Health and environmental matters
- Process operations
- Instrument, electrical, and mechanical design

This team verifies and ensures that the plant is built according to Basell safety design criteria and assists the licensee with any safety, health, or environmental concerns regarding the process and related facilities.

The use of high-mileage catalysts, the absence of undesired by-products from the reaction, and a closed-loop hexane system in the polymerization contribute to the process being clean and having minimal environmental impact.

Intrinsic Safety of the Process. An advantage of the *Hostalen* process is the self-stabilization and a safe behavior even during and after an emergency shutdown. In case of power failure, no uncontrolled polymerization reaction and no heat production will take place. All operations are controlled from a control room equipped with sophisticated control devices (distributed control system with fully electronic instrumentation).

A fully redundant emergency shutdown (ESD) system and an interlock system are provided to ensure automatic intervention in case of emergency to put the plant in safe conditions without delay due to the operator decision. In addition, the design of each *Hostalen* process plant includes a number of safety features, such as

- Uninterruptible power supply (UPS) for computer control and critical instrumentation control
- Instrument air emergency buffer
- Emergency blowdown system
- Automatic fire protection systems

After an emergency shutdown, the *Hostalen* process plant can be started up without emptying or cleaning of the system and without loss of feed materials. This is an important consideration when the reliability of the electrical supply to the plant is not optimal.

Process Capability

Basell's *Hostalen* slurry process is an excellent PE polymerization technology. Its advantages include the easy and accurate reproducibility of polymerization and the high heat transfer (no "hot spots") together with a unique cooling system. Because of this cooling system, there is theoretically no limitation in the polymerization capacity of the reactor system.

The continuous improvement in development of the catalytic low-pressure slurry process leads to the manufacture of HDPE with bimodal molecular weight distribution. The properties of polyolefins depend mainly on polymer structure. The main relationships are shown in Fig. 14.6.3. The flexibility of the *Hostalen* process facilitates product adjustments and the manufacture of tailored products for different markets.

Unimodal/Bimodal Swing Plant. As described earlier, the *Hostalen* process is able to run different setups of the two reactors, depending on the product needs. For unimodal HDPE, the reactors are operated in parallel; for bimodal HDPE, they are operated in series (cascade) (Fig. 14.6.4). Both modes can achieve the same capacity and can be switched from one to the other in a short transition time.

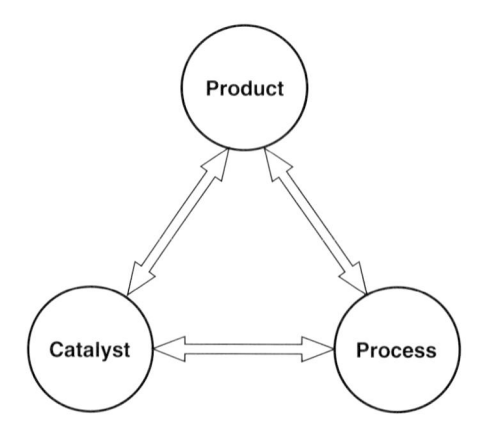

FIGURE 14.6.3 Catalyst-process-product relationships.

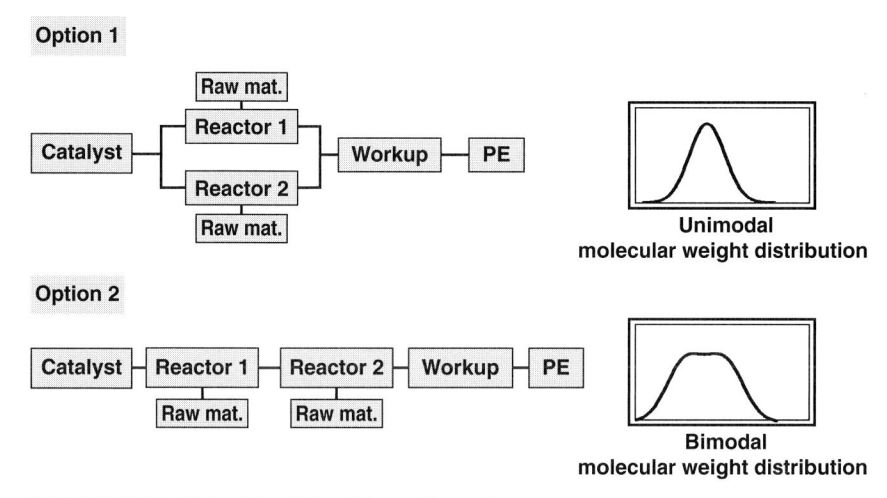

FIGURE 14.6.4 Unimodal and bimodal operation modes.

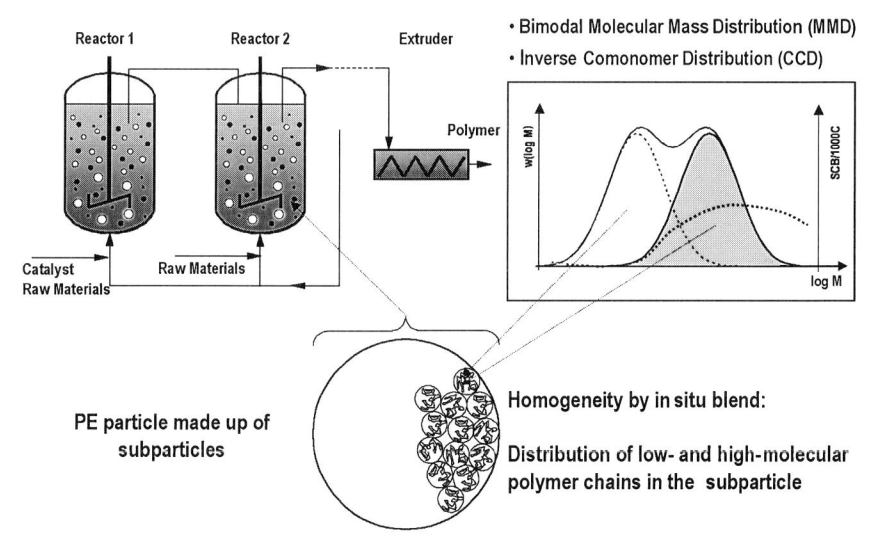

FIGURE 14.6.5 In situ blending of bimodality.

Excellent Homogeneity Based on Reactor in Situ Blend. The development of the catalysts for the *Hostalen* process, especially the third-generation catalyst, provides the opportunity for an in situ blending in the polymerization reactor with excellent homogeneity, which is needed to meet the requirements of outstanding HDPE resins for pipe and film applications (Fig. 14.6.5).

Tailored Product Range. All HDPE grades ranging from low-molecular-mass injection-molding products to high-molecular-mass film, pipe, and blow-molding grades can be produced with molecular mass related densities between 0.940 and 0.965 g/cm^3. *Hostalen* process plants are designed to meet the unique requirements of individual licensees. As a result of the cascade configuration, the process has the flexibility to adjust products according to the licensee HDPE market needs. The optimized balance of properties for a grade can be achieved by selecting the suitable process conditions in polymerization. This gives the licensee an opportunity to follow future market improvements and changes easily.

Plant Size and Scaleup Potential. The *Hostalen* process can be easily scaled up due to its macroscale homogeneity. Reactor size has to be selected to keep the residence time at an appropriate level, as given by the space-time yield. *Hostalen* process facilities have been designed for 40- to 350-kMTA single-line polymerization and single-line extrusion based on 8000 operating hours per year and a standard-grade slate. The limitation on a single-line plant capacity is linked to the extrusion capacity.

PRODUCT RANGE AND APPLICATIONS

Overview

The characteristic properties of HDPE products are

- Low density compared with other plastics (<0.965 g/cm^3)
- High resistance to stress cracking, depending on the grade
- High toughness and elongation at break
- Very good electrical and dielectric properties
- Very good chemical resistance
- Good processing and machining properties

Basic HDPE Structures and Properties

The properties of HDPE are determined primarily by density, molecular weight distribution, and comonomer distribution. With increasing density (higher crystallinity), the following properties also increase:

- Tensile strength
- Elastic modulus (rigidity)
- Hardness
- Resistance to chemicals
- Impermeability to gases and vapors

Increasing molecular weight gives rise to higher

- Impact strength
- Notched impact strength
- Tensile strength
- Elongation at break
- Environmental stress cracking resistance (ESCR)

For unimodal HDPE with increasing density, impact strength and stress cracking resistance will decrease.

BASELL *HOSTALEN* TECHNOLOGY FOR BIMODAL HDPE PRODUCTION   14.83

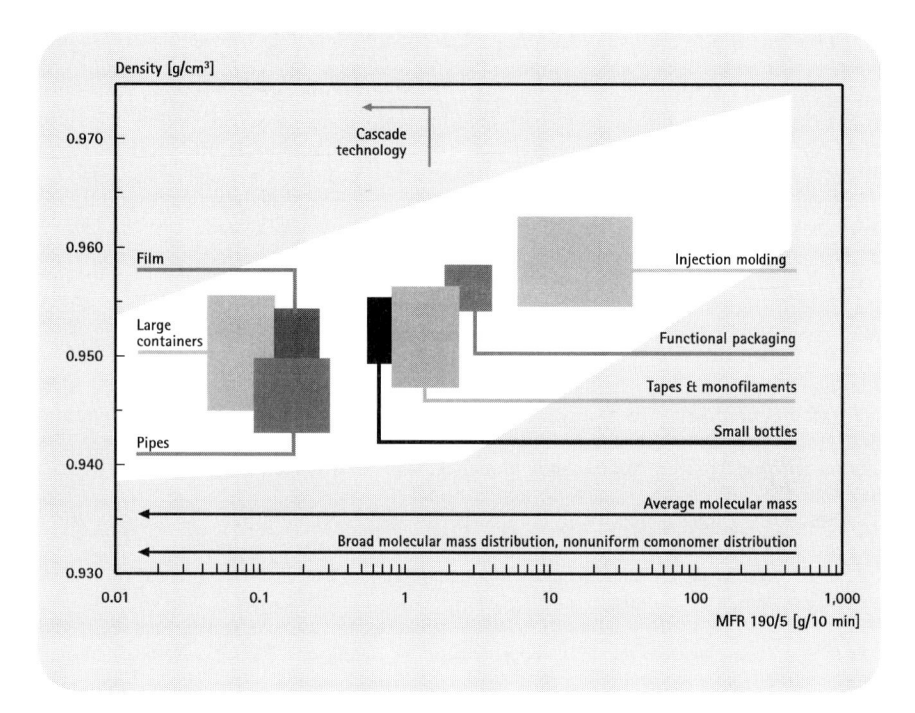

FIGURE 14.6.6 Product range and applications.

The mechanical properties of bimodal *Hostalen* grades are improved significantly as compared with the unimodal ones, e.g., the relation between stiffness and stress crack resistance. PE molecules are branched. The degree of branching in the molecular chains and the length of the side chains have an important effect on the properties of polyethylene. Products from the *Hostalen* process are "linear polyethylene." The HDPE grades produced in the process have a low degree of branching with only short side chains.

The *Hostalen* process offers a broad product portfolio in the density range of approximately 0.940 to 0.965 g/cm^3 (Fig. 14.6.6). The melt flow rate (MFR in g/10 min) varies between 2.2 for MFR 190/21.6 (high-molecular-weight product for large blow-molding grade) to almost 60 for MFR 190/5 (injection-molding grade).

Applications of *Hostalen* Products

Bimodal Pipe Products. Outstanding resin and final part properties can be achieved with bimodal HDPE. The molecular mass distribution of such a grade shows two peaks, with the low-molecular-weight part mainly forming the crystalline and the high-molecular-weight polyethylene mainly forming the amorphous part with high numbers of tie molecules per volume between the crystalline lamellae. If the comonomer (1-butene) is mainly incorporated in the long chains, the bimodal material has a higher density than unimodal HDPE.

These intrinsic "polymer alloys" or "reactor blends" feature a very high stiffness combined with an unprecedented toughness because of the very efficient comonomer incorporation in the long chains. The level of stress cracking resistance increases by several orders of magnitude compared with unimodal grades.

Hostalen PE 80 pipe resins have been used by customers for decades. This leading position was strengthened when bimodal PE 100 pipe resins were commercialized in the mid-1990s by a Basell predecessor company. They fulfill the required criteria easily and with a high safety margin: creep rupture strength, stress crack resistance, and resistance to rapid crack propagation. For the benefit of converters and end users, the prestigious PE 100+ Association approved *Hostalen* CRP 100 as part of its Positive List of Materials.

Hostalen CRP 100, according to Basell's customers, is very resistant to slow and rapid crack propagation. The S^4 (small-scale steady-state) test demonstrates that the critical pressure for bimodal pipes is more than an order of magnitude improved as compared with conventional unimodal pipe resins.

Another important pipe property, the creep behavior under internal pressure, can be pushed beyond the typical limits with a bimodal grade such as *Hostalen* CRP 100.

Bimodal Film Products. Bimodal HDPE resins are used widely to produce thin, stiff, and paper-like films. They have a very low melt flow rate (MFR), high average molecular mass, broad bimodal molecular mass distribution, and nearly linear polymer chains with only a few short-chain branches. The HDPE film thickness is less than half that of LDPE film for the same application.

The HDPE resins typically are converted into thin films by a blowing process. After extruding the polymer melt through a ring die, the tube is stretched and blown up to form the film. Whereas inside the extruder shear properties are most important, behavior under strain is crucial after passing the die.

To achieve high throughput and acceptable backpressure in the extruder, both shear viscosity and elasticity must be low under processing conditions. This favorable rheologic profile can be achieved and tailored by the bimodal molecular design.

In addition to favorable mechanics, bimodal film resins offer converters processing advantages such as bubble stability and extreme downgauging. At higher pressures, the throughput increases by approximately 20 percent or—reverse—at the same output level the machine pressure may be reduced significantly compared with unimodal film grades.

One of the characteristic mechanical film properties is the dart drop (DD) value. Although it depends on surface roughness and other parameters, the DD value indicates the optimal processing window in terms of blowup ratio, throughput, or neck length.

Medium-Molecular-Weight Film Applications. In close cooperation with a film producer, a special medium-molecular-weight HDPE grade was developed for candy wrap. It constitutes the main part of a twist film that is covered like a sandwich in a coextrusion cast-film process. A special stretching process yields a tailored film with good foldability and wrapability and acceptable "deadfold" behavior. On top of these mechanical properties, the resin shows good flowability, low backpressure, high output, and a good film surface.

In comparison with the unimodal predecessor grade, the new bimodal product demonstrates the advantages of the *Hostalen* BM technology and exceeds the complex market requirements regarding processability, stretchability, toughness, fish eyes, and barrier properties. Current applications go beyond cast film into blown film (short neck, blowup ratio <4) and also small blow molding with favorable ESCR at high-duty machines.

Another bimodal film grade was developed for high-performance food packaging films that have to provide an excellent moisture barrier. It is defined by a lower density (0.949 compared with 0.952 g/cm^3) and a more narrow molecular mass distribution (MFR 21.6/MFR 5 is 11 compared with 21). Films of this grade are relatively flexible and transparent with a smooth surface. These films are preferred for cereal packaging at form-fill-seal machines at high speed.

Blow-Molding Products. Blow-molding grades of the *Hostalen* process have been applied successfully over decades from small bottles to Jerry cans, drums, and intermediate bulk containers or heating oil storage tanks. In many dangerous goods packaging materials, *Hostalen* grades were tested superior with respect to cold impact strength, ESCR, and chemical resistance to filling agents.

General requirements for a suitable HPDE grade reflect the two principal steps of the blow-molding process: parison extrusion and blowing/welding. The first step needs excellent processability under shear and stress, as well as a high melt strength; good extensibility and weldability are important in the second.

In other applications, the molecular weight distribution (MWD) is a key element in tailoring blow-molding grades. For smaller bottles, it may be almost unimodal. However, for larger containers, a second mode that can be related to a particular catalyst/cocatalyst system is shown to be relevant. Systematic trials in a cascade operation lead to significant improvements in the stress cracking resistance measured by the full notch creep test (FNCT), whereas most of the other parameters can be kept almost unchanged.

Tapes and Monofilaments. Special *Hostalen* grades for tapes are suitable to run on high-performance film-stretching machines with stretch ratios of up to 8:1. The tapes feature an extremely high mechanical strength and are used for fruit and vegetable packaging or as woven fabrics for agricultural or construction purposes.

Hostalen monofilament grades are used for ropes and yarns in the fishing industry and for geotextiles.

Injection-Molding Products. Customers use *Hostalen* injection-molding grades for bottles, fish crates, bins, caps, and closures and for many other household articles.

PROCESS ECONOMICS

The *Hostalen* process is a reliable and robust stirred-tank slurry process. Compared with other bimodal HDPE processes, the operating costs of *Hostalen* process plants are among the lowest in the industry as a result of

- High plant availability
- High conversion rate of up to 99.5 percent for ethylene
- Rapid grade change
- Short catalyst change downtime
- High percentage of premium grades
- High catalyst yields (e.g., *Avant* Z 501)
- High lot-to-lot consistency
- Low maintenance costs due to low-pressure equipment

Low Catalyst Cost

Licensees usually get the right to produce the Ziegler-type THE, THT, and THB catalysts on site. The simple in-plant fabrication of these catalysts based on commercial raw materials results in low catalyst costs.

TABLE 14.6.1 Typical Specific Consumption/Cost (per 1000 kg of PE Produced)

	Unit	HDPE
Monomer*	kg	1017
Hydrogen	kg	0.7
Catalyst + cocatalyst†	€	2–6
Hexane‡	kg	4 + 3 = 7
Stabilizers†,§	€	4–25
Electric power*	kWh	450–550
Steam	kg	500
Cooling water¶	m^3	180

*Average value, depending on grade slate.
†Depending on grade, cost Germany 2002.
‡Process + catalyst preparation.
§UV and special stabilizer only, carbon black/dyes excluded.
¶At $\Delta T = 10°C$.

Raw Materials

Main feed components into the *Hostalen* slurry process are ethylene, comonomers, and hydrogen. Ethylene from a state-of-the-art cracker can be used in the *Hostalen* slurry process without purification.

Consumption and Utility Figures

Plant design is based on the assumption that the specified raw and auxiliary materials and utilities are available continuously at the battery limit. The consumption figures in Table 14.6.1 are for polymerization and extrusion.

CHAPTER 14.7
BASELL *LUPOTECH* G TECHNOLOGY FOR HDPE AND MDPE PRODUCTION

Cyrus Ahmadzade and Gijs ten Berge
Basell Polyolefine GmbH
Frankfurt, Germany

GENERAL PROCESS DESCRIPTION

The *Lupotech* G process technology is designed to produce from ethylene (and 1-hexene as comonomer) high-density polyethylene (HDPE) or medium-density polyethylene (MDPE) polymer of excellent quality. Single-line capacity of 350,000 metric tons per annum (MTA) is achievable, and process reliability has proved to be very high.

The process consists of a single fluidized-bed gas-phase polymerization reactor. Downstream from the reaction section is a product finishing section, additivation, and extrusion (Fig. 14.7.1).

PROCESS CHEMISTRY AND THERMODYNAMICS

By catalytic polymerization of gaseous ethylene, polyethylene powder, which has a crystalline structure and a melting point of approximately 125°C depending on the type of product, is formed. The polymerization is an exothermic reaction:

$$nC_2H_4(g) \rightarrow 1/n(C_2H_4)_n(s)$$

The polymerization takes place at 20 to 25 bar(g) and about 85 to 116°C depending of the type of polyethylene produced. Copolymerization with 1-hexene can be applied.

For the manufacture of MDPE and HDPE in the *Lupotech* G process, the catalyst is Cr-based and requires a cocatalyst (aluminum–organic component). In the *Lupotech* G process, the catalyst is fed directly into the reactor without using a prepolymerization step.

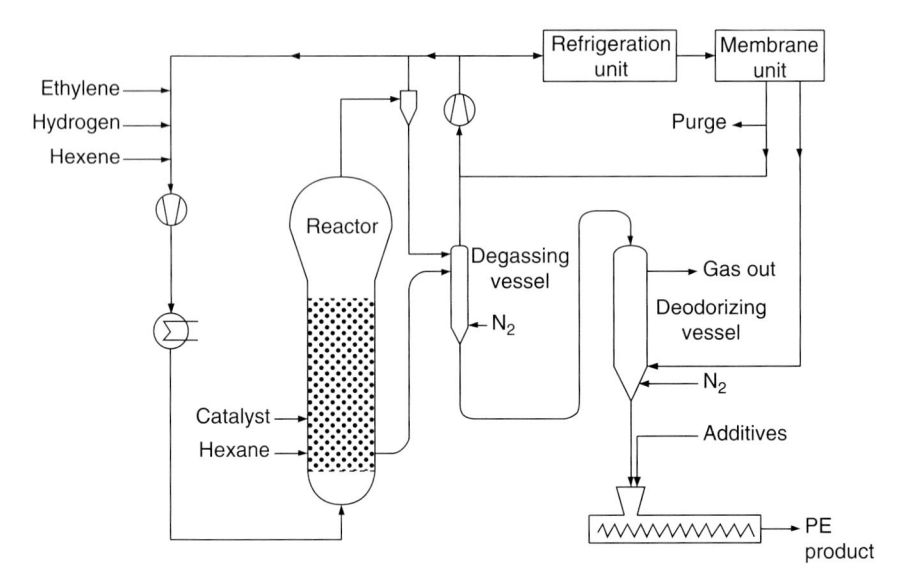

FIGURE 14.7.1 Simplified process flow diagram.

Polymerization occurs at the interface between the solid catalyst and the polymer matrix, which is swollen with monomers during polymerization. From gas-state monomer to solid-state polymer, ethylene experiences a physicochemical transition within a very short time. The polymerization environment changes with the catalyst, reactant composition of the reaction loop, and reactor operating conditions.

The Basell chromium catalyst used in the polymerization process consists of chromium(VI) trioxide chemically fixed on silica gel. This catalyst is produced under oxidizing conditions by heating the silica gels that have been doped with chromium(III) salts (precursor or precatalyst). During this heat treatment, the chromium(III) oxidizes to chromium(VI), the chromium(VI) is fixed, and the silica gel's hydroxyl group is eliminated as water (Fig. 14.7.2).

The catalyst activity is increased significantly by adding small quantities of Al-alkyls (cocatalyst). The time of response decreases, thus leading to a reduction in residence time. Additionally, the catalyst productivity rises (Fig. 14.7.3).

The *Lupotech* G process operates for Cr catalyst at high reactor temperatures near the melting point of the product to improve the driving force for heat removal and to reduce the amount of fines in the bed. The reactor temperature is kept within the darker, rainbow-shaped area, which corresponds to the center explanatory box (Fig. 14.7.4).

In this range, the amount of fines is small, and the bulk density is high. A decrease in temperature would lead to an increased level of fines because the particles become more brittle and more porous. An increase in temperature would cause a loss of fluidization because the PE particles would start to melt and then stick together, causing a shutdown of the process.

LUPOTECH G PROCESS PERSPECTIVE

The *Lupotech* G process was developed during the 1980s by a Basell predecessor company; the first commercial unit was started up in 1992. A second world-scale unit was start-

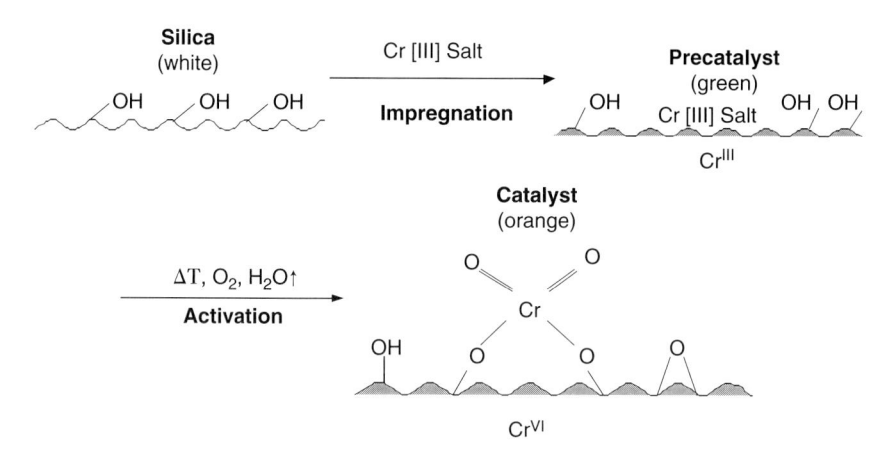

FIGURE 14.7.2 Chromium catalyst activation chemistry.

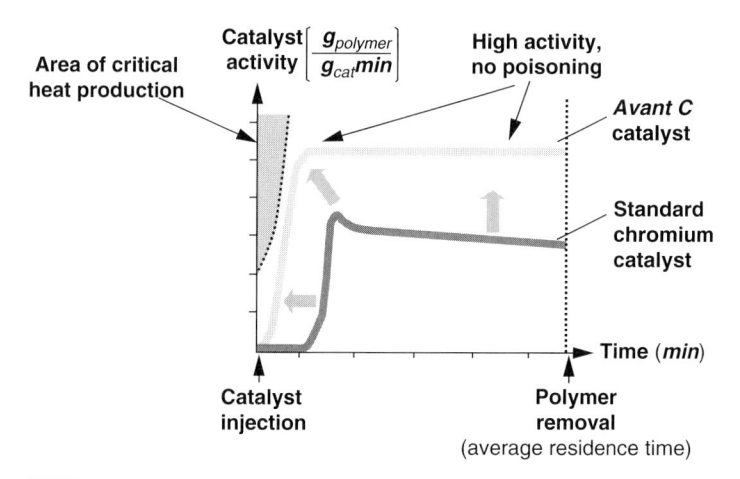

FIGURE 14.7.3 Basell *Avant* C catalyst benefit in operation.

ed up in 2000 (Fig. 14.7.5). The *Lupotech* G process has been available for license since 2000, and two licenses have been sold. The product slate has been increased and improved continuously and includes all MDPE and HDPE grades that can be achieved with a single-reactor gas-phase PE technology. The quality of especially the Cr-based products is very high owing to the excellent *Avant* C catalyst's performance and characteristics.

PROCESS DESCRIPTION

Purification and Dosing of Feedstock

Purification is required if the feedstock cannot meet the quality requirements for the polymerization. The main raw materials used in the polymerization are

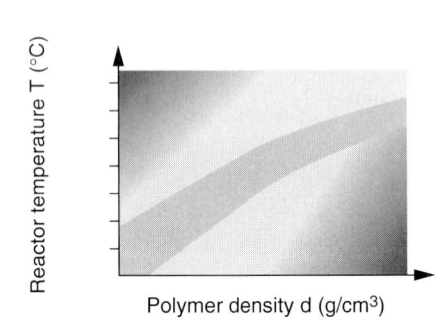

FIGURE 14.7.4 Optimal operating window of Cr catalyst.

FIGURE 14.7.5 Production unit with polymerization section, monomer preparation and recovery unit, and deodorizing unit.

- Ethylene and 1-hexene as monomers
- Catalysts and cocatalyst
- n-Hexane to reduce formation of dust in the fluidized bed and as a solvent for the cocatalyst, antistatic agents
- Antistatic agents to prevent deposits in the cycle gas system

Where needed, 1-hexene is first treated to remove CO; hexane and 1-hexene are purified on a molecular sieve to remove mainly water; and ethylene is purified in several stages using catalysts and a molecular sieve to remove mainly CO, acetylene, and water by physical adsorption, hydrogenation, and oxidation reaction.

Polymerization and Powder Handling

The fluidized-bed reactor is a vertical pressure vessel. A fluidized bed of polymer particles in a gaseous mixture is maintained by a cycle gas compressor. The cycle gas enters the reactor through a distributor plate at the bottom to achieve an even gas flow over the entire cross section. In the upper part of the conical reactor, the gas velocity decreases with increasing diameter of the reactor to keep the particles in the fluidized bed.

The distribution plate at the bottom of the reactor plays an important role in operation of the reactor. To achieve an even gas distribution, the pressure drop across the distribution plate should be in a certain ratio to the pressure drop through the fluidized bed. Since the polymer particles are hot and active, settling must be prevented to avoid agglomeration. Maintaining a sufficient cycle gas flow rate through the distributor plate to achieve fluidization at the base of the bed is very important in fluidized-bed polymerization reactor operation.

The unreacted gas leaves the reactor at the top. Entrained particles are removed by a cyclone to keep the cycle gas loop clean. The coarse fraction is routed back to the degassing vessel. The reaction heat flow plus the compressor power are removed by a cycle gas cooler, and the gas is then routed back to the bottom reactor inlet via the cycle gas compressor.

The raw materials ethylene, 1-hexene, catalysts, hydrogen (if any), and n-hexane, along with cocatalyst and antistatic agents, are fed continuously into the reactor or cycle gas.

Ethane as an inert component is accumulated in the cycle gas/recycle gas system because the ethylene stream fed to the cycle gas loop contains ethane from the cracker. Additional ethane is always generated in small quantities in the ethylene purification column by hydrogenation of ethylene.

A gas-polyethylene mixture is discharged from the reactor via discharge pipes and depressurized in a degassing vessel. Polyethylene powder contains adsorbed HC, mainly C_6. To remove most of the C_6, a continuous stream of nitrogen is sent to the degassing vessel. The gas phase consisting of carrier gas from the reactor and cyclone discharge, stripping N_2 and stripped C_6, is drawn off at the top as recovery gas and compressed in a multistage compressor and recycled partly to the reactor. The other part is sent to a treatment section to condense the C_6 and to separate nitrogen from ethylene to improve ethylene usage. A bleed stream is sent to the flare to avoid any buildup of ethane or other inert components.

Recovery Gas Treatment and HC Recovery

The gas stream drawn off at the top of the degassing vessel enters the recovery gas compressor, which is a multistage compressor. The recovery gas is compressed to reaction pressure and recycled back under pressure control into the reactor cycle gas loop after the cycle gas cyclone after being cooled down in the aftercooler. The condensate is recycled back to the process.

Offgas Treatment

To prevent accumulation of nitrogen gases in the system, a side stream is taken off from the recovery gas. It is refrigerated to recover condensables, which are recycled back to the process. The vapor fraction is sent to a membrane unit to make the separation of nitrogen from the hydrocarbons, mainly ethylene.

Installation of the membrane unit minimizes the specific consumption of ethylene, and by recycling the nitrogen, the consumption of fresh nitrogen is minimized. The hydrocarbon fraction (permeate), mainly ethylene and ethane, penetrates the membrane owing to its selective permeability and is cycled back as feed to the recovery gas compressor.

The other fraction (retentate) consists mainly of nitrogen. The retentate flow is controlled by the concentration of nitrogen in the cycle gas loop. Nitrogen is used as a stripping medium either for the deodorizing vessel or for the degassing vessel if the purity with respect to the level of hydrocarbons is sufficient.

Deodorization

The PE powder from the degassing vessel is conveyed to the deodorizing silos. Before the PE powder enters the deodorizing silos, the conveying nitrogen is separated from the powder.

In the deodorizing silos, any residual hydrocarbons are removed by purging with a mixture of hot nitrogen (and steam, when necessary) after a sufficient residence time. The powder is discharged at the bottom of the deodorizing silos to the extrusion section.

PRODUCT SPECIFICATIONS

The physical and mechanical properties of polyethylene material depend on density, crystallinity, molecular weight and its distribution, and other characteristics, such as swelling ratio. These variables, in turn, are controlled by polymerization conditions and are not mutually independent.

These parameters can be influenced in different ways during polymerization. A measure of crystallinity is density, which, in turn, is determined by the branching of the polyethylene. One general possibility for influencing the density is presented by copolymerization with α-olefins, e.g., 1-hexene, which increases the degree of branching (Fig. 14.7.6).

In commercial copolymers, high-molecular-weight fractions generally contain fewer comonomer units and fewer branch chains, whereas low-molecular-weight fractions contain two to four times the comonomer content of the high-molecular-weight fractions.

The use of different catalysts makes it possible to produce products with varying distributions of molar mass and molecular weight. With Cr catalyst types, the molecular weight depends on the catalyst activation temperature. The higher the temperature, the lower is the molecular weight, and the narrower is the molecular weight distribution.

As a general characteristic, the chromium catalyst gives a broader molecular weight distribution than a Ziegler-Natta catalyst.

Commercial resins are specified mainly by density and molecular weight (melt index). The processing and end-use properties are governed mainly by these variables. The density range for MDPE and HDPE is 0.93 to 0.96 g/cm^3. The applications of products from the *Lupotech* G process are film extrusion and small and large blow-molding items (bottles, Jerry cans, drums, and intermediate bulk containers) (Table 14.7.1).

HDPE
0.93-0.97 g/mL

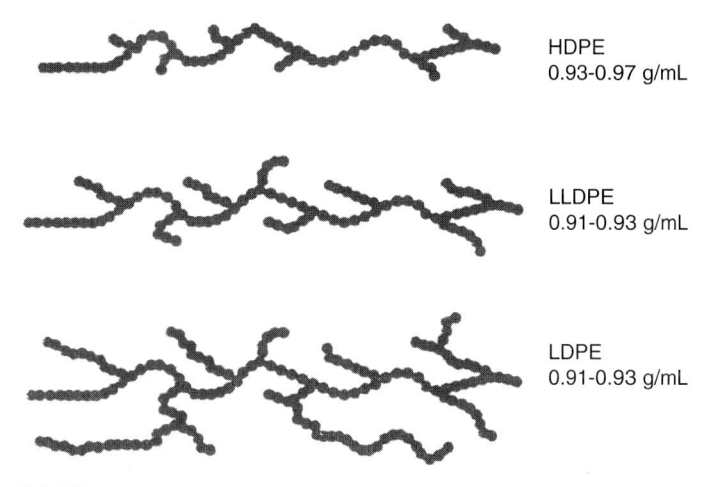

LLDPE
0.91-0.93 g/mL

LDPE
0.91-0.93 g/mL

FIGURE 14.7.6 Polyethylene chain branching simplified.

TABLE 14.7.1 Physical Properties and Applications

Application	MFI 21.6/190 (g/10 min)	Density (kg/m^3)
Film	13	933
Film	13	937
Containers, IBCs	6	945
Bottles	0.25*	950
Bottles	0.25*	955
Jerry cans, drums	10	955

*MFI 2.16/190.

TABLE 14.7.2 Specific Consumption

	Consumption
Ethylene and 1-hexene, kg	1008–1010
Electricity, kWh	370
Steam, kg	90–130
Cooling water ($\Delta T = 10K$), t	150

PROCESS ECONOMICS

Owing to the concept of the single fluidized-bed gas-phase polymerization reactor and related simple equipment, the investment figures are very competitive to similar gas-phase PE process technologies. Typical net consumption per metric ton of PE (depending on product mix) is low (Table 14.7.2).

CHAPTER 14.8
BASELL *LUPOTECH* T TECHNOLOGY FOR LDPE AND EVA-COPOLYMER PRODUCTION

André-Armand Finette and Gijs ten Berge
Basell Polyolefine GmbH
Frankfurt, Germany

GENERAL PROCESS DESCRIPTION

The *Lupotech* T process technology is designed to produce from ethylene low-density polyethylene (LDPE) polymer or ethylene vinyl acetate (EVA) copolymers by copolymerization with comonomers such as vinyl acetate or acrylates. Single-line capacity of 400,000 metric tons per annum (MTA) can be offered.

The polymerization, after initiation by feeding of small amounts of initiator such as peroxide or oxygen, takes place at high pressure in a tubular reactor. The tubular reactor system can be offered in a single-feed configuration or a multiple-feed configuration. The polymer produced in the reactor is separated from the residual ethylene. The polymer is additivated and extruded to solid pellets, and the monomer is recycled and recompressed to operating pressure. Flow schemes for *Lupotech* TS and TM are given below.

PROCESS CHEMISTRY AND THERMODYNAMICS

Reaction Mechanism

The free-radical polymerization process for LDPE manufacture can be illustrated in a detailed kinetic scheme. This scheme allows the calculation of structural properties such as molecular weight distribution and branching frequencies. Therefore, it distinguishes several reaction steps, e.g., initiator decomposition, radical chain propagation, chain transfer to monomer and to modifier, intra- and intermolecular chain transfer, ß-scission of secondary radicals, and chain termination.

The overall reaction takes place in the following steps:

$$\text{Initiation} \rightarrow \text{propagation} + \text{chain transfer} \rightarrow \text{termination}$$

Initiation. The polymerization of ethylene under high pressure can be initiated by

- Compounds that decompose into free radicals (e.g., peroxides)
- Oxygen

The simplest type of initiation takes place by decomposition of peroxide into two radicals:

$$\text{R-O-O-R'} \xrightarrow{k_d} \text{R-O*} + \text{*O-R'}$$

For initiation and molecular weight control in free-radical polymerization of ethylene, a combination of several substances can be applied. Typically used free-radical initiators belong to the classes of dialkyl peroxides, peroxyalkyl esters, peroxycarbonates, and diacyl peroxides. The choice of the initiator mainly depends on its half-life at application temperature. To generate a more or less constant radical concentration level over a bigger range of temperatures (e.g., 150 to 300°C), a combination of different initiators is commonly applied. A typical mixture consists of a low- and a high-temperature decomposing peroxide dissolved in hydrocarbons.

Propagation. The polymer is formed by multiple addition of the monomer to the free-radical end of a growing polymer chain:

$$\text{-R*} + \text{CH}_2\text{=CH}_2 \xrightarrow{k_p} \text{-R-CH}_2\text{-CH}_2\text{*'}$$

$$\text{-R-CH}_2\text{-CH}_2\text{*} + n(\text{CH}_2\text{=CH}_2) \rightarrow \text{-R-(CH}_2\text{-CH}_2)_n\text{-CH}_2\text{-CH}_2\text{*}$$

The heat generated during the polymerization reaction is about 3600 kJ/kg polymer.

When the reaction mixture contains not only ethylene but also other so-called comonomers such as propylene, vinyl acetate, or acrylates, the reaction is more complex:

$$\text{-CH}_2\text{-CH}_2\text{*} + \text{CH}_2\text{=CH}_2 \xrightarrow{k_{11}} \text{-CH}_2\text{-CH}_2\text{*}$$
$$\text{-CH}_2\text{-CH}_2\text{*} + \text{CH}_2\text{=CRH} \xrightarrow{k_{12}} \text{-CH}_2\text{-CRH*}$$
$$\text{-CH}_2\text{-CRH*} + \text{CH}_2\text{=CRH} \xrightarrow{k_{22}} \text{-CH}_2\text{-CRH*}$$
$$\text{-CH}_2\text{-CRH*} + \text{CH}_2\text{=CH}_2 \xrightarrow{k_{21}} \text{-CH}_2\text{-CH}_2\text{*}$$

Each of these reactions is characterized by its rate constant $k_{i,j}$. The composition of the polymer is determined by the copolymerization parameters $r_1 = k_{11}/k_{12}$ and $r_2 = k_{22}/k_{21}$. If $r_1 = r_2 = 1$, the comonomer is randomly distributed in the polymer chain. This is largely the case when vinyl acetate is copolymerized with ethylene.

Chain Transfer. Besides reacting with ethylene or a comonomer, a growing polymer chain also may react with transfer agents (modifiers). Modifiers are chemical substances that easily transfer an H atom to the free-radical end of a growing polymer chain. By this reaction, the modifier becomes a radical itself. This radical can start a new polymer chain while the growth of the polymer chain to which the H atom is transferred is stopped.

$$-CH_2-CH_2* + HX \xrightarrow{k_{tr}} -CH_2-CH_3 + X*$$

$$X* + n(CH_2=CH_2) \xrightarrow{k_p} X-CH_2-CH_2*$$

The free radical on the growing polymer chain is not eliminated from the reaction but just transferred to a new molecule.

The effectiveness of a modifier depends on its chemical structure, its concentration, temperature, and pressure. A concentration-independent measure of its effectiveness is the *chain-transfer constant,* defined as the ratio of kinetic coefficients for the transfer reaction to this substance and radical chain propagation reaction. Usually the effectiveness of chain-transfer agents is increased with rising temperature and with reduced pressure. The chain-transfer constant of modifiers falls from aldehydes, which are more effective than ketones or esters, to hydrocarbons. Unsaturated hydrocarbons typically have higher transfer constants than saturated hydrocarbons, and a strong effect on polymer density must be considered because of the ability to copolymerize giving a higher frequency of short-chain branches in the polymer.

Even the polymer itself can react as a chain-transfer agent. In this case, one has to distinguish between intramolecular and intermolecular chain transfer.

Intramolecular chain transfer leads to short-chain branching with mainly butyl groups at the branches:

Intermolecular chain transfer forms long chain branches:

$$-CH_2-CH_2* + H-CR'R''H \xrightarrow{k_{LCB}} -CH_2-CH_3 + *CR'R''H$$

Termination. The growth of polymer chain radicals can be stopped by various reactions. For example, two chain radicals may combine with saturation of the free valences:

$$R'-CH_2-CH_2* + *CH_2-CH_2-R'' \xrightarrow{k_t} R'-CH_2-CH_2-CH_2-CH_2-R''$$

This reaction forms very large molecules. Also, two chain radicals may form a terminal double bond (disproportionation):

$$R'-CH_2-CH_2* + *CH_2-CH_2-R'' \xrightarrow{k_t'} R'-CH=CH_2 + CH_3-CH_2-R''$$

The chain length is unaffected by this reaction. Finally, a chain radical may react with an initiator radical:

$$R'-CH_2-CH_2* + *I \xrightarrow{k_t''} R'-CH_2-CH_2-I$$

In all three cases, two free radicals disappear from the reaction mixture. In order to keep the concentration of radicals in the reaction mixture constant, a new initiator molecule must decompose into free radicals.

Reaction Kinetics. All the preceding reactions take place simultaneously in a polymerization reactor. However, the rates of the individual reactions depend on the concentration of the reactants involved in the reaction and the individual reaction rate constants k_i. These rate constants are largely dependent on pressure and temperature:

$$k_i = A_i \exp\left[-\frac{{}^A E_i + \Delta V_i (P - P_0)}{RT}\right]$$

Each rate constant k_i is characterized by a particular value of the factor A_i, the activation energy ${}^A E_i$, and the activation volume ΔV_i.

For polymerization at constant pressures, we can write

$$k_i = A_i' \exp(-{}^A E_i/RT)$$

Copolymerization

High-pressure conditions can be applied to polymerize ethylene with a large variety of chemically different monomers via a free-radical kinetic mechanism. Polar comonomers such as acrylic and methacrylic acid and their esters, as well as vinyl acetate, are used in large-scale production. The product properties in these copolymers are very different from those of homopolyethylene (LDPE) and depend strongly on the type of comonomer and the built-in ratio of the monomers. The fundamental technologies of high-pressure manufacturing processes have been described already in detail above. Depending on the nature of the comonomers, the process is selected with regard to its back-mixing capability. The residence-time distribution of the reaction mixture and the applied polymerization concept have a strong impact on polymer structure, product homogeneity, and process operability.

In general, three different cases of copolymerization behavior have to be distinguished: the copolymerization of ethylene with comonomers with higher reactivity (e.g., acrylates, methacrylates, styrene), with monomers of equal reactivity (e.g., vinylacetate), and with monomers of minor reactivity (e.g., vinylethers, propylene, and higher olefins). Copolymerization with vinylacetate proceeds nearly ideally, which means that the polymer has nearly the same composition as the comonomer mixture at any time during the polymerization. On the other hand, acrylic and methacrylic acid and their esters polymerize much more rapidly than ethylene, so a polymer with high comonomer content can be formed from a monomer mixture with only small quantities of the comonomer.

Aside from the composition and comonomer distribution along the copolymer chain, the molecular weight distribution (MWD) and the degree of branching are of special importance for their application. Thus, for example, copolymer products of ethylene and acrylic acid with higher degrees of polymerization have improved adhesion-promoting properties in steel pipe coating applications. Molecular weight can be controlled by adding modifiers, e.g., aldehydes or ketones, despite the fact that most comonomers themselves have a high radical chain-transfer capability, restricting the highest achievable polymer molecular weight. Chain-transfer-rate coefficients of comonomers must be considered in the polymerization process, although these values often are uncertain for appropriate reaction conditions.

Polymerization Techniques

Heat Control. During the polymerization of ethylene, a large amount of heat is generated. For a smoothly running process it is essential to control the heat balance of the polymerization reaction very closely. The thermodynamic stability of the ethylene demands that the peak temperature of the reaction mixture does not exceed 350°C. Therefore, more than half the heat generated must be removed from the reaction mixture by heat transfer through the wall of the tubes.

The overall heat transfer is determined by the temperature differential between the reaction mixture and the cooling medium in the jackets and by the heat transfer

1. From the reaction mixture to the inner surface of the reactor tubes
2. From the reactor tubes to the hot water in the jackets

For a good heat transfer to the inner surface of the reactor tubes, it is important to maintain a turbulent flow in the tubes. Therefore, it is necessary to increase the velocity in the tubes from time to time by "kicking" (reducing the reactor pressure some 100 bar). To remove the heat from the tubes, it is important to avoid any vaporization of the cooling medium in the jackets. Therefore, the flow of the cooling medium in the jackets must be throttled to build up a reasonable pressure at the outlet of each cooling circle.

Depending on the reaction pressure, an oxygen-initiated reaction starts at a temperature of about 170 to 180°C. Low-temperature-decomposing peroxides may initiate the reaction at temperatures of 140 to 150°C.

Control of Product Properties

Density and Melt-Flow Index. The correlation between density D of a polymer, on the one hand, and the peak temperature T and the reaction pressure P, on the other hand, can be described by

$$D = a_0 - a_1 \times T + a_2 \times P$$

In this equation, a_1 and a_2 are constants, whereas a_0 is a function of both the melt-flow index and the type of modifier.

The reason for the decrease in density with an increasing peak temperature is the increase in short-chain branching due to intramolecular chain transfer with increasing temperature. However, as the pressure increases, the short-chain branching is suppressed in favor of the chain growth. This is the reason why the density increases with increasing pressure at a constant peak temperature.

A correlation between density and conversion can be derived qualitatively from the preceding equation. Because the conversion is largely determined by the difference between starting temperature and peak temperature, the conversion must increase with an increasing temperature peak. The consequence of this is that

1. Increasing pressure increases conversion at the same density.
2. At constant pressure, conversion is higher for low-density materials than for high-density materials.

The melt-flow index (MFI) is controlled by the modifier used during polymerization. However, there is an important difference between so-called strong modifiers and weak modifiers such as propane and propylene.

The chain-transfer activity of a strong modifier such as aldehydes is only slightly influenced by the temperature. The MFI of the product is determined almost exclusively by the concentration of the modifier in the reaction mixture; the MFI increases with the amount of modifier fed into the reactor per hour.

The behavior of a weak modifier such as hydrocarbons is quite different from that of a strong one. The chain-transfer activity of a weak modifier depends mainly on the reaction temperature, whereas changes in concentration have only a slight effect on the MFI.

Therefore, in order to change the MFI with a weak modifier, it is necessary to change the peak temperature. This is done normally by changing the initiator concentration (varying the peroxide flow) to the reactor, but changing temperature also will change density. To keep density constant in this case, one has to adjust the pressure, the peak temperature, and the modifier flow at the same time.

Controlling Conversion. The conversion in a tubular reactor is determined by the heat balance. The equation for the overall heat balance is

$$Q = G \times H \times dU = G \times C_p \times dT_g + \pi \times D \times k \times (T_g - T_w) \times dL$$

where G = mass flow through tubes
H = enthalpy
U = conversion
C_p = specific heat of the reaction mixture
T_g = temperature of the reaction mixture
Q = generated heat
D = inner diameter of the reactor tubes
k = overall heat-transfer coefficient
T_w = water temperature in the jackets
L = effective length of reaction zone

From this equation one can easily derive that the conversion U will increase

- With an increasing dT_g of the reaction mixture
- With a larger heat-transfer coefficient k
- With an increasing difference between reaction temperature T_g and water temperature T_w
- With a larger effective length of the reaction zone

The terms G, H, C_p, and D are constant for a given design of a reactor and cannot be changed.

The increase in dT_g is limited on the one hand by the starting temperature of the initiator and on the other hand by the peak temperature, which is determined by the required product density. The heat-transfer coefficient k can be influenced by opening the pressure control valve of the reactor periodically. This will increase the flow rate of the reaction mixture and improve the turbulent heat transfer to the wall of the tubes. However, because "kicking" causes very strong stress loads on the tube material, it is recommend to minimize the amplitude and frequency of these pulses.

At a given peak temperature T_g the temperature difference $T_g - T_w$ can be influenced only by the jacket temperature T_w. Lowering T_w not only will increase the temperature difference $T_g - T_w$ but also will shift the temperature profile downstream and increase the effective length L of the actual reaction area. If the jacket temperature is lowered too much, the heat-transfer coefficient k will be decreased because of increasing thickness of the wall layer.

Controlling Temperature Profile. The temperature profile is further controlled by the temperature of the reactor preheater, by the reaction pressure P, and by initiator concentration. The ranges of all these controlling measures are given by the intended density and MFI of the product and by the stability of the reaction.

Effect of Reaction Conditions on Product Properties. The effect of pressure, peak temperature, and type of modifier on the polymer density has already been described. The influence of peak temperature and type and concentration of modifier on the MFI also has been explained. Besides these correlations, pressure, peak temperature,

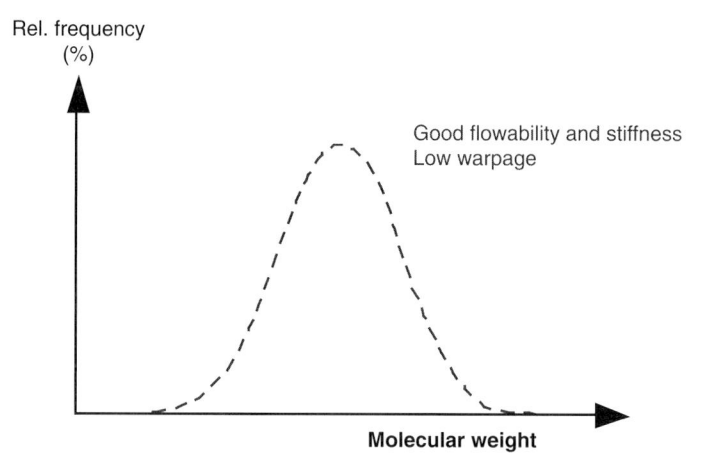

FIGURE 14.8.1 Narrow MWD for injection-molded parts.

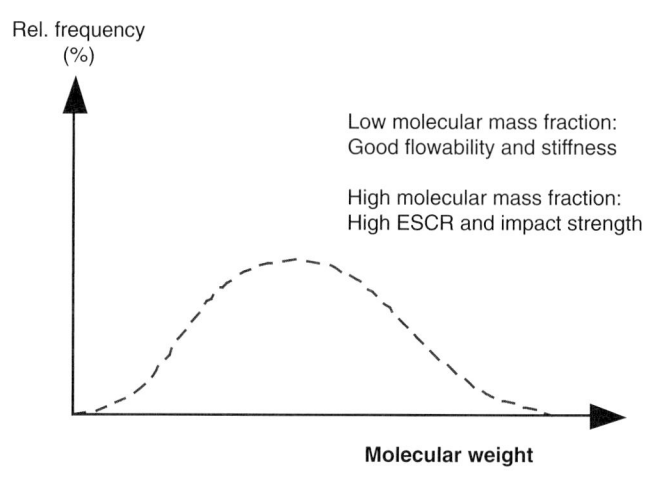

FIGURE 14.8.2 Broad MWD for blow molding and film.

initiator concentration, type and concentration of modifier have an influence on the molecular structure of the polymer (Figs. 14.8.1 and 14.8.2).

Increasing temperature will narrow the molecular mass distribution (MMD), which will result in improved film properties. Increasing peak temperature and initiator concentration will broaden the MMD and deteriorate film properties. High modifier concentration will give a narrow MMD (Table 14.8.1).

It was mentioned earlier that the properties of LDPE depend originally on the MWD and the distribution of chain branches regarding their length and frequency. Size-exclusion chromatography (SEC) may be the most important method for analyzing the MWD of polymer material because of its simplicity and applicability over a wide range of molecular weights. The original value analyzed by this method is the quantity of polymer molecules fractionated with respect to their hydrodynamic volume. This hydrodynamic volume depends not only of the molecular weight but also on the branching structure of the molecule (especially on long-chain branching). This means that the SEC hydrodynamic volume

TABLE 14.8.1 Correlation between Reaction Conditions and Molecular Structure*

	SCB	LCB	Mn	Mw/Mn
Pressure ↑	↘	↘	↗	↘
T_{peak} ↑	↑	↗	↘	↗
Initiator ↑	→	↗	↓	↗
Modifier ↑	→	↓	↓	↘
Conversion ↑	→	↗	→	↑

*SCB = short-chain branching; LCB = long-chain branching; Mn = number molecular mass; Mn/Mw = molecular mass distribution.

must be correlated with the molecular weight of polyethylene (PE) samples determined, for example, by light scattering or ultracentrifugal methods and with known branching structure. Because this correlation cannot be made for every sample, a certain error arises from individual structures.

A very powerful tool for a quantitative analysis of branching structure is ^{13}C nuclear magnetic resonance (NMR) spectroscopy. Today, chain lengths up to 10 carbon atoms can be distinguished, and the relative accuracy with respect to more frequently present short chains, such as methyl-, ethyl-, butyl-, and 2-ethylhexyl-, is around 10 percent.

Structural characteristics are presented mostly in the form of averaged values, e.g., the number or weight average of molecular weight or the average number of short chains per 1000 carbon atoms. To obtain a more detailed picture of the molecular structure, different analytical methods can be applied in combination. One example is the combination of SEC and temperature-rising elution fractionation (TREF). The TREF method classifies the polymer sample into fractions with different tendencies to crystallize, which correlates very well with the short-chain branching frequency. After application of this preparative TREF method, the fractions can be analyzed with SEC and, for example, with ^{13}C NMR to obtain more detailed information about the chain-length distribution of the main chain and the branches.

LUPOTECH T PROCESS PERSPECTIVE

The *Lupotech* T process for production in tubular reactors of low-density polyethylene (LDPE) has been licensed successfully all over the world. Over 50 years of manufacturing experience, combined with modern tools such as process simulation, advanced control, polymer research, and material/equipment optimization, make the *Lupotech* T process the most attractive technology with tubular reactors for LDPE production.

The *Lupotech* T process is being improved and upscaled continuously, and capacities up to 400 kMTA per single line are now feasible. *Lupotech* T products cover the whole manufacturing range and can be used for all LDPE applications, including a wide range of vinyl-acetate copolymer applications. Basell's current *Lupotech* T process is the result of decades of development based on a strong research and development (R&D) program focusing on all key features of the technology (process, equipment, products, and applications).

An Overview of *Lupotech* T Development

LDPE was discovered in 1933 and attained commercial importance because of its electrical insulation and optical and mechanical properties. The first commercial plants based on Basell technology became operational in the 1950s. Since then, through R&D, significant advances have made the *Lupotech* T process a world-class technology.

In the 1970s, a major breakthrough in terms of process safety, plant reliability, and operating windows was the introduction of modified high-pressure steels with increased toughness as base material for the reactor piping. This resulted in superior plant integrity and also enabled higher operating pressures, which expanded the LDPE product portfolio with higher-density LDPE grades and an increase in film quality.

In the 1980s, organic peroxides were introduced as radical initiators, leading to both higher conversion rates and more efficient process control. In the 1990s, R&D refocused on process upscaling, process simulation and modeling, and advanced control.

The introduction of *Luposim* T, Basell's proprietary simulation program for high-pressure tubular reactors, created a powerful tool for process and product development/improvement. In terms of scale-up, *Lupotech* T technology achieved a breakthrough in January 2001 with the startup of the world's largest LDPE plant at Aubette, France, with a single-line capacity of 320 kMTA. The successful operation of this unique plant is a showcase for the process technology.

Based on advanced process simulation, which has been validated in existing plants and has proven its capabilities with the scale-up of the 320-kMTA LDPE plant, Basell is now in the position to offer plant capacities of up to 400 kMTA single line with dedicated customer design.

Key Characteristics

The *Lupotech* T process is a high-pressure tubular reactor technology developed for the whole range of LDPE products, including hydrolyzed EVA (HEVA) copolymers.

Safety and Loss Prevention. Polymerization of ethylene under high pressure in a tubular reactor takes place in supercritical ethylene as diluent. Therefore, no additional solvent is required.

Plants using the *Lupotech* T process can meet all environmental and safety requirements. Around the world, over 50 million metric tons have been produced with *Lupotech* T technology without major incidents.

Reducing Resource Intensity. In the *Lupotech* T process, unreacted monomers are recovered and recycled constantly. If necessary, discontinuous hydrocarbon purges can be sent for "thermal recovery" or to a flare system. Since hydrocarbons as solvent are not needed, there are no major effluent streams.

Design Capability/Flexibility. Single-line capacities up to 400 kMTA (based on 8000 operating hours per year) can be provided, covering the whole range of LDPE products.

Process Versatility. Related to specific process features, *Lupotech* T technology can produce the whole range of LDPE products, including HEVA and *N*-butylacrylate-modified LDPE. Fast grade changes ensure that the amount of off-grade material is kept very low, allowing a broad grade slate to be produced in one reactor.

Product Quality. *Lupotech* T technology delivers a very homogeneous product quality with a low gel level, which means blending/homogenizing silos are not required.

Economical. Capital costs for the *Lupotech* T process are low as a result of highly efficient design. Based on low raw materials and energy consumption figures,

Lupotech T technology is characterized by competitive operating costs, whereas the final product is sold at a premium over other PE products. The combination of low investment and low operating costs makes the *Lupotech* T process an attractive high-pressure PE process.

PROCESS DESCRIPTION

In the *Lupotech* T process, ethylene is polymerized to LDPE at temperatures above 150°C and pressures between 2000 and 3100 bar. For the production of copolymers, specific comonomers such as vinylacetate or butylacrylate are used.

The overall manufacturing process can be divided into the following steps/process units:

- Precompression of ethylene
- Compression to reaction conditions
- Polymerization reaction
- Polymer-gas separation
- Recycling of unreacted gases
- Extrusion, pelletizing
- Degassing
- Storage and packaging

High-pressure tubular reactors are offered in two generally different forms, as shown in Figs. 14.8.3 and 14.8.4.

In a TS reactor, the total ethylene flow from the hypercompressor is preheated to 150 to 180°C and fed to the inlet of the first reaction zone. Reaction is initiated by injection of organic peroxides. As the reaction mixture cools after the first reaction peak, additional peroxide initiator is added to start a second reaction zone. There can be further peroxide injection points, giving a total of three to five reaction zones.

In a TM reactor (multiple cold gas injection), the compressed ethylene is split into several streams and fed into the reactor at a number of positions. In addition, the ethylene can be dosed with oxygen prior to compression. Approximately 40 to 70 percent of the total ethylene is preheated to 150 to 180°C before addition of peroxide, which, together with the oxygen, sets off the first reaction zone. The other 30 to 60 percent of the ethylene is cooled and further split into more or less equal quantities, which are injected at different locations along the reactor. Together with addition of initiator, this results in the desired number of reaction zones. Additional reaction zones can be created by dosing peroxide directly to the reactor.

Reaction heat in this reactor type can be absorbed by the cold gas feeds, having the disadvantage that this part of the gas feed has only a reduced residence time in the reactor. The balance of these two effects can result in higher conversion rates, depending on the product grade. The TM reactor can have a stepwise increase in reactor diameter, and the reaction mixture flow velocity can change significantly owing to the additional gas feeds. This reactor is equipped with a similar pressure-control valve as a TS reactor.

LDPE Process Sections

1. *Precompression of ethylene.* Polymer-grade ethylene and recycled ethylene are compressed by a primary compressor to approximately 300 bar. A modifier (e.g., hydrocarbons, ketones, or aldehydes) to control the molecular weight is added.

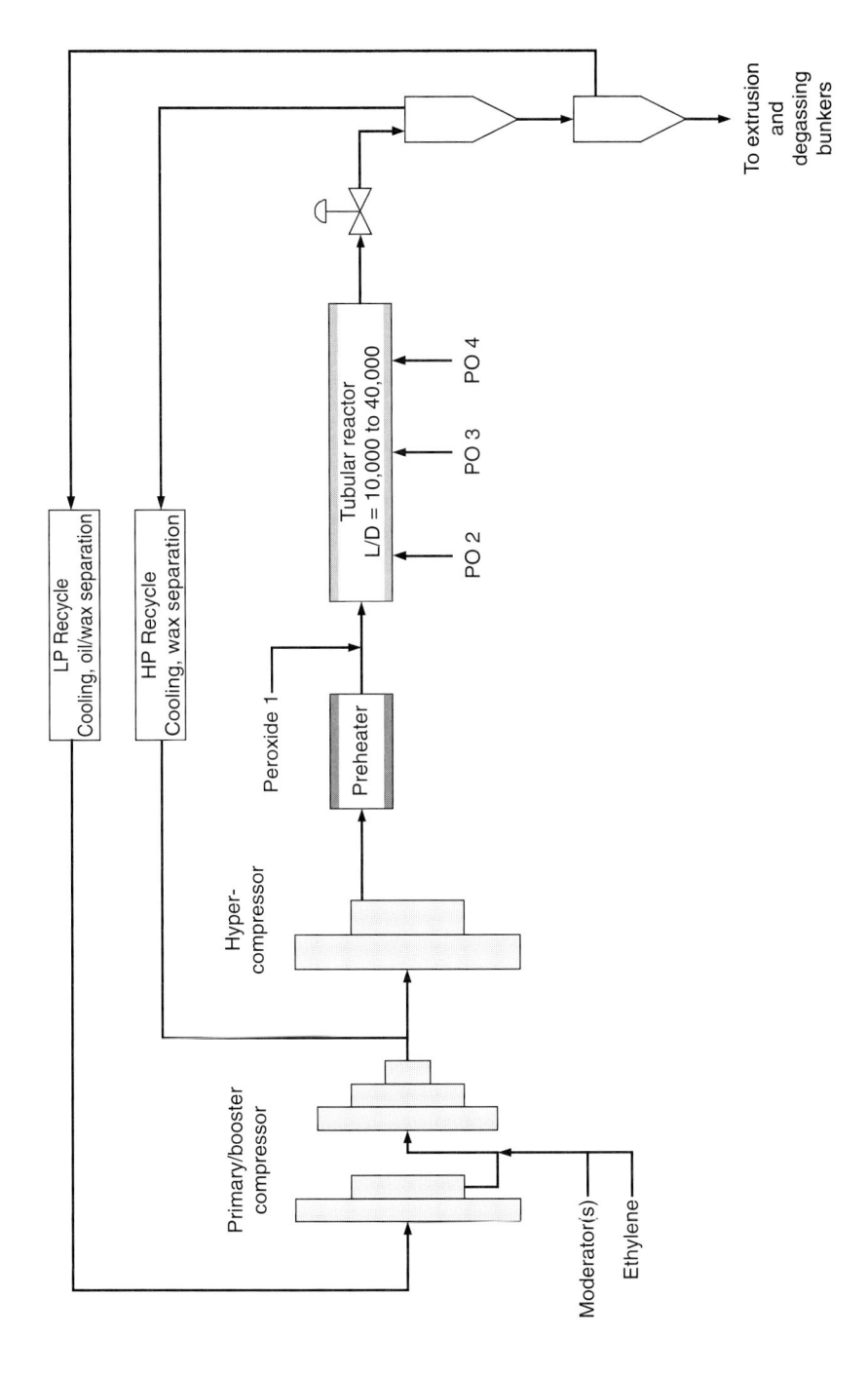

FIGURE 14.8.3 *Lupotech* TS process.

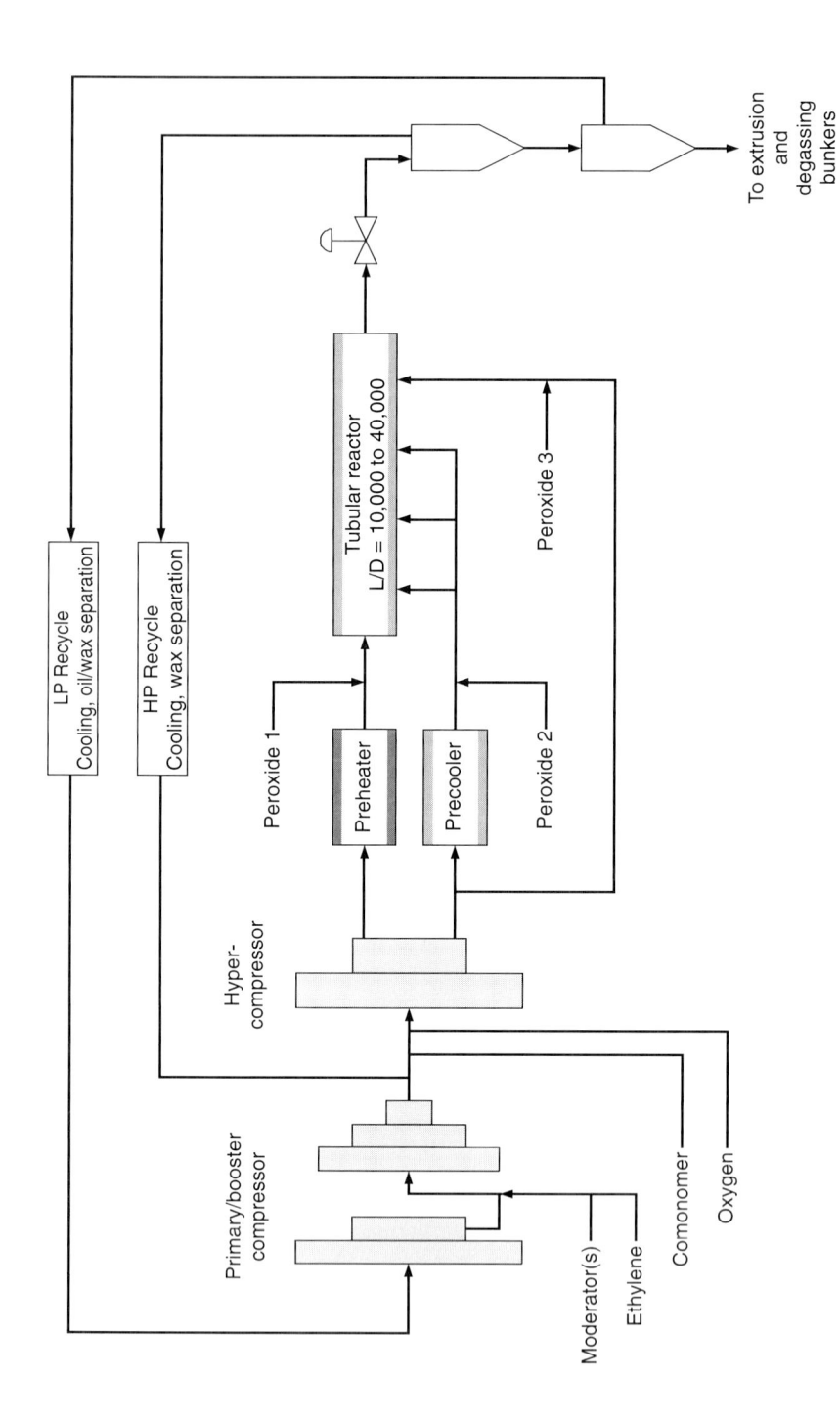

FIGURE 14.8.4 Lupotech TM process.

2. *Compression to reaction conditions.* The ethylene stream, together with the recycled ethylene out of the high-pressure recycle loop, is further compressed by the hypercompressor up to 3100 bar. The hypercompressor is a symmetric two-stage compressor in two lines (streams A and B). In the first stage, the compressor feed gas is compressed from about 270 bar(g) to approximately 1250 bar(g). Cooling of the gas is carried out after stage 1 only. The gas in both streams is compressed in the second stage to pressures up to 3200 bar. After each stage, the gas passes through a discharge pulsation dampener. The ethylene finally discharged has to be further heated for the reaction to take place.

3. *Polymerization reaction.* The reaction mixture leaving the hypercompressor is fed to a tubular reactor consisting of three different segments. In the preheater, the reaction mixture is heated to the initiator temperature of about 150 to 180°C, followed by a sequence of reaction and cooling zones. The reactor consists of a large number of jacketed tubes with a total length-to-diameter ratio between 10,000 to 40,000. The inner diameter of the high-pressure tubes can vary between 10 and 100 mm, having a monomer throughput of between 10 and 190 t/h.

Polymerization pressure is typically between 2000 and 3200 bar, and maximum temperatures are below 350°C. Under polymerization conditions, the polymer is dissolved in the supercritical ethylene, resulting in a one-phase reaction mixture.

In the reaction zones, the polymerization is initiated by adding the polymerization initiator (oxygen or peroxide). Owing to the exothermic nature of ethylene polymerization, the reaction mixture heats up, and the resulting polyethylene remains dissolved in the supercritical ethylene. Temperature is controlled by the concentration and type of added initiators, as well as by the hot water circuit. Side streams of cooled fresh ethylene can be used for a fast cooldown of the reaction mixture and restart of the initiation. The conversion of more than 35 percent per pass can be achieved by this procedure.

4. *Polymer-gas separation.* At the end of the reactor, the mixture of polymer and unreacted ethylene is depressurized, passes a postreactor cooler, and enters the high-pressure product separator. The unreacted ethylene is separated from the polymer melt at approximately 300 bar. The polymer melt is let down through the product valve into the low-pressure product separator (approximately 0.2 to 2.0 bar) to remove the remaining ethylene from the resin.

5. *Recycling of unreacted gases.* Ethylene streams from both separators are recycled to the respective compressors by stepwise cooling. The high-pressure stream is recycled to the hypercompressor and the low-pressure stream to the booster stage of the primary compressor. To avoid accumulation of impurities and inerts in the loops, a minor quantity of ethylene is withdrawn as a purge gas stream and sent to the ethylene recovery at battery limits.

6. *Extrusion, pelletizing.* The melt, still containing minor quantities of ethylene, is passed from the low-pressure product separator into the extruder for further degassing, for the incorporation of additives, if demanded, and for pelletizing by an underwater face-cutter system. The extruder can be a single-screw machine or a twin-screw machine.

7. *Degassing.* After passing through the extruder, the pellets still contain traces of ethylene and are sent to the degassing silos, which are purged by a constant flow of air. Within a certain period of time, the remaining hydrocarbons diffuse out of the pellets. Airflow is designed to keep the concentration level of ethylene safe.

8. *Storage, bagging.* Pellets are conveyed pneumatically into storage silos; from there the products can be shipped in bulk or bagged.

Process Safety

Basell has a safety record unmatched in the industry. By 2003, combined Basell technologies have achieved nearly 7 million operating hours without a major incident.

Specific Safety Features of Lupotech T

1. Fast-acting safety chain (thermocouples, pressure transmitters, hydraulic system)
2. Fast depressurization of the reactor system in case of a reactor safety shutdown (RSD) program) to safe levels
3. Autofrettage of high-pressure reactor equipment manufactured of K 10 X-steel material (result: leak before break)
4. TV cameras for an overall view of reactor and plant area
5. Inertization system for pellet degassing silos in case of elevated ethylene concentrations in the silo atmosphere or additional spare blowers for air

Automatic Reactor Safety Relief System. The polymerization process is secured by up-to-date safety systems installed independently in addition to the distributed control system. This safety system is hardwired and therefore extremely reliable. It fulfills all standards for such production plants.

This reactor shutdown (RSD) program transfers the reactor unit into a safe condition by depressurizing the compressors and the reactor. Shortly after, the reactor system is ready for another startup.

The LDPE plant is fitted with "conventional" mechanically operated relief devices to protect against overpressure. The high-pressure reaction and high-pressure recycle sections also contain emergency expansion valves to rapidly depressurize the plant in the event of abnormal operating conditions occurring, and these are activated via trip systems.

Active depressurizing systems are critical to the safety of high-pressure tubular reactors. Requirements are

- Early detection of a reactor exotherm
- Quick-acting response to rapidly depressurize the reactor
- Reactor contents discharged safely
- Polymer removed to reduce pollution from dust
- Key system components that are reliable, effective, and affordable

  The relief systems consist of

- Hydraulically operated emergency expansion valves on the reactor and high-pressure recycle systems that open in response to initiation of safeguarding system trips. The valves open automatically in the event of hydraulic oil pressure failure.
- High-pressure spring-loaded relief valves or bursting disks in the reactor and high-pressure recycle systems.
- Normal spring-loaded relief valves or bursting disks on parts of the process operating below 300 bar(g).

  The overall philosophy is that hydrocarbon releases to atmosphere are avoided wherever possible. To this end,

- Relief valves on hydrocarbon duty are routed to flare (however, bursting disks are routed to atmosphere).
- For those sections working above 20 bar(g), and in particular at 325 or 1600 bar, very cold temperatures may occur downstream of the relief valve, and the relief/flare system must be designed accordingly.

- Vents, which may contain polymer or waxes, are steam-jacketed where necessary to protect against blocking with polymer. These vents are checked routinely for possible blockages by purging with nitrogen.

In vent systems, low points should be avoided where practicable, with those remaining being fitted with a boot to collect liquid with level alarms at the low points to warn of any liquid accumulation.

Emergency Shutdown. The main hazard affecting the reaction section is relevant to the decomposition reaction of ethylene. Thermal decomposition of ethylene (as well as the polymerization reaction for this process) is a radical reaction, leading to the formation of carbon (soot), hydrogen, and methane. The reaction is favored by high pressure and high temperature and is strongly exothermic, although other factors (impurities, iron oxides, or other metal oxides) are supposed to facilitate initiation of this reaction. Decomposition reactions will take place at approximately 3100 bar and 355°C.

Ethylene decomposition is strongly exothermic and favored by high pressure. The associated reaction heat causes a simultaneous temperature and pressure increase inside the reactor, resulting in self-acceleration of the decomposition reaction. This can lead to extremely high pressures and temperatures in a few seconds. Whenever an abnormal situation is detected in the reactor section, the plant must be shut down and depressurized by an intrinsic safe system.

Apart from the ethylene decomposition reaction, there are other hazards related to the high-pressure part of the LDPE plant:

- Major damage to the hypercompressor
- Major gas releases in the compressor, reactor, or high-pressure recycle sections

The ultimate safeguard for these hazards is also plant shutdown by rapid depressurization. As explained earlier, this is the only way to immediately mitigate the hazards related to the high operating pressure. Figure 14.8.5 shows the polymerization unit of a 320-kMTA plant.

FIGURE 14.8.5 Polymerization unit, 320-kMTA plant.

PRODUCT SPECIFICATIONS

Overview

By means of variations in process conditions, a wide range of products can be obtained, ranging from standard LDPE grades to EVA copolymers or N-butylacrylate-modified copolymer. The process has no limitations to the number of reactor grades, and the product mix can be adjusted to match market demand and economical product ranges (Fig. 14.8.6). Reactor grades from MFR 0.15 to 50 and from density 0.917 to 0.934 g/cm^3 can be produced. For copolymers, the fraction of comonomer may range from 3 percent up to 28 percent and has the most important influence on the key product parameters, especially rubber-like properties for HEVA grades. Commonly available additives, as used worldwide in the LDPE business, can be easily incorporated directly into the product to yield the final product.

Lupolen LDPE is the trade name for the Basell line of polyethylene low-density resins, which includes homopolymers and copolymers. The range of applications consists of products for film, injection moldings, blow moldings, sheathings for cables and wires and steel pipes, profiles, and sheets. Special features of *Lupolen* LDPE are the large scope, easy processing, good resistance to environmental stress cracking, well-balanced property combination of good mechanics/opticals/drawdown, and an overall excellent product consistency.

Applications of LDPE

Depending on the kind of application and demanded property mix, the specific *Lupolen* grade can be selected, including an appropriate additive mix (e.g., slip, antiblock, and stabilizers for specific applications). EVA grades are especially suited for low-temperature applications. Other outstanding properties are enhanced stress crack resistance, flexibility, and transparency.

Film. The wide range of products for both homopolymers and copolymers enables a wide range of film applications, including shrink and heavy-duty packing films, films for laminates, carrier bags, composite and crack films, deep freeze bags, and agricultural films such as mulch and greenhouse films.

Injection Molding and Blow Molding. Both homopolymers and copolymers are used for many standard applications. For parts in the medical and pharmaceutical sector, *Lupolen* LDPE homopolymers are especially suitable owing to their high purity. The product line matches the technical properties required, namely, flowability, rigidity, and stress cracking resistance. The range of grades includes specialty products for pharmaceutical applications and products with good organoleptic properties.

Pipe Coating. *Lupolen* grades for coating steel pipes provide effective corrosion protection with a well-balanced property profile, including good long-term properties, excellent processing, high mechanical strength, and good chemical resistance.

Cable and Wire. *Lupolen* LDPE homopolymers cover the area of cable applications from power cables, insulation for coaxial cables, and telephone cable cores to sheathing material with very high resistance to environmental stress cracking.

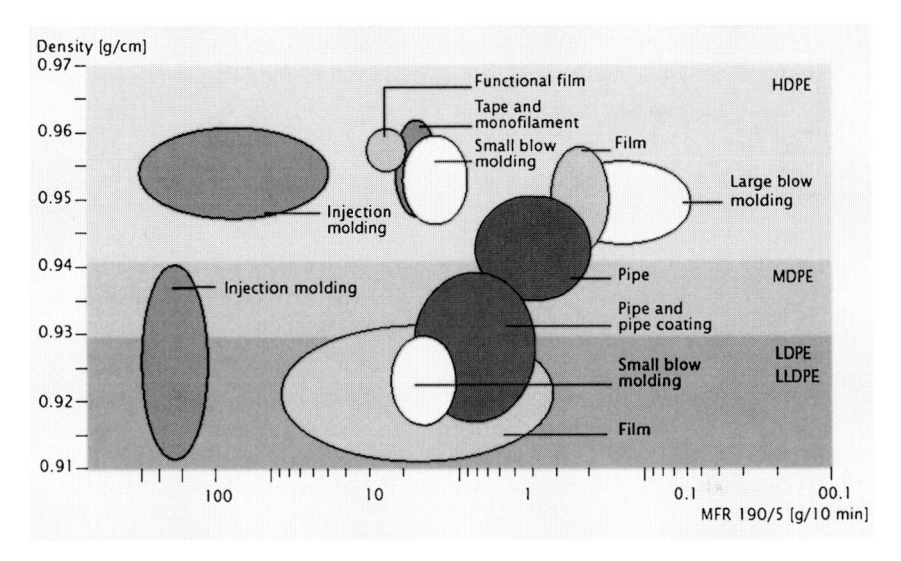

FIGURE 14.8.6 Product families of polyethylene.

PROCESS ECONOMICS

Lupotech T process plant capital and operating costs are competitive and very well rewarded as a result of

- Efficient reactor design and plant layout
- High conversation rates
- Low monomer consumption
- Very short transition time
- High process reliability
- Waste heat integration concept (Table 14.8.2)

TABLE 14.8.2 Typical Specific Consumption (per 1000 kg of PE Produced)

	LDPE
Monomer, kg	1010
Initiator, kg	0.25
Product additives, kg*	2.5
Electric power, kWh*†	700–1000
Steam, kg*†	−1200 (= export)
Cooling water, m^3	75–100
Boiler feed water, m^3*†	1.4
Nitrogen, $N \cdot m^3$	4
Instrument air, $N \cdot m^3$	25

*Typical; may vary depending on grade slate.
†Extrusion included, depending on grade slate and plant capacity.

CHAPTER 14.9
UNIPOL™ PE GAS-PHASE PROCESS: DELIVERING VALUE TO THE PE INDUSTRY

Mardee McCown Kaus
Univation Technologies LLC
Houston, Texas

INTRODUCTION

Univation Technologies is a joint venture between the Dow Chemical Company and ExxonMobil Chemical Company and a partnership to drive a renewal of polyethylene (PE) technology around the world and to deliver value to the PE industry.

Union Carbide, a subsidiary of the Dow Chemical Company, was the original developer of the UNIPOL™ PE process, the premier gas-phase PE manufacturing technology and the worldwide leader in licensing PE technology. In the course of building this position, the industry's preeminent technology licensing and postlicense support organization was created and resides within Univation today.

ExxonMobil Chemical was first to develop breakthrough metallocene catalyst systems for the UNIPOL PE gas-phase process. ExxonMobil Chemical also developed Super-Condensed Mode Technology (SCM-T), which can dramatically increase the production capacity of existing UNIPOL polyethylene reaction systems.

The joint venture consolidates the most advanced PE catalysis, process technology, and product and applications know-how—focused on the UNIPOL gas-phase process. The mission of Univation Technologies is to build on a base of the industry's best process, catalyst, and product technologies to deliver

- UNIPOL gas-phase process and catalyst technology
- UCAT™ conventional catalysts, including Ziegler-Natta and chrome
- XCAT™ metallocene catalysts for the production of high-performance (HPR) and easy-processing (EZP) mLLDPE

- PRODIGY™ bimodal catalyst technology for bimodal product property advantages produced with single-reactor economics
- Capacity expansion technologies (CET), including Super-Condensed Mode Technology (SCM-T)

Univation has catalyst manufacturing plants at Mont Belvieu and Seadrift, Texas, and St. Charles, Louisiana. The facilities at Seadrift and St. Charles offer licensees state-of-the-art Ziegler-Natta and chromium-based catalyst systems. Univation's metallocene catalyst manufacturing facility at Mont Belvieu, Texas, has increased its capacity to meet future global demand and can provide catalyst for the manufacturing equivalent of over 2.5 million tons of PE per year.

The UNIPOL PE process for new plants includes

- State-of-the-art UNIPOL PE technology
- Ziegler-Natta and chromium catalyst systems and product capability
- Metallocene catalyst systems and product capability
- Bimodal catalyst systems and product capability
- Advanced process control

The UNIPOL PE process for retrofits of existing plants include

- Metallocene and bimodal catalyst systems and product capability
- Capacity expansion technology (CET), including
  - Complete systems analysis
  - Condensed-mode operation
  - Super-condensed-mode operation
  - Advanced process control

PE licensing technology packages from Univation offer the following advantages:

- Access to developments from the world's largest research and development (R&D) activity focused on the gas-phase PE process and advanced catalysis
- Ability to produce the broadest and most versatile product line
- Low investment and operating costs
- More than 30 years of licensee service experience, including engineering, R&D, and manufacturing support
- Metallocene, bimodal HDPE, and other advanced catalyst technologies with proven performance in the UNIPOL PE process
- Reliable supply of Ziegler-Natta, metallocene, chrome, and bimodal HDPE catalysts
- Opportunities to improve the capacity and thus the value of existing licensees' PE businesses with retrofits using condensed-mode and super-condensed-mode technology innovations

HISTORY

Univation's Experience in Gas-Phase PE

For nearly 40 years, the UNIPOL PE process has been actively developed and enhanced with numerous catalyst, product, and process improvements (Fig. 14.9.1). And the UNIPOL PE process has achieved an impressive licensing position in worldwide markets.

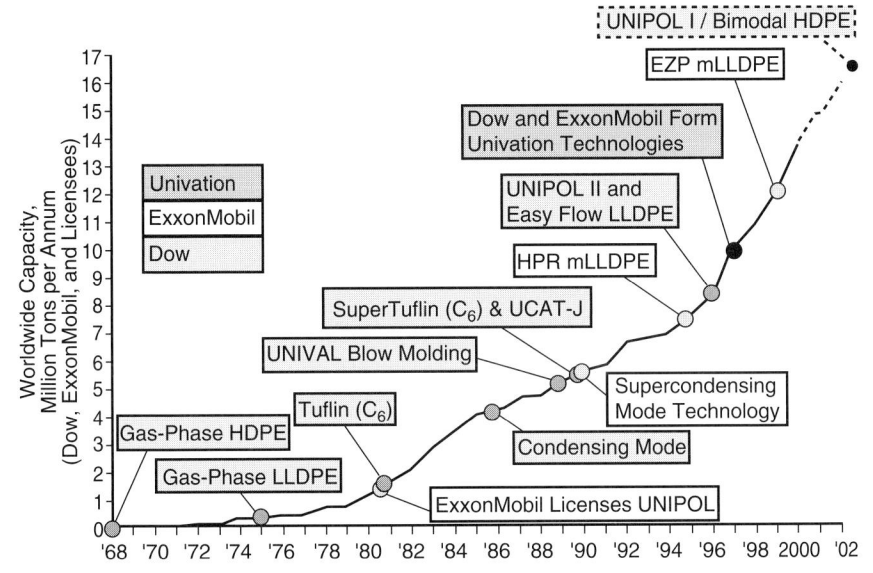

FIGURE 14.9.1 History of UNIPOL PE technology.

UNIPOL PE technology was first commercialized in 1968 and from the beginning represented a step-change improvement in the versatility and economics of polyethylene production. It has become the most widely chosen process worldwide, and licensees operate facilities in 25 different countries. With over 1500 reactor-years of proven, safe, and successful operation, the credentials of this technology are well established. Today there are 90 operating UNIPOL PE reactor lines worldwide. Total UNIPOL PE operating capacity is nearly 15 million tons per year. UNIPOL licensees have ranged from new producers to established leaders in the polyolefins business. Many licensees have reaffirmed their choice by building additional UNIPOL plants.

UNIPOL PE technology is supported by a commitment from Dow and ExxonMobil to keep the platform at the forefront of value-added PE technology development and to continue to deliver value to the PE industry.

GENERAL PROCESS DESCRIPTION

The UNIPOL PE process produces a broad range of polyethylene products. Using conventional, metallocene, and newly engineered bimodal catalyst systems, PE products can be produced for all significant market segments (Fig. 14.9.2).

At the core of the UNIPOL PE process is a gas-phase fluid bed providing throughput to produce a very uniform product quality and leading to reduced blending and handling facilities compared with other polymer processes.

Raw Materials Purification and Ethylene Purification

All polymerization catalysts are sensitive to certain impurities in the raw materials feed streams. The sensitivity depends on the catalyst in question and the particular impurity involved. Normally, molecular sieves are used for water removal from all major raw mate-

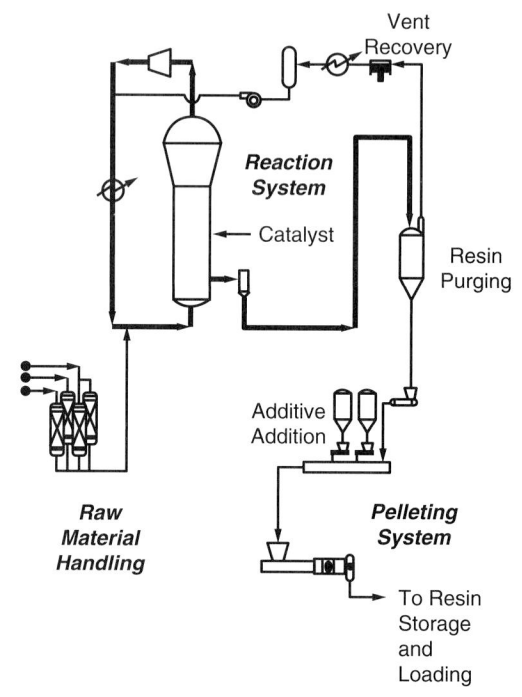

FIGURE 14.9.2 UNIPOL PE process summary.

rials. Depending on the quality of the various materials being used, it also may be appropriate to include one or more purification steps for selected raw materials to remove other impurities.

The purification steps included in the process typically are required only as guard beds to protect against impurity spikes. Bed life between regenerations is relatively long (measured in months, not days).

The polymerization catalysts also are sensitive to traditional poisons such as sulfur, carbonyls, etc.; however, these materials are rarely present in sufficient quantity to be of concern.

Catalyst

All polymerization catalysts purchased from Univation are shipped and stored in special shipping containers under a nitrogen blanket. They are shipped in ready-to-use form.

Reaction

The reaction system can be subdivided into several sections: reaction loop, catalyst feeding, and product discharge and separation (Fig. 14.9.3). These parts will be discussed individually.

Reaction Loop

- The reaction loop consists of a reactor, a cycle gas compressor, and a cycle gas cooler. The reactor is a continuous backmixed fluid-bed reactor that contains a fluidized bed of

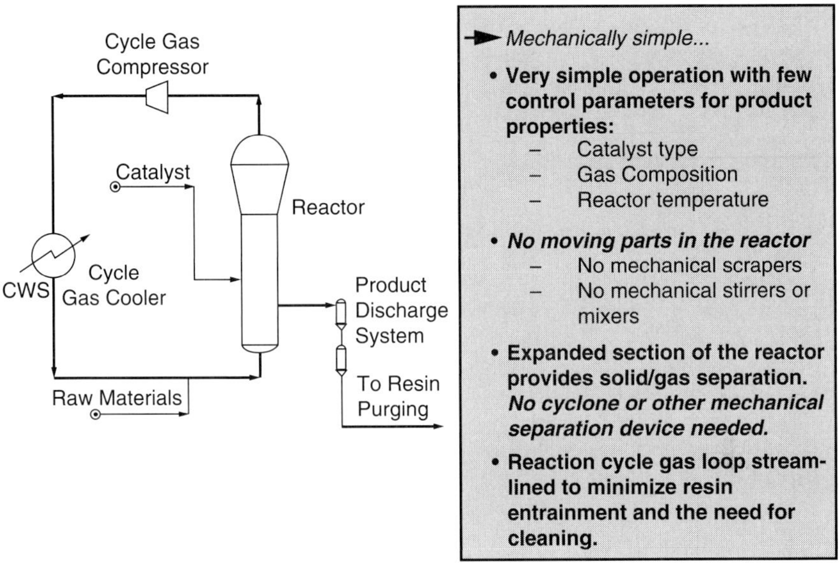

FIGURE 14.9.3 UNIPOL PE reaction system.

granular PE resin with a trace of catalyst. The cycle compressor circulates reaction gas through the bed, providing the agitation required for good backmixing, supplying raw materials to the reaction sites, and carrying away heat of reaction.

- No significant temperature or composition gradients exist within the fluid bed. Reaction heat is removed from the circulating gas by an external heat exchanger. Depending on the reaction conditions, a percentage of the gas can condense into a liquid phase that enhances heat removal from the reactor. Capitalizing on Univation's condensed-mode and super-condensed-mode technologies, operation at high space/time (i.e., tons of production per unit volume) can be achieved easily.
- Product properties are controlled by adjusting reaction conditions. Computer models are used to determine the required reaction conditions based on the catalyst type and the specific product being produced.
- Gas composition is controlled via analyzers, and system pressure is held constant by adjusting the raw materials feed rate. The reaction bed temperature is mild and is controlled at a constant value for each product by adjusting the temperature of cycle gas returned to the reactor. Molecular weight distribution can be varied by changing catalyst type.
- The polymerization reaction can be stopped or slowed down with a "kill" system to protect against emergencies.
- The reaction loop is readily designed to accommodate all catalyst systems, ranging from conventional Ziegler-Natta and chromium systems to metallocene systems, with little or no modifications.

Catalyst Feeder. Polymerization catalysts are supplied in ready-to-use form. Ziegler-Natta and bimodal HDPE catalysts come either as a dry powder or as slurry depending on the catalyst family. Chromium and metallocene catalysts are supplied as dry powders. All are fed directly from the shipping containers into the reaction system catalyst feeder.

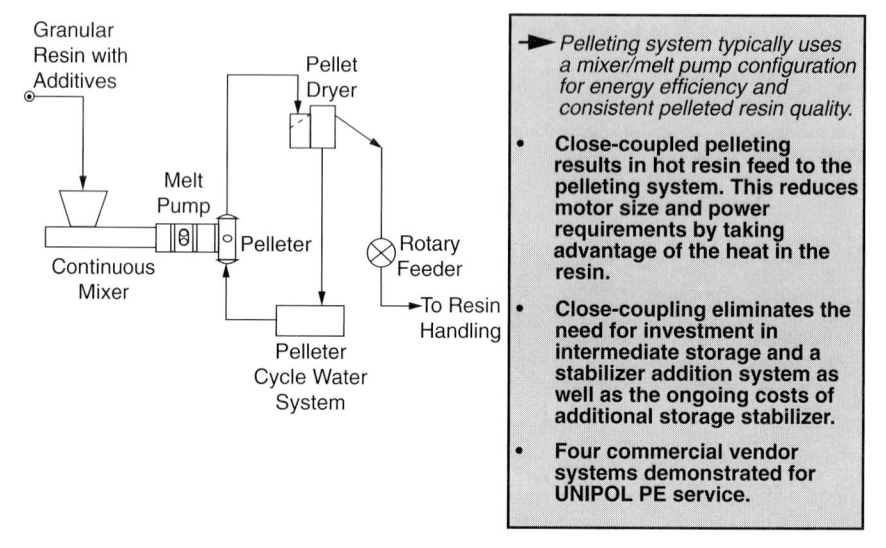

FIGURE 14.9.4 Pelleting.

Resin Additive Handling. Facilities are provided for handling both solid and liquid additives. Solid additives are metered into the main stream of granular resin. Liquid additive is pumped directly from the shipping container into the pelleting system.

Pelleting. The UNIPOL reactor produces a granular resin with a consistent particle size distribution that is readily pelletized. The stability and reliability of the UNIPOL PE process allow the pelletizing operation to be accomplished in the most economical fashion by "close coupling" the mixer and pelletizing operations to the reaction system. Resin and additives flow by gravity directly to a low-energy pelleting system that typically uses a mixer/melt-pump configuration. The pelleting system produces pellets that are air-conveyed to storage (Fig. 14.9.4).

Typical Resin Handling, Blending, Storage, and Shipping

Resin handling, blending, storage, and shipping systems are based on conventional, commercially available equipment and use air conveying throughout. UNIPOL PE facilities in operation use a variety of resin handling systems based on owner preferences for inventory storage, operating schedule of distribution/packaging facilities, etc.

Control Systems

The entire process is controlled by conventional instrumentation, usually a distributed control system (DCS). An enhanced version of advanced process control (APC+) performs special control calculations to optimize operating conditions and expedite product transitions, including closed-loop supervisory control of the conventional instrumentation. Online analyzers are provided to periodically monitor impurities in the feed streams. Continuous online analyzers also provide closed-loop control of the reactor gas composition for resin property control based on model predictions. Models have been developed based on first principles for each catalyst system to control resin properties.

Advanced Process Control (APC+) Software

APC+ software is a new Univation innovation and is a proven integrated process control package that can enhance the performance of your UNIPOL PE process unit dramatically. Engineered specifically for UNIPOL PE process, APC+ is the ideal tool for ensuring that plants run at optimal efficiency and productivity—day in and day out.

APC+ software enhances plant performance by

- Maximizing production rate
- Reducing off-grade product
- Increasing aim-grade consistency
- Improving system reliability

APC+ was developed using engineering first principles of polymerization technology specifically for the UNIPOL PE reactor configurations and unique catalyst technologies. The technology has demonstrated high service factor and low maintenance over many reactor years of UNIPOL PE operation. APC+ was implemented on a robust, open architecture platform for simplified expansion and broad compatibility with the UNIPOL platform.

With APC+, you can count on

- Increased production rates and rate consistency
- Fewer reactor upsets and less downtime
- Better control at higher production rates
- Consistent transitions
- Increased aim-grade production
- Enhanced resin quality and consistency

Proven Results.   In commercial operations, APC+ has demonstrated improvements that translate to millions of dollars in savings each year:

- 5 to 9 percent increase in production rate
- 5 to 7 percent reduction in non-aim-grade product
- 1 to 2 percent reduction in unplanned downtime

Backed by Unrivaled UNIPOL Experience.   Operators have choices when it comes to process control software. But only APC+ incorporates demonstrated UNIPOL best practices and uses engineering first principles for gas-phase polymerization and deep knowledge of the UNIPOL PE reaction system—not empirical models or other generic approaches. And only with APC+ will you get the benefit of Univation's extensive commercial operating experience in gas composition control, production rate control and maximization, and automated resin property control.

PROCESS PERSPECTIVE

Figure 14.9.5 shows the locations of existing plants and locations of plants under construction.

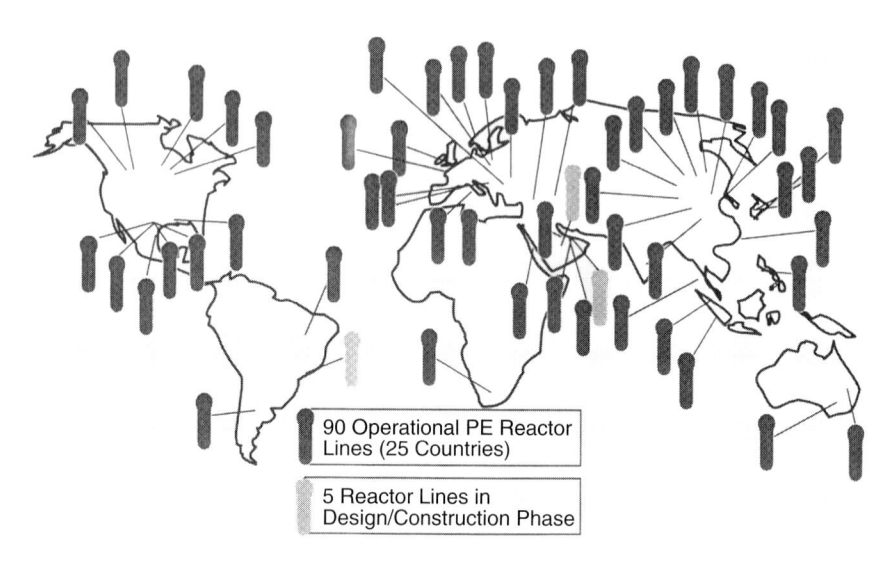

FIGURE 14.9.5 UNIPOL PE production locations around the world.

PRODUCT AND BY-PRODUCT SPECIFICATIONS

UNIPOL PE Products

The UNIPOL PE gas-phase fluidized-bed reaction system has the ability to make a wider range of products than any other PE technology. Combining the UNIPOL PE process with Univation's advanced catalyst systems further expands the product capabilities of the resin producer. Commercial products span the ranges of melt indices from less than 0.01 to greater than 150 dg/min, density from 0.910 to 0.970 g/cm^3 and cover narrow, intermediate, broad, and bimodal molecular weight distributions—the latter currently for HDPE only. This range of properties is not available with competing gas-phase, slurry, or solution-phase processes. As a result of these broad capabilities, products from UNIPOL PE process plants compete in all significant market applications (Fig. 14.9.6).

Hundreds of commercial PE resins are made around the world in the UNIPOL PE process with UCAT Ziegler-Natta and chromium conventional catalysts produced by Univation Technologies. Using XCAT metallocene catalysts and PRODIGY bimodal catalysts can further enhance this broad product capability.

PE resins made by the UNIPOL PE process are marketed worldwide by Univation's licensees under such trademarks as ExxonMobil Polyethylene (ExxonMobil), Ladene (Saudi Basic Industries Corp.), Novapol (Novacor Chemicals, Ltd.), Aecithene (Polifin, Ltd.), Alkatoff (Qenos Australia), Asrene (Chandra Asri), Flexirene, Clearfex, and Eraclene (Polimeri Europa), Liten (Chemopetrol), Titanex (Titan), and Noratec (Mitsubishi Petrochemical Co.).

PE covers a broad range of applications owing to its versatility, product properties, and economics. Since the introduction of LDPE in the 1930s, the demand for PE has grown rapidly, and the evolution has placed ever-increasing demands on the performance of PE products.

PE is the largest volume polymer consumed in the world (Fig. 14.9.7). It is a versatile polymer that offers high performance relative to other polymers and alternative materials

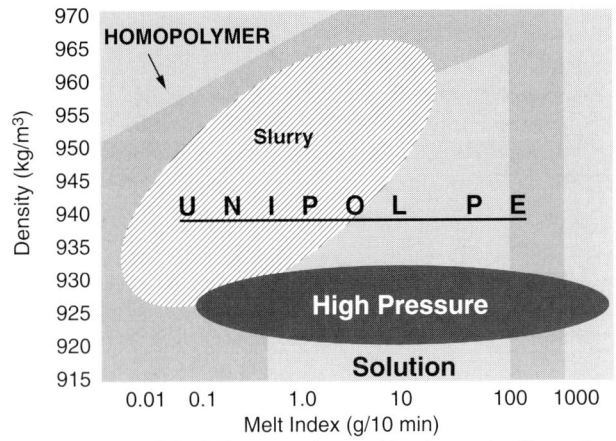

FIGURE 14.9.6 UNIPOL PE process: broad product capabilities.

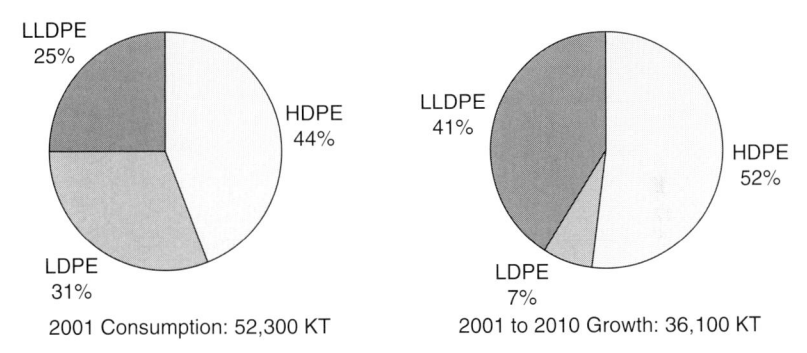

- 2001 Consumption: 52,300 KT
- 2001 to 2010 Growth: 36,100 KT

- Consumption grows at a rate of 6% pa between 2001 and 2010

- Growth rates by product are:
  6.8% pa for HDPE, 1.6% pa for LDPE and 8.8% pa for LLDPE

- LLDPE penetration rate into LDPE moves from 45% in 2001 to 60% in 2010

Source: Univation Technologies

FIGURE 14.9.7 World consumption by PE product.

such as glass, metal, or paper. Univation Technologies offers its licensees the ability to manufacture the industry's broadest PE product portfolio.

Linear low-density polyethylene (LLDPE) is the fastest growing resin due to its penetration in the film market, offering a balance of toughness and stiffness (Fig. 14.9.8). Its out-

Pie Charts

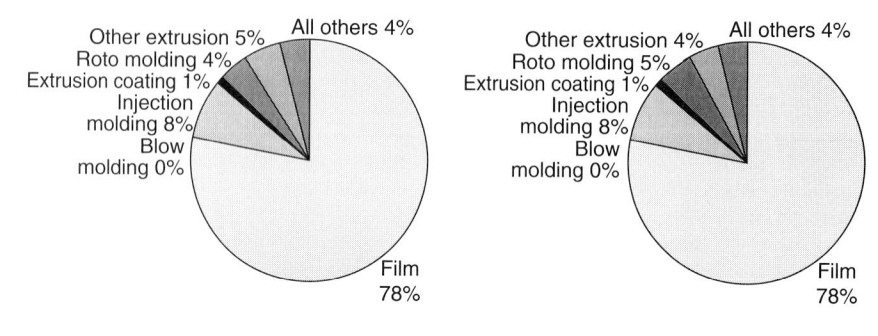

Left chart (2001 Consumption: 13,143 KT):
- Other extrusion 5%
- Roto molding 4%
- Extrusion coating 1%
- Injection molding 8%
- Blow molding 0%
- All others 4%
- Film 78%

Right chart (2001 to 2010 Growth: 14,820 KT):
- Other extrusion 4%
- Roto molding 5%
- Extrusion coating 1%
- Injection molding 8%
- Blow molding 0%
- All others 4%
- Film 78%

- LLDPE consumption grows by 8.8% pa between 2001 and 2010

- Segments with above-average growth rates are: extrusion coating (12.9% pa) and roto molding (9.5% pa)

- Film has an on-average growth rate

Source: Nexant/ChemSystems

FIGURE 14.9.8 World LLDPE consumption by segment.

standing toughness relative to other products has opened new application areas in addition to facilitating material savings through downgauging in existing applications. LLDPE is used either pure or in rich blends with LDPE on extrusion equipment designed to optimize output.

Low-density polyethylene (LDPE) has been one of the most useful and versatile polyolefins ever produced (Fig. 14.9.9). Both autoclave and tubular high-pressure technology offer an excellent combination of low manufacturing costs and product performance, which makes LDPE a key ingredient in thousands of applications.

High-molecular-weight high-density polyethylene (HMW) HDPE is used in a variety of film applications, including institutional and container can liners, grocery sacks, and merchandise bags (Fig. 14.9.10). As industry continues to reduce material costs through downgauging, there has been a need to compensate with stiffer products. The range of HMW HDPE resins has been designed for easy extrusion and is ideal for a wide variety of applications.

WASTES AND EMISSIONS

Product Discharge System (PDS)

As catalyst and raw materials are fed to the reactor, PE is produced and accumulates in the fluidized bed contained within the reactor, causing the bed level to rise. When the bed level reaches a predetermined height, a product-discharge sequence is initiated automatically. A discharge valve opens, allowing a portion of the fluid bed to pass into the discharge system. The discharge system is designed to maximize the quantity of resin exiting the reactor while minimizing the amount of reactant gas that exits. The PE then flows to a transfer system that moves the resin to the purging system. The bed level is controlled by varying the product discharge frequency.

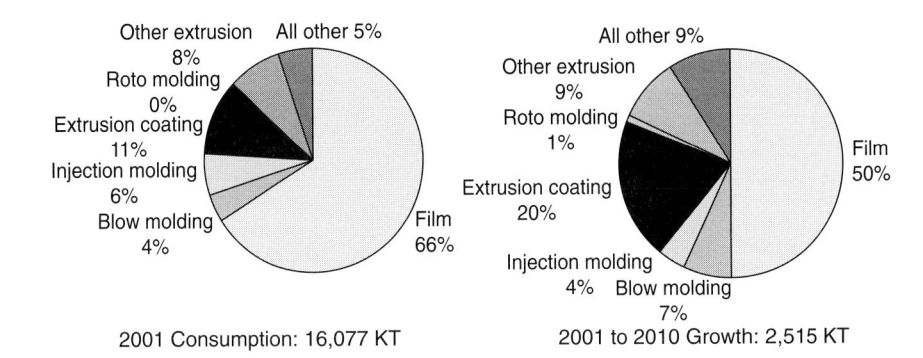

2001 Consumption: 16,077 KT
- Other extrusion 8%
- All other 5%
- Roto molding 0%
- Extrusion coating 11%
- Injection molding 6%
- Blow molding 4%
- Film 66%

2001 to 2010 Growth: 2,515 KT
- All other 9%
- Other extrusion 9%
- Roto molding 1%
- Extrusion coating 20%
- Injection molding 4%
- Blow molding 7%
- Film 50%

- LDPE consumption grows by 1.6% pa between 2001 and 2010

- Segments with above-average growth rates are:
  blow molding (2.8% pa), extrusion coating (2.7% pa) and roto molding (2.7% pa)

Source: Nexant/ChemSystems

FIGURE 14.9.9 World LDPE consumption by segment.

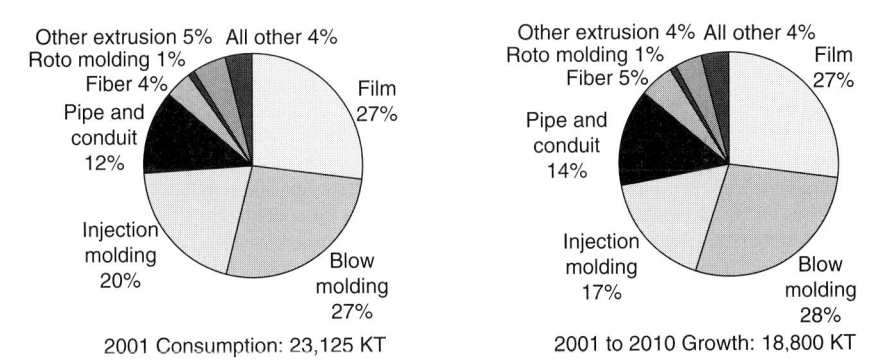

2001 Consumption: 23,125 KT
- Other extrusion 5%
- All other 4%
- Roto molding 1%
- Fiber 4%
- Pipe and conduit 12%
- Injection molding 20%
- Blow molding 27%
- Film 27%

2001 to 2010 Growth: 18,800 KT
- Other extrusion 4%
- All other 4%
- Roto molding 1%
- Fiber 5%
- Pipe and conduit 14%
- Injection molding 17%
- Blow molding 28%
- Film 27%

- HPDE consumption grows by 6.8% pa between 2001 and 2010

- Segments with above-average growth rates are:
  pipe & conduit (7.8% pa), fiber (7.4% pa) and film (7% pa)

- Blow molding and roto molding have on-average growth rates

Source: Nexant/ChemSystems

FIGURE 14.9.10 World HDPE consumption by segment.

Resin Degassing and Vent Recovery

Resin leaving the reactor contains absorbed hydrocarbons. These hydrocarbons are purged and recovered from the resin to meet safety, environmental, monomer utilization, and product quality standards. Resin that has been purged of absorbed hydrocarbons satisfies all safety and environmental requirements.

PROCESS ECONOMICS

UNIPOL PE Process Economics

Based on competitive analyses, Univation is confident that the UNIPOL PE process provides the lowest investment option for the production of PE. Univation can most reliably estimate the inside battery limits (ISBL) portion of a UNIPOL PE facility built on the U.S. Gulf Coast. The method used in the preparation of this estimate is the same as that used by Univation's parent companies for their own UNIPOL PE facilities.

A UNIPOL PE bidding contractor can provide more rigorous investment estimates that are based on the actual plant location and client situation. An estimate of this type would be reliable enough to use for securing financing, project approval, and so on.

Manufacturing Cost Elements

The utility usages that include cooling water, steam, power, nitrogen, and air for the UNIPOL PE process are low and essentially independent of plant capacity on a per-ton-of-resin basis.

The principal raw materials for the UNIPOL PE process are ethylene, butene, hexene, hydrogen, and catalyst. Raw materials requirements per pound of resin are also independent of reactor line capacity and operating rates. In addition, the conversion rate of monomer (ethylene) and comonomer (butene and hexene) to PE is very high in the UNIPOL PE process—98 to 99 percent.

UNIPOL PE reactors have broad turndown capability. Within the limits of control system tuning parameters, the system can be operated virtually at any production rate without affecting product properties. High-capacity reactor lines can be operated at reduced rates with economics comparable with those of smaller reactors of other technologies. Since the investment cost for the UNIPOL PE process enjoys very favorable economies of scale, it is therefore attractive to consider installation of a large(r) reactor line if warranted by the expected timing of future market growth.

Low staffing requirements, including operation and maintenance of a PE facility, are key to minimizing plant period or fixed costs. UNIPOL PE plant staffing is quite low relative to plant capacity. The ISBL area of a UNIPOL PE unit requires only four shift operator positions and one shift laboratory position. No permanent shift maintenance is required; hence UNIPOL PE has very modest fixed costs on a per-ton-of-resin basis. Skill levels of the operators, maintenance personnel, and laboratory analysts are typical of that found throughout the petrochemical industry. No unique skills are required. Because of the simplicity of the different jobs, the operators can easily learn the skills for all the jobs and rotate between all positions at regular intervals.

Catalyst Overview

The Univation catalyst business has the manufacturing and distribution infrastructure in place to deliver PE catalyst all over the world. The Univation catalyst business is the number one producer of catalysts for the PE industry with a catalyst portfolio that includes Ziegler-Natta catalysts, chromium catalysts, metallocene catalysts, and engineered bimodal catalysts.

Catalyst Evolution

Over the years, PE catalysts have undergone dramatic evolution (Fig. 14.9.11). Starting in the 1930s and 1940s, the first PEs were produced in high-pressure reactors using free-radical catalysts that produced some long-chain branching and an assortment of short-chain branches for low-density PE. In the 1950s, Ziegler-Natta (Z-N) and chromium-based catalysts were introduced to make high-density PEs in lower-pressure liquid-phase processes with improvements evolving through the 1970s.

In the late 1960s, Union Carbide (now a subsidiary of the Dow Chemical Company) learned how to use the Z-N and chromium types of catalyst in the gas phase to produce HDPE, and in the late 1970s, it learned how to extend these catalysts to make linear low-density polyethylenes (LLDPE). Gas-phase improvements with these catalysts continue even today. In the latter half of the 1990s, ExxonMobil developed the first commercial single-site metallocene catalysts for use in the gas phase, using their licensed-in UNIPOL PE technology as the development platform.

In early 2000s, Univation began focusing on commercializing a new generation of catalyst capable of producing bimodal products in a single reactor. Traditionally, if producers wanted PE with toughness, stiffness, and impact resistance combined with easy processability, they needed two reactors to achieve those goals.

HDPE resins made using Univation's breakthrough PRODIGY bimodal catalyst offer the best of both worlds: good processability and enhanced performance properties. Univation is currently commercializing bimodal catalyst technology to make bimodal HDPE with these attributes in one UNIPOL PE reactor.

Initiators	Phillips Catalysts	Ziegler-Natta Catalysts	Metallocene Catalysts	Engineered Catalysts
•Low-density polyethylene	•Linear PEs •Primarily high density •Low and high density	•Linear PEs •Low and high density	•Precise control over molecular architecture of highly tailored polymers •Improved processability and properties	•Product performance equivalent to benchmarked multiple-reactor HDPEs •Capital investment savings for PE producers compared to existing staged processes
1930s	1950s	1970s	1990s	2000s

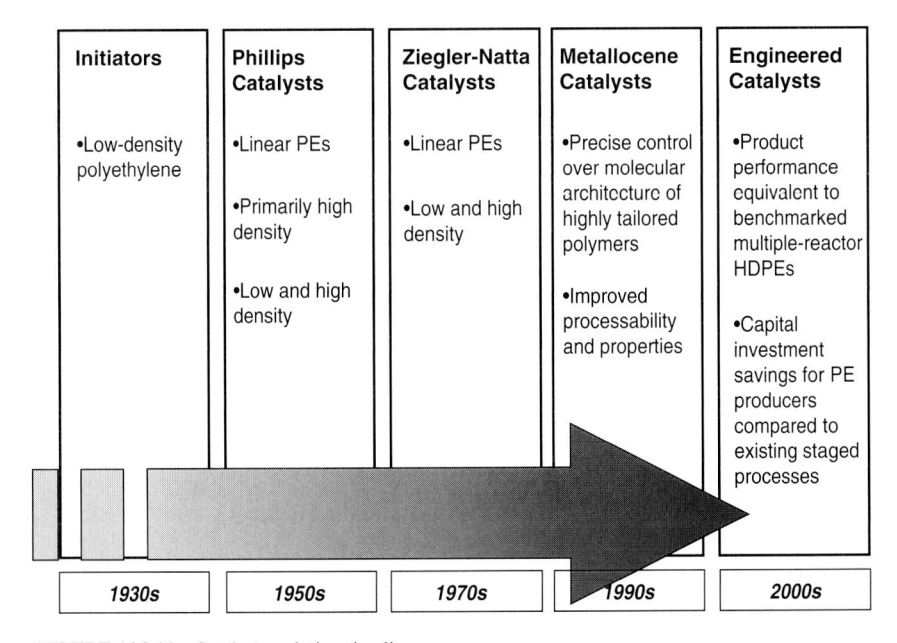

FIGURE 14.9.11 Catalyst evolution timeline.

UNIPOL PE Market Applications

With a UNIPOL PE license, the whole portfolio of catalysts is available—chromium, Ziegler-Natta, metallocene, and bimodal. The combination of this array of catalysts with the UNIPOL PE process allows production of the full range of PE products—high and low density, high and low molecular weight, and narrow to broad molecular weight distributions.

Catalyst Design

A good polymerization catalyst must have high productivity to provide low residues without the need for ash removal at an attractive cost per unit of PE. It must provide controllable particle morphology to achieve good mixing characteristics, high bulk density, and trouble-free conveying. It must have the right kinetic profile to achieve good operability and on-stream time and to ensure uniform product quality. It must have a good response to chain-transfer mechanisms and to comonomer incorporation to provide good control over molecular weight and density. And it must provide the right balance of processability and product properties. With the extensive experience of Dow and ExxonMobil, Univation has the tools and insights needed to develop these catalysts. The results are evident in both the flexible product capability and the outstanding process operability available with UNIPOL PE technology (Table 14.9.1).

XCAT Metallocene Catalyst. Metallocene catalysis brings important process technology benefits to gas-phase PE production that can be used to enhance manufacturing efficiencies with the UNIPOL PE process. With specific metallocene catalyst systems, high catalyst activities, unique process response characteristics, and control of polymer molecular architecture can be combined synergistically in the UNIPOL PE process for improved manufacturing efficiencies. Univation's XCAT metallocene catalysts produce resins that have few low-molecular-weight, high-comonomer-content molecules, resulting in low extractables, low film blocking, and low odor and taste properties. They also have few high-molecular-weight, low-comonomer-content molecules, resulting in excellent clarity, controllable peak melting points, lower heat-seal initiation temperatures, and a broad sealing range. They have narrow compositional distributions, resulting in superior toughness, high hot tack, and faster line speeds. The very narrow molecular weight distributions of XCAT high-performance (HPR) mLLDPE and mVLDPE resins provide higher drawdown capabilities with tough, balanced films. The broader molecular weight distribution of the XCAT EZP mLLDPE resins provides controlled melt strength, good bubble stability, and excellent drawdown capabilities with greatly improved processability and clarity (Fig. 14.9.12).

Along with the traditional advantages of UNIPOL PE technology, metallocene technology provides expanded process capability for producing improved products for conventional markets and novel products for new markets not accessible with older catalysts. These metallocene catalysts also reduce comonomer and hydrogen usage, improve bulk density, reduce resin stickiness, and for low-density products, provide the opportunity to run at higher reactor outputs (Table 14.9.2).

PRODIGY Bimodal Catalysts. Univation's PRODIGY bimodal catalysts are a new generation of catalyst that is capable of producing bimodal products in single UNIPOL PE reactor. While the current focus is on commercialization of bimodal HDPE film and PE-100 pipe product, the catalyst family is, additionally, capable of producing bimodal products suitable for other applications like advanced blow-molding products (Table 14.9.3).

TABLE 14.9.1 UCAT Conventional Catalysts

Ziegler-Natta		Chromium	
\[PE resin\]			
UCAT-J	UCAT-A	UCAT-G	UCAT-B
LLDPE	LLDPE	MDPE	LLDPE
MDPE	MDPE	HDPE	MDPE
HDPE	HDPE		HDPE
Typical applications			
Linear low-density blown and cast films for trash bags, industrial liners, heavy-duty sacks, stretch film, and multilayer constructions.	Linear low-density blown and cast films for trash bags, industrial liners, heavy-duty sacks, stretch film, and multilayer constructions.	High-density film, pipe, and large-part blow molding	High-density blow molding for intermediate-size containers and hose and tubing applications
Linear low density and high density for injection molding and rotomolding	Linear low density and high density for injection molding and rotomolding		

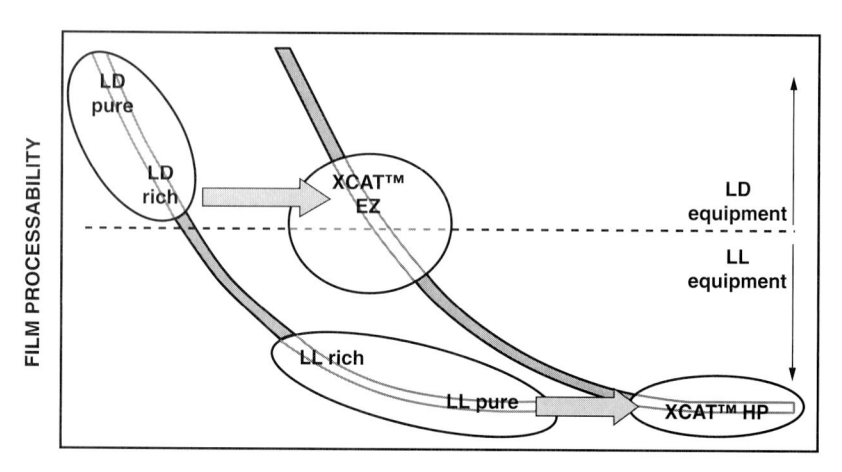

FIGURE 14.9.12 XCAT metallocene LLDPE catalysts; LD and LLDPE film market structure and trends.

TABLE 14.9.2 XCAT Metallocene Catalysts

High-performance metallocene	Easy processing metallocene
\multicolumn{2}{c}{PE resin}	
HPR mVLDPE HPR mLLDPE	EZP mLLDPE
\multicolumn{2}{c}{Typical applications}	
Metallocene very low-density and metallocene linear low-density packaging film for food and nonfood; single-layer or multilayer film; laminates for frozen food, meat, cheese, and poultry Industrial film Heavy-duty sacks Cast and blown stretch film	Metallocene linear low-density packaging film for food and nonfood; single-layer or multilayer film; laminates for frozen food, meat, cheese, and poultry; collation shrink warp Industrial film Heavy-duty sacks Cast and blown stretch film Pallet shrink wrap

TABLE 14.9.3 PRODIGY Bimodal Catalysts

\multicolumn{2}{c}{Engineered bimodal}	
\multicolumn{2}{c}{PE resin}	
Bimodal HDPE BMC-100	Bimodal HDPE BMC-200
\multicolumn{2}{c}{Typical applications}	
Bimodal high-density films—high-strength grocery sacks, shopping bags, trash can liners Blow molding—small bottles, HIC bottles, industrial drums; pressure pipes for potable water and municipal gas distribution	Pressure pipes for potable water, municipal gas distribution, and other demanding applications where PE-100 pipe products are used

		Process Capabilities			Typical Applications				Delivery Mode		Global Availability
		Gas-Phase	Slurry	Other	VLDPE	LLDPE	MDPE	HDPE	Dry	Slurry	2004
Conventional Catalyst (Ziegler-Natta)	UCAT™-A	◆		▲		◆	◆	◆	◆		◆
Conventional Catalyst (Ziegler-Natta)	UCAT-J	◆	▲	▲		◆	◆	◆		◆	◆
Conventional Catalyst (Chromium)	UCAT-B	◆					◆	◆			◆
Conventional Catalyst (Chromium)	UCAT-G	◆				◆	◆	◆			◆
Metallocene Catalyst (Easy Processing)	XCAT™ EZ	◆	◆	▲		◆	▲		◆	▲	◆
Metallocene Catalyst (High Performance)	XCAT HP			▲	◆		▲		◆	▲	◆
Engineered Catalyst (Bimodal/Film)	PRODIGY™ 100	▲				▲		▲	▲		▲
Engineered Catalyst (Bimodal/Pipe)	PRODIGY 200	▲		▲				▲	▲		▲

◆ Commercially Available
▲ Development Progressing

FIGURE 14.9.13 Univation PE catalyst portfolio.

Catalyst Summary

The catalysts available for use with the UNIPOL PE process provide licensees with the ability to manufacture the broadest product mix in the industry. Close to 40 years of catalyst development experience, coupled with the continuing commitment to catalyst R&D, ensures an ongoing stream of new catalyst introductions (Fig. 14.9.13).

UNIPOL PE Process Summary

Mechanically simple, technologically elegant, refined, and robust:

- No moving parts in the reactor
- No cyclone or other mechanical gas/solid separation equipment
- Few pieces of equipment
- Mild process conditions
- No solvents to handle or recover

A single UNIPOL PE unit can produce an unmatched range of products:

- HDPE
- LLDPE
- mLLDPE

Product properties are easily controlled for excellent consistency. Standard raw materials and utilities are used. Excellent conversion costs result from

- Simple equipment
- Condensing mode
- Advanced process control
- Recovery and reuse of vent streams
- Gravity flow of resin and close-coupled pelleting

UNIPOL PE process means leadership in PE manufacturing technology for new plants:

- State-of-the-art UNIPOL PE technology
- Ziegler-Natta and chromium catalyst systems and product capability
- Metallocene catalyst systems and product capability
- Bimodal catalyst systems and product capability
- Advanced process control

Retrofits

- Metallocene and bimodal catalyst systems and product capability
- Capacity expansion technology (CET), including
  - Complete systems analysis
  - Condensed-mode operation
  - Super-condensed-mode operation
  - Advanced process control

Capability to deliver value to expanding traditional markets as well as new applications to meet specific customer needs.

CHAPTER 14.10
NOVA CHEMICALS SCLAIRTECH™ LLDPE/HDPE SWING TECHNOLOGY

Keith Wiseman
NOVA Chemicals Corporation
Calgary, Alberta, Canada

INTRODUCTION

Based on a number of significant process technology developments that started in the late 1930s and continue today, polyethylene (PE) in its many forms enjoys a ranking as the world's largest-volume commodity polymer. This position has been achieved in large part owing to a combination of a diverse range of performance properties and attractive economics of manufacture and use, resulting in the ability to displace a number of more traditional materials (notably glass, metal, and wood) and to create novel applications in market segments across a range of industries. These factors have driven historical PE growth rates to exceed that of gross domestic product on a global basis, resulting in an annual demand approaching 60,000 kt (132 billion lb) by 2004.

Invented by Imperial Chemical Industries in the late 1930s using high temperature and pressures in combination with free-radical initiators to polymerize ethylene molecules into highly branched chains, the first version of this polymer became known as *high-pressure, low-density polyethylene* (LDPE). Produced typically in a density range of 0.915 to 0.930 g/cm^3 and based mainly on performance characteristics of flexibility, optics, toughness, and processability, this material has found acceptance for many years in such areas as agricultural film and a wide range of flexible packaging and coating applications. Recent process developments allowing world-scale plant capacities and improved investment economics are ensuring this material's significant ongoing role and profile in the PE industry.

In the 1950s, the discovery of transition metal–based catalysts and their ability to polymerize ethylene under more moderate conditions than that of high-pressure LDPE resulted in a new form of PE. Initial process developments by Phillips Petroleum in the United States (using chromium-based catalysts) and soon after Hoechst in Europe (using titanium-based Ziegler catalysts) resulted in the invention of a "linear" PE with little or no branching and substantially higher crystallinity than its LDPE predecessor. Produced in a typical density range of 0.947 to 0.965 g/cm^3, this material became known as *high-density polyethylene*

(HDPE). With the ability to now deliver performance characteristics of higher stiffness, tensile strength, hardness, and temperature resistance, these resins found use in new markets such as rigid packaging (bottles, materials-handling containers) and pressure pipe, resulting in new growth opportunities for the PE industry.

Building on the application of Ziegler-Natta catalysts by its U.S. parent company to the production of HDPE, DuPont Canada was the first to extend the envelope into lower-density polymers in a solution process through the copolymerization of linear α-olefins (1-butene and 1-octene) with ethylene. Produced in a density range of 0.915 to 0.940 g/cm^3 with melt indices initially from approximately 0.5 to 10, these new materials represented the invention of what would later be known as *linear low-density polyethylene* (LLDPE) and *medium-density polyethylene* (LMDPE). The capability of this solution process was later expanded further to include densities down to 0.905 (very-low-density PE, or VLDPE) and melt indices from about 0.2 to in excess of 150. These linear copolymers exhibited a unique array of improved performance features, such as toughness, stress crack resistance, excellent sealability, and improved film optics. The ability to combine enhanced performance with tailored processability characteristics resulted in significant growth, particularly in the flexible packaging arena, in addition to such applications as cable jacketing, caps and closures, industrial shipping containers, and storage tanks. As well as producing LLDPE and LMDPE, this new process also was capable of efficiently producing HDPE to 0.965-g/cm^3 density on the same production line—a capability that eventually would be described as linear "swing" PE and be offered globally for license under the name SCLAIRTECH™ linear polyethylene technology. The combination of product versatility and process flexibility allows PE businesses based on the SCLAIRTECH process to satisfy the needs of a diverse range of markets and applications through a varied offering of PE performance features.

NOVA Chemicals Corporation purchased the SCLAIRTECH PE technology in 1994 and immediately invested resources in its growth and development. In a short period of time, several significant developments by the company in the area of catalysis resulted in the invention of an advanced Ziegler-Natta catalyst. Soon after, a unique family of single-site catalysts also emerged. Combining these new catalysts with novel reactor engineering and a series of process enhancements, an evolved process known as the Advanced SCLAIRTECH™ PE Technology was announced in 1998 and first commercialized in Joffre, Alberta, Canada, in 2001 (Fig. 14.10.1). Like its predecessor, this swing technology also produces resins across the full density range, melt indices from fractional to in excess of 150, and a wide range of molecular weight distribution characteristics, now allowing the production of resins tailored to the most demanding customer requirements in the global PE marketplace.

CHEMISTRY AND CATALYSIS

In the SCLAIRTECH process, ethylene, in the presence of a solvent and optionally in combination with an α-olefin comonomer (1-butene or 1-octene), is polymerized by a catalyst at a temperature above the melting point of the resulting polymer. Under these conditions, the monomers and polymer are soluble in the solvent, creating a single phase in the reactor. In the absence of an α-olefin comonomer, high-density PE is produced. This is illustrated for ethylene by the following general reaction:

Homopolymer reaction

$$CH_2=CH_2 + n[CH_2=CH_2] \rightarrow \text{-}[CH_2\text{-}CH_2\text{-}CH_2\text{-}CH_2]_{n+1}\text{-}$$

As increasing levels of comonomer are introduced to the reactor, the comonomer is incorporated into the growing polymer chain, generating short-chain branches. These

FIGURE 14.10.1 NOVA Chemicals' two SCLAIRTECH technology plants in Ontario, Canada.

short-chain branches reduce the ability of the polymer to crystallize in the solid state, consequently reducing the density of the final product. This is illustrated for ethylene/1-butene copolymerization by the following general reaction:

Copolymer reaction

$$CH_2=CH\text{-}CH_2\text{-}CH_3 + n[CH_2=CH_2] \rightarrow \text{-}[CH_2\text{-}CH(C_2H_5)\text{-}CH_2\text{-}CH_2]_{n+1}\text{-}$$

The type and level of comonomer are used to control the final density of the product in a range of 0.905 to 0.965 g/cm^3. Different comonomers result in different physical properties, even at the same density. Typically within a range up to 8 carbons, the longer the α-olefin chain in the comonomer, the better are the physical properties of the resulting polymer. For example, in the SCLAIRTECH process, the use of 1-octene creates a resin of superior mechanical performance than an otherwise similar grade produced with 1-butene comonomer.

Hydrogen is also used in many grades as a molecular weight control agent. When added to the polymerization process, hydrogen terminates the growing polymer chain and generates a transition metal–hydrogen bond into which ethylene can reinsert, initiating a new polymer molecule.

The polymerization of ethylene and the comonomer is extremely exothermic, and the reactors in the SCLAIRTECH process are operated adiabatically (i.e., none of the heat of reaction is removed by cooling). The required reactor temperature varies with the polymer grade, catalyst system, and specific process technology. The reactor temperature, along with other factors such as the hydrogen level, must be controlled to generate the desired polymer chain length and resulting molecular weight. This is achieved by cooling the reactor feed stream and controlling both the ethylene concentration (in the feed) and the mass fraction of ethylene converted. In this manner, the heat of reaction provides most of the energy requirement for the subsequent separation of polymer and solvent.

Typically, a very high (95 percent or greater) per-pass ethylene conversion is achieved. The comonomer conversion is much lower owing to its low reactivity relative to ethylene, the value depending on a number of factors such as the catalyst and comonomer type and comonomer-to-ethylene feed ratio. Unreacted comonomer is separated from the solvent, purified, and returned to the reactor.

The residence time of the reactors is very short (on the order of a few minutes), and extremely active catalysts are employed to deliver high ethylene conversions without high levels of residual catalyst in the product.

A range of proprietary catalyst systems of both the conventional Ziegler-Natta and single-site types is used in the SCLAIRTECH and Advanced SCLAIRTECH processes. The catalysts typically are generated on site from simple commercially available starting materials. These components are pumped as solutions and are mixed under the appropriate conditions of time, temperature, and molar ratios to generate the active catalyst species online in a simple arrangement upstream of the polymerization reactor(s). The use of a Ziegler-Natta catalyst results in a resin exhibiting a relatively broad polymer molecular weight distribution and a short-chain branch distribution in which the lower-molecular-weight molecules have higher comonomer content. The use of a single-site catalyst results in a very narrow molecular weight distribution and a more uniform short-chain branch distribution over the range of molecular weights. For both catalyst types, these factors can be manipulated through the use of multiple reactor designs operated alone or in series.

In the SCLAIRTECH process, the catalysts are based on mixtures of vanadium and titanium compounds used in conjunction with various aluminum alkyls as cocatalysts and impurity scavengers. In the new Advanced SCLAIRTECH process, an advanced Ziegler-Natta–type catalyst, based on magnesium/titanium chemistry, is employed. This catalyst is extremely active and results in low levels of residual catalyst in the resin. NOVA Chemicals also has recently commercialized a proprietary single-site catalyst system that is used in the dual-reactor Advanced SCLAIRTECH process. Unlike many other single-site catalysts, this catalyst system has proven to be very stable and reliable in commercial operation.

Of particular note in the SCLAIRTECH technology is the ability for a single catalyst system (whether Ziegler-Natta or single site) to produce resins across a range of densities, molecular weights, and molecular weight distributions in order to service a wide array of market applications. This is unlike most competitive PE processes that require several unique catalysts to effectively manipulate specific resin characteristics (notably molecular weight distribution). These different catalyst types often are not compatible, and therefore, significant time and effort can be required to change from one to the other during a resin transition in the production cycle. The SCLAIRTECH and Advanced SCLAIRTECH technologies suffer no such transition issues related to catalyst change in either Ziegler-Natta or single-site catalyst families.

PROCESS OVERVIEW

A manufacturing facility based on the SCLAIRTECH technology can be divided into three main process areas:

- Reaction
- Distillation
- Finishing

A reference diagram is shown in Fig. 14.10.2. A brief description of each section is also presented below.

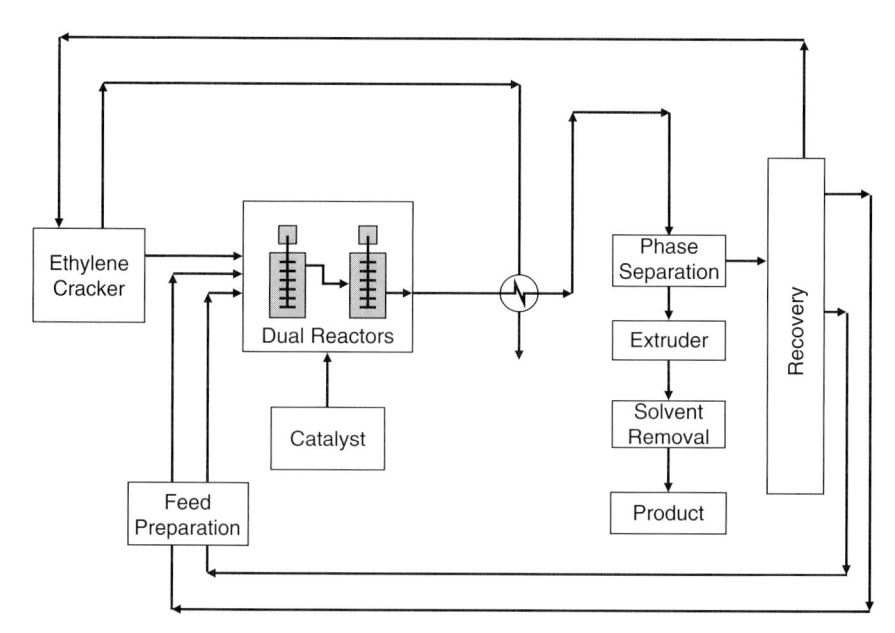

FIGURE 14.10.2 Process schematic for SCLAIRTECH PE technology.

Reaction Area

In order to remove traces of water, oxygen, and other polar impurities that are detrimental to catalyst activity, the solvent and comonomer streams are purified with the use of conventional fixed-bed adsorption systems. Ethylene is fed from outside the battery limits into the reaction system, often without the need for any further purification. Owing to the high conversion of ethylene, typically greater than 95 percent, unreacted ethylene need not be recycled within the plant but rather can be returned to a cracker facility. In cases where there is not a cracker convenient to the PE plant, the required equipment is installed to recycle the ethylene within plant operations.

A solution of ethylene, solvent, and when required, a comonomer is made online and pumped up to reaction pressure to ensure that the polymer will stay in solution throughout the reactor system. Comonomers such as 1-buene and 1-octene are used commonly. Terpolymer products, using a mixture of 1-butene and 1-octene, also can be produced. This feature provides the unique capability to optimize product performance and variable cost.

In the SCLAIRTECH process, the reactors are run completely full (i.e., there is no vapor space), and the contents of the reactors are homogenized by intensive mixing, or alternately, zoned behavior can be introduced.

Two reactor configurations are used. In the original SCLAIRTECH technology, the reactor system consists of a tubular reactor in series or parallel with a continuous-stirred-tank reactor (CSTR). Varying the amount of polymer produced in each reactor, as well as the conditions in each reactor, allows for the broad range of products that can be manufactured. In the newly developed Advanced SCLAIRTECH process, two CSTRs are used in place of the tubular reactor and single CSTR. A single reactor can be operated, or both reactors can be operated in series.

The use of multiple reactors is an important feature of this technology. The flexibility of the reactor configuration allows for a high degree of control over the polymer architecture and is

14.136 POLYETHYLENE

used to modify the polymer molecular weight, molecular weight distribution, and short-chain branch distribution to optimize the physical properties and processability of the final resin.

The final stage in the reaction system is phase separation. Here, the pressure is reduced so that the hydrocarbons, consisting of the solvent, unreacted comonomer, and unreacted ethylene, are removed as vapor to the distillation section, and the molten polymer is fed directly to the extruder as the first step in the finishing operation.

Distillation Area

A simple distillation system is used primarily to separate the solvent, unreacted comonomer, and unreacted ethylene (Fig. 14.10.3). In addition, any impurities or side

FIGURE 14.10.3 Distillation area.

products produced in the reaction area are removed. The solvent and comonomer are then recycled directly to the reaction area. The ethylene is either reclaimed in a cracker unit or recycled internally.

Finishing Area

The molten polymer from the reaction area is further processed in an extruder. Proprietary extruder designs have allowed SCLAIRTECH technology facilities to maximize production rates. In addition, since the polymer is fed to the extruder in a molten state, a relatively small motor is required compared with other process technologies, where material in solid form must be melted during the extrusion step. The extruder system is used to incorporate the required additives, as well as to pelletize the polymer.

The pellets are then stripped with low-pressure steam in a proprietary-design plug-flow vessel to remove any residual solvent. The design of the vessel promotes the diffusion of solvent from the pellet and minimizes the transition losses between different polymer grades (Fig. 14.10.4).

FIGURE 14.10.4 Vessel for removal of residual solvent.

ADVANTAGES OF THE SCLAIRTECH TECHNOLOGY PLATFORM

Advantages of the Dual-Reactor Configuration

The dual-reactor configuration, whether with two CSTRs or a tubular reactor with a single CSTR, provides the ability to control the shape of the molecular weight distribution of resins that are produced. The ability to control conditions in each reactor independently allows a SCLAIRTECH process operator to tailor polymer properties for specific end-use application requirements.

This is particularly true in the Advanced SCLAIRTECH configuration. Using two CSTRs allows for the manufacture of a precise component in each reactor. This provides for the manufacture of products that can be distinctly bimodal in both molecular weight distribution and comonomer distribution. An example of a representative molecular weight distribution for a dual-reactor grade is shown in Fig. 14.10.5.

Minimal Transition Times

All polyethylene plants produce a range of products. The use of reactors exhibiting very short residence time means that grade-to-grade transitions are accomplished more efficiently than competitive PE technologies. The result is less off-specification transition material and greater plant versatility. In recent years, it has been recognized that minimizing the transition time between grades is key to being able to profitably respond to market needs.

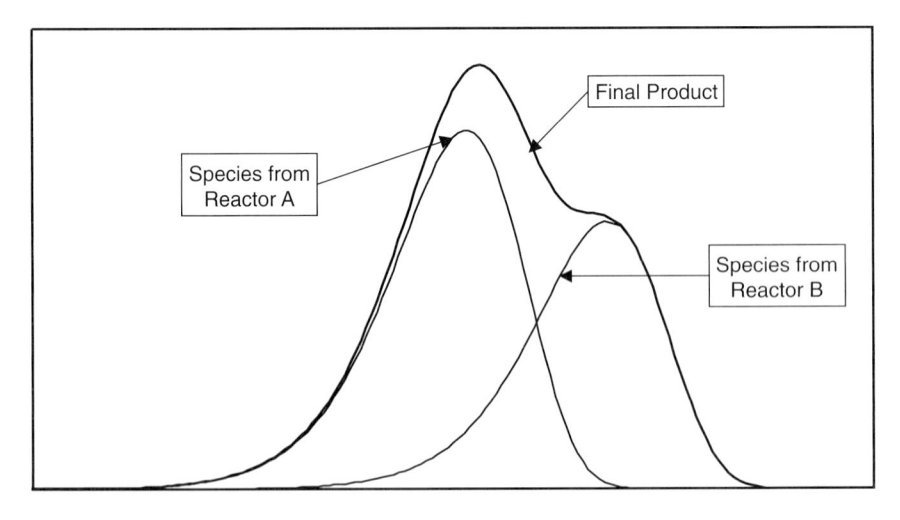

FIGURE 14.10.5 Representative molecular weight distribution of a dual-reactor grade.

In the SCLAIRTECH process, the polymer is maintained in solution throughout the reaction stream. In some competitive PE technologies, process temperature must be maintained below the sticking temperature of the product. This limits the amount of ethylene that can be reacted, resulting in the need for a relatively large holdup of polymer in the reactor. Beyond the negative effect on transition times, this has an impact on the stability of the reactor, as discussed below.

Since the retention time in the reactor is very short in the SCLAIRTECH technology, its contents can be turned over very quickly, typically in the space of several minutes. This allows grade changes to be performed at full operating rates with minimal transition time. Often the specifications for consecutive products will overlap, resulting in no transition losses in the reactor.

Another unique feature of the SCLAIRTECH process is the ability to produce the full range of products with a single catalyst system. This eliminates penalties resulting from the need to shut down, purge the reactor, and restart with a new catalyst. This operation is required in some competitive PE manufacturing technologies, mainly for transitions between narrow and broad molecular weight distribution grades, where catalyst incompatibility can be an issue.

World-Class Safety Record

The SCLAIRTECH technology enjoys an enviable safety record. DuPont Canada and NOVA Chemicals have a combined operating experience of over 71 reactor-years without a process-related lost-time accident.

One of the contributors to this excellent safety record is high single-pass ethylene conversion in the reactor. In technologies where there is a relatively low conversion of ethylene during each pass, a considerable amount of free ethylene may be present in the reactor. In this case, any disturbance that would increase the reaction rate could result in a runaway reaction. These disturbances could include uncontrolled increases to the feed temperature.

In the SCLAIRTECH process, conversion can be maximized to an economic optimum with no safety concerns. Since the per-pass conversion of ethylene is very high, there is little free ethylene available in the reactor. In the situation mentioned earlier, an unexpected increase in feed temperature therefore would result in the evolution of minimal additional heat.

For economic reasons, the reactors are run at or near the maximum operating temperature for the catalyst. The reaction therefore is self-stabilizing. Any upset that unexpectedly increases the reaction temperature is counteracted by slightly lower catalyst activity.

Low Operating Costs

A unique feature of the SCLAIRTECH platform is its ability to employ ethylene and butene of lower purity than is normally considered "polymer grade." The facility therefore can employ a lower-cost feedstock than is possible with other technologies.

Longer term, this can have a greater benefit for the overall petrochemical complex. There may not be a need to debottleneck other plants in the complex as polymer production increases. Rather, the SCLAIRTECH technology facilities can be shifted to an available lower-purity feedstock, freeing up capacity in other units of the complex.

Ease of Capacity Expansion

A series of technology innovations has allowed operators to successfully increase the capacity of SCLAIRTECH technology plants by 50 percent or more with minimal capital investment. This is largely due to the ability to increase the polymer concentration in the

TABLE 14.10.1 Comparison of Economic Factors

	SCLAIRTECH	Competitive low-pressure technologies
Capital cost	Higher	Lower
Operating cost	Similar	Similar
Grade-transition costs	Lower	Higher
Product inventory	Lower	Higher
High-value products	Higher	Lower
Product slate sophistication	Higher	Lower

reactor without a comparative increase in the solvent rate—thereby eliminating the need for changes to the distillation area and reactor feed systems.

ECONOMICS

Assessing business success for a PE venture requires consideration of both cost and revenue factors. Table 14.10.1 compares several relevant economic factors between a business based on a SCLAIRTECH technology facility and that of a typical competitive low-pressure PE technology.

While the investment cost for a plant based on the SCLAIRTECH technology is incrementally higher than that of some comparable low-pressure technologies, such costs can be more than offset by the value realized in the marketplace. This value stems from a combination of specialty, premium-priced resins and a broad slate of products tailored to specific end-use requirements. An owner of a SCLAIRTECH facility, using the full capability of the technology, can realize significantly higher value from the asset.

PRODUCT CAPABILITY

End-use performance characteristics and processability can be designed by manipulation of the four key resin parameters that typically are used to describe a PE resin (i.e., density, melt index, molecular weight distribution, and comonomer distribution). Additives that may be incorporated into any grade, either by the resin manufacturer or by the processor, also confer important characteristics to a resin.

PE resins across the complete density range from VLDPE to homopolymer HDPE are produced on the SCLAIRTECH technology. The typical density range available covers 0.905 to 0.965 g/cm^3. Molecular weight capability (I_2 melt index, as measured by ASTM D1238) ranges from fractional to in excess of 150 dg/min. By manipulating the multiple-reactor configuration characteristic of the technology, molecular weight distribution also can be varied from narrow to broad using the same catalyst system. Copolymers may be manufactured by reacting ethylene with either butene or octene comonomer, or both may be used simultaneously to produce ethylene-butene-octene terpolymers. Manipulation of the relative monomer composition of these terpolymers permits the unique capability to optimize the performance/economics relationship of a resin to the specific requirements of a customer or market application.

Resins produced by the SCLAIRTECH technology deliver an excellent combination of processability and physical performance. They also exhibit very low residuals, low extractables, and excellent film optical properties (as measured by haze, gloss, and gel levels)—characteristics that make them particularly attractive in food packaging applications. The resins have been marketed in North America since 1960 and globally for over 30 years, earning an excellent reputation for quality in demanding, premium-priced applications.

A process such as the Advanced SCLAIRTECH technology that uses multiple CSTRs allows unique species to be produced in each reactor—and combined in situ into the final product. In this manner, a targeted balance of characteristics can be achieved without the need for costly postreactor blending that otherwise would be necessary to realize such performance. In addition to being able to be designed with varied amounts of the components from each reactor, such resins can be truly bicompositional in comonomer content and molecular weight. Resins therefore can be designed with an almost infinite number of combinations to meet the needs of a targeted customer or market.

The Advanced SCLAIRTECH technology also offers a choice of catalyst system to further differentiate product capability. Resins that are produced using the company's advanced Ziegler-Natta catalyst deliver an excellent combination of performance and processability for demanding applications. Where exceptional toughness, optics, and sealing performance are required, NOVA Chemicals' proprietary single-site catalyst is employed. The ability to control rheology using multiple reactors allows these exceptional performance characteristics to be realized without the deterioration in processability commonly associated with resins based on single-site catalysts.

In many cases, unique performance characteristics and suitability for a market application derive from the specific additives used in a resin. The SCLAIRTECH technology offers considerable freedom to tailor performance characteristics through the manipulation of additive formulation. Depending on the processing technique to be used and end-use qualities desired, additives used may include process aids and thermal stabilizers; slip, antistatic and antiblock agents; and ultraviolet light stabilizers.

Combining this inherent versatility with unique process flexibility for rapid, economical grade changes allows the tailoring of resins for a broad range of PE market end uses. Figure 14.10.6 illustrates the broad product capability of the SCLAIRTECH technologies.

Table 14.10.2 illustrates a brief selection from the over 100 distinct grades that have been produced on the SCLAIRTECH PE process by either NOVA Chemicals or its licensees around the world.

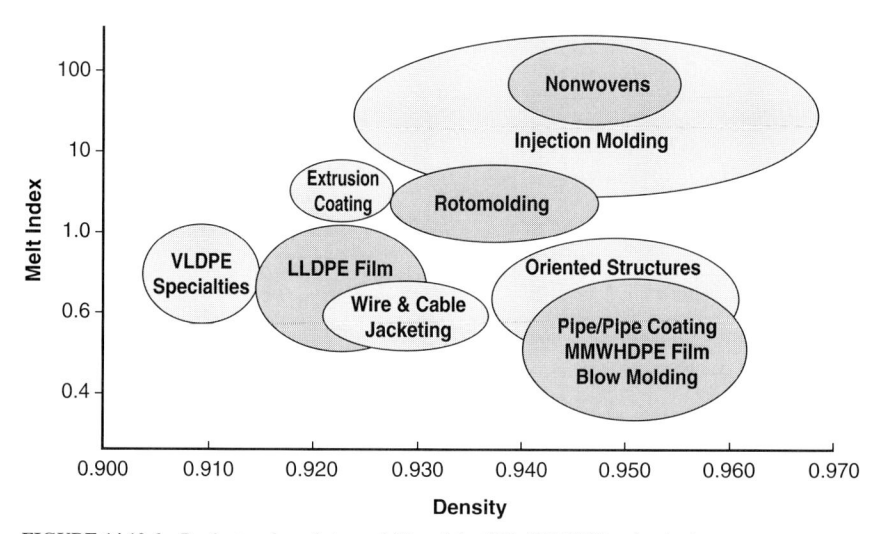

FIGURE 14.10.6 Product and market capability of the SCLAIRTECH technologies.

TABLE 14.10.2 Selection of Resins from SCLAIRTECH Technologies

	Density (g/cm³)	Melt index (dg/min)	Molecular weight distribution*	Comonomer	Market application
Film	0.912	0.5	N	C4/C8	Biaxially oriented shrink
	0.912	0.9	N	C8	VLDPE sealant
	0.917	1.0	N	C8	Clarity sealant
	0.917	4.0	N	C8	Cast stretch
	0.919	0.75	N	C4	Automatic packaging
	0.919	0.75	N	C8	Lamination
	0.920	0.72	N	C4	Liquid packaging
	0.920	1.0	N	C8	Performance packaging
	0.926	0.8	N	C8	Heavy-duty bags
	0.934	2.7	N	C8	Hygiene film
	0.936	0.72	B	C4	Coextrusion
	0.942	0.72	N	C8	Paper overwrap
	0.960	1.2	B	Homopolymer	Moisture barrier film
Extrusion coating	0.919	5.3	N	C8	Paper coating
	0.963	18.2	N	Homopolymer	Woven fabric coating
Injection molding	0.924	19	N	C4	Housewares
	0.926	35	N	C4	Housewares
	0.932	150	N	C8	Thin-wall lids
	0.947	4.8	N	C4	Refuse containers
	0.951	51	N	C4	Ice cream containers
	0.953	7.0	N	C8	Industrial pails
	0.952	62	N	C4	Thin-wall containers
	0.964	10	N	Homopolymer	Crates/totes
Rotomolding	0.924	19	N	C4	Toys
	0.925	4.8	N	C4	Chemical containers
	0.937	5.2	N	C8	Shipping containers
	0.942	1.7	N	C8	Large storage tanks
Blow molding	0.939	1.8	N	C4	Injection blow molding
	0.947	0.34	B	C4	Portable fuel cans
	0.950	0.67	B	C4	Detergent bottles
	0.957	0.41	B	C4	Household chemicals
Pipe and pipe coating	0.942	0.32	B	C4	General-purpose pipe
	0.941	0.38	B	C4	Steel pipe coating
	0.944	3.3	N	C4	Cross-link tubing base
Wire and cable	0.961	0.38	B	Homopolymer	Water bottles
	0.922	11.5	N	C4	Silane cross-link base
	0.936	0.72	B	C4	Cable jacketing base
	0.942	0.62	B	C4	Flame-retardant base
Oriented structures	0.931	0.95	N	C8	Extruded netting
	0.936	1.77	N	C4	Extruded netting
	0.951	31.5	N	C4	Nonwovens
	0.961	0.55	B	Homopolymer	Woven fabric

*N = narrow; B = broad.

TABLE 14.10.3 SCLAIRTECH PE Global Installations and Capacities

Licensee	Location	Capacity (t/yr)	Startup date
NOVA Chemicals (A-line)	Corunna, Canada	125,000	1960
NOVA Chemicals (B-line)	Corunna, Canada	180,000	1977
Polimeri Europa	Priolo, Italy	185,000	1987
SK Corporation	Ulsan, South Korea	190,000	1990
Fushun Ethylene Company	Fushun, China	100,000	1992
Politeno Indústria e Comércio S.A.	Camacari, Brazil	195,000	1992
Reliance Industries, Ltd. (first plant)	Hazira, India	220,000	1992
Reliance Industries, Ltd. (second plant)	Hazira, India	220,000	1997
Polinter	El Tablazo, Venezuela	190,000	1994
Eleme Petrochemicals Limited	Port Harcourt, Nigeria	270,000	1995
GAIL (India) Limited	Etawa, India	160,000	1999
Uzbekneftegaz	Qarshi, Uzbekistan	125,000	2002
NOVA Chemicals	Joffre, Canada*	385,000	2001

*New Advanced SCLAIRTECH technology facility.

COMMERCIAL INSTALLATIONS

The versatility and flexibility of the SCLAIRTECH technology allow a very sophisticated product slate of high-performance resins to be manufactured economically to meet the needs of a broad cross section of the PE marketplace. Commercial experience extends back to 1960, and since the early 1970s, these resins have been marketed globally from Canada into more than 60 countries, earning a solid reputation for lot-to-lot uniformity, quality, and performance. This process has been demonstrated commercially at an annual capacity of 385,000 t and in principle can be employed economically in single-train plants ranging to greater than 400,000 t/yr.

The combination of a sophisticated product slate with attractive economics and a global reputation for quality resins has made the SCLAIRTECH technology an attractive option in the linear polyethylene licensing business. Table 14.10.3 lists the commercial installations of NOVA Chemicals and its global licensees as of the end of 2003. The relative ease with which the SCLAIRTECH process lends itself to economical increases in capacity has resulted in most of these facilities having been significantly debottlenecked from their initial design capacity.

SUMMARY

NOVA Chemicals' SCLAIRTECH linear PE technology is a true swing process capable of economically producing a sophisticated slate of resins across a wide range of densities, melt indexes, and molecular weight distributions. Fast process response and a choice from NOVA Chemicals' proprietary family of advanced Ziegler-Natta or single-site catalysts allow specific resins to be tailored to the individual needs of targeted customers or markets depending on the business strategy of the individual operator. The technology has a long history of safe and reliable operation in many countries around the world. In addition, the resins maintain an excellent global reputation for quality and performance, making SCLAIRTECH technology an attractive and popular choice in the global PE licensing arena (Fig. 14.10.7).

FIGURE 14.10.7 NOVA Chemicals Advanced SCLAIRTECH technology plant in Alberta, Canada.

ACKNOWLEDGMENT

The author would like to thank Dave Houser and Dr. Steve Brown for their contributions to the content of this chapter, and Jamie Nielsen and Rob Jarron for their assistance in its preparation.

DISCLAIMER

Copyright © 2004 NOVA Chemicals Corporation. This information is furnished without warranty, representation, condition, inducement, or license of any kind. All implied warranties and conditions (including warranties and conditions of quality, merchantability, and fitness for a particular purpose) are specifically excluded. No freedom from infringement of a patent owned by NOVA Chemicals or others is to be inferred.

PART 15

POLYETHYLENE TEREPHTHALATE

CHAPTER 15.1
UOP SINCO SOLID-STATE POLYMERIZATION PROCESS FOR THE PRODUCTION OF PET RESIN AND TECHNICAL FIBERS

Stephen M. Metro and James F. McGehee
UOP LLC
Des Plaines, Illinois

INTRODUCTION

The history of thermoplastic polyesters goes back to 1929 with the pioneering work of W. H. Carothers, who synthesized the first polyesters from adipic acid and ethylene glycol. Polyesters only became of industrial interest in 1941, with the synthesis of high-melting-point (high-molecular-weight) products based on terephthalic acid. In the 1950s, Du Pont and Imperial Chemical Industries, Ltd., began manufacture of fiber melt-spun synthetic polyester products. Yet it wasn't until the mid-1960s that thermoplastic polyesters were employed as construction materials. These were fast-crystallizing grades of polyethylene terephthalate (PET) containing additives that gave uniform and controlled morphology. A short time later, copolyesters were developed with enhanced properties. PET containers formed by stretch blow molding were introduced in the United States by Du Pont in the mid-1970s. Since that time, the market for PET has grown at 10 percent per year, with initial growth spawned by direct replacement of glass and metal, as well as the introduction of new products such as heat-set containers.

Today, PET is used to produce a variety of products, including filament and staple fiber, film, tire cord, technical yarns, and packaging resins. PET typically is produced by combining ethylene glycol with either purified terephthalic acid (PTA) or dimethyl terephthalate (DMT) in the presence of a metal catalyst. PTA has been the preferred feedstock since the development of high-purity PTA process in the 1960s. PET is produced in a melt-phase polymerization process unit, a topic for another chapter. The basic building block for PET

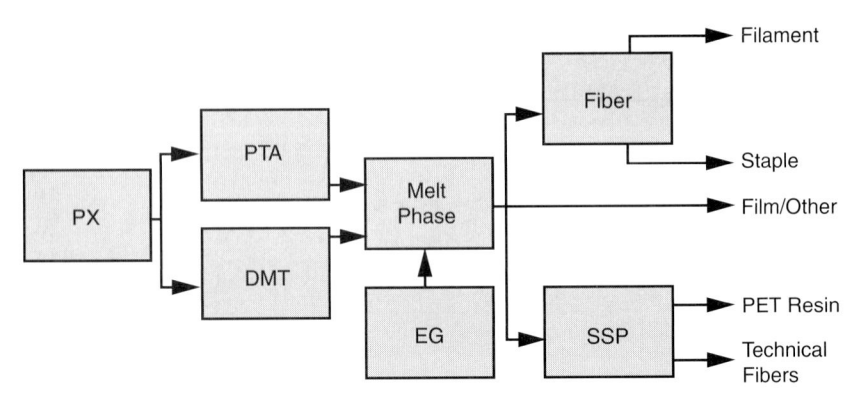

FIGURE 15.1.1 The polyester chain.

is *p*-xylene, produced by UOP's Parex™ process, covered in Chap. 13.3. Refer to Fig. 15.1.1 for an overview of the polyester chain.

According to industry reports for 2004, PET has a worldwide production exceeding 30 million metric tons per annum (MTA). Most of it is homopolymer, which is used primarily for the manufacture of staple and filament fibers. Approximately one-third of all PET produced today is "solid stated" using a solid-state polymerization (SSP) process. SSP was first introduced for the production of industrial technical fibers such as tire cord, which require higher molecular weights to produce the desired mechanical strength and tenacity. SSP, by definition, is the further processing of a polymer in the "solid state" (more correctly, semisolid, or "glassy," state above the glass transition temperature) to increase the molecular weight to much higher levels than can be accomplished easily in the melt phase. The main objective of the SSP process is to increase the polymer's molecular weight, density, and viscosity, thus improving the physical and mechanical properties for select applications.

The most common measurement of the polymer's molecular weight is the intrinsic viscosity (IV). Table 15.1.1 summarizes a few of the major applications and corresponding IVs and molecular weights for solid-stated PET products.

The largest quantity of PET solid stated is for bottle-grade PET resin, exceeding 95 percent of all material produced. Bottle-grade PET resin applications include mineral water (0.72 to 0.78 IV) and carbonated soft drink (CSD) (0.78 to 0.85 IV) bottles. The SSP process increases the tensile strength of the PET resin, permitting high-speed preform and stretch blow molding, producing bottles with excellent clarity and high resistance to stress cracking. The process also reduces the amount of contaminants in the resin, such as acetaldehyde (AA), which imparts a bad taste to foods and beverages. Bottle-grade PET

TABLE 15.1.1 Common Applications for Solid-Stated PET Products

Application	IV (dL/g)	Mn (g/mol)
Textile fibers*	0.57–0.65	17,000–21,000
Film*	0.60–0.65	18,500–21,000
Bottles	0.72–0.85	24,000–31,000
Frozen food trays	0.85–0.95	31,000–36,000
Technical fibers	0.95–1.05	36,000–42,000

*SSP not required.

resin is typically a copolymer containing 2 to 4 percent of isophthalic acid (IPA). High-tenacity technical fibers such as industrial yarns and tire cords are also produced by solid-stating homopolymer PET (no IPA) with product IVs up to 1.05 dL/g.

The solid-state polycondensation process can be either batch or continuous. Batch SSP processing is used for producing smaller quantities of solid-stated specialty products using a large tumble dryer operating under vacuum. Continuous SSP processing is more suitable for production quantities of commodity resins greater than 30 metric tons per day (MTD; 10,000 MTA). The UOP Sinco SSP process is a continuous process with capacities ranging from 1 to 800 MTD. SSP unit capacity limits continue to increase based on market demands for increased production quantities of commodity resins at lower costs.

The SSP process is not limited to only upgrading the PET polymer but is also used commercially to solid-state polybutylene terephthalate (PBT), polyethylene naphthalate (PEN), polytrimethylene terephthalate (PTT), and other polymers such as polyamides. SSP is also used for upgrading postconsumer PET bottles (R-PET recycle applications). This chapter focuses on upgrading virgin PET because it is by far the most common application of the SSP process.

MELT-PHASE POLYMERIZATION

As mentioned earlier, one route to PET production is the reaction of purified terephthalic acid (PTA) with mono-ethylene glycol (EG). The overall chemistry is represented in Fig. 15.1.2.

As shown in Fig. 15.1.1, production of PET resin for packaging and technical fibers typically proceeds through two stages. The first stage involves production of a lower-molecular-weight PET polymer in a continuous melt polymerization process unit. The second stage employs a separate SSP process unit to further increase the polymer's molecular weight for more demanding applications. While the focus of this chapter is the SSP process, the melt polymerization process is described below to provide additional background. It should be noted that SSP is required because it is not possible or practical to produce the high-molecular-weight polymer in a melt polymerization process unit alone (which operates at higher temperatures) primarily due to thermal degradation of the product.

The continuous melt polymerization process, as used for manufacturing food-grade PET resin, is shown schematically in Fig. 15.1.3. PET is a chain-growth polymer that undergoes condensation polymerization in the presence of a catalyst. The melt-phase continuous polymerization (CP) line consists of three to five reactors that are staged as follows:

1. Esterification of EG, DEG, and the mixture of PTA and IPA to produce an equilibrium mixture of cyclic and linear oligomers having degree of polymerization (DP) of 1 to 7.

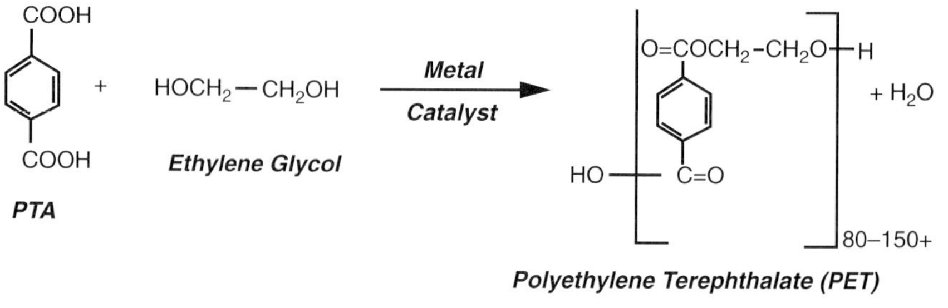

FIGURE 15.1.2 Overall process chemistry: PTA to PET.

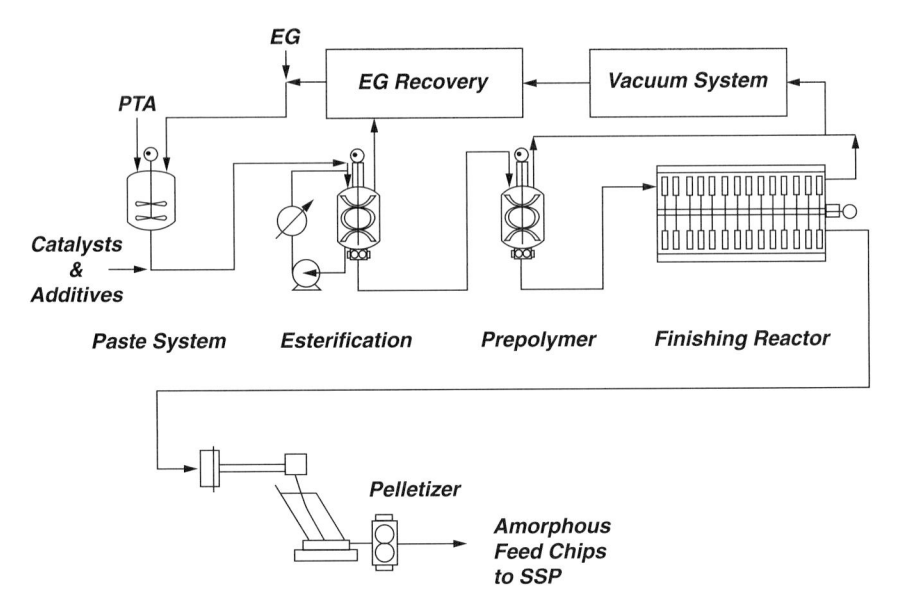

FIGURE 15.1.3 Continuous melt polymerization process.

This process is primarily autocatalytic. Heat is applied to the slurry of reactants, which are not mutually soluble. The reaction is driven forward as the oligomers dissolve the incoming phthalic acids.

2. In the prepolymer stage, antimony or titanium catalysts and other additives are added to the reaction mixture to further increase the molecular weight of the polymer being formed. Vacuum and heat are applied during this stage to remove EG and H_2O to shift the equilibrium toward increased polymerization; the excess EG and H_2O are driven off and subsequently recovered.

3. A finishing reactor is employed in the last stage to increase the degree of polymerization to greater than 30. The finishing reactor can be of a number of proprietary designs; it typically pulls sheets or films of highly viscous melt using techniques that encourage mass transfer. Deep vacuum, approximately 3 mmHg, is used to create the strong driving force for devolution of excess EG and H_2O.

The melt-phase process is concluded by extruding and cooling the molten polymer and subsequently cutting the solidified strands into cylindrical shapes 2 to 3 mm in diameter and length using an underwater strand pelletizer. More recently, some producers have been employing underwater melt cutting equipment to produce PET spheres. The amorphous PET pellets, or chips, are subsequently screened, classified, dried, and sent to intermediate storage silos prior to being fed to the SSP unit.

SSP PROCESS CHEMISTRY

The SSP process is accomplished by heating the polymer in solid-chip form at a temperature above the polymer's glass transition temperature (80°C) but below its melting point

(~250°C). The SSP process typically is conducted in an inert-gas environment such as nitrogen to prevent degradation of the polymer at the elevated temperature.

In solid stating, similar reactions take place to those in the melt-phase polymerization process. The molecular weight of PET is increased by *condensation polymerization* (chain growth). Chain growth can be represented as follows:

$$(\text{Polymer DP} = m) + (\text{polymer DP} = n) = \rightarrow (\text{polymer DP} = m + n)$$

where DP refers to the degree of polymerization, or number of repeating units. Other polymers, e.g., polyethylene or polypropylene, are *step-growth polymers,* which add repeating units one at a time. PET is a linear polymer, for practical purposes, unless branching agents are introduced.

Molecular weight increase (joining two chains together) can take place by either of two primary reactions, *polycondensation* or *esterification.* Polycondensation eliminates one molecule of EG per molecule of PET, and esterification eliminates one molecule of water per molecule of PET.

Polycondensation

$$\text{PET-COO-CH}_2\text{-CH}_2\text{-OH} + \text{HO-CH}_2\text{-CH}_2\text{-OOC-PET}$$
$$K_{1,\text{reverse}} \uparrow \downarrow K_{1,\text{forward}} \qquad (1)$$
$$\text{PET-COO-CH}_2\text{-CH}_2\text{-OOC-PET} + \text{HO-CH}_2\text{-CH}_2\text{-OH}\uparrow$$

Esterification

$$\text{PET-COOH} + \text{HO-CH}_2\text{-CH}_2\text{-OOC-PET}$$
$$K_{2,\text{reverse}} \uparrow \downarrow K_{2,\text{forward}} \qquad (2)$$
$$\text{PET-COO-CH}_2\text{-CH}_2\text{-OOC-PET} + \text{H}_2\text{O}\uparrow$$

Polycondensation and esterification are reversible reactions. In the melt phase, which is at higher temperature and the PET is in liquid form, the K_{forward} and K_{reverse} reaction rates become equal after a certain time in any stage. This is called *chemical equilibrium.* However, in solid stating, conditions are maintained to remove the reaction products rapidly using dry, organic-free gas. It is generally assumed that the K_{forward} reaction rate is much higher than the K_{reverse} rate under good operating conditions.

PET is considered to have "living" end groups, meaning that the polymer is not "dead," or unreactive. Any PET can go on to increase or decrease molecular weight under the right conditions. For example, this can happen under heating with or without moisture, as happens in the drying and extrusion molding of the resin.

Side Reactions

Other important reactions influencing PET quality are those side reactions which arise from thermal or catalytic effects, called *degradation reactions.* The major ones are *acetaldehyde-forming reactions, chain scission (breakage),* and *color formation.*

Acetaldehyde Formation. In conventional PET melt-phase processes, the acetaldehyde content can be 20 to 100 ppm, which is dissolved in the solid feed chip. The mechanism of its formation is from thermal reactions in the melt-phase process finisher that form a vinylester end group (VEG) by dehydration of a terminal ethylene glycol. The VEG subsequently reacts by polycondensation to form a PET chain and one molecule of acetaldehyde:

$$\text{PET-COO-CH}_2\text{-CH}_2\text{OOC-PET}$$
$$\downarrow K_3 \quad (3)$$
$$\text{PET-COOH} + \text{CH}_2\text{=CH-OOC-PET}$$

$$\text{PET-COO-CH=CH}_2 + \text{HO-CH}_2\text{-CH}_2\text{-OOC-PET}$$
Vinylester end

$$\downarrow K_4 \quad (4)$$
$$\text{PET-COO-CH}_2\text{-CH}_2\text{-OOC-PET} + \text{CH}_3\text{CHO}\uparrow$$
Acetaldehyde

or

$$\text{PET-COO-CH=CH}_2 + \text{H}_2\text{O}$$
Vinylester end

$$\downarrow K_5 \quad (5)$$
$$\text{PET-COOH} + \text{CH}_3\text{CHO}\uparrow$$

Water is especially detrimental in solid stating because it reverses the molecular weight increase by the $K_{2,\text{reverse}}$ rate in reaction (2). Any reaction causing chain scission leads to a molecular weight and IV reduction.

Color Formation. PET can form an undesirable yellow color in the presence of oxygen above about 180°C. The reactions are nonspecific but believed to be oxygen-catalyzed dehydration reactions:

$$\text{PET-CO[O-CH}_2\text{-CH}_2\text{OCH}_2\text{-CH}_2\text{O]OC-PET-COO-CH}_2\text{-CH}_2\text{OOC-PET}$$
Diethylene glycol group

$$\downarrow \text{heat, time, O}_2 \quad (6)$$
$$\text{PET-CO[O-CH}_2\text{-C=CH}_2\text{-CH}_2\text{O]OC-PET-COO-CH}_2\text{-CH}_2\text{OOC-PET} + \text{H}_2\text{O}\uparrow$$
Internal double bond

These double-bonded molecules can proceed to condense or cross-link to other structures, which tend to appear yellow in reflected light or fluoresce under ultraviolet light. Solid stating in an all-nitrogen environment in the UOP Sinco SSP process largely minimizes these reactions.

Oligomer Loss. Oligomers of PET are those linear and cyclic polymers having degree of polymerization from 1 to approximately 6. Oligomers generally react quickly in SSP to form higher-molecular-weight PET. A small fraction, however, can vaporize and enter the nitrogen, reacting to form a fine dust that is removed in a gas filter. There is also a small amount of free terephthalic acid that can be found in this dust.

CRYSTALLIZATION OF PET

PET, like polyamides, exists in either an amorphous or a semicrystalline state. The form of the crystals depends on whether the material is oriented (mechanically strain-induced crystallinity) or produced by heating the amorphous material above its glass transition temperature. In the first case, the crystal domains are less than the wavelength of light, and the material remains clear. An example is the oriented sidewall of a PET blow-mold-

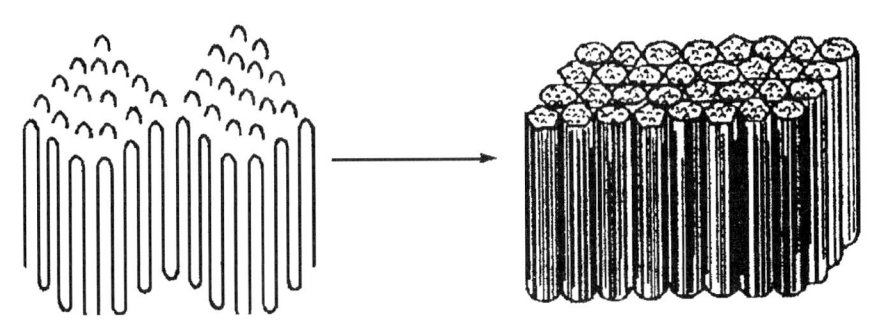

FIGURE 15.1.4 Formation of the internal structure of a PET crystallite.

ed container. In the second case, the crystal domains are larger, and the PET material becomes opaque. This occurs in the typical rapid primary crystallization step at 180 to 200°C in an SSP unit.

The crystal domains appear as round shapes under polarized light and are called *spherulites*. They may not be strictly spherical; using modern techniques such as atomic force microscopy (AFM), a correct picture has begun to appear only recently. According to recent work,[7] the crystallization process in linear PET occurs when the molecular chains in the amorphous phase are aligned and folded to form straight molecular chains at the nanometer scale, and small crystallites are formed. This is represented in Fig. 15.1.4. The crystallites aggregate and align together into polygon-like rod-shaped crystallites when observed at the submicron scale. Finally, large micron-sized crystallites are formed and appear on the edge area of the inside cross section of a PET pellet when divided in half. The structures are not uniform in dimension and contain much amorphous and poorly organized material.

The picture of secondary crystallization is less clear. However, a high perfection of crystal structures is known to be undesirable for good SSP reactivity because buried chain end groups are difficult to react, lowering the overall reaction rate.[8] It is for this reason that the surface material in solid stating is often found to have higher molecular weight than the inside of the particle.

As stated earlier, the main objective of the SSP process is to increase the polymer's molecular weight to upgrade the material for more demanding applications and finished products. The SSP process is conducted at an elevated temperature, typically in the 195 to 220°C temperature range, as described in the preceding section. The feed chips, having an amorphous molecular structure, will stick together or agglomerate if directly polymerized inside a hopper (SSP reactor) at elevated temperatures. To prevent agglomeration, the amorphous feed chips are first crystallized, typically to a crystalline level ranging from 30 to 50 percent. During the crystallization stage, the particles must be sufficiently agitated to prevent sustained interparticle contact, which would lead to sticking under these conditions. The partially crystallized chips become free-flowing after this pretreatment step to permit further processing.

By heating amorphous chips after the softening of the PET at 80°C (glass transition temperature), primary crystallization occurs. The primary crystals are formed rapidly, particularly on the chip's surface and are formed mainly at low molecular weight. When the PET is heated further, secondary crystallization occurs, and the degree of crystallization increases. The secondary crystallization is the result of rearrangement of the primary crystals as they become more "perfect." The upgrading of chips without any sticking problem can be achieved only with sufficiently crystallized chips.

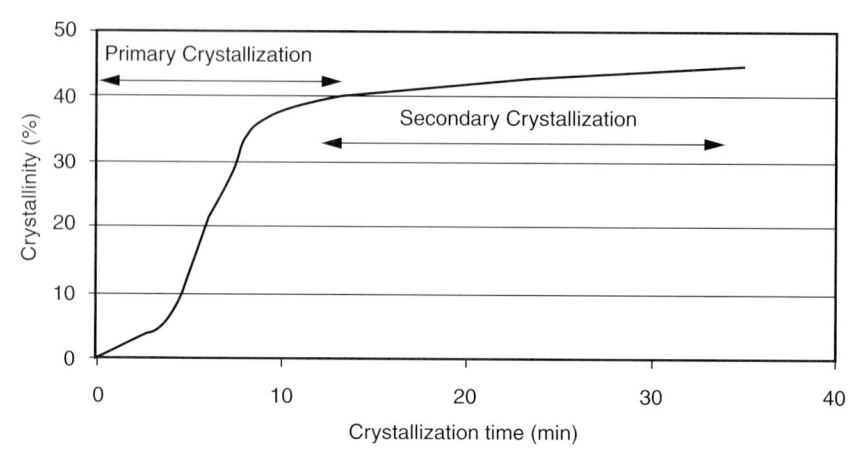

FIGURE 15.1.5 Crystallization of PET.

The UOP Sinco SSP process performs crystallization in two steps: primary crystallization in a fluid-bed precrystallizer and secondary crystallization in a paddle-shaft crystallizer, later described in the detailed process description (Fig. 15.1.5).

STICKING TENDENCY OF PET

Several factors contribute to PET chips' sticking tendency during processing, including chip temperature, chip velocity, chip-to-chip contact time, chip crystallinity, chip pretreatment before reaction, additives/modification levels, SSP reaction time, the consolidation pressure of chips in the SSP reactor, chip shape/geometry/surface area, chip surface roughness, and oligomer and fines content.

The three main factors related to the sticking or agglomeration tendency are

1. PET chip isophthalic (IPA) content; agglomeration tendency increases as IPA content increases.
2. SSP reaction temperature; agglomeration tendency increases as temperature increases.
3. Polymer pretreatment (crystallization); variable effect.

DETAILED PROCESS DESCRIPTION

Most commercial continuous SSP processes are subject to the following processing steps:

1. Precrystallization
2. Crystallization
3. SSP reaction (polycondensation)
4. Product chip cooling
5. Nitrogen purification

Please refer to the simplified process flow scheme contained in Fig. 15.1.6 for reference to the process description.

UOP SINCO SOLID-STATE POLYMERIZATION PROCESS 15.11

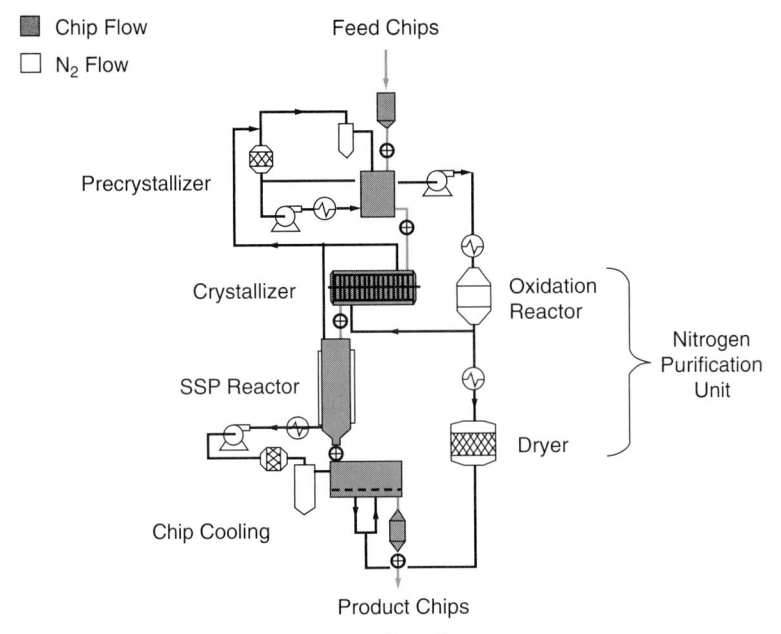

FIGURE 15.1.6 UOP Sinco SSP process flow scheme.

The UOP Sinco SSP process is conducted in a nitrogen atmosphere and follows the preceding steps as follows.

Precrystallization

The amorphous PET chips are transported from raw material storage silos by a dense-phase conveying system to the chip feeding section of the SSP process unit. Amorphous PET feed chips are loaded continuously into the precrystallizer through a rotary valve. The rotary valve serves to isolate the conveying system from the SSP process unit, thus preventing air from entering the precrystallizer.

Chips entering the precrystallizer are immediately subject to a high gas flow at a temperature of ~190°C (374°F), keeping the particles in a high degree of motion. The fluidization action serves to prevent chips from sticking together as the polymer moves through the glass transition temperature (80°C). Nitrogen flowing in a closed circuit is used to fluidize the chips inside the precrystallizer, a fluid-bed-type heat exchanger. The UOP Sinco precrystallizer employs two distinct internal fluidization zones operating at different conditions to optimize the precrystallization process.

The amorphous feed chips entering the SSP process unit typically have PET dust on the surfaces of the chips, which must be removed. The precrystallizer also performs a dedusting operation, removing very small particles that have a negative effect on processing.

The precrystallizer gas loop consists of a blower, heater, cyclone, and filter. The heater can be a heat exchanger operating with electricity, steam, or hot diathermic oil. Diathermic oil is often the heating medium of choice owing to availability from the continuous melt polymerization unit that often accompanies the SSP process unit.

The chips exit the hot dedusting and precrystallization section through a rotary valve positioned above the crystallizer.

Crystallization Section

The dedusted, precrystallized PET chips are fed to the UOP Sinco crystallizer through a rotary valve. The crystallizer is a horizontal mechanical heat exchanger employing motor-driven paddle-type screw conveyers inside to heat and gently agitate the chips. Diathermic oil, also known as *heat-transfer material* (HTM), flows through the inside of the paddles, as well as the crystallizer's external jacket. The design of the machine optimizes heat transfer and crystallization of the PET chips.

HTM is circulated through the crystallizer via a pump. Chips flow from inlet to outlet of the horizontal machine in near plug-flow movement.

Nitrogen flowing countercurrent to the flow of chips inside the crystallizer is used to sweep and remove acetaldehyde and the other volatile compounds released during the heating and crystallization process. UOP Sinco employs other methodology to enhance crystallization, developed after several years of applied research.

The crystallizer heats the chips from ~190°C (374°F) to ~220°C (428°F) and increases the crystallinity to a minimum level required to permit free flow of chips inside the SSP reactor. The crystallizer sets the SSP reaction temperature, and no further heat is added nor removed. The total pretreatment time before SSP reaction is about 1 hour when considering both the precrystallization and crystallization steps.

Solid-State Polycondensation (SSP) Reaction Section

Hot PET chips originating from the crystallizer enter the top of the SSP reactor through a rotary valve located at the reactor's inlet. The SSP reaction takes place inside the SSP reactor, which is a long cylindrical hopper-type vessel. The chips flow down through the tube-type reactor by gravity.

The postpolycondensation reaction typically is conducted in the 205 to 220°C (401 to 428°F) temperature range, but is dependent on polymer properties. The IV is built over a 10- to 15-hour period (typical for bottle-grade PET resin) by controlling the residence time of the moving bed of chips. High-tenacity technical fibers such as tire cord PET resin require longer residence times, as well as different operating temperatures. The required residence time depends on feed and product properties, reactor temperature, reactor volume, etc. The product chips reach a crystallinity level in the 50 to 60 percent range, which depends on a number of variables.

Nitrogen introduced into the bottom of the reactor flows upward, countercurrent to chip flow. The nitrogen stream removes volatile impurities such as EG, water, olygomers, and acetaldehyde, which are released by diffusion from the solid chips during SSP. Removal of EG and water is important because the reaction kinetics are determined by the partial pressure of EG and water, as well as the reaction temperature. The process is operated at a low gas-to-solids ratio, one of the patented features of the UOP Sinco SSP process.

The hot PET chips are discharged from the reactor to the chip cooling and dedusting section through a rotary valve.

Product Cooling Section

The hot PET product chips are cooled down below 60°C (140°F), as required, using the UOP Sinco fluidized-bed cooler. The fluid-bed cooler also performs final dust removal. A shell and tube–type heat exchanger (chip static cooler) also can be used in conjunction with the fluid-bed cooler for chip cooling to decrease electrical energy consumption.

The product chips exit the cooling section though a rotary valve and typically are conveyed via air to product storage silos, bagging operations, or directly to injection-molding machines for perform production. For production of technical fibers (tire cord, etc.), hot product chips can be conveyed directly to spinning machines, eliminating the cooling and reheating steps and thus conserving overall heat energy.

Nitrogen Purification Section

As discussed earlier, volatile organic hydrocarbons and water are liberated during the crystallization and reaction processing steps and must be removed to shift the equilibrium reaction to favor increasing polymer chain length (IV) and to minimize undesirable side reactions. The UOP Sinco SSP process employs a nitrogen purification unit (NPU) to purify the circulating nitrogen. The NPU is another patented feature of the UOP Sinco SSP process. A catalytic combustion process removes the hydrocarbons; the resulting moisture generated during the combustion is removed using dryers containing molecular-sieve adsorbent.

Hydrocarbons are removed by catalytic combustion inside the reactor using a UOP precious-metal catalyst. A precisely controlled amount of air is introduced into the oxidation reactor to permit near-stoichiometric reaction conditions. Oxygen and hydrocarbon analyzers provide feedback to the combustion control loop. The control of oxygen concentration exiting the reactor is very important because higher concentrations can react with the polymer being processed, having a negative impact on product color properties.

The moisture-rich hot nitrogen stream, free from hydrocarbons, passes through an exchanger to partially cool it before being conducted to the molecular-sieve-type dryers. The exchanger picks up heat and transfers it to the gas stream entering the oxidation reactor, thus conserving total heat energy in the system.

Most of the moisture-rich hot gas stream originating from the oxidation reactor is fed to a pair of dryer vessels through an exchanger operating with cooling water. The dryers employ a high-capacity, long-life UOP molecular-sieve adsorbent bed. The dryers operate on a timed cycle. While one dries the process stream, the other is being regenerated. The dryer circuit completely removes water adsorbed by the molecular sieves during the previous operation cycle.

The purified, cooled gas stream is sent to the product cooling section and the SSP reaction section. To compensate for any leakage in the SSP process unit, a small amount of makeup nitrogen is introduced to prevent any air from entering the system.

REACTIONS OF THE CATALYTIC NITROGEN PURIFICATION SYSTEM

Oxidation Unit

In this section the hydrocarbons (EG, acetaldehyde, and traces of heavier hydrocarbons) are oxidized in a catalytic bed reactor using instrument air in near-stoichiometric conditions. The hydrocarbons are oxidized completely to form carbon dioxide and water:

$$2CH_3CHO + 5O_2 \xrightarrow{\text{Catalyst}} 4CO_2 + 4H_2O \quad \text{Acetaldehyde} \tag{7}$$

$$2C_2H_6O_2 + 5O_2 \xrightarrow{\text{Catalyst}} 4CO_2 + 6H_2O \quad \text{Ethylene glycol} \tag{8}$$

Drying Unit

In this section the water is adsorbed on molecular sieves. This is a physical binding of the water to a desiccant, called a *zeolite*, and is not a chemical reaction. The water adsorbed is the sum of the water physically present in the feed, the water produced from the esterification reaction in SSP, and the water of combustion from the preceding reactions in the oxidation unit.

OXIDATION OF PET

Also, at the elevated process temperatures in which the SSP process is conducted (e.g., above 180°C), the PET polymer will react with oxygen and degrade, resulting in poor "b" color and optical clarity for bottle-grade PET resin applications. To prevent oxidation, the pretreatment and reaction processes are conducted in an inert environment. High-purity nitrogen typically is used in commercial SSP processes. After the PET polymer is preconditioned (crystallized), it can be charged to the SSP reactor to increase the polymer IV to the desired level and then is subsequently cooled. Cooling product chips at temperatures below 180°C generally will not have a measurable effect on product color or other quality parameters.

PROCESS VARIABLES

As described earlier, the main variables that are controlled in the UOP Sinco SSP process are the chip residence time and the temperature inside the SSP reactor. Both have direct control on the product chip IV, the main product quality parameter.

As illustrated in Fig. 15.1.7, the reaction rate increases in a fairly linear manner with time for the IV range required for bottle-grade PET resin but decreases at IVs above 0.90, as required for high-tenacity technical yarns (0.95 to 1.05 IV).

As mentioned in the process chemistry section of this chapter, polycondensation and esterification are reversible reactions. The polycondensation reaction is controlled by mass

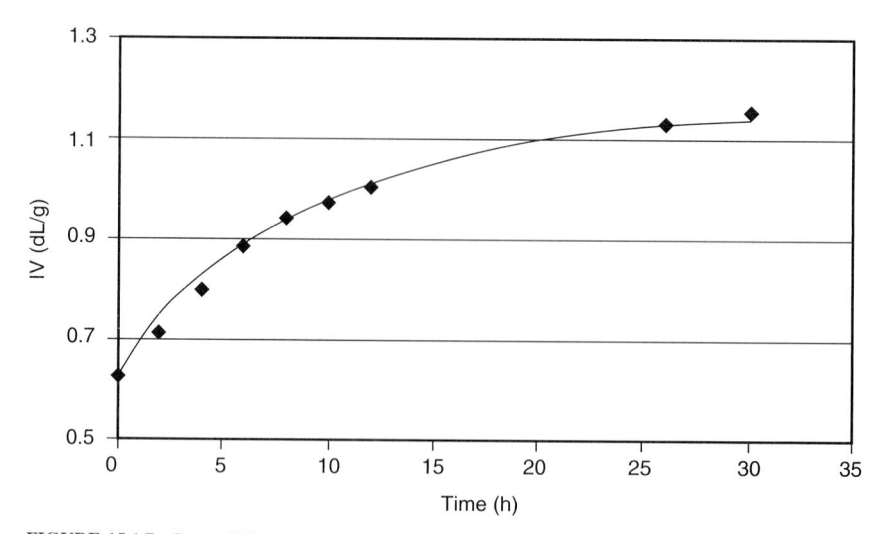

FIGURE 15.1.7 Rate of PET postpolycondensation in the solid state.

transport—molecular diffusion of EG and water in the solid matrix of the PET granules, as well as at the particle-gas interface. Conditions during solid stating are maintained to remove the reaction products rapidly using dry, organic-free gas and thus drive the reaction forward to produce the higher-molecular-weight polymer.

The overall reactivity depends on several feed characteristics, including catalyst type and amount, OH/COOH end-group balance, chip geometry, and comonomer type and amount. The following is a summary of methods for increasing the reaction rate:

- Increase reaction temperature
- Increase catalyst amount (e.g., Sb)
- Optimize (COOH) end-group content
- Increase comonomer content
- Decrease chip size
- Increase gas-to-solids ratio (decrease EG and water content)

FEED PROPERTIES

Table 15.1.2 contains properties for a typical bottle-grade PET amorphous feed.

The amorphous PET feed to SSP units vary among various bottle-grade PET resin producers, including

Catalyst type. The most widely used catalyst is antimony; however, some commercial PET resin producers (TPA-based polymer) use titanium, germanium, phosphorous, etc. Zinc and manganese typically are used for DMT-based melt polymerization processes.

Comonomers. DEG and IPA are most common, but CHDM also has been used. Comonomer content typically ranges from 2 to 6 percent for bottle-grade PET resin. Comonomers such as IPA reduce the polymer melting point, crystallization rate from the melt, and stress cracking propensity. Comonomers permit manufacture of thicker preforms and unique container shapes. Homopolymers are used for production of industrial yarns and technical fibers.

TABLE 15.1.2 Typical Bottle-Grade PET Amorphous Feed

Property	Value
Intrinsic viscosity, dL/g	0.60
Diethylene glycol, % m/m	1.5
Isophthalic acid, % m/m	2.0
Carboxyl end group, mEq/kg	35–45
Antimony (Sb), ppm	250
Cobalt (Co), ppm	10
Phosphorous (P), ppm	20
Acetaldehyde, ppm	60
Moisture content, wt %	0.4
Weight of chips (each), mg	18
Chip dimensions, mm	$2 \times 2.5 \times 3$
Black specks/foreign particles	None
Dust content, ppm	300 (max)
Melting point, °C	~250
Chip temperature, °C	Ambient

TABLE 15.1.3 Typical Bottle-Grade PET Resin Product

Property	Value
Intrinsic viscosity, dL/g	0.85
Intrinsic viscosity lift, dL/g	0.25
Carboxyl end group, mg/kg	Less than base grade resin
Acetaldehyde, ppm	1
Crystallinity, %	>50
Dust (fines), ppm	100 max
Black specks/foreign particles	None
Chip temperature, °C	60 max

Notes: (1) L color and b color are other key quality parameters. O_2 in the process can have a negative affect on b color. L color is controlled by the CP process and is not affected by the SSP. (2) Refer to Table 15.1.1 for other solid-stated products produced and their corresponding IVs.

Chip geometry. Feed chips can be produced as pellets (produced from a strand pelletizer), pastilles (half-spheres produced on a rotoformer), spheres (from underwater melt cutter), powder, or flakes (postconsumer flakes to be recycled). The most common form is the standard cylindrical pellet produced from a strand pelletizer. Different PET resin producers produce different size chips in addition to the type of chips just described.

Initial IV. Typically, this is in the 0.52 to 0.64 range but can be as low as 0.25 in the case of some recent new-generation PET resin technologies.

PRODUCT PROPERTIES

Table 15.1.3 lists the properties of a typical bottle-grade PET resin product.

PRODUCT YIELD

Typically, 1.007 kg of amorphous feed is required to produce 1.000 kg of product. Feed consumed includes dust on incoming feed chips, moisture in feed chips, EG, AA, and H_2O released during the postpolycondensation reaction, as well as a small amount of dust generated from processing the solid chips.

WASTES AND EMISSIONS

The only waste generated from the UOP Sinco SSP process is a small amount of water condensate formed in the nitrogen purification unit. The water condensate is clear and contains a small quantity of dissolved CO_2. There are no waste gas streams or emissions.

UTILITIES

Utilities vary based on process unit capacity and configuration. Required utilities include electric, HTM, cooling water, nitrogen, and air.

FIGURE 15.1.8 UOP Sinco SSP process unit.

EQUIPMENT CONSIDERATIONS

Because the SSP process operates under relatively mild operating conditions, special materials of construction are not required. Process equipment and piping typically are constructed from type 304 stainless steel, ensuring contamination-free operation for food-contact applications. The simplicity of the process design and use of conventional metallurgy result in low capital investment and maintenance expenses for the SSP process.

The UOP Sinco SSP process employs field-proven specialized process equipment developed for SSP technology. In conjunction with the process license and basic design, UOP Sinco supplies key process equipment, including the precrystallizer, crystallizer, and fluid-bed cooler. Molecular sieves and catalysts also have been developed for the SSP application by UOP and are included with the design, license, and equipment package (Fig. 15.1.8).

COMMERCIAL EXPERIENCE

UOP Sinco (formerly Sinco Engineering of the M&G Group) developed and built their first commercial SSP process unit in 1986, having a production capacity of 70 MTD of bottle-grade PET resin. The first commercial SSP process unit was licensed by UOP Sinco in 1992, having a capacity of 90 MTD. UOP Sinco has been at the forefront of many technical advances, including commercializing the largest-capacity SSP units, starting with a 300-MTD unit in 1993. Based on market demands for higher production capacities and lower production costs, UOP Sinco has pushed capacities over 500 MTD for a single-train SSP unit and over 1000 MTD for a dual-process-train SSP unit currently in operation. UOP Sinco's first PBT and tire cord units were introduced in 1996.

UOP Sinco has licensed more than 70 SSP process units for a total installed capacity of over 4.5 million MTA. UOP Sinco has licensed over 40 percent of the SSP capacity worldwide since 2000 after UOP acquired the Sinco SSP business. UOP will continue to innovate and introduce improvements to maximize value to PET resin producers.

REFERENCES

1. *Ullman's Encyclopedia of Industrial Chemistry*, 5th ed., Vol. 28. New York: Wiley VCH, 1997.
2. *Encyclopedia of Chemical Terminology*, 4th ed., Vol. 19. New York: Wiley, 1995.
3. McGehee, J. F., and Witt, A. R., "Future Directions in Solid State Polymerization of Polyethylene Terephthalate," presented at AIChE Spring Meeting, March 2003, New Orleans, LA.
4. Bhatt, G. M., and Fike, L., "Effect of Copolymer Content on the Tackiness of PET Chips During Continuous SSP," presented at the Polyester 2000 World Congress, Zurich, Switzerland.
5. "Polyethylene Terephthalate," Report 02/03-6, Nexant Chem Systems, White Plains, NY, Sept. 2000.
6. Rieckmann, T., and Volker, S., "Solid State Polycondensation of PET," Conference Paper, August 2001.
7. Lu, W., Debelak, K. A., Witt, A. R., et al., "Structural Features of Crystallized Poly(ethylene terphthalate) Polymers." *J Polymer Sci [B] Polymer Physics* 40:245–254, 2002.
8. Hayes, N. W., et al., "Crystallization of PET from the Amorphous State: Observation of Different Rates for Surface and Bulk Using XPS and FTIR." *Surface and Interface Analysis* 24(10):723–728, 1996.
9. Giordano, D., "Approach to Crystallization for the Optimization of the SSP Process," presented at ACHEMA, Frankfurt, Germany, 1997.
10. Mallon, F. K., and Ray, W. H., "Modelling of Solid-State Polycondensation: I. Particle Models." *J Appl Polymer Sci* 69:1233–1250, 1998.
11. Jacobson, L. L., and Ray, W. H., "Analysis and Design of Melt and Solution Polycondensation Processes." *AIChE J* 38:911–925, 1992.
12. James, D. E., and Packer, L. G., "Effect of Reaction Time on Polyethylene Terephthalate Properties." *Ind Eng Chem Res* 34:4049–4057, 1995.
13. Crystallization Kinetics," supplement in *Concise Encyclopedia of Polymer Science and Engineering*. New York: Wiley, 1990.
14. Kang, C. K., "Modeling of Solid-State Polymerization of Polyethylene Terephthalate." *J Appl Polymer Sci* 68:837–846, 1998.
15. Huang, B., and Walsh, J. J., "Solid-Phase Polymerization Mechanism of Polyethylene Terephthalate Affected by Gas Flow Velocity and Particle Size." *Polymer* 39:6991–6999, 1998.
16. Jabarin, S. A., and Lofgren, E. A., "Solid State Polymerization of Polyethylene Terephthalate: Kinetic and Property Parameters." *J Appl Polymer Sci* 32:5315–5335, 1986.
17. Bensoin, J. M., and Choi, K. Y., "Identification and Characterization of Reaction By-products in the Polymerization of Polyethylene Terephthalate." *J Macromol Sci Rev Macromol Chem Phys* C29(1):55–81, 1989.
18. U.S. Patent No. 4238593, to Duh, 1980.

PART 16

POLYPROPYLENE

CHAPTER 16.1
BASELL *SPHERIPOL* TECHNOLOGY FOR PP PRODUCTION

Maurizio Dorini and Gijs ten Berge
Basell Polyolefine GmbH
Frankfurt, Germany

GENERAL PROCESS DESCRIPTION

The *Spheripol* process technology is designed to produce from propylene a wide range of high-quality polypropylene (PP) homopolymers, random copolymers, and heterophasic impact copolymers by copolymerization with comonomers such as ethylene and butene. Single-line polymerization capacity of more than 450,000 metric tons per annum (MTA) is achievable. The process consists of a polymerization unit with the typical loop reactor(s) in which a slurry of liquid propylene and solid PP is circulated; in heterophasic impact copolymer production, this is followed by a gas-phase fluidized-bed reactor. Downstream from the reaction section is a product finishing section, additivation, and extrusion.

PROCESS CHEMISTRY AND THERMODYNAMICS

Homopolymer

The stereospecific polymerization of propylene by Professor Giulio Natta in 1954 was a scientific breakthrough with immediate industrial success. By obtaining alignment of propylene molecules, a crystalline polymer isotactic polypropylene was synthesized at high yields in a manner that could be exploited commercially. This achievement, which spurred interest in organometallic chemistry, as well as in polymerization, was recognized by award of the 1963 Nobel Prize in chemistry to Natta and Ziegler, the latter having discovered the basic catalyst system for polymerizing ethylene.

Polymerizing propylene with various heterogeneous catalyst systems in hydrocarbon diluent bulk liquid phase or gas phase may make isotactic PP.

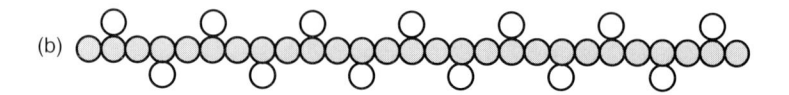

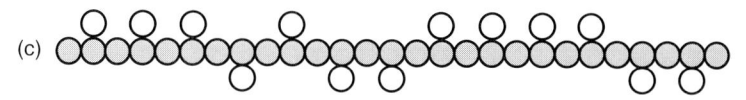

FIGURE 16.1.1 Stereospecificity of PP polymer chain: (*a*) isotactic; (*b*) syndiotactic; (*c*) atactic.

Polymerization Reaction. Propylene can polymerize into three distinct structural chains: isotactic, syndiotactic, and atactic. Isotactic PP occurs when all the methyl groups are arranged on the same side of the carbon chain attached to every other carbon atom, as shown in Fig. 16.1.1*a*. Syndiotactic PP occurs when the methyl groups are attached to every other carbon atom in the chain but are located on alternating sides of the chain, as shown in the Fig. 16.1.1*b*. Atactic PP occurs when the methyl groups are randomly dispersed along the carbon chain, as shown in the Fig. 16.1.1*c*.

In PP homopolymers, the balance between atactic and isotactic fractions has a direct impact on several critical physical properties such as stiffness and resistance to impact and is of primary interest when discussing polymer chemistry. Production of syndiotactic PP requires specialized catalyst and donor systems not yet fully commercially employed.

Role of the Catalyst System Components. The catalyst systems are composed of three basic parts: (1) a solid catalyst, generally $TiCl_4$ supported on $MgCl_2$; (2) stereoregulating agents, an internal and external Lewis base; and (3) an aluminium alkyl.

Catalyst Function. Catalyst is composed of two main elements: a transition metal salt and an inert support structure. The $MgCl_2$ support structure of the catalyst has several roles. It creates a highly disorganized crystalline structure. Thus the active centers, where polymerization takes place, are greater in number and reactivity than those of other conventional catalysts. This explains the high activity of these catalyst systems. $MgCl_2$ contributes indirectly to the stereospecificity owing to interactions with the donor. Finally, owing to its chemical-physical properties, $MgCl_2$ can be converted into particles with controlled shapes and sizes. These characteristics are also transferred to the catalyst and to the polymer.

The active part of the catalyst is titanium tetrachloride ($TiCl_4$). In this form, the catalyst will not actively polymerize and requires "activation" by an aluminium alkyl and a Lewis base.

Catalyst Evolution. The rapid and successful commercialization of PP manufacturing is due primarily to the continuous development of new and improved catalysts. Each evolution or generation of catalysts both reduced the complexity of the manufacturing process and expanded control over resin morphology. A brief summary of PP catalyst evolution is shown in Table 16.1.1.

TABLE 16.1.1 Performance of Different Catalyst Generations

Generation	Catalyst composition	Yield, kg PP/g cat.	Isotactic index, wt %	Morphology control	Process requirements
1	TiCl$_3$/AlCl$_3$ + DEAC	1	90–94	Not possible	Deashing + atactic removal
2	TiCl$_3$ + DEAC	10–15	94–97	Possible	Deashing
3	TiCl$_4$/ester/MgCl$_2$ + AlR$_3$/ester	15–30	90–95	Possible	Atactic removal
4	TiCl$_4$/diester/MgCl$_2$ + TEA/silane (HY/HS)	30–60	95–99	Possible	—
5	TiCl$_4$/diether/MgCl$_2$ + TEA	70–120	95–99	Possible	—
6	Zirconocene + MAO	—	90–99	—	—

Note: DEAC = diethyl aluminum chloride; TEA = triethyl aluminum; MAO = methylaluminoxane.

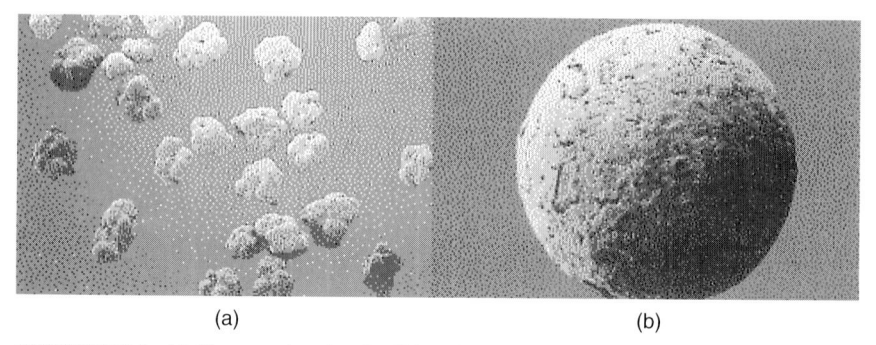

FIGURE 16.1.2 (*a*) Close-up of catalyst for flake. (*b*) Close-up of spherical catalyst.

Catalyst Size/Shape. The polymer flake size and shape are a large copy of the size and shape of the catalyst particle. Average diameter of the polymer particles depends on the average diameter of the catalyst and on the polymerization mileage.

Catalyst Activity. The activity of a catalyst is its tendency and ability to polymerize propylene (and other comonomers) under standard conditions. Activity is measured in units called Zieglers. One Ziegler is defined as one gram of polymer per millimole of $TiCl_3$ per atmosphere of propylene pressure per hour.

Catalyst Types. Several catalysts can be used in the *Spheripol* technology, each tailored according to the final application of the polymer. The catalyst can be granular or spherical form; it can be tailored to have a very high isotactic index, very high mileage, etc. Depicted in Fig. 16.1.2 are microscopic photos of granular and spherical catalyst particles.

Replication Factor. The particle size distribution (PSD) of the PP flake is a replica of the PSD of the catalyst. In the industrial process, a broader PSD flake is obtained than in batch polymerization. More fine particles are produced because of friction in the reactors, flash line, and other processing equipment. Also, fines are produced owing to a residence time distribution that is different in batch polymerization compared with the continuous plant process. In batch polymerization, all the catalyst particles have the same time to grow and make polymer. In the loop reactor, there is a distribution of residence times. There are particles that stay in the reactor for only a few minutes and others that stay for a few hours.

The ratio between the average diameter of the flake and the diameter of the catalyst is called the *replication factor*. The replication factor is proportional to the cube root mileage. The higher the mileage, the larger is the particle of polymer produced. The replication factor generally is calculated by the formula

$$\text{Replication factor} = [(g\ PP/g\ cat.) \times (2/0.9)]^{1/3}$$

Aluminium Alkyl. Polymerization of olefins with Ziegler-Natta catalysts, it is generally understood, involves a stepwise insertion of the monomer into a transition metal-carbon bond as follows:

$$\text{Mt-R} + n\text{C=C} \rightarrow \text{Mt-(C-C)}_n\text{-R}$$

Catalysis of propylene requires the presence of an aluminum alkyl. For the polymerization of propylene, the most widely used alkyl is triethyl aluminum, $(C_2H_5)_3Al$, or TEAL for short. The activation reaction is as follows:

$$TiCl_4 + (C_2H_5)_3Al\text{-} \rightarrow TiCl_3 + (C_2H_5)_2 AlCl + (C_2H_5)^{\cdot}$$

The reaction is exothermic with a ΔH of 131.6 kcal/kg $TiCl_4$.

In addition to assisting in the polymerization reaction, TEAL acts as a scavenger for certain monomer poisons, most notably water, which would seriously impair or completely stop the reaction. Therefore, TEAL is added to the reaction in amounts greater than the stoichiometric amount required to activate the catalyst. The amount of excess TEAL depends mainly on monomer purity.

Stereoregulation and Isotacticity: Donor Function. During polymerization, olefins can be inserted into the carbon-metal bond in two different orientations. Random insertion will produce a material with a high-irregularity, or atactic, structure. Atactic PP is soft and sticky and of little value as a material. The inclusion of a Lewis base or electron donor into the polymerization reaction greatly increases the regularity at which the olefin molecules are inserted into the polymer chain, producing a highly usable stereoregular (isotactic) polymer. The electron donors are called *donor* for short. Along with the evolution of catalysts, there has been a parallel evolution of donors.

The degree to which propylene is polymerized in an orderly or stereoregular manner is measured by the amount of crystalline material produced. This is called the *isotactic index* and is measured as percent by weight. In general, homopolymer PP is produced with an isotactic index in the 92 to 99 percent range. Increasing or decreasing the amount of donor controls the isotactic index of the polymer.

Measurement of isotactic index is a fairly complex test using extraction with boiling heptane. This test generally has been replaced with measurement of the atactic portion, which uses xylene and, more recently, magnetic resonance technology. Most plants refer to the atactic portion of the polymer as *xylene solubles* (XS). While not exactly the same, xylene solubles and isotactic index have been used interchangeably.

Polymer Chain Length Control. The length of the polymer chain has a significant impact on its performance, particularly its flowability. This is a critical parameter for downstream use of the resin. Direct measurement of the polymer chain is difficult and not suited to plant laboratories. Instead, polymer chain length is measured indirectly in several ways. For many years, the intrinsic viscosity (IV) of the polymer was measured. IV results were directly proportional to polymer chain length. That is, the higher the IV, the longer is the average chain length.

While still used to some extent during the production of impact copolymers, IV has been replaced with the much quicker and more repeatable melt-flow rate (MFR). As the term implies, MFR is the weight of melted polymer that can flow through a specific orifice under a standard load for a given temperature and time.

Standard load = 2.16 kg

Standard temperature = 230°C (for very high MFR products, a lower temperature is used)

Standard time = 10 min

As would be expected, MFR results are the inverse of chain length. That is, polymers with long chain length will have a low MFR.

With the development of automated testing machines in the late 1970s, polymer MFR can be measured quickly with high accuracy and repeatability.

Hydrogen is added to the polymerization reaction to control the molecular weight (i.e., chain length) of the polymer by acting as a chain-transfer agent. Hydrogen increases the activity of the catalyst.

Molecular Weight of Polymer versus MFR. Polymer chain length (or molecular weight) is directly related to MFR. Laboratory work has shown the relationship to be closely approximated by the following formula:

$$\log \text{MW} = -0.2773 \times \log \text{MFR} + 5.7243 \qquad (r^2 = 0.9780)$$

Molecular Weight Distribution (MWD) of the Polymer. As with all reactions, the polymer chain lengths produced are not equal. However, for a given catalyst-donor system, the distribution of chain lengths is constant. Changing reaction temperature, residence time, etc., will not significantly affect the distribution.

A fundamental measurement of MWD is *gel permeation chromatography* (GPC), currently known as *size-exclusion chromatography,* which responds to the physical size of the different molecules in dilute solution. In this technique, the polymer solution is passed over a porous medium from which the different-sized molecules elute at different times. The curve of eluted material versus time becomes the MWD. From that curve it is possible to calculate the weight-average MW (M_w) and the number-average MN (M_n). The ratio of these two, M_w/M_n, the *polydispersity,* is a widely used term to describe MWD.

Mathematically,

$$M_n = \frac{\Sigma n_i M_i}{\Sigma n_i} \qquad \text{and} \qquad M_w = \frac{\Sigma n_i M_i^2}{\Sigma n_i M_i}$$

where n_i = number of moles of fraction i
M_i = molecular weight of molecules in fraction i

Because the rheology of PP responds strongly to the MWD, a correlation between the polydispersity from GPC and a rheologic parameter called the *polydispersity index* (PI) has been established. The PI is considerably easier to determine and consequently is growing in use. Die swell (from the melt-flow rate measuring instrument) also corresponds directly with PI, providing a relatively simple method for measuring PI.

The MWD of the polymer plays an important role in the performance of the resin. A narrow MWD is beneficial for fiber applications, whereas a broad MWD improves resin performance in oriented films.

In the *Spheripol* process, broadening of the MW may be obtained by polymerizing widely different MWs (i.e., different melt-flow rates) in separate reactor stages, but this approach is limited by the degree to which the different viscosity may be homogenized in the extrusion process. This method of polymerization is called *bimodal.* In bimodal polymerization, it is important that the high-molecular-weight component be at least 50 percent of the total. Low-molecular-weight material can be blended into the high-molecular-weight material without problem.

Conversely, the addition of alkyl peroxides to the polymer during the extrusion process will cause breaking of longer-chain molecules, thus narrowing the MWD. This process is called *visbreaking* or *cracking.*

Mileage. While catalyst performance or yield has been clearly and precisely defined, the unit Ziegler generally is not used in manufacturing plants. The more generic and slightly less precise term mileage is commonly accepted as the unit for measuring and comparing performance. Mileage is defined as the kilograms of PP produced per gram of catalyst.

Propane concentration has an effect on catalyst mileage. Increasing propane concentration reduces mileage because the concentration of propylene at the catalyst's active sites is decreased. Comparison of mileage results always must be made based on results using the same polymerization basis.

Catalyst Yield: Impact of Residence Time. Catalyst mileage is directly affected by residence time—the period of time in which the catalyst is in the reactors. From its kinetics, the long life of the catalyst allows for a higher residence time and thus the possibility for higher mileage.

All other things being equal, slurry density in loop reactors should be maintained as high as possible to maximize the residence time, hence the catalyst mileage. However, there is an upper limit. To have good processability of the slurry, the proper amount of unreacted monomer is needed to keep the polymer well suspended. The parameter that defines the limit of slurry concentration is the power consumption of the reactor circulating pumps.

Catalyst Yield: Impact of MFR. As discussed previously, the MFR of the polymer is proportional to the amount of hydrogen added to the reaction. Since hydrogen acts to increase the reactivity of the catalyst, catalyst mileage therefore is proportional to MFR. When comparing catalyst mileage, it must be done at the same MFR.

Catalyst Yield: Impact of Propane Concentration. Another important factor affecting catalyst performance is the propane concentration in the reactors. A small amount of propane enters the reactors as part of the propylene feed stream (generally 0.2 to 0.7 percent) entering the plant battery limits. Plants with propane splitters operate with as much as 6 percent propane in the supply propylene. Obviously, the higher the propane concentration in the reactor, the lower is the propylene concentration. Since polymerization rate is a direct function of the active center concentration of the catalyst and the monomer concentration, the higher the propane concentration, the lower is the reactivity.

Propane entering with the feed monomer does not react. Approximately 50 percent of the monomer entering the reactors is polymerized, with the remaining being separated, recovered, and recycled. Thus propane concentration in the reactor builds up with time. The increasing propane level reduces catalyst mileage. In order to prevent a significant loss in mileage, a small stream of propane must be purged continuously. Increasing propane concentration reduces the cost of purged monomer but increases catalyst cost (i.e., lower mileage). As such, there is an economic balance point between catalyst mileage and propane level in the reactors. This balance depends on the cost of monomer and the cost of catalyst.

Monomer Quality. The quality of the monomer in the reactors has a great impact on both catalyst mileage and resin quality. Monomer contaminants fall into several general categories depending on how they affect the reaction. The first group of contaminants reacts with the aluminum-alkyl cocatalyst and greatly reduces catalyst yield when present at the parts per million level. Examples of these materials would be CO_2, H_2S and H_2O. Water is by far the most common poison of the group.

The second group of contaminants reacts with the active sites of the catalyst and seriously reduces catalyst yield when present at the parts per billion level. Examples of these poisons are CO, arsine, phosphine, etc.

The third type of contaminants is hydrogenating compounds, which have a strong effect on the MFR or the polymer and, consequently, on the yield. Some of these also will affect the stereospecificity of the polymer—generally increasing the atactic level.

Random Copolymer

Random Factor. Adding ethylene directly to the polymerization reactor(s) produces random copolymer. Ethylene molecules are added to the polymer chain at random locations. The higher the concentration of ethylene in the feed, the higher is the chance of having two or more ethylene molecules bound between the propylene molecules. The random factor is the ratio between the percentage of single-bound ethylene and the total percentage of bound ethylene. The higher the ethylene concentration in the feed, the lower is the random factor. In other words, increasing the ethylene concentration in the copolymerization reaction increases the probability of having short-chain polyethylene between two molecules of propylene. No parameters have been developed to control the random factor. The catalyst inherently produces the random factor.

Polymerization Reaction. To produce random copolymer, ethylene is added directly to the polymerization reactors. The addition of ethylene increases the polymerization rate significantly. This behavior may be accounted for by assuming that propylene insertion in the Ti-carbon bond is easier when an ethylene unit is bonded to the transition metal or that ethylene insertion after a regioirregular propylene insertion reactivates a "dormant" site. The reactivity ratio is the ethylene-propylene mole ratio in the polymer divided by the ethylene-propylene ratio in the liquid monomer circulating in the reactor.

Polymer Tacticity. Adding ethylene to the polymer increases the xylene soluble fraction. On average, having 1 percent ethylene bound in the copolymer gives an increase in xylene soluble fraction of about 1 percent. Therefore, when making homopolymer with 3 percent xylene soluble, the corresponding 3.5 percent random copolymer (RACO) would have a total xylene soluble level of 6.5 percent. This increases the stickiness of the polymer and contributes to handling and processing problems.

Ethylene does not affect the hydrogen-propylene relationship. That is, the same amount of hydrogen is required to meet a specific MFR for both homopolymer and random copolymer.

The addition of ethylene to the polymer chain lowers the melting point significantly. With 3 to 4 percent atactic fraction, homopolymer polypropylene has a melting point of approximately 165°C, whereas a random copolymer with 2 percent ethylene bonded has a melting point of about 150°C.

Physical Characteristics of RACOs. The addition of ethylene to the homopolymer chain modifies the physical properties of the resin as follows:

- Increases clarity (beneficial for injection-molding applications, such as the type of resins used in food containers)
- Reduces melt temperature (beneficial for some film applications)
- Reduced stiffness
- Increased impact resistance (at moderate temperatures)

Heterophasic Copolymers

Polymer Structure. Heterophasic copolymers (commonly called *impact copolymers*, or HECOs) are an intimate mixture of homopolymer and bipolymers. Bipolymers are mainly amorphous materials produced through polymerizing ethylene and propylene in the gas-phase reactor (GPR) installed downstream of the loop reactors. Homopolymer is first produced in the loop reactors. The flake flows to the GPR, where ethylene and propylene are fed in a predetermined ratio. In the GPR, bipolymerization occurs, and the bipolymer grows inside the homopolymer sphere. The result is a semicrystalline matrix (homopolymer) with a nearly amorphous elastomeric component (bipolymer) dispersed within it. The final product is heterophasic in that it is a mixture of two different products: homopolymer and bipolymer. In fact, by fractionation, it is possible to separate the two materials from each other.

Catalyst Requirement. To produce a polymer with the desired characteristics, a proper amount of rubber must be produced in the copolymerization step, and consequently, a long-life catalyst structure showing a high residual activity after the homopolymerization step is needed. Furthermore, the rubber must be retained inside the flake granule and homogeneously dispersed in the homopolymer matrix to prevent undesired reactor fouling. Consequently, a proper catalyst architecture is required. This polymerization technique, in fact, behaves in such a way that the rubber grows around the homopolymer microgranules. The internal voids of the homopolymer granule are occupied by the growing rubber, which, after the pores are filled, migrates to the particle surface. A high porosity thus is required to bear the desired amount of rubber inside the granule and prevent its migration to the surface with consequent reactor fouling and operability problems.

Reactivity Ratio. The bipolymer that is produced with gas of a certain composition of ethylene and propylene will have a different composition than the gas because of the different reactivities of the monomer with the catalyst. Ethylene is more reactive than propylene in the gas phase. Knowing the reactivity ratio and the gas composition, it is possible to calculate the composition of the bipolymer. As a result, a specific bipolymer composition can be produced by controlling the gas ratio in the GPR.

The gas ratio $(C_2^=)/(C_2^= + C_3^=)$ is controlled carefully in order to produce the correct bipolymer composition. Bipolymers should have good elastomeric properties, thus improving the impact resistance of the copolymer, a highly desirable characteristic property.

Control of Physical Properties. With homopolymers, where physical properties play a less important role, only the molecular weight of the polymer (i.e., MFR) and the amount of atactic material (xylene solubles) are controlled. Hydrogen is used to control the MFR, and the TEAL-donor ratio is used to control the level of xylene solubles.

With heterophasic copolymers, the amount and composition of the bipolymer fraction play critical roles in determining the physical properties of the polymer. Physical properties are controlled as follows:

Physical property/parameter	Is controlled by adjusting
Catalyst mileage/activity	Slurry density in reactor
MFR of homopolymer	Hydrogen concentration in (loop) reactor
XS of homopolymer	T/D ratio in the (loop) reactor
$C_2^=$ content of the copolymer	Total $C_2^=$ fed to the GPR
$C_2^=$ content in biopolymer	Gas ratio $(C_2/C_2 + C_3)$ in the GPR
IV of the biopolymer	Hydrogen ratio (H_2/C_2) in the GPR
Final MFR of the copolymer	Hydrogen ratio (H_2/C_2) in the GPR

Balance of Physical Properties. The critical physical properties of copolymers are MFR, flexural modulus, tensile strength, Izod impact and dropped-weight impact, and HDT. In polymeric materials there is a balance between stiffness and resistance to impact. Stiffness generally must be reduced in order to gain impact strength. This relationship holds true within any given copolymer structure. However, it may be possible to somewhat increase both the impact and stiffness by changing either or both the catalyst and donor systems.

Table 16.1.2 shows how physical properties are affected by changes in polymer structure. This table should be used as a general guideline only because the additive package used can influence the final properties of the resin.

TABLE 16.1.2 Heterophasic Copolymers: Standard Parameters versus Mechanical and Optical Characteristics

Action ↑	Unit	Flex modulus	Izod	Ball drop Trans. temp.	Ball drop Energy	Stress whitening	Tensile yield	Gloss	Hardness	HDT
Homopolymer										
Melt-flow rate	g/10 min	←	→	←	→	→	←	→	←	←
Molecular weight distribution	—	←	→	x	←		←		←	←
Isotactic index	%	←	x		x					
Bipolymer										
Total amount of biopolymer C_3-C_2	% wt	→	←	→	←	→	→	→	→	→
Xylene soluble of biopolymer	% wt	→	←	→	←	→	→	x	→	→
Intrinsic viscosity of xylene soluble	dL/g	→	←	→*	←	←	→			
Ethylene content in xylene soluble	% wt	→*	←*	→*	←		→*	→*	→*	→*
Xylene insoluble (crystalline biopolymer)	% wt	→*	←*	→*	←	←*	→*	→*	→*	→*
Intrinsic viscosity of xylene insoluble	dL/g	→				←				
Ethylene content in xylene insoluble	% wt		←*	→*	←*	←*	→*	→*	→*	→*
Polyethylene	% wt	→*								

Note: When the action increases (↑), the parameters of the left column, the mechanical and optical properties, increase (↑), decrease (↓), or remain unchanged (x). If the increase of this fraction means a reduction of the amorphous one, these correlations are inverted.
*At constant amorphous fraction content.

SPHERIPOL PROCESS PERSPECTIVE

Today's *Spheripol* process is the result of 40 years of continual improvement. However, to truly appreciate the unique capabilities of this technology, it is helpful to understand the evolution of the PP industry and the breakthroughs that led to discovery of the *Spheripol* process.

In the 1960s, PP processes employed first-generation low-yield catalysts (<1000 kg PP/kg catalyst) in mechanically stirred reactors filled with an inert hydrocarbon diluent. Polymer produced with these catalysts had unacceptably high residual metals and contained 10 percent atactic PP, which required separation. Removal of catalyst residues and atactic PP involved treatment of the polymer with alcohol, multiple organic and/or water washings, multistage drying, and elaborate solvent, amorphous, and catalyst separation systems. These processes were costly and difficult to operate; they also required extensive water treatment facilities and catalyst residue disposal systems.

An intermediate step was reached with the second-generation catalyst, increasing yield (6000 to 15,000 kg PP/kg catalyst) and isotacticity but not yet at a level that allowed simplification of the production process.

In the 1970s, the discovery of third-generation high-yield catalysts (15,000 to 30,000 kg PP/kg catalyst) eliminated the need for catalyst residue removal, but atacticity was still unacceptably high. This simplified the washing but did not eliminate the atactic recovery steps.

In the 1980s, fourth-generation high-yield, high-selectivity (HY/HS) catalysts (30,000 kg PP/kg catalyst) eliminated the need for catalyst and atactic removal. This further simplified the process and improved product quality. Other breakthroughs occurred in the process design through the refinement of gas-phase and bulk polymerization reactors that led to the development of *Spheripol* technology in 1982.

Today, the capabilities of the *Spheripol* process are further enhanced by the current catalyst generation, which has the ability to produce new families of reactor-based products with improved properties. They offer even greater control over morphology, isotacticity, and molecular weight and continually challenge new frontiers in the development of propylene polymers.

By mid-2003, 91 *Spheripol* PP lines had been licensed for a capacity of over 16 million t/yr. *Spheripol* technology accounts for 45 percent of all high-yield, high-selectivity PP technologies available.

PROCESS DESCRIPTION

The *Spheripol* process, using HY/HS catalysts supplied by Basell, has the unique ability to produce polymer spheres directly in the reactor. Spherical PP differs considerably from the small, irregularly shaped, granular particles produced with some other technologies and provides significant advantages in terms of process reliability.

The *Spheripol* process (Fig. 16.1.3) is a modular technology. In its most widely adopted configuration, the polymerization section involves the following main units:

- Catalyst feeding
- Polymerization
  - Prepolymerization
  - Bulk polymerization (homopolymer/random copolymer and terpolymer)
  - Gas-phase polymerization (heterophasic impact and speciality copolymer), optional (gas-phase copolymer unit can be added at a later stage without affecting initial plant configuration or involving significant implementation costs)
- Finishing

16.14 POLYPROPYLENE

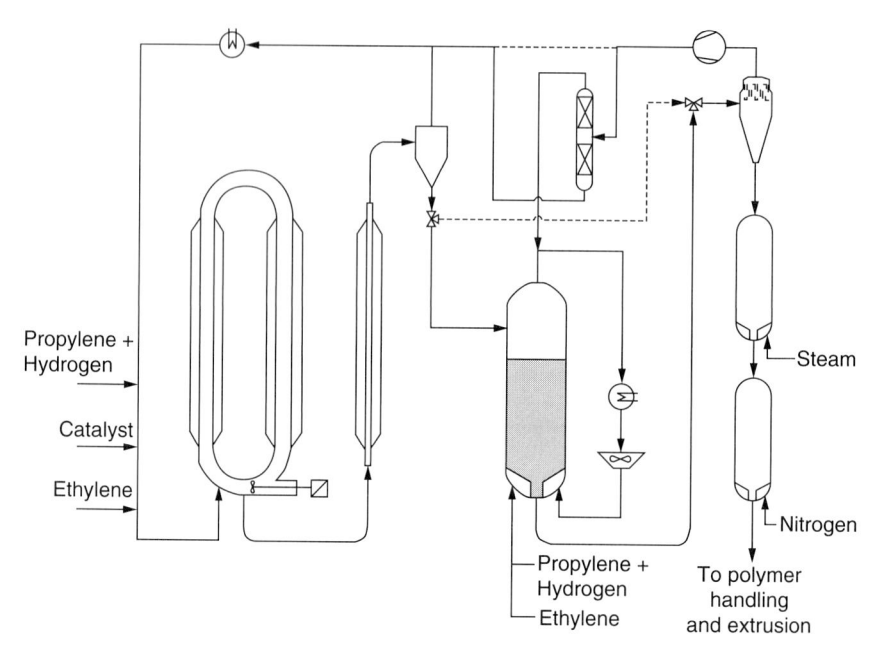

FIGURE 16.1.3 Simplified process scheme of *Spheripol* process.

The catalytic system has three components: the solid catalyst, an aluminium-alkyl used to activate the catalyst, and a Lewis base used to control the crystallinity of the homopolymer grade. The three components are received separately in the plant, unloaded in small drums, and metered to the polymerization section. Monomers fed to the reactors can be treated to remove poisons.

Bulk polymerization employs jacketed tubular loop reactors completely filled with liquid propylene to produce homopolymer, random copolymer, and terpolymer. The catalyst, liquid propylene, and hydrogen for molecular weight control are fed continuously into the loop reactor. Residence time in the reactor is lower than other technologies because of the high monomer density and increased catalyst activity. The polymerization reaction is exothermic; the heat of reaction is removed by means of cooling water circulating into the reactor jackets. The loop reactor is used because it offers low cost and high heat transfer and maintains uniform temperature, pressure, and catalyst distribution. The low residence time also results in short transitions during grade changes, whereas the complete filling of the reactors eliminates any risk of contamination between different grades owing to the presence of an interface between the actual reaction volume and disengagement.

A homogeneous mixture of porous PP spheres (where active catalyst resides) suspended in liquid propylene is circulated inside the reactor loop by means of a dedicated pump. If the production of random copolymer or terpolymer is desired, ethylene and/or butene-1 are fed—in defined ratio with the plant capacity—into the loop reactor. This process achieves very high solid concentration (at least 50 percent by weight) and excellent heat removal (by water circulation in the reactor jacket) and temperature control (no hot spots). The resulting polymer is discharged continuously from the reactor through a flash heater into a first-stage degassing vessel. Unreacted propylene from the flash vessel is recovered, condensed, and pumped back into the loop reactor.

For the production of impact and speciality impact copolymers, polymer from the first flash vessel is fed to a gas-phase fluidized-bed reactor that operates in series with the loop reactors. (This gas-phase reactor can be bypassed when homopolymer or random copolymer is produced.) In this reactor, an elastomer (ethylene/propylene rubber) polymerizes within the homopolymer matrix that resulted from the first reaction stage. The carefully developed pores inside the homopolymer particle allow the rubber phase to grow inside without showing the sticky nature of the rubber to upset operations by forming agglomerates.

Fluidization is maintained by adequate recirculation of reacting gas; reaction heat is removed from the recycle gas by a cooler, and then the cooled gas is recycled back to the bottom of the gas-phase reactor for fluidization. This type of gas-phase reactor is efficient because it maintains a high degree of turbulence in order to enhance monomer diffusion and reaction rates and offers an efficient heat-removal system.

In impact copolymer production, at least 60 percent of the final product is produced in the first-stage loop reactor. In addition, since ethylene is more reactive than propylene, the gas-phase reactors are smaller than would be required if this design were to be used for homopolymer production. Spherical morphology ensures high reliability and elimination of fouling phenomena, which frequently disrupt other gas-phase systems.

Polymer discharged from the reactors flows to a low-pressure separator and subsequently to a steam treatment vessel where catalyst residues are neutralized and the dissolved monomer is removed, recovered, and recycled back to the reactor system.

From the steamer, polymer is discharged into a small fluidized-bed dryer with a hot nitrogen closed-loop system to remove the moisture. The final product is conveyed to an extrusion unit, where it is mixed with additives and extruded to pellets.

Spheripol Key Characteristics

The *Spheripol* process offers licensees a simple and economical method of producing a wide range of PP products of the highest quality. Today, more companies are using the *Spheripol* process than the technologies of the three closest competitors combined.

Safety and Environment

Basell has a safety record unmatched in the industry. By 2003, Basell technologies had achieved nearly 7 million operating hours without any major incident.

Intrinsic Safety of the Process. The design of each *Spheripol* process plant includes a number of safety features, such as

- Proprietary catalyst deactivation system, which immediately stops all reaction
- Interlocks that help to prevent operator error
- Computer-controlled emergency shutdown systems
- Uninterruptible power supply (UPS) for computer control and critical instrumentation control
- Instrument air emergency buffer
- Emergency blowdown system to empty the plant quickly in the event of an emergency
- Gas detectors that instantly determine and highlight (on a graphic easy-to-read board) the source of any hydrocarbons in the event of leakage into the atmosphere
- Automatic fire protection systems

Depending on the severity of the situation, the plant can be shut down manually in a step-by-step, controlled fashion more rapidly by both manual and computer control or by instant automatic shutdown.

Reducing Resource Intensity. Features of the *Spheripol* process help to reduce both resource consumption and emissions. These include use of high-yield, highly stereospecific catalysts; recovery and recycle of unreacted monomers; the absence of undesired by-products from the reaction; and low energy consumption.

Process Capability

Design Flexibility. *Spheripol* process plants are designed to meet the particular requirements of individual licensees, yet they are flexible enough to be expanded easily to meet future needs as the business develops. Two critical design elements that are expanded easily include capacity and product range.

A range of single-line capacities from 40 to 450 kt/yr is available for homopolymer, random copolymer, or heterophasic impact copolymer either using polymer (99.5 percent typical assay) or chemical-grade (94 percent typical assay) propylene. This wide capacity range and the modular installation approach allow, within limits, easy debottlenecking. This minimizes initial capital costs and allows new capacity to be added later when required.

The product range also can be expanded easily. Often new entrants to the PP business will build a plant to produce only homopolymer and random copolymer products because these are the least expensive, are easy to operate, and their products account for 75 percent of all PP sold in the world. A basic homopolymer plant can be expanded easily at a later date to produce heterophasic impact and speciality impact copolymers. Impact copolymers are more specialized products, which require additional capital investment and technical support.

The simple design of a *Spheripol* process plant does not require mandatory equipment or instrument/electrical vendor lists. There is also a list of suitable suppliers and designs for critical equipment, enabling purchasers to benefit from the most economical pricing available.

Versatility. In comparison with any wholly gas-phase technology, a *Spheripol* process plant offers on a single polymerization line the widest range of homopolymers, random copolymers, and terpolymers, as well as heterophasic impact and specialty impact copolymers covering all PP application fields. Intense efforts in product application development for all the major market areas in the world ensures that Basell's PP products keep a leading position in most profitable market segments, with excellent results in PP specialties and "high quality" demanding application introduction and market position.

Key to this versatility is the application of Basell's high-yield, high-selectivity *Avant* catalysts:

- High polymerization activity (mileage over 40,000 kg PP/kg catalyst), resulting in extremely high polymer purity
- Stereospecificity control of polymer
- Morphology control of particle size, shape, and distribution
- Molecular weight distribution control
- Use of polymer or chemical-grade monomer
- Homopolymer, random copolymers and terpolymers, heterophasic impact, and specialty

PROCESS ECONOMICS

Reliability and Operability

The *Spheripol* process has proven to be extremely reliable. In use since 1982, it has been continuously refined and optimized. Yearly worldwide surveys of operating *Spheripol* process lines reveal an average on-stream operability of around 98 percent. Of an average 2 percent downtime, less than 1 percent is due to process features.

Key contributors to this high operability include

- A simple, straightforward process design with simple and reliable equipment
- Easy online product change
- Rapid restart after shutdown
- No scheduled maintenance downtime for cleaning or inspection purposes

Consumptions

Spheripol process plant capital and operating costs are among the lowest in the industry as a result of

- Smaller reactor volumes (completely full of reacting liquid, no dead space)
- Minimum transition time (due to very low residence time)
- Lowest monomer consumption (complete recovery of the unreacted monomer)
- Use of low-cost chemical-grade or polymer-grade monomer
- Low steam and electric power consumption (Table 16.1.3)

PRODUCTS AND APPLICATIONS

Isotactic PP is well suited for a variety of end uses, ranging from flexible and rigid packaging to fibers and large molded parts for automotive and consumer products. PP is recyclable, an important consideration in many packaging and automotive applications, and it can be incinerated without toxic emissions. It can be processed using most methods, including extrusion, extrusion coating, blow molding and stretch blow molding, injection molding, and thermoforming. Its physical properties can be enhanced easily through the addition of fillers such as calcium carbonate or talc. Polypropylene has excellent chemical resistance and electrical insulating properties.

Typical applications for PP products produced in the *Spheripol* process include fibers and filaments, oriented and cast film, injection-molding items, blow-molded bottles and parts, and thermoformed containers.

Random copolymers can be produced with excellent optical properties and sealing initiation temperatures in compliance with the U.S. Food and Drug Administration (FDA) regulations for food contact.

Heterophasic copolymers with outstanding low-temperature behavior, high impact strength, and enhanced stiffness can be obtained in the widest range of melt viscosities.

As a global average, homopolymer accounts for 65 to 70 percent of all PP; random copolymer and terpolymer account for 10 to 15 percent and heterophasic copolymer for 15 to 20 percent.

TABLE 16.1.3 Typical Specific Consumption (per 1000 kg of PP Produced)

	Homopolymer	Impact copolymer
Monomers, kg*	1002	1002
Hydrogen, $N \cdot m^3$†	0.01–0.5	0.01–0.5
Catalysts and chemicals, $ (typical general-purpose stabilization)	22–25	23–26
Electric power, kWh‡	250–270	280–300
Steam, kg§	280	280
Cooling water, m^3¶	110	120
Nitrogen, $N \cdot m^3$	15	15

*Net consumption of 100% monomers/comonomers included.
†Depending on reactor gas composition, according to polymer MI.
‡Extrusion included, consumption depending on product MI.
§Low-pressure steam and high-pressure steam consumption for extruder die plate or barrel heating is excluded.
¶At $\Delta T = 10°C$.

Film and fiber are the two largest segments in the global PP market, but injection/blow molding and extrusion account for significant quantities. Typical PP fabrication processes and end uses include the following major fields:

Fabrication process	Markets/end uses
Film extrusion	BOPP film/WQB film/cast film (flexible packaging for textiles, confectionery, bakery and cigarette wrap)
Multifilaments	Woven sacs (raffia)/fibrillated tape for carpet backing, geotextiles, rope and twine, upholstery, and cigarette tow
Nonwovens (melt-blown and spun-bonded)	Geotextiles, medical application
Injection molding	Automotive, appliances, housewares, furniture, consumer products, packaging (crates, cases, caps and closures, and thin-walled and transparent containers)
Blow molding	Packaging
Profile extrusion	Pipes, conduits, corrugated sheet, wire and cable extrusion coating, lamination

Basell *Avant* catalysts are capable of manufacturing products with varied combinations of physical properties (see Fig. 16.1.4 and Table 16.1.4) to meet the increasing demands of customers in new and more challenging applications. In comparison with any gas-phase technology, a *Spheripol* process plant is making available on a single polymerization line the widest range of PP products, including homopolymers, random copolymers, and terpolymers, as well as heterophasic impact and specialty impact copolymers covering all PP application fields. Extensive product application development has enabled *Spheripol* process products to keep a leading position in most market segments worldwide, and particularly in PP specialties and applications demanding high quality levels, where excellent results have been achieved.

For better evaluation of the capabilities of the *Spheripol* process and to demonstrate its performance versus competing technologies, listed below are some of the *Spheripol* process products from Basell in the specialties and high-performance PP market segments:

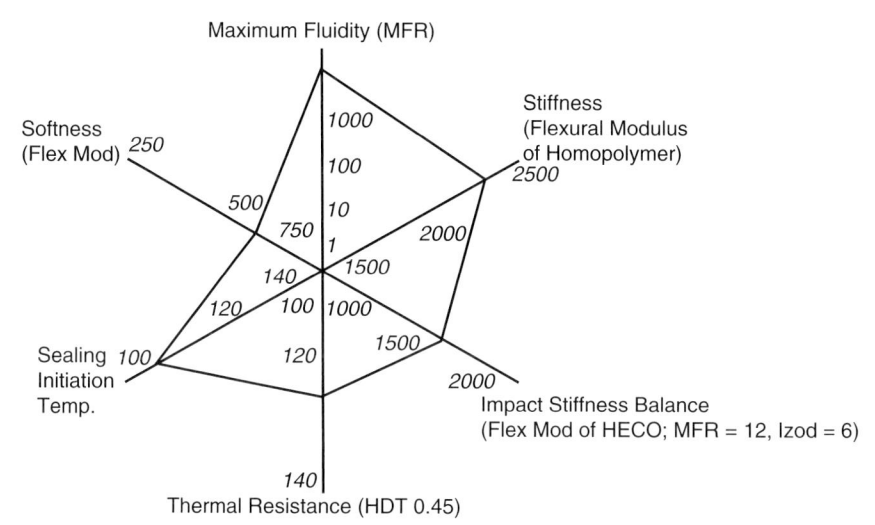

BASELL *SPHERIPOL* TECHNOLOGY FOR PP PRODUCTION 16.19

[Figure: radar chart with axes — Maximum Fluidity (MFR); Stiffness (Flexural Modulus of Homopolymer); Impact Stiffness Balance (Flex Mod of HECO; MFR = 12, Izod = 6); Thermal Resistance (HDT 0.45); Sealing Initiation Temp.; Softness (Flex Mod)]

FIGURE 16.1.4 Impact copolymer production.

TABLE 16.1.4 *Spheripol* Process Capability

MFR, g/10 min	0.3 to >1600
Xylene insolubles, %	90 to 99%
Particle size, mm	0.3 to 5.0
Melting point, °C	130 to 165

Homopolymers

- Very high processability BOPP, for application on very fast tenter machines (over 300 m/min)
- Single- and multilayer cast film produced on high-speed lines using new grades particularly suitable for thin gauges
- Fine denier continuous filament and nonwoven fabric products for disposable clothing, lining, medical hygienic, feminine care, diapers, crop protection in agriculture
- High-clarity gamma-ray-resistant material for injection-grade syringes

In addition, a number of new commercial grades have been introduced successfully to the market for very-high-stiffness homopolymers (flexural modulus higher than 2300 MPa) for injection and thermoforming.

Random Copolymers and Terpolymers

- Terpolymer grades for BOPP coextrusion with very low SIT (115°C), good hot tack, and excellent optical properties are also suitable for shrinkable film production on tubular BOPP lines
- High MFR random copolymers for injection and medical article blow molding
- Random copolymer for stretch blow-molding grades for transparent bottles with excellent gloss for table water or other noncarbonated beverages (PVC or PET replacement),

and also for injection-molded containers with good clarity and gloss for foodstuffs, housewares, cosmetic lids, closures, and caps articles

A new family of very-high-clarity random copolymers of propylene and butene-1 with less hexane extractables (very important for food contact) is under commercial development. Expected uses include high-performance cast film applications with different stabilization packages, as well as for extrusion applications, such as thermoforming, blow molding, sheet extrusion, and blown film.

Heterophasic Impact Copolymers

- Improved impact/stiffness balance copolymers for corrugated pipe, injection, and thin-wall injection. (These grades have been introduced recently on a commercial scale and made available to *Spheripol* process licensees.)
- Very high MFR impact copolymers for thin-wall injection. (MFR 70 and 100 g/10 min pelletized commercial grades are regularly produced in Basell's *Spheripol* process plants.)
- Specialty impact heterophasic copolymers for bumpers, presenting extremely high impact resistance even at very low temperatures. (These reactor grades were developed in the *Spheripol* process to the requirements of major car manufacturers in Europe.)
- High impact for special applications such as a low MFR grade free of fish eyes for film and tape or extrusion and blow molding.

In addition, heterophasic copolymers with very high creep resistance are now available for pipe applications. There is also a family of high-rigidity "ultra" grades for fast injection applications, as well as high-clarity impact grades for cosmetic packaging and housewares.

CHAPTER 16.2
BASELL *SPHERIZONE* TECHNOLOGY FOR PP PRODUCTION

Riccardo Rinaldi and Gijs ten Berge
Basell Polyolefine GmbH
Frankfurt, Germany

GENERAL PROCESS DESCRIPTION

The new *Spherizone* process technology is designed to produce from propylene an extremely wide range of high-quality polypropylene homopolymers, random copolymers, and heterophasic impact copolymers by copolymerization with comonomers such as ethylene and butene-1. Single-line capacity of 400,000 metric tons per annum (MTA) is achievable right now; higher capacities will be designed for in the near future.

The process consists of a polymerization unit with the revolutionary multizone circulating reactor (MZCR). In case of heterophasic impact copolymer production, the MZCR is followed by a gas-phase fluidized-bed reactor. Downstream of the reaction section is the product finishing section, additivation, and extrusion.

PROCESS CHEMISTRY AND THERMODYNAMICS

Homopolymer

The stereospecific polymerization of propylene by Natta in 1954 was a scientific breakthrough with immediate industrial success. By obtaining alignment of propylene molecules, a crystalline polymer of isotactic PP was synthesized at high yields in a manner that could be exploited commercially. This achievement, which spurred interest in organometallic chemistry, as well as in polymerization, was recognized by the award of the 1963 Nobel Prize in chemistry to Natta and Ziegler, the latter having discovered the basic catalyst system for polymerizing ethylene.

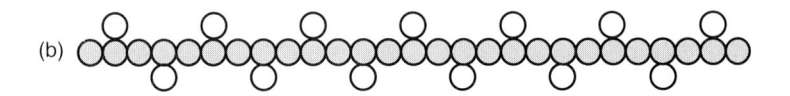

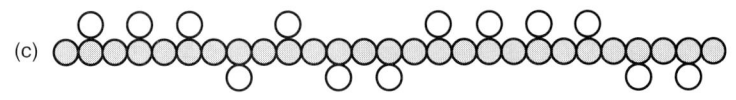

FIGURE 16.2.1 Stereospecificity of PP polymer chain: (*a*) isotactic; (*b*) syndiotactic; (*c*) atactic.

Polymerizing propylene with various heterogeneous catalyst systems in hydrocarbon diluent bulk liquid phase or gas phase may make isotactic PP.

Polymerization Reaction. Propylene can polymerize into three distinct structural chains: isotactic, syndiotactic, and atactic. Isotactic PP occurs when all the methyl groups are arranged on the same side of the carbon chain, attached to every other carbon atom, as shown in Fig. 16.2.1*a*. Syndiotactic PP occurs when the methyl groups are attached to every other carbon atom in the chain but are located on alternating sides of the chain, as shown in Fig. 16.2.1*b*. Atactic PP occurs when the methyl groups are dispersed randomly along the carbon chain, as shown in Fig. 16.2.1*c*.

In PP homopolymers, the balance between atactic and isotactic fractions has a direct impact on several critical physical properties such as stiffness and resistance to impact and is of primary interest when discussing polymer chemistry. Production of syndiotactic PP requires specialized catalyst and donor systems not yet fully commercially used.

Role of the Catalyst System Components. The catalyst systems are composed of three basic parts: (1) a solid catalyst, generally $TiCl_4$ supported on $MgCl_2$; (2) stereoregulating agents, an internal and external Lewis base; and (3) an aluminiumalkyl.

Catalyst Function. Catalyst is composed of two main elements: a transition metal salt and an inert support structure. The $MgCl_2$ support structure of the catalyst serves several roles. It creates a highly disorganized crystalline structure. Thus the active centers, where polymerization takes place, are greater in number and reactivity than those of other conventional catalysts. This explains the high activity of these catalyst systems. $MgCl_2$ contributes indirectly to the stereospecificity owing to interactions with the donor. Finally, owing to its chemical-physical properties, $MgCl_2$ can be converted into particles with controlled shapes and sizes. These characteristics are also transferred to the catalyst and the polymer.

The active part of the catalyst is titanium tetrachloride ($TiCl_4$). In this form, the catalyst will not actively polymerize and requires "activation" by an aluminum alkyl and a Lewis base.

Catalyst Evolution. The rapid and successful commercialization of PP manufacturing is due primarily to the continuous development of new and improved catalysts. Each evolution or generation of catalysts both reduced the complexity of the manufacturing process and expanded the control over resin morphology. A brief summary of PP catalyst evolution is shown in Table 16.2.1.

Catalyst Size/Shape. The polymer flake size and shape constitute a large copy of the size and shape of the catalyst particle. Average diameter of the polymer particles depends on the average diameter of the catalyst and on the polymerization mileage.

Catalyst Activity. The activity of a catalyst is its tendency and capability to polymerize propylene (and other comonomers) under standard conditions. Activity is measured in units called Zieglers. One Ziegler is defined as one gram of polymer per millimole of $TiCl_3$ per atmosphere of propylene pressure per hour.

Catalyst Types. Several catalysts can be used in the *Spherizone* technology; each of them is tailored according to the final application of the polymer. The catalyst can be granular or spherical form; it can be tailored to have a very high isotactic index, very high mileage, and so on. Figure 16.2.2 presents microscopic images of granular and spherical catalyst particles.

Replication Factor. The particle size distribution of the PP flake is a replica of the particle size distribution of the catalyst. In the industrial process, a broader PSD flake is obtained compared with batch polymerization because more fine particles are produced as a result of friction in the reactors and other processing equipment. Also, fines are produced owing to a residence time distribution that is different in the batch polymerization compared with the continuous plant process. In the batch polymerization, all the catalyst particles have the same time to grow and make polymer. In the multizone circulating reactor, there is a distribution of residence times. There are particles that stay in the reactor for only a few minutes and others that stay for a few hours.

The ratio between the average diameter of the flake and the diameter of the catalyst is called the *replication factor*. The replication factor is proportional to the cube root mileage. The higher the mileage, the larger is the particle of polymer produced. The replication factor generally is calculated by

$$\text{Replication factor} = [(g\ PP/g\ cat.) \times (2/0.9)]^{1/3}$$

Aluminum Alkyl. It is generally understood that polymerization of olefins with Ziegler-Natta catalysts involves a stepwise insertion of the monomer into a transition metal-carbon bond as follows:

$$Mt\text{-}R + n C{=}C \rightarrow Mt\text{-}(C\text{-}C)_n\text{-}R$$

Catalysis of propylene requires the presence of an aluminum alkyl. For the polymerization of propylene, the most widely used alkyl is triethyl aluminum, $(C_2H_5)_3Al$, or TEAL for short. The activation reaction is as follows:

$$TiCl_4 + (C_2H_5)_3Al\text{-} \rightarrow TiCl_3 + (C_2H_5)_2\ AlCl + (C_2H_5)^{\cdot}$$

The reaction is exothermic with a $\Delta H = 131.6$ kcal/kg $TiCl_4$.

In addition to assisting in the polymerization reaction, TEAL acts as a scavenger for certain monomer poisons, most notably water, which would seriously impair or completely stop the reaction. Therefore, TEAL is added to the reaction in amounts greater than the stoichiometric amount required to activate the catalyst. The amount of excess TEAL depends mainly by monomer purity.

TABLE 16.2.1 Performance of Different Catalyst Generations

Generation	Catalyst composition	Yield, kg PP/g cat.	Isotactic index, wt %	Morphology control	Process requirements
1	TiCl$_3$/AlCl$_3$ + DEAC	1	90–94	Not possible	Deashing + atactic removal
2	TiCl$_3$ + DEAC	10–15	94–97	Possible	Deashing
3	TiCl$_4$/ester/MgCl$_2$ + AlR$_3$/ester	15–30	90–95	Possible	Atactic removal
4	TiCl$_4$/diester/MgCl$_2$ + TEA/silane (HY/HS)	30–60	95–99	Possible	—
5	TiCl$_4$/diether/MgCl$_2$ + TEA	70–120	95–99	Possible	—
6	Zirconocene + MAO	—	90–99	—	—

Note: DEAC = diethyl aluminum chloride; TEA = triethyl aluminum; MAO = methylaluminoxane.

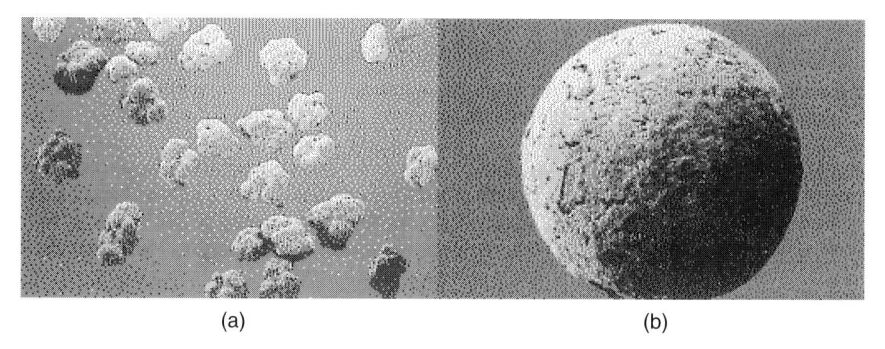

FIGURE 16.2.2 (*a*) Close-up of catalyst for flake. (*b*) Close-up of spherical catalyst.

Stereoregulation and Isotacticity: Donor Function. During polymerization, olefins can be inserted into the carbon-metal bond in two different orientations. Random insertion will produce a material with a high-irregularity, or atactic, structure. Atactic PP is soft and sticky and of little value as a material. The inclusion of a Lewis base or electron donor into the polymerization reaction greatly increases the regularity at which the olefin molecules are inserted into the polymer chain, producing a highly usable, stereoregular, or isotactic, polymer. The electron donors are called *donor* for short. Along with the evolution of catalysts, there has been a parallel evolution of donors.

The degree to which propylene is polymerized in an orderly or stereoregular manner is measured by the amount of crystalline material produced. This is called the *isotactic index* and is measured as percent by weight. In general, homopolymer PP is produced with an isotactic index in the 92 to 99 percent range. Increasing or decreasing the amount of donor controls the isotactic index of the polymer.

Measurement of isotactic index is a fairly complex test using extraction with boiling heptane. This test generally has been replaced with measurement of the atactic portion, which uses xylene and, more recently, magnetic resonance technology. Most plants refer to the atactic portion of the polymer as *xylene solubles* (XS). While not exactly the same, xylene solubles and isotactic index have been used interchangeably.

Polymer Chain Length Control. The length of the polymer chain has a significant impact on its performance, particularly its flowability. This is a critical parameter for downstream use of the resin. Direct measurement of the polymer chain is difficult and not suited to plant laboratories. Instead, polymer chain length is measured indirectly in several ways. For many years, the intrinsic viscosity (IV) of the polymer was measured. IV results were directly proportional to polymer chain length. That is, the higher the IV, the longer is the average chain length.

While still used to some extent during the production of impact copolymers, IV has been replaced with the much quicker and more repeatable *melt-flow rate* (MFR). As the term implies, MFR is the weight of melted polymer that can flow through a specific orifice under a standard load for a given temperature and time.

Standard load = 2.16 kg

Standard temperature = 230°C (for very high MFR, a lower temperature is used)

Standard time = 10 min

As would be expected, MFR results are the inverse of chain length. That is, polymers with long chain length will have a low MFR.

With the development of automated testing machines in the late 1970s, polymer MFR can be measured quickly with high accuracy and repeatability.

Hydrogen is added to the polymerization reaction to control the molecular weight (i.e., chain length) of the polymer by acting as a chain-transfer agent. Hydrogen increases the activity of the catalyst.

Molecular Weight of Polymer versus MFR. Polymer chain length (or molecular weight) is directly related to MFR. Laboratory work has shown the relationship to be closely approximated by the following formula:

$$\log \text{MW} = -0.2773 \times \log \text{MFR} + 5.7243 \qquad (r^2 = 0.9780)$$

Molecular Weight Distribution (MWD) of the Polymer. As with all reactions, the polymer chain lengths produced are not equal. However, for a given catalyst-donor system, the distribution of chain lengths is constant. Changing reaction temperature, residence time, and so on will not affect the distribution significantly.

A fundamental measurement of MWD is *gel permeation chromatography,* currently known as *size-exclusion chromatography,* which responds to the physical size of the different molecules in dilute solution. In this technique, the polymer solution is passed over a porous medium from which the different-sized molecules elute at different times. The curve of eluted material versus time becomes the molecular weight distribution (MWD). From that curve it is possible to calculate the *weight-average molecular weight* (M_w) and the *number-average molecular number* (M_n). The ratio of these two, M_w/M_n, the *polydispersity,* is a widely used term to describe MWD.

Mathematically:

$$M_n = \frac{\Sigma n_i M_i}{\Sigma n_i} \qquad \text{and} \qquad M_w = \frac{\Sigma n_i M_i^2}{\Sigma n_i M_i}$$

where n_i = number of moles of fraction i
M_i = molecular weight of molecules in fraction i

Because the rheology of PP responds strongly to the MWD, a correlation between the polydispersity from GPC and a rheologic parameter called the *polydispersity index* (PI) has been established. The PI is considerably easier to determine and consequently is growing in use. Die swell (from the MFR measuring instrument) also corresponds directly with PI, providing plants with a relatively simple method for measuring PI.

The MWD of the polymer plays an important role in the performance of the resin. A narrow MWD is beneficial for fiber applications, whereas a broad MWD improves resin performance in oriented films.

In the *Spherizone* process, broadening of the molecular weight may be obtained by polymerizing widely different molecular weights (i.e., different MFRs) in the separate reactor stages. This method of polymerization is called *bimodal.* In bimodal polymerization, it is important that the high-molecular-weight component be at least 50 percent of the total. Low-molecular-weight material can be blended into the high-molecular-weight material without problem.

Conversely, the additional of alkyl peroxides to the polymer during the extrusion process will cause a breaking of longer-chain molecules, thus narrowing the MWD. This process is called *visbreaking* or *cracking.*

Mileage. While catalyst performance or yield has been defined clearly and precisely, the unit Ziegler generally is not used in manufacturing plants. The more generic and slightly less precise term mileage is commonly accepted as the unit for measuring and comparing performance. Mileage is defined as the kilograms of PP produced per gram of catalyst.

Propane concentration has an effect on catalyst mileage. Increasing propane concentration reduces mileage because the concentration of propylene at the catalyst's active sites is decreased. Comparison of mileage results always must be made based on results using the same polymerization basis.

Catalyst Yield: Impact of Residence Time. Catalyst mileage is affected directly by residence time—the period of time in which the catalyst is in the reactors. From its kinetics, the long life of the catalyst allows for a higher residence time and thus the possibility for higher mileage.

Catalyst Yield: Impact of MFR. As discussed previously, the MFR of the polymer is proportional to the amount of hydrogen added to the reaction. Since hydrogen acts to increase the reactivity of the catalyst, catalyst mileage therefore is proportional to MFR. When comparing catalyst mileage, it must be done at the same MFR.

Catalyst Yield: Impact of Propane Concentration. Another important factor affecting catalyst performance is the propane concentration in the reactor. A small amount of propane enters the reactors as part of the propylene feed stream (generally 0.2 to 0.7 percent) entering the plant battery limits. Plants with propane splitters operate with as much as 6 percent propane in the supply propylene. Obviously, the higher the propane concentration in the reactor, the lower is the propylene concentration. Since polymerization rate is a direct function of the active center concentration of the catalyst and the monomer concentration, the higher the propane concentration, the lower the reactivity.

Propane entering with the feed monomer does not react; thus propane concentration in the reactor builds up over time. The increasing propane level reduces catalyst mileage. In order to prevent a significant loss in mileage, a small stream of propane must be purged continuously. Increasing propane concentration reduces the cost of purged monomer but increases catalyst cost (i.e., lower mileage). It can be seen easily that there is an economic balance between catalyst mileage and propane level in the reactors. This balance depends on the cost of monomer and the cost of catalyst.

Monomer Quality. The quality of the monomer in the reactors has a great impact on both catalyst mileage and resin quality. Monomer contaminants, or poisons, fall into several general categories depending on how they affect the reaction.

The first group of contaminants reacts with the aluminum-alkyl cocatalyst and greatly reduces catalyst yield when present at the parts per million level. Examples of these materials are CO_2, H_2S, and H_2O. Water is by far the most common poison of the group. The second group of contaminants reacts with the active sites of the catalyst and seriously reduces catalyst yield when present at the parts per billion level. Examples of these poisons are CO, arsine, and phosphine. The third group of contaminants consists of hydrogenating compounds, which have a strong effect on the MFR or the polymer and, consequently, on the yield. Some of these poisons also will affect the stereospecificity of the polymer—generally increasing the atactic level.

Random Copolymer

Random Factor. Adding ethylene directly to the polymerization reactor produces random copolymer. Ethylene molecules are added to the polymer chain at random locations. The higher the concentration of ethylene in the feed, the higher is the chance of having two or more ethylene molecules bound between the propylene molecules. The random factor is the ratio between the percentage of single-bound ethylene and the total percentage of bound ethylene. The higher the ethylene concentration in the feed, the lower will be the random factor. In other words, increasing the ethylene concentration in the copolymerization reaction increases the probability of having short-chain polyethylene between two molecules of propylene. No parameters have been developed to control the random factor. The catalyst inherently produces the random factor.

Polymerization Reaction. To produce random copolymer, ethylene is added directly to the polymerization reactor. The addition of ethylene increases the polymerization rate significantly. This behavior may be accounted for by assuming that propylene insertion in the Ti-carbon bond is easier when an ethylene unit is bonded to the transition metal or that the ethylene insertion after a regioirregular propylene insertion reactivates a "dormant" site. The reactivity ratio is the ethylene-propylene mole ratio in the polymer divided by the ethylene-propylene ratio in the liquid monomer circulating in the reactor.

Polymer Tacticity. Adding ethylene to the polymer increase the xylene soluble fraction. On average, having 1 percent ethylene bound in the copolymer gives an increase in xylene soluble of about 1 percent. Therefore, when making homopolymer with 3 percent xylene soluble, the corresponding 3.5 percent random copolymer (RACO) would have a total xylene soluble level of 6.5 percent. This increases the stickiness of the polymer and contributes to handling and processing problems.

Ethylene does not affect the hydrogen-propylene relationship. That is, the same amount of hydrogen is required to meet a specific MFR for both homopolymer and random copolymer.

The addition of ethylene to the polymer chain lowers the melting point significantly. With a 3 to 4 percent atactic fraction, homopolymer polypropylene has a melting point of approximately 165°C, whereas a random copolymer with 2 percent ethylene bonded has a melting point of about 150°C.

Physical Characteristics of RACOs. The addition of ethylene to the homopolymer chain modifies the physical properties of the resin as follows:

- It increases clarity (beneficial for injection-molded applications, such as the type of resins used in food containers).
- It reduces melt temperature (beneficial for some film applications).
- It reduces stiffness.
- It increases impact resistance (at moderate temperatures).

Heterophasic Copolymers

Polymer Structure. Heterophasic copolymers (commonly called *impact copolymers*, or HECOs) are an intimate mixture of homopolymers and bipolymers. Bipolymers are mainly amorphous materials produced through polymerizing ethylene and propylene in the gas-phase reactor (GPR) installed downstream of the MZCR. Homopolymer is first produced in the MZCR. The flake flows to the GPR, where ethylene and propylene are fed in a predetermined ratio. In the GPR, bipolymerization occurs, and the bipolymer grows inside the homopolymer sphere. The result is a semicrystalline matrix (homopolymer) with a nearly amorphous elastomeric component (bipolymer) dispersed within it. The final product is heterophasic in that it is a mixture of two different products: homopolymer and bipolymer. In fact, by fractionation, it is possible to separate the two materials from each other.

Catalyst Requirement. To produce a polymer with the desired characteristics, a proper amount of rubber must be produced in the copolymerization step, and consequently, a long-life catalyst structure showing a high residual activity after the homopolymerization step is needed. Furthermore, the rubber must be retained inside the flake granule and be dispersed homogeneously in the homopolymer matrix to prevent undesired reactor fouling. Consequently, proper catalyst architecture is required. This polymerization technique, in fact, behaves in such a way that the rubber grows around the homopolymer microgranules. The internal voids of the homopolymer granule are occupied by the growing rubber, which, after the pores are filled, migrates to the particle surface. A high porosity is thus required to bear the desired amount of rubber inside the granule and prevent its migration to the surface with consequent reactor fouling.

Reactivity Ratio. The bipolymer that is produced with gas of a certain composition of ethylene and propylene will have a different composition than the gas because of the different reactivities of the monomer with the catalyst. Ethylene is more reactive than propylene in the gas phase. Knowing the reactivity ratio and the gas composition, it is possible to calculate the composition of the bipolymer. Conversely, a specific bipolymer composition can be produced by controlling the gas ratio in the GPR.

The gas ratio $(C_2^=)/(C_2^= + C_3^=)$ is controlled carefully to produce the correct bipolymer composition. Bipolymer should have good elastomeric properties, thus improving the impact resistance of the copolymer, a highly desirable characteristic property.

Control of Physical Properties. With homopolymers, where physical properties play a less important role, only the molecular weight of the polymer (i.e., MFR) and the amount of atactic material (xylene solubles) are controlled. Hydrogen is used to control the MFR, and the TEAL-donor ratio is used to control the level of xylene solubles.

With heterophasic copolymers, the amount and composition of the bipolymer fraction play critical roles in determining the physical properties of the polymer. Physical properties are controlled as follows:

Physical property/parameter	Is controlled by adjusting
Catalyst mileage/activity	Residence time in reactor
MFR of homopolymer	Hydrogen concentration in reactor
XS of homopolymer	TEAL-donor ratio in the reactor
$C_2^=$ content of the copolymer	Total $C_2^=$ fed to the GPR
$C_2^=$ content in biopolymer	Gas ratio $(C_2/C_2 + C_3)$ in the GPR
IV of the biopolymer	Hydrogen ratio (H_2/C_2) in the GPR
Final MFR of the copolymer	Hydrogen ratio (H_2/C_2) in the GPR

Balance of Physical Properties. The critical physical properties of copolymers are MFR, flexural modulus, tensile strength, Izod impact, and dropped weight impact and heat deflection versus temperature. In polymeric materials, there is a balance between stiffness and resistance to impact. The *Spherizone* process greatly improves the stiffness/impact balance compared with existing PP resins by the nature of its polymerization conditions. Stiffness generally will be reduced in order to gain impact strength. This relationship holds true within any given copolymer structure. However, it may be possible to increase both the impact and stiffness somewhat further by changing either or both the catalyst and donor systems.

Table 16.2.2 shows how physical properties are affected by changes in polymer structure. This table should be used as a general guideline only because the additive package used can influence the final properties of the resin.

SPHERIZONE PROCESS PERSPECTIVE

Development of the *Spherizone* technology was started in 1995 and subsequently was scaled up from pilot plant to commercial size in 2002, when the new multizone circulating reactor was installed at Basell's Brindisi, Italy, plant. The plant has been running very well since startup.

Even in the early stages in the development and commercialization of the new process, the Brindisi plant is running reliably. Since many parts of the process are structurally identical to the equivalent parts in the *Spheripol* process (see Chap. 16.1) with its proven high reliability, such consistency in performance is expected for future plants as well.

Since the process was made avaiable for licensing in October 2003, several licenses have been granted.

PROCESS DESCRIPTION

The *Spherizone* process, together with Basell's high-yield, high-selectivity catalysts, has the ability to produce spherical polymer particles directly in the reactor. Spherical PP differs considerably from the small, irregularly shaped granular particles obtained by some other technologies and provides some significant advantages in terms of process reliability.

The *Spherizone* process is a modular technology and typically is composed of the following sections:

- Catalyst feeding
- Polymerization
  - Prepolymerization
  - Polymerization in the MZCR (homopolymer grades, medium wide, very wide MWD, random copolymers and terpolymers, two-composition polymers, homopolymer/random copolymer–twin random copolymers, random/heterophasic copolymers) (Fig. 16.2.3)
  - Gas-phase polymerization in a fluidized-bed reactor (heterophasic impact and speciality copolymer) as a further option; the gas-phase copolymer unit can be added at a later stage without affecting initial plant configuration or involving significant implementation costs (Fig. 16.2.4)
- Finishing

The catalyst, a spherical solid having a controlled particle size, is dispersed in a mixture of paraffinic oil and grease. The catalyst is fed continuously to the multizone circulating reactor (MZCR), which is the core of this new technology. This loop reactor consists of two (or more) distinct reaction zones, each operating under its own peculiar fluid-dynamic regime.

In the so-called riser (see also Figs. 16.2.3 and 16.2.4), the polymer particles are entrained upward by the monomer flow in a fast-fluidization regime. Gas superficial velocities are maintained at much higher values than the average particle terminal velocities so that a highly turbulent flow regime ensues. This generates an optimal heat-exchange coefficient between the single particles and the surrounding gas and ensures that the reaction

TABLE 16.2.2 Heterophasic Copolymers: Standard Parameters versus Mechanical and Optical Characteristics

Action ↑	Unit	Flex modulus	Izod	Ball drop Trans. temp.	Ball drop Energy	Stress whitening	Tensile yield	Gloss	Hardness	HDT
Homopolymer
Melt-flow rate	g/10 min	↑	→	↑	→	→	↑	→	↑	↑
Molecular weight distribution	—	↑	→	x						
Isotactic index	%	↑	x		x		↑		↑	↑
Bipolymer
Total amount of biopolymer C$_3$-C$_2$	% wt	→	↑	→	↑	→	→	→	→	→
Xylene soluble of biopolymer	% wt	→	↑	→	↑	→	→	x	→	→
Intrinsic viscosity of xylene soluble	dL/g	→	↑	→*	↑	↑				
Ethylene content in xylene soluble	% wt	→	↑	→*	↑					
Xylene insoluble (crystalline biopolymer)	% wt	→*	↑*	→*	↑	↑*	→*	→*	↓*	↓*
Intrinsic viscosity of xylene insoluble	dL/g	→	↑	→	↑	↑		→		
Ethylene content in xylene insoluble	% wt							→*		
Polyethylene	% wt	↓*	↑*	↓*	↑*	↑*	↓*	↓*	↓*	↓*

Note: When the action increases (↑), the parameters of the left column, the mechanical and optical properties, increase (↑), decrease (↓), or remain unchanged (x).
*At constant amorphous fraction content. If the increase of this fraction means a reduction of the amorphous one, these correlations are inverted.

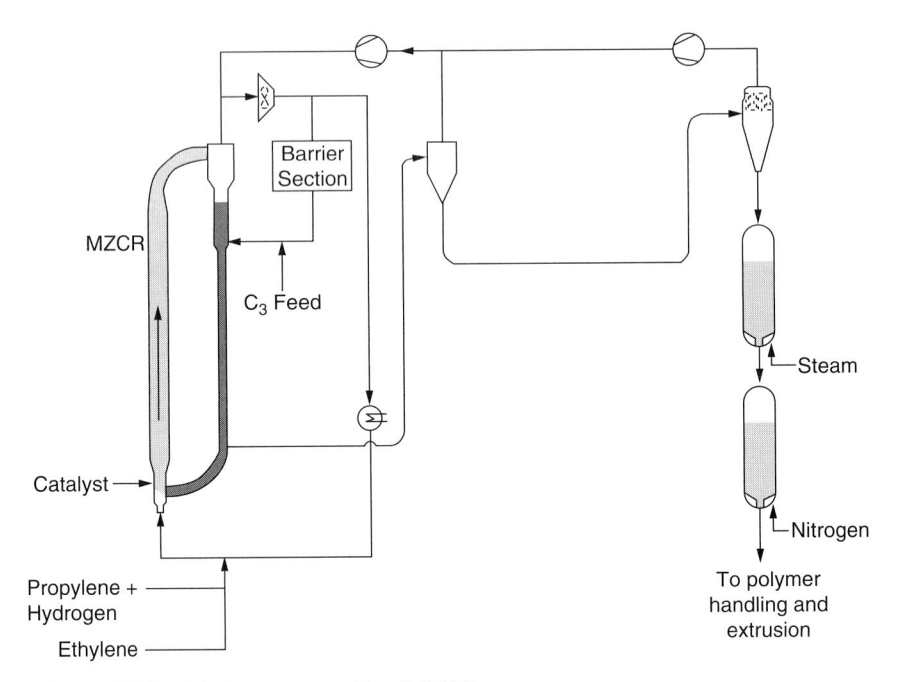

FIGURE 16.2.3 *Spherizone* process with only MZCR.

temperature is kept constant along the reaction bed. Head losses in this area are comparable with those along a fluidized bed of the same solids holdup while maintaining a high bed voidage, as typical of fast-fluidized beds.

In the top of the reactor, the riser gas is then separated from the solids, which enter the so-called downcomer. This section operates as a moving packed bed, with the polymer flowing downward. Since it operates adiabatically, the reaction heat will increase the temperature of the solid bed as the polymer descends. Therefore, care is taken to recirculate enough polymer to prevent the formation of hot spots and generally excessively high temperatures along the bed, which may jeopardize the flowability of the polymer and the recirculation itself.

The polymer's loop circulation is set up and defined by the pressure balance between the two polymerization zones. As it flows down under gravity, the downcomer polymer bed "pumps" the gas downward and recovers the head losses developed in the riser, the gas-polymer separator, and all other sections of the reactor. Actually, the moving packed bed develops a pressure profile increasing from top to bottom of the reactor as per Ergun's law, according to the pressure head necessary for recirculation. The overall pressure balance determines the required pressure head, which is at all times maintained lower than the maximum attainable in conditions of incipient fluidization.

The pressure head, in turn, will determine the differential (slip) velocity between the gas that flows down with the solids and the polymer itself. This is only a function of the reactor's fluodynamics, i.e., of the solids flow and the pressure balance to be maintained in the reactor, and is in all cases independent of reaction conditions. The actual fluid-dynamic conditions of the reactor are such that the gas is in all cases flowing down with the polymer. Since the pressure head is determined by the differential in velocities between solids and gas, it is apparent that for the same overall head losses, and therefore the same slip velocity, higher polymer flows in the downcomer will significantly increase the flow of gas entrained with the polymer. This plays an important part in the bimodal operation of the reactor.

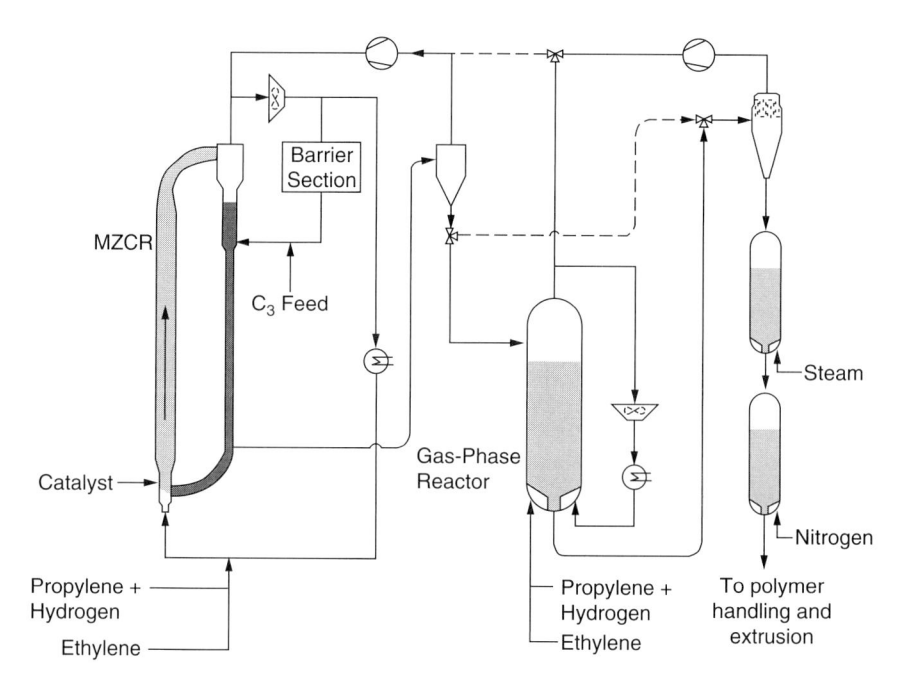

FIGURE 16.2.4 *Spherizone* process with MZCR + 1 GPR.

At the bottom of the downcomer, the polymer particles are recirculated into the riser via a J-valve-like piece of equipment. Furthermore, any suitable valve also may be installed at the bottom of the downcomer as an additional device to further control polymer flow.

The packed-bed fluid dynamics of the downcomer are essential to bimodal operation. The intergranular gas normally would have the same composition as the riser gas, resulting in a monomodal polymer. However, the two sections of the reactor can be operated at different compositions of hydrogen (used as chain-transfer agent) and comonomer, allowing for the development of a bimodal polymer structure (in terms of MFR and/or comonomer concentration/type) at a macromolecular level. This can be accomplished by the introduction of a liquid or gas propylene stream on the top of the downer just below the polymer level so that the riser gas is replaced by a gas with a different composition (Fig. 16.2.5). Typically, low-molecular-weight polymers are produced in the riser, whereas the injection of a hydrogen-poor monomer stream in the liquid or vapor phase allows the production of higher-molecular-weight polymers in the downcomer.

The ratio between the barrier flow rate and that of the recirculating solid can be very low (<0.1 wt). However, the propylene requirements for the barrier flow generally are independent of, and usually higher than, the production rate. Therefore, the net incoming propylene flow is not necessarily enough to guarantee the stripping requirements, and the difference should be recovered in a dedicated section and then recycled to the reactor. There is, therefore, a pressing need to minimize the requirements for the barrier flow rate and the energy consumption involved in its recovery.

The polymer recirculation rate has a direct influence on the amount of gas flowing down the downcomer, which results in the requirement to limit such polymer flow to the minimum allowable by the thermal balance of the downcomer.

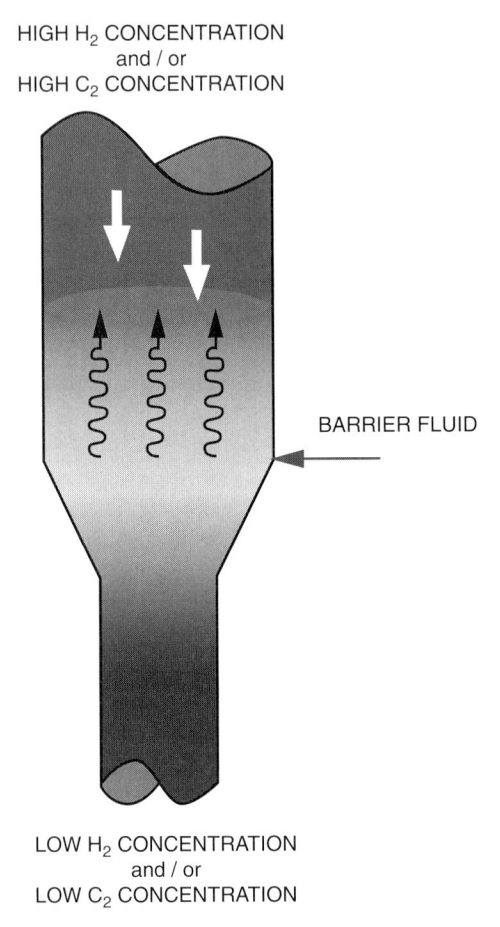

FIGURE 16.2.5 "Stripping" section in downcomer of MZCR.

In this, the use of a liquid barrier stream is particularly useful because its latent heat helps to remove the heat of reaction developed in the downcomer. The barrier stream is fed and easily distributed in the moving packed bed. The reaction heat and the solid sensible heat vaporize the barrier liquid that is fed slightly in excess of the theoretical intergranular amount of gases moving downward so that a net gas flow upward is created in the upper point of the downcomer. Thus the polymer level above the injection point forms a "seal" that prevents contamination between the two reaction zones as long as the packed-bed flow regime is maintained.

Obviously, the reactor also can yield monomodal homopolymer and random copolymer products by operating the sections under the very same conditions. In this case, the lack of the cooling effect of the evaporating barrier stream forces the use of higher solids recirculation rates to keep the polymer temperatures all along the downcomer well below the softening point.

The continuous, massive recirculation of the polymer particles between the two zones makes the residence time per pass in each zone one order of magnitude smaller that the overall residence time. In bimodal operation, this allows an intimate mixing of the different polymers being produced, giving a very good homogeneity of the final product. Moreover, the holdup and residence times in each leg can be adjusted to vary the split of production between the two areas to suit different requirements.

From the top of the reactor, the unreacted monomer is recirculated to the bottom of the riser by a centrifugal compressor. A reactor cooler is positioned in that circuit to remove the heat of reaction and to ensure that the desired operating temperature is maintained. The gas flow through the circuit can be varied according to the actual solids recirculation in order to maintain the desired conditions of fast fluidization and solids concentration in the riser.

The pressure in the system is maintained by controlling the fresh monomer flow rates equal to those reacting inside the reactor. From the delivery of the circulation compressor, a side stream is withdrawn and sent to a monomer distillation section, where the desired barrier flow rate is recovered. The top, hydrogen-rich gas is recycled to the riser of the reactor.

The spherical product is continuously withdrawn from the reactor and separated from the unreacted monomer gas in a separator operating at intermediate pressure. The gas is then compressed and recycled back to the MZCR via a small, one-stage reciprocating compressor.

As an option, the polymer can then be fed to a fluidized-bed gas-phase reactor operated in series with the MZCR, where additional copolymerization can take place to yield high-impact copolymer PP. This gas-phase reactor may be bypassed when homopolymer or random copolymers are produced. In this reactor, an elastomer (ethylene/propylene rubber) formed by the introduction of ethylene is allowed to polymerize within the homopolymer matrix that resulted from the first reaction stage. The pores developed inside the polymer particle in the MZCR upstream allow the rubber phase to develop without the formation of agglomerates resulting from the sticky nature of the rubber.

The fluidization in the reactor's polymer bed is maintained by adequate recirculation of reacting gas. The reaction heat is removed from the recycle gas by a cooler, and the cooled gas is recycled back to the bottom of the gas-phase reactor for fluidization. This gas-phase reactor maintains a high degree of turbulence, enhances monomer diffusion and reaction rates, and ensures an efficient particle heat removal.

Depending on the adopted configuration, from the intermediate pressure separator or the fluidized-bed reactor, the product is discharged to a low-pressure filter, where the unreacted monomer gas is recovered. The polymer is then steam stripped of any residual dissolved monomers in an additional vessel. The same unit also neutralizes the residual active catalyst. The residual hydrocarbons are removed and recovered and can be sent back to the reactor system, whereas the polymer is dried by a closed-loop nitrogen system in a small fluidized bed.

The resulting polymer, free from volatiles, is transported pneumatically by nitrogen to the extrusion unit, where it is mixed with additives and extruded to pellets, typically by a single extrusion line. The product then enters a homogenization silo before it is stored, bagged, or shipped.

The *Spherizone* technology can be operated using either polymer- or chemical-grade propylene.

Environmental Considerations

Features of the *Spherizone* process help to reduce both resource consumption and emissions. These include use of high-yield, highly stereospecific catalysts; recovery and recycle of unreacted monomers; the absence of undesired by-products from the reaction; and low energy consumption. Owing to the MZCR concept, in the polymerization section of the process the overall energy consumption may be reduced by 0 to 30 percent, depending on the type of polymer produced.

Modular Approach

The design can be adjusted for homopolymer, random copolymer, or heterophasic impact copolymer. The modular structure of the process is such that a *Spherizone* process plant can be upscaled in successive stages, starting with a basic setup for monomodal operations and later adding the barrier recovery section and a fluid-bed gas-phase reactor for the production of bimodal and impact copolymer, respectively.

Therefore, the *Spherizone* process can be designed to meet the particular requirements of individual licensees, yet it is flexible enough to be expanded easily to meet future needs as business develops. New entrants to the PP business may want to build a plant producing only homopolymers and random copolymers because these are the least expensive and are easy to operate, and their products account for 75 percent of all PP sold in the world. Such a basic homopolymer plant can be expanded easily at a later date to produce heterophasic impact and specialty impact copolymers. Impact copolymers are more specialized products that require additional capital investment and technical support.

Versatility

Product-wise, the versatility of the *Spherizone* process is demonstrated by the high-quality product range that includes all standard polypropylene grades, as well as many unique, special products (Fig. 16.2.6). One key to this versatility is, as mentioned earlier, the unique design and operation of the MZCR, allowing for many kinds of intimately mixed polymer compositions to be produced. Another key is the application of the high-yield, high-selectivity catalysts provided by Basell as *Avant* catalysts:

- High polymerization activity (mileage over 40,000 kg PP/kg catalyst), resulting in extremely high polymer purity
- Stereospecificity control of polymer
- Morphology control of particle size, shape, and distribution

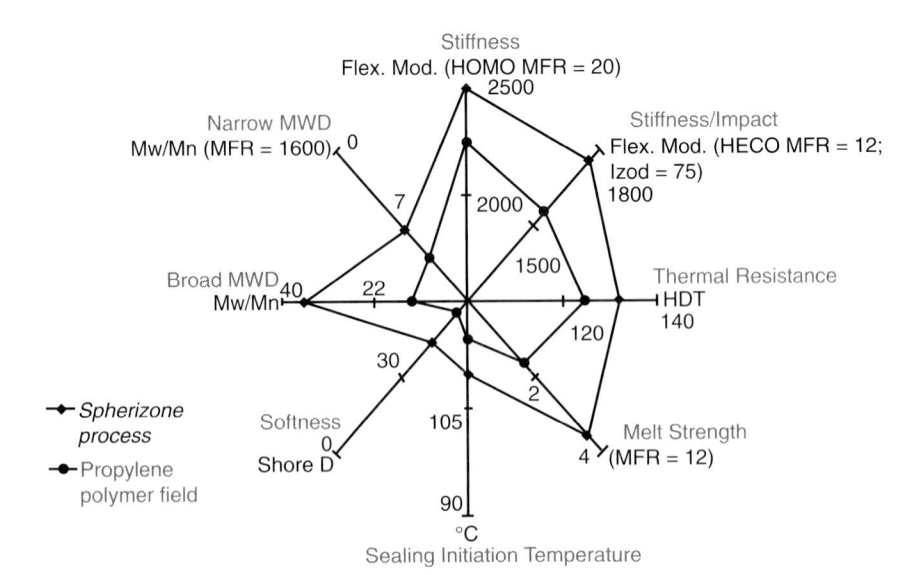

FIGURE 16.2.6 Expanding PP properties through *Spherizone* process technology.

- Molecular weight distribution control
- Use of polymer- or chemical-grade monomer
- Homopolymer, random copolymers, high-clarity random copolymers and terpolymers, two compositions (homopolymer/random copolymers, twin-random copolymers, random/heterophasic copolymers) in the MZCR only, and heterophasic impact and speciality impact copolymer production with an additional gas-phase reactor

ECONOMICS

Reliability and Operability

The *Spherizone* process has been running consistently well at the first commercial plant in Brindisi, Italy, after a scale-up from pilot plant size to 160 kMTA. The new process builds on the extensive knowledge gained from the *Spheripol* technology and offers further operability advantages:

- A simple, straightforward process design with simple and reliable equipment
- Easy online product change
- Rapid restart after shutdown
- No scheduled maintenance downtime for cleaning or inspection purposes

Consumptions

Typical specific consumptions (per 1000 kg of PP produced) are shown in Table 16.2.3.

PRODUCTS AND APPLICATIONS

Products from the *Spherizone* process are superior in quality owing to the optimal mixing of the different polymer properties at a macromolecular level. As a result, depending on the product family produced, the following properties are improved over the excellent quality of products from the *Spheripol* process:

Homopolymer/homo/random copolymer:

- Stiffness
- Fluidity
- Heat resistance
- Homogeneity
- Processability
- Controlled crystallization

Twin random copolymer:

- Homogeneity
- Opticals
- Impact
- Seal initiation temperature (SIT)

TABLE 16.2.3 Typical Specific Consumptions

	Homopolymer	Impact copolymer
Monomers, kg*	1002	1002
Catalyst, kg	0.025	0.025
Electric power, kWh†	110	135
Steam, kg‡	80	120
Cooling water, m^3§	85	85

*Net consumption of 100% monomers/comonomers included.
†Extrusion not included, consumption depending on product MI.
‡Low-pressure steam, high-pressure steam consumption for extruder die plate or barrels heating is excluded.
§At $\Delta T = 10°C$.

Random/heterophasic copolymer:

- Softness
- Opticals
- Low-temperature resistance

Heterophasic copolymer:

- Fluidity
- Impact/stiffness balance
- Heat resistance

Products produced with the *Spherizone* process maintain consistently high quality from one run to the next owing to the good control of reaction conditions and consistent catalyst quality. The *Spherizone* process can produce a range of products from high-quality homopolymer PP to various special polyolefin products.

General. Isotactic PP is well suited for a variety of end uses, ranging from flexible and rigid packaging to fibers and large molded parts for automotive and consumer products. PP is recyclable, an important consideration in many packaging and automotive applications, and it can be incinerated without toxic emissions. PP can be processed via extrusion, extrusion coating, extrusion and injection-stretch blow molding, injection molding, and thermoforming. Its physical properties can be enhanced easily through the addition of fillers such as calcium carbonate and talc. PP has excellent chemical resistance and electrical insulating properties.

Typical applications for PP products include fibers and filaments, oriented and cast film, injection-molding items, blow-molded bottles and parts, and thermoformed containers. Random copolymers can be produced with excellent optical properties and sealing initiation temperatures in compliance with the U.S. Food and Drug Administration (FDA) regulations for food contact. Heterophasic copolymers with outstanding low-temperature behavior, high impact strength, and enhanced stiffness can be obtained in the widest range of melt viscosities.

Worldwide, homopolymer accounts for 65 to 70 percent of all PP, random copolymer and terpolymer for 10 to 15 percent, and heterophasic copolymer for 15 to 20 percent. Film and fiber are the two largest segments in the global PP market, but injection/blow molding and extrusion account for significant quantities. Typical PP fabrication processes and end uses include the major fields shown in Table 16.2.4.

TABLE 16.2.4 Typical PP Processes and End Uses

Fabrication process	Markets/end uses
Film extrusion	WQBF/cast film/BOPP film (flexible packaging for textiles, confectionery, bakery, and cigarette wrap)
Multifilaments	Woven sacs (raffia)/fibrillated tape for carpet backing, geotextiles, rope and twine, upholstery, and cigarette tow
Nonwovens (melt blown and spun bonded)	Geotextiles, medical applications
Injection molding	Automotive, appliances, housewares, furniture, consumer products, packaging (crates, cases, caps and closures, thin-walled and transparent containers)
Blow molding	Packaging
Profiles extrusion	Pipes, conduits, corrugated sheet, wire and cable extrusion coating, and lamination

Owing to the excellence in the quality of the products produced by the *Spherizone* process, and considering the effect of these parameters on the various processing techniques or application fields of the final product, it is apparent that basically all markets can benefit from these products. However, initially, owing to the limited availability of the *Spherizone* process, the main markets will be the specialities in each sector, where premiums are paid for top-quality products.

Spherizone technology, when used with Basell's *Avant* catalysts, is capable of manufacturing PP products with virtually unlimited combinations of physical properties to meet the increasing demands of customers in new and more challenging applications. Products from the *Spherizone* process, owing to their improved product properties, will be another driving factor for material substitution, further enlarging the market for PP and specialty applications.

CHAPTER 16.3
BORSTAR POLYPROPYLENE TECHNOLOGY

Jouni Kivelä, Helge Grande, and Tarja Korvenoja
Borealis Polymers O/Y
Porvoo, Finland

INTRODUCTION

The Borstar polypropylene (PP) process is a versatile technology. Homopolymers, random copolymers, heterophasic copolymers, and very-high-rubber-content heterophasic copolymers can be produced. The process is modular, the basic module consisting of a loop reactor–gas-phase reactor combination. A typical plant is shown in Fig. 16.3.1.

PP with a melt-flow rate (MFR) ranging from 0.1 to 1200 can be produced with the Borstar PP process. Currently, Ziegler-Natta catalyst is used, but single-site catalysts can be used in the future. When producing homopolymers and random copolymers, the process consists of a loop reactor and a gas-phase reactor in series. One or two gas-phase reactors (GPRs) are combined with this arrangement when heterophasic copolymers are produced.

Propylene, catalyst, cocatalyst, donor, hydrogen, and comonomer (in the case of random copolymers) are fed into the loop reactor, where propylene is used as polymerization medium (bulk polymerization). The loop reactor, which is designed for supercritical conditions, typically is operated at 80 to 100°C and 50 to 60 bar. The propylene and polymer mixture coming from the loop reactor enters a fluidized-bed gas-phase reactor, where propylene is consumed in polymerization. The reactor typically is operated at 80 to 100°C and 22 to 35 bar. Fresh propylene, hydrogen, and comonomer (in the case of random copolymers) are fed into the reactor. After removing hydrocarbon residuals, the polymer powder is transferred to extrusion.

In the case of heterophasic copolymers, the polymer from the gas-phase reactor is transferred into another, smaller gas-phase reactor where the rubbery copolymer is made. After this step, hydrocarbon residuals are removed, and the powder is transferred to extrusion.

The basic module, the loop–gas-phase reactor combination, enables high once-through conversion (minimized recycle) because the unreacted monomer from the loop reactor is consumed in the gas-phase reactor. Polymerization conditions in each reactor

FIGURE 16.3.1 Borstar PP plant, Austria.

can be controlled independently, enabling production of both standard unimodal and broad-molecular-weight multimodal grades. The production rate ratio between the reactors can be adjusted to meet the targeted product properties. The reactor combination, the high operating temperatures, the Borealis nucleation technology, and the proprietary catalyst together offer

- Very high catalyst productivity (60 to 80 kg PP/g catalyst depending on grade produced)
- Competitive electricity, steam, and monomer usage
- Extremely wide product window for product tailoring in molecular weight and comonomer distribution
- Possibility for very high stiffness products
- Possibility for very soft products
- Low nucleation cost

The Borstar PP technology is based on proprietary knowledge, and it is covered by a broad range of patents. The first commercial plant went on stream in May 2000 in Austria, and it produces a wide range of added-value polymers that already are recognized by customers to be at the leading edge of modern PP technology.

The product scope of the Borstar process is exceptionally broad, as demonstrated in the following operating window:

Flexural modulus   200 to 2300 MPa
$MFR_{(2.16)}$ range   0.1 to 1200 g/min
MWD range (M_w/M_n)   3.5–20
Isotacticity range   95–99.2%
Comonomer content   Random copolymer up to 10 wt % ethylene
Comonomer distribution   Accurately controlled across the molecular weight range
Rubber content   25 wt % rubber (for typical module 2 plant)

FEATURES OF THE BORSTAR PP PROCESS TECHNOLOGY

Modular Process

The Borstar PP process is available as a modular process. The first module (1) contains two reactors in series and is suitable for making a very wide range of homopolymers and random copolymers. The second module can be installed directly or as a later investment. This module (module 2) uses a third reactor in series with the first two that allows the production of a wide range of heterophasic (block) copolymers.

Module 1: Homopolymers and Random Copolymers. The Borstar PP process is based on the same technology concept as Borstar PE, featuring a slurry loop and one gas-phase reactor running in series, with the possibility to control each reactor independently. Approximately 50 percent of the polymer product is made in each reactor, but this can be varied with the grade being produced. The unique combination of reactors provides complete control over a wide range of molecular weight distributions, from unimodal to highly bimodal, for homopolymers and random copolymers; in addition, the comonomer distribution in the random copolymers also may be controlled to provide an optional balance of properties.

The slurry loop reactor has been designed for a high reactor temperature right up to and including *supercritical* propylene as the slurry diluent. The high temperature strongly enhances the catalyst activity, thereby improving the production economics.

Module 2: Heterophasic (Block) Copolymers. In module 2, a second, smaller gas-phase reactor is added to produce heterophasic copolymers with up to 25 percent rubber incorporated. These products offer a superior stiffness/impact balance compared with competing processes because the polymer matrix is made in the unique first step with a loop and a gas-phase reactor combination, where very high stiffness can be made.

Process Benefits: Why Use a Loop–Gas-Phase Reactor Combination?

The heart of the process, the loop–gas-phase reactor combination, offers significant possibilities for polymer tailoring in homopolymers, random copolymers, and a matrix of heterophasic copolymers:

- Tailoring of molecular weight distribution
- Tailoring of isotacticity
- Tailoring of comonomer content

The loop reactor operates in a high-temperature area up to supercritical conditions. Loop reactor pressure is always kept above the critical pressure of the reaction mixture. High reactor temperature is beneficial for a number of reasons:

- Higher catalyst activity
- Higher heat-transfer capability (allowing more output), owing to the large temperature difference between the cooling water and the reactor
- Low density of the reaction liquid, allowing higher solids concentration (= conversion)
- Better randomness (comonomer distribution)
- Higher isotacticity

High pressure in the loop reactor, of course, allows very high hydrogen concentrations to be used, enabling very high MFRs to be produced.

The first gas-phase reactor also operates in a high-temperature area to maximize catalyst productivity. There are no inert components added to the first GPR; it operates in a full monomer environment. In the case of bimodal product, the low-molecular-weight fraction usually is produced in the first gas-phase reactor. The GPR was chosen as the second step for following reasons:

- Very high hydrogen concentrations are possible.
- Production split between loop and first GPR is easy to control.
- Very high comonomer contents are possible.
- Gas-phase reactor consumes the unreacted monomer from the loop reactor, making the once-through conversion over the loop–first GPR combination very high.
- Part of first GPR cooling comes from vaporizing the monomer coming from the loop reactor.
- No steam thus is needed for vaporizing the monomer in the loop outlet.

PROCESS DESCRIPTION

The reactor part of the Borstar PP process consists of a cascade of a slurry loop reactor followed by a gas-phase fluidized-bed reactor using proprietary Ziegler-Natta catalysts. The process is suitable for producing a wide range of both bimodal and unimodal PP grades.

A Borstar PP plant consists of the following areas:

- Feedstock preparation area
- Loop reactor area
- First gas-phase reactor area
- Second gas-phase reactor (when module 2 for block copolymers is required)
- Recovery area
- Pelletizing and blending area
- Bagging and storage area (OSBL)

Wet End (Fig. 16.3.2)

The Borstar PP process consists of three reactors and a prepolymerizer in series. The prepolymerizer, loop reactor, and first gas-phase reactor are used for production of

BORSTAR POLYPROPYLENE TECHNOLOGY   16.45

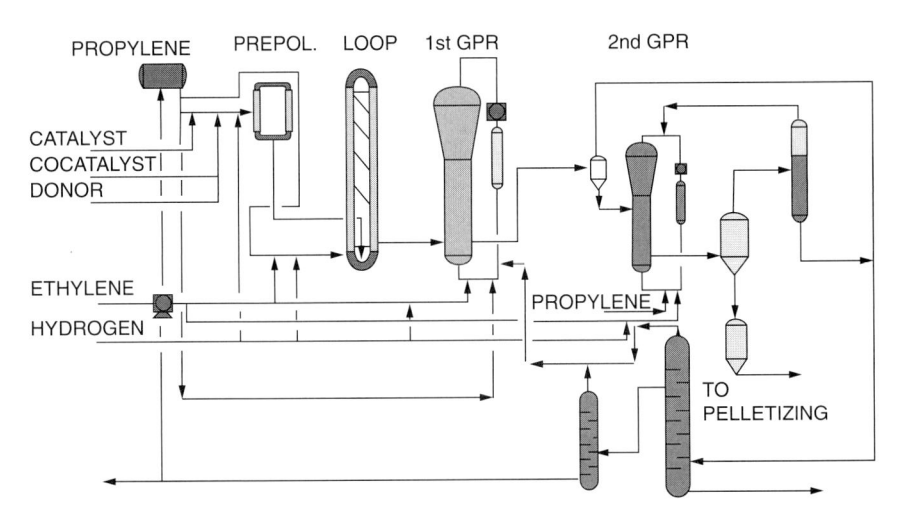

FIGURE 16.3.2 Borstar PP process, wet end.

homopolymers, random copolymers, and the matrix of heterophasic copolymers, whereas the second GPR is used for production of rubber for heterophasic copolymers.

The main raw material is propylene. Ethylene is used as a comonomer. The design has been made especially for operation with Borealis' proprietary Ziegler-Natta catalyst; triethyl aluminum-alkyl (TEAL) is used as cocatalyst. An external donor is used for control of isotacticity. Hydrogen is used for control of molecular weight.

Feed Preparation. The feed area consists of a propylene feed tank and adjacent pumps and compressors for feeding liquid propylene, hydrogen, and ethylene. Depending on the quality of the raw materials, the reactor feeds may have to be purified to remove catalyst poisons. Typical catalyst poisons are water, oxygen, carbon monoxide, carbon dioxide, sulfur components, arsine, and methyl acetylene. The purification is carried out in beds filled with catalysts or absorbents dedicated for each poisonous component using separate purification trains for each feed stream. The type of catalyst used in the beds is available from catalyst and absorbent vendors. Each bed is designed for a certain lifetime depending on the amount of impurities in the feed, after which the bed is regenerated for reuse.

Catalyst Handling. The catalyst is delivered in dry form under nitrogen blanket in catalyst transport drums. The transport drums are emptied into catalyst tanks. The catalyst area includes batchwise preparation of catalyst in carrier, catalyst feed pumps and vessels, and pumps for feeding donor. The Ziegler-Natta catalyst suspended in a carrier is fed from the catalyst feed tanks into the prepolymerization reactor with catalyst feed pumps. The cocatalyst, TEAL, is stored under nitrogen pressure and is fed into the reactors with a metering pump.

Loop Reactor Including Prepolymerizer. The prepolymerization reactor is a small loop reactor operated at 50 to 60 bar(g) and low temperature. The prepolymerized catalyst, propylene, hydrogen, and ethylene (when producing random copolymers) are fed to the loop reactor. The polymerization in the loop reactor takes place in bulk propylene. The heat of reaction is removed by cooling water in the reactor jacket. The loop reactor is operated at 50 to 60 bar(g) and 80 to 100°C.

In the loop reactor, PP forms discrete particles on each grain of catalyst. The resulting polymer slurry is circulated in the reactor at high velocity by an axial-flow pump. The slurry concentration in the reactor is 40 to 60 wt %.

Pressure control of the loop reactor is based on the product outlet. The pressure in the loop reactor controls slurry withdrawal from the reactor.

The loop reactor outlet is operated in a continuous mode. The slurry from the loop reactor is fed directly to the first GPR.

First Gas-Phase Reactor. Polymer is transferred to the first GPR for further polymerization without addition of new catalyst. The first GPR produces more homopolymer or random copolymer as desired. The reactor is a fluid-bed reactor operating at 22 to 35 bar(g) and 80 to 100°C.

In the GPR, fluidizing gas, circulated with a circulation gas compressor, is cooled in the circulation gas cooler to remove the heat of reaction and compression. The required amount of cooling depends on the current production rate and product grade. Reactor temperature is controlled by adjusting the inlet temperature of the circulation gas. Reactor pressure is controlled with fresh propylene feed or by routing excess circulation gas to the recovery section. The reactor bed level is controlled by product takeoff.

The completed homopolymer/random copolymer is directed toward the polymer degassing process step.

Second Gas-Phase Reactor. When producing block copolymers, the powder from the first GPR is fed via a powder-transfer system to the second GPR. The rubber part of the heterophasic copolymer is produced in this reactor.

The reactor is a fluid-bed reactor operating at 15 to 25 bar(g) and 75 to 90°C. Polymer powder from second GPR enters the polymer degassing step.

Polymer Degassing. Polymer powder from first or second GPR enters a low-pressure flash tank, where the polymer powder and hydrocarbon gases are separated. The separated gaseous hydrocarbons are fed to the recovery area. The powder flows by gravity to a purge bin, where it is treated with hot nitrogen and steam to remove residual hydrocarbons and to deactivate the catalyst. The purged powder is transferred with a closed-loop nitrogen conveying system to one of the powder bins in the dry-end section.

Recovery Section. The recovery section is designed to recover propylene, ethylene, and hydrogen and to recycle the components back to the reactors. Other functions of this section are to remove light impurities, which are catalyst poisons, and light inert components and heavy hydrocarbons and cocatalyst residues. The recovery is operated differently depending on whether the second GPR is in operation or not.

Dry End (Fig. 16.3.3)

Pelletizing Building and Blending Area. The dry-end part of the process starts after the purge bin. From the purge bin, the polymer powder is transferred pneumatically with nitrogen to the powder silos. These powder silos serve as a buffer between the reaction section and pelletizing. From the powder silos, the polymer powder is fed via a powder roof bin to the continuous mixer.

Solid additives are transferred from big bags or from solid additive drums on the ground level to feed bins by a vacuum conveying system. From the feed bins, the solid additives are fed to the continuous mixer with loss-in-weight feeders. In the continuous mixer, polymer powder and additives—solid and in some cases liquid—are mixed before feeding into the extruder, where the mixture is melted, homogenized, and pelletized.

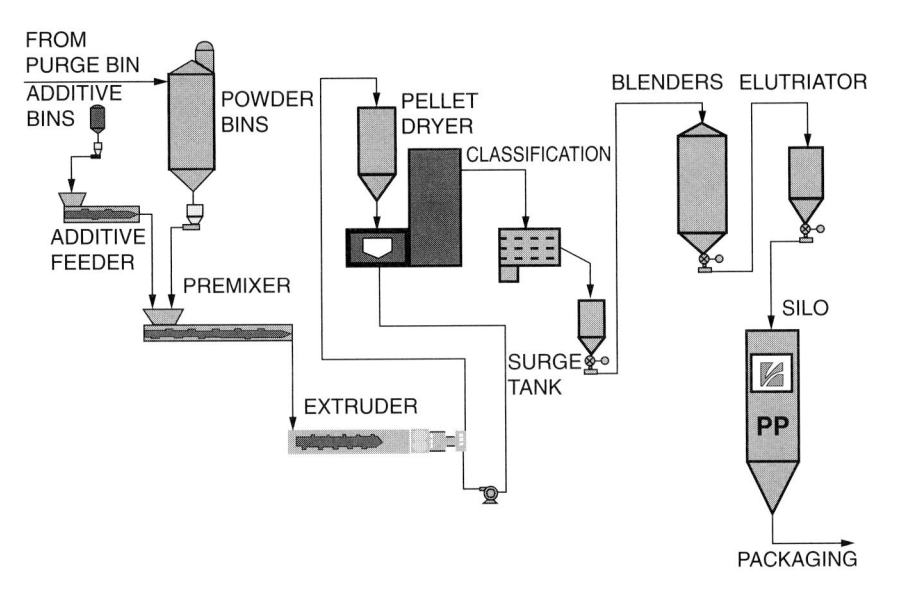

FIGURE 16.3.3 Borstar PP process, dry end.

The system upstream of the extruder is slightly pressurized by nitrogen to ensure oxygen-free material feeding to the extruder. Downstream of the extruder, the process consists of a pellet water system and an air conveying system that transfers the pellets into pellet blenders.

The pellet water system, consisting of pellet water circulation equipment, a centrifugal dryer, and a product screen, is used for cooling, drying, and classifying the polymer pellets.

Dried pellets are transported pneumatically from pelletizing to blenders located in the blender area. Homogenization in the blenders is done by circulating the polymer pellets with a pneumatic conveying system.

Bagging and Storage. From the blenders, the polymer pellets are transferred pneumatically to pellet bulk silos and further to bagging via elutriators, where dust and streamers are separated from the pellets.

A typical bagging line consists of a feed bin, bagging machine, palletizing machine, and wrapper. The polymer pellets also can be routed from the storage silos to a truck loading station, where manual big bag loading also is possible.

PRODUCTION CYCLE AND GRADE TRANSITIONS

The length of the production cycle depends on marketing requirements. The recommended production cycle is typically 6 weeks. The expected amount of transition material is as low as 0.5 percent of total production.

Grade transitions are controlled easily, and as with all processes, the relative actual transition quantities depend on the specific grade change being made. Experience from the Borstar PP plant in Austria and the pilot plant in Finland shows that the multimodal production technique may be compared readily with other common PP processes when it comes to transition times and amounts produced for a given type of transition.

Transition time in the Borstar PP process is kept to a minimum because the Borstar PP process uses one catalyst family. This reduces transitions imposed by catalyst changes in some other processes. The use of one catalyst is possible because of superior reactor design and configuration, allowing molecular tailoring by the process to give the properties and not by changing the catalyst (especially for block copolymers).

The transitions in addition are made all the more simple to manage by the advanced process control system developed especially for the Borstar process.

ADVANCED PROCESS CONTROL

Introduction

BorAPC is the proprietary advanced process control system developed by Borealis for the control of polyolefin plants. The purpose of BorAPC is to

- Stabilize production in the reactors
- Improve first-time-right yields of prime-grade polymer product through online closed-loop control of the polymer properties
- Maximize use of the production capacity by operating with full and stable control near the maximum capability while still respecting the constraints of the plant
- Improve plant performance during grade transitions

BorAPC helps the process operator to monitor and control the polymerization reactors and the key quality properties of the polymers produced and to optimize the performance of the plant.

There are two main functions:

- *Supervisory control* in the form of nonlinear multivariable predictive controllers that will stabilize the conditions in the reactors and the polymer quality properties. This is closed-loop real-time process control. The multivariable controller is called *OnSpot*.
- *Basic calculations* are online process calculations that calculate and predict important variables that cannot be measured directly. This information is used both for the online process control to assist the operator and for offline use by plant engineers in reporting and follow-up.

BorAPC was developed based on Borealis' operational experience and already has been installed in several polymerization plants. It can be installed together with any modern DCS control system.

Advanced Process Control System

BorAPC has multivariable model-based predictive controls that are especially suitable to handle control of polymerization processes. Examples of controlled variables are

- Concentrations of catalyst, hydrogen, and comonomer in the reactors
- Amount of polymer solids in the reactors
- Reaction rates in each reactor
- The balance ("split") between reaction rates in the prepolymerization and the loop and gas-phase reactors

BorAPC has a fundamental physical nonlinear process model that uses measurement information from the plant to estimate key process variables and predict their future behavior. The predictions are used to assist the operator and for closed-loop multivariable control.

The BorAPC control application is implemented in a separate computer with an interface to the DCS control system. An estimator uses online information from process measurements and analysis data from the laboratory to update selected parameters in the prediction models to ensure that the prediction models reflect the real plant as closely as possible.

The basic calculation package provides real-time property calculations with accuracy at least as good as conventional steady-state process simulation tools. It can be implemented in a separate computer or be configured as a set of equations in the DCS system.

CATALYST

The product range produced with Borstar technology is based on the use of a specially developed proprietary Ziegler-Natta catalyst family that maintains a high level of activity throughout all the reactors (the catalyst is only injected at one point—the prepolymerizer).

ENVIRONMENT

Emissions to Atmosphere

All hydrocarbon-containing purges are routed to the flare. Only nitrogen purges and conveying air are vented to atmosphere.

VOC Emissions

The main sources of VOCs (volatile organic compounds) emissions are fugitive sources. Fugitive emissions emanate from a variety of point sources in process operations, primarily in the form of leakage. These leaks occur from various process and piping equipment such as pumps, flanges, compressors, vents, and drains. The main components are

| Ethylene | 10 to 15 wt % |
| Propylene | 85 to 90 wt % |

Liquid Effluents

Wastewaters. Total wastewater sources are

- Washing and flushing waters from the process areas (discontinuous)
- Washing waters from the silos (discontinuous—varies significantly with the number of silos required)
- Overflow of pellet water tank (about 2000 m^3/yr)
- Cooling water analysis (continuous)

Other Liquid Effluents. Liquid effluents and waste streams are

- Flushing oil from TEAL handling
- Heavy hydrocarbons (HCs) from diluent recovery

TABLE 16.3.1 Typical Raw Materials and Utilities Consumption

Raw material/ utility	Unit per ton PP	Consumption/unit/ton PP Homo/random, typical	Heterophasic, typical
Total monomer	ton	1.010	1.010
Catalyst	kg	0.015	0.015
Cocatalyst	kg	0.20	0.20
Steam	kg	170	170
Donor	kg	0.01–0.05	0.01–0.05
Hydrogen	kg	0.1	0.2
Cooling water		100	100
Electricity, wet end	kWh	160	180
Electricity, dry end (grade dependent)	kWh	160	160

Wastes

Solid waste produced or formed in the process consists of

- Purification bed material
- Waste polymer
- Pelletizing additive spills

OPERATING REQUIREMENTS

Utilities, Additives, and Ancillaries Consumption

Typical consumption of raw materials and utilities for a Borstar PP plant for a broad product range from homopolymer via random to heterophasic copolymers are given in Table 16.3.1.

PRODUCTS

The product property window of the Borstar PP process is broader than in other technologies owing to the unique combination of advanced reactor technology and the Borealis enhanced-properties catalyst (Fig. 16.3.4). This means in practice that Borstar, in addition to allowing for the production of all the conventional PP grades, also provides the option to produce enhanced products both to established and to completely new applications. These improved properties already have been proved by customers' testing materials produced commercially.

Creep and heat deflection temperature (HDT) translate into savings potential either by replacement of more expensive engineering plastics or in direct weight reduction. Benefits and opportunities have been identified in applications ranging from pipe extrusion and molding of durable goods to automotive and construction products.

The Borstar soft PP product range adds a completely new feature to a "normal" PP plant. This product line will offer the market new possibilities to create "all-PP" solutions. A whole new soft PP family combining good optical properties with low temperature impact has been tested in different cast film and molding applications.

BOREALIS

Borstar PP broadens the product performance window

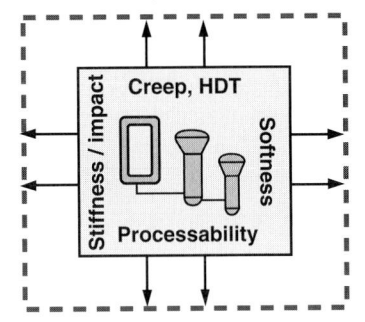

[Diagram: central box labeled "Creep, HDT" / "Processability" with "Stiffness / impact" on left and "Softness" on right, arrows pointing outward in all directions]

→ Market advantages in terms of broader properties window
→ Cost savings by increased output and downgauging possibilities

FIGURE 16.3.4 Product performance window.

Excellent processability usually can be transferred immediately into an improved cost position at every converter. Higher output at fiber spinning, in-line thermoforming, cast film production, and several molding applications already has been verified by Borstar advanced grades. Ongoing tests also show very encouraging results regarding improved fiber bonding index and cast film optics owing to the very consistent isotacticity distribution and the excellent randomness achieved. Another advantage of the improved flow is experienced as a reduction in flow marks in the surface when molding complicated articles with long and narrow flow channels and in the extrusion of high-molecular-weight pipe grades (Fig. 16.3.5).

Heterophasic Copolymers

In all heterophasic copolymer applications, improvement in the stiffness/impact balance is a constant challenge for PP development engineers. With the new options available from the Borstar PP technology, a significant lift in this ratio has been achieved (Fig. 16.3.6).

One exciting example of this improvement shift is a new injection-molding grade (MFR = 12) targeted for transport packaging and thin-wall packaging applications (Fig. 16.3.7).

Compared to the reference, a considerable increase in the stiffness, falling weight impact, top load, and the creep properties has been achieved. This new balance of properties will open up several new options in the next round:

- Higher performance in critical applications, i.e., new market opportunities
- Reduced article weight while maintaining the mechanical properties of the reference
- Concentrating the improvement into increased flow (MFR), which immediately transforms into increased output at the molder

The same improvement in mechanical properties also has been achieved at higher MFRs (Fig. 16.3.8).

For the low-MFR range, heterophasic copolymers intended for heavy-duty applications such as transport packaging (pallets, bins, crates, boxes) and luggage, improved creep

BOREALIS

Borstar PP gives direct end-use benefits

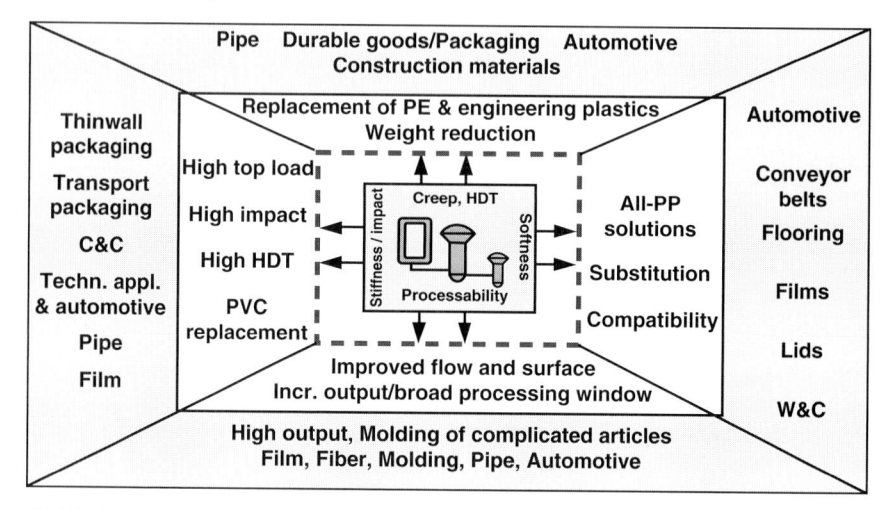

```
                    Pipe   Durable goods/Packaging   Automotive
                              Construction materials

Thinwall          Replacement of PE & engineering plastics       Automotive
packaging                    Weight reduction
                 High top load                                   Conveyor
Transport                            Creep, HDT    All-PP        belts
packaging        High impact                       solutions
C&C                                                              Flooring
                 High HDT                          Substitution
Techn. appl.                      Processability                 Films
& automotive     PVC                               Compatibility
Pipe             replacement                                     Lids
Film                Improved flow and surface
                 Incr. output/broad processing window            W&C

              High output, Molding of complicated articles
                  Film, Fiber, Molding, Pipe, Automotive
```

FIGURE 16.3.5 End-use benefits.

BOREALIS

Borstar PP gives optimum impact / stiffness balance for a broad stiffness range

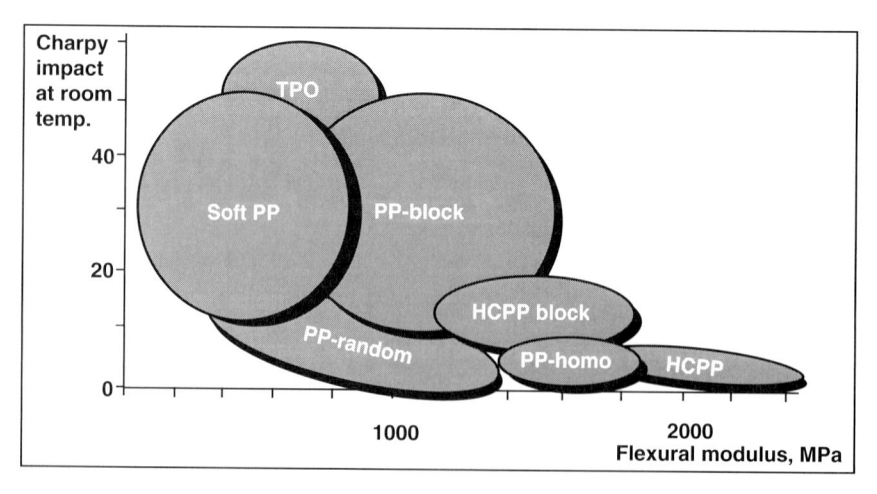

[Chart: Charpy impact at room temp. vs Flexural modulus, MPa; showing regions TPO, Soft PP, PP-block, PP-random, HCPP block, PP-homo, HCPP]

FIGURE 16.3.6 Impact/stiffness balance.

BORSTAR POLYPROPYLENE TECHNOLOGY 16.53

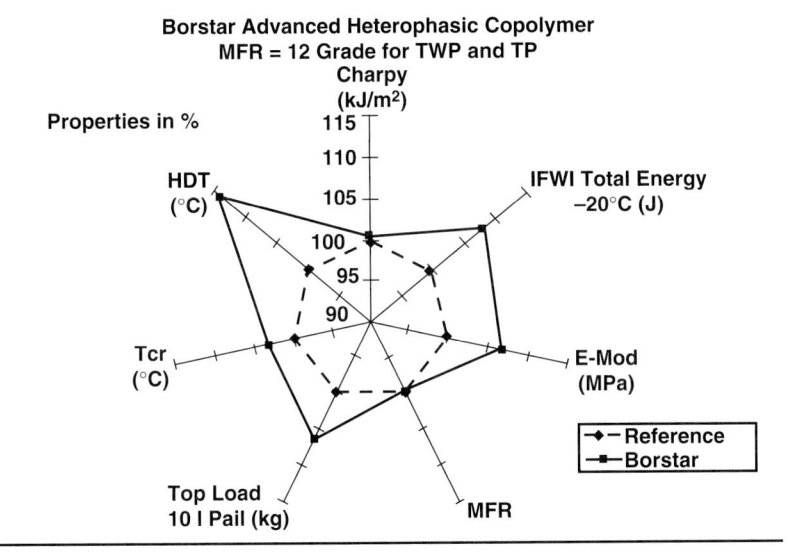

FIGURE 16.3.7 Borstar advanced heterophasic copolymer.

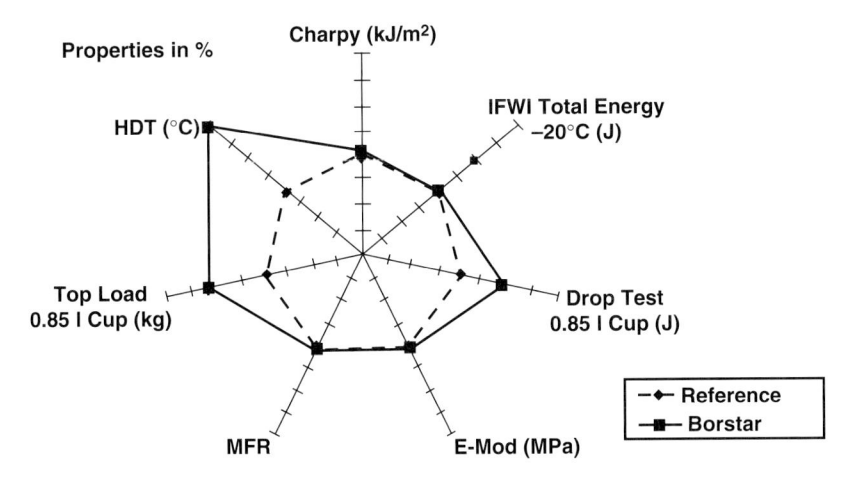

FIGURE 16.3.8 Higher-MFR advanced heterophasic copolymer.

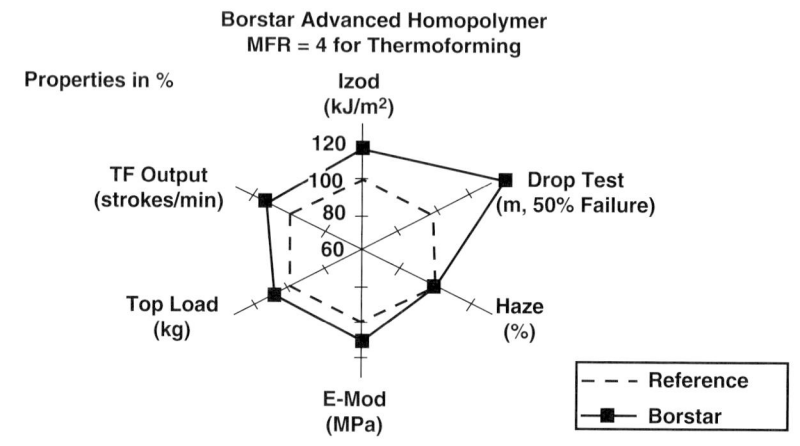

FIGURE 16.3.9 Advanced homopolymer.

properties and processability have been demonstrated while still maintaining the established stiffness/impact balance.

Homopolymers

The Borstar enhanced product performance window also extends into homopolymers and random copolymers. A concrete example is a new homopolymer for thermoforming where a 10 to 15 percent output increase has been proven (Fig. 16.3.9).

In comparison with the reference, this Borstar advanced grade shows improvement in all essential TF properties and therefore will, in addition to the higher output, also offer possibilities for downgauging. The higher isotacticity and easier crystallization also will show up as improvements in properties such as HDT, WVTR, and surface gloss.

The same tendency is seen on cast film properties (Fig. 16.3.10). The favorable polymer characteristics, such as a broad and/or a bimodal MWD in combination with high stereoregularity/isotacticity and low extractables, opens up for downgauging and new applications such as hot fill and for substitution of higher heat-resistant polymer solutions. The Borstar option of broadening the MWD in a controlled manner also will improve processability, which is a prerequisite for the success of the oncoming high-output cast film lines, where polymer flow and shear behavior will be critical factors.

Random Copolymers

Polymerization kinetics for random copolymers tend to concentrate the ethylene comonomer in the low-molecular-weight fraction. The Borstar reactor configuration allows for a narrower comonomer distribution, leading to a more homogeneous polymer composition with

- Lower amount of volatiles
- No odor/smoke during processing

BOREALIS

Borstar PP impact on cast film properties

Polymer characteristics
- Broad and/or bimodal MWD
- High stereoregularity/isotacticity
- Low extractables

Application benefits
- Further downgauging
- Entrance to hot-fill applications
- Substitution potential

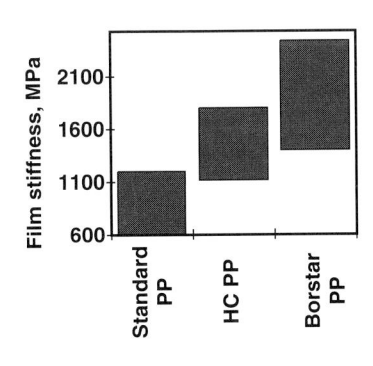

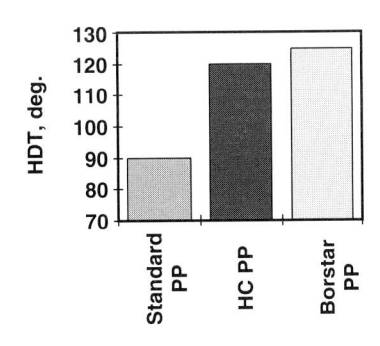

FIGURE 16.3.10 Cast film properties.

BOREALIS

Borstar randoms: Good opticals with low extractables

- Narrow comonomer distribution by incorporating different amounts of comonomer in different molecular weight fractions

→ No blooming

→ Food contact approval

→ Medical applications

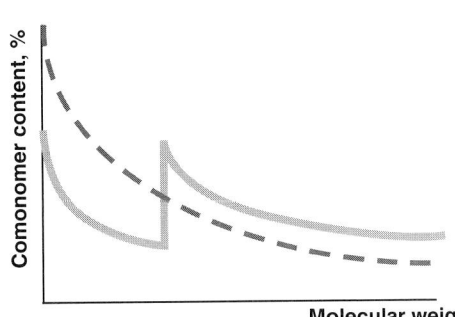

FIGURE 16.3.11 Borstar grades with good opticals and low extractables.

- Better clarity
- Improved film shrink force control
- Higher comonomer content (Fig. 16.3.11)

See Fig. 16.3.12.

BOREALIS

The Borstar product perfomance window is based on

- Very flexible process configuration with proprietary Advanced Process Control system
    - High-temperature operation
    - Predictive system based on theoretical and empirical model
    - Improved product consistency

- Bi- and multimodal process configuration
    - Wide operating window allowing for tailormaking of desired properties
    - Cost-saving potential in terms of reduced cycle time and downgauging

- Proprietary catalyst with enhanced properties
    - High isotacticity when required
    - Controlled crystallization and crystallinity
    - Improved control of isotactic sequence lengths
    - High content of comonomer and excellent randomness
    - High activity (low ash)

FIGURE 16.3.12 Borstar performance window.

CHAPTER 16.4
UNIPOL™ POLYPROPYLENE PROCESS TECHNOLOGY

Barry R. Engle
Dow Chemical Company
Danbury, Connecticut

GENERAL UNIPOL PP PROCESS DESCRIPTION

Overview

The UNIPOL™ PP process for polypropylene (PP) is thoroughly proven technology developed by the world leader in gas-phase polyolefins process technology. It is simple, flexible, economical, safe, and superior to alternative technologies. The process simplicity derives from advanced gas-phase fluidized-bed technology and advanced catalyst technology. Fluidized-bed technology brings many economic advantages over the competing processes, which use prepolymerization, multiple reactors in series (up to four), or mechanical stirrers. Advanced catalyst technology permits the elimination of process steps for catalyst passivation, ash removal, and atactic polymer removal and disposal. The result is a very stable but flexible system that uses simple equipment, produces a full range of commercial products, provides broad turndown with uniform product properties, and is easy to operate. The simple equipment used provides high reliability and reduces the need for specialized maintenance capabilities.

The high safety standards for UNIPOL process technology are validated by over 1700 reactor-years of safe operation (end of 2003). The process is designed with a focus on safety and, as a gas-phase process, offers many safety advantages over alternative technologies. The reaction system is designed to shut down gently without outside intervention in the event of loss of utilities or automatic controls. The process requires no reaction solvents and thereby avoids both the cost of the associated recovery systems and any potential hazards involved with handling large inventories of flammable liquids, such as risks associated with liquid pool fires. The dynamics of the fluid-bed reaction are inherently self-limiting, eliminating the risks associated with overpressurization of equipment in the event of control failure. All these design features make compliance with environmental, health, and safety regulations readily achievable through proper design and operation.

16.58   POLYPROPYLENE

The UNIPOL PP process, which includes all process facilities from receipt of raw materials through pelleting, is described in this section. A brief description of simplified resin blending and handling systems downstream of pelleting is also provided. The gas-phase fluidized bed's complete backmixing and unparalleled stability produce a very uniform product quality and permit a significant reduction in blending and handling facilities compared with the requirements for other polymer processes.

Process Description

A simplified flow diagram is presented in Fig. 16.4.1.

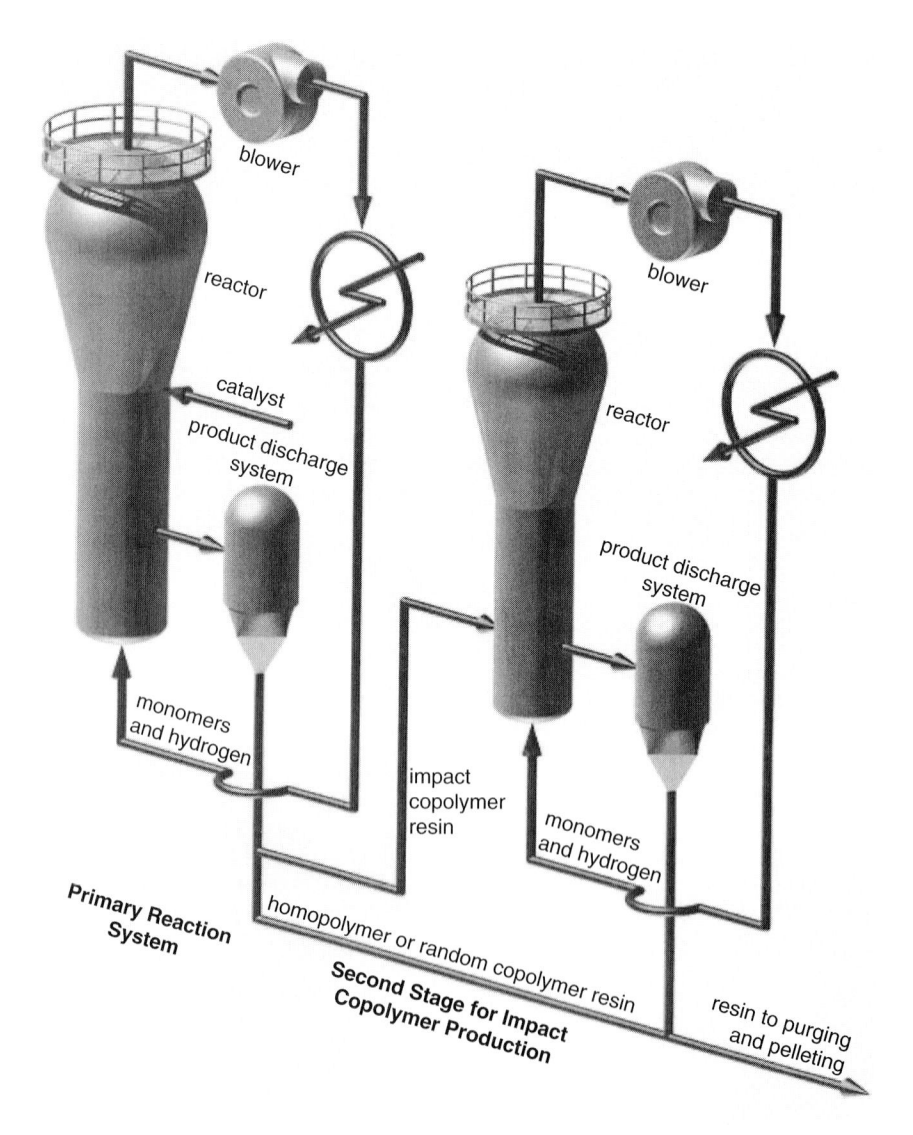

FIGURE 16.4.1   UNIPOL PP process flowsheet.

Raw Materials Receipt and Storage. The standard design for a facility employing the UNIPOL PP process uses polymerization-grade propylene (99.5 wt % propylene). The process can be adapted to use lower-purity propylene, if desired. Propylene received at the battery limits is treated to remove poisons that could affect catalyst performance.

Ethylene (used for copolymers), hydrogen, and nitrogen meeting standard commodity specifications are used in the reaction system without further treatment. Other miscellaneous materials are used simply as received, with no further treatment.

Polymerization Catalyst. SHAC™ PP polymerization catalyst purchased from Dow is delivered ready to use. Catalyst is stored and shipped under a nitrogen blanket.

Reaction. Reaction system 1, used for producing homopolymer and random copolymers, consists of a fluidized-bed reactor, cycle gas compressor, cycle gas cooler, catalyst feeder, and product removal system. The cycle gas compressor circulates reaction gas upward through the bed, providing the agitation required for fluidization, backmixing, and heat removal. No mechanical stirrers or agitators are needed in UNIPOL process reactors. The cycle gas exiting overhead from the reactor passes through the cooler that removes the heat of reaction. The catalyst feeder continuously feeds catalyst to the reactor while product is removed using a special, high-efficiency product removal system.

Reaction system 2 operates in series with reaction system 1 to produce impact copolymers. The impact reaction system is a smaller replica of reaction system 1 and includes a reactor, cycle gas compressor, cycle gas cooler, and a product removal system. For a homopolymer and random copolymer only plant (H/R-only plant), the second gas-phase reaction system is not required.

The reaction system is very stable and easy to operate and control with only five independent process variables. As a result, the UNIPOL PP process system produces a reliably uniform product. The five process control variables are

Reaction temperature: Constant for each resin grade

Reaction pressure: Constant

Gas composition: Constant for each resin grade

Bed level: Constant

Reaction rate: Controlled by catalyst feed rate

Resin properties, controlled by the gas composition and reaction temperature, are independent of reaction rate. The process computer performs the calculations needed to quantify certain parameters that cannot be measured directly and optimizes reaction rate, product grade changes, and so on. Each reaction system is optimized with regard to monomer and energy consumption and represents the most advanced technology available.

Resin Degassing and Vent Recovery. Granular resin from the product discharge system is conveyed to a vessel for purging with nitrogen to remove residual hydrocarbons. The vent gas is processed to recover hydrocarbons and nitrogen purge gas, which is returned to the process. The purged resin is combined with additives and then flows to pelleting.

Pelleting. Resin leaves the degassing system in the form of free-flowing granules, which flow directly to a low-energy pelleting system. The pelleting system produces uniform pellets, which are air-conveyed to storage.

Typical Resin Handling. These systems are based on conventional, commercially available equipment. Standard pneumatic conveying is used throughout the resin handling system. In typical plants operating the UNIPOL process, two or three combination storage/continuous-blending bins per reactor line are provided for surge capacity and product change/off-grade recovery flexibility. Shipment normally is made in bulk containers (van boxes, hopper cars, etc.) or small packages such as 25-kg bags. Shipping containers are also used for inventory storage.

Product Yield

The conversion rate of monomer (propylene) and comonomer (ethylene) to PP is essentially 99 percent for the UNIPOL PP process. Raw material requirements per pound of resin are also independent of reactor line capacity and operating rates.

PROCESS CHEMISTRY

Process Chemistry and Historical Development

The UNIPOL fluid-bed process was developed in the 1960s and commercialized in 1968. The reaction system, used for producing homopolymers, random copolymers, and impact copolymers, consists of a fluidized-bed reactor, cycle gas compressor, cycle gas cooler, catalyst feeder, and product removal system. The cycle gas compressor circulates reaction gas upward through the bed, providing the agitation required for fluidization, backmixing, and heat removal. No mechanical stirrers or agitators are needed in UNIPOL process reactors.

Catalysts available from Dow are high-activity Ziegler-Natta-based catalysts supported on magnesium chloride. An aluminum-alkyl is used as a cocatalyst. Internal and external electron donors are used to vary isotacticity and molecular weight distribution; a broad range of donors can be used, providing the ability to modify the product capabilities of a single catalyst system simply by varying the donor. The ratio of hydrogen to monomer controls product molecular weight (melt flow).

Third-generation catalyst systems (e.g., SHAC 310 catalyst) typically contain an aromatic monoester internal donor, whereas a chemically similar ester is used as the external selectivity control agent (SCA). This external SCA can be varied at the reactor to control product xylene solubles. Third-generation catalysts are characterized by their relatively fast rate of polymerization decay, as well as an inverse relationship between productivity and isotactic index. They generally have broader molecular weight distributions than fourth-generation materials, which can provide benefits in certain product applications.

In fourth-generation catalysts (e.g., SHAC 205, 320, and 330 catalysts), alkoxy silanes are used for selectivity control in conjunction with organic ester internal donors. Fourth-generation catalysts offer a slower rate of polymerization decay; thus while the initial polymerization rate is below a third-generation system, fourth-generation catalysts offer significantly higher net productivity over the course of the UNIPOL PP process reactor residence time. Fourth-generation catalysts also generally are capable of significantly lower xylene soluble products (higher selectivity to isotactic PP) without the loss in productivity seen for third-generation catalysts; they have other product advantages in areas of oligomers and organoleptics (odor and taste). Molecular weight distribution also can be tailored to a specific product application via selection of an appropriate SCA.

The catalyst support, which is $MgCl_2$ based, also varies in morphology (shape) from different catalyst systems and can provide advantaged properties in terms of resin powder bulk density, process operability, and a broadened window of product capability. The history and attributes of the SHAC catalyst families are summarized in Table 16.4.1.

TABLE 16.4.1 Catalyst Families and Benefits

SHAC catalysts	Primary area of advantage	Isotactic index	Polymer morphology	Benefits delivered
SHAC 103	Process and product	90–98	Irregular	Sufficiently high activity that de-ashing no longer required
SHAC 201	Product	90–99	Irregular	Low oligomers and low odor; narrower MWD, able to reach lower XS
SHAC 205	Process	90–99	Regular	Resin bulk density increase, measurable throughput increase for random copolymers
SHAC 310	Process	90–99	Spherical	Significant increase in resin bulk density giving opportunity for throughput increase while maintaining easy operating SHAC 103 catalyst kinetic profile
SHAC 320	Process	90–99	Spherical	Resin bulk density increase, expansion of product grade range
SHAC 330	Process and product	90–99	Spherical	Lower XS while maintaining high activity

Typical Productivity and Selectivity

In bulk PP laboratory autoclave reactors that are run under standard conditions of 1 hour and 67°C, productivity will vary with the SHAC catalyst and external donor (SCA) type selected. For SHAC 310 catalyst, 22 kg PP/g catalyst is given at approximately 4.0 wt % XS. For a prepolymerized autoclave run, SHAC 205 and 320 catalysts offer 50, 72, or 84 kg PP/g catalyst, depending on SCA selection. Commercial UNIPOL PP process results will vary.

PROCESS PERSPECTIVE

Technology Development

Dow is committed to the rapid development of new, advanced technology to meet the ever-increasing needs of PP producers in regional markets worldwide. As a global leader in polyolefins production and technology, Dow has a wide array of laboratory facilities at work on the discovery, development, and scale-up of new technologies, catalysts, and the resulting polymer products. Facilities include extensive small-scale catalyst preparation laboratories, catalyst and polypropylene pilot plants, and propylene applications laboratories. Major research and development (R&D) locations are

- Houston, Texas (catalyst preparation, polypropylene product applications)
- South Charleston, West Virginia (gas-phase polymerization pilot plants)
- Bound Brook, New Jersey (polymer rheology testing)
- Midland, Michigan (catalyst preparation, polymer material science and characterization)
- Freeport, Texas (polymer material science and characterization)
- External collaborations (new catalysts, new catalyst precursors, organic ligand synthesis)

Driven to continually advance its standing as the leading science and technology company in the chemical industry, Dow is leveraging its wide-ranging R&D organization to create new and powerful catalyst technology. Through the application of polymer materials science and advanced chemical screening technologies such as combinatorial and computational chemistry, Dow's significant investment in R&D programs to support the UNIPOL PP process will ensure a steady stream of novel heterogeneous and single-site catalysts designed to create generations of market-leading PP products.

Licensed Plants in Operation

Dow has 35 years of UNIPOL fluid-bed, gas-phase process experience starting with the first commercial polyethylene plant in 1968 and extending the UNIPOL technology to UNIPOL PP in 1986. The UNIPOL process has proven its durability, flexibility, and environmental qualities in over 120 polyolefin reactor lines worldwide comprising 1700 reactor-years of operating experience (at the end of 2003).

Aside from Dow's own UNIPOL PP facilities, there are 23 licensees operating 34 reaction lines in 15 countries on six continents (Fig. 16.4.2). Approximately 6 million tons of PP are produced annually using the UNIPOL PP process. For a complete list of operating commercial facilities, see Table 16.4.2.

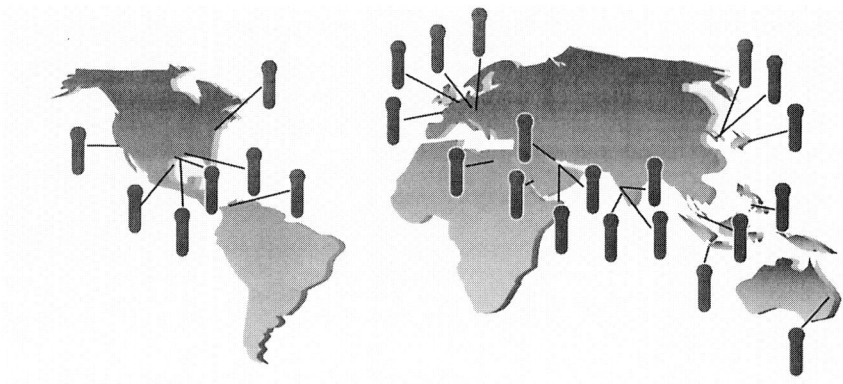

- 23 operating licensees
- Plants in 15 countries on 6 continents
- Over 6 million tons annual production in 36 operating lines

FIGURE 16.4.2 UNIPOL PP world presence.

Technology Delivery

The UNIPOL PP process technology is thoroughly documented in manuals that cover the underlying scientific knowledge, typical plant operating procedures, and a comprehensive process design package that provides the basic engineering information from which an engineering contractor can work. Experienced personnel work closely with licensees throughout the process of turning the basic technology into a new plant.

Licensees are trained in all facets of UNIPOL PP technology—from process theory to practical, hands-on experience. Initial orientation courses covering both process and product are held for technical staff and the project team. A thorough operations training program for plant staff is provided in advance of startup. In this more comprehensive program, licensee personnel from plant operations, maintenance, and laboratory groups spend weeks at the UNIPOL process technology training center in Houston, Texas, in a university-like program, which already has produced more than 1500 graduates.

The trainees receive expert instruction in small study groups. Course work in the classroom is supplemented by training time spent on a state-of-the-art computerized plant simulator that reproduces the behavior and control room displays of a real plant. This training tool compresses years of experience into weeks. Instructors can program unexpected changes into the simulator, and trainees' responses can be replayed for study and correction. Scale models are used to teach plant layout and maintenance techniques, and licensees are provided all-inclusive training manuals to keep for continuing reference.

An experienced Dow team, including a chief engineer and several operating specialists, provide on-site assistance during the commissioning and startup of each licensee's new plant. After the plant begins operation, licensees can participate in an exchange program to receive technical improvements.

Bidding Contractor Program

A group of several bidding contractors skilled in working with UNIPOL PP process technology offers extensive support to the client prior to license signing. These bidding con-

TABLE 16.4.2 UNIPOL PP Units in Operation

Company	Country	Startup year	Reactor lines	Product capability
Dow P-1	United States	1985	1	H/R/I
L. G. Caltex	Korea	1988	1	H/R/I
Huntsman 1	United States	1989	1	H/R/I
Polychim	France	1989	1	H/R
Propilco	Colombia	1990	1	H/R/I
Epsilon Polymers 1	United States	1991	1	H/R/I
Basell	Australia	1991	1	H/R/I
Dow Wesseling	Germany	1991	1	H/R/I
Huntsman 2	United States	1991	1	H/R
TPI 1	Indonesia	1992	2	H/R
DSM (now SABIC)	Germany	1992	2	H/R
Japan Polychem	Japan	1992	1	H/R/I
Solvay (now BP) 1	Belgium	1992	1	H/R/I
Polypropylene Malaysia	Malaysia	1992	1	H/R
SABIC 1, Ibn Zahr	Saudi Arabia	1993	1	H/R/I
TPI 2	Indonesia	1995	1	H/R/I
Reliance 1	India	1996	2	H/R/I
Epsilon Polymers 2	United States	1996	1	H/R
Solvay (now BP) 2	United States	1996	1	H/R/I
Hyosung T&C, Ltd.	Korea	1996	1	H/R/I
PIC	Kuwait	1997	1	H/R/I
J. G. Summit	Philippines	1998	1	H/R
Reliance 2	India	1999	2	H/R/I
ARCO	United States	1999	1	H/R
Pinnacle Polymers	United States	1999	2	H/R/I
Reliance 3	India	1999	1	H/R/I
SABIC 2, Yanpet	Saudi Arabia	2000	1	H/R
OPC	Egypt	2001	1	H/R
SABIC 3, Ibn Zahr 2	Saudi Arabia	2001	1	H/R
Conoco Phillips	United States	2003	2	H/R/I

tractors have access to up-to-date confidential technical information so that they can accurately estimate the cost of a UNIPOL PP facility and provide a fixed-price bid before a license is signed. The bidding contractors have comprehensive knowledge of and experience in implementing projects employing UNIPOL PP process technology. They receive regular training and updates on technology advancements and work closely with Dow during project development, scoping, and implementation. UNIPOL bidding contractors as a group have been responsible for the design and construction of over 75 percent of all the UNIPOL production lines currently in operation.

PRODUCTS AND BY-PRODUCTS

The UNIPOL PP process offers PP manufacturers the broadest range of homopolymers, random copolymers, and impact copolymers available today. PP resins produced by the UNIPOL PP process have wide-ranging applications in many market areas. They possess excellent uniformity and cover an extensive range of melt flows, xylene solubles, and comonomer content. These products include CEFOR™ specialty random copolymers with exceptional optical properties and outstanding heat-sealing behavior and high-impact copolymers with an excellent balance of impact strength and stiffness.

Approximately 200 commercial products are made from the UNIPOL PP process. Dow can provide product research support to UNIPOL PP licensees to develop specific products for their local markets.

Homopolymers

The UNIPOL PP process can produce commercially proven homopolymers for all applications. These include fibers, filaments, carpet backing, woven and nonwoven fabrics, tubular water quench, flexible and rigid packaging, and injection molding for all types of parts.

Random Copolymers

Random copolymers (copolymers of propylene and ethylene or butene made in one reactor) provide superior clarity, flexibility, and heat-seal performance and lower crystallization temperature than homopolymers. The UNIPOL PP process provides random copolymers for use in blow molding, injection molding, and monolayer and coextruded films.

Impact Copolymers

Impact copolymers are not true copolymers. Rather, they are formed as an intimate dispersion of homopolymer and ethylene-propylene rubber phases created directly in two sequentially operated UNIPOL PP reaction systems. Impact copolymers provide excellent low-temperature impact strength. Major end uses include film, automotive applications, appliances, furniture, and rigid packaging.

Controlled Rheology

Postreactor controlled rheology technology (i.e., peroxide cracking) offers the flexibility to make a number of products with different properties while using the same polymerization reactor conditions. These cracked products have higher melt flows and narrower molecular weight distributions than the reactor base resins produced.

UNIPOL PP PRODUCT ATTRIBUTES SUMMARY

Film Grades

BOPP Film. A key attribute of the UNIPOL PP process is the ability to easily tailor the xylene solubles content to allow production of resins with high xylene solubles without the sacrifice in production rates often experienced by users of competing technologies. Higher xylene solubles yield significantly improved polymer stretchability and processibility. UNIPOL biaxially oriented polypropylene (BOPP) grades are designed to run at high speeds (without web break) on Brückner, Mitsubishi, and Dornier lines. One of the most popular UNIPOL PP BOPP grades,

5D98, has been tested at more than 470 m/min line speed on a state-of-the-art Brückner line. This grade has been certified as one of only two worldwide BOPP film grades to run at this speed (2003).

High polymer stretchability and excellent processability also result from the specific combination of SHAC catalyst, cocatalyst, and SCA used to manufacture this grade. The catalyst system contributes to an engineered polymer crystallinity that yields significantly advantaged stretchability in UNIPOL BOPP products.

These two properties—high xylene solubles and unique homopolymer crystallinity—make the UNIPOL PP BOPP grades superior to the competitive grades in processability and line speed.

Despite the higher xylene solubles, UNIPOL PP grades meet all required tensile properties for all BOPP applications.

Cast Film. UNIPOL PP random and homopolymer cast film grades are market leaders in North America. Random copolymer cast film grades offer better clarity and coefficient of friction (COF) when compared with other competitive products. These superior properties are the result of unique comonomer addition properties of the SHAC catalyst and the specific external donor. The combination of UNIPOL process backmixing and SHAC catalyst technology provides an enhanced level of comonomer randomness in the polymer chain backbone. In turn, this reduced crystallinity yields better clarity and a fast migration path for slip additives to quickly provide surface COF properties.

TWQ Film. UNIPOL tubular water quenched (TWC) grades contain an optimized amount of xylene solubles to promote slip-additive migration to the surface. This provides superior film opening characteristics in TWQ grades. TWQ formulations contain the best balance of antiblock and slip agents, selected through our substantial international market experience and internal TWQ-focused research.

Heat-Seal Film. The UNIPOL PP process is adept at producing heat-seal random copolymers with a very high ethylene content. A high-ethylene random copolymer provides desired low heat-seal temperatures and excellent hot-tack properties. The incorporation of high ethylene levels into random copolymers is a unique property of the fluidized-bed gas-phase process; competitive liquid-phase processes cannot achieve high ethylene content without reductions in manufacturing rates and efficiency.

The UNIPOL PP product slate also offers CEFOR® *butene* random copolymers—superb heat-seal products with exceptionally high clarity.

Textile Grades

Spunbond. UNIPOL PP offers outstanding homopolymers and random copolymers for the spunbond market. UNIPOL PP spunbond grades are designed to provide narrower molecular weight distributions (MWDs) than those of competing resins. This narrow MWD, generated by a unique combination of SHAC catalyst, cocatalyst, and SCA, coupled with controlled rheology, leads to higher line speeds. The SHAC catalyst advanced platforms also provide very low oligomer content for UNIPOL PP spunbond grades—a critical property for this high-surface-area, high-throughput application. Additionally, UNIPOL PP spunbond grades are industry leaders for their no-gas-fading additive package.

Bulk Continuous Filament (BCF). UNIPOL PP BCF grades contain a specially designed additive package to promote excellent color, even in white-carpet applications. UNIPOL PP has many formulation options, including resins tailored to ultraviolet light (UV)–stabilized and no-gas-fading applications. The optimized MWD of our BCF grades provides better drawing performances and higher fiber strength than competitive grades. Low xylene solubles and high melt-flow BCF grades result in superior spinning and heat setting performance for carpet face yarn applications requiring high resilience.

Tape (Raffia). UNIPOL PP raffia grades are strong (high tenacity) grades for the tape market. UNIPOL PP tape grades are designed to reduce the water carryover for tape processes. Grade formulations address the needs of all raffia applications including UV stabilization. The narrower MWD of the UNIPOL tape grades leads to higher extruder throughput and higher line speeds when compared with competitors' grades.

Molding and Extrusion

Homopolymers. Homopolymer MFRs range from 0.5 to 100 g/10 min and are suitable for a wide variety of complicated mold applications and applications requiring short cycle times. These nucleated and nonnucleated grades are formulated to contain different additives, including antistatic and mold-release agents for all types of injection-molding applications.

Random Copolymers. UNIPOL PP clarified random copolymers with their broad melt-flow ranges are exceptionally successful in the marketplace. These copolymers are carefully formulated to improve clarity and provide excellent gloss for variety of applications. Such grades also present superior processability and shorter cycle times for the demanding applications of random copolymers in this market.

Impact Copolymers. Medium- and high-impact copolymers provide a wide range of melt-flow capabilities with excellent stiffness to address a variety of applications. UNIPOL PP impact copolymers are recognized with excellent molding properties for injection-molding applications without warpage or mold lines and grades with high blush resistance.

The new SHAC catalyst system capability provides ultrahigh stiffness and high impact strength IMPPAX copolymers. These new UNIPOL PP impact copolymers are gaining excellent market recognition. Commercial examples include exceptionally high-impact ductile-failure grades at $-8°C$ and 30 MF. UNIPOL PP IMPPAX technology also includes a commercially produced range of thin-wall injection-molding (TWIM) grades from 35 to 130 MF that combines superior molding cycle throughput, high impact strength, and outstanding stiffness (ASTM D790/D4101 1% Secant Flexural Modulus above 200,000 psi) while using low-cost nucleating agents.

WASTES AND EMISSIONS

The UNIPOL PP technology is designed to emit essentially no polluting effluents, which enables the process to meet regulations for air, water, and solid wastes in many different countries. The small number of process steps and the absence of solvent facilitate compliance with hydrocarbon emissions standards. Plants have been permitted and constructed in 15 countries on 6 continents, including countries with very stringent environmental requirements.

Hydrocarbon discharges from occasional routine depressurization of the reactor and sieve regenerations are flared. All but the last traces of hydrocarbons removed from the resin granules are recovered or flared. Polymer particulate emission to the air is very low, with dust controlled by filters on gas exhausted from conveying stations. Surface drainage is collected and screened to remove polymer spills and sold as off-grade. Solid waste consists of a minimal amount of inactive catalyst and treatment beds, which can be landfilled. There are no aqueous process effluent streams whatsoever from the gas-phase process.

PROCESS ECONOMICS

The UNIPOL PP process offers PP manufacturers a cost-effective and versatile method of producing a growing range of end products. With a full range of high-performance SHAC polymerization catalysts and an all-gas-phase reactor, there is no catalyst residue removal, atactic resin extraction, or liquid handling stage. This process simplicity offers significant economic advantages over other commercial PP processes, with capital cost savings upfront and low operating costs over the life of the plant.

Investment

The simplicity of the UNIPOL PP process results in the lowest inside battery limits (ISBL) investment costs of the competing technology licensors. Several attributes of the UNIPOL PP process contribute to its inherently low total installed cost (TIC):

- No mechanical agitators in the reactor
- No solvent to handle, purify, recover, or recycle
- Most raw materials meeting commercial purchase specifications can be used with no pretreatment
- No catalyst pretreatment or prepolymerization required
- No intermediate resin storage owing to close-coupled pelleting
- No product batch blending owing to excellent product uniformity
- No special design requirements for unusual safety or health considerations

Similarly, the efficiency of the UNIPOL PP process results in minimal outside battery limits (OSBL) facilities costs:

- Efficient use of utilities results in lowest investment for utilities supply and distribution.
- The lack of continuous liquid and solid wastes, coupled with the high polymerization efficiency, results in low waste treatment and emissions control facilities.

Finally, the UNIPOL PP process has a very favorable capacity scaling exponent. Coupling this fact with the UNIPOL PP process' industry-leading single-reaction-line capacity means that significant incremental capacity can be achieved for very little upfront investment.

While total installed cost (TIC) is significantly dependent on project scope, location, business climate, and so on, Table 16.4.3 provides some idea of the TIC for various UNIPOL PP facilities.

TABLE 16.4.3 UNIPOL PP Investment, 2003 USGC Basis

Product mix	Homopolymer		Homo/random/impact	
Capacity, kMTA	300	415	300	415
Investment, million USD				
ISBL	54.4	59.6	69.0	78.1
OSBL	29.4	34.3	30.4	35.2
Total	83.8	93.6	99.4	113.3
$/t/yr	279	226	331	273

ISBL includes all process facilities from raw materials receipt through pelleting. OSBL basis is the construction of the plant in an existing petrochemical complex with basic utilities supply available. OSBL includes utilities distribution, storm water collection and filtering, flare, product handling, small packaging, and a warehouse capable of holding 3 weeks' production.

Dow is continually working to lower the installed cost of the UNIPOL PP process facilities. For up-to-date investment information, contact Dow's UNIPOL PP licensing group.

Operating Cost

High monomer usage efficiency and low utilities consumption enable UNIPOL PP licensees to be the lowest-cost producers of PP resins. Monomer and utilities usage are independent of facility nameplate capacity on a per ton of resin basis. The conversion efficiency of monomer (propylene) and comonomer (ethylene) to PP is 99 percent across essentially the complete range of UNIPOL PP products.

	Monomer usage,* kg monomer/kg product
Homopolymer	1.009
Random copolymer	1.009
Impact copolymer	1.009–1.013

*On a contained monomer basis.

Utilities costs (cooling water, steam, power, nitrogen, and plant air) are very low, totaling about $20/t for homopolymer and random copolymer products and $25/t for impact copolymer products (2003 USGC basis).

Catalyst and chemicals costs vary significantly with product mix. For details, contact Dow's UNIPOL PP licensing group.

Staffing

Low staffing requirements are key to minimizing plant period costs. UNIPOL PP operations and maintenance staffing is quite low relative to plant capacity. Regardless of capacity, the ISBL area of a UNIPOL process unit requires only four shift operator positions and one shift laboratory position. No permanent shift maintenance is required.

Skill levels of the operators, maintenance personnel, and laboratory analysts are typical of that found throughout the petrochemical industry. No unique skills are required. Because of the simplicity of the different jobs, the operators can easily learn the skills for all the jobs and rotate among all positions at regular intervals.

CHAPTER 16.5
CHISSO GAS-PHASE POLYPROPYLENE PROCESS

Takeshi Shiraishi
Chisso Corporation
Tokyo, Japan

TECHNOLOGY BACKGROUND AND HISTORY

The Chisso gas-phase polypropylene (PP) process is the most up-to-date version of a joint development between BP Chemicals (formerly Amoco Chemical Company) and Chisso Corporation. This gas-phase process was developed originally by Amoco for homopolymer in the mid-1970s and was expanded by Chisso to include ethylene-propylene impact copolymer production in 1987.

The Amoco/Chisso process was modified and improved by the two companies over a considerable period of time during the existence of a contractual relationship between the two companies involving both technology information exchange and licensing of the PP process. However, Chisso and Amoco decided to end their relationship amicably in the spring of 1995. Since then, each company has been independently carrying out licensing activities. Currently, six PP plants are in commercial operation using Amoco/Chisso gas-phase PP process technology, and four Chisso technology gas-phase PP process plants are in operation, under construction, or planned (see Table 16.5.3). The Chisso gas-phase PP process technology incorporates catalyst modifications as well as modifications for each step of the process section for copolymer production.

Recently, Chisso and Japan Polychem Corporation (a subsidiary of Mitsubishi Chemical Corporation of Japan) established a new PP joint venture, Japan Polypropylene Corporation. The rights to license this technology also were given to the new JV company.

POLYMERIZATION MECHANISM AND POLYMER TYPE

The following Ziegler-Natta type of catalyst system is used widely for the commercial production of PP:

Catalyst	TiCl$_4$/diester internal donor/MgCl$_2$
Cocatalyst	Triethyl aluminum (TEA)
Stereomodifier	External silane donor

Many theories have been presented in the literature concerning the propylene polymerization mechanism. Most theories treat the polymerization as proceeding by a coordinated anionic mechanism with an active growing organotitanium species. The polymerization of propylene can be indicated as follows:

$$\text{Ti-Et} + n(\text{CH}_2\text{=CH})\underset{\overset{|}{\text{CH}_3}}{} \rightarrow \text{Ti-(CH}_2\text{-CH)}_n\text{-Et}\underset{\overset{|}{\text{CH}_3}}{}$$

The first step involves the interaction of TEA with the catalytically active Ti centers on the surface of the supported catalyst. The ethyl group (Et), which is coordinated with both Ti and Al, is the point of initiation of an isotactic polymer molecule. The initiation occurs when a polarized propylene molecule inserts itself between the Ti and Et.

Homopolymer

Homopolymer is produced with only propylene as the monomer. It is known that heterogeneous Ziegler-Natta catalysts produce a mixture of PPs having different stereoregularities, i.e., isotactic, syndiotactic, and atactic. The two most common types of polypropylene that can be produced are shown below.

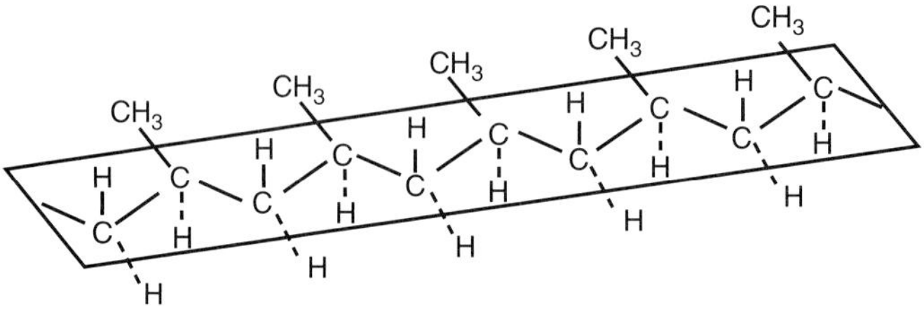

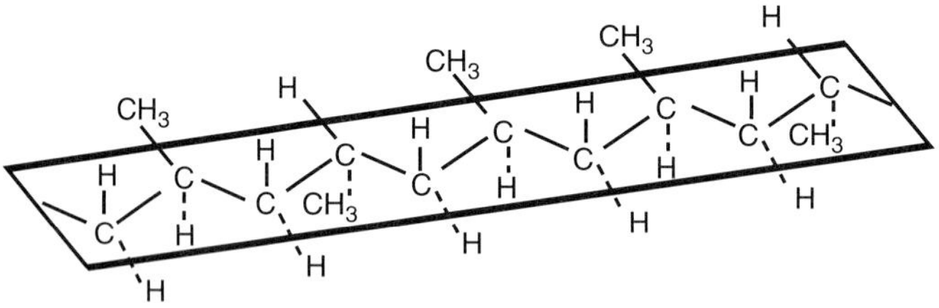

The methyl groups in isotactic PP are oriented in the same direction, resulting in a highly crystalline polymer. On the other hand, the methyl groups in atactic PP are oriented randomly, which results in an amorphous polymer. The crystalline material is the commercially desirable product because the higher-crystallinity polymer has excellent physical and mechanical properties owing to its higher stiffness. However, this crystallinity can be tailored as necessary for each particular product application. The catalyst itself, as well as the type of external silane donor used as a stereomodifier, is very important for the control of the PP polymer stereoregularity/crystallinity.

Random Copolymer

PP homopolymer is known for its high melting point (approximately 165°C) and high stiffness. However, for some applications, such as films or blow-molding (bottles) applications, a lower melting point, higher transparency, and greater softness are required. In order to improve these properties, a random ethylene-propylene copolymerization is carried out. PP random copolymer is produced by polymerizing a small amount of ethylene comonomer with propylene monomer. Ethylene units become randomly distributed in the main propylene polymer chain, which disrupts the stereoregularity of the isotactic polymer chain and results in lowering the crystallinity of the polymer.

Impact Copolymer

To improve the impact strength of homopolymer, "block" copolymerization is used commonly by means of an ethylene-propylene random copolymerization step after the initial homopolymerization step. Impact copolymer production usually is carried out using two or more reactors in series. PP homopolymer is produced in the first reactor, and an ethylene-propylene random copolymer is produced in the second and subsequent reactors. The lifetime of an individual active reacting polymer chain is very short; therefore, the homopolymer chains produced in the first reactor are no longer considered as "living" when they enter the second reactor. The random copolymer chains therefore are newly initiated chains that are separate and distinct from the homopolymer chains. Thus the impact copolymer is an intimate blend of homopolymer and random copolymer chains (Fig. 16.5.1).

The ethylene-propylene rubber produced in the second, or copolymer, stage of polymerization is an amorphous and sticky polymer material. Therefore, optimal morphology (size, shape, and porosity) is required based on catalyst performance so as to achieve ease of polymer powder flow during production of impact copolymer powders.

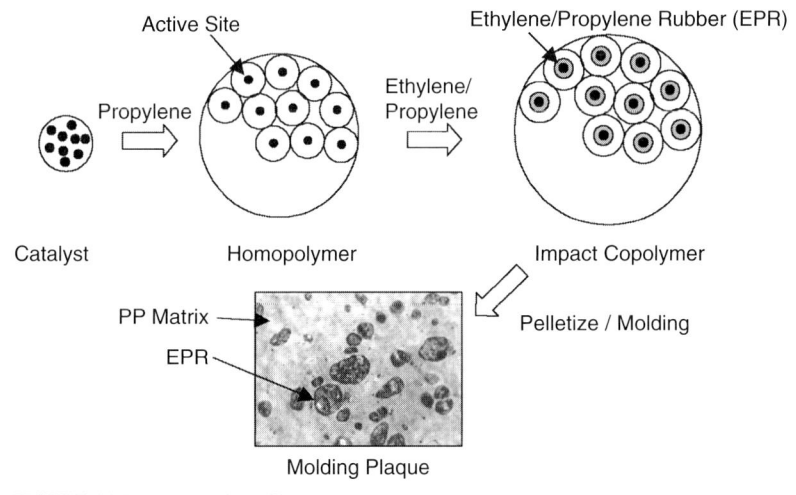

FIGURE 16.5.1 Production of impact copolymer.

PROCESS FEATURES

The Chisso PP gas-phase process features the use of a horizontal agitated reactor with the high-performance Toho Corporation of Japan THC-C–supported catalyst. This process and catalyst combination provides PP producers with the advantages of low energy consumption, superior ethylene-propylene impact copolymer product properties, production of a minimum quantity of transition products, high polymer throughput, a high process unit operating factor, and low operational costs. Every process step has been optimized/simplified, resulting in a low initial capital investment requirement and reduced manufacturing costs while enhancing product uniformity and providing excellent product quality control.

Reactor Technology: Unique Reactor Design

One of the features of the Chisso process is its innovative reactor design. Polymerization takes place in a horizontal reactor containing a mechanical agitator, with catalyst system injection near one end of the reactor and powder removal at the other end. The residence time distribution (RTD) approaches that of plug-flow movement in the reactor, comparable with that obtained from the use of more than three continuous stirred-tank reactors (CSTRs). With the minimal short circuiting of catalyst and polymer particles, high-performance impact copolymer products can be produced using only two Chisso reactors in series. In addition, this reactor design is estimated to have one of the highest space-time yields (production rate per unit reactor volume, kg/h · m^3) of the newer PP technologies, thus contributing to a low competitive capital cost.

Minimal Transition Products

With a plug-flow reactor, product transitions are quick and accomplished readily with a minimal offgrade production penalty. The production of offgrade/transition melt-flow-rate product material is estimated to be about 40 percent less than that produced by an equivalent CSTR in a series reactor system.

Energy Efficiency

The Chisso gas-phase PP process is designed to be energy efficient. Ample polymer powder mixing is carried out by mild mechanical agitation. Unlike a hydrocarbon diluent slurry or propylene bulk process, steam consumption is minimized because there is no need to flash a polymer slurry to separate the product powder. Similarly, owing to lower reactor pressures, there are less dissolved and trapped volatiles to be removed from the polymer powder.

Catalyst

The Chisso gas-phase PP process employs a new advanced Toho THC-C–supported catalyst having controlled particle morphology, very high activity, and very high stereoselectivity for polymer production. The high activity facilitates high reactor throughput, as well as process simplifications, which leads to low capital investment requirements, low operating costs, and superior product properties. The yield of Toho THC-C–supported catalyst is in the range of 25,000 to 45,000 kg PP/kg catalyst depending on the product being produced. The stereoselectivity of polymer can be adjusted easily by proper process variable changes to meet different market product requirements. The Toho THC-C–supported catalyst system has the ability to produce powder products of controlled morphology that are suitable for use in the Chisso gas-phase process. The resulting polymer powder is characterized by a narrow particle size distribution containing a very low amount of polymer fines. Excellent powder flow and handling are achieved for copolymer production while maintaining a wide operational window for comonomer content in the product. With use of the THC-C catalyst, the process unit design and operating procedures are simplified. With the presence of less polymer fines, plant operation is very reliable, very stable, and requires minimal mechanical maintenance.

Safety and Environmental Cleanliness

More than a decade of operating experience has demonstrated that the Chisso gas-phase process has inherently safe operational characteristics owing to the absence of large quantities of process hydrocarbon liquids. The process is also environmentally clean; the reaction section operates without the release of any liquids to the environment, and a minimum quantity of offgases are vented and burned in a plant flare stack. There is complete control of the polymerization reactor and its cooling system because the exothermic polymerization reaction can be stopped smoothly and easily by stopping catalyst feed to the reactor. In an emergency situation, such as an electrical power failure, a thermal runaway can be prevented by "killing" the exothermic polymerization reaction and rapidly reducing the reactor pressure. After either a "planned" or emergency shutdown, the polymerization reaction is restored rapidly by repressurizing the reactor and resuming catalyst feed. Process shutdowns resulting from fouling of the reactor or its auxiliaries have not been encountered during routine operating conditions. The design of the reactor and its associated cooling system has proven to be very reliable.

Versatile Product Capability

Owing to the flexibility of the Chisso process, wide-ranging polymer product designs, such as melt-flow rates (MFRs), polymer stereoregularity/crystallinity, comonomer content in random copolymer, and rubber content in impact copolymer, are possible, as shown in Table 16.5.1.

Using this technology, licensees are able to produce a wide range of products that serve many applications, including injection molding, blown film, cast film, biaxially oriented film (BOPP), blow molding, sheet thermoforming, and fibers. In particular, impact copolymer products exhibit a superior balance of stiffness and impact resistance over a broad temperature range owing to the use of THC-C high-performance catalyst and near plug flow of the catalyst and polymer powder in the reactors.

TABLE 16.5.1 Typical Property Ranges of License Grades

	Homopolymer	Random copolymer	Impact copolymer
MFR range,* g/10 min	0.5–40	1.8–35	0.65–50
Tensile strength, MPa	31–37	21–33	20–28
Flexural modulus, MPa	1100–1800	590–1200	720–1400
Heat deflection temp., °C (at 0.45 MPa)	105–135	85–105	87–125
Izod impact at 23°C, kJ/m^2, 3.2-mm notched bar	2–20	3.5–7.5	7–N.B.† (>70)
Dupont impact at −20°C (J)	—	—	10–N.B. (>40)

*Process not limited to 50 MFR. Controlled rheology (CR) grades are included.
†N.B. = nonbreak.

Competitive Capital and Operating Costs

Owing to the simplicity of its process design, energy efficiency, high product throughput, and minimum offgrade production, the Chisso process is fully cost competitive with the most recent leading PP manufacturing technologies.

PROCESS DESCRIPTION

Figure 16.5.2 is a simplified process block flow diagram for the Chisso gas-phase PP process. Monomer vapor in the reactor reacts continuously in the presence of the high-activity catalyst system to form solid particles of crystalline PP powder. After vapor separation and catalyst system residue deactivation, the stabilized polymer powder is pelletized and blended as necessary for shipment. The key sections of the process are monomer purification (as required by the monomer source), catalyst preparation, polymerization, powder deactivation, and product pelletization, as summarized below.

Monomer Purification (as Required by the Monomer Source)

Either chemical- or polymer-grade propylene can be used with the Chisso process, provided known catalyst poisons do not exceed process/catalyst specifications. Otherwise, in order to achieve the highest activity potential of the Toho THC-C–supported catalyst, these poisons are removed from the monomer stream using an appropriate monomer purification system. In this manner, specific catalyst poisons such as moisture, oxygen and oxygenated compounds, and sulfur-containing compounds are reduced to acceptable levels. The purification system is regenerated as needed.

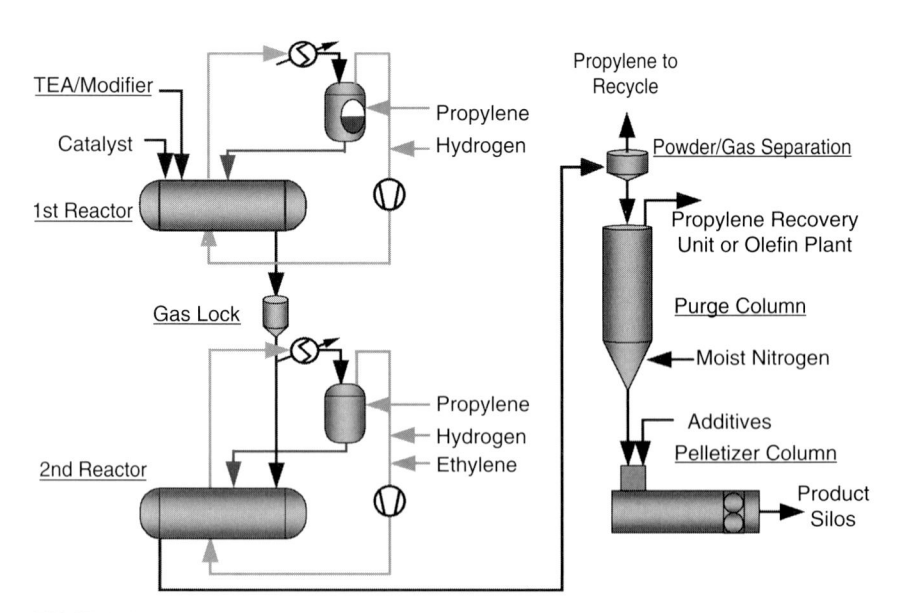

FIGURE 16.5.2 Simple flow diagram of the Chisso process.

Catalyst Preparation

Catalyst preparation and catalyst feeding equipment have been simplified. Catalyst, which is usually in a hydrocarbon slurry, is metered and fed to the reactor from the catalyst hold drum. Triethyl aluminum (TEA) as cocatalyst and the silane external catalyst modifier/donor, both in neat form, are also metered respectively and fed directly to the reactor. The feed rate of the catalyst is accurately controlled to achieve the desired production throughput by maintaining a ratio of cocatalyst and catalyst modifier/donor compared with the catalyst feed rate.

Polymerization

Chisso's reactor operates at a relatively low pressure of 2.0 to 2.5 MPa and 60 to 80°C. In the reactor, particles of crystalline PP are formed continuously by gas-phase polymerization of monomer in the presence of the catalyst system. The evaporated monomer leaving the reactor is partially condensed and recycled to the reactor. The liquid monomer is sprayed onto the agitated powder bed to provide cooling for the exothermic polymerization reaction. The flow of liquid monomer is controlled to achieve the desired reactor temperature profile. Fresh makeup monomer is added to the reactor as necessary. The uncondensed gas is returned to the bottom of reactor via a recycle compressor. Hydrogen is added to this stream to control the molecular weight of the polymer.

For homopolymer production, propylene is the only monomer fed to the system. For random copolymer production, a relatively small quantity and controlled ratio of ethylene is added to the propylene.

For impact copolymer production, a second reactor in series is required for the sequential polymerization. In the Chisso process, the first reactor, or homopolymer reactor, is physically elevated above the copolymer reactor. The homopolymer polymer powder is transferred by gravity through a simple gas-lock system, which then feeds the second, or copolymer, reactor. The reliable and effective gas-lock system allows the simplified transfer of powder to the copolymer reactor from the homopolymer reactor and prevents leakage of any of the reactants between reactors. This is critically important to produce the highest-quality copolymer products.

In almost all respects, the operation of the second reactor system is similar to that of the first reactor, except that ethylene as well as propylene is fed to the second reactor. The ethylene-to-propylene ratio is accurately controlled to produce the desired ethylene-propylene copolymer product composition.

Powder Deactivation

For either homopolymer or copolymer product production, powder from the last polymerization reactor is released periodically to a gas-powder separation system and into a purge column. The release of pressure on the powder to near-ambient conditions facilitates the removal and recovery of olefin monomers for recycling. Countercurrent moist nitrogen gas flow in the purge column deactivates the catalyst system residues. The monomer in the purge column offgas stream is optionally fed to a propylene recovery unit or sent to a nearby olefin plant for recovery. The powder is then conveyed to a powder weigh bin for gravimetric metering to the pelletizing equipment.

Product Pelletization

In the product pelletization operation, polymer powder is converted into a wide variety of pelletized resins, each tailored for specific market applications. The extrusion-pelletization system is designed to accurately meter the additives required for the highest product grade performance and consistency. Polymer powder and additives are fed continuously to an energy-efficient mixer–pump system and then fed to an underwater die-face cutter and formed into uniform polymer pellets. These pellets are cooled and dried. After blending, the pellets may be sent directly to a shipping silo and/or storage silo.

SAFETY AND ENVIRONMENTAL CONSIDERATIONS

The Chisso gas-phase process is inherently safer than a liquid-phase PP process because, as mentioned previously, the flammable hydrocarbon holdup in the Chisso reactor is maintained at a minimum. The operating pressure for the Chisso gas-phase reactor is the lowest of the competing gas-phase PP technologies. In an emergency, the reactor can be depressured safely and quickly to a plant flare system.

The Chisso gas-phase PP process is an environmentally clean process. Any releases of emissions and waste effluents are minimal. Also, special considerations have been given to the proper handling of any release of waste materials.

PRODUCT CAPABILITIES

Product Properties of License Grades

The typical product property ranges of the licensed grades are shown in Table 16.5.1. Chisso's standard licensed grades are designed to meet with the maximum number of customer requirements.

Application and Markets

- *Homopolymer.* Injection molding, blow molding, thermoforming, sheet, tape (raffia), fiber, cast and BOPP films, profile extrusion.
- *Random copolymer.* Thin-walled injection molding, low-heat-seal and high-transparency films, blow molding, medical, packaging, parts.
- *Impact copolymer.* Automotive parts, appliances, housewares, rigid packaging, injection molding, thermoforming.

Product Capability: Wide Window for Polymer Design

Some experiences with polymers from the Chisso process are shown below:

Homopolymer	
Xylene insolubles*	94–99.5 wt %; reactor-made MFR, fractional, 300 g/10 min
Random copolymer	
Ethylene content	Up to 6 wt %
Impact copolymer	
Reactor-made MFR	Fractional, 100 g/10 min
Reactor-made rubber Content†	Up to 60 wt %

*Higher value shows high stereoregularity or crystallinity of polymer, resulting in the production of a high-stiffness product.
†The content of ethylene-propylene rubber produced in second reactor.

TABLE 16.5.2 Typical Unit Consumption (Unit/t of PP Pellet)

	Homopolymer	Impact copolymer
Propylene and ethylene, kg	1005	1005
Electricity, kWh	327	354
Steam, kg	85	100
Cooling water, t	117	127
Nitrogen, $N \cdot m^3$	35	38

ECONOMICS

The typical unit that can produce not only homopolymer but also impact copolymers would contain only two reactors. Owing to the simplicity and efficiency of the Chisso process, the required capital investment is estimated by Chisso to be less than those of competitive PP processes.

A gas-phase process is more economical to operate compared with liquid-phase PP processes. With liquid-phase PP processes, a large amount of steam energy is required to evaporate the liquid that leaves the reactor with the polymer. In addition, the Chisso process is highly energy efficient owing to mild reactor mechanical agitation, efficient polymerization heat removal by the liquid monomer, and simplified operation for each process step.

The typical unit consumption figures based on a 300,000-MTA single-line production process are shown in Table 16.5.2.

REFERENCE PLANTS

Table 16.5.3 on p. 16.80 gives reference plants.

TABLE 16.5.3 Reference Plants

Location	Startup date	Design capacity (MTA)	Current capacity (MTA)
BP (former Amoco Chemical Company)			
Chocolate Bayou, Tex. (no. 2)*	Dec. 1979	110,000	218,000
Chocolate Bayou, Tex. (no. 3)*	June 1992	135,000	135,000
Geel, Belgium*	Sep. 1996	200,000	280,000
Japan Polypropylene (former Chisso Petrochemical Corp.)			
Goi, Japan (no. 4)*	June 1987	30,000	65,000
—, Japan	— 2006	(300,000)†	—
Yokkaichi, Japan*	Dec. 1990	40,000	77,700
SABIC EuroPetrochemicals (former DSM)			
Geleen, The Netherlands*	Nov. 1996	150,000	250,000
FCFC			
Mailiao, Taiwan	June 2000	300,000	300,000
Mailiao, Taiwan†	June 2004	(384,000)†	—
FPC/USA			
Point Comfort, Tex.	Mar. 2001	300,000	(360,000)†
CURRENT CAPACITY TOTAL			1,625,700 (2,309,700)‡

*Joint license of Amoco/Chisso Technology.
†Under construction or plan.
‡Including the capacity under construction or plan.

PART · 17

POLYSTYRENE

CHAPTER 17.1
BP/LUMMUS TECHNOLOGY FOR THE PRODUCTION OF EXPANDABLE POLYSTYRENE

Robert Stepanian
ABB Lummus Global
Bloomfield, New Jersey

INTRODUCTION

Expandable polystyrene (EPS) beads are tiny spheres of general-purpose polystyrene (see Chap. 17.2) impregnated with a physical blowing agent (usually pentane) that expands on further processing (i.e., molding, extrusion) to produce the characteristic "beaded" nature of EPS packaging and insulation. EPS has good thermal insulation and shock-absorbing properties, high compressive strength, and very low density and is resistant to moisture.

Current EPS demand is estimated at over 3 million metric tons per annum (MTA).[1] The two major end uses for EPS—packaging (meat trays, plates, containers, cups, shipping containers) and insulation/construction (extruded foam board, insulation board for roofing)—together account for over 85 percent of EPS demand. Other uses include shipping containers, furniture, ice chests, and helmet liners.

OPERATING PLANTS

As shown in Table 17.1.1, BP plants using this technology are in operation in France and Germany, with a total capacity of approximately 170,000 MTA of EPS. A 40,000-MTA EPS unit was licensed to Petrochina and is operating in China.

PROCESS CHEMISTRY

Polymerization consists of three steps: initiation, propagation, and termination. It is the control of these steps (the amount of free radicals formed, the length of the polymerization chain, etc.) that governs the physical properties of the polymer produced.

TABLE 17.1.1 Plants Using BP/Lummus Expandable Polystyrene Technology

Owner	Location	Total capacity at site (MTA)
BP	Wingles, France	90,000
BP	Marl, Germany	80,000
Petrochina	Dalian, China	40,000

In the initiation step, the double bond on the carbon atom is broken to form a free radical. This is accomplished by application of heat or a chemical initiator:

$$M \rightarrow M\cdot$$

During propagation, the highly unstable free radical combines with another styrene molecule, forming a new free radical. This process continues and the polymer chain grows:

$$M\cdot + M \rightarrow MM\cdot + M \rightarrow MMM\cdot + M \rightarrow -\{MMMM\cdot\}$$

In the termination reaction, two free radicals can combine to terminate a chain, or other chemicals can be used to terminate a chain:

$$-\{MMM\cdot\} + \{\cdot MMM\} \rightarrow -\{MMMMMM\}-$$

PROCESS DESCRIPTION

The BP/Lummus styrene polymerization technology for the manufacture of expandable polystyrene (EPS) is a batch suspension process followed by continuous dewatering, drying, and size classification. The following description references the processing steps identified in Fig. 17.1.1.

Depending on the formulation, styrene monomer, water, initiators, suspending agents, nucleating agents, and other minor ingredients are added to the reactor (1). The contents are then subjected to a time-temperature profile under agitation. The combination of suspending agent and agitation disperses the monomer to form beads. At the appropriate time, a premeasured quantity of pentane is introduced into the reactor. Polymerization is then continued to essentially 100 percent conversion. After cooling, the polystyrene beads and water are discharged to a slurry tank (2).

From this point, the process becomes continuous. The bead-water slurry is centrifuged (3) so that most of the "mother liquor" is removed. The beads enter a pneumatic dryer (4), where the surface moisture is removed.

The dry beads are then screened (5), yielding as many as four product cuts. External lubricants are added in a proprietary blending operation (6), and the finished product is conveyed to shipping containers.

FEEDSTOCK/PRODUCT SPECIFICATIONS

Feedstock

Styrene monomer feed typically has the specifications shown in Table 17.1.2.

Product Properties and Applications

Table 17.1.3 shows typical EPS product properties and end uses of these products.

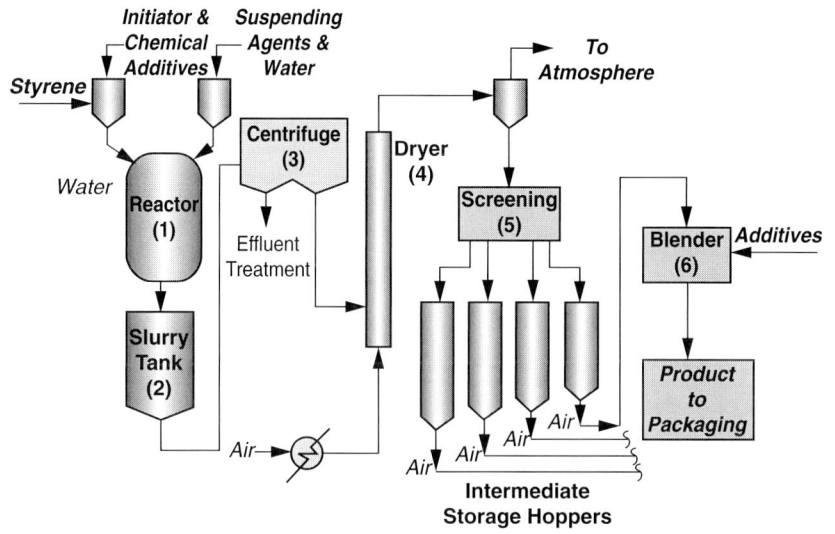

FIGURE 17.1.1 Process flow schematic.

TABLE 17.1.2 Typical Feedstock Specifications

Monomer content	99.8 wt % min
Aldehydes as benzaldehyde	200 wppm max
Peroxides as hydrogen peroxide	100 wppm max
Sulfur	30 wppm max
Chlorides as Cl	10 wppm max
Polymer content	Negative methanol or 10 wppm max
Polymer solubility	Complete in benzene
Inhibitor (p-tert-butyl catechol)	10–15 wppm
Color (APHA)	10 max

WASTE AND EMISSIONS

The design of a BP/Lummus EPS plant incorporates process features and equipment to minimize emissions, effluent, and waste from process operation. Gaseous effluents from an EPS plant are controlled to meet the requirements of U.S. federal regulations for environment protection.

The major sources of volatile organic carbon (VOC) emission are the atmospheric vents from feed mix drums, reactors, and the slurry tank. Table 17.1.4 lists the various emissions from an EPS plant.

PROCESS ECONOMICS

Table 17.1.5 provides raw materials consumption, and Table 17.1.6 provides utilities consumption per metric ton of expandable polystyrene.

TABLE 17.1.3 Typical EPS Product Properties and End Uses

Type	Bead size (mm)	Comment
Regular	0.9–1.4	Blocks with good mechanical and thermal properties, with improved cycle time
	0.7–0.9	Blocks or moldings for construction and packaging
	0.7–0.9	Molded packaging with fast processing
	0.4–0.7	Thin-walled molded packaging with fast processing
Flame retardant	1.4–2.0	Very-low-density blocks for impact sound deadening
	0.9–1.4	Blocks with good mechanical behavior for construction
	0.7–0.9	High-quality blocks or molded parts for construction
	0.4–0.7	Ceiling tiles and decorative articles, molded parts with very good surface aspect

Thermal properties

Density, kg/m^3	Thermal conductivity, W/m·K
15	0.036–0.038
20	0.034–0.036
30	0.032–0.034

Mechanical properties

Density, kg/m^3	Flexural strength, N/mm^2 [DIN 53423]	Tensile strength, N/mm^2 [DIN 43430]
15	0.24–0.28	0.15–0.20
20	0.29–0.34	0.19–0.26
30	0.45–0.52	0.30–0.45

TABLE 17.1.4 Typical Emissions per Metric Ton of Product

Type of waste emissions	Amount
Gaseous emission: pentane, t	0.0023
Wastewater (for water treatment), t	2.0
Solid waste (EPS), kg	5.0

TABLE 17.1.5 Raw Materials per Metric Ton of Product

Styrene and pentane	1000–1015 kg
Process chemicals	25–45 kg
Demineralized H$_2$O	1000 kg

TABLE 17.1.6 Utilities per Metric Ton of Product

Electricity, kWh	150
Cooling water, m^3	120
LP steam, kg	420

SUMMARY OF PROCESS FEATURES

The main characteristics and advantages of the BP/Lummus expandable polystyrene process include

- Consistently high-quality product
- Narrow bead size distribution
- High reactor productivity
- Fines recycle, which minimizes raw material consumption

REFERENCE

1. *CMAI News*, April 1999.

CHAPTER 17.2
BP/LUMMUS TECHNOLOGY FOR THE PRODUCTION OF GENERAL-PURPOSE AND HIGH-IMPACT POLYSTYRENES

Robert Stepanian
ABB Lummus Global
Bloomfield, New Jersey

INTRODUCTION

Polystyrene is a versatile polymer with a broad range of applications. Approximately 40 percent of polystyrene is used in packaging and single-use applications (film and sheet) such as cups, lids, cutlery, and rigid packaging. About 30 percent is used in electronics and appliances, including refrigerator parts, television cabinets, computer housings, CD cases, and floppy disk and videocassette housings. The balance is used for kitchen, laboratory, office, and medical goods and furniture and in building and construction.[1] In 2002, global capacity was approximately 14.8 million metric tons per annum (MTA).[2]

General-purpose (also known as crystal) polystyrene (GPPS), a styrene homopolymer, has high clarity and fairly good mechanical properties and is easy to process into most types of objects. However, it is relatively brittle. It is used where light transmission properties are important.

High-impact polystyrene (HIPS) contains an elastomer (usually polybutadiene rubber) to improve impact strength and therefore is milky or opaque in appearance. It is tougher than general-purpose polystyrene but still has good processability.

Both types have reasonably good environmental resistance. They are produced in a variety of grades having various physical properties, such as tensile properties, melt-flow rate, and impact strength. Additives are used to impart specific properties, such as lubricants to aid processing, antistatic agents, ultraviolet light (UV) stabilizers, and bluing agents.

Polystyrene can be produced by suspension, solution, or mass (bulk) polymerization. The BP/Lummus polystyrene process uses a bulk continuous process to produce a wide range of general-purpose and high-impact polystyrenes.

TABLE 17.2.1 Plants Using BP/Lummus Polystyrene Technology

Owner	Location	Total capacity at site (MTA)
BP	Wingles, France	200,000
BP	Marl, Germany	180,000
BP	Trelleborg, Sweden	80,000
Shanghai SECCO (BP/SPC JV)	Caojing, China	300,000
FCFC	Ningbo, China	200,000
Salavatnefteorgsyntez	Salavat, Russia	70,000

OPERATING PLANTS

As shown in Table 17.2.1, BP plants using this technology are in operation in France (Fig. 17.2.1), Germany, and Sweden, with a total capacity of approximately 460,000 MTA of GPPS and HIPS. A 300,000-MTA GPPS/HIPS unit will start up in China in 2005.

PROCESS CHEMISTRY

Polymerization consists of three steps: initiation, propagation, and termination. It is the control of these steps (the amount of free radicals formed, the length of the polymerization chain, etc.) that governs the physical properties of the polymer produced.

In the initiation step, the double bond on the carbon atom is broken to form a free radical. This is accomplished by application of heat or by a chemical initiator:

$$M \rightarrow M\cdot$$

During propagation, the highly unstable free radical combines with another styrene molecule, forming a new free radical. This process continues and the polymer chain grows:

$$M\cdot + M \rightarrow MM\cdot + M \rightarrow MMM\cdot + M \rightarrow -\{MMMM\cdot\}$$

In the termination reaction, two free radicals can combine to terminate a chain, or other chemicals can be used to terminate a chain:

$$-\{MMM\cdot\} + \{\cdot MMM\} \rightarrow -\{MMMMMM\} -$$

PROCESS DESCRIPTION

The production of general-purpose and high-impact polystyrenes is essentially the same with the exception of the initial rubber-dissolution step for high-impact polystyrene. The following description references the processing steps identified in Fig. 17.2.2.

The production of high-impact polystyrene begins with the granulation and dissolving of rubber and other additives in styrene monomer (1) and then transfer of the rubber solution to storage tanks (2). From this point on, the production steps for general-purpose polystyrene and high-impact polystyrene are the same. The feed mixture is preheated (3) and fed continuously to the prepolymerizer (4).

Prepolymerization may be initiated thermally or chemically depending on the product desired. For HIPS, this is a critical step in the process because the rubber morphology and physical properties of the resulting product are controlled during prepolymerization.

BP/LUMMUS TECHNOLOGY FOR GPPS AND HIPS   17.11

FIGURE 17.2.1 BP's polystyrene facility in Wingles, France.

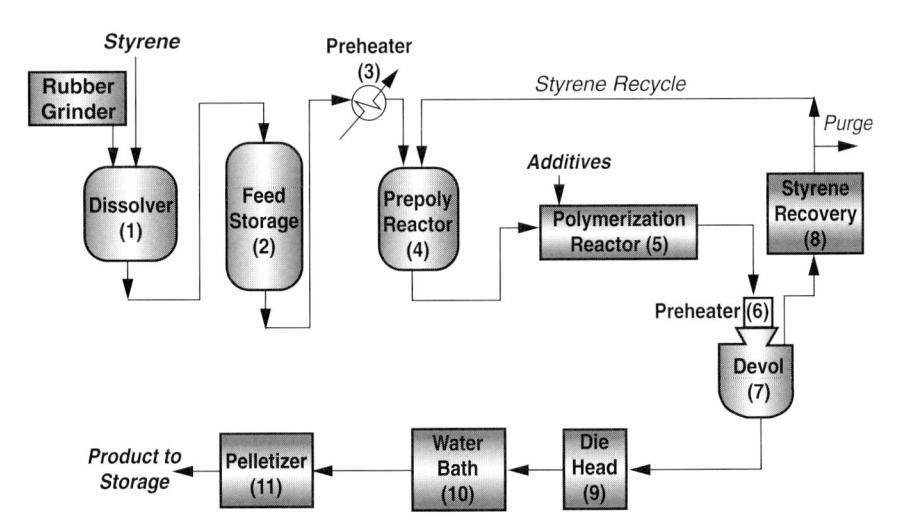

FIGURE 17.2.2 Process flow schematic.

Following prepolymerization, the polymer mixture is pumped through a polymerization reactor system (5) of proprietary design. At the exit of the reactor, the polymerization is essentially complete. The polymer mixture is preheated (6) in preparation for devolatilization.

The devolatilizer system (7) is held under a very high vacuum to remove unreacted monomer and solvent from the polymer melt. The residuals are sent to a styrene recovery system (8), and recovered styrene is recycled back to the prepolymerizer. The polymer melt is then pumped through a die head (9) to form strands, a water bath (10) to cool the strands, and a pelletizer (11) to form pellets, and then is screened to remove large pellets and fines. An external lubricant may be added at this time. The resulting product is air-conveyed to inspection hoppers prior to bulk storage and packaging facilities.

Product Grade Changes

The BP/Lummus process has extremely high productivity and high single-train capacity, which are attributed to the proprietary reactor design. The mass flow rate per unit of reactor volume is one of the highest in the industry. Because of this, typically only 2 to 6 hours are required for transition from one comparable grade to another, or switching from a GPPS grade to a HIPS grade, without sacrificing product quality. However, careful attention is required when switching from a HIPS grade to a GPPS grade. During GPPS production, any remnants of HIPS or rubber-containing material in the production line may have an impact on GPPS product quality.

FEEDSTOCK AND PRODUCT SPECIFICATIONS

Feedstock

Styrene monomer feed typically has the specifications shown in Table 17.2.2.

Product Properties and End Uses

Tables 17.2.3 and 17.2.4 show typical GPPS and HIPS product properties and their end uses, respectively. Figure 17.2.3 shows typical product applications.

TABLE 17.2.2 Typical Feedstock Specifications

Monomer content	99.8 wt % min
Aldehydes as benzaldehyde	200 wppm max
Peroxides as hydrogen peroxide	100 wppm max
Sulfur	30 wppm max
Chlorides as Cl	10 wppm max
Polymer content	Negative methanol or 10 wppm max
Polymer solubility	Complete in benzene
Inhibitor (p-tert-butyl catechol)	10–15 wppm
Color (APHA)	10 max

TABLE 17.2.3 Typical GPPS Product Properties and End Uses

Property	Units	Very easy MF,* standard HR	Easy MF, improved HR	Easy MF, high HR	Exceptional MS, improved HR	High HR, good MS	Exceptional MS, improved HR	Exceptional MS, high HR
Melt flow	g/10 min	23	10.0	7.4	3.8	2.5	3.1	1.4
Vicat	°C	85	88	102	93	101	90	102
Deflection temperature (under load)	°C	75	78	90	84	90	81	90
Charpy impact strength	kJ/m²	9	10	8	15	14	16	16
Tensile stress	MPa	40	50	52	55	55	55	58
Nominal strain at break	%	1.5	2	2	3	3	2	3
Flexural strength	MPa	70	80	75	100	100	100	100
Tensile modulus	MPa	3100	3200	3300	3300	3300	3300	3300
Specific gravity	g/cm³	1.04	1.04	1.05	1.04	1.05	1.04	1.05
Applications		Petri dishes, medical packaging, coextrusion	Medical tubes, pipettes, food packaging, paper substitute	Foamed insulation boards	CD cases, drinking cups, egg cartons, in-line extrusion, packaging	Shower enclosures, insulating boards, audio cassettes, injection blown packaging	Shower enclosures, CD cases, medical articles	Biaxially oriented sheet (OPS), foamed insulation boards

*MF = melt flow; HR = heat resistance; MS = mechanical strength.

TABLE 17.2.4 Typical HIPS Product Properties and End Uses

Property	Units	Easy MF,* high rigidity, medium impact	High HR, medium impact	Easy MF, high impact	Very high impact, improved HR	Improved ESCR, very high impact	High HR, very high impact
Melt flow	g/10 min	22	5.7	10	5.5	4.3	2.5
Vicat	°C	80	94	85	90	89	96
Deflection temperature (under load)	°C	71	85	74	80	80	85
Charpy notched impact strength	kJ/m^2	5	7	8	13	10	10
Yield stress	MPa	30	24	23	24	24	27
Nominal strain at break	%	35	45	50	45	65	40
Tensile modulus	MPa	2500	2450	2100	1900	1800	1900
Specific gravity	g/cm^3	1.04	1.04	1.03	1.03	1.03	1.03
Applications		Household articles, plant pots, small cases	Packaging, vending cups, CD trays	Closures, cups, packaging, storage boxes	Food packaging, electronics, profiles for furniture	Refrigerator liners, fatty food packaging	Electrical equipment, TV backs, video cassettes

*MF = melt flow; HR = heat resistance; ESCR = environmental stress crack resistance.

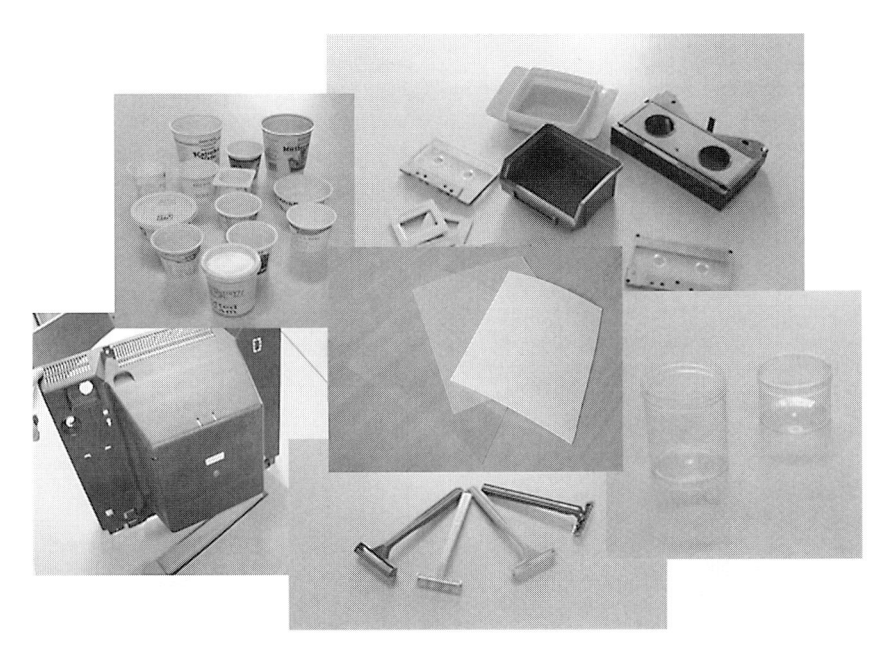

FIGURE 17.2.3 Typical product applications.

WASTE AND EMISSIONS

The design of the BP/Lummus plant incorporates process features and equipment to minimize emissions, effluent, and waste from process operation. Gaseous effluents from a PS plant are controlled to meet the requirements of U.S. Federal regulations for environment protection.

There are no flue gas stacks or air emissions with the exception of the hot oil heater. Table 17.2.5 shows a typical flue gas analysis per metric ton of product. Table 17.2.6 lists all the waste and emissions from a PS plant.

PROCESS ECONOMICS

Table 17.2.7 provides raw materials consumption per metric ton of polystyrene. Table 17.2.8 provides utilities consumption per metric ton of polystyrene.

SUMMARY OF PROCESS FEATURES

The main characteristics of the BP/Lummus polystyrene process are

- Large single-train capacity of up to 130,000 MTA
- Consistently high product quality

TABLE 17.2.5 Typical Hot Oil Heater Flue Gas Analysis

Compound	kg per MT of product
CO	2.0
VOC	0.18
SO_2	0.3

TABLE 17.2.6 Waste Emissions per Metric Ton of Product

Types of waste emissions	GPPS	HIPS
Gaseous emission, t	None*	None*
Liquid emission, t	0.03†	
Wastewater, t	0.07‡	
Solids emission		
Spent alumina, kg	0.2	
Fines/overs, kg	1.0	

*Vents from PS plant are a negligible source of VOC emissions due to the low vapor pressure of the respective chemicals.
†Organic waste is generated from the die fume mist eliminator system.
‡0.02 t/t of product of wastewater is sent to wastewater treatment and 0.05 t/t of product is sent to rain/stormwater separator.

TABLE 17.2.7 Raw Materials per Metric Ton of Product

	GPPS	HIPS
Styrene and mineral oil, kg	1011	937
Rubber, kg	—	73
Additives, kg	1	2

TABLE 17.2.8 Utilities per Metric Ton of Product

	GPPS	HIPS
Electricity, kWh	97	110
Fuel, 10^3 kcal	127	127
Cooling water, m^3	46	26
LP steam, kg	6	6

- Broad product line covering all major applications
- High reactor productivity, better plug-flow characteristics with excellent temperature control
- Chemically initiated polymerization, which improves polymer properties; polymer-to-rubber grafting is enhanced and oligomer by-products are reduced
- Efficient devolatilization system that results in products with low residual monomer content, suitable for food-grade applications

- Low capital cost
- Unit flexibility allowing for quick product grade changes
- Better turndown capability without affecting the physical properties of the polymer

REFERENCES

1. *CMAI News*, April 1999.
2. Source: Citigroup Smith Barney (New York).

CHAPTER 17.3
POLIMERI EUROPA GENERAL-PURPOSE POLYSTYRENE PROCESS TECHNOLOGY

Francesco Pasquali and Riccardo Inglese
"C. Buonerba" Research Centre
Mantova, Italy

INTRODUCTION

Within the range of industrial monomers, styrene is probably one of the most important and extensively studied. Its propensity to generate free radicals even at relatively low temperature and thus to propagate the polymer chain has been well known since the middle of 1800.

Since it is polymerizable with all the most common mechanisms, such as anionic, cationic, and Ziegler-Natta catalysts, a lot of effort has been made in the last century to individuate and develop the most efficient process for producing crystal polystyrene, commonly known as *general-purpose polystyrene* (GPPS).

The first commercial GPPS was produced by IG Farben (1931) by means of a bulk polymerization batch process. After the outbreak of World War II, the need for styrene for synthetic rubber production caused a sudden slowdown in investment in the field of polystyrene. For the same reason, styrene production was increased significantly. During the postwar period, the requirement of a process optimized from a productivity/cost point of view led to the almost simultaneous development of the Dow and BASF (Germany) continuous processes. Compared with the previous batch technology, the continuous process has many advantages in terms of investment and operating costs, environmental impact, residual monomer content in the product, and uniformity of final product quality parameters.

In 1971, Polimeri Europa (at that time Montedison and then EniChem) began producing GPPS using the continuous-mass process technology developed at the Mantova Research Centre. The main features are as follows:

- Proprietary, accurate process and mechanical design of key equipment (reactor, devolatilizer)

- Simple process scheme and easy process control
- Flexible technology allowing a tailor-made solution for specific needs in terms of plant capacity and product range
- Minimum quantity of foreign materials introduced in the process

Even though GPPS production technology can be considered to be well established, especially in the last decades, the market requirements, in terms of quality and the environmental impact of GPPS, pushed Polimeri Europa research and development (R&D) to continuously improve its technology and its product portfolio by innovating its key proprietary equipment and optimizing the process cycle. The results of this effort make Edistir® GPPS, with its wide product portfolio, a benchmark within the European scenario.

PROCESS CHEMISTRY

The GPPS process is based on the radical polymerization of styrene. From the point of view of a single radical, the sequence of events by which styrene monomer is converted to a high-molecular-weight polymer can be divided into four distinct steps: initiating of radical formation, initiation of chain formation, propagation, and termination of the chains.

Even though styrene is almost unique in the extent of its ability to undergo spontaneous polymerization simply by heating the monomer, the most advanced polystyrene production processes are based on an alternative method of initiating styrene polymerization, the addition of a known free-radical generator.

Thermal Decomposition of Initiators

Several molecules have been found to be very efficient in promoting the polymerization reaction initiation step. The various catalysts are used at different temperatures depending on their rates of decomposition, but only the peroxides find extensive use as radical sources because other classes of compounds usually are either not readily available or not stable enough.

Anyway, they generally act by a mechanism that works as follows: The peroxidic initiator thermally decomposes via homolytic cleavage of the O—O bond:

$$R-O-O-R' \rightleftharpoons R-O\cdot + \cdot O-R'$$

Polymer Chain Initiation

The radical formed by peroxide scission can react with styrene, promoting the formation of a styrenic radical:

Propagation

The initiated monomer radical thus can propagate according to the following scheme:

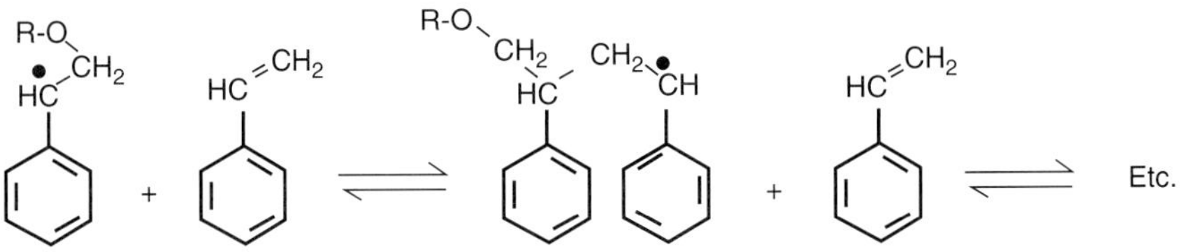

[reaction scheme showing propagation of styrene radical polymerization] Etc.

The final polymer product will have the following arrangement of monomer units:

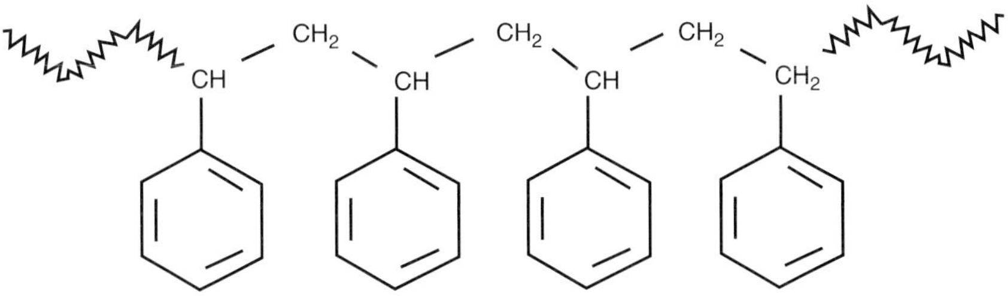

[structural diagram of polystyrene chain showing head-to-tail placement of phenyl groups]

This type of arrangement is usually referred to as *head-to-tail placement*.

Termination

A basic requirement for the production of long chains is the high reactivity of the initiating and polymer free radicals. Since these radicals are uncharged, any pair of radicals can react together extremely rapidly, in most cases requiring little or no activation energy.

Two mechanisms have been proposed for the mutual destruction of radicals: (1) combination termination or coupling

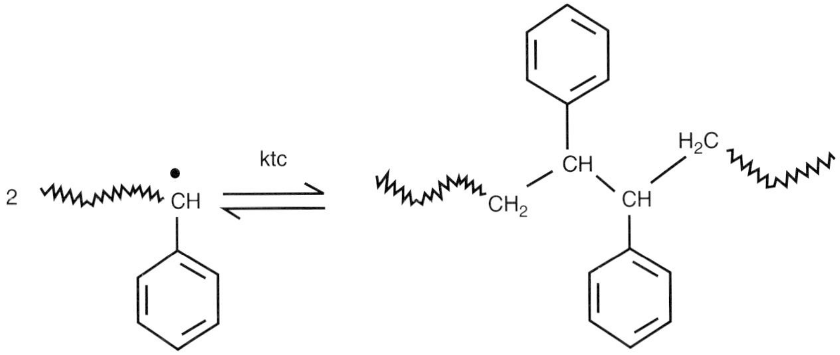

[reaction scheme showing combination termination of two styrene radicals with rate constant ktc]

and (2) disproportion termination

17.22 POLYSTYRENE

In the specific case of two polystyrene radicals at temperatures up to 80°C, termination occurs almost exclusively by combination. In addition, at temperatures above 80°C, kinetic models have shown that molecular weight data are best fit by using combination over disproportion as the principal mode of termination, although this has not been proven to occur exclusively. There has been no good evidence to support a significant amount of the disproportion reaction for polystyrene termination.

Side Reactions: Oligomers

Even if a peroxide initiator is used to promote the polymerization reaction, a significant number of styrene units undergo thermal reaction, following the mechanism described below. The thermal initiation starts with a Diels-Alder reaction between two monomer molecules, providing free radicals for the polymerization reaction:

[1-phenyltetraline]

This reaction, together with consecutive rearrangements and coupling, competes with polymer chain propagation and leads to the formation of dimers (such as the 1-phenyltetraline reported in the example) and trimers. The presence of residual oligomers in the polymer bulk can have a serious effect on its final properties and its behavior during processing at the end user in terms of either performance or environmental impact (release of bad-smelling volatile matter, soiling of process equipment, etc.).

A typical thermally initiated commercial polystyrene could contain up to 0.1 wt % total dimer and 1.0 wt % total trimer. For this reason, Polimeri Europa process facilities, especially the reactor and devolatilizers, have been designed in such a way as to maximize the peroxide-initiated polymerization performances. In this way it is possible to produce GPPS with a very low oligomer content without affecting production rates.

DESCRIPTION OF THE PROCESS FLOW

Polimeri Europa GPPS technology is based on a continuous-mass, peroxide-initiated polymerization (Fig. 17.3.1). GPPS is obtained by polymerization of styrene, added with chemicals in a mixing section. The mass reaction occurs in the presence of solvent (ethyl-

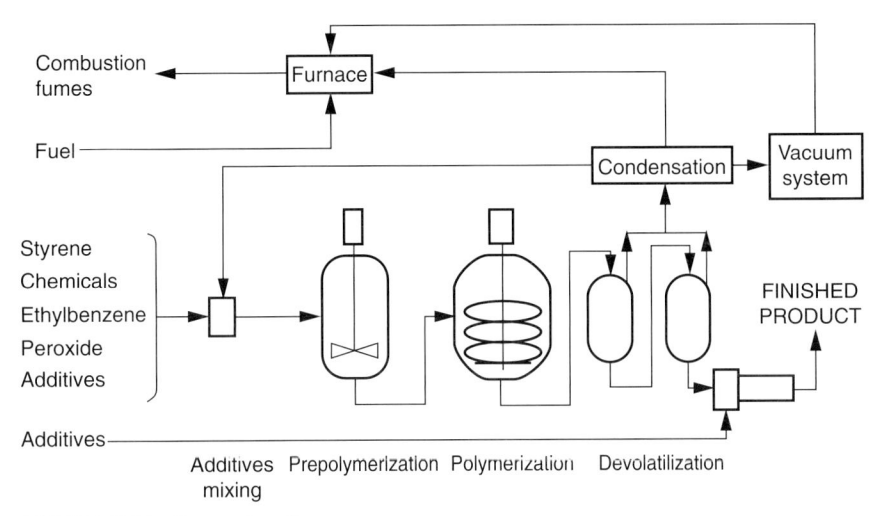

FIGURE 17.3.1 Process flow diagram.

benzene, typically 5 to 10 percent). At startup, a preliminary amount of solvent is fed to the reactor; during the normal run, the initial ethyllbenzene is substituted gradually by non-polymerizable recycled organic matter.

This mixture is thus fed, together with peroxidic initiator and process aids, to the polymerization section, generally composed by a sequence of two continuously stirred tank reactors (CSTRs); the reaction heat is removed by evaporation of the volatile matter. Other reaction section arrangements are also possible in order to fit specific requirements.

At the end of the reaction train, the polymer solution is sent to a devolatilizing unit, under vacuum, in two stages in sequence. The monomer and low-boiling compounds are removed from the polymer, which is finally sent to the pelletizing unit. The heat is provided by the diathermic oil system.

The vapor mixture, after condensation, is constantly recycled to the mixing section. Noncondensed vapors/inert gas from the vacuum system and liquid organic purge from the condensation section are recovered as fuel in a furnace, where diathermic oil for the process is heated.

PROCESS ADVANCED DESIGN FEATURES

Even if the process scheme basically calls to mind the most common current technologies, the Polimeri Europa GPPS process is provided with the following proprietary advanced design features:

- *Polymerization section.* Generally, the main items are two continuous stirred tank reactors (CSTRs) with optimized design of the stirrer in such a way as to ensure the better compromise between thermal control and polymer production rate. By ensuring a good fluodynamic and thermal homogeneity, the Polimeri Europa CSTR reactor provides maximum control of molecular weight growth and distribution.

- *Devolatilization section.* This is a two-stage section with high heat and mass transfer rates, at very low residence times. This combination of factors leads to very efficient monomer and other organic residual matter removal even at relatively low temperature (where polymer chain degradation is minimized) and without the addition of water or other stripping agents.

- *Condensation section.* This section allows customization of the balance between product purification from low-molecular-weight components (waxes) and specific raw materials consumption. In any case, the waxes removed can be thermally oxidized in the process furnace together with inert gases and the noncondensable vapor stream, leading to minimal environmental impact of the whole plant and energy recovery.
- *Raw materials feed section.* The raw materials are fed to a purification section designed to selectively remove impurities (mainly organic matters) that can affect the polymerization kinetics, process fluctuations, and finally, product characteristics. The chemical feeding system and mixing facilities provide the most efficient balance between additive concentration and their effect on polymer properties. Additives, such as internal lubricants and mold-release agents, also can be fed in downstream, directly before the pelletizer, in order to maximize the efficiency of the chemicals and of the reaction volume.
- *Plant layout.* The integrated process and the mechanical design of the equipment around the reaction and devolatilization sections maximize the efficiency of the plant. The compact layout also leads to a production unit having the best compromise between investment cost and manufacturing/maintenance operability.
- *Flexibility.* As a consequence of its optimized plant design and the efficiency of its proprietary equipment, the Polimeri Europa GPPS technology can match any requirements in terms of product portfolio, keeping the polymer quality at the top level.

PROCESS PERFORMANCE

For all the reasons just mentioned, the Polimeri Europa GPPS technology minimizes the effects of raw material and chemical impurities either on the process or on the product structural parameters. Nevertheless, the low purity of the feed stocks, mainly styrene, may cause a significant increase in plant specific consumption. Assuming 99.9 percent of styrene purity, the typical raw materials and utilities consumption per metric ton of polymer is reported in Table 17.3.1.

PLANT CAPACITY

Polimeri Europa typically suggests the process scheme described earlier for a production rate of about 100,000 metric tons per annum (MTA). The versatility of the Polimeri Europa technology, however, makes it easily possible to provide for larger capacities, even up to 200 kMTA. Tailor-made process design also can be carried out in order to fit specific requirements, such as lower plant capacity (i.e., 50 kMTA), special grades, and/or specific product ranges.

TABLE 17.3.1 Specific Consumption for a Typical GPPS Unit

Raw materials, kg	1005–1008
Electric power, kWh	100
Fuel gas, 10^4 kcal	15–20
Steam, kg	0

COMMERCIAL EXPERIENCE

Polimeri Europa GPPS units, based on proprietary technology, are on-stream in Italy (1971, 75 kMTA) and in Belgium (1988, 80 kMTA), making Polimeri Europa one of the major European producers of general-purpose polystyrene. One GPPS unit (80 kMTA), licensed by Polimeri Europa, has been on-stream in Hong Kong since the early 1990s. A second one (50 kMTA) was started up in Brazil in 2000.

THE EDISTIR GPPS PRODUCT PORTFOLIO

Polimeri Europa GPPS products are characterized by a unique balance between key properties:

- Very low residual monomer and oligomer content
- Low amounts of chemical consumption for the same property balance
- Fine-tuned macromolecular structure (molecular weight and its distribution)

This set of characteristics allows the Edistir product portfolio to cover even the most challenging fields of application.

Packaging

- Thermoformed and injection-molded cups and food containers
- Injection-molded disposable items and cutlery

Direct Gassing

- Insulation panels (XPS)
- Expanded trays

Others

- Injection molding of CD boxes, housewares, medical articles, and toys
- Injection molding of refrigerator clear internal components
- Extrusion of clear panels and shower boxes
- Oriented and biaxially oriented extruded films

CHAPTER 17.4
POLIMERI EUROPA EXPANDABLE POLYSTYRENE PROCESS TECHNOLOGY

Dario Ghidoni and Riccardo Inglese
"C. Buonerba" Research Centre
Mantova, Italy

INTRODUCTION

Polystyrene has been produced commercially by IG Farben since 1931 by means of a batch polymerization process. However, a great deal of further work has been carried out in order to develop an efficient process to produce expandable polystyrene (EPS), and only during World War II were polystyrene foams first made in Great Britain. The product was made available commercially by means of a batch suspension polymerization process only in the United States by the Dow Chemical Company; it was used for the construction of life rafts on troop transports for floating equipment to shore.

Since 1955 significant innovations in both the process and the product have been made. Nowadays, EPS is produced mainly with a batch suspension polymerization process into foamable polystyrene beads. The blowing agent, a low-boiling hydrocarbon, can be incorporated either during the polymerization reaction or later in a separate impregnation step.

An alternative method involves a continuous process for the production of expandable pellets. The blowing agent is added during extrusion of PS, and the final pellets are cooled under pressure.

In 1974, Polimeri Europa (at that time Montedison and then EniChem) started the production of EPS in Italy (Mantova) and in Belgium (Feluy) using a batch suspension polymerization technology developed at the Mantova Research Centre. The suspending agent was an organic salt. The blowing agent was incorporated during the polymerization. In 1981, a new process with an inorganic suspending agent was developed at Mantova and since has been implemented in both the plants.

The main features are as follows:

- An inorganic suspending agent, which leads to a narrow bead size distribution
- Single-step technology (impregnation of the beads with a prefoaming agent is done during the polymerization step)
- Very good reactor filling and no need for solvent washing of the reactors
- Easy availability in the world market of the chemicals employed in the process
- Flexible and wide product range developed and fine-tuned for any market requirement

Even though the EPS production process is well established, especially in the last decades, market needs in terms of both quality and environmental impact pushed Polimeri Europa research and development (R&D) to continuously improve its proprietary equipment and optimize its process cycle. The result of this effort makes Extir® EPS, with its wide product portfolio, a benchmark within the European scenario.

PROCESS CHEMISTRY

The EPS process is based on a batch water suspension polymerization. Owing to the presence of a suitable suspending agent, styrene is dispersed in small droplets inside the water phase in an agitated reactor. The radical polymerization, peroxide initiated, occurs in the styrene droplets. It can be divided into four distinct steps: initiation of radical formation, initiation of chain formation, propagation, and termination of the chains.

Thermal Decomposition of Initiator

Various initiators are used at different temperatures depending on their rates of decomposition, but only the peroxides find extensive use as radical sources because other classes of compounds usually are either not readily available or not stable enough. The peroxide initiator thermally decomposes via homolytic cleavage of the O—O bond:

$$R-O-O-R' \rightleftharpoons R-O\cdot + \cdot O-R'$$

Polymer Chain Initiation

The radicals formed by peroxide scission can react with styrene, promoting the formation of a styrenic radical:

Propagation

The initiated monomer radical thus can propagate according to the following scheme:

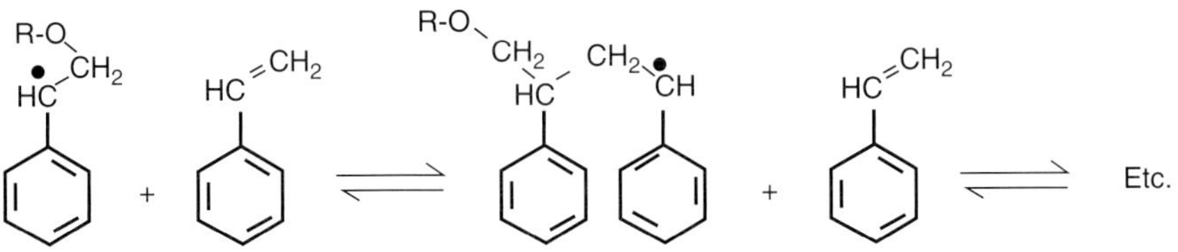

The final polymer product will have the following arrangement of monomer units:

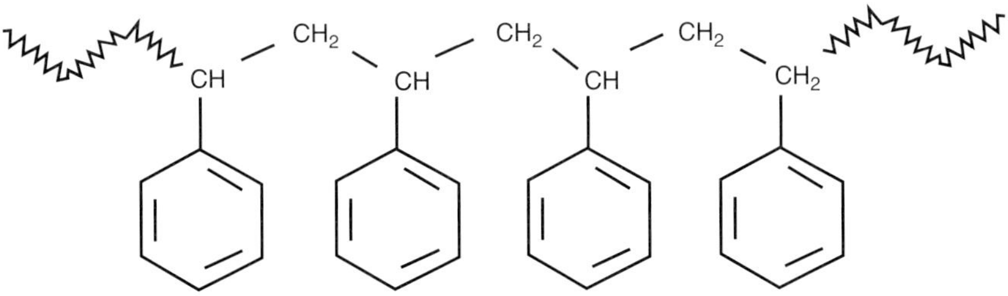

This type of arrangement is usually referred to as *head-to-tail placement*.

Termination

A basic requirement for the production of long chains is the high reactivity of the initiating and polymer free radicals. Since these radicals are uncharged, any pair of radicals can react together extremely rapidly, in most cases requiring little or no activation energy.

Two mechanisms have been proposed for the mutual destruction of radicals: (1) combination termination or coupling

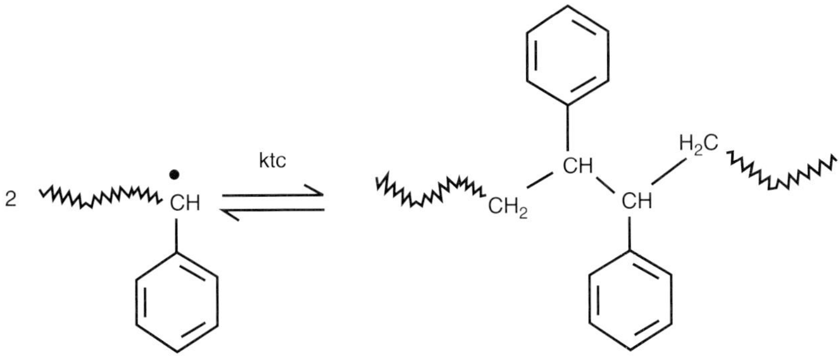

and (2) disproportion termination

$$2 \;\underset{}{\overset{CH_2 \cdot}{\underset{CH}{\text{\textasciitilde}}}}\text{-Ph} \;\underset{}{\overset{ktd}{\rightleftharpoons}} \; \text{\textasciitilde}CH_2\text{-}CH(Ph)\text{-}CH_2\text{\textasciitilde} \;+\; \text{\textasciitilde}CH=CH\text{-}Ph$$

In the specific case of two polystyrene radicals at temperatures of up to 80°C, termination occurs almost exclusively by combination.

In addition, at temperatures above 80°C, kinetic models have shown that molecular weight data fit better by using combination over disproportion as the principal mode of termination, even if that seems not to occur exclusively. There has been no good evidence to support a significant amount of the disproportion reaction for polystyrene termination.

At an appropriate time, the blowing agent is added under pressure and incorporated into the polystyrene.

Residual Volatile Content

An important target is to reach a very low amount of residual volatile content in EPS beads; this is mostly related to two main conditions:

- High styrene purity (hopefully >99.9 percent)
- Use of an efficient peroxide to achieve almost total styrene monomer conversion

Side reactions are less effective in such conditions, where an intensive use of peroxide initiator is made at a relatively low reaction temperature, and the final result is a typical oligomer content below 0.1 percent.

Coating Technology

Special care must be taken in the coating technology, which greatly influences final EPS performance during both processing and final use. Each EPS fraction has a proper lubrication recipe, according to the final application; the main additives are as follows:

- Mixture of glycerol esters as controller of the interaction between the steam and the surface of the beads
- Stearic salts of some metals as antilumping agents

DESCRIPTION OF PROCESS FLOW

The Polimeri Europa EPS technology is a typical batch suspension polymerization with an inorganic suspending agent (Fig. 17.4.1).

Internal chemicals are dissolved in the styrene. This solution, preheated to a certain temperature, is then transferred into a reactor that is prefilled with demiwater and a suspending agent under agitation. Polymerization starts and continues until a certain polymer content is achieved, and then the blowing agent is added. The reaction continues until the residual styrene reaches the target value based on a proper temperature

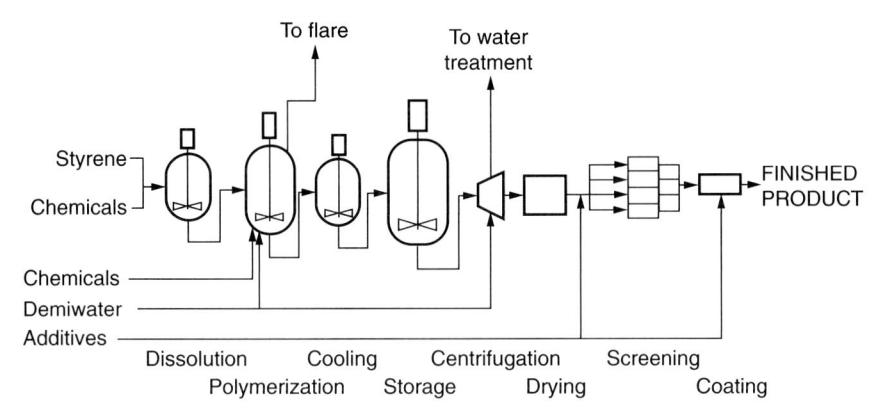

FIGURE 17.4.1 Process flow diagram.

cycle and recipe. Final bead size is controlled chemically during the polymerization up to the desired diameter. Blowing agent escaping from the polymerization section is sent to a flare.

After polymerization, the EPS bead slurry is cooled to an intermediate temperature and then transferred to pressurized cooling tanks, where it reaches the final discharge temperature. The slurry is then transferred into a bigger slurry tank that can contain several batches in order to get good homogeneity and to continuously feed the centrifugation section. Wet beads are then still continuously fed to a drier and to the screening section to be divided into desired grades. All the wastewater can be treated by a common biologic plant. Single fractions are stored temporarily in silos and batchwise lubricated with a proper coating recipe. All the residual amounts of the coated fractions (e.g., excess weights of the lubricating section) can be recovered in the slurry tank section.

PROCESS ADVANCED DESIGN FEATURES

Even if the process scheme basically calls to mind the most common current technologies, the Polimeri Europa EPS technology is unique in that it is provided with the following proprietary advanced design features:

- *Polymerization section.* Proprietary reactor design, able to give narrow bead size distributions; high reactor filling and no need for chemical washing. Proprietary suspension system, based on an inorganic agent, able to
  - Get an optimal control on bead size distribution and bead internal cellular structure
  - Stabilize the small styrene droplets, avoiding the risk of agglomeration during the whole polymerization cycle

  Single-step technology; the impregnation of the beads with a prefoaming agent is done during the polymerization step, thus avoiding a further and expensive steeping phase. Wide bead size range from 0.2 to 3 mm; a typical narrow size distribution (σ/x) of 0.18 can be obtained for any bead size. Easy availability in the world market of the chemicals used in the process.

- *Coating section.* Highly developed coating deposition technology, able to fully tailor EPS grades to the desired customer needs. Easy availability in the world market of the chemicals employed in the process.

TABLE 17.4.1 Specific Consumption for Typical EPS Unit

Raw materials, kg	1010–1015
Electric power, kWh	150
Nitrogen, N·m^3	80
Demineralized water, m^3	1.2
Steam, kg	400–450

- *Flexibility.* As a consequence of its optimized plant design and of the efficiency of its proprietary equipment, the Polimeri Europa EPS technology can match any possible requirement in terms of product portfolio, keeping the polymer quality at the top level.

PROCESS PERFORMANCE

The Polimeri Europa EPS technology minimizes the effects of raw materials and chemical impurities either on the process or on the product structural parameters. Nevertheless, the low purity of the feed stocks, especially styrene or chemicals, may cause a significant worsening of the bead size distribution or suspension polymerization instability. Assuming 99.9 percent styrene purity, the typical raw materials and utilities consumption per metric ton of polymer is reported in Table 17.4.1.

PLANT CAPACITY

Polimeri Europa typically suggests the process scheme described earlier for a production rate ranging from 30 to 50 kMTA. Based on such technology, it is possible to provide convenient solutions for either larger or smaller capacities.

COMMERCIAL EXPERIENCE

Polimeri Europa EPS units are on-stream in Italy (1976, 35 kMTA), Belgium (1983, 35 kMTA), and Hungary (1991, 40 kMTA), making Polimeri Europa one of the major European producers of expandable polystyrene.

THE EXTIR EPS PRODUCT PORTFOLIO

The Polimeri Europa EPS technology allows producers to obtain a high-quality product range that satisfies all the needs of the most qualified customers. Within the Extir product portfolio it is possible to find a suitable grade for all the current processing technologies (molding, transfer, vacuum, or blocks), which can cover the following main applications.

General Purpose (Normal Grades)

- Boxes and containers for marble and food
- Packaging

- Lightened concrete
- Light and heavy blocks

Fire-Resistance Grades

- Industrial packaging
- Sheets and blocks for insulation in the building industry
- Blocks for external insulation
- Wall padding sheets
- Floor bodies
- Disposable forms
- Lightened mortars and bricks

CHAPTER 17.5
POLIMERI EUROPA HIGH-IMPACT POLYSTYRENE PROCESS TECHNOLOGY

Francesco Pasquali and Franco Balestri
"C. Buonerba" Research Centre
Mantova, Italy

INTRODUCTION

First commercialized during the 1930s, general-purpose polystyrene (GPPS) immediately met with worldwide success. Nevertheless, despite its clear success, a strong limitation for many applications was found immediately because of its brittleness. Many attempts were made to increase the impact resistance of PS by adding rubber, but only at the end of the 1950s was a rubber-modified polystyrene production process patented by Monsanto and Dow.

At the end of the 1970s, Polimeri Europa (at that time Montedison and then EniChem) started the production of high-impact polystyrene (HIPS) via a continuous mass process in two plants located in Italy and Belgium. During the following decade, the Italian plant was modified by the introduction of new equipment developed at the Mantova Research Centre.

Within a few years, two new units, acknowledging the improved proprietary technology developed meanwhile by Polimeri Europa research and development (R&D), were started up in Italy. The main features are as follows:

- Proprietary, accurate process and mechanical design of key equipment (reactor, devolatilizer)
- Simple process scheme and easy process control
- Flexible technology allowing tailor-made solutions for specific needs in terms of plant capacity and product range
- Minimum quantity of foreign materials introduced in the process

Even though HIPS production technology can be considered to be well established, especially in the last decades, the market requirements in terms of quality and environmental impact of HIPS pushed Polimeri Europa R&D to continuously improve its technology and product portfolio by innovating its key proprietary equipment and optimizing the process cycle. The results of this effort make Edistir® HIPS, with its wide product portfolio, a benchmark within the European scenario.

PROCESS CHEMISTRY

The most common method for manufacturing HIPS consists of dissolving polybutadiene rubber (PBU) in styrene monomer and polymerizing the mixture in a first reaction step in the presence of peroxide initiators, chain-transfer agents, and antioxidants until a conversion of about 30 percent is reached.

The Phase Inversion

The rubber-styrene solution, which is homogeneous at the beginning of the reaction, separates into two phases at a PS concentration of about 1 percent because PS has a very limited miscibility with PBU. At this stage, the mixture consists of two solutions:

- A solution of PBU in styrene as a continuous phase
- A solution of PS in styrene as a disperse phase

As the polymerization proceeds, the quantity of PS in the solution increases, and the ratio of rubber solution to PS solution decreases. When the volume of the PS-styrene solution becomes greater than the other phase solution, the so-called phase inversion occurs: The PS-styrene solution becomes the continuous phase, and the PBU-styrene solution becomes the disperse phase, occurring as droplets suspended in the continuous phase.

Phase inversion is accompanied by a sharp decrease in solution viscosity, the viscosity of the new continuous phase (PS-styrene) being significantly lower than that of the previous one. After having completed the phase inversion, the viscosity increases as polymerization continues.

During the early phase of the reaction, one key chemical reaction occurs, the chemical grafting of the rubber phase, which, combined with the above-mentioned phase inversion process, leads to the formation of rubber particles.

The Grafting Reaction

Grafting generally is believed to start with a primary radical attack on the rubber backbone; most authors describe this reaction as hydrogen abstraction from the allylic position in the polybutadiene molecule:

$$R\cdot + -CH_2-CH=CH-CH_2- \rightleftarrows RH + -\cdot CH-CH=CH-CH_2-$$

As the separation starts, the PS and PBU chains of the graft copolymer participate in the process, forming an interface at the phase boundary. A graft copolymer acts as a polymeric emulsifier, stabilizing the droplets and reducing droplet size.

Since the thermal reaction can occur even at relatively mild conditions, the presence of peroxide initiators significantly enhances the efficiency of grafting of PS chains onto the rubber molecules. In terms of the grafting of styrene, the following reactions are believed to be of prime importance:

(1) IN· + PBU → PBU· + S → graft
(2) IN· + S → homo-PS
(3) PS· + PBU → PS + PBU → graft
(4) PS· + 1,2-vinyl → graft
(5) IN + 1,2-vinyl → graft

where IN is initiator, PBU is polybutadiene, S is styrene, and homo-PS is polystyrene homopolymer.

Reactions (3) and (4) represent thermal grafting. Reactions (1) and (5) represent catalytic grafting. The grafting reaction has to be associated with a very efficient stirring of the reaction mixture.

As a matter of fact, in the absence of appropriate stirring, phase inversion does not take place, and the product consists of a continuous rubber phase containing a very high concentration of embedded PS particles.

Matrix molecular weight, which affects mainly mechanics and melt flow, is usually regulated by process conditions (temperature, initiator concentration, solvent content) and by the addition of a suitable chain-transfer agent. Moving along the reaction train, the propagation of polymer chain growth goes on until a proper conversion is achieved (typically 65 to 75 percent of solid content). The polymer mixture is then fed to the devolatilization unit, where another unique chemical reaction takes place.

The Cross-Linking Mechanism

In the final stage of the manufacture of HIPS, the polymer mixture is transferred to the devolatilization section. At the end of this unit, the grafting and homopolymerization reactions, which have prevailed up to this point, cannot continue because the monomer concentration is close to zero. Thus the addition of the free radicals to the polybutadiene double bond can occur, leading to a cross-linking reaction.

It is very clear that cross-links are created in the devolatilization step of the continuous process; this step is time- and temperature-dependent, and the process conditions have to be adapted to the desired cross-link density.

From one side, cross-linking prevents particle breakdown during fabrication processes. As a matter of fact, if cross-linking is too low, the original physical properties may be lost as a result of rubber particle breakdown during high-shear injection-molding or extrusion processes. On the other hand, if the level of cross-linking is too high, the efficiency of the rubber phase toughening mechanism may drop, affecting physical properties such as elongation at break and impact strength in a very serious way. In this sense, an optimal temperature and residence time balance is crucial to lead to efficient monomer removal without affecting final product properties.

DESCRIPTION OF PROCESS FLOW

The Polimeri Europa HIPS technology is based on a continuous mass peroxide-initiated polymerization (Fig. 17.5.1). HIPS is obtained by polymerization of styrene in a rubber-styrene solution.

Rubber, after being ground in a mill, is dissolved in styrene in a proper section and then added with chemicals in a mixing section. The mass reaction occurs in the presence of solvent (ethylbenzene, typically 5 to 10 percent). At startup, a preliminary amount of ethylbenzene is fed; during a normal run, the initial ethylbenzene is gradually substituted by nonpolymerizable recycled organic matter.

17.38 POLYSTYRENE

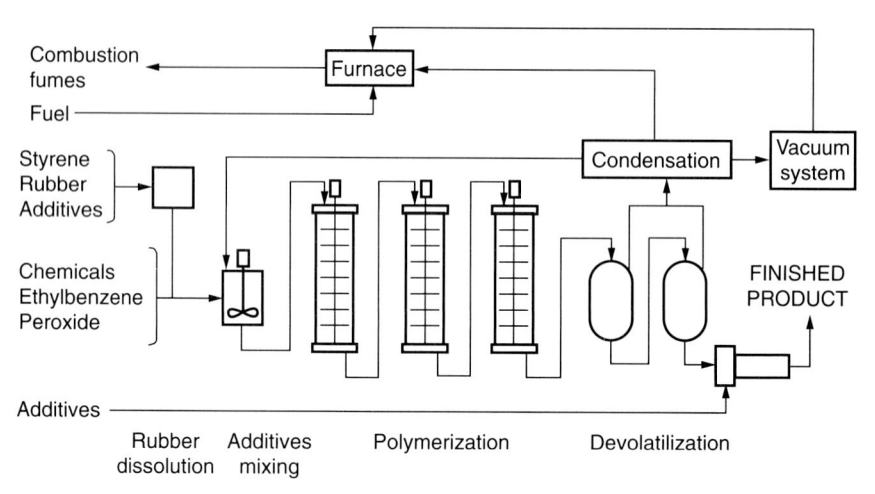

FIGURE 17.5.1 Process flow diagram.

This mixture is thus fed, together with the peroxide initiator and process aids, to the polymerization section, generally composed of a sequence of three plug-flow reactors; the reaction thermal profile is controlled by diathermic oil circulating inside internal coils. Other reaction section arrangements are also possible in order to meet specific requirements.

At the end of reaction train, the polymer solution is sent to a devolatilizing unit, under vacuum, in two stages in sequence. The monomer and low-boiling compounds are removed from the polymer, which is finally sent to the pelletizing unit. The heat is provided by the diathermic oil system.

The vapor mixture, after condensation, is constantly recycled to the mixing section. Noncondensed vapors/inert gases from the vacuum system and liquid organic purge from the condensation section are recovered as fuel in a furnace, where diathermic oil for the process is heated.

PROCESS ADVANCED DESIGN FEATURES

Similar to what was discussed in Chap. 17.3 for GPPS, even if the process scheme basically calls to mind the most common current technologies, the Polimeri Europa HIPS technology is unique, being provided with the following proprietary advanced design features:

- *Polymerization section.* The main items are generally three full plug-flow reactors (PFRs); thanks to agitation and a high specific thermal exchange surface area, they are characterized by very precise control of the thermal reaction profile. In this way, it is possible to achieve maximum control of the morphology of the disperse phase, together with good efficiency of the catalytic grafting reaction. This synergy leads to an excellent balance between production rate and polymer quality.
- *Devolatilization section.* This involves two-stage operation with high heat and mass transfer rates at very low residence times. This combination of factors leads to a very efficient monomer and organic matter removal even at relatively low temperature (where polymer chain degradation and cross-linking of the rubber phase are minimized) without the addition of water or other stripping agents.
- *Condensation section.* This allows producers to customize the balance between

product purification from low-molecular-weight components (waxes) and raw materials specific consumption. In any case, the waxes removed can be thermally oxidized in the process furnace together with inert gas and the noncondensed vapor stream, leading to minimal environmental impact of the whole plant and energy recovery.

- *Raw materials feed section.* The chemical feeding system and the mixing facilities are such that they provide the most efficient balance between additive concentration and the effect on polymer properties. Additives, such as internal lubricants and mold-release agents, also can be fed in downstream, directly before the pelletizer, to maximize the efficiency of the chemicals and the reaction volume. An optional section can be designed to feed acrylonitrile, allowing the production of very high environmental stress cracking resistant HIPS grades for special uses in the refrigeration industry.
- *Plant layout.* The integrated process and mechanical design of equipment around the reaction and devolatilization sections maximize the efficiency of the plant. The compact layout leads to a production unit having the best compromise between investment cost and manufacturing/maintenance operability.
- *Flexibility.* As a consequence of its optimized plant design and the efficiency of its proprietary equipment, the Polimeri Europa HIPS technology can match any possible requirement in terms of product portfolio, keeping the polymer quality at the top level.

PROCESS PERFORMANCE

For all the reasons mentioned earlier, the Polimeri Europa HIPS technology minimizes the effects of raw materials and chemical impurities on the process and product structural parameters. Nevertheless, the low purity of feed stocks, mainly styrene, may cause a significant increase in plant specific consumption. Assuming 99.9 percent styrene purity, the typical raw materials and utilities consumption per metric ton of polymer is reported in Table 17.5.1.

PLANT CAPACITY

Polimeri Europa typically suggests the process scheme described earlier for a production rate of about 100 kMTA. The versatility of Polimeri Europa technology, however, makes it easily possible to also provide convenient solutions for larger capacities, even up to 200 kMTA. Tailor-made process design also can be carried out in order to meet specific requirements, such as lower plant capacity (i.e., 50 kMTA), special grades, and/or unique product ranges.

TABLE 17.5.1 Specific Consumption for a Typical HIPS Unit

Raw materials, kg	1005–1008
Electric power, kWh	130
Fuel gas, 10^4 kcal	15–20
Steam, kg/t	0–30

COMMERCIAL EXPERIENCE

Polimeri Europa HIPS units, based on proprietary technology, are on-stream in Italy (1981, 75 kMTA; 1992, 40 kMTA), in Hungary (1991, 75 kMTA), and in Belgium (1979, 80 kMTA), making Polimeri Europa one of the major European producers of high-impact polystyrene. One HIPS unit (80 kMTA) licensed by Polimeri Europa has been on-stream in Hong Kong since the early 1990s. A second one (70 kMTA) was started up in Brazil in 2000.

THE EDISTIR HIPS PRODUCT PORTFOLIO

Polimeri Europa HIPS products are characterized by a unique balance between key properties:

- Very low residual monomer and oligomer content
- Very high rubber phase efficiency (reduced rubber consumption)
- Minimized rubber cross-linking and polymer degradation

This set of characteristics allows the Edistir product portfolio to cover even the most challenging fields of application.

Refrigeration Industry

- Environmental stress cracking resistant (ESCR) grades for highly chemical-resistant inner liners
- Very high gloss grades to enhance the aesthetic properties of the internal cabinet
- Medium-impact grades combining high flowability and stiffness for injection-molded internal parts

Telectronic

- A wide range of grades having a very good balance among toughness, stiffness, gloss, and flowability for injection molding of technical parts, housings, and covers
- High-impact grades with good flowability and high stiffness
- High-impact grades, with good stiffness and high thermal resistance

Packaging

- High-performance grades for extrusion and thermoforming of disposable tumblers, flatware, cups, lids, and containers for dairy products and frozen food

Other Applications

- High-flowability grades for very fast injection molding of toys and housewares

PART 18

VINYL CHLORIDE AND POLYVINYL CHLORIDE

CHAPTER 18.1
VINNOLIT VINYL CHLORIDE AND SUSPENSION POLYVINYL CHLORIDE TECHNOLOGIES

Ulrich Woike and Peter Kammerhofer
Vinnolit GmbH & Co. KG
Ismaning, Germany

COMPANY INTRODUCTION

Vinnolit was formed in 1993 by a merger of the polyvinyl chloride (PVC) activities of Hoechst AG and Wacker-Chemie GmbH. The new company drew on the experience of its two founders, both active in the vinyl sector for almost 60 years. Today Vinnolit is the leading producer in Germany, one of the most important manufacturers in Europe, and worldwide is among the top 10 in the vinyl industry—with capacities of 650,000 metric tons per annum (MTA) of PVC and 630,000 MTA of vinyl chloride (VCM). Vinnolit produces and markets a wide range of PVC grades covering all kinds of PVC applications, as well as electrolysis products, 1,2-dichloroethane (EDC), and VCM.

The know-how of Vinnolit in EDC, VCM, and suspension PVC (S-PVC) is being licensed worldwide by VinTec, the technology center of Vinnolit. The licensing activities were started in 1965, and since then, many plants all around the globe have been designed and constructed in close cooperation with our engineering partner Uhde, formerly part of Hoechst AG and now part of the Thyssen Krupp Group. Both partners are able to offer a highly competitive technology for the whole production chain from chloralkali electrolysis through EDC/VCM production to S-PVC.

The partnership is a long-term synergistic combination of Vinnolit's modern, proven, and highly competitive technologies with Uhde's international engineering and service capabilities.

PROCESS PERSPECTIVE

The VCM and S-PVC industry was developed in the 1930s on acetylene and hydrochloride (HCl) feedstock. Beginning in the 1950s, the feedstocks changed to ethylene and chlorine. The share of the ethylene-based plants in 2002 was about 94 percent of annual production. The world annual VCM and PVC consumption in 2003 was 27 million MTA. The annual average growth forecast for 2003–2008 is about 3.8 percent.[1] Plant sizes have changed from approximately 100,000 MTA in the 1960s to 350,000 MTA in 2003. Such capacities typically are installed as single-stream VCM monomer plants with two cracking furnaces and in S-PVC plants as two production lines with four to six polymerization reactors.

VINNOLIT VINYL CHLORIDE MONOMER (VCM) PROCESS

General Process Description

A balanced vinyl chloride plant consists of the following process units (Fig. 18.1.1):

- Direct chlorination
- Oxychlorination
- EDC distillation
- EDC cracking
- VCM distillation
- Chloroprene and benzene chlorination

The production process for vinyl chloride monomer (VCM) from ethylene and chlorine is based on two different routes. In the *direct chlorination* and the *oxychlorination* processes, EDC is produced. Both reactions proceed exothermally.

Because of its better quality, the EDC produced by direct chlorination can be used directly for cracking, whereas the EDC from the oxychlorination process has to pass a purification stage (EDC distillation) before cracking. The EDC distillation unit has three distillation columns. In the first column, the heads column, water and low-boiling by-products are removed from the top of the column. The bottom product of the heads column is combined with the unconverted recycled EDC from the cracking furnace and fed to the high-boil column, where high-boiling by-products are separated. The column overhead product is pumped to the furnace feed EDC tank, and the bottom product is transferred to the vacuum column for further concentration of the high-boiling by-products. To minimize reboiler fouling, this column is operated under vacuum. The overhead product of the vacuum column is used as feed EDC, whereas the bottom product with a high-boil content of greater than 95 percent is fed to by-product incineration.

In the EDC cracking unit, EDC is cracked under formation of VCM and HCl. These products and the unconverted EDC are separated in the VCM distillation unit. In the first distillation column, the HCl column, the overhead product HCl is fed to the oxychlorination unit as raw material. The bottom product, a mixture of EDC and VCM, is separated in the VCM column. Traces of HCl are removed in the VCM stripper and recycled to the HCl column. The final product, VCM, is sent to VCM storage for shipment or for further use in a PVC plant. The low-boiling by-products in the unconverted EDC (coming from the bottom of the VCM column) are chlorinated in the benzene and chloroprene chlorina-

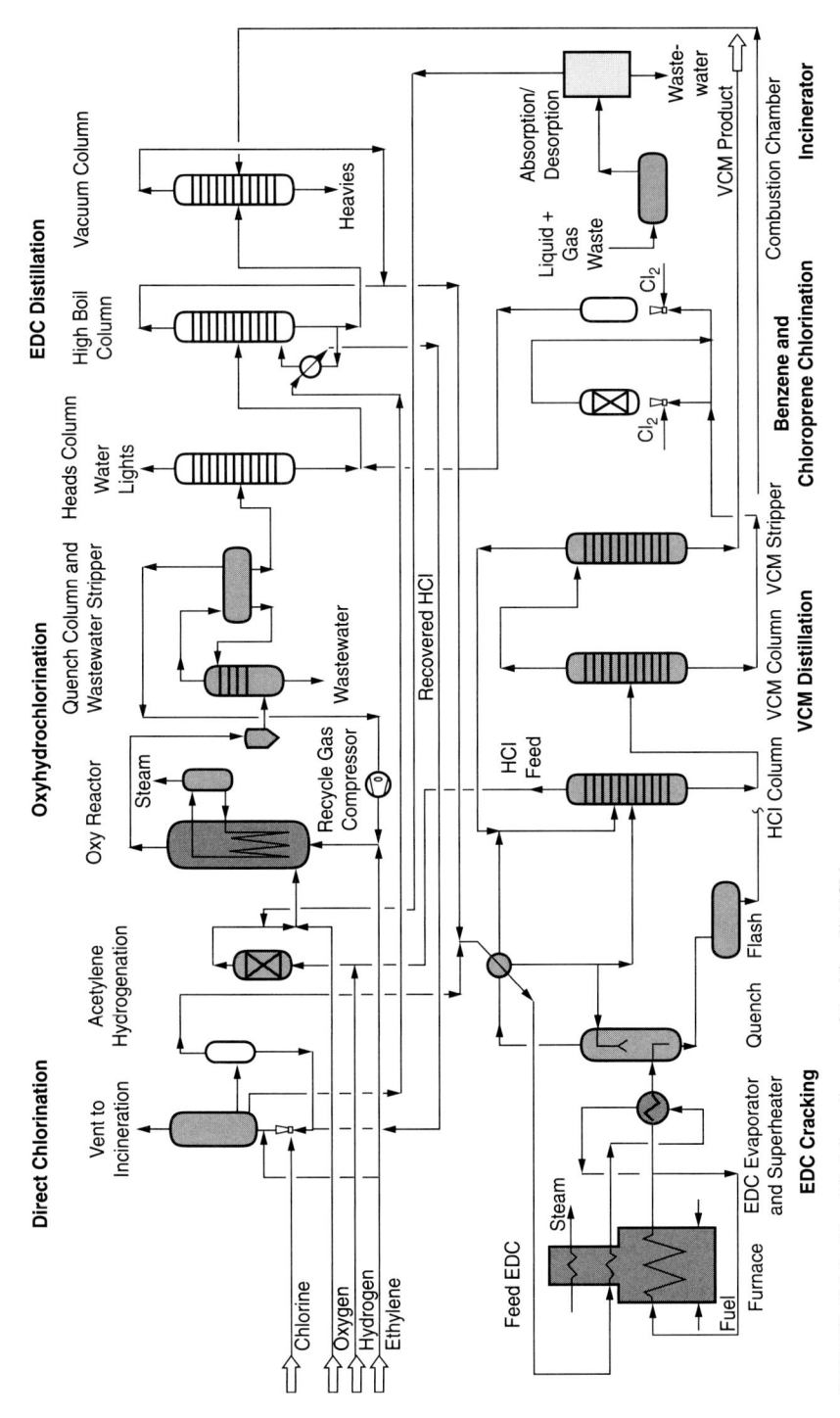

FIGURE 18.1.1 Simplified flow diagram of the Vinnolit VCM process.

18.5

tion unit to high-boiling by-products that are then removed in the EDC distillation unit.
The main reactions are

Direct chlorination:
$$C_2H_4 + Cl_2 \rightarrow C_2H_4Cl_2 - 218 \text{ kJ/mol}$$

Oxychlorination:
$$C_2H_4 + 2HCl + \tfrac{1}{2}O_2 \rightarrow C_2H_4Cl_2 + H_2O - 270 \text{ kJ/mol}$$

EDC cracking:
$$C_2H_4Cl_2 \rightarrow C_2H_3Cl + HCl + 71 \text{ kJ/mol}$$

Water obtained from this process is treated in appropriate water purification plants. Waste gases and liquid by-products are fed to the HCl recovery unit and converted to HCl. The recovered HCl is reused either for the production of hydrochloric acid or in the oxychlorination process. This leads to a complete use of the input chlorine.

Product Specification

VCM as feedstock for the PVC production has to fulfill a very stringent specification to ensure the required quality for the broad range of many PVC products in different applications (Table 18.1.1).

Wastes and Emissions

The major by-product formation takes place in the EDC cracking unit, a smaller amount in the oxychlorination unit, and a negligible amount in the direct chlorination unit. All liquid by-products are removed from the intermediate products in the distillation units. All high-boiling liquid by-products (boiling point higher than EDC, >84°C at atmospheric pressure) are removed at the bottom of a vacuum column; the low-boiling liquid by-products (boiling point lower than EDC, <84°C at atmosphere pressure) are separated and removed in the heads column (column for the separation of water and lights) (Table 18.1.2).

The vent gases are collected in two headers, a dry header and a wet vent header, and routed to an incineration unit for gaseous and liquid residues (Table 18.1.3).

The quantity and quality of the wastewater leaving the plant plays an important role in a modern chemical process (Table 18.1.4). The only source of process water in a modern vinyl chloride plant is the reaction water in the oxychlorination process (see below).

Economics

Table 18.1.5 shows the unit economics of direct chlorination, oxychlorination, and a total VCM plant.

Commercial Experience

Since 1965, Vinnolit has licensed its modern EDC/VCM process with an installed capacity of around 9.3 million MTA EDC and 4 million MTA VCM in plants worldwide (Table 18.1.6).

TABLE 18.1.1 Typical Vinyl Chloride Quality

Hydrochloride	<0.5 ppm (w/w)
Iron	Not detectable
Suspended solids	Not detectable
Propene	2 ppm (w/w)
Acetylene	Not detectable
1,3-Butadiene	6.5 ppm (w/w)
Methyl chloride	60 ppm (w/w)
Ethyl chloride	2 ppm (w/w)
Monovinyl acetylene	1.5 ppm (w/w)
Vinyl chloride	>99.99% (w/w)

TABLE 18.1.2 Expected Liquid By-Product Formation in a VCM Plant in Mass Percent Based on VCM Produced

Total amount of high-boiling by-products	1.5
Total amount of low-boiling by-products	0.5

TABLE 18.1.3 Expected Vent Gas Composition Fed to the Incineration Unit in Volume Percent

Ethylene	3
1,2-Dichloroethane	1
Carbon dioxide	40
Carbon monoxide	4
Nitrogen	Balanced

TABLE 18.1.4 Expected Wastewater Composition Outlet of the VCM Plant to Biologic Treatment (Values in mg/L Water)

Dissolved copper	<0.5
Dissolved solids	15
Chlorinated hydrocarbons	<1
Chemical oxygen demand (COD)	2000–3000

VINNOLIT DIRECT CHLORINATION PROCESS

Process Description

1,2-Dichloroethane (EDC) is produced by an exothermic reaction from the feedstocks ethylene and chlorine; the raw material chlorine is manufactured by either membrane, diaphragm, or mercury processes.

Vinnolit offers two different reactor types:

1. *Liquid-phase reactor with forced draft EDC circulation,* which operates at temperatures between 80 and 120°C depending on battery limit, product, and heat recovery requirements. Gaseous ethylene and chlorine are dissolved in the EDC circulation stream by a special mixing device. The exothermic reaction takes place in the liquid phase and is

TABLE 18.1.5 Unit Economics of Direct Chlorination, Oxychlorination, and Complete VCM Plant

	Direct chlorination*		Oxychlorination*	VCM plant† including DC and OC
	Stand-alone	Integrated in VCM plant		
Consumption figures in kg/1000 kg of product				
Ethylene	284.5	283.1	289	456
Chlorine	718.6	718.6	—	584
Hydrochloride	—	—	744	—
Oxygen	—	—	174	139
Yields in mass percent				
Ethylene	99.65	99.9	98.1	98.5
Chlorine	99.8	99.8	—	98.5
Hydrochloride	—	—	99	—
Oxygen	—	—	93	92.4
Utility consumption per 1000 kg of product				
Electrical power, kWh	10	15	45	100
Steam consumption, kg	0	0	0	0
Steam generation, kg	800	800	1000	0
Cooling water, m³ (10°C temperature difference)	25	25	25	130
Fuel gas, MJ	—	—	—	2720
Catalyst consumption, g	Only for initial filling	Only for initial filling	20	16‡
Erection costs built on U.S. Gulf Coast in 2003, excluding investments outside battery limit	For 320,000-MTA EDC: US$15 million	For 320,000-MTA EDC: US$15 million	For 320,000-MTA EDC: US$40 million	For 400,000-MTA VCM: US$150 million§

*All figures are based on EDC.
†All figures are based on VCM.
‡Oxychlorination catalyst consumption based on VCM.
§Costs include direct and oxychlorination units.

TABLE 18.1.6 VCM Plants Using Vinnolit Technology

	Plant capacity, MTA			Year of agreement
	DC	OC	VCM	
Total installed capacity up to 1988	1,200,000	1,200,000	1,600,000	1965–1988
AECI, South Africa	80,000	—	—	1988
SNEP, Morocco	24,000	—	30,000	1989
Pequiven, Venezuela	105,000	105,000	130,000	1989
	—	105,000	—	2000
Finolex, India	—	105,000	130,000	1989
Chlorvinyl, Ukraine	280,000	280,000	350,000	1991
Beijing No. 2, China	70,000	—	—	1992
SCAC, China	—	—	100,000	1993
	—	115,000	—	1996
	160,000	—	—	2000
	—	—	100,000	2002
Sasol Polymers, South Africa	130,000	130,000	165,000	1994
	164,000	164,000	205,000	2002
Hydro Plast, Sweden	260,000	—	—	1994
BSL/Dow, Germany	288,000	145,000	180,000	1996
Dow, United States, Canada, Germany	Confidential	—	—	1996
BorsodChem, Hungary	—	225,000	—	2002
Panjin, Liaohe, China	—	65,000	80,000	1996
	70,000	—	—	1998
EPC, Egypt	—	80,000	—	1997
	—	80,000	—	1997
Petkim, Turkey	—	120,000	—	1997
	140,000	—	152,000	2000
Sinopec Qilu, China	—	80,000	—	1997
	295,000	295,000	370,000	2002
SADAF, Saudi Arabia	260,000	—	—	1998
QVC, Qatar	355,000	175,000	220,000	1998
Vinnolit, Germany	250,000	280,000	330,000	2000
Uhde, Middle East	540,000	272,000	340,000	2003

catalyzed by the proprietary Vinnolit direct chlorination catalyst at a reaction temperature of between 80 and 120°C in a loop reactor. The reaction heat of the low-temperature chlorination (LTC) at 80°C can be removed by an air or water cooler; the reaction heat of the high-temperature chlorination (HTC) at 120°C can be recovered in a reboiler of the EDC distillation unit. The EDC produced is removed by evaporation, which keeps the catalyst in the reactor loop.

Without distillation, the EDC is fed directly to the EDC cracking unit. The vent gas stream from the reactor overhead either can be fed to an incineration unit or can be used in the oxychlorination unit. This process is extremely friendly to the environment because no further product and wastewater treatment is required, and no catalyst is discharged with the product.

2. *Liquid-phase boiling reactor with natural draft,* which operates at reaction temperatures of between 90 and 120°C with or without heat recovery.

Vinnolit's new high-temperature direct chlorination (HTC) reactor provides an energy-efficient technology for the production of furnace feed EDC without distillation. The liquid-phase reaction of ethylene and chlorine releases 218 kJ/mol of EDC. In a simple carbon steel U-shaped loop reactor, chlorine and ethylene are separately solved in EDC before the reaction takes place. In combination with the special Vinnolit catalyst, this method significantly minimizes by-product formation.

Downstream of the reaction zone, the reactor content boils owing to the lower static pressure and the thermosyphon effect of circulation. The EDC vapor leaves the horizontal vessel and either enters the reboiler of a column (e.g., reboiler of high-boil, heads, and/or vacuum column) or a heat exchanger that condenses the EDC vapor. The reaction heat is transferred to the column indirectly. A fraction of the condensed EDC is fed back to the reactor; the rest is fed directly to the EDC cracker without further distillation.

Because of the high yields, the Vinnolit DC reactor can be operated without a tailgas reactor in (1) stand-alone mode with vents fed to an incineration unit or (2) as part of a complete VCM plant, where the vent gas streams can be fed to the oxychlorination reactor in order to recover the remaining ethylene quantities. If sales EDC specification is the target, only a small stripper column has to be installed to eliminate traces of HCl.

Process Chemistry and Operating Data

The direct chlorination of ethylene proceeds via a polar reaction mechanism in a polar solvent (1,2-dichloroethane). The proprietary Vinnolit catalyst, an inorganic complex, polarizes chlorine, which then acts as an electrophilic reagent to attack the ethylene double bond, thus facilitating the addition reaction.

Main Reaction

$$\underset{\text{Ethylene}}{C_2H_4} + \underset{\text{Chlorine}}{Cl_2} \rightarrow \underset{\text{1,2-Dichloroethane + Energy}}{C_2H_4Cl_2} - 218 \text{ kJ/mol}$$

Side Reactions. The main by-product of this reaction is 1,1,2-trichloroethane formed by substitution chlorination of EDC accompanied by the release of HCl. The reaction probably occurs by a free-radical mechanism. The formation of 1,1,2-trichloroethane can be reduced by the addition of radical scavengers such as oxygen. The HCl coming from the 1,1,2-trichloroethane formation then reacts with ethylene to give chloroethane:

$$\underset{\text{Ethylene}}{C_2H_4} + \underset{\text{Hydrochloric acid}}{HCl} \rightarrow \underset{\text{Chloroethane (ethyl chloride)}}{C_2H_5Cl}$$

$$\underset{\text{1,2-Dichloroethane}}{C_2H_4Cl_2} + \underset{\text{Chlorine}}{Cl_2} \rightarrow \underset{\text{1,1,2-Trichloroethane}}{C_2H_3Cl_3} + HCl$$

Operating Parameters

Operating temperature: 80–120°C
Operating pressure: 0.5–2.5 bar

Detailed Process Description for Boiling Reactor (Fig. 18.1.2)

Process Feeds and Reactor. Gaseous chlorine from a membrane, diaphragm, or mercury plant is dissolved in an EDC stream and mixed with an ethylene-containing EDC stream. The feeds are ratio controlled to ensure a small ethylene excess in the reaction zone. EDC circulates in the reactor loop by the density difference between the liquid-vapor mixture in the upstream pipe and sole liquid in the downstream leg. The reaction heat is removed by evaporation of EDC. The vapor-liquid separation takes places in the horizontal separation drum.

Product Condensation and Heat Recovery. For energy recovery, the product vapor is either condensed in the reboilers of the EDC distillation unit or in a water or air

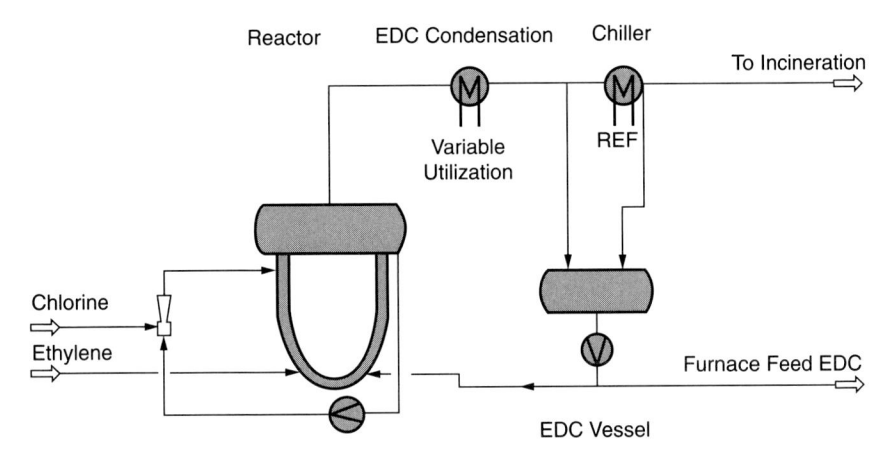

FIGURE 18.1.2 Simplified flow diagram of the Vinnolit direct chlorination boiling reactor.

condenser. The EDC produced is pumped into an EDC storage tank. The product can be used as feed in the EDC cracker without further purification. The bigger portion of the EDC, which is used only to remove the reaction heat from the reactor, is recycled into the reactor loop.

Vent Gas System and Reactor Control. The reactor vent gas is compressed by a liquid ring compressor and fed to the oxychlorination reactor. To prevent a flammable gas mixture (ethylene and oxygen) in the reactor overhead system, ethylene is added as a safening gas. This ethylene-containing vent gas stream is fed into the oxychlorination reactor. The operating mode of the vent gas system can be changed from ethylene as a safening gas to nitrogen. In the later case, the vent gas is fed to an incineration unit.

Important Process Features

- *Low manufacturing costs.* Carbon steel as the main construction material, together with high raw material yields (99.9 percent for ethylene and 99.8 percent for chlorine) and a product quality that requires no further treatment, ensures a highly competitive process with low production costs. The HTC (high temperature chlorination) boiling reactor is simple because no EDC washing, wastewater treatment, and EDC distillation facilities are necessary. A catalyst makeup is not required.
- *Energy savings.* Vinnolit's DC process significantly reduces the steam consumption in a balanced EDC/VCM plant. The savings in steam are approximately 800 kg for each ton of EDC produced. The reaction heat preferably can be used in EDC distillation.
- *Low capital costs.* A simple design with a small amount of equipment results in low unit investment costs.
- *No environmental effluents.* The closed catalyst cycle, the vent gas recovery in the oxychlorination unit, and an EDC quality that requires no further treatment generate *no* liquid effluents.
- *Operability and maintainability.* A corrosion-inhibiting catalyst system and simple equipment without major moving parts keep the maintenance costs low.

TABLE 18.1.7 Typical DC EDC Quality (Feed to EDC Cracking Unit without Further Purification)

1,1,2-Trichloroethane	<400 ppm (w/w)
Hydrogen chloride	<100 ppm (w/w)
Chloroethane	<20 ppm (w/w)
Ethene	<100 ppm (w/w)
Total oxygenated chlorinated compounds such as 2-chloroethanol and chloral	<60 ppm (w/w)
1,2-Dichloroethane	>99.93 % (w/w)

TABLE 18.1.8 Expected Vent Composition Gas in Case of Using Nitrogen as Safening Gas in Volume Percent (Vent Stream Is Fed to the By-Product Incineration Unit)

Oxygen	4.5
Ethylene	3
Carbon dioxide	0.6
1,2-Dichloroethane	0.8
Hydrogen	0.3
Nitrogen	Balance

Feed and Product Specification

Polymerization-grade ethylene and chlorine from different sources such as membrane, diaphragm, and mercury plants can be used in the Vinnolit direct chlorination process. In case of a direct chlorination vent gas recovery in the oxychlorination unit, there is no limitation for oxygen and inerts affecting the yields.

Wastes and Emissions

High yields lead to negligible by-product formation of less than 0.07 percent of the EDC produced. In the case of using the EDC in the EDC cracking unit, no EDC purification is necessary, and therefore, no liquid by-product stream will be generated (Table 18.1.7). In the case of vent gas utilization in the oxychlorination unit, all the ethylene in the vent stream can be recovered (Table 18.1.8).

VINNOLIT OXYCHLORINATION PROCESS (FIG. 18.1.3)

Process Description

In the Vinnolit oxychlorination process, EDC is produced by an exothermic reaction from the feedstocks ethylene, anhydrous hydrogen chloride (HCl), and oxygen. The exothermic reaction is catalyzed by a copper chloride catalyst in a single-step fluidized-bed reactor at a reaction temperature of 200 to 225°C. The reaction heat is either recovered by the production of 10-bar steam at a constant pressure level or by heating of other heat-transfer fluids.

FIGURE 18.1.3 Oxychlorination plant using Vinnolit technology.

The separation of the catalyst fines takes place in a newly developed hot gas catalyst filter or, alternatively, by wastewater treatment meeting even the most stringent regulations for copper, dioxins, and furanes. The environmentally friendly process uses recycle gas, which is fed back to the reactor after quenching and condensing EDC and water. After EDC removal, the process water meets even the most stringent environmental requirements.

After removal of carbon dioxide and chloral/chloroethanol, the crude EDC is purified in the EDC distillation unit and can be used as furnace feed or sales EDC.

Process Chemistry and Operating Data

Main Reaction. This can be summarized in one equation:

$$2C_2H_4 + 4HCl + O_2 \rightarrow 2C_2H_4Cl_2 + 2H_2O - 270 \text{ kJ/mol}$$
Ethylene   Hydrogen chloride   Oxygen   EDC   Water   + Energy

Catalytic Reaction. The preceding main reaction can be divided into the following steps:

1. Reduction of cupric chloride to cuprous chloride by ethylene under formation of 1,2-dichloroethane:

$$2CuCl_2 + C_2H_4 \rightarrow C_2H_4Cl_2 + Cu_2Cl_2$$
Cupric chloride   Ethylene   1,2-EDC   Cuprous chloride

2. Oxydation of cuprous chloride to cupric oxide and cupric chloride by oxygen:

$$2Cu_2Cl_2 + O_2 \rightarrow 2CuO \cdot CuCl_2$$
Cuprous chloride   Oxygen   Cupric oxide/cupric chloride

3. Reaction of cupric oxide/cupric chloride with hydrogen chloride, forming cupric chloride and water:

$$CuO \cdot CuCl_2 + 2HCl \rightarrow 2CuCl_2 + H_2O$$
Cupric oxide/cupric chloride   Hydrogen chloride   Cupric chloride   Water

Side Reactions. The side reactions can be summarized as follows:

Oxidation reactions (i.e., ethylene is partly oxidized to carbon dioxide and water):

$$C_2H_4 + 3O_2 \rightarrow 2CO_2 + 2H_2O + \text{energy}$$
Ethylene   Oxygen   Carbon dioxide   Water   Energy

Substitution reactions (i.e., EDC is chlorinated, yielding 1,1,2-trichloroethane):

$$2C_2H_4Cl_2 + 2HCl + O_2 \rightarrow 2C_2H_3Cl_3 + 2H_2O$$
EDC   Hydrogen chloride   Oxygen   1,1,2-Trichloroethane   Water

Operating Parameters

Operating temperature: 20–225°C
Operating pressure: 2.5–4.0 bar(g)
Steam pressure: 10 bar(g)

Detailed Process Description (Fig. 18.1.4)

Process Feeds. Hydrogen chloride from the EDC cracking process is preheated and treated in a hydrogenation reactor, where acetylene and hydrogen are reacted to ethylene in the presence of a noble metal fixed-bed catalyst. The purpose of the partial hydrogenation of the acetylene is to increase the yields by preventing the formation of by-products such as trichloroethylene and tetrachloroethylene.

Oxygen is preheated as well and mixed with hydrogen chloride before entering the oxychlorination reactor. Ethylene is fed into the recycle gas stream and heated before entering the oxychlorination reactor.

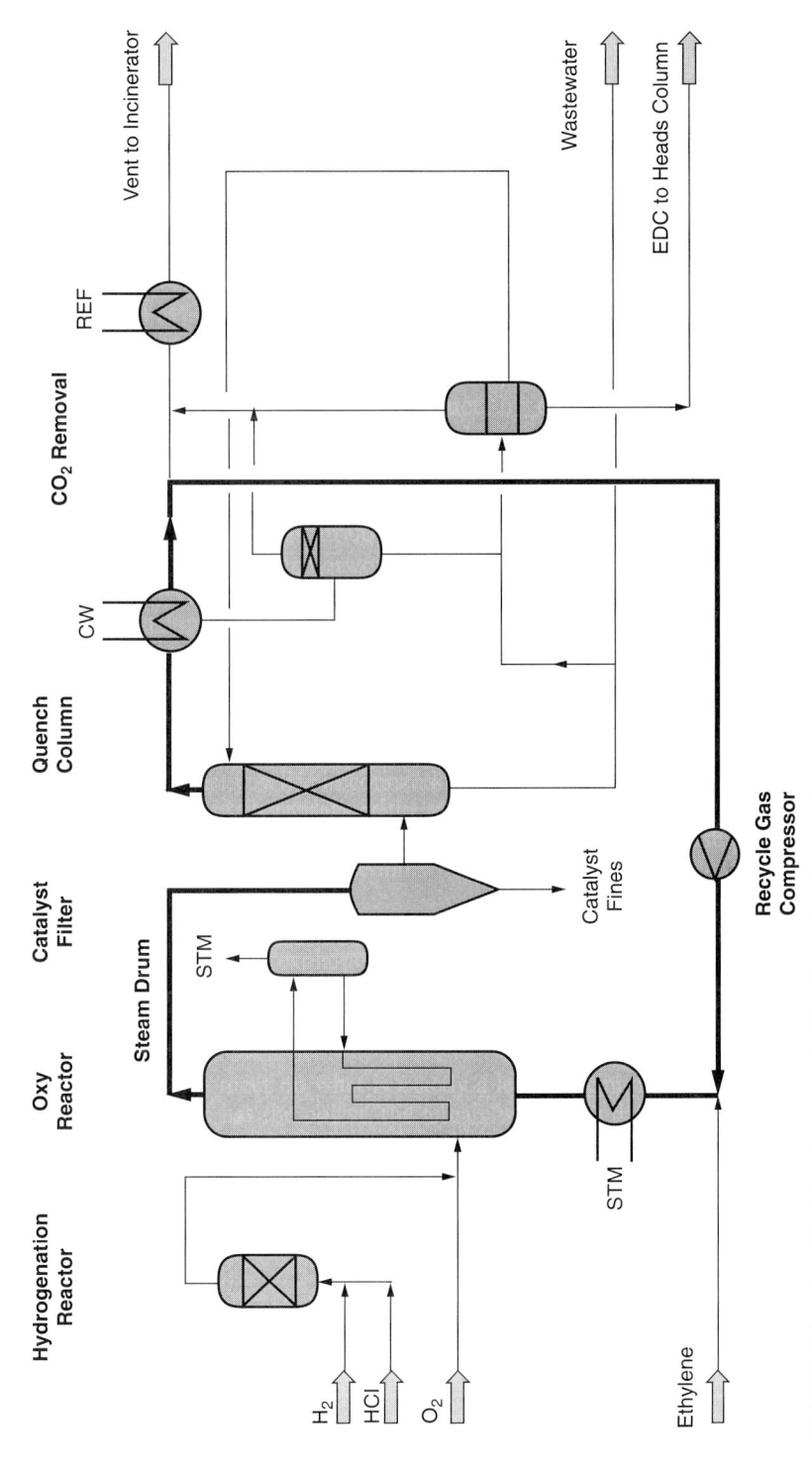

FIGURE 18.1.4 Simplified flow diagram of the Vinnolit oxychlorination unit.

Reaction and Hot Gas Filtration. The oxychlorination reactor is a fluidized-bed reactor that provides intensive contact of process feed gases with the catalyst. The main reactor parts are the gas distributor, the cooling coils, and the cyclones for separation of the catalyst from the reaction gases. The catalyst has a particle size distribution of 20 to 80 μm with flow characteristics similar to liquids. The catalyst lifetime is unlimited, and only small catalyst quantities, which pass the reactor cyclones, have to be replaced by makeup catalyst approximately once a year. The generation of fines (catalyst attrition) is mainly influenced by the gas distributor design. The newly developed Vinnolit gas distributor reduces the fines formation to a minimum. The heat produced by the reaction is removed by direct generation of steam in vertical tubes immersed in the catalyst bed. The recycle gas with the EDC produced and water leave the oxychlorination reactor and enter the catalyst filtration for removal of the fine catalyst particles from the hot reactor effluent gas. The purpose of this further catalyst recovery is to prevent contamination of the oxychlorination wastewater with dioxins, furanes, and copper from the oxycatalyst. The separated catalyst fines are periodically dumped into drums.

Quench Column and Product Condensation. The separation of reaction water and EDC with recycle gas takes place in the quench column, where the gases are scrubbed with recycled process water and caustic soda. Gaseous EDC with vaporized water and recycle gas leave the column at the top, whereas the condensed water is discharged to the neutralization drum and further to the wastewater stripper for the separation of chlorinated hydrocarbons such as EDC and chloroform.

Water and EDC are separated from the recycle gas by condensation in the crude EDC condenser. In the crude EDC decanter, the water phase is separated from the EDC phase by gravity. After a caustic treatment, the crude EDC is pumped to the EDC distillation unit, whereas the water is recycled as scrubbing liquid back to the quench column.

Recycle Gas System and Reactor Control. The preheated recycle gas is compressed by a gas compressor, very often a centrifugal compressor, and recycled to the oxychlorination reactor. To keep the system pressure constant, a small quantity of recycle gas is purged to an incineration unit.

In order to control the OC reactor, the ethylene and oxygen content in the recycle gas are analyzed online. Typical recycle gas composition is shown in Table 18.1.9.

Catalyst Storage. During maintenance shutdown, which has to occur every 2 years, the catalyst is transferred into the catalyst hopper (not shown in Fig. 18.1.4). For the pneumatic transfer of the catalyst, recycle gas or nitrogen can be used. The catalyst hopper is also used to for the addition of the small catalyst makeup quantities that can be charged into the reactor during operation.

TABLE 18.1.9 Expected Vent Gas Composition in Volume Percent

Ethylene	3
1,2-Dichloroethane	2–3
Carbon dioxide	40
Carbon monoxide	4
Nitrogen	Balanced

Note: 1,2-Dichloroethane content depends on vent gas condensation temperature.

Important Process Features

- *Reliability.* A stable temperature control combined with an excellent heat transfer and a uniform temperature profile (no hot spots) in the fluidized bed guarantees an on-stream time of more than 99 percent per year. A specially designed raw materials sparger system allows for operation spans of more than 2 years without maintenance. Vinnolit oxychlorination reactors are made of carbon steel and have been in operation for almost 30 years. Compared with competitive processes, the increased heat-transfer area allows for a higher steam temperature and steam pressure in the cooling coils, which gives a much wider safety margin to the critical surface temperature with the danger of hydrochloric acid dew point corrosion occurring. All common oxychlorination catalysts available on the market can be used.
- *Safety.* The oxygen is mixed with anhydrous hydrogen chloride outside the reactor and is fed into the fluidized bed *independently* of the ethylene. The oxygen concentration in the recycle stream is approximately 0.5 percent by volume, which is far from the explosive range.
- *Flexibility.* A turndown ratio as low as 20 percent capacity utilization can be achieved. The load of the oxychlorination reactor can be increased from 20 to 100 percent within 30 minutes, and vice versa. In addition, an immediate startup after a shutdown makes operation very easy.
- *Low manufacturing costs.* The unlimited catalyst lifetime, the low losses due to the highly efficient cyclone system (less than 20 g catalyst per ton of produced EDC), the high raw materials yields (98.0 percent ethylene, 99 percent anhydrous hydrochloride, and 93 percent oxygen), and the possibility to use inexpensive oxygen from PSA units ensure a highly economical process.

Product and By-Product Specification

Not only polymerization-grade ethylene and hydrogen chloride from the EDC cracking process but also ethylene and/or anhydrous hydrogen chloride containing offgas streams can be used in the Vinnolit oxychlorination process (Table 18.1.10). For example, the following vent gas streams or low-purity feeds are recovered in commercial oxychlorination plants from following processes:

- Ethylene-containing vent gas streams from direct chlorination and acetaldehyde and monochloroacetic production
- Hydrochloride from the chlorinated by-product incineration unit, chlorinated methanes, chlorinated ethanes, epichlorohydrin, and isocyanates plants [e.g., methylene diphenyldiisocyanate (MDI), toluene diisocyanate (TDI), and hexamethylen diisocyanate (HDI)]
- Low-cost oxygen from pressure swing adsorption (PSA) units

TABLE 18.1.10 Typical Oxycrude EDC Quality Excluding Water (Feed to Distillation Unit)

1,1,2-Trichloroethane	<2000 ppm (w/w)
Chloroform	<2000 ppm (w/w)
Carbon tetrachloride	<1000 ppm (w/w)
Chloroethane	<200 ppm (w/w)
Trichloroethanol (chloral)	< 10 ppm (w/w)
2-Chloroethanol	<150 ppm (w/w)
1,2-Dichloroethane	>99.4 % (w/w)

Wastes and Emissions

High yields are the key for a low by-product formation of less than 0.6 percent of the produced EDC and a vent gas stream of only 100N·m³/100,000 MTA EDC capacity. A highly efficient hot gas filtration system for the small quantities of catalyst fines or a simple wastewater treatment ensures extremely low copper as well as dioxin and furane concentrations in the oxychlorination wastewater. No other wastewater treatment than the EDC removal via steam stripping is required, so the charter for European Council for Vinyl Manufacturers (ECVM) is met easily (emissions to the environment: EDC <5 g/t of EDC purification capacity, copper <1 g/t of oxychlorination capacity, dioxin-like components <1 μg TEQ/t of oxychlorination capacity).

VINNOLIT THERMAL CRACKING PROCESS OF 1,2-DICHLOROETHANE TO VINYL CHLORIDE

Process Description (Fig. 18.1.5)

Liquid EDC is preheated in three steps to 210°C and evaporated in an external EDC evaporator heated by the cracking gas leaving the furnace. In the cracking furnace, pure and dry EDC is partly cracked to vinyl chloride (VCM) and hydrogen chloride (HCl) in an endothermic reaction at approximately 490°C. The reaction heat (HR = 71 kJ/mol) to the cracking furnace supplies radiant wall burners.

The conversion rate is 55 to 60 percent. The cracked gas leaving the furnaces is first cooled to 270°C in the tubes of the EDC evaporator and then cooled to 160°C in the downstream quench system. The different gaseous and liquid streams from the quench system are fed to the HCl column.

Process Chemistry and Operating Data

Main Reaction. The purified EDC is cracked in the following endothermic reaction:

$$C_2H_4Cl_2 \rightarrow C_2H_3Cl + HCl + 71 \text{ kJ/mol}$$
$$\text{EDC} \quad \text{Vinyl chloride} \quad \text{Hydrogen chloride} - \text{Energy}$$

Side Reactions. The main side reactions are the formation of acetylene by dehydrochlorination of vinyl chloride, the benzene and mono-vinyl-acetylene formation from acetylene, and the addition reaction of acetylene and mono-vinyl-acetylene to chloroprene.

Formation of acetylene:

$$C_2H_3Cl \rightarrow C_2H_2 + HCl$$
$$\text{Vinyl chloride} \quad \text{Acetylene} \quad \text{Hydrogen chloride}$$

Formation of benzene:

$$3C_2H_2 \rightarrow C_6H_6$$
$$\text{Acetylene} \quad \text{Benzene}$$

Formation of mono-vinyl-acetylene:

$$2C_2H_2 \rightarrow C_4H_4$$
$$\text{Acetylene} \quad \text{Mono-vinyl-acetylene}$$

FIGURE 18.1.5 Vinyl chloride plant using Vinnolit technology.

Formation of chloroprene:

$$C_4H_4 + HCl \rightarrow C_4H_5Cl$$
Mono-vinyl-acetylene　　Hydrogen chloride　　2-Chloro-butadiene (chloroprene)

$$C_2H_2 + C_2H_3Cl \rightarrow C_4H_5Cl$$
Acetylene　　Vinyl chloride　　2-Chloro-butadiene (chloroprene)

Operating Parameters

Operating temperature: 480–500°C

Operating pressure: 16–24 bar

Detailed Process Description (Fig. 18.1.6)

Feed Preheating and Evaporation. In the first step, EDC (furnace feed grade) is preheated with process gas and steam before entering the convection section of the

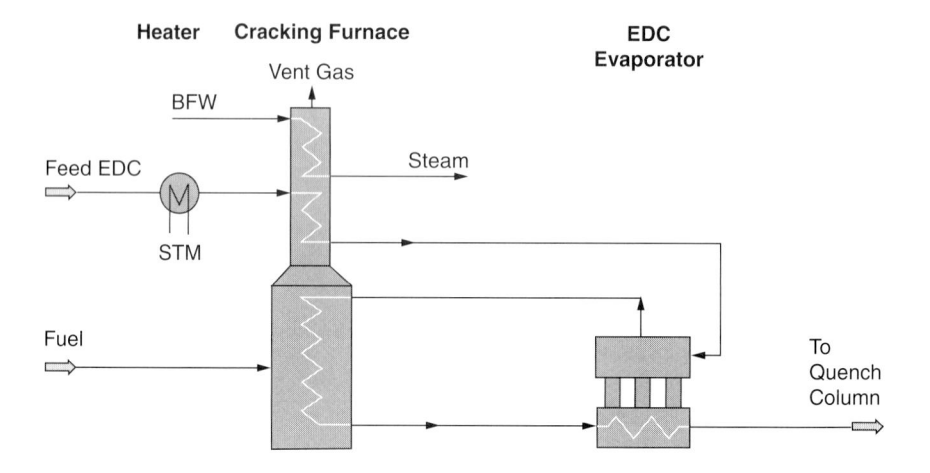

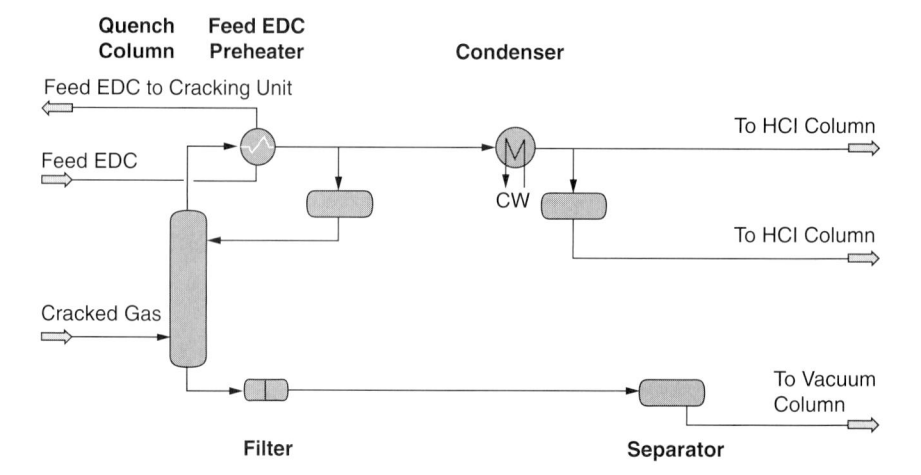

FIGURE 18.1.6 Simplified flow diagram of the Vinnolit cracking unit.

cracking furnace. At temperatures close to boiling conditions, the EDC is fed into the evaporator. The EDC evaporator of the Vinnolit process is heated with cracking gas coming directly from the cracking furnace, which reduces the fuel consumption significantly. Gaseous EDC is fed into the cracking furnace under flow control.

EDC Cracking Furnace. The cracking furnace is a box-type furnace, and its radiant section is equipped with wall-mounted radiant burners. The burners may be fired with natural gas, hydrogen, mixtures of gaseous hydrocarbons, or light oil. The burners typically are forced-draft burners using combustion air from a radial blower or natural draft burner. The excess air (oxygen) in the furnace is measured and controlled by an online oxygen analyzer. In the radiant section, the EDC vapor flow is superheated to approximately 490°C in countercurrent with the flue gas and cracked to VCM and HCl at a cracking rate of approximately 55 percent.

The upper part (convection section) of the cracking furnace is used for EDC preheating and for steam generation. The EDC cracking products—a mixture of nonconverted EDC, VCM, HCl, and by-products—flow from the cracking furnace to the EDC evaporator for energy recovery.

A typical furnace operation conversion rate is 55 percent. In principle, this rate of 55 percent can be increased up to 65 percent, accompanied by a pronounced increase in the formation of by-products and fouling.

Furnace Decoking. During the operating period, coke deposits inside the tubes of the cracking furnace. Coking is monitored by the increase in pressure drop. The following cleaning methods typically are used:

- Thermal decoking with a mixture of steam and air. The coke layer is removed by thermal oxidation at temperatures above 620°C.
- Decoking with steel pellets in a high-velocity nitrogen flow.
- Pigging of the coils with water as the carrier liquid.

Quench System. The cracked gases are cooled in the cracker quench column and processed in the quench column overhead and bottom system. The cracked gases are cooled mainly by condensed EDC and VCM evaporation. The design and operation of the quench column ensure that no solids are carried over to the downstream VCM distillation unit.

The quench overhead vapors, which consist primarily of HCl, VCM, and EDC, are condensed in several steps, such as a steam generator, EDC preheater, and water condenser or air cooler. The condensed liquid is partly used as reflux for the quench column. The gaseous and vapor streams from the quench section are sent to the VCM distillation unit.

The quench column bottoms effluent is routed to a flash drum under flow control in order to separate the solid coke particles from the liquid stream. The flash overhead stream is sent to the VCM distillation, and the coke-containing bottom stream is sent to the vacuum column for EDC recovery.

Important Process Features

- *Energy saving.* Recovery of primary and secondary energy is done by using the energy from the cracking gas for external EDC evaporation and steam generation. Minimized electrical power consumption in the HCl column occurs by operation at energy- and yield-optimized cracker pressure.
- *Low manufacturing costs.* The long cracking furnace operating cycles between cleaning shutdowns of up to 2 years, the high raw materials yield (99.5 percent EDC), and the possibility to adapt the heat recovery to the customer requirements guarantee low production costs.

VINNOLIT SUSPENSION POLYVINYL CHLORIDE (S-PVC) PROCESS

Process Description

A schematic overview of the production steps is given in Fig. 18.1.7.

The process is divided into three main steps. Raw materials, DM water, and chemical process materials are fed to the *polymerization unit,* where vinyl chloride monomer is converted to PVC in pressurized reactors.

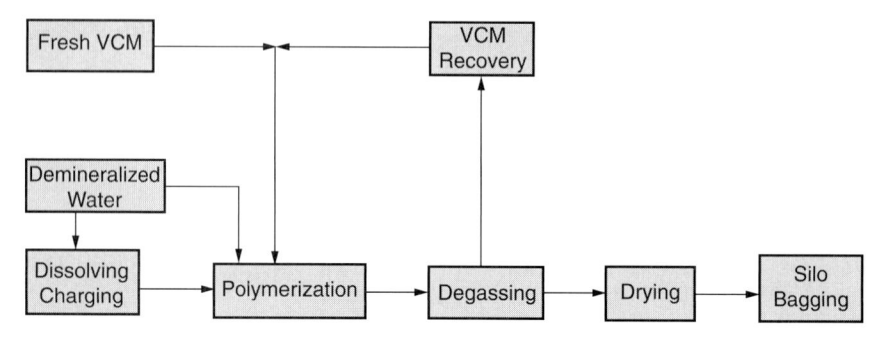

FIGURE 18.1.7 Suspension PVC process.

In the *degassing unit,* unreacted VCM is removed and sent to VCM recovery, where the recovered VCM is condensed, stored, and reused.

After mechanical separation of the water from the PVC slurry in the *drying unit,* the wet cake enters the dryer, where the residual water is evaporated. The dry PVC powder is screened and conveyed to the silo station, where it is bagged or loaded into containers or trucks for shipment.

Process Chemistry and Operating Data

Chemical Reaction. The production of suspension PVC can be described by the following equation:

$$n\underset{\text{Vinyl chloride}}{CH_2=CHCl} \rightarrow \underset{\text{Polyvinyl chloride}}{[-CH_2-CHCL-]_n} + \underset{\text{Enthalpy}}{\text{approx. 1600 kJ/kg}}$$

The reaction takes place in droplets of liquid VCM, which are dispersed in the water phase. On addition of initiators, the double bonds are broken, and reaction heat in the range of 1500 to 1750 kJ/kg is generated. The length of the resulting polymer chain depends on the polymerization temperature and defines the molecular weight of the PVC obtained. The molecular weight is determined by measuring the viscosity of a dilute polymer solution. The resulting viscosity numbers are mathematically transformed into K-values, which are used for the specification of the different PVC types.

Operating Parameters

Operating temperature: 50–65°C
Operating pressure: 7–11 bar(g)
Conversion rate: 80–94%
pH value: <7

Detailed Process Description (Fig. 18.1.8)

The polymerization is carried out in pressurized reactors. Before a batch start, the internal walls of the empty and flushed reactor are coated with an antifouling agent supplied by means of a steam injection nozzle. The coated reactor walls have to be rinsed again with demineralized water. The remaining coating film prevents polymer scaleup on the internal surfaces of the reactor.

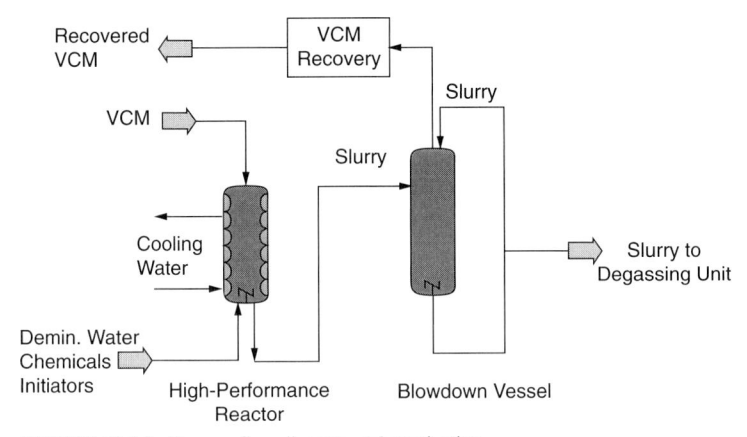

FIGURE 18.1.8 Process flow diagram: polymerization.

According to the recipe, demineralized water, dispersing agents, additives, and activators are fed into the reactors via metering devices and controlled by a DCS system. The common chemicals that are used for S-PVC polymerization are

Initiator
- Various peroxide compounds
- 40–50 percent emulsion in water
- 1.5–5 percent active oxygen content

Dispersing agent
- Partly saponified polyvinylacetate
- Cellulose ester

Additives
- Caustic soda
- Sodium nitrite
- Silicon-based defoaming agents

Appropriate quantities of VCM and hot demineralized water are fed simultaneously into the reactor so that the reactor content achieves the polymerization temperature immediately after charging. After polymerization ignition, external cooling is started for an isotherm temperature control during the total polymerization period. During polymerization, a certain quantity of demineralized water is fed continuously into the reactor to compensate for the volume shrinkage.

For safety reasons, the reactor is equipped with an automatic short-stopping system. In case of abnormal high reactor pressure, the stopper solution is fed automatically into the reactor by means of nitrogen pressure via a rupture disk.

As soon as the nominal VCM/PVC conversion rate in the range of 85 to 90 percent is reached, the reactor content is discharged into a blowdown vessel (Fig. 18.1.9).

In order to achieve high productivity, excellent product quality, and very good environmental and safety conditions, the design and operating conditions of the polymerization process must fulfill the following requirements:

- Short nonreaction time (preparation of the reactor, charge, discharge)
- Short reaction time (conversion VCM → PVC)
- Optimized recipes (quality, economics)
- High safety standards

FIGURE 18.1.9 High-performance reactors, with 115 m^3 reactor volume, at Vinnolit's production site in Knapsack, near Cologne, Germany.

Nonreaction Time (NRT). In Vinnolit's S-PVC process, the NRT is less than 100 minutes for large reactors up to 150 m^3 of volume owing to the closed and clean reactor technology and the simultaneous hot water and VCM charging (SHWC).

Clean Reactor Technology. The polymer buildup on the reactor walls during polymerization, with its detrimental effect on heat transfer and product quality, is a major problem of the PVC process. Use of buildup-suppressing agents (antifouling agents) is helpful but not sufficient. After a certain number of batches, the reactors have to be opened and cleaned with high-pressure water. The removal of VCM from the reactor before opening, high-pressure cleaning, and preparations for the restart of the reactor (remove of air/oxygen) take time and reduce productivity.

Using Vinnolit's clean reactor technology, it is possible to run a reactor for 300 batches or more without any high-pressure cleaning procedure (Fig. 18.1.10). After discharge of a batch, the reactor is rinsed with normal-pressure water to wash out the remaining PVC grains from the walls and internals. No time for opening and hydroblasting of the reactor is lost. Vinnolit's clean reactor technology is characterized by

- Optimization of design and construction of the reactor and the internals
- Application of a highly effective antifouling agent
- Adapted recipes

Vinnolit's charging procedure is described as simultaneous hot water and VCM charge (SHWC). Appropriate quantities of hot water and VCM at ambient temperature are charged into the reactor. Polymerization temperature is reached at the end of the charging

FIGURE 18.1.10 Clean reactor after 300 batches.

procedure. With adapted recipes, this procedure reduces the time compared with sequential charging of water and VCM, saving the time that is usually required for the heatup of the mixture.

Reaction. Owing to the exothermal polymerization reaction, the requirements for heat transfer and agitator system increase with reactor size. In conventional reactors, with heat removal via the water phase, productivity is mainly restricted by the cooling water supply temperature, as well as by design of the reactor, agitation, and internals (Fig. 18.1.11).

A higher heat flow rate Q out of the reactor can be realized by different technical solutions:

- Increased temperature difference (cooling water − polymerization temperature)
- Increased cooling surface area
- Higher heat-transfer coefficient (lower wall thickness)

Increase in temperature difference can be realized only by using chilled water, which is only economical at sites with extremely cheap energy for the preparation of chilled water. Enlargement of the cooling surface area is achieved mostly either by an additional external cooler (e.g., reflux condenser) or cooled internals (e.g., chilled baffles). This approach is realized in various technologies.

Vinnolit's high-performance reactor (HP reactor) is characterized by an arrangement of cooling halfpipes on the inner shell of the reactor. Improvement in the heat-transfer coefficient is possible by reducing reactor wall thickness. A design comparison of conventional reactors and the Vinnolit's HP reactor is given in Fig. 18.1.12.

In a conventional reactor, the polymerization heat has to be transferred through an over 30-mm-thick reactor wall; the cooling halfpipes inside Vinnolit's HP reactor are 3 to 4 mm thick. The performance comparison of different reactor systems in Fig. 18.1.13 shows that the maximum heat flow rate of the HP reactor is more than 200 percent higher than in conventional reactors. The enlargement of the cooling surface area, the significant decrease in the wall thickness, and the increase of the Reynold's number (higher turbulence at the structured cooling wall) all lead to this improvement.

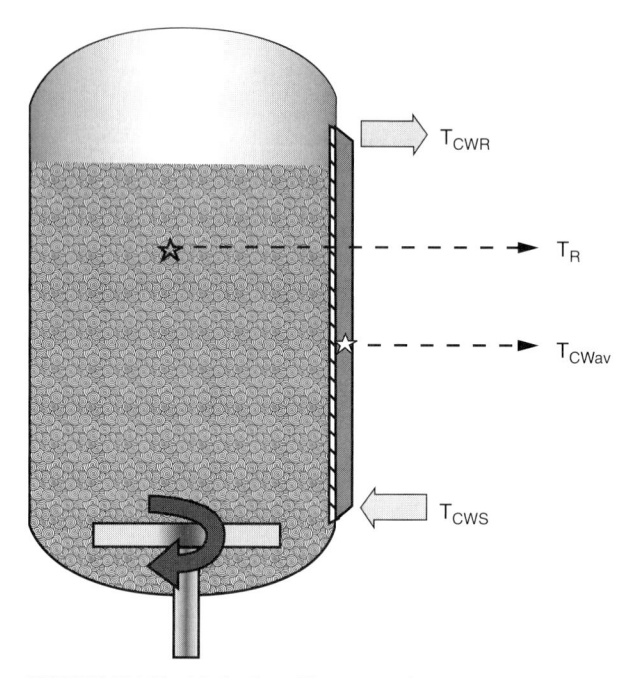

FIGURE 18.1.11 Mechanism of heat removal. T_{CWS} = cooling water supply temperature, T_{CWR} = cooling water return temperature, T_{CWav} = cooling water average temperature, T_R = polymerization temperature, A = cooling surface area, k = heat-transfer coefficient; $\dot{Q} = k \cdot A \cdot \Delta(T_R - T_{CWav})$.

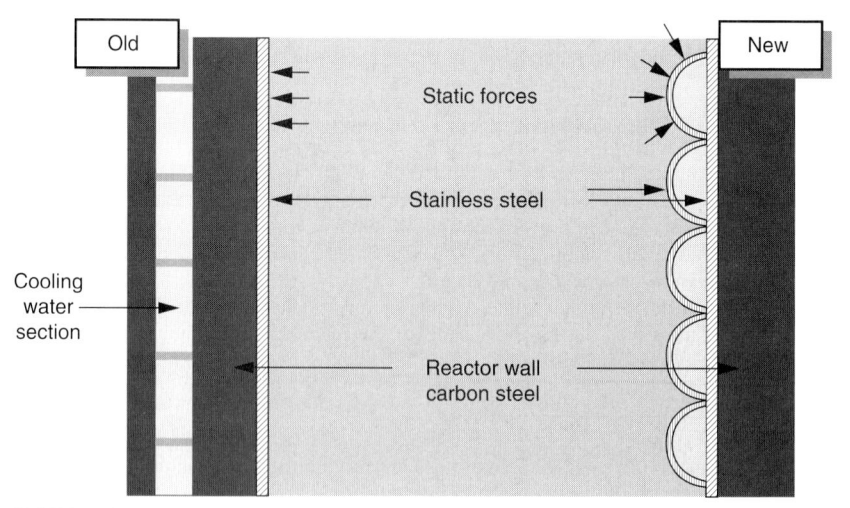

FIGURE 18.1.12 Comparison of conventional reactor versus Vinnolit's HP reactor.

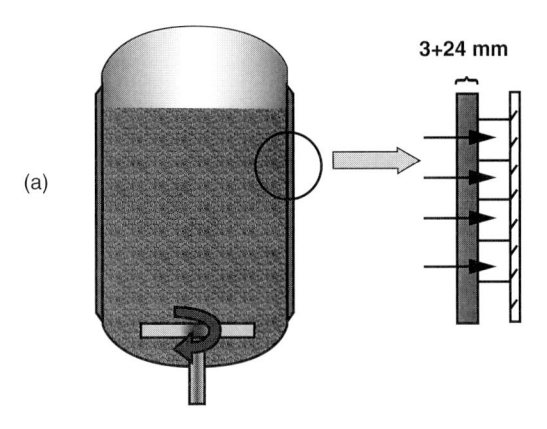

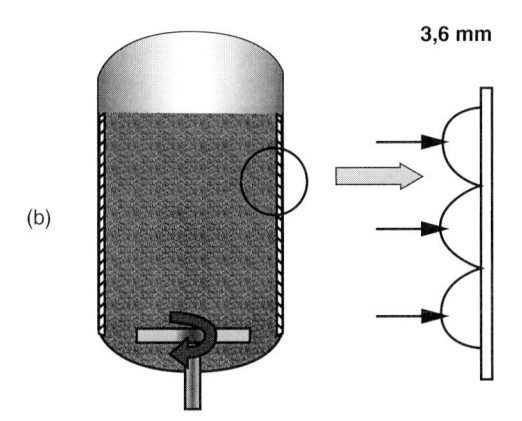

FIGURE 18.1.13 Comparison of different reactor systems. (*a*) Conventional reactor: cooling surface, 100%; heat transmission, 100%; heat-transfer coefficient, 100%; performance, 100%. (*b*) High-performance reactor: cooling surface, 122%; heat transmission, 283%; heat-transfer coefficient, 175%; performance, 213%.

Degassing. The PVC slurry containing unreacted VCM is discharged from the polymerization reactors into the blowdown vessel (Fig. 18.1.14). This vessel functions as a buffer tank between the discontinuously operated polymerization and the continuously operated degassing and drying unit. Because of the pressure drop between the HP reactor and the blowdown vessel, most of the unreacted VCM is flashed out and recovered in the VCM recovery unit by compression and condensation. In the blowdown vessel, the VCM content in the slurry can vary between 20,000 and 40,000 ppm.

For energy saving reasons, the suspension stream to the degassing column is preheated with the already degassed slurry from the degassing column. In the predegassing vessel, the suspension is heated up to 95°C by condensation of the offgas from the degassing column and by steam injection.

18.28 VINYL CHLORIDE AND POLYVINYL CHLORIDE

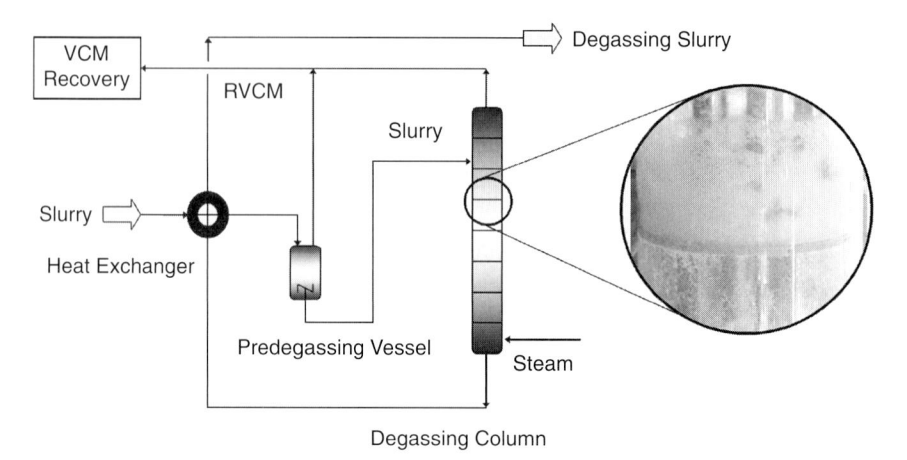

FIGURE 18.1.14 Process flow diagram: degassing.

Under flow control, the suspension is transferred to the sieve-tray degassing column. The unreacted VCM is stripped with steam. The slurry is heated by introducing the stripping steam at the bottom of the column. The temperatures are 100 to 110°C at the bottom and 95 to 105°C at the top. The column is operated close to atmospheric pressure. Opening and cleaning of the column are not necessary because of the self-cleaning design. Unreacted VCM (continuous stream at constant pressure) flows to the VCM recovery unit.

Drying. The slurry is dewatered by means of a solid-bowl decantation centrifuge yielding a wet PVC cake with a residual moisture content of 20 to 33 percent depending on the resin grade (Fig. 18.1.15). The water from the centrifuges is sent through the fresh air heater in order to recover the sensitive heat and subsequently fed to the wastewater treatment.

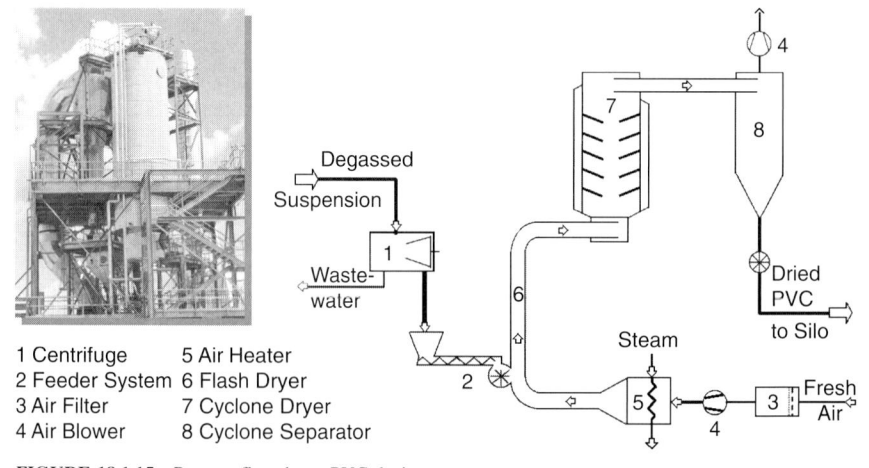

1 Centrifuge   5 Air Heater
2 Feeder System   6 Flash Dryer
3 Air Filter   7 Cyclone Dryer
4 Air Blower   8 Cyclone Separator

FIGURE 18.1.15 Process flowsheet: PVC drying.

The wet PVC cake is conveyed into the flash dryer by means of product vibrating feeders and product paddle feeders. The product paddle feeder distributes PVC in the hot air stream. The fresh atmospheric air is filtered, compressed by a fresh air blower, and heated in the fresh air heater.

The residual moisture of PVC grains leaving the flash dryer is in the range of 3 to 5 percent. The air and the PVC particles then enter the bottom drying chamber of the cyclone dryer. Here the solids are dried to a residual moisture content of 0.2 to 0.3 percent during a residence time of 10 to 20 minutes. The temperature of the exhaust air leaving the top of the cyclone dryer is approximately 50 to 70°C depending on product grade. Both air and solids have the same temperature.

Cyclone Dryer: Function and Construction. The cyclone dryer is a cylindrical vessel with several chambers. Each chamber is separated from the next one by a horizontal annular conical plate. Hot gas and wet solids are introduced tangentially into the lowest chamber. Owing to centrifugal forces, the solid particles are separated from the gas phase so that a ring of PVC dust rotates in the first chamber with high velocity.

With a spinning stream, the gas leaves the first chamber and enters the second chamber. Solid particles from the dust-laden rotating ring in the first chamber are redispersed into the hot gas before they enter the next chamber, first the small and light particles and then the larger and heavier particles.

Having passed the annular baffle plate, the spinning solid stream expands, and the solid particles flow to the wall of the dryer, where they decelerate by friction. At the dryer wall, the particles reverse and return back into the center, where they are redispersed into the spinning stream again. In this way, the chambers are successively filled with solid particles, one after the other, starting from the bottom of the dryer. After a while, all the chambers are filled with the same solid concentration. At the top of the dryer, the solid particles and the hot gas leave the dryer.

The drying energy is supplied by the hot gas and by introducing hot water into the jacket. The temperature of the hot water should not be below the calculated dew point temperature to prevent condensation in the dryer.

The combination of the spinning stream and the annular baffle plates causes a permanent separation and redispersion of gas and solid particles. The separation is caused by the centrifugal forces, whereas the redispersion is caused by gravitation. This leads to a high-velocity difference between the gas and the solid particles that causes an intensive mass and heat exchange. During residence in the dryer, the particles are separated by size and weight: The small and light particles follow the gas flow, which means a short residence time. The larger and heavier the particles are, the longer is the residence time. Owing to the separation and redispersion, the PVC particles have different residence times in the cyclone dryer, between 30 minutes and 5 seconds, depending on particle size and weight.

This selective effect is the main characteristic feature of Vinnolit's cyclone dryer and results in all particles leaving the cyclone dryer with the same moisture content irrespective of their size. By this selective drying, thermal degradation is prevented, and overdrying of the drying goods is avoided. Owing to the rotational flow, the cyclone dryer is self-cleaning. No encrustation or deposits form, not even in the case of heat-sensitive or cohesive solids. Changing the product type is very easy and takes only 20 to 30 minutes. Owing to the absence of mechanical or rotary elements, minimum maintenance is required.

Application of the Cyclone Dryer System. Cyclone dryers are also used for various other products besides polymers, such as pulp, sesame seeds, inorganic chemicals, and others. An overview is given in Table 18.1.11.

Up to 2003, more than 30 cyclone dryers have been installed and have been in operation around the globe for many years.

TABLE 18.1.11 Overview of Cyclone Dryer System

Product	Moisture content Before drying	Moisture content After drying	Drying temperature		Max. residence time
Suspension PVC	30%	0.1%	165°C	55°C	20 min
Emulsion PVC	45%	0.2%	160°C	55°C	13.8 min
Sesame seeds	40%	4%	114°C	70°C	5.6 min
Polytetraflourethylene	50%	0.01%	250°C	150°C	9.8 min
Pinewood chips*	54%	12.8%	312°C	71°C	10.2 min
Grass protein*	35%	9%	206°C	91°C	5.3 min
Dextrin*	20.8%	2%	143°C	70°C	3.6 min
Municipal solid waste*	31.5%	6%	186°C	71°C	1.5 min
Polysaccharide*	60%	4.8%	196°C	96°C	2.6 min
Sodium bicarbonate*	8%	<0.5%	262°C	115°C	6.6 min
Potassium nitrate*	2.8%	0.08%	375°C	139°C	6.0 min
Pentaaerithrol*	24%	0.2%	112°C	86°C	Not determined
Polyvinyl buthyralen*	35%	<1%	90°C	46°C	14.2 min

*Product data from Babcock-BSH GmbH.

Important Process Features

Vinnolit and its parent companies Hoechst and Wacker have been active in the field of S-PVC production since 1938 and have spent a lot of effort to develop their processes and to meet the market requirements as well as environmental and regulation requirements. Vinnolit is approved according to DIN ISO 9001 (quality management) and according to DIN ISO 14001 (environmental).

Polymerization. In order to shorten the cycle time of polymerization, reactor productivity has been optimized. The cycle time (CT) covers the reaction time (RT) and the nonreaction time (NRT). The high-performance reactor technology minimizes the reaction time and improves the efficiency of the reactor without using chilled water and/or a reflux condenser.

Optimization of the Reaction Profile. The short nonreaction time is realized by

- Clean and closed reactor technology. Eliminating the opening and the high-pressure cleaning after each discharge of a batch by using an efficient antifouling agent to cover the reactor wall and internals.
- Simultaneous hot water/VCM charging (SHWC), i.e., no heat-up time after charging, no heating device necessary, and thus, reduced charging time.
- Optimized design of reactor and internals.
- Adapted recipes.

All these parameters, together with an automatic and computer-controlled charging system, allow the reactor to be run under clean and closed conditions over long production periods without mechanical or high-pressure water cleaning.

Reactor volumes of up to 150 m^3 have been realized. High performance of the reactor at up to 600 MTA/m^3 can be achieved.

Advanced Reactor Safety Philosophy. To ensure safe and proper production and to control the reaction parameters even in cases of unexpected or abnormal incidents, e.g., due to power failure, the following means or equipment items can be provided:

- Redundant instrumentation and an emergency short-stopper system
- Predictive measures to avoid hazardous operating conditions
- Graduated operational steps in cases of failure to be carried out depending on the current situation or the status of the batch.

Degassing. The Vinnolit degassing column is a proprietary process for degassing of the slurry; it leads to less than 1 ppm VCM in the final product. Consequently, the residual VCM content in the decanter wastewater and in the dryer exhaust is extremely low (less than 1 ppm).

Low Steam Consumption (≤140 kg/t PVC). Low residual VCM is seen in the PVC powder (less than 1 ppm for general-purpose PVC, less than 50 ppb for grades used in medical application). Gentle degassing conditions result in no thermal degradation of the product.

Drying. The PVC powder is dried by the modern and approved Vinnolit cyclone:

- Gentle drying at a low temperature level
- Intensive heat and mass exchange due to the high relative velocity between air and solids
- Uniform final moisture of all particles, and no electrostatic effect due to overheating or overdrying
- Grade change without opening and cleaning
- No moving parts, easy to operate, and easy to clean
- Low investment cost and low maintenance cost
- Small holdup of PVC
- Minor PVC losses during grade changes

Feed and Product Specification

- VCM (see Table 18.1.1)
- Demineralized water
- Conductivity: ≤1 µS/cm
- SiO_2-: ≤0.02 mg/L
- pH value: 6.0–7.0
- Color: Transparent, colorless
- PVC product grades for rigid application are shown in Tables 18.1.12 and 18.1.13.

Waste and Emissions

Wastewater. The process wastewater leaving the plant has the following specifications:

Normal flow rate	2.5 m^3/t PVC
Temperature	40–50°C
VCM content	Below 0.5 mg/m^3 water
Suspended solids	Below 20 mg/kg water
pH value	6–9

18.32 VINYL CHLORIDE AND POLYVINYL CHLORIDE

TABLE 18.1.12 PVC Product Grades

Parameter	Unit	Test method ISO	Grade A	Grade B	Grade C
K-value			56–58	59–61	66–68
Bulk density	g/L	60	570–610	550–600	550–600
Sieve analysis					
>63 µm	%	4610	>95	>95	>95
>250 µm		4610	≤1	≤1	≤5
Porosity	%	4608	15–19	16–20	19–23
Volatile matter	%	1269	<0.3	<0.3	<0.3
Flowability	s/150 g	6186 10-mm nozzle	<25	<25	<25
Residual VCM	ppm	6401	≤1	≤1	≤1
Impurities	PT/250 g	VN F/5 6.2.3*	≤10	≤10	≤10
Thermostability	Min	VN J/8*	>30	>30	>40

*Vinnolit standard.

TABLE 18.1.13 PVC Product Grades for Flexible Applications

Parameter	Unit	Test method ISO	Grade E	Grade F	Grade G
K-value			64–66	69–71	74–76
Bulk density	g/L	60	520–560	470–510	460–500
Sieve analysis					
>63 µm	%	4610	>95	>95	>95
>250 µm		4610	≤1	≤1	≤1
Porosity	%	4608	25–29	31–35	31–35
Volatile matter	%	1269	<0.3	<0.3	<0.3
Flowability	s/150 g	6186 10-mm nozzle	≤35	≤35	≤35
Residual VCM	ppm	6401	≤1	≤1	≤1
Impurities	PT/250 g	VN F/5 6.2.3*	≤10	≤10	≤10
Thermostability	Min	VN J/8*	>35	>40	>40

*Vinnolit standard.

Gaseous Effluents. The exhaust air from the drying unit contains:

Solids (PVC dust)	10 mg/m^3
Vinyl chloride	0.1 mg/m^3

Total losses of VCM and PVC are as follows:

VCM losses:

Wastewater	5 g/t PVC
Offair from drying unit	7 g/t PVC
Leakage	4 g/t PVC
Offgas to incineration	410 g/t PVC

PVC losses:

Offair from drying unit	140 g/t PVC
Settling basin	200 g/t PVC
Wastewater	50 g/t PVC
Off grade	170 g/t PVC

Maximum total VCM and PVC losses are 986.0 g/t PVC.

Economics

The erection costs of a suspension PVC unit with a capacity to produce 400,000 MTA of S-PVC are about US$180 million (U.S. Gulf Coast 2003). This sum excludes outside battery limits.

Consumption Figures.   Expected consumption figures per metric ton of S-PVC, K-value 67, rigid are as follows:

Raw materials and auxiliaries
    Vinyl chloride (100%)   1001 kg
    Demineralized water   2.3 m^3
Utilities
Steam   800 kg
Electric power   170 kWh
Cooling water (dT = 4 K)   192 m^3

Commercial Experience

Vinnolit has licensed the modern suspension PVC process since 1973 with an installed capacity of around 1,440,000 MTA in 20 plants worldwide, as listed in Table 18.1.14 on p. 18.34.

ABBREVIATIONS AND ACRONYMS

EDC	1,2-dichloroethane
VCM	vinyl chloride monomer
MDI	methylene diisocyanate
TDI	toluene diisocyanate
HDI	hexamethylene diisocyanate
HCl	hydrogen chloride
DC	direct chlorination
OC	oxychlorination
COD	chemical oxygen demand
HTC	high-temperature chlorination
LTC	low-temperature chlorination

TABLE 18.1.14 S-PVC Worldwide Capacity

Licensee	Country	Capacity, MTA	Unit Polymerization	Unit Degassing	Unit Drying	Year of agreement
Georgia Gulf	United States	100,000	X	X	X	1973
Shell	Netherlands	30,000		X		1978
Buna	Germany	60,000	X	X	X	1979
Petroquimica Col.	Colombia	42,000	X	X	X	1979
Formosa Plastics	Taiwan	25,000			X	1984
Westlake	United States	70,000			X	1985
Formosa Plastics	Taiwan/United States	100,000			X	1985
Indupa	Argentina	45,000	X	X	X	1985
Denka	Japan	24,000			X	1989
SNEP	Morocco	35,000			X	1992
Finolex	India	80,000	X	X	X	1994
Georgia Gulf	United States	2 × 80,000			XX	1994
EKO	Greece	38,000			X	1995
Hydro Polymers	United Kingdom	100,000			X	1996
Condea-Vista	United States	100,000			X	1996
SNEP	Morocco	45,000			X	1997
Condea-Vista	United States	100,000		X	X	1997
SCAC	China	120,000			X	1998
Royal Polymers	Canada	80,000			X	2000
Uhde	Middle East	300,000	X	X	X	2003
Finolex	India	150,000	X	X	X	2004

S-PVC	suspension polyvinyl chloride
NRT	nonreaction time
SHWC	simultaneous hot water and VCM charging
HP	high-performance reactor
STM	medium-pressure steam
CW	cooling water
REF	refrigerant
BFW	boiler feed water

REFERENCES

1. CMAI, *2004 World Vinyls Analysis*.
2. Christoph Heinze, *Chem Eng* 7:274–279, 1984.

CHAPTER 18.2
CHISSO POLYVINYL CHLORIDE SUSPENSION PROCESS TECHNOLOGY AND VINYL CHLORIDE MONOMER REMOVAL TECHNOLOGY

Seiichi Uchida
Chisso Corporation
Tokyo, Japan

CHISSO POLYVINYL CHLORIDE SUSPENSION PROCESS TECHNOLOGY

Introduction

This general information introduces the polyvinyl chloride (PVC) suspension process technology that was developed and commercialized by Chisso Corporation of Japan. In addition to Chisso's own PVC plants, Chisso has licensed to many companies the suspension PVC process technology and the proprietary continuous process to remove unreacted vinyl chloride monomer (VCM) from aqueous PVC slurries (see Table 18.2.6).

The Chisso process can produce a wide range of PVC grades with homopolymer K-values (product characterization parameter) that range from 58 to 74, vinyl acetate copolymers, as well as PVC specialty grades.

Chisso Corporation is a pioneer in the Japanese PVC industry and has developed a great deal of experience in the areas of PVC production, compounding, and polymer processing.

Process Description

The Advantages of the Chisso PVC Production Process. The important characteristics of the Chisso process may be summarized as follows:

   1. *High-quality products.* The Chisso process produces high-quality products, each of which has unique physical properties such as excellent transparency, high heat stability, excellent weatherability, superior electrical properties, superior processability, and excellent gelation. The Chisso PVC process technology has consistently satisfied all its worldwide customers' process and product requirements.

   2. *Low production cost.* By achieving the exhaustive recovery of vinyl chloride monomer (VCM) with substantial energy savings, the Chisso process minimizes the consumption of raw materials and utilities. With the incorporation of a highly automated control system (DCS), the Chisso process can be operated by a minimum number of trained technicians. This results in a low production cost for PVC.

   3. *A safe and nonpolluting process.* The Chisso process plant is designed for operation with a high degree of safety and meets all waste effluent standards, as has been demonstrated based on Chisso's licensees' long and satisfactory experience. The entire PVC production and monomer recover process is virtually a closed system.

   4. *Low investment requirement.* The Chisso process has undergone continuous improvement so that higher productivity is accomplished by employing compact, reliable, nonfouling production facilities.

A Chisso PVC technology process plant is designed with unique equipment that has been skillfully and carefully engineered by Chisso Engineering Co., Ltd., of Japan using its chemical engineering expertise, which has been accumulating for over 50 years of practical project engineering. It is this experience that ensures that a Chisso process PVC plant can be established with a relatively low investment requirement.

Process Description (Fig. 18.2.1)

   1. *Polymerization section.* Purified VCM, pure water, and other chemicals, including a polymerization initiator, are charged into a reactor. The reactor contents are stirred vigorously, keeping good PVC slurry particle suspension conditions. At the elevated reactor temperature, the VCM reacts to form the PVC particles in the presence of the initiator. The PVC polymer particles produced by the reaction form a concentrated, dense polymer slurry in the reactor. When the reaction reaches a final stage, the unreacted VCM is recovered into a VCM gas holder, and the polymer slurry is discharged from the reactor to a slurry hold tank.

   2. *VCM recovery section.* The unreacted VCM gas in the gas holder is compressed and condensed in the process monomer recovery section of the plant. The recovered liquid VCM is stored in a vessel and is recycled as feedstock for the succeeding polymerization reaction.

   3. *VCM removal section.* The PVC slurry is pumped from the slurry hold tank to the continuous VCM removal process section of the plant, where residual VCM is effectively "stripped" out from the slurry with steam and is recovered to the VCM recovery process section of the plant. The VCM removal process section is designed and engineered using the well-proven Chisso VCM removal technology.

   4. *Drying section.* The PVC slurry from the VCM removal process section is fed into a continuously operating centrifuge that separates the wet PVC particle cake from water. The wet PVC cake, discharged from the centrifuge, is then fed into a fluidized-bed dryer, where moisture is removed from the PVC cake and powder particles. The dried PVC powder then passes through an oscillating screen to eliminate oversized particles and is then

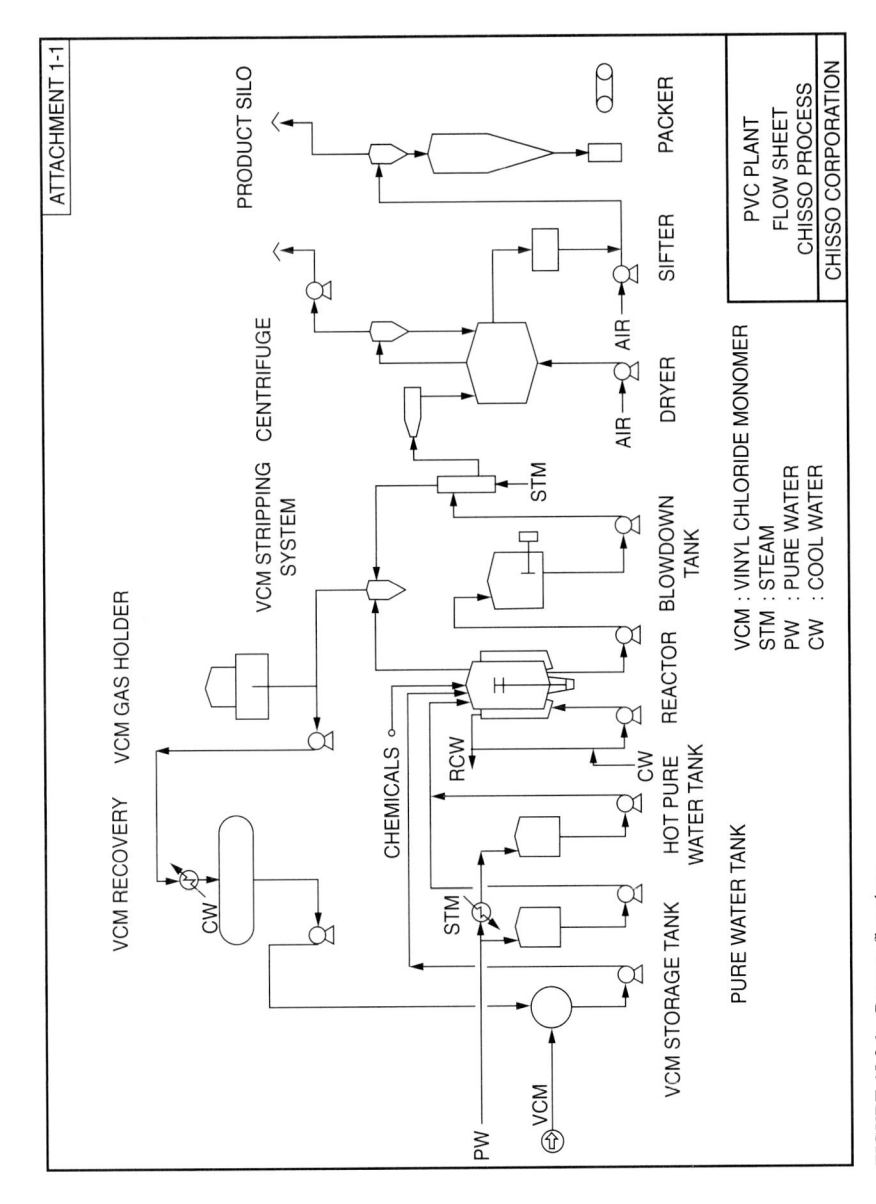

FIGURE 18.2.1 Process flowsheet.

conveyed pneumatically to product powder silos. The PVC product powder has excellent uniform quality and is ready for shipping.

5. *PVC product handling.* The PVC powder product is stored temporarily in product powder silos and then bagged. The powder bagging machine automatically carries out the weighing and powder packaging work. The filled bag is then automatically transferred from the bagging machine to a palletizer via a conveyor and stacked onto pallets. Alternately, the PVC powder can be conveyed from the product powder storage silos and loaded onto trucks for bulk polymer shipment.

TABLE 18.2.1 Typical Raw Materials and Utilities Consumption per Metric Ton of PVC

Vinyl chloride monomer, kg	<1003
Electric power, kWh	<160
Steam, kg	<700
Catalyst and chemicals, US$	<12

TABLE 18.2.2 Typical Grades of Chisso PVC Products

	SM homopolymer	SL-P homopolymer	SL-K homopolymer	SR homopolymer
Appearance	White powder	White powder	White powder	White powder
K-value	72	66	66	60
Bulk density, g/cm^3	0.49	0.51	0.56	0.54
Volatile matter, %	Max. 0.3	Max. 0.3	Max. 0.3	Max. 0.3
Particle size (pass %, 42 mesh screen)	100	100	100	100

TABLE 18.2.3 Rigid Applications

Style	Service	Processing	SM	SL-KSR
Film, sheet	Packaging, building materials, vacuum lamination	Calendaring Extrusion	0	0
Pipe	Industrial uses Water pipes Electrical uses	Extrusion	0	0
Board	Industrial uses Building materials	Extrusion	0	0
Injection moldings	Joints, fishing floats, fans, sundries	Extrusion	0	0
Bottle, vessel	Food use General purpose	Blow molding		0
Extrusion	Curtain rails, moldings, eaves, gutters, blinds	Extrusion	0	0

TABLE 18.2.4 Plasticized Applications

Style	Service	Processing	SM	SL-P	SR
Film	Agricultural use, packaging, general purpose, toys, houseware	Calendaring Extrusion	0	0	0
Sheet	General purpose, packaging, tape, personal ornaments, wall coverings	Calendaring Extrusion	0	0	
Leather	Personal ornaments, furniture, handbags, clothes, car seats	Calendaring	0	0	
Belt	Industrial use, personal ornaments	Calendaring Extrusion	0	0	
Hose, tube	Garden hoses General tubes	Extrusion	0	0	
Wire coating	Insulation Sheaths	Extrusion	0	0	
Injection moldings	Shoes Sandals	Injection	0	0	
Profile extrusion	Door packing	Extrusion	0	0	
Tile	Floors, walls	Calendaring	0	0	

Process Features

Typical consumption figures of raw materials and utilities for the standard Chisso PVC product grades are given in Table 18.2.1.

Typical Properties of Chisso PVC Products.   Table 18.2.2 lists typical grades of Chisso PVC products.

The S series is the standard Chisso PVC product. SM, SL-P, SL-K, and SR are considered commodity grades. SL-K and SR are used for rigid applications, whereas SM and SL-P are mainly used for plasticized applications.

Applications of Chisso's PVC products are given in Tables 18.2.3 and 18.2.4.

Plant Requirements

The following figures are based on a PVC plant with a production capacity of 100,000 metric tons per annum (MTA).

Required Process Technicians

Foremen	1 person/shift
Operators	2 persons/shift

Note: These numbers of workers are based on normal operation under Japanese conditions, and the labor requirements for product handling, maintenance work, and utility facilities operation are not included.

18.42 VINYL CHLORIDE AND POLYVINYL CHLORIDE

Required Plant Site Area. Plant site area within battery limit: approximately 8000 m^2.

Required Plant Investment. Process plant within battery limit: approximately US$300 per metric ton of production capacity on a turnkey basis, depending on plant site and process requirements.

Conditions

1. Net price as of September 1996, Southeast Asia base
2. The plant shall be installed on a prepared site.
3. All utilities and effluents shall be delivered to the battery limit of the plant site.

Scope of Estimate

1. The plant cost includes the following items:
    - Engineering cost and fee
    - Supervision service cost and fee
    - Machinery and equipment
    - Piping
    - Instruments
    - Electricals
    - Steel structure
    - Civil engineering
    - Control and switch rooms
    - Erection work
    - Insulation and painting
    - Product packing (25-kg bags) and bulk truck-loading facility
2. The following items are excluded from the plant cost:
    - License and know-how fee
    - Land purchase and site preparation
    - Utility supply facility
    - Water treatment facility
    - Firefighting facility
    - Product warehouse
    - VCM storage tank facility

Process Design Basis

1. Annual production capacity, 100,000 MTA; operating hours, 8000 h/yr.
2. Major equipment requirement (Table 18.2.5).

TABLE 18.2.5 Major Equipment Requirements

Section	Capacity	Number per production train	Production trains
Polymerization	100 m^3	3	1
VCM recovery	Approximately 1500 m^3/h	1	1
VCM stripping	15 t/h	1	1
Drying	15 t/h	1	1
Product handling	—	1	1

TABLE 18.2.6 Licensed Chisso Process PVC Plants

Name of company	Location	Capacity, MTA	Date of first startup
Chisso	Minamata, Japan	50,000	1941
Chisso Petrochemical	Goi, Japan	60,000	1963
Chisso	Mizushima, Japan	70,000	1970
CGPC	Taiwan	60,000	1981
Asahimas Chemical	Indonesia	67,000	1989
Asahimas Chemical	Indonesia	70,000	1992
Asahimas Chemical	Indonesia	80,000	1996
Dagu Chemical Factory	Tianjin, China	40,000	1997
Apex Petrochemical	Thailand	120,000	1997
Asahimas Chemical	Indonesia	Revamping 56,000	1997
Cangzhou	Hebei, China	150,000	1999
Cangzhou	Hebei, China	Expansion 80,000	1999
Tianjin Dagu Chemical	Tianjin, China	64,000	2001
CGPC	Chung Shan, China	170,000	(2005)
Shanghai Chloralkali Chemical	Shanghai, China	Debottle 70,000	(2004)
Tianjin Dagu Chemical	Tianjin, China	160,000	(2004)

Chisso Licensed Commercial Installations

Licensed PVC plants employing Chisso process technology are listed in Table 18.2.6.

CHISSO VINYL CHLORIDE MONOMER REMOVAL PROCESS TECHNOLOGY

Introduction

A superior and unique process to remove vinyl chloride monomer (VCM) from an aqueous polyvinyl chloride (PVC) slurry has been developed and commercialized by Chisso Corporation of Japan. The first commercial unit using the process, with a capacity of 2.5 t dry PVC/h, was installed in 1977 at the Chisso Minamata, Japan, plant (see Table 18.2.6). Through commercial operational experience from the first unit, Chisso found that the performance of the monomer removal process was far better than originally expected in terms of VCM removal efficiency, lower energy requirement, PVC product quality, and operability. Since then, several units have been added to other Chisso PVC manufacturing facilities with various improvements and modifications. The process also has been adopted under license from Chisso by other PVC producers in Japan, Asia, Europe, and North and South America. At present, more than 60 (70 at present) commercial units worldwide are in satisfactory operation. In addition to these installations, since 1993, latex stripping of paste PVC (microsuspension PVC and emulsion PVC) has been developed by Chisso. Two commercial latex stripping columns are currently being operated satisfactorily.

Principles of the VCM Removal Process

The Chisso VCM removal process is based on the principles of steam stripping of the residual VCM out of the aqueous PVC slurry that is obtained through the suspension polymerization operation. Although the principle is relatively common technology and by no means new for a degassing operation, great attention was paid by Chisso engineers during the course of process development and subsequent equipment design so as to prevent any deterioration in product quality due to heat degradation of the PVC and to avoid operational problems in the PVC slurry handling.

Investigations of the VCM diffusion rate in PVC particles, as well as the characteristics of PVC slurry, were coupled with Chisso's engineering expertise. This led to the development of a unique and compact design for the stripping column. The steam stripping column's unique internal design allows the achievement of plug flow for the PVC slurry. This results in a highly efficient VCM stripping operation without degradation of the PVC product. This stripping operational efficiency was achieved at a low operational cost requirement. The internal surfaces of the steam stripping column are kept clean by using a special washing device that allows reliable, continuous equipment operation.

Process Features

Although the VCM stripping efficiency is highly dependent on the nature of PVC product, it is to some extent modifiable by proper selection of operating conditions. Excellent commercial-scale stripping results have been obtained in operational commercial-scale units that compare closely with those experienced with laboratory and pilot-plant equipment.

Typical commercial-scale operation reduces the VCM content in PVC slurry for product having K-values of 58 to 72 from approximately 30,000 ppm down to less than 1 ppm on a dry PVC powder base. The stripping process is continuous and can be fully automated so that no additional operational technician time is required. The stripping process also can be incorporated easily into most the DCS systems at existing PVC manufacturing plants.

The Chisso VCM stripping process also can be applied successfully not only to suspension homopolymers having a wide range of molecular weights, copolymers such as vinyl chloride/vinyl acetate or vinyl chloride/vinyl ether, but also to paste-type polymers.

Typical commercial PVC stripping operational data are as shown in Table 18.2.7.

As shown in this table, PVC slurry with a comparatively high VCM content is treated very effectively in the stripping column. Thus, depending on product requirements, an increase in productivity can be expected for an entire PVC plant.

The vinyl chloride monomer stripped out of the aqueous slurry is recovered for reuse in succeeding polymerization batches. The steam consumption is projected to be well below 130 kg/t dry PVC in ordinary cases, although it will depend on the properties of the PVC particles and slurry feed conditions, such as temperature, solid PVC content in the slurry, etc.

Investment Requirements

The stripping column, which is the major component of the VCM recovery process, is very compact, and the installation cost of the total VCM removal facility can be minimized as long as the recovery process step can be combined efficiently with other existing PVC manufacturing facilities. Although most of the equipment is of stainless steel construction to avoid contamination problems in the PVC product, the total investment

TABLE 18.2.7 VCM Content in PVC (ppm, Dry PVC Base)

	Inlet slurry	PVC product
PVC homopolymers (K-value, 58–87)	20,000–30,000	<1

for the Chisso VCM removal process is considerably less than that required for other commercially available VCM monomer recovery processes. A typical VCM removal facility with a capacity of 15 t dry PVC per hour requires a site area of approximately 100 m² (10 m × 10 m).

Patents and License

The patents for the VCM removal process and its apparatus have been issued or assigned to Chisso Corporation in more than 10 countries, and a license for the process and patents is available to PVC producers worldwide. The process is particularly recommended for PVC manufacturing plants that are operated under VCM emission control regulations similar to those issued by the Environmental Protection Agency (EPA) in the United States. This is so because of the remarkable reduction of VCM content in the PVC-water slurries and in the very low amount of VCM experienced in the exhaust gases from PVC dryers.

Chisso is fully prepared to carry out preliminary VCM stripping tests on PVC resins for potential licensees. Actual experimental tests are advisable so as to optimize the design of VCM stripping facilities because the VCM stripping rate is closely related to actual polymerization recipes and operating conditions.

Commercially Licensed Installations

VCM stripping systems employing the Chisso process technology are listed in Table 18.2.8.

TABLE 18.2.8 Chisso Process Technology VCM Stripping Systems

Company	Capacity, t/h	Startup
Chisso Corp.	2.5	1977
Chisso Corp.	4	1979
Chisso Corp.	4.6	1981
Chisso Corp.	6	1986
Chisso Corp.	7	1986
Chisso Corp.	8	1987
Japanese Company (A)	3, 5	1978
Japanese Company (A)	5	1994
Japanese Company (B)	3	1980
Japanese Company (B)	8	1981
Japanese Company (B)	6	1982
Japanese Company (B)	4	1985
Japanese Company (B)	3	1988
Japanese Company (B)	11	1989
Japanese Company (C)	5	1984
Japanese Company (D)	4	1986
Japanese Company (D)	5	1988
Japanese Company (E)	6	1987
Japanese Company (E)	7	1988
Japanese Company (E)	4, 8	1991
Japanese Company (E)	18	2003
Japanese Company (F)	7, 10	1994
Taiwanese Company (G)	20, 10, 5	1985

(Continued)

TABLE 18.2.8 Chisso Process Technology VCM Stripping Systems (*Continued*)

Company	Capacity, t/h	Startup
Taiwanese Company (G)	5 (×2)	1998
Taiwanese Company (G)	5	1998
Taiwanese Company (G)	20	1999
Taiwanese Company (G)	10 (×3)	1999
Taiwanese Company (G)	20 (×2)	1999
Taiwanese Company (G)	5	2001
Taiwanese Company (H)	23	(20
European Company (I)	5.5 (×2)	1989
European Company (J)	6	1991
European Company (J)	6	1994
European Company (J)	6 (×3)	1996
European Company (J)	19	1999
European Company (J)	19	2000
European Company (J)	19	2000
European Company (K)	5, 10	1995
Indonesian Company (L)	10	1989
Indonesian Company (L)	10	1992
Indonesian Company (L)	12	1996
Indonesian Company (L)	5	1997
Indonesian Company (M)	3	1989
Korean Company (N)	5 (×2)	1990
Korean Company (N)	14	1995
Korean Company (N)	2.5 (×2)	2004
Korean Company (N)	4	2004
Korean Company (O)	12	1996
Korean Company (O)	12	1996
Korean Company (O)	20	2000
American Company (P)	20	1996
American Company (P)	20	2003
Thai Company (Q)	15	1997
Chinese Company (R)	6	1997
Chinese Company (S)	15	1998
Chinese Company (S)	15	1999
Chinese Company (S)	15	2003
Chinese Company (T)	20	1999
Chinese Company (T)	14	1999
Chinese Company (T)	8	1999
Chinese Company (U)	8	2000
Chinese Company (V)	13 (×2)	2003
Chinese Company (W)	12.5 (×2)	(2004)
Chinese Company (W)	20 (×2)	(2004)

INDEX

A

ABB Lummus Global cumene production via CD*Cumene* technology, **4.3–4.9**
  process chemistry, **4.4**
  process description, **4.4–4.7**
  process economics, **4.7–4.9**
  process perspective, **4.4**
  summary of process features, **4.9**
ABB Lummus Global propylene production via olefins conversion technology, **10.35–10.41**
  conclusion, **10.41**
  development and commercial history, **10.36**
  process chemistry, **10.37**
  process description, **10.37–10.38**
  process economics, **10.38–10.40**
  summary of process features, **10.40–10.41**
ABB Lummus Global SRT cracking technology for production of ethylene, **6.3–6.20**
  commercial operations, **6.19**
  cracking heater, **6.8–6.11**
  economic aspects, **6.19–6.20**
  process chemistry, **6.5**
  process flow schematic, **6.11–6.15**
  recent technology advances, **6.16–6.19**
  refinery and ethylene plant integration, **6.15–6.16**
  worldwide feed slate, **6.4–6.5**
ACETICA process, **1.3–1.13**
Acetone:
  from isopropanol, **9.31**
  via KBR phenol process, **9.31–9.50**
  via Polimeri Europa process, **9.8–9.12**
  via Sunoco/UOP phenol process, **9.15–9.25**
Acetone and phenol:
  conversion to bisphenol A, **9.51–9.59**
  from cumene, **4.11**
Acetylene, removal from butadiene streams, **3.12**
Acetylene and hydrochloric acid, vinyl chloride process, **18.4**
Acrylates, comonomers for polyethylene production, **14.95**, **14.98**
Adsorptive separation, for p-xylene production (*see* UOP Parex process for p-xylene production)
Advanced MHAI process, **13.16**, **13.18**, **13.21**
Advanced process control (APC+) software, **14.119**
Advanced recovery system (ARS), **6.28**, **6.42**
Alkylation of benzene, **4.11**
Alkylation reaction of ethylene and benzene, **5.13–5.21**
AlliedSignal/UOP phenol process, **9.13**
Ammonolysis of phenol, **2.3**
Amoco Chemical Company, development of polypropylene process, **16.71**
Aniline:
  market, **2.4**
  production history, **2.3–2.4**
  worldwide end use of, **2.4**
AO Orgsteklo (Russia) cumene plant, **4.5**
Aromatics complex, **13.8–13.9**
  Parex unit, **13.23–13.24**
Asahi Kasei (Japan) "classic" SM project, **11.4**
ATOFINA Petrochemical [joint licenser of Stone & Webster (Badger) styrene technology], **5.23**, **5.32**, **11.13**
Axens' Eluxyl process, **13.8**

B

Badger Licensing LLC, **5.23**, **11.13**
Badger styrene technology [*see* Stone & Webster (Badger) styrene technology]
Basell *Avant* catalysts, **16.18**
Basell *Hostalen* technology for bimodal HDPE production, **14.71–14.86**
  general process description, **14.71**

1

Basell *Hostalen* technology for bimodal HDPE production (*Cont.*):
  Hostalen process perspective, **14.**74–**14.**75
  process chemistry, **14.**71–**14.**73
  process description, **14.**75–**14.**82
  process economics, **14.**85–**14.**86
  product range and applications, **14.**82–**14.**85
Basell *Lupotech* G technology for HDPE and MDPE production, **14.**87–**14.**93
  general process description, **14.**87
  Lupotech G process perspective, **14.**88–**14.**89
  process chemistry and thermodynamics, **14.**87–**14.**88
  process description, **14.**89–**14.**92
  process economics, **14.**93
  product specifications, **14.**92–**14.**93
Basell *Lupotech* T technology for LDPE and EVA-copolymer production, **14.**95–**14.**111
  general process description, **14.**95
  Lupotech T process perspective, **14.**102–**14.**104
  process chemistry and thermodynamics, **14.**95–**14.**102
  process description, **14.**104–**14.**109
  process economics, **14.**111
  product specifications, **14.**110–**14.**111
Basell *Spherilene* technology for LLDPE and HDPE production, **14.**3–**14.**13
  general process description, **14.**3
  process chemistry and thermodynamics, **14.**3–**14.**7
  process description, **14.**8–**14.**9
  process economics, **14.**12–**14.**13
  products and applications, **14.**10–**14.**12
  Spherilene process perspective, **14.**7–**14.**8
Basell *Spheripol* technology for PP production, **16.**3–**16.**20
  general process description, **16.**3
  process chemistry and thermodynamics, **16.**3–**16.**13
  process description, **16.**13–**16.**17
  process economics, **16.**17
  products and applications, **16.**17–**16.**20
  Spheripol process perspective, **16.**13
Basell *Spherizone* technology for PP production, **16.**21–**16.**39
  economics, **16.**37
  general process description, **16.**21
  process chemistry and thermodynamics, **16.**21–**16.**30
  process description, **16.**30–**16.**37
  products and applications, **16.**37–**16.**39
  Spherizone process perspective, **16.**30

BASF butadiene extraction technology, **3.**3–**3.**9
  economics, **3.**7
  environmental considerations, **3.**8
  process description, **3.**4
  process features, **3.**8–**3.**9
  process perspective, **3.**3
BASF Ludwigshafen methanol synthesis plant, **7.**3
Bechamp process, **2.**3
Benzene:
  sources and uses, **5.**25
  typical specification for, **5.**25
Benzene feed specifications, for ethylbenzene production, **5.**11
Biaxially oriented polypropylene (BOPP), **16.**65–**16.**66
Bimodal film products, produced by *Hostalen* process, **14.**84
Bimodal high-density polyethylene (HDPE), **14.**4
  Basell *Hostalen* process for, **14.**71–**14.**86
  UNIPOL process for, **14.**114
Bimodal polyethylene, via Borstar technology, **14.**15–**14.**29
Bisphenol A process (*see* QBIS process for high-purity bisphenol A)
Blow molding:
  of polyethylenes, **14.**101, **14.**110–**14.**111
  polyethylene for, **14.**92–**14.**93
  resins from SCLAIRTECH technologies, **14.**142
Blow-molding products, produced by *Hostalen* process, **14.**85
BorAPC, **14.**21–**14.**22, **16.**48
Borstar LLDPE and HDPE technology, **14.**15–**14.**29
  advanced process control, **14.**21–**14.**23
  Borstar PE products, **14.**23–**14.**28
  process description, **14.**15–**14.**20
  process economics, **14.**29
Borstar polypropylene technology, **16.**41–**16.**56
  advanced process control, **16.**48–**16.**49
  catalyst, **16.**49
  environment, **16.**49–**16.**50
  features of Borstar PP process technology, **16.**43–**16.**44
  operating requirements, **16.**50
  process description, **16.**44–**16.**47
  production cycle and grade transitions, **16.**47–**16.**48
  products, **16.**50–**16.**56
BP Chemicals:
  development of polypropylene process, **16.**71
  phenol process, **9.**3, **9.**8, **9.**31

INDEX

BP/Lummus technology for production of expandable polystyrene, **17**.3–**17**.7
  feedstock/product specifications, **17**.4–**17**.5
  operating plants, **17**.3
  process chemistry, **17**.3–**17**.4
  process description, **17**.4
  process economics, **17**.5–**17**.6
  summary of process features, **17**.7
  waste and emissions, **17**.5
BP/Lummus technology for production of general-purpose and high-impact polystyrenes, **17**.9–**17**.17
  feedstock and product specifications, **17**.12–**17**.15
  operating plants, **17**.10
  process chemistry, **17**.10
  process description, **17**.10–**17**.12
  process economics, **17**.15
  summary of process features, **17**.15–**17**.17
  waste and emissions, **17**.15
BP-SPC (SECCO) "classic" SM project, **11**.4–**11**.5
Brückner film production line, **16**.65
BSL Petrochemieverb. butadiene plant, **3**.4
Bulk continuous process, to produce general-purpose and high-impact polystyrenes, **17**.9–**17**.17
Butadiene production, CATADIENE process for, **10**.43
1-Butene, comonomer in polyethylene, **14**.132, **14**.139
1-Butene, comonomer to control PE density, **14**.16, **14**.18, **14**.23
Butene, copolymerization with propylene via *Spheripol* technology, **16**.3
Butene-1, copolymerization with propylene via *Spherizone* technology, **16**.21
Butenes, as ethylene comonomer, **14**.71
Butenes and ethylene, conversion to propylene (*see* ABB Lummus Global propylene production via olefins conversion technology)
Butylbenzene, **14**.16
BYPC (China) "classic" SM project, **11**.4–**11**.5

C

"C. Buonerba" Polimeri Europa Styrenics Research Centre, **5**.20
C. F. Braun, **6**.51
Cable and wire, applications of polyethylene, **14**.110–**14**.111
CATADIENE process, **10**.43
Catalyst and chemical consumption, **10**.49
Catalytic alkylation of benzene, **4**.3

CATOFIN dehydrogenation process (*see* Propylene via CATOFIN propane dehydrogenation technology)
CCR, **10**.28–**10**.29
CCR-based re-formate, **13**.8
CCR platforming, aromatics product, **13**.23–**13**.24
CD*Cumene* technology (*see* ABB Lummus Global cumene production via CD*Cumene* technology)
CD*Hydro* catalytic distillation hydrogenation technology, **6**.16–**6**.17, **10**.37
CDTECH *EB* process, **4**.3
  (*See also* Lummus/UOP liquid-phase EB*One* process and CDTECH *EB* process)
CEFOR:
  butene random polypropylene copolymers, **16**.66
  specialty random copolymers, **16**.54–**16**.65
Chemical Research & Licensing, **4**.3
Chevron Phillips Chemical "classic" SM project, **11**.4–**11**.5
Chevron Phillips slurry-loop-reactor process for polymerizng linear polyethylene, **14**.31–**14**.44
  history, **14**.31–**14**.32
  polymer finishing and packaging, **14**.37
  process description, **14**.32–**14**.35
  slurry-loop reactor, **14**.35–**14**.37
  summary, **14**.44
  technical advantages of Chevron Phillips slurry-loop process for PE, **14**.38–**14**.44
  utilities, **14**.37–**14**.38
Chiba styrene monomer, **5**.4
Chisso gas-phase polypropylene process, **16**.71–**16**.80
  economics, **16**.79
  polymerization mechanism and polymer type, **16**.71–**16**.73
  process description, **16**.76–**16**.78
  process features, **16**.74–**16**.76
  product capabilities, **16**.78
  reference plants, **16**.79–**16**.80
  safety and environmental considerations, **16**.78
  technology background and history, **16**.71
Chisso polyvinyl chloride suspension process technology and vinyl chloride monomer removal technology, **18**.37–**18**.46
  Chisso polyvinyl chloride suspension process technology, **18**.37–**18**.43
  Chisso vinyl chloride monomer removal process technology, **18**.43–**18**.46
Chiyoda acetic acid process ACETICA, **1**.3–**1**.13
  chemistry, **1**.4–**1**.6

Chiyoda acetic acid process ACETICA (*Cont.*):
  economics of Chiyoda ACETICA technology, **1.**11
  experience, **1.**12
  process description, **1.**7
  process features, **1.**7
  process yield and emissions, **1.**11
  product specifications, **1.**11
  scope of Chiyoda's package of services, **1.**12
Chlorine, manufacture, **18.**7
Chlorine and ethylene, for production of vinyl chloride, **18.**4–**18.**21
Chrome silica catalysts, in Chevron Phillips process, **14.**34–**14.**35, **14.**41
"Classic" styrene technology (*see* Lummus/UOP "classic" styrene technology and Lummus/UOP SMART styrene technology)
Clearflex, **14.**61
CMC International Tendering Company, **1.**13
Coke, in Polimeri Europa ethylbenzene process, **5.**15–**5.**16
Coke formation and deposition, in ABB Lummus ethylene process, **6.**8–**6.**9
Comonomer content in polyethylenes, **14.**4
Continuous-stirred-tank reactor (CSTR):
  in *Hostalen* process, **14.**77
  in SCLAIRTECH process, **14.**135–**14.**136
CPChem slurry-loop process (*see* Chevron Phillips slurry-loop-reactor process for polymerizing linear polyethylene)
Cracking, of hydrocarbons, **6.**51–**6.**63
Crystal polystyrene, **17.**9
Cumene:
  impurities, **4.**16
  typical purchase specification, **9.**41–**9.**42
  value chain, **9.**32
Cumene and toluene, from ethylbenzene, **11.**17
Cumene hydroperoxide, **9.**8–**9.**9
  route to produce phenol, **9.**13–**9.**29
Cumene-phenol process (*see* Polimeri Europa cumene-phenol processes)
Cumene production:
  ABB Lummus Global process, **4.**3–**4.**9
  history, **4.**3
  Polimeri Europa process, **9.**3–**9.**7
Cymere, **4.**15–**4.**16

D

Deethanizer, **6.**58–**6.**60
Dehydrogenation:
  of ethylbenzene, **11.**16–**11.**24
    with Lummus/UOP process, **11.**3–**11.**11
  of isobutane, **10.**43
  of propane, **10.**43–**10.**50

Dehydrogenation reaction of ethylbenzene, **11.**25–**11.**34
Demethanizer, **6.**51
Dephenolation of water, **9.**56
Dephlegmators, **6.**51
Depropanizer, **6.**60–**6.**61
1,2-Dichloroethane (EDC):
  thermal cracking, **18.**18–**18.**21
  via direct chlorination, **18.**3–**18.**12
  via oxychlorination, **18.**6, **18.**8, **18.**12–**18.**18
Diethylbenzene, **5.**6, **5.**14–**5.**15, **5.**29
Dimethylether, conversion to propylene, **10.**3, **10.**6–**10.**9
Diphenyl propane, **4.**14
Diphenylethane, **5.**29
Direct chlorination, route to vinyl chloride monomer, **18.**4–**18.**6
Distillers Co. phenol process, **9.**31
Divinylbenzenes, from ethylbenzene in SM reactors, **5.**37
Dornier film production line, **16.**65
Dow Chemical Company, **3.**12
  E-3 ethylene plant, **6.**21, **6.**46–**6.**47
  gas-phase polyethylene process, **14.**113, **14.**115
  KLP process, **3.**12
  polystyrene foam, **17.**27
  SHAC catalyst system, **16.**59–**16.**63
DuPont/Canada, low-density polyethylene process, **14.**132
DuPont/KBR aniline process, **2.**3–**2.**8
  operating requirements, **2.**7
  process chemistry, **2.**5
  process description, **2.**5–**2.**6
  product quality, **2.**7–**2.**8
  technology features, **2.**6
  wastes and emissions, **2.**7–**2.**8
DuPont polyesters, **15.**3

E

E PTA: Lurgi/Eastman/SK process, **12.**3–**12.**12
  chemistry overview and product specification, **12.**4
  economics of E PTA technology, **12.**10–**12.**11
  highlights and benefits of E PTA technology, **12.**9–**12.**10
  process description, **12.**4–**12.**8
Eastman terephthalic acid process (*see* E PTA: Lurgi/Eastman/SK process)
EBMax process:
  catalyst supercage structure, **5.**27
  (*See also* ExxonMobil/Badger ethylbenzene technology)
EB*One* process (*see* Lummus/UOP liquid-phase EB*One* process and CDTECH *EB* process)

Edistir GPPS, **17.**20, **17.**25, **17.**32, **17.**40
Elastomers:
　in polystyrene, **17.**9
　production with Basell *Spherizone*
　　technology, **16.**29
EniChem:
　polystyrene technology, **17.**9, **17.**27, **17.**35
　styrene production, **11.**25
　(*See also* Polimeri Europa cumene-phenol
　　processes)
Environmental emissions, **10.**49
Ether complex, **10.**30–**10.**31
Ethyl tertiary butyl ether (ETBE), **10.**30
Ethylbenzene, **4.**15–**4.**16
　boiling points of pure, **5.**26
　conversion in xylene isomerization, **13.**16–**13.**18
　conversion to benzene and ethylene, **13.**16
　conversion to styrene, **11.**3–**11.**34
　feed specifications for Lummus/UOP
　　process, **11.**10
　history, **5.**3, **5.**13, **5.**23–**5.**24
　integration with dehydrotechnology SM
　　plants, **5.**25
　physical properties, **5.**26
　production:
　　by Polimeri Europa process, **5.**13–**5.**21
　　(*See also* ExxonMobil/Badger ethylben-
　　　zene technology)
Ethylbenzene product specifications, for
　Lummus/UOP process, **5.**11
Ethylene:
　comonomer in Borstar process, **16.**45
　copolymerization with propylene via
　　Spheripol technology, **16.**3, **16.**9–**16.**11
　copolymerization with propylene via
　　Spherizone technology, **16.**21,
　　16.27–**16.**28
　feed to ethylbenzene unit specification, **5.**10
　production capacity, **6.**3
　specification for polymer-grade, **5.**24
Ethylene and butene, conversion to propylene
　(*see* ABB Lummus Global propylene pro-
　　duction via olefins conversion technology)
Ethylene and chlorine, for production of vinyl
　chloride, **18.**4–**18.**21
Ethylene copolymerization:
　with acrylic acid, **14.**61
　with alpha olefins, **14.**61
　with methyl acrylate, **14.**61
　with vinyl acetate, **14.**59, **14.**61, **14.**63
Ethylene production technology, Stone &
　Webster, **6.**21–**6.**48
Ethylene-propylene copolymer, produced by
　Chisso technology, **16.**76–**16.**78

Ethylene-propylene impact copolymer, **16.**71,
　16.73
Ethylene vinyl acetate copolymer, **14.**95,
　14.98
Ethyltoluene, **5.**29
EVA copolymers, in polyethylene chains,
　14.7
Expandable polystyrene (EPS):
　demand, **17.**3
　history, **17.**27
　process, **17.**3–**17.**7
　production via Polimeri Europa process,
　　17.23–**17.**33
　properties, **17.**4, **17.**6
Extrusion coating, resins from SCLAIRTECH
　technologies, **14.**142
ExxonMobil/Badger ethylbenzene technology,
　5.23–**5.**38
　catalyst requirements, **5.**37
　EBMax plant design, **5.**38
　EBMax process catalysts, **5.**26
　EBMax process designs for dilute ethylene
　　feedstocks, **5.**33
　ethylbenzene manufacturing, **5.**23–**5.**24
　process chemistry and EBMax catalyst
　　performance, **5.**28
　process design customization and
　　optimization, **5.**33
　properties of ethylbenzene, **5.**25–**5.**26
　raw materials and utilities consumption,
　　5.37
　technology conversion and capacity
　　expansion with EBMax, **5.**34
ExxonMobil Chemical Company, **6.**51–**6.**52
　ethylbenzene technology, **5.**23
　gas-phase polyethylene process, **14.**113,
　　14.115
ExxonMobil ethylene technology (*see* KBR
　SCORE ethylene technology)
ExxonMobil high-pressure process technology
　for LDPE, **14.**45–**14.**58
　LDPE markets, **14.**54–**14.**56
　LDPE vs. LLDPE, **14.**53
　process overview/description, **14.**48–**14.**53
　product capability/grade slate, **14.**54
　reaction mechanism, **14.**46–**14.**48
　strengths of ExxonMobil technology,
　　14.56
　summary, **14.**57
ExxonMobil PxMax *p*-xylene from toluene,
　13.3–**13.**13
　aromatics complex and PxMax unit
　　description, **13.**8
　case I: grassroots PxMax unit, **13.**8–**13.**10

ExxonMobil PxMax *p*-xylene from toluene (*Cont.*):
  case II: retrofit of selective TDP to PxMax, **13**.11–**13**.12
  case III: retrofit of nonselective TDP to PxMax, **13**.12–**13**.13
  conclusion, **13**.13
  operating performance, **13**.5–**13**.7
  process chemistry, **13**.4–**13**.5
  process description, **13**.5
  PxMax retrofit and debottleneck applications, **13**.7–**13**.8
ExxonMobil XyMax xylene isomerization, **13**.15–**13**.21
  commercial experience, **13**.21
  operating performance, **13**.18–**13**.20
  process chemistry, **13**.16–**13**.17
  process description, **13**.17–**13**.18
  XyMax cycle length, **13**.20–**13**.21

F

Feedstock and utility consumption, **10**.47
  feedstock considerations, **13**.26
Film:
  applications of polyethylene, **14**.110–**14**.111
  resins from SCLAIRTECH technologies, **14**.142
Film extrusion, polyethylene for, **14**.92–**14**.93
Film packaging, applications of polyethylene in, **14**.24
Fischer-Tropsch catalysts, **10**.15
Flex, **14**.60
  end-use applications for, **14**.56
Flexirene, **14**.60
Fluidized-bed technology, for polypropylene production, **16**.57–**16**.69
Formosa Chemicals and Fibre Corp. (Taiwan), cumene plant, **4**.5
Formosa Plastics (Taiwan), ethylene furnaces, **6**.45–**6**.46
Free-radical polymerization, for low-density polyethylene production, **14**.46–**14**.47
Friedel-Crafts alkylation, **4**.12
Friedel-Crafts reaction, in cumene production, **9**.4

G

"G. Donegani" Polimeri Europa corporate Research Centre, **5**.20
Gas-to-liquids (GTL) technology, **10**.15–**10**.16
Gas-to-polyolefin (GTP) technology, **10**.16
Gasoline, from methanol, **10**.4–**10**.5, **10**.9
General-purpose polystyrene (GPPS):
  process designation, **17**.10–**17**.12

General-purpose polystyrene (*Cont.*):
  properties, **17**.9, **17**.13
  via Polimeri Europa technology, **17**.19–**17**.25
Guizhou Crystal Organic Chemical Group Co., **1**.13

H

Haldia Petrochemicals butadiene plant, **3**.4
Hercules Inc. phenol process, **9**.3, **9**.8, **9**.31–**9**.32
Heterophasic copolymers:
  produced by Borstar process, **16**.51–**16**.53
  produced by *Spherizone* technology, **16**.30–**16**.31
Heterophasic impact copolymer production, **16**.21
1-Hexene, comonomer for polyethylene production, **14**.87, **14**.91–**14**.92
High-density high-molecular-weight PE (HMWPE), **14**.4–**14**.5
High-density polyethylene (HDPE):
  Basell *Lupotech* G process for, **14**.87–**14**.93
  Basell *Spherilene* process for, **14**.3–**14**.13
  Chevron Phillips process for, **14**.39
  NOVA Chemicals process for, **14**.131–**14**.144
  UNIPOL process for, **14**.113–**14**.130
High flux heat-exchanger tubing, **13**.28–**13**.29
High-impact polystyrene (HIPS):
  process description, **17**.10–**17**.12
  product properties and end uses, **17**.14–**17**.15
  properties, **17**.9
High-performance polyethylenes, **14**.12
Hoechst AG:
  Basell predecessor company, **14**.74
  linear polyethylene process, **14**.131
  polyvinyl chloride activities, **18**.3
Homopolymers, produced by Borstar process, **16**.54
Hostalen process technology (*see* Basell *Hostalen* technology for bimodal HDPE production)
Hüls selective hydrogenation process (SHP), **3**.11, **10**.29
HY/HS catalysts, **16**.13
Hydrocarbons, solubility, **3**.3–**3**.4
Hydrochloric acid and acetylene, vinyl chloride process, **18**.4
Hydroformylation (oxo reaction), **8**.1–**8**.6
Hydrogen, as molecular weight controller, **14**.73
Hydrogenation, of aldehydes, **8**.6
Hydroperoxidation, of cumene, **9**.3, **9**.7–**9**.12
HySorbXP process, **13**.25

I

Imperial Chemical Industries (ICI):
  high-pressure oxo process (*see* Johnson Matthey Oxo Alcohols Process)

Imperial Chemical Industries (ICI) (*Cont.*):
  invention of polyethylene, **14.**131
  polyesters, **15.**3
Injection molding:
  of polyethylenes, **14.**110–**14.**111
  resins from SCLAIRTECH technologies, **14.**142
Injection-molding products produced by *Hostalen* process, **14.**85
Isobutane:
  dehydrogenation plants, **10.**43
  diluent for polyethylene production, **14.**32–**14.**36
Isobutylene:
  process (*see* UOP Oleflex process for light olefin production)
  production, **10.**43
Isomar unit, production of xylenes, **13.**23–**13.**24
Isopropylbenzene (*see* Cumene)

J
Japan Polychem Corporation, **16.**71
Japan Polypropylene Corporation (licenser of Chisso polypropylene technology), **16.**71
Johnson Matthey Oxo Alcohols Process, **8.**3–**8.**13
  benefits of Johnson Matthey technology, **8.**9
  capital costs, **8.**12
  feed specifications, **8.**10–**8.**11
  operational experience, **8.**13
  process description, **8.**3–**8.**7
  process economics, **8.**12
  process flowsheet, **8.**7–**8.**8

K
Kaucuk (Czech Republic) "classic" SM project, **11.**4–**11.**5
KBR/DuPont aniline process (*see* DuPont/KBR aniline process)
KBR phenol process, **9.**31–**9.**50
  acetone netback, **9.**48–**9.**49
  environmental features, **9.**44–**9.**45
  feedstock and product properties, **9.**41–**9.**43
  history, **9.**31–**9.**32
  investment/economies of scale, **9.**46–**9.**48
  markets, **9.**32–**9.**33
  operating economics, **9.**46
  process chemistry, **9.**34–**9.**36
  process description, **9.**36–**9.**41
  product storage and shipping, **9.**43–**9.**44
  production yields, **9.**43
  safety, **9.**45–**9.**46
  technology advantages, **9.**49
  utility requirements, **9.**43

KBR SCORE ethylene technology, **6.**51–**6.**63
  development and history, **6.**51–**6.**52
  future developments, **6.**63
  optimal recovery-section design, **6.**57–**6.**63
  selective cracking furnace technology, **6.**52–**6.**57
Kinetic severity function (KSF), **6.**24–**6.**25
KLP process (*see* UOP KLP 1,3-butadiene from acetylene process)

L
Latex stripping, of paste PVC, **18.**43
Linear high-density polyethylene (HDPE), chrome-based catalyst for, **14.**31
  Basell *Spherilene* process for, **14.**3–**14.**13
  Chevron Phillips process for, **14.**39–**14.**40
  consumption by industrial segment, **14.**121–**14.**122
  NOVA Chemicals process for, **14.**131–**14.**144
  Polimeri Europa process for, **14.**59–**14.**70
  UNIPOL process for, **14.**113–**14.**130
Linear polyethylene, Chevron Phillips process for, **14.**31–**14.**44
Low-cost ethylene technology (LCET), **6.**51
Low-density polyethylene (LDPE):
  Basell *Lupotech* T process for, **14.**95–**14.**111
  discovery of, **14.**45
  ExxonMobil process for, **14.**35–**14.**58
  Polimeri Europa process for, **14.**59–**14.**70
Low-NO$_x$ burner concepts, **6.**26
  in thylene production, **6.**18–**6.**19
LP methanol process, Lurgi technology, **7.**3–**7.**17
Lummus:
  expandable polystyrene technology, **17.**3–**17.**7
  general-purpose polystyrene technology, **17.**9–**17.**17
Lummus/UOP "classic" styrene technology and Lummus/UOP SMART styrene technology, **11.**3–**11.**11
  economics, **11.**9–**11.**10
  process chemistry, **11.**5–**11.**6
  process descriptions, **11.**6–**11.**9
  process perspective, **11.**4–**11.**5
  summary of process features, **11.**11
Lummus/UOP liquid-phase EB*One* process and CDTECH *EB* process, **5.**3–**5.**12
  economics, **5.**9
  process chemistry, **5.**4
  process description, **5.**5
  process features, **5.**12
  process perspective, **5.**4
Lupotech G technology (*see* Basell *Lupotech* G technology for HDPE and MDPE production)

Lupotech T technology (*see* Basell *Lupotech* T technology for LDPE and EVA-copolymer production)
Lupotech TS and TM, **14.**95, **14.**105–**14.**106
Lurgi, Fisher-Tropsch synthesis, **7.**3
Lurgi low-pressure process (*see* Lurgi MegaMethanol technology)
Lurgi MegaMethanol technology, **7.**3–**7.**17
  process description, **7.**5–**7.**17
Lurgi MTP technology, **10.**3–**10.**13
  detailed process description, **10.**4–**10.**9
  process economics, **10.**11–**10.**12
  process overview, **10.**3–**10.**4
  products, by-products, wastes, and emissions, **10.**9–**10.**10
  technical and commercial status, **10.**10–**10.**11
Lurgi terephthalic acid process (*see* E PTA: Lurgi/Eastman/SK process)

M

M&G Group, **15.**17
M. W. Kellogg Company, **6.**51
Mantova Research Centre, polystyrene technology, **17.**27, **17.**35
Medium-density polyethylene (MDPE):
  Basell *Lupotech* G process for, **14.**87–**14.**93
  Basell *Spherilene* process for, **14.**3–**14.**13
MegaMethanol process, **10.**3
  (*See also* Lurgi MegaMethanol technology)
Melt-flow rate (MFR), defined, **16.**7, **16.**25
Metallocene catalyst, for UNIPOL polyethylene process, **14.**113
Metallocenes, in Chevron Phillips process, **14.**34, **14.**41
Metathesis chemistry, for propylene production, **6.**16, **10.**37
Methanol:
  carbonylation, **1.**3–**1.**13
  conversion to dimethylether, **10.**3–**10.**6
  conversion to ethylene and propylene (*see* UOP/Hydro MTO process)
  production costs, **10.**17
Methanol synthesis, Lurgi MegaMethanol technology, **7.**3–**7.**17
Methyl iodide, in Chiyoda process, **1.**4–**1.**6
Methyl tertiary butyl ether (MTBE), **10.**30
α-Methylstyrene (AMS):
  from cumene, **11.**17
  recovered by-product in phenol plant, **9.**43
Mitsubishi Chemical Corporation of Japan, **16.**71
  film production line, **16.**65
Mitsui Chemicals, **5.**4
Mixed xylenes, definition, **13.**23

Mobil High Activity Isomerization (MHAI) process, **13.**15–**13.**16, **13.**18–**13.**20
Mobil High Temperature Isomerization (MHTI) process, **13.**15–**13.**16
Mobil MTG (methanol-to-gasoline) process, **10.**15
Mobil Research and Development Corporation, alkylation process, **5.**24
Mobil Selective Toluene Disproportionation (MSTDP) process, **13.**3, **13.**11–**13.**12
Mobil Toluene Disproportionation (MTDP) process, **13.**3, **13.**12–**13.**13
Mobil Vapor Phase Isomerization (MVPI) process, **13.**15–**13.**16
Molecular weight distribution, measurement of, **16.**8, **16.**26
Molecular weight of polymer vs. melt-flow rate, **16.**7, **16.**26
Monomodal high-density polyethylene, Basell *Hostalen* process for, **14.**71–**14.**86
Mononitrobenzene (MNB) production, **2.**5
Montedison:
  catalyst discoveries, **14.**7
  polystyrene technology, **17.**9, **17.**27, **17.**35
  styrene production, **11.**25
MTO process (*see* UOP/Hydro MTO process)
MTP technology (*see* Lurgi MTP technology)
Multizone circulating reactor (MZCR), **16.**21, **16.**30, **16.**33, **16.**35, **16.**37

N

Nan Ya Plastics Corp., bisphenol A plant, **9.**57
Naphtha steam cracker, material balances, **10.**39
Natta, Giulio, invention of stereospecific polymerization, **16.**3, **16.**21
Natural gas conversion, liquids and polyolefins, **10.**16–**10.**17
Nippon Styrene Monomer Company (Japan), **5.**4
Nitration of benzene, **2.**3–**2.**8
NOVA Chemicals SCLAIRTECH LLDPE/HDPE swing technology, **14.**131–**14.**144
  advantages of SCLAIRTECH technology platform, **14.**138–**14.**140
  chemistry and catalysis, **14.**132–**14.**134
  economics, **14.**140
  process overview, **14.**134–**14.**138
  product capability, **14.**140–**14.**143
  summary, **14.**143–**14.**144

O

OCT (*see* ABB Lummus Global propylene production via olefins conversion technology)
1-Octene, comonomer in polyethylene, **14.**132
Octenes feedstock, to oxo alcohols plant, **8.**11
Olefins, conversion to alcohols, **8.**3–**8.**13

Oleflex process (*see* UOP Oleflex process for light olefin production)
On-purpose acetone, **9.**31
On-purpose phenol, **9.**31
Organic peroxides, for low-density polyethylene production, **14.**46–**14.**47
Organoleptics, **14.**35
Oxo alcohols process (*see* Johnson Matthey Oxo Alcohols Process)
Oxo reaction, **8.**3–**8.**6
Oxychlorination, route to vinyl chloride monomer, **18.**4–**18.**6

P
Parex process (*see* UOP Parex process for *p*-xylene production)
Phenol:
   demand by end-use application, **9.**33
   demand by region, **9.**32–**9.**33
   demand growth by region and Asia by application, **9.**33
   via toluene oxidation process, **9.**32
Phenol and acetone:
   conversion to bisphenol A, **9.**51–**9.**59
   for polycarbonates and nylon, **4.**11
   from cumene, **4.**11
Phenylacetylene reduction technology, **11.**13
Phillips Petroleum Company:
   low-density polyethylene process, **14.**131
   slurry-loop-reactor design, **14.**31
Pipe coating:
   resins from SCLAIRTECH technologies, **14.**142
   with polyethylene, **14.**110–**14.**111
Plate-fin heat exchangers, **6.**51
Polimeri Europa cumene-phenol processes, **9.**3–**9.**12
   cumene technology, **9.**3–**9.**7
   phenol technology, **9.**8–**9.**12
Polimeri Europa ethylbenzene process, **5.**13–**5.**21
   experience, **5.**21
   features, **5.**20
   performance, **5.**21
   process chemistry, **5.**14–**5.**15
   process flow, **5.**16
Polimeri Europa expandable polystyrene process technology, **17.**27–**17.**33
   commercial experience, **17.**32
   description of process flow, **17.**30–**17.**31
   Extir EPS product portfolio, **17.**32–**17.**33
   plant capacity, **17.**32
   process advanced design features, **17.**31–**17.**32
   process chemistry, **17.**28–**17.**30
   process performance, **17.**32

Polimeri Europa general-purpose polystyrene process technology, **17.**19–**17.**25
   commercial experience, **17.**25
   description of process flow, **17.**22–**17.**23
   Edistir GPPS product portfolio, **17.**25
   plant capacity, **17.**24
   process advanced design features, **17.**23–**17.**24
   process chemistry, **17.**20–**17.**22
   process performance, **17.**24
Polimeri Europa high-impact polystyrene process technology, **17.**35–**17.**40
   commercial experience, **17.**40
   description of process flow, **17.**37–**17.**38
   Edistir HIPS product portfolio, **17.**40
   plant capacity, **17.**39
   process advanced design features, **17.**38–**17.**39
   process chemistry, **17.**36–**17.**37
   process performance, **17.**39
Polimeri Europa polyethylene high-pressure technologies, **14.**59–**14.**70
   chemistry and thermodynamics, **14.**61–**14.**63
   detailed process description, **14.**65–**14.**67
   high-pressure reactor technologies, **14.**63–**14.**65
   plant battery limits, **14.**69–**14.**70
   process performance, **14.**69
   reactor safety discharge system, **14.**67–**14.**68
Polimeri Europa styrene process technology, **11.**25–**11.**34
   description of process flow, **11.**27–**11.**32
   process and mechanical design advanced features, **11.**32–**11.**33
   process chemistry, **11.**26–**11.**27
   process performance, **11.**33
Polyamides, via solid-state polymerization, **15.**5
Polybutadiene rubber, in polystyrene, **17.**36–**17.**38
Polybutylene terephthalate (PBT), **15.**5
Polyester:
   applications by molecular weight, **15.**4
   history, **15.**3–**15.**4
   product chain, **15.**4
   production, **15.**3–**15.**18
Polyester fibers, SK Chemicals, **12.**3
Polyethylbenzene, **5.**14–**5.**15, **5.**32
Polyethylene:
   Basell *Hostalen* process for, **14.**71–**14.**86
   Basell *Lupotech* G process for, **14.**87–**14.**93
   Basell *Lupotech* T process for, **14.**95–**14.**111
   Basell *Spherilene* process for, **14.**3–**14.**13

Polyethylene (*Cont.*):
   Borstar process for, **14.**15–**14.**29
   Chevron Phillips process for, **14.**31–**14.**44
   ExxonMobil process for, **14.**45–**14.**58
   NOVA Chemicals SCLAIRTECH process for, **14.**131–**14.**144
   pipe applications of, **14.**26–**14.**27
   Polimeri Europa process for, **14.**59–**14.**70
   UNIPOL PE gas-phase process for, **14.**113–**14.**130
   wire and cable applications of, **14.**2
Polyethylene naphthalate (PEN), **15.**5
Polyethylene terephthalate (PET):
   applications by molecular weight, **15.**4
   bottle-grade product properties, **15.**16
   chemistry, **15.**5–**15.**8
   crystallite, **15.**9
   crystallization, **15.**8–**15.**10
   production process, **15.**3–**15.**18
   worldwide production figures, **15.**4
Polymer-grade terephthalic acid (*see* E PTA: Lurgi/Eastman/SK process)
Polypropylene (PP):
   atactic, **16.**4, **16.**22
   Basell *Spheripol* process for, **16.**3–**16.**20
   Basell *Spherizone* process for, **16.**30–**16.**39
   Borstar process for, **16.**41–**16.**56
   Chisso process for, **16.**71–**16.**80
   chrome-based catalyst for, **14.**31
   heterophasic copolymers, **16.**10, **16.**12
   impact copolymers, **16.**10, **16.**12
   isotactic, **16.**3–**16.**4, **16.**7, **16.**22, **16.**25, **16.**38
   syndiotactic, **16.**4, **16.**22
   terpolymers, **16.**18
   UNIPOL process for, **16.**57–**16.**69
Polystyrene:
   utilization, **17.**9
   via BP/Lummus expandable polystyrene technology, **17.**3–**17.**7
   via BP/Lummus general-purpose polystyrene technology, **17.**9–**17.**17
   via Polimeri Europa expandable polystyrene technology, **17.**27–**17.**33
   via Polimeri Europa general-purpose polystyrene technology, **17.**19–**17.**25
   via Polimeri Europa high-impact polystyrene technology, **17.**35–**17.**40
Polytrimethylene terephthalate (PTT), **15.**5
Polyvinyl chloride (PVC):
   Vinnolit process for, **18.**3–**18.**4, **18.**21–**18.**35
   world consumption and growth, **18.**4
Polyvinyl chloride slurry, vinyl chloride monomer removal, **18.**43

PRODIGY bimodal catalyst technology, **14.**114, **14.**120, **14.**125–**14.**126, **14.**128
Propane, dehydrogenation, **10.**43–**10.**50
n-Propylbenzene, **4.**16–**4.**17
Propylene:
   ABB Lummus Global propylene production via olefins conversion technology, **10.**35–**10.**41
   as by-product of ethylene production or transportation fuels, **10.**35
   Basell *Spheripol* process for polymerization of, **16.**3–**16.**20
   Basell *Spherizone* process for polymerization of, **16.**30–**16.**39
   Borstar technology for polymerization of, **16.**41–**16.**56
   from methanol, **10.**3–**10.**13
   stereospecific polymerization, **16.**3
   UOP Oleflex process for light olefin production, **10.**27–**10.**34
   world production sources, **10.**35–**10.**36
Propylene oxide/styrene monomer unit, **5.**33
Propylene via CATOFIN propane dehydrogenation technology, **10.**43–**10.**50
   process chemistry, **10.**44
   process description, **10.**44–**10.**47
   process economics, **10.**47
   product quality and by-products, **10.**47–**10.**48
PxMax process (*see* ExxonMobil PxMax *p*-xylene from toluene)
Pyrolysis Yield Prediction System (PYPS), **6.**7

Q

Q-Max process (*see* UOP Q-Max process)
QBIS process for high-purity bisphenol A, **9.**51–**9.**59
   commercial experience, **9.**56–**9.**57
   overview, **9.**51–**9.**56
   wastes and emissions: expected performance, **9.**57–**9.**59

R

Radiant coil material, **6.**11
Random copolymers, produced by Borstar process, **16.**54
REI (Reaction Engineering, Inc.), **6.**32
Rhodium carbonylation, catalytic cycle for, **1.**5
Riblene, **14.**60
Rotomolding, resins from SCLAIRTECH technologies, **14.**142
Rubber:
   in polystyrene, **17.**36–**17.**38
   production with Basell *Spherizone* technology, **16.**29

Rubber (*Cont.*):
 production with Borstar PP process technology, **16.**43

S

SABINA (Shell/BASF/ATOFINA) butadiene plant, **3.**4
SCLAIRTECH process (*see* NOVA Chemicals SCLAIRTECH LLDPE/HDPE swing technology)
SCORE ethylene technology (*see* KBR SCORE ethylene technology)
SECCO (BP/SPC) butadiene plant, **3.**4
Selective catalytic reduction (SCR), **6.**27
Selective linear exchanger (SLE), **6.**25
SHC PP polymerization catalyst, **16.**59–**16.**62
Shell ethylene plant, **6.**51
Simulated-moving-bed (SMB) separation, **13.**23
Sinco Engineering, **15.**17
SK Chemicals terephthalic acid technology (*see* E PTA: Lurgi/Eastman/SK process)
Slurry-cascade process, **14.**74
Slurry-loop-reactor process (*see* Chevron Phillips slurry-loop-reactor process for polymerizing linear polyethylene)
SMART styrene technology (*see* Lummus/UOP "classic" styrene technology and Lummus/UOP SMART styrene technology)
Solid-state polymerization (SSP) process, for polyethylene terephthalate, **15.**3–**15.**18
Solubility, of hydrocarbons, **3.**3–**3.**4
Sorbex process, **13.**26, **13.**28
Spherical polypropylene, **16.**30
Spherilene technology (*see* Basell *Spherilene* technology for LLDPE and HDPE production
Spheripol technology (*see* Basell *Spheripol* technology for PP production)
Spherizone technology (*see* Basell *Spherizone* technology for PP production)
SRT cracking technology (*see* ABB Lummus Global SRT cracking technology for production of ethylene)
SSP process, for polyethylene terephthalate, **15.**3–**15.**18
Stone & Webster (Badger) styrene technology, **11.**13–**11.**24
 operating economics, **11.**23–**11.**24
 process chemistry, **11.**16–**11.**18
 process description, **11.**18–**11.**23
 product specification, **11.**23
 properties, **11.**15
 styrene industry, **11.**13–**11.**14

Stone & Webster (Badger) styrene technology (*Cont.*):
 styrene manufacturing, **11.**15–**11.**16
 use of styrene monomer, **11.**14–**11.**15
Stone & Webster ethylbenzene technology, **5.**23
Stone & Webster ethylene technology, **6.**21–**6.**49
 development history: pyrolysis, **6.**23–**6.**27
 development history: recovery, **6.**27–**6.**29
 economic drivers, **6.**21–**6.**23
 megaplant design issues, **6.**44–**6.**46
 process description, **6.**29–**6.**44
 project execution aspects, **6.**47–**6.**49
Styrene:
 boiling points, **11.**15
 by-product of propylene oxide manufacture, **5.**23
 comonomer for polyethylene production, **14.**98
 demand, **11.**13
 history, **11.**25
 market growth, **11.**13
 producers, **11.**14
 properties, **11.**15
 uses, **11.**14–**11.**15
 via BP/Lummus expandable polystyrene technology, **17.**3–**17.**7
 via BP/Lummus general-purpose polystyrene technology, **17.**9–**17.**17
 via Lummus/UOP "classic" and SMART technologies, **11.**3–**11.**11
 via Polimeri Europa expandable polystyrene technology, **17.**27–**17.**33
 via Polimeri Europa general-purpose polystyrene technology, **17.**19–**17.**25
 via Polimeri Europa high-impact polystyrene technology, **17.**35–**17.**40
 via Stone & Webster (Badger) styrene technology, **11.**13–**11.**24
Sunoco/UOP phenol process, **9.**13–**9.**29
 cumene production, **9.**13–**9.**14
 overall process description/chemistry, **9.**15–**9.**16
 phenol production, **9.**14–**9.**15
 process flow and recent technology advances, **9.**16–**9.**29
 Sunoco/UOP cumene peroxidation route to phenol production, **9.**15
Super-Condensed Mode Technology (SCM-T), **14.**114
Suspension polyvinyl chloride:
 Chisso process, **18.**37–**18.**46
 Vinnolit process, **18.**3–**18.**4, **18.**21–**18.**35

Syngas:
  Lurgi MegaMethanol technology, **7.4–7.12**
  plant technology, **10.**16
  to methanol, **10.**16–**10.**17

T
Tapes and monofilaments, produced by *Hostalen* process, **14.**85
Tatoray unit, production of xylenes, **13.**23–**13.**24
Tertiary amyl ethyl ether (TAEE), **10.**30
Tertiary amyl methyl ether (TAME), **10.**30
Thermal cracking:
  of 1,2-dichloroethane to vinyl chloride, **18.**18–**18.**21
  of hydrocarbons via ABB Lummus technology, **6.**3–**6.**20
Thermoplastic polyesters, **15.**3–**15.**18
Thyssen Krupp Group (licenser of PVC production chain), **18.**3
Titanium magnesium catalysts, in Chevron Phillips process, **14.**35, **14.**41
Toho Corporation of Japan, THC-C-supported catalyst, **16.**74–**16.**75
Toluene, **5.**29
  conversion to *p*-xylene by PxMax process, **13.**3–**13.**13
  in EBMax reaction systems, **5.**36
Toluene and cumene, from ethylbenzene, **11.**17–**11.**18
Transalkylation:
  in ethylbenzene production, **5.**14, **5.**16, **5.**18
  of DIPB (diisopropylbenzene), **4.**13–**4.**14
  of polyethylbenzenes with benzene, **5.**28
TransPlus process, **13.**4
Triethyl aluminum (TEAL), cocatalyst in Borstar process, **14.**17
Triisopropylbenzene, **4.**14

U
UCAT conventional catalysts, **14.**113, **14.**126, **14.**128
Uhde (Vinnolit engineering partner), **18.**3
Ultra-low-density polyethylene (ULDPE), **14.**59, **14.**61
Ultra Selective Conversion (USC) coil design, **6.**23–**6.**24, **6.**31
Ultrahigh-molecular-weight PE (UHMWPE), **14.**4–**14.**5
Union Carbide, gas-phase polyethylene process, **14.**113
UNIPOL PE gas-phase process, **14.**113–**14.**130
  general process description, **14.**115–**14.**119
  history, **14.**114–**14.**115

UNIPOL PE gas-phase process (*Cont.*):
  process economics, **14.**124–**14.**130
  process perspective, **14.**119–**14.**120
  product and by-product specifications, **14.**120–**14.**122
  wastes and emissions, **14.**122–**14.**124
UNIPOL polypropylene process technology, **16.**57–**16.**69
  general UNIPOL PP process description, **16.**57–**16.**60
  process chemistry, **16.**60–**16.**62
  process economics, **16.**68–**16.**69
  process perspective, **16.**62–**16.**64
  product attributes summary, **16.**65–**16.**67
  products and by-products, **16.**64–**16.**65
  wastes and emissions, **16.**67–**16.**68
UNIPOL PP IMPPAX, **16.**67
Univation Technologies, gas-phase polyethylene process, **14.**113
UOP catalytic condensation process, **4.**11
UOP cumene technology, **9.**3
UOP ethylbenzene process (*see* Lummus/UOP liquid-phase EB*One* process and CDTECH *EB* process)
UOP/Hydro MTO process, **10.**15–**10.**26
  conclusions, **10.**26
  economic basis, **10.**20
  economic comparisons, **10.**23–**10.**25
  economic sensitivity, **10.**25–**10.**26
  investment estimates, **10.**20–**10.**22
  MTO technology, **10.**18
UOP KLP 1,3-butadiene from acetylene process, **3.**11–**3.**14
  butadiene, **3.**11–**3.**12
  commercial experience, **3.**13
  economics and operating costs, **3.**13–**3.**14
  KLP process, **3.**12
  process chemistry, **3.**12–**3.**13
UOP Oleflex process for light olefin production, **10.**27–**10.**34
  dehydrogenation plants, **10.**29–**10.**31
  process description, **10.**27–**10.**29
  propylene production economics, **10.**31–**10.**34
UOP Parex process for *p*-xylene production, **13.**23–**13.**30
  case study, **13.**29
  commercial experience, **13.**29
  description of process flow, **13.**26–**13.**28
  equipment considerations, **13.**28
  process performance, **13.**26
  vs. crystallization, **13.**23–**13.**26
UOP phenol process (*see* Sunoco/UOP phenol process)

UOP platforming process, **10.**28
UOP Q-Max process, **4.**11–**4.**19
   case study, **4.**18–**4.**19
   commercial experience, **4.**19
   description of process flow, **4.**14–**4.**15
   feedstock considerations, **4.**15
   process chemistry, **4.**12
   process performance, **4.**18
UOP Sinco, history of, **15.**17–**15.**18
UOP Sinco solid-state polymerization process for production of PET resin and technical fibers, **15.**3–**15.**18
   commercial experience, **15.**17–**15.**18
   crystallization of PET, **15.**8–**15.**10
   detailed process description, **15.**10–**15.**13
   equipment considerations, **15.**17
   feed properties, **15.**14–**15.**16
   melt-phase polymerization, **15.**5–**15.**6
   oxidation of PET, **15.**14
   process variables, **15.**14
   product properties, **15.**16
   product yield, **15.**16
   reactions of catalytic nitrogen purification system, **15.**13–**15.**14
   SSP process chemistry, **15.**5–**15.**8
   sticking tendency of PET, **15.**10
   utilities, **15.**16
   wastes and emissions, **15.**16
USX (Ultra Selective Exchanger), **6.**25–**6.**26, **6.**32

V

Very-low-density polyethylene (VLDPE):
   Basell *Spherilene* process for, **14.**3–**14.**13
   Polimeri Europa process for, **14.**59–**14.**70
   properties of, **14.**132
Vinnolit vinyl chloride and suspension polyvinyl chloride technologies, **18.**3–**18.**35
   process perspective, **18.**4
   Vinnolit direct chlorination process, **18.**7–**18.**12
   Vinnolit oxychlorination process, **18.**12
   Vinnolit S-PVC process, **18.**21–**18.**35
   Vinnolit VCM process, **18.**4–**18.**7

VinTec (licenser of Vinnolit technology), **18.**3
Vinyl acetate, comonomer for polyethylene production, **14.**95, **14.**98
Vinyl chloride:
   Chisso process for polymerization of, **18.**37–**18.**43
   removal process for, **18.**43–**18.**46
   Vinnolit process for polymerization of, **18.**3–**18.**4, **18.**21–**18.**35
   world consumption and growth, **18.**4
Vinylethers, comonomers for polyethylene production, **14.**98

W

Wacker-Chemie GmbH, polyvinyl chloride activities, **18.**3
Wire and cable, resins from SCLAIRTECH technologies, **14.**142

X

XCAT metallocene catalysts, **14.**113
p-Xylene:
   by fractional crystallization, **13.**23
   conversion to terephthalic acid, **12.**4–**12.**12
   from toluene by PxMax process, **13.**3–**13.**13
   production via UOP Parex process, **13.**23–**13.**30
   selectivity in zeolite, **13.**4–**13.**5
Xylene isomerization via XyMax process (*see* ExxonMobil XyMax xylene isomerization)
XyMax (*see* ExxonMobil XyMax xylene isomerization)

Z

Zeigler-Natta catalysts:
   high-pressure polyethylene process using, **14.**59, **14.**61, **14.**63, **14.**69
   in Basell *Hostalen* process, **14.**71–**14.**76
   in Chevron Phillips process, **14.**35, **14.**41

ABOUT THE EDITOR-IN-CHIEF

ROBERT A. MEYERS earned his Ph.D. in organic chemistry at UCLA, was a post-doctoral fellow at Cal Tech, and was manager of chemical processes for TRW. He has published in *Science* and the *Journal of the American Chemical Society* as well as in *Chemical Engineering*. He has written or edited 12 scientific books and holds more than 20 chemical patents. His research has been reviewed in *The New York Times, The Wall Street Journal,* and *Chemical Engineering*. A more-detailed biography of Dr. Meyers appears in *Who's Who in the World.*